Y0-BDT-720

Progress in Mathematics
Volume 275

Series Editors

H. Bass
J. Oesterlé
A. Weinstein

Andreas Juhl

Families of Conformally Covariant Differential Operators, Q-Curvature and Holography

Birkhäuser
Basel · Boston · Berlin

Author:

Andreas Juhl
Humboldt-Universität
Institut für Mathematik
10099 Berlin
Germany
e-mail: ajuhl@math.hu-berlin.de

and

Universitet Uppsala
Matematiska Institutionen
Box-480
75106 Uppsala
Sweden
e-mail: andreasj@math.uu.se

2000 Mathematics Subject Classification 53BXX (several: 20, 25, 50), 53C25, 53C80, 58 J50, 81T50

Library of Congress Control Number: 2009924478

Bibliographic information published by Die Deutsche Bibliothek.
Die Deutsche Bibliothek lists this publication in the Deutsche Nationalbibliografie; detailed bibliographic data is available in the Internet at http://dnb.ddb.de

ISBN 978-3-7643-9899-6 Birkhäuser Verlag AG, Basel · Boston · Berlin

© 2009 Birkhäuser Verlag AG
Basel · Boston · Berlin
P.O. Box 133, CH-4010 Basel, Switzerland
Part of Springer Science+Business Media
Printed on acid-free paper produced from chlorine-free pulp. TCF ∞
Printed in Germany

ISBN978-3-7643-9899-6   e-ISBN 978-3-7643-9900-9

9 8 7 6 5 4 3 2 1   www.birkhauser.ch

Contents

Preface ix

1 Introduction

    1.1   Hyperbolic geometry and conformal dynamics   2

    1.2   Automorphic distributions and intertwining families   6

    1.3   Asymptotically hyperbolic Einstein metrics.
            Conformally covariant powers of the Laplacian   9

    1.4   Intertwining families 11

    1.5   The residue method for the hemisphere 17

    1.6 Q-curvature, holography and residue families 20

    1.7   Factorization of residue families. Recursive relations 32

    1.8   Families of conformally covariant differential operators 42

    1.9   Curved translation and tractor families 46

    1.10 Holographic duality. Extrinsic Q-curvature.
           Odd order Q-curvature 50

    1.11 Review of the contents 55

    1.12 Some further perspectives 58

2 Spaces, Actions, Representations and Curvature

    2.1   Lie groups, Lie algebras, spaces and actions 63

    2.2   Stereographic projection 67

    2.3   Poisson transformations and spherical principal series 71

    2.4   The Nayatani metric 81

    2.5   Riemannian curvature and conformal change 82

**3 Conformally Covariant Powers of the Laplacian, Q-curvature
and Scattering Theory**

    3.1   GJMS-operators and Q-curvature 87

    3.2   Scattering theory 91

4   Paneitz Operator and Paneitz Curvature

4.1 P_4, Q_4 and their transformation properties   106

4.2   The fundamental identity for the Paneitz curvature   108

4.3 Q_4 and v_4   114

5   Intertwining Families

5.1   The algebraic theory   117

    5.1.1   Even order families $\mathcal{D}_{2N}(\lambda)$ 117

    5.1.2   Odd order families $\mathcal{D}_{2N+1}(\lambda)$ 127

    5.1.3 $\mathcal{D}_N(\lambda)$ as homomorphism of Verma modules   129

5.2   Induced families   131

    5.2.1   Induction   131

    5.2.2   Even order families: $D_{2N}^{nc}(\lambda)$ and $D_{2N}^{c}(\lambda)$ 139

    5.2.3   Odd order families: $D_{2N+1}^{nc}(\lambda)$ and $D_{2N+1}^{c}(\lambda)$ 148

    5.2.4   Eigenfunctions of $\Delta_{\mathbb{H}^n}$ and the families $D_N^{nc}(\lambda)$ 154

5.3   Some low order examples   161

5.4   Families for $(\mathbb{R}^n, S^{n-1})$ 165

    5.4.1   The families $D_N^b(\lambda)$ 165

    5.4.2 $D_1^b(\lambda)$, $D_2^b(\lambda)$ and $D_3^b(\lambda)$ 172

    5.4.3 $D_3^b(0)$ for $n = 4$ and (P_3, T) for $(\mathbb{B}^4, S^3)$ 176

5.5   Automorphic distributions   178

6   Conformally Covariant Families

6.1   Fundamental pairs and critical families   190

6.2   The family $D_1(g; \lambda)$ 194

6.3 $D_2(g; \lambda)$ for a surface in a 3-manifold   195

6.4   Second-order families. General case   201

6.5   Families and the asymptotics of eigenfunctions   208

6.6   Residue families and holographic formulas for Q-curvature   214

6.7 $D_2(g; \lambda)$ as a residue family   235

6.8 $D_3^{\text{res}}(h; \lambda)$ 236

6.9   The holographic coefficients v_2, v_4 and v_6   239

6.10   The holographic formula for Q_6   254

6.11   Factorization identities for residue families.
       Recursive relations   264

6.12   A recursive formula for P_6. Universality   318

6.13   Recursive formulas for Q_8 and P_8   325

6.14   Holographic formula for conformally flat metrics   329

6.15 v_4 as a conformal index density   339

6.16   The holographic formula for Einstein metrics   343

6.17   Semi-holonomic Verma modules and their role   356

6.18   Zuckerman translation and $\mathcal{D}_N(\lambda)$ 360

6.19   From Verma modules to tractors 381

6.20   Some elements of tractor calculus 388

6.21   The tractor families $D_N^T(M, \Sigma; g; \lambda)$ 403

6.22   Some results on tractor families 418

6.23 $\mathcal{J}$ and Fialkow's fundamental forms 445

6.24 $D_2(g; \lambda)$ as a tractor family 450

6.25   The family $D_3^T(M, \Sigma; g; \lambda)$ 455

6.26   The pair (P_3, Q_3) 463

Bibliography 469

Index 485

Preface

A basic problem in geometry is to find canonical metrics on smooth manifolds. Such metrics can be specified, for instance, by curvature conditions or extremality properties, and are expected to contain basic information on the topology of the underlying manifold. Constant curvature metrics on surfaces are such canonical metrics. Their distinguished role is emphasized by classical uniformization theory. A more recent characterization of these metrics describes them as critical points of the determinant functional for the Laplacian. The key tool here is Polyakov's variational formula for the determinant. In higher dimensions, however, it is necessary to further restrict the problem, for instance, to the search for canonical metrics in conformal classes. Here two metrics are considered to belong to the same conformal class if they differ by a nowhere vanishing factor. A typical question in that direction is the Yamabe problem ([165]), which asks for constant scalar curvature metrics in conformal classes.

In connection with the problem of understanding the structure of Polyakov type formulas for the determinants of conformally covariant differential operators in higher dimensions, Branson ([31]) discovered a remarkable curvature quantity which now is called Branson's Q-curvature. It is one of the main objects in this book.

Q-curvature is a scalar local Riemannian curvature invariant on manifolds of even dimension. On surfaces it coincides with Gauß curvature. On four-manifolds it first appeared in connection with the conformally covariant Paneitz operator. In this case, it is a certain linear combination of squared scalar curvature, the squared norm of Ricci curvature and the Laplacian of scalar curvature. On a manifold of dimension n, Q-curvature is an nth-order curvature invariant. One of its remarkable properties is that its behaviour under conformal changes of the metric is governed by an nth-order *linear* conformally covariant differential operator. In dimensions two and four, the respective operators are the Laplacian and the Paneitz operator. In higher dimensions, these operators are replaced by certain conformally covariant powers of the Laplacian (GJMS-operators) ([124]).

Besides their significance in conformal geometry, GJMS-operators also play an important role in physics. This is due to the fact that their definition extends to Lorentzian manifolds. They are common generalizations of the Yamabe operator and the conformally covariant powers of the wave operator on Minkowski space.

In recent years, new connections of Q-curvature with other parts of mathematics and theoretical physics have been discovered. Probably the most remarkable one is the relation to geometric scattering on asymptotically hyperbolic Einstein spaces. This is a relation in the spirit of the conjectural AdS/CFT-duality ([171]), which connects gravitation with gauge field theory, and which led to an outburst of activities in theoretical physics (for reviews see [1], [82]). In geometric analysis, recent efforts are being directed towards an understanding of the geometric significance of Q-curvature in dimension four, for instance, by studying Yamabe type problems and Q-curvature flows (for a review see [170]). Here one of the problems is to characterize the conformal classes which contain a metric with constant Q-curvature. Much less is known in higher dimensions.

Although Q-curvature is an intrinsic Riemannian curvature invariant, all known conceptual definitions in general dimension take one or another extrinsic point of view. The situation somewhat resembles Weyl's formula for the volume of a tube ([130]). This formula shows that the volume of a tube of a closed submanifold of Euclidean space is a polynomial in its radius, and the coefficients depend only on the intrinsic curvature of the submanifold. In particular, the Euler characteristic of the submanifold appears in the leading coefficient.

In the present book, we develop a new extrinsic point of view towards Q-curvature with the emphasis on general structural results. The guiding idea is to associate to a hypersurface $i : \Sigma \hookrightarrow M$ and a general background metric g on M certain one-parameter families of conformally covariant local operators which map functions on M to functions on Σ. Q-curvature and the GJMS-operator of the submanifold $(\Sigma, i^*(g))$ appear in the respective linear and constant coefficients of these families, and the fundamental transformation law of Q-curvature is a direct consequence of the covariance of the family. In particular, we introduce two specific constructions of conformally covariant families with such properties: the residue families and the tractor families.

The setting of residue families is more restricted, however. Here Σ is the boundary of M, and the background metric on M is a canonical extension of a given metric on Σ. Such situations arise in connection with conformally compact Einstein metrics and the Fefferman-Graham construction of an ambient metric. The closely related Poincaré-Einstein metrics associate to any conformal class on Σ a diffeomorphism class of conformally compact Einstein metrics on M with the given class as conformal infinity ([199]). The method of the ambient metric was introduced in [99] as a fundamental systematic construction of conformal invariants. During the last two decades the ambient metric had a major influence on the subject of conformal geometry. For full details see [96].

Poincaré-Einstein metrics are used in theoretical physics in connection with the speculative holographic principle in quantum gravity ([29], [227]). The bulk space/boundary duality between superstring theory on AdS-space and supersymmetric Yang-Mills theory on Minkowski space is regarded as a concrete manifestation of the principle.

In a pure gravity setting with homogeneous metrics, a related bulk/boundary duality is Helgason's well-known theory of Poisson transformations in harmonic analysis on symmetric spaces ([140]). Versions of that transform for conformally compact Einstein spaces play an important role in attempts to establish rigorous statements in the AdS/CFT-duality.

The residue families are defined by a certain residue construction, which has its origin in the spectral theory of Kleinian manifolds. This explains the name. These families can be regarded as local counterparts of the global scattering operator. They naturally lead to an understanding of Q-curvature of a metric on the boundary at infinity as part of a hologram of the associated Poincaré-Einstein metric in one higher dimension. More precisely, the *holographic formulas* describe Q-curvature in terms of holographic coefficients of the Poincaré-Einstein metric and its harmonic functions. Combining that relation between Q-curvature and residue families with structural properties of residue families (factorization relations), uncovers recursive structures among Q-curvatures and GJMS-operators. It is here where the lower order relatives Q_{2N} ($2N < n$) of Branson's Q-curvature Q_n become important. All in all, the residue families are an effective tool for the systematic study of the interplay between the asymptotic geometry of Poincaré-Einstein metrics on bulk space and GJMS-operators (and Q-curvatures) of their conformal infinities.

The theory of the tractor families is an attempt to take a wider perspective. Here the conformal compactifications of Poincaré-Einstein metrics are replaced by arbitrary background metrics, and we extract the intrinsic Q-curvature of the submanifold using an appropriate extrinsic construction near Σ. That perspective leads to the notions of extrinsic and odd order Q-curvatures, which relate the subject of Q-curvature with conformal submanifold theory. The tractor families are defined in terms of the conformally invariant tractor calculus ([17]). A closely related construction was used in [40] in a different connection.

For certain classes of metrics, residue families and tractor families coincide. Such relations imply tractor formulas for GJMS-operators and Q-curvature, and will be termed *holographic duality*.

The new approach to Q-curvature grew out of results which relate the divisors of Selberg zeta functions to automorphic distributions. Such results are related to the dream of an interpretation of the Riemann-Weil explicit formula in analytic number theory as a version of a Lefschetz fixed point formula. The hope is that a cohomological interpretation may also contain a key to the Riemann hypothesis ([78]). In the same spirit, it was shown in [151] that, using Osborne's character formula, the Selberg trace formula can be regarded as a Lefschetz formula for the geodesic flow. This leads to characterizations of the divisors of Selberg zeta functions in terms of cohomology of Anosov foliations and representation theory. The basic principle is that the complex numbers which appear as zeros or poles of a zeta function are characterized by the non-vanishing of the Euler characteristics of associated complexes. Moreover, the values of the corresponding Euler characteristics yield the multiplicities. Equivariant Poisson transformations translate these results into characterizations in terms of group cohomology with values in distributions on the geodesic boundary of rank one symmetric spaces. The latter result can be regarded as a version of holography: the divisor, i.e., zeros and poles with multiplicities, of a zeta function, which is defined by the lengths of closed geodesics of a hyperbolic manifold, is completely characterized in terms of a theory which is formulated on a manifold of one dimension less. The cohomological objects which correspond to the zeros of the Riemann zeta function remain to be found, however.

The fascination of Q-curvature stems from its central role in the complex web of ideas outlined above. In this framework, we observe how classical and modern differential geometry, geometric analysis, harmonic analysis and theoretical physics meet each other.

Although in four dimensions the geometric meaning of Q-curvature has been studied intensively in recent years, there are only few results in higher dimensions. It would be pleasing if the perspectives and the structural insights presented here help to enter this unexplored field. Presently, the future role of Q-curvature is hard to predict, and it seems that we are now taking only the first steps towards its comprehension.

The reader will easily notice that the theory in this book has open ends on different levels. In addition to a number of explicitly formulated conjectures, there are results that are derived under conditions which certainly can be relaxed, and the full consequences of some arguments and constructions are not yet predictable. Moreover, the basic ideas should apply also in different contexts. We hope that this will motivate further investigations.

At first glance, it might seem that the text contains a jungle of complicated formulas. To some extent, this is typical for the subject. On the other hand, we believe that the ambitious reader finally will be delighted by the ways in which complex but beautiful formulas emerge from simple principles, albeit sometimes through non-trivial calculations. The disclosure of some of the hidden structures is one of the aims of this work.

First and foremost, the book is a research monograph presenting a new theory. On the other hand, we have attempted of a self-contained presentation of the material so that it should be accessible for non-specialists. Although we strictly concentrate on the development of new ideas, we necessarily touch upon many of the recent developments in conformal differential geometry. Therefore, the text may also serve as an informal introduction to the subject. We hope that we have succeeded in finding some balance between the presentation of structural ideas and the discussion of full (calculational) details. In particular, we also included proofs of some results which might be considered as well-known by specialists. But since the various fields which are touched upon here do not have a common folklore, proofs are given if required for the sake of a coherent presentation. Also, due to varying conventions, it was sometimes easier to supply proofs than to refer to the literature. The list of references is not representative for any of the numerous fields linked with the subject.

The early phases of the work were financed by a grant of the Swedish Research Council (VR) at Uppsala University. Since 2005 Sonderforschungsbereich 647 "Space-Time-Matter" at Humboldt-University, Berlin supported the research as part of a project initiated by H. Baum. Special thanks go to the participants of my courses and the seminars at Humboldt-University on the subject during the years 2005–2008. Over the years, I benefited a lot from discussions with H. Baum, T. Branson, A. Čap, R. Gover, F. Leitner, T. Leistner, M. Olbrich, B. Ørsted, P. Somberg, V. Souček, and from the stimulating annual conferences in Srni. In a series of lectures in Srni 2005, I had the privilege of presenting part of the results. In later stages of the project, discussions with R. Graham influenced the shape of the theory. Finally, I am grateful to the reviewers for valuable hints.

Berlin, autumn 2008

Chapter 1

Introduction

In the present book we develop a new approach towards Branson's Q-curvature. The central idea is to relate it to certain one-parameter families of conformally covariant differential operators which are associated to hypersurfaces in Riemannian manifolds.

In this chapter we describe the main ideas, review the central results and discuss related perspectives. Full details are given in the later chapters.

In the first two sections, we start with a brief discussion of the origin of the new method in the spectral geometry of Riemannian manifolds of constant negative curvature (Kleinian manifolds). This is an extremely rich subject in its own which connects hyperbolic geometry and hyperbolic dynamics with conformal geometry and conformal dynamics. We use Selberg zeta functions of Kleinian manifolds to illustrate these relations. In particular, we relate zeros of Selberg zeta functions to automorphic distributions. Everything here is, basically, a consequence of the fact that the same Lie group appears as a group of isometries and as a conformal group. The geometric framework of the AdS/CFT-correspondence ([82]) is a Lorentzian analog of this situation. In that case, too, there is one Lie group which appears both as an isometry group of a space and as a conformal group of a suitable completion of it. This is the source of the links of the present theory to theoretical physics.

Some observations on Selberg zeta functions and automorphic distributions (Section 1.2) naturally suggest the existence of one-parameter families of differential intertwining operators for principal series representations of the conformal group. These families are the flat model cases of analogous families in conformal geometry which will dominate the investigation of Q-curvature.

There are two main constructions of such families which will be called the residue families (Section 1.6) and the tractor families (Section 1.9). Both constructions generalize different aspects of the flat models, and coincide for conformally flat metrics (Section 1.10).

The main results of the book do not depend on Selberg zeta functions. In fact, their role can be ignored by those readers who are primarily interested in the emerging perspective towards the structure of Q-curvature. On the other hand, the suggestive power of the zeta functions has not been exhausted, and similar arguments in other contexts are expected to have interesting consequences.

1.1   Hyperbolic geometry and conformal dynamics

Let $\mathbb{B}^{n+1} = \{x \in \mathbb{R}^{n+1} \,|\, |x| < 1\}$ be the unit ball in $\mathbb{R}^{n+1}$ with the Riemannian metric

$$g = \frac{4}{(1-|x|^2)^2} \sum_{i=1}^{n+1} dx_i^2.$$

g has constant sectional curvature -1. $(\mathbb{B}^{n+1}, g)$ is the ball model of hyperbolic geometry. The connected component $G^{n+1} = SO(1, n+1)^\circ$ of the identity in $SO(1, n+1)$ acts isometrically on $\mathbb{B}^{n+1}$, and $\mathbb{B}^{n+1}$ can be identified with the homogeneous space G^{n+1}/K^{n+1}, where $K^{n+1} = SO(n+1)$ is the maximal compact subgroup which fixes the origin. For background material on hyperbolic geometry see [206].

The non-compact Lie group G^{n+1} acts transitively also on the geodesic boundary S^n of $\mathbb{B}^{n+1}$, and S^n can be identified with a homogeneous space G/P, where P is the isotropy group of a point. The action of G^{n+1} on S^n has an additional metric aspect: G^{n+1} acts by conformal diffeomorphisms. Here a diffeomorphism κ of S^n is called *conformal* with respect to the round metric g_c, if

$$\kappa_*(g_c) \quad \text{and} \quad g_c$$

are conformally equivalent, i.e., differ by a nowhere-vanishing function

$$\kappa_*(g_c) = e^{2\Phi(\kappa)} g_c, \quad \Phi(\kappa) \in C^\infty(S^n).$$

The action of G^{n+1} on S^n gives rise to the spherical principal series representations

$$\pi_\lambda^c : g \mapsto \left(\frac{g_*(\mathrm{vol}(g_c))}{\mathrm{vol}(g_c)} \right)^{-\frac{\lambda}{n}} \circ g_*, \quad \lambda \in \mathbb{C}$$

on $C^\infty(S^n)$. Here g_* denotes push-forward and $\mathrm{vol}(g_c)$ is the Riemannian volume form of g_c. Alternatively, π_λ^c can be identified with a representation which is induced by a character of the minimal parabolic subgroup P. For the role and the structure of principal series representations in harmonic analysis we refer to [156].

The fact that the same Lie group is an isometry group and a conformal group has numerous consequences. One of these is that the intertwining operator for spherical principal series π_λ^c can be interpreted as a scattering operator of hyperbolic space. More precisely, for $\Re(\lambda) < -\frac{n}{2}$ the integral operator

$$I_\lambda : \omega \mapsto \int_{S^n} \frac{1}{|x-y|^{2\lambda+2n}} \omega(y) \, \mathrm{vol}(g_c)$$

is well defined and satisfies the intertwining relation

$$I_\lambda \circ \pi_\lambda^c(g) = \pi_{-\lambda-n}^c(g) \circ I_\lambda \qquad (1.1.1)$$

for all $g \in G^{n+1}$. I_λ admits a meromorphic continuation to $\mathbb{C}$ with simple poles in $-\frac{n}{2} + \mathbb{N}_0$. Its residue

$$R_{2N} = \mathrm{Res}_{\lambda=-\frac{n}{2}+N}(I_\lambda), \ N \in \mathbb{N}_0$$

is a differential operator which satisfies

$$R_{2N} \circ \pi_{-\frac{n}{2}+N}^c(g) = \pi_{-\frac{n}{2}-N}^c(g) \circ R_{2N}, \ g \in G^{n+1}.$$

More explicitly, R_{2N} is a constant multiple of the operator

$$\prod_{j=\frac{n}{2}}^{\frac{n}{2}+N-1} (\Delta_{g_c} - j(n-1-j)). \qquad (1.1.2)$$

On the other hand, an eigenfunction $u \in C^\infty(\mathbb{B}^{n+1})$ of the Laplace-Beltrami operator of the hyperbolic metric g, i.e., a solution of

$$-\Delta_{\mathbb{B}^{n+1}} u = \lambda(n-\lambda)u, \ \Re(\lambda) = \frac{n}{2}, \ \lambda \neq \frac{n}{2},$$

has an asymptotics of the form

$$u(r,b) \sim (1-r^2)^\lambda \mathcal{S}(\lambda)(\omega)(b) + (1-r^2)^{n-\lambda}\omega(b) + \cdots, \ b \in S^n$$

for $r \to 1$. u is determined by its boundary value ω. The quality of ω (smooth function, distribution, hyperfunction) reflects the growth of u for $r \to 1$ ([140], [152], [233]). $\mathcal{S}(\lambda)$ is the scattering operator of the space $\mathbb{B}^{n+1}$. It is equal to the product $(\Gamma(\lambda)/\Gamma(\lambda - \frac{n}{2}))I_{\lambda-n}$, up to a non-vanishing factor.

Next, we discuss some aspects of spectral theory on Kleinian manifolds. This illustrates the rich relations between analysis on hyperbolic spaces and conformal invariants on spheres, and serves as a preparation of the considerations in Section 1.2. Let $\Gamma \subset G^{n+1}$ be a discrete subgroup without torsion, i.e., without non-trivial elements which are conjugate to elements of the maximal compact subgroup $SO(n+1)$. Γ acts properly discontinuously on $\mathbb{B}^{n+1}$, and the quotient $X^{n+1} = \Gamma \backslash \mathbb{B}^{n+1}$ is a smooth hyperbolic manifold (Kleinian manifold). For $n = 2$, Γ can be viewed as a discrete subgroup of $PSL(2, \mathbb{C})$, i.e., Γ is a Kleinian group in the classical sense.

The dynamics of discrete groups acting on the sphere by conformal transformations is very complex, in general. S^n admits a Γ-invariant decomposition

$$S^n = \Omega(\Gamma) \cup \Lambda(\Gamma)$$

into the closed limit set $\Lambda(\Gamma)$ and its complement $\Omega(\Gamma)$. Γ operates properly dis-continuously on the proper set $\Omega(\Gamma)$. The limit set $\Lambda(\Gamma)$ is a fractal-like set. Γ is called cocompact if X is compact. In that case, $\Omega(\Gamma) = \emptyset$ and $\Lambda(\Gamma) = S^n$. Γ is called convex-cocompact if it acts with a compact quotient on the convex hull of $\Lambda(\Gamma)$. Then $\Gamma\backslash\Omega(\Gamma)$ is compact, and the hyperbolic manifold $X = \Gamma\backslash\mathbb{B}^{n+1}$ can be compactified by adding $\Gamma\backslash\Omega(\Gamma)$. For an account of the complexity and astonishing beauty of limit sets of Schottky groups acting on S^2 we refer to [184].

Associated to any convex-cocompact discrete group Γ there are two closely related objects: the Selberg zeta function $Z_\Gamma(\lambda)$ and the scattering operator $S_\Gamma(\lambda)$. The zeta function Z_Γ is defined by an Euler product over the prime closed oriented geodesics in X, i.e., by the prime periodic orbits of the geodesic flow Φ_t of X on the unit tangent bundle SX. Let

$$Z_\Gamma(\lambda) = \prod_{\text{p.p.o.}\, c} \prod_{N \geq 0} \det\left(\text{id} - S^N(P_c^-)e^{-\lambda|c|}\right), \ \Re(\lambda) > \delta(\Gamma). \qquad (1.1.3)$$

In (1.1.3), the first product runs over the prime periodic orbits c of Φ_t. For any periodic orbit c, P_c^- denotes the corresponding linear Poincaré map on the stable tangent bundle $T^-(SX) \subset T(SX)$. $S^N(\cdot)$ denotes the induced linear map on symmetric tensors. $\delta(\Gamma)$ is the critical exponent of Γ (topological entropy of Φ_t). We refer to [153] for the notions of dynamics used here. In the special case of a cocompact $\Gamma \subset PSL(2, \mathbb{R})$ $(n = 2)$, the above definition reduces to

$$Z_\Gamma(\lambda) = \prod_{\text{p.p.o.}\, c} \prod_{N \geq 0} (1 - e^{-(\lambda+N)|c|}), \ \Re(\lambda) > 1 = \delta(\Gamma).$$

Selberg ([213]) introduced and studied this function for cofinite Γ. For the details we refer to [139].

In (1.1.3), the zeta function is regarded as an invariant of the geodesic flow on SX. Alternatively, Z_Γ can be viewed as being associated to the action of Γ on S^n. In fact, prime closed orbits of Φ_t correspond to attracting fixed points, and the eigenvalues of P_c^- and the quantity $e^{-|c|}$ can be read off from the linearization of the action of Γ at its fixed points.

Z_Γ is a relative of Ruelle's zeta function ([104], [209], [151])

$$Z_\Gamma^R(\lambda) = \prod_{\text{p.p.o.}\, c} (1 - e^{-\lambda|c|})^{-1}, \ \Re(\lambda) > \delta(\Gamma),$$

which, in turn, should be viewed as an analog of Riemann's zeta function ([230], [197])

$$\zeta(\lambda) = \prod_p (1 - p^{-\lambda})^{-1}, \ \Re(\lambda) > 1.$$

For a cocompact Γ, we have $\delta(\Gamma) = n$, and Z_Γ admits a meromorphic contin-uation to $\mathbb{C}$. All poles are real (if there are any). Z_Γ has infinitely many zeros in the critical strip $0 < \Re(\lambda) < n$. All non-real zeros are on the line $\Re(\lambda) = \frac{n}{2}$. These zeros have an interpretation in terms of hyperbolic geometry as well as in terms of conformal dynamics. In fact, $\lambda_0 \notin \mathbb{R}$ is a zero of Z_Γ iff $\lambda_0(n-\lambda_0)$ is an eigenvalue of the Laplacian $-\Delta_X$. Since X is compact, the spectrum of $-\Delta_X$ consists only of eigenvalues of finite multiplicity. On the other hand, $\lambda_0 \notin \mathbb{R}$ is a zero of Z_Γ iff the space

$$\{\omega \in C^{-\infty}(S^n) \mid \pi^c_{\lambda_0-n}(\gamma)\omega = \omega, \ \gamma \in \Gamma\}$$

of Γ-automorphic distributions is non-trivial. Here π^c_λ denotes the spherical principal series representation on distributions. These characterizations also concern multiplicities. The equivalence of both descriptions of non-real zeros follows from Helgason's theory of Poisson transformations. The notion of automorphic distributions (for cofinite discrete groups) was coined in [210]. For beautiful applications to the functional equations of L-functions we refer to [182].

For a convex-cocompact Γ, the volume of X is infinite (if Γ is not cocompact). $-\Delta_X$ has a continuous spectrum $[n^2/4, \infty)$ of infinite multiplicity and a finite discrete spectrum in $[0, n^2/4]$. The scattering operator $\mathcal{S}_\Gamma(\lambda)$ arises by averaging $I_{\lambda-n}$ over Γ. It describes the asymptotics of generalized eigenfunctions (Eisenstein series). $\mathcal{S}_\Gamma(\lambda)$ is a meromorphic family of pseudo-differential operators which acts on functional spaces on the compact manifold $\Gamma\backslash\Omega(\Gamma)$. The poles of $I_{\lambda-n}$ in $\frac{n}{2}+\mathbb{N}_0$ lead to poles of $\mathcal{S}_\Gamma$ in $\frac{n}{2}+\mathbb{N}$.

Z_Γ admits a meromorphic continuation to $\mathbb{C}$. The zeros of Z_Γ can be characterized in two different manners. In order to simplify the discussion, we restrict here to non-real zeros. On the one hand, such zeros of Z_Γ are related to non-real poles of the scattering operator $\mathcal{S}_\Gamma$. Such a characterization is natural if one regards Z_Γ as an object in the spectral geometry of the Kleinian manifold $\Gamma\backslash\mathbb{B}^{n+1}$. The following description, however, is more natural from the point of view of the dynamics of the action of Γ on the sphere S^n: zeros of Z_Γ correspond to non-trivial spaces

$$\{\omega \in C^{-\infty}(\Lambda(\Gamma)) \mid \pi^c_{\lambda-n}(\gamma)\omega = \omega, \ \gamma \in \Gamma\} \tag{1.1.4}$$

of Γ-automorphic distributions on S^n with support in the limit set $\Lambda(\Gamma)$. In fact, the limit set $\Lambda(\Gamma)$ is the closure of the set of all (attracting) fixed points, and the local factors of the zeta function are defined by quantities which describe the behaviour of the action at these fixed points. The celebrated Patterson-Sullivan measure μ_{PS} is a canonical Γ-automorphic distribution with support in $\Lambda(\Gamma)$. It satisfies the conformal invariance

$$\pi^c_{\delta(\Gamma)-n}(\gamma)\mu_{PS} = \mu_{PS}, \ \gamma \in \Gamma,$$

and is responsible for the simple zero of Z_Γ at $\delta(\Gamma)$.

For a convex-cocompact Kleinian group Γ, the set of all $\lambda \in \mathbb{C}$ for which the space (1.1.4) is non-trivial, is discrete, and defines a notion of a spectrum of such a Kleinian group Γ, viewed as an object of dynamics on the sphere.

For background information and details we refer to [50], [51], [26], [28], [148], [151], [163], [188], [196], [195], [201], [206], [224], [225], [226] and the references therein.

The relation between divisors of zeta functions and automorphic distributions will be important in Section 1.2.

Note that the attempts to extend the relations of hyperbolic geometry of Kleinian manifolds to conformally invariant structures on the sphere, to the setting of rational iteration on the Riemann sphere, are known under the name of Sullivan's dictionary ([183], Chapter 5). In particular, the analogs of the Kleinian limit sets are the Julia sets of rational maps, Patterson-Sullivan measures (see [196]) are replaced by conformal measures (see [232] and the references therein), and hyperbolic laminations serve ([169]) as analogs of the hyperbolic 3-space.

In the present book, the spectral geometry of Kleinian manifolds is used as a source of inspiration in another direction. We forget the dynamics but retain the geometry of Kleinian manifolds at infinity, at least asymptotically. Convex-cocompact Kleinian groups Γ give rise to Kleinian manifolds which can be seen as asymptotically hyperbolic Einstein manifolds with natural conformal compactifications (in the sense of [199]). The fundamental role of such spaces for conformal geometry was discovered in [99] (see also [96]).

More recently, the constructions in [99] proved to be important also in connection with the so-called AdS/CFT-correspondence [171]. The original geometric setting of the AdS/CFT-correspondence was the conformal completion of the Lorentzian analog AdS_{n+1} of the real hyperbolic space $\mathbb{B}^{n+1}$. The universal covering of the anti de Sitter space AdS_{n+1} is a simply connected complete Lorentzian manifold of curvature -1 which has Einstein's static universe

$$Ein_n = (\mathbb{R} \times S^{n-1}, -dr^2 + g_{S^{n-1}})$$

as a conformal boundary. $O(2, n)$ operates by isometries on AdS_{n+1} and by conformal diffeomorphisms on the quotient $S^1 \times S^{n-1}$ with the metric $-g_{S^1} + g_{S^{n-1}}$. Geometric questions concerning the dynamics of discrete groups $\Gamma \subset O(2, n)$ and the conformal completions of other anti de Sitter space times were the subject of recent studies (see [102], [103]).

1.2 Automorphic distributions and intertwining families

Let H^+ be the upper hemisphere of $S^n \subset \mathbb{R}^{n+1}$ (with its standard metric g_c of curvature 1) with boundary S^{n-1}. In the present section, we use the relation between automorphic distributions and zeros of Selberg zeta functions to suggest the existence of certain polynomial families of differential intertwining operators

$$C^\infty(S^n) \to C^\infty(S^{n-1}).$$

Let $\Gamma \subset G^n$ be a discrete torsion-free cocompact subgroup acting on the hyperbolic space $\mathbb{B}^n = \mathbb{H}^n$. The action of G^n on $\mathbb{H}^n$ extends to an action on the compactification $\overline{\mathbb{H}^n} = \mathbb{H}^n \cup S^{n-1}$. Here $S^{n-1} = \partial_\infty(\mathbb{H}^n)$ is the geodesic boundary of $\mathbb{H}^n$. G^n acts by conformal transformations on (S^{n-1}, g_c) (see the discussion in 1.1). By cocompactness of Γ, the limit set $\Lambda(\Gamma)$ of its action on $\overline{\mathbb{H}^n}$ coincides with S^{n-1}.

Now consider S^{n-1} and $\mathbb{H}^n$ as the respective submanifolds of S^n and $\mathbb{H}^{n+1}$ defined by the intersection with the plane $x_{n+1} = 0$. Then G^n is the subgroup of G^{n+1} which preserves the subsphere S^{n-1}. If Γ is regarded as a subgroup of G^{n+1}, it operates with an infinite volume quotient space X^{n+1} on $\mathbb{H}^{n+1}$. Γ is then convex-cocompact. The limit set of its action on $\overline{\mathbb{H}^{n+1}}$ coincides with the limit set of its action on $\overline{\mathbb{H}^n}$. The sphere S^n decomposes as

$$S^n = \Lambda(\Gamma) \cup \Omega(\Gamma) = S^{n-1} \cup \Omega^+(\Gamma) \cup \Omega^-(\Gamma),$$

where Γ operates properly discontinuously on the proper set $\Omega(\Gamma)$. The quotient manifold $X^{n+1} = \Gamma \backslash \mathbb{H}^{n+1}$ is a hyperbolic cylinder with cross section $X^n = \Gamma \backslash \mathbb{H}^n$.

We consider the geodesic flows of X^n and X^{n+1} on the corresponding sphere bundles. The geodesic flow of a Riemannian manifold X is Hamiltonian on the cotangent bundle T^*X and hyperbolic (or Anosov) on the sphere bundle SX of X if the curvature is negative. All periodic orbits of the geodesic flow of X^{n+1} are periodic orbits of the flow of X^n. In both cases, the periodic orbits give rise to a Selberg zeta function

$$Z^m(\lambda) = \prod_{\text{p.p.o. } c} \prod_{N \geq 0} \det(\mathrm{id} - S^N((P_c^{m-1})^-)e^{-\lambda|c|}), \quad \Re(\lambda) > \delta(\Gamma)$$

($m = n+1$ and $m = n$). Although both zeta functions Z^{n+1} and Z^n are defined by the same set of prime periodic orbits, they do *not* coincide since the local factors of the Euler products contain the differing contracting parts $(P_c^{n-1})^-$ and $(P_c^n)^-$ of the corresponding linear Poincaré maps P_c^{n-1} and P_c^n of periodic orbits on the sphere bundles. Here $n-1$ and n indicate the dimensions of the corresponding stable tangent bundles. However, the identity

$$\left(\frac{Z^m}{Z^m}\right)'(\lambda) = \sum_{\text{p.o. } c} |c_0| \frac{1}{\det(\mathrm{id} - (P_c^{m-1})^-)} e^{-\lambda|c|}, \quad \Re(\lambda) > \delta(\Gamma) \qquad (1.2.1)$$

(here the sum runs over all periodic orbits, and c_0 denotes the prime periodic orbit of c) and the relation

$$(P_c^n)^- = (P_c^{n-1})^- \oplus e^{-|c|} \mathrm{id}$$

imply the factorization

$$Z^{n+1}(\lambda) = \prod_{N \geq 0} Z^n(\lambda + N). \qquad (1.2.2)$$

Now Z^n admits a meromorphic continuation to $\mathbb{C}$. A streamlined proof of that fundamental fact can be built on a version of an Atiyah-Bott-Lefschetz fixed point formula for the geodesic flow of X^n ([151]) by interpreting the right-hand side of (1.2.1) as an analog of the right-hand side of the Atiyah-Bott-Lefschetz formula ([13], [14]) for a diffeomorphism of a closed manifold. In that argument, Osborne's fundamental character formula ([138]) plays a central role.

Poles of Z^m may appear only on the real line (at non-positive integers). The zeros of Z^n off the real line are on the line $\Re(\lambda) = \frac{n-1}{2}$ (Riemann hypothesis). They are of spectral nature: $\frac{n-1}{2} + is$ is a zero of Z^n iff $s^2 + (\frac{n-1}{2})^2$ is an eigenvalue of the non-negative Laplacian $-\Delta_{X^n}$. (1.2.2) implies that Z^{n+1} also admits a meromorphic continuation to $\mathbb{C}$. That is actually only a very special case of the same result for a general convex-cocompact Γ ([195]). More precisely, any non-real zero $\frac{n-1}{2} + is$ of Z^n generates a ladder

$$\frac{n-1}{2} + is - \mathbb{N}_0$$

of zeros for Z^{n+1}. These zeros can be interpreted as poles of the scattering operator (scattering resonances) of the infinite-volume space X^{n+1}. From the dynamical point of view ([151]), it is more natural to regard the occurrence of a (non-real) zero λ of Z^{n+1} as a consequence of the existence of a non-trivial space

$$C^{-\infty}(\Lambda(\Gamma))^{\Gamma}_{\lambda-n} = \left\{\omega \in C^{-\infty}(S^n) \mid \operatorname{supp}(\omega) \subset \Lambda(\Gamma),\ \pi^c_{\lambda-n}(\gamma)\omega = \omega,\ \gamma \in \Gamma\right\}$$

of Γ-automorphic distributions on S^n which are supported on the limit set of Γ (as opposed to a space of (generalized) eigenfunctions of the Laplacian on X^{n+1}). An analogous complete characterization of the divisor of Z^{n+1} (i.e., including a description of the divisor on the real line) requires us to consider Γ-cohomology with values in spaces of distributions ([50], [151]). The above correspondence between non-real zeros of Z^{n+1} and automorphic distributions on the limit set $\Lambda(\Gamma) = i(S^{n-1})$, of course, generalizes a correspondence between non-real zeros of Z^n and Γ-automorphic distributions on S^{n-1}.

Now the ladder structure of the set of non-real zeros of Z^{n+1} suggests that the Γ-automorphic distributions on $\Lambda(\Gamma) = i(S^{n-1})$ arise as the result of the application of a series of universal, i.e., Γ-independent, G^n-equivariant families of local operators

$$C^{-\infty}(S^{n-1})_\lambda \to C^{-\infty}(\Lambda(\Gamma))_{\lambda-N-1},\ N \geq 0$$

to the Γ-automorphic distributions on S^{n-1}. Here the subscripts refer to the respective module structures defined by the spherical principal series representations π^c_λ on S^n and S^{n-1}. Moreover, by taking adjoints, the existence of G^n-equivariant operator families

$$D^c_N(\lambda) : C^\infty(S^n)_\lambda \to C^\infty(S^{n-1})_{\lambda-N}$$

which are compositions of differential operators on S^n and S^{n+1} (and the restriction i^*), is suggested.

The structure of the intertwining families $D_N^c(\lambda)$ will be clarified in Section 1.4. It will be an important observation that these families can be described in terms of the intertwining operators $P_{2N}(g_c)$ (see (1.1.2)) on both spheres S^n and S^{n-1}. In order to put this into the appropriate perspective, and as a preparation for the later sections, we continue in Section 1.3 with the introduction of the conformally covariant powers of the Laplacian (GJMS-operators) in full generality.

1.3 Asymptotically hyperbolic Einstein metrics. Conformally covariant powers of the Laplacian

For a convex-cocompact Γ, the hyperbolic metric on $X^{n+1} = \Gamma\backslash\mathbb{B}^{n+1}$ has a second-order pole at infinity in the following sense. The complement of a compact convex core X_0 of X is isometric to $(0, \infty) \times \Sigma$ ($\Sigma \subset X$ a suitable hypersurface) with a metric of the form

$$g = dt^2 + e^{-2t}h_- + h_0 + e^{2t}h_+.$$

Here h_+ is a metric on Σ. It can be viewed as a metric on $\Gamma\backslash\Omega(\Gamma)$. Substituting $r = e^{-t}$, we find that $r^2 g$ extends smoothly to $r = 0$, and pulls back to a conformally flat metric on $\Gamma\backslash\Omega(\Gamma)$. In other words, the metric g admits a conformal compactification.

The Fefferman-Graham construction of the Poincaré-Einstein metric ([99], [96]) can be regarded as kind of an inverse. It associates to a given Riemannian manifold (Σ, h) a metric $g = r^{-2}(dr^2 + h_r)$ with a second-order pole on $(0, \varepsilon) \times \Sigma$ so that h arises by conformal compactification of g: $i^*(r^2 g) = h$. Moreover, conformally equivalent metrics on Σ lead to diffeomorphic Poincaré-Einstein (and ambient) metrics. Note that here we pass from the consideration of conformal diffeomorphisms (as in the previous sections) to the consideration of conformal classes $[h]$ of metrics.

The Poincaré-Einstein metric and the equivalent Lorentzian Ricci flat ambient metric are basic tools for systematic constructions of conformally invariant objects. One such construction is that of the conformally covariant powers of the Laplacian in [124]. For further results see [121].

In the case $\Sigma = S^n$ with the conformal class $[h_c]$ of the round metric, the ambient metric is the usual Lorentzian metric

$$-dx_0^2 + dx_1^2 + \cdots + dx_{n+1}^2$$

on $\mathbb{R}^{1,n+1}$ which pulls back to h_c on

$$S^n = \left\{ x \in \mathbb{R}^{1,n+1} \mid -x_0^2 + \cdots + x_{n+1}^2 = 0, \ x_0 = 1 \right\}.$$

The Lorentzian metric induces a metric on the hypersurface $-x_0^2 + x_1^2 + \cdots + x_{n+1}^2 = -1$, and the latter space is isometric to the ball model of hyperbolic geometry.

In the general case, the Lorentzian metric is replaced by the ambient metric, and the analog of the hyperbolic metric is the Poincaré-Einstein metric

$$g_E = r^{-2}(dr^2 + h_r)$$

on the space $(0, \varepsilon) \times \Sigma$. It is a solution of the Einstein equation $\mathrm{Ric}(g) + ng = 0$ for small r. Here h_r denotes a path of metrics on Σ so that h_0 is the given metric h. g_E is asymptotically hyperbolic in the sense that its sectional curvatures tend to -1 for $r \to 0$.

The basic facts on the spectrum of the Laplacian on general asymptotically hyperbolic manifolds were established in the pioneering works [179], [180] (see [181] for an overview of geometric scattering theory).

For Poincaré-Einstein metrics, geometric scattering theory yields a scattering operator $\mathcal{S}(h; \lambda)$ on $C^\infty(\Sigma)$ which displays the transformation formula

$$\mathcal{S}(e^{2\varphi}h; \lambda) = e^{-\lambda\varphi} \circ \mathcal{S}(h; \lambda) \circ e^{(n-\lambda)\varphi} \tag{1.3.1}$$

under conformal changes $h \mapsto e^{2\varphi}h$ of the metric. $\mathcal{S}(h; \lambda)$ generalizes the scattering operator $\mathcal{S}_\Gamma(\lambda)$ of convex-cocompact Kleinian groups (Section 1.1). Similarly as $\mathcal{S}_\Gamma$, for even n, the scattering operator $\mathcal{S}(h; \lambda)$ has simple poles in the set $\{\frac{n}{2}, \frac{n}{2}+1, \ldots, n\}$ (under some mild assumption), and its residues are constant multiples of conformally covariant powers $P_{2N}(h)$ of the Laplacian on Σ. (1.3.1) implies that

$$P_{2N}(e^{2\varphi}h) = e^{-(\frac{n}{2}+N)\varphi} \circ P_{2N}(h) \circ e^{(\frac{n}{2}-N)\varphi}. \tag{1.3.2}$$

(1.3.1) is a far-reaching generalization of the intertwining relation (1.1.1). In particular, the operator $P_{2N}(h)$ generalizes (1.1.2), and is a modification of the power Δ_g^N of the Laplacian by lower order terms, i.e., $P_{2N}(h)$ is of the form $\Delta_h^N + LOT$.

Here we use the following terminology. A *natural* differential operator on functions on a smooth manifold M is a rule $g \mapsto D(M; g)$ that associates to any Riemannian metric g on M a linear differential operator on $C^\infty(M)$ which is given locally by a universal polynomial formula in terms of the metric g, its inverse, the Levi-Civita connection and the curvature tensor using tensor products and contraction. A natural differential operator $D(M; g)$ on $C^\infty(M)$ is called *conformally covariant* of weight (a, b) if

$$D(M; e^{2\varphi}g) = e^{-b\varphi} \circ D(M; g) \circ e^{a\varphi}, \tag{1.3.3}$$

for all $\varphi \in C^\infty(M)$ and certain conformal weights a, b. In that sense, the GJMS-operator P_{2N} is conformally covariant of weight $(\frac{n}{2} - N, \frac{n}{2} + N)$.

The original construction ([124]) of the operators P_{2N} rests on the Fefferman-Graham ambient metric. In the following, these operators will be called GJMS-operators. The above relation to scattering theory was established in [128]. For more details we refer to Chapter 3.

In a sense, the most important element in the sequence of GJMS-operators is the *critical* GJMS-operator P_n on M^n. Its order coincides with the dimension of the underlying manifold, and its significance comes from the fact that it appears in the transformation formula for the critical Q-curvature (Section 1.6).

The first two GJMS-operators P_2 and P_4 are well known. P_2 is the *conformal Laplacian* or *Yamabe operator*

$$Y = P_2 = \Delta - \frac{n-2}{4(n-1)}\tau \qquad (1.3.4)$$

and P_4 is also known as the *Paneitz operator*

$$P_4 = \Delta^2 + \delta\left((n-2)\mathsf{J}g - 4\mathsf{P}\right)\#d + \frac{n-4}{2}\left(\frac{n}{2}\mathsf{J}^2 - 2|\mathsf{P}|^2 - \Delta\mathsf{J}\right). \qquad (1.3.5)$$

For the notation we refer to Section 2.5.

For even n, and general metrics, the operators P_{2N} exist for $2N \leq n$. However, the construction is obstructed at $2N = n$. This reflects the fact that the very *existence* of conformally covariant operators of the form $\Delta^N + LOT$ is obstructed at $2N = n$. In fact, Graham ([118]) showed that there exists no natural conformally covariant modification of Δ_g^3 on manifolds of dimension 4. Later Gover and Hirachi ([112]) completed the picture by showing an analogous non-existence theorem for $2N > n$. However, for specific conformal classes (as for locally conformally flat metrics) GJMS-operators may exist for all orders ([96], [120]). On the other hand, there are no such obstructions for odd n.

1.4 Intertwining families

Chapter 5 will be devoted to a detailed study of polynomial families of differential intertwining operators $C^\infty(S^n) \to C^\infty(S^{n-1})$. A profound understanding of their structure will be the basis of the later discussion of Q-curvature. The structure of the intertwining families $D_N^c(\lambda) : C^\infty(S^n) \to C^\infty(S^{n-1})$ is influenced by an underlying algebraic structure and by the curvature of the background metrics. In order to eliminate the influence of the curvature, we first study the analogous intertwining families in the flat case $r^{n-1} \hookrightarrow \mathbb{R}^n$. We shall describe such families in terms of homomorphisms of generalized *Verma modules*. This will be the key to the construction of curved analogs by applying a version of the curved translation principle. The latter method will yield the so-called *tractor families* (see Section 1.9). On the other hand, an interpretation of the families in the cases $(\mathbb{R}^n, \mathbb{R}^{n-1})$ and (S^n, S^{n-1}) in terms of the asymptotic analysis of eigenfunctions of Laplacians of associated conformally compact metrics will lead to the *residue families* (see Section 1.5).

The following result describes the intertwining families $D_N^c(\lambda)$ in terms of GJMS-operators on S^n and S^{n-1} (for the round metric g_c). On the round sphere

(S^n, g_c), the GJMS-operators P_{2N} are given by the explicit formulas

$$P_{2N}(g_c) = \prod_{j=\frac{n}{2}}^{\frac{n}{2}+N-1} (\Delta_{g_c} - j(n-1-j)) \tag{1.4.1}$$

for all $N \geq 1$ (if n is even) ([23], [31], [32], [110], [120]).

In Section 1.1, we have seen that these operators are intertwining operators for spherical principal series representations of G^{n+1}. We briefly recall the relation between the conformal covariance and the intertwining property. For $g \in G^{n+1}$, we have $g_*(g_c) = \Phi_g^2 g_c$ for some non-vanishing $\Phi_g \in C^\infty(S^n)$. The conformal covariance of P_{2N} states, in particular, that

$$(\Phi_g)^{\frac{n}{2}+N} \circ P_{2N}(g_*(g_c)) \circ (\Phi_g)^{-\frac{n}{2}+N} = P_{2N}(g_c),$$

i.e.,

$$(\Phi_g)^{-\frac{n}{2}-N} \circ P_{2N}(g_c) \circ (\Phi_g)^{\frac{n}{2}-N} = g_* \circ P_{2N}(g_c) \circ g^*,$$

i.e.,

$$P_{2N}(g_c) \circ ((\Phi_g)^{\frac{n}{2}-N} g_*) = ((\Phi_g)^{\frac{n}{2}+N} g_*) \circ P_{2N}(g_c).$$

The latter identity can be written in the form

$$P_{2N}(g_c) \circ \pi^c_{-\frac{n}{2}+N}(g) = \pi^c_{-\frac{n}{2}-N}(g) \circ P_{2N}(g_c), \tag{1.4.2}$$

where

$$\pi^c_\lambda(g) = \left(\frac{g_*(\mathrm{vol}(g_c))}{\mathrm{vol}(g_c)}\right)^{-\frac{\lambda}{n}} \circ g_* = \Phi_g^{-\lambda} \circ g_*.$$

The main result on the intertwining families $D^c_N(\lambda)$ is

Theorem 1.4.1.

(i) *For any $N \in \mathbb{N}_0$, there exists a polynomial family*

$$D^c_{2N}(\lambda) = (-1)^N \lambda^{2N} i^* + A_{2N-1}\lambda^{2N-1} + \cdots + A_1\lambda + A_0 \tag{1.4.3}$$

of differential operators $C^\infty(S^n) \to C^\infty(S^{n-1})$ of order $2N$ which satisfies the intertwining relations

$$\pi^c_{\lambda-2N}(g) \circ D^c_{2N}(\lambda) = D^c_{2N}(\lambda) \circ \pi^c_\lambda(g), \ g \in G^n, \ \lambda \in \mathbb{C}$$

for the spherical principal series on S^n and S^{n-1}. The operator coefficients in (1.4.3) are compositions of the GJMS-operators $P_{2j}(S^n)$ and $P_{2j}(S^{n-1})$ for $j = 1, \ldots, N$, and the restriction i^*. For

$$\lambda \notin \left\{-\frac{n-1}{2}+N, \ldots, -\frac{n-1}{2}+2N-1\right\},$$

the normal order of $D_{2N}^c(\lambda)$ is $2N$. If

$$\lambda = -\frac{n-1}{2}+N+k-1, \; k = 1,\ldots,N,$$

then the normal order of $D_{2N}^c(\lambda)$ is $2k-2$. In particular, $D_{2N}^c(-\frac{n-1}{2}+N)$ has normal order 0, i.e., is tangential to S^{n-1}.

(ii) *For any $N \in \mathbb{N}_0$, there exists a polynomial family*

$$D_{2N+1}^c(\lambda) = (-1)^N\lambda^{2N+1}D_1^c + B_{2N}\lambda^{2N} + \cdots + B_1\lambda + B_0 \qquad (1.4.4)$$

of differential operators $C^\infty(S^n) \to C^\infty(S^{n-1})$ of order $2N+1$ which satisfy the intertwining relations

$$\pi_{\lambda-2N-1}^c(g) \circ D_{2N+1}^c(\lambda) = D_{2N+1}^c(\lambda) \circ \pi_\lambda^c(g), \; g \in G^n.$$

Here $D_1^c(\lambda) = D_1(S^n, S^{n-1}; g_c; \lambda) = i^*\nabla_{N(g_c)}$. The operator coefficients in (1.4.4) are compositions of the GJMS-operators $P_{2j}(S^n)$, $P_{2j}(S^{n-1})$ for $j = 1,\ldots,N$ and D_1^c. For

$$\lambda \notin \left\{-\frac{n-1}{2}+N+1,\ldots,-\frac{n-1}{2}+2N\right\},$$

the normal order of $D_{2N+1}^c(\lambda)$ is $2N+1$. If

$$\lambda = -\frac{n-1}{2}+N+k, \; k = 1,\ldots,N,$$

then the normal order of $D_{2N+1}^c(\lambda)$ is $2k-1$. In particular, $D_{2N+1}^c(-\frac{n-3}{2}+N)$ has normal order 1.

Note that $D_0^c(\lambda) = i^*$ and $D_1^c(\lambda) = D_1(S^n, S^{n-1}; g_c; \lambda) = i^*\nabla_{N(g_c)}$. From a more general point of view, this is a special case of (1.8.7) since $S^{n-1} \hookrightarrow S^n$ is a totally geodesic embedding, and the second fundamental form of the submanifolds S^{n-1} vanishes.

We emphasize again that in Theorem 1.4.1 equivariance is to be understood with respect to G^n and not with respect to G^{n+1}.

The proof of Theorem 1.4.1 uses, among other things, the following result. It provides a set of factorization identities which can be used to determine the families recursively. Analogous arguments will play a central role in the theory of residue families (see Section 1.7).

Theorem 1.4.2. *The families $D_{2N}^c(\lambda)$ of even order satisfy the identities*

$$D_{2N}^c\left(-\frac{n}{2}+j\right) = D_{2N-2j}^c\left(-\frac{n}{2}-j\right)P_{2j}(S^n) \qquad (1.4.5)$$

and

$$D_{2N}^c\left(2N-j-\frac{n-1}{2}\right) = P_{2j}(S^{n-1})D_{2N-2j}^c\left(2N-j-\frac{n-1}{2}\right) \qquad (1.4.6)$$

for $j = 0, \ldots, N$. In particular, for $j = N$, we have the relations

$$D^c_{2N}\left(-\frac{n}{2}+N\right) = i^* P_{2N}(S^n),$$

$$D^c_{2N}\left(-\frac{n-1}{2}+N\right) = P_{2N}(S^{n-1})i^*.$$

Similarly, the families $D^c_{2N+1}(\lambda)$ of odd order satisfy the identities

$$D^c_{2N+1}\left(-\frac{n}{2}+j\right) = D^c_{2N-2j+1}\left(-\frac{n}{2}-j\right) P_{2j}(S^n)$$

and

$$D^c_{2N+1}\left(2N+1-j-\frac{n-1}{2}\right) = P_{2j}(S^{n-1})D^c_{2N+1-2j}\left(2N+1-j-\frac{n-1}{2}\right)$$

for $j = 0, \ldots, N$. In particular, we have the relations

$$D^c_{2N+1}\left(-\frac{n}{2}+N\right) = D^c_1 P_{2N}(S^n),$$

$$D^c_{2N+1}\left(-\frac{n-3}{2}+N\right) = P_{2N}(S^{n-1})D^c_1.$$

Note that, for even order families, the special values for which the families factorize belong to the two non-intersecting intervals

$$\left[-\frac{n}{2}, -\frac{n}{2}+N\right] \quad \text{and} \quad \left[-\frac{n-1}{2}+N, -\frac{n-1}{2}+2N\right].$$

Similarly, for odd order families, these special values are contained in the disjoint intervals

$$\left[-\frac{n}{2}, -\frac{n}{2}+N\right] \quad \text{and} \quad \left[-\frac{n-3}{2}+N, -\frac{n-3}{2}+2N\right].$$

Theorem 1.4.2 can be applied as follows. In order to find an explicit formula for $D^c_{2N}(\lambda)$, we use the fact that it is given by a polynomial of degree $2N$ with leading term $(-1)^N \lambda^{2N} i^*$ (Theorem 1.4.1/(i)). Therefore, it is enough to determine the remaining $2N$ operator coefficients $A_0, \ldots, A_{2N-1}$. Now the $2N$ non-trivial identities (1.4.5), (1.4.6) can be used to determine $D^c_{2N}(\lambda)$ by using the known families $D^c_2(\lambda), \ldots, D^c_{2N-2}(\lambda)$. A similar procedure applies to the odd order families.

For low orders, it is easy to derive explicit formulas. Here we give such formulas for the families $D^c_N(\lambda)$ of order ≤ 3. The families of even order 0 and 2 are $D^c_0(\lambda) = i^*$ and

$$D^c_2(\lambda) = -(2\lambda+n-3)i^* P_2(S^n) + (2\lambda+n-2)P_2(S^{n-1})i^*$$
$$-\left(\lambda+\frac{n-2}{2}\right)\left(\lambda+\frac{n-3}{2}\right)i^*. \quad (1.4.7)$$

This formula for $D^c_2(\lambda)$ follows from the recursive relations.

For odd orders 1 and 3, we have $D_1^c(\lambda) = i^* \operatorname{grad}(\mathcal{H}_0) = i^* \partial/\partial n$ (this operator does not depend on λ since $H = 0$) and

$$D_3^c(\lambda) = -\frac{1}{3}(2\lambda + n - 5)D_1^c P_2(S^n) + \frac{1}{3}(2\lambda + n - 2)P_2(S^{n-1})D_1^c$$
$$- \left(\lambda + \frac{n-5}{2}\right)\left(\lambda + \frac{n-2}{2}\right)D_1^c. \quad (1.4.8)$$

Theorem 1.4.1 and Theorem 1.4.2 both follow from algebraic results using an induction construction. In fact, all operators are induced by homomorphism of generalized Verma modules. In order to formulate the result, we introduce a bit more notation. Let $\mathfrak{g}_m$ be the Lie algebra of $G^m = SO(1, m)^\circ$. Let $\mathfrak{p}_m \subset \mathfrak{g}_m$ be a parabolic subalgebra. We have the decompositions

$$\mathfrak{g}_m = \mathfrak{n}_m^- \oplus \mathfrak{m}_m \oplus \mathfrak{a} \oplus \mathfrak{n}_m^+ \quad \text{(triangle decomposition)},$$
$$\mathfrak{p}_m = \mathfrak{m}_m \oplus \mathfrak{a} \oplus \mathfrak{n}_m^+ \quad \text{(Langlands decomposition)}.$$

Here $\mathfrak{a}$ is abelian and 1-dimensional. The abelian subalgebras $\mathfrak{n}_m^\pm$ have dimension $m - 1$. $\mathfrak{a}$ operates by $\operatorname{ad}(X) = \pm\alpha(X)$ on $\mathfrak{n}^\pm$. Each $\lambda \in \mathbb{C}$ defines a character ξ_λ of $\mathfrak{p}_m$ on $\mathbb{C}(\lambda)$ by $\xi_\lambda(X) = \lambda\alpha(X)$, $X \in \mathfrak{a}$ (ξ_λ operates by 0 on $\mathfrak{m} \oplus \mathfrak{n}^+$). Let $\mathcal{U}(\mathfrak{g}_m)$ be the universal enveloping algebra of $\mathfrak{g}_m$. Now let $I_\lambda(\mathfrak{g}_m) \subset \mathcal{U}(\mathfrak{g}_m) \otimes \mathbb{C}(\lambda)$ be the left $\mathcal{U}(\mathfrak{g}_m)$-ideal which is generated by the elements

$$X \otimes 1 - 1 \otimes \xi_\lambda(X)1 \in \mathcal{U}(\mathfrak{g}_m) \otimes \mathbb{C}(\lambda), \ X \in \mathfrak{p}_m.$$

We consider the left $\mathcal{U}(\mathfrak{g}_m)$-module

$$\mathcal{M}_\lambda(\mathfrak{g}_m) = (\mathcal{U}(\mathfrak{g}_m) \otimes \mathbb{C}(\lambda))/I_\lambda(\mathfrak{g}_m).$$

$\mathcal{M}_\lambda(\mathfrak{g}_m)$ is a *generalized Verma module* for $\mathfrak{g}_m$. It is generated by the highest weight $1 \otimes 1$.

Theorem 1.4.3. *For any $N \in \mathbb{N}_0$, there exists a polynomial family of elements $\mathcal{D}_N(\lambda) \in \mathcal{U}(\mathfrak{n}_{n+1}^-)$ so that the map*

$$\mathcal{U}(\mathfrak{g}_n) \otimes \mathbb{C}(\lambda - N) \ni T \otimes 1 \mapsto i(T)\mathcal{D}_N(\lambda) \otimes 1 \in \mathcal{U}(\mathfrak{g}_{n+1}) \otimes \mathbb{C}(\lambda)$$

induces a family of homomorphisms

$$\mathcal{D}_N(\lambda) : \mathcal{M}_{\lambda-N}(\mathfrak{g}_n) \to \mathcal{M}_\lambda(\mathfrak{g}_{n+1})$$

of $\mathcal{U}(\mathfrak{g}_n)$-modules. $\mathcal{D}_N(\lambda)$ induces the family $D_N^c(\lambda)$. Here $i : \mathcal{U}(\mathfrak{g}_n) \hookrightarrow \mathcal{U}(\mathfrak{g}_{n+1})$ is induced by the canonical inclusion $i : \mathfrak{g}_n \hookrightarrow \mathfrak{g}_{n+1}$.

The latter result (Theorem 5.1.5) is not only an existence result. Its proof yields explicit formulas for the families $\mathcal{D}_N(\lambda)$. In terms of the basis $\{Y_j^-\}$ of $\mathfrak{n}_{n+1}^-$, we have the explicit formula

$$\mathcal{D}_{2N}(\lambda) = a_0(\lambda)(Y_n^-)^{2N} + a_1(\lambda)\Delta_{n-1}^-(Y_n^-)^{2N-2} + \cdots + (\Delta_{n-1}^-)^N,$$

where $\Delta_{n-1}^- = \sum_{j=1}^{n-1}(Y_j^-)^2$ and

$$a_j(\lambda) = \frac{N!}{j!(2N-2j)!}(-2)^{N-j}\prod_{k=j}^{N-1}(2\lambda-4N+2k+n+1), \ j = 0,\ldots,N-1. \quad (1.4.9)$$

The coefficients of $\mathcal{D}_{2N}(\lambda)$ can be recorded in an associated polynomial which turns out to be a classical orthogonal polynomial. In fact, the polynomial

$$a_0(\lambda)x^{2N} - a_1(\lambda)x^{2N-2} \pm \cdots + (-1)^N x^0$$

in x coincides with the *Gegenbauer polynomial* $C_{2N}^{-\lambda-\frac{n-1}{2}}(x)$, up to a rational coefficient in λ ([21], 3.15). Similarly, for odd N, we find

$$\mathcal{D}_{2N+1}(\lambda) = b_0(\lambda)(Y_n^-)^{2N+1} + b_1(\lambda)\Delta_{n-1}^-(Y_n^-)^{2N-1} + \cdots + (\Delta_{n-1}^-)^N Y_n^-,$$

where

$$b_j(\lambda) = \frac{N!}{j!(2N-2j+1)!}(-2)^{N-j}\prod_{k=j}^{N-1}(2\lambda-4N+2k+n-1), \ j = 0,\ldots,N-1. \quad (1.4.10)$$

In that case, the associated polynomial

$$b_0(\lambda)x^{2N+1} - b_1(\lambda)x^{2N-1} \pm \cdots + (-1)^N x$$

is related to the Gegenbauer polynomial $C_{2N+1}^{-\lambda-\frac{n-1}{2}}(x)$.

The intertwining maps $\mathcal{D}_N(\lambda)$ identify the $\mathcal{U}(\mathfrak{g}_n)$-module $\mathcal{M}_{\lambda-N}(\mathfrak{g}_n)$ with a submodule of $\mathcal{M}_\lambda(\mathfrak{g}_{n+1})$. This result suggests that we ask for a description of the decomposition of $\mathcal{M}_\lambda(\mathfrak{g}_{n+1})$ under the action of $\mathcal{U}(\mathfrak{g}_n)$. Here it seems natural to expect the multiplicity-free branching rule

$$\mathcal{M}_\lambda(\mathfrak{g}_{n+1}) \simeq \bigoplus_{N\geq 0} \mathcal{M}_{\lambda-N}(\mathfrak{g}_n). \quad (1.4.11)$$

The homomorphisms in Theorem 1.4.3 induce families of differential operators $C^\infty(\mathbb{R}^n) \to C^\infty(\mathbb{R}^{n-1})$ which intertwine the non-compact models of the spherical principal series on $\mathbb{R}^n$ and $\mathbb{R}^{n-1}$. The even order families are given by the formula

$$D_{2N}^{nc}(\lambda) = \sum_{j=0}^N a_j(\lambda)\Delta_{\mathbb{R}^{n-1}}^j i^* (\partial/\partial x_n)^{2N-2j}, \quad (1.4.12)$$

where $i : \mathbb{R}^{n-1} \ni x' \mapsto (x',0) \in \mathbb{R}^n$ and $\Delta_{\mathbb{R}^{n-1}} = \sum_{j=1}^{n-1} \partial^2/\partial x_i^2$. In that case the background metric is flat.

The polynomial degree of these families coincides with the polynomial degree of the families $\mathcal{D}_N(\lambda)$. It equals N for $\mathcal{D}_{2N}(\lambda)$. This is in contrast to the case of the hemisphere, where the corresponding polynomial degree is $2N$. We interpret the higher order of the polynomials in the latter case as an effect of the non-trivial curvatures of the metrics of the two-sphere.

1.5 The residue method for the hemisphere

In Section 1.4, we have seen that certain families of homomorphisms of Verma modules are responsible for the existence of intertwining families of differential operators. In Section 1.9, this will be the starting point for an application of a version of Eastwood's curved translation principle. It will enable us to construct curved analogs of the intertwining families.

In the present section, we describe a second construction of the intertwining families $D_N^c(\lambda) : C^\infty(S^n) \to C^\infty(S^{n-1})$. It rests on the asymptotic analysis of eigenfunctions of Laplacians for the hyperbolic metric. That method will find a far-reaching generalization in Section 1.6 in terms of the residue families. Although the present situation does not fall into the framework of residue families, the discussion here serves as an illustration of the idea of the construction of residue families.

The first step is to recognize that there is a canonical G^n-invariant metric g_N on $\Omega(\Gamma) = H^+ \cup H^-$ so that $(\Omega(\Gamma), g_N)$ is isometric to two copies of hyperbolic n-space. The intrinsic definition of the Γ-invariant metric g_N on $\Omega(\Gamma)$ for any geometrically finite Kleinian groups Γ is due to Nayatani ([185]). Its G^n-invariance and the identification of $\Omega(\Gamma)$ with (two copies of) hyperbolic space is a consequence of the special situation here. An alternative (but not intrinsic) definition of g_N is

$$g_N = |\mathcal{H}_0|^{-2} g_c,$$

where $\mathcal{H}_0 \in C^\infty(S^n)$ is the restriction of the height function x_{n+1}. $\mathcal{H}_0$ is a defining function for $S^{n-1} \subset S^n$. Then g_N has constant negative curvature, and we have an isometry

$$(\Omega(\Gamma), g_N) \simeq S^0 \times (\mathbb{H}^n, g_c) \simeq \{\pm\} \times (\mathbb{H}^n, g_c).$$

$(\Omega^\pm(\Gamma), g_N)$ then are conformally compact manifolds, and g_N is a Poincaré metric with the round metric g_c on the boundary $S^{n-1} = \partial\Omega^\pm$ as conformal infinity (see [128] and Chapter 3).

Now we consider eigenfunctions for the Laplacian Δ_N of g_N. For simplicity we restrict attention to $H^+ = \Omega^+(\Gamma)$. Eigenfunctions for Δ_N on H^+ correspond to hyperfunctions on S^{n-1} via Helgason's theory of Poisson transformations ([140], [152]). In particular, any $f \in C^\infty(S^{n-1})$ gives rise to an eigenfunction

$$-\Delta_N u = \mu(n-1-\mu)u, \; u \in C^\infty(H^+)$$

on H^+. The boundary function f can be recovered from u via the leading coefficients in its (radial) asymptotics near the boundary. More precisely, the asymptotics of u near the boundary has two leading terms, and all lower order terms are given by certain families of differential operators on S^{n-1} acting on these leading terms; the family parameter is the spectral parameter μ.

These differential operators can be determined (at least in principle) by an iterative procedure. For certain values of μ, they reduce to multiples of GJMS-operators on S^{n-1}. In order to construct the families $D_N^c(\lambda)$, we shall combine these families with normal derivatives in a suitable way.

Before we describe the method to find these compositions, we take a closer look at the resulting families for the subsphere S^{n-1}. We introduce coordinates $(0,1) \times S^{n-1} \ni (\rho, x) \mapsto (\sqrt{1-\rho^2}x, \rho) \in H^+ \subset S^n$. ρ is the height function. In these coordinates,

$$g_N = \frac{1}{\rho^2}\left(\frac{d\rho^2}{1-\rho^2} + (1-\rho^2)g_{S^{n-1}}\right),$$

and the Laplacian Δ_N reads

$$(1-\rho^2)\rho^2\frac{\partial^2}{\partial\rho^2} - 2\rho^3\frac{\partial}{\partial\rho} - (n-2)\rho\frac{\partial}{\partial\rho} + \frac{\rho^2}{1-\rho^2}\Delta_{S^{n-1}}.$$

We try to find a *formal* eigenfunction

$$u(\rho, x) \sim \sum_{N\geq 0} \rho^{\mu+N} a_N(f)(x), \ x \in S^{n-1}.$$

Note that the latter form of the asymptotics of the eigenfunction differs from the usual radial asymptotics often used in harmonic analysis. The condition $-\Delta_N u = \mu(n-1-\mu)u$ is equivalent to

$$(1-\rho^2)\rho^2 \sum_{N\geq 0}(\mu+N)(\mu+N-1)\rho^{\mu+N-2}a_N(f)$$

$$-2\sum_{N\geq 0}(\mu+N)\rho^{\mu+N+2}a_N(f) - (n-2)\sum_{N\geq 0}(\mu+N)\rho^{\mu+N}a_N(f)$$

$$+\left(\sum_{m\geq 1}\rho^{2m}\right)\left(\sum_{N\geq 0}\rho^{\mu+N}\Delta_{S^{n-1}}(a_N(f))\right) = -\mu(n-1-\mu)\sum_{N\geq 0}\rho^{\mu+N}a_N(f).$$

A comparison of the coefficients of the powers of ρ shows that $a_0(f)$ is free, $a_{odd}(f) = 0$, and

$$N(2\mu+N-(n-1))a_N(f)$$
$$= (\mu+N-2)(\mu+N-1)a_{N-2}(f) - \Delta_{S^{n-1}}(a_{N-2}(f) + \cdots + a_0(f)) \quad (1.5.1)$$

for even $N \geq 2$. Now assume that $\Re(\mu) = \frac{n-1}{2}$. Since

$$2\mu+N-(n-1) \neq 0$$

for $N \geq 2$, the latter formula implies that all coefficients a_{even} can be determined from $a_0(f)$ by applying a differential operator of order $2N$. It is a polynomial in the Laplacian $\Delta_{S^{n-1}}$. Notice that the recursion formula (1.5.1) says that, in order to find $a_N(f)$, one needs to know *all* terms with smaller indices.

Now let $u \in C^\infty(H^+)$ be a *genuine* eigenfunction of Δ_N with eigenvalue $\mu(n-1-\mu)$, $\Re(\mu) = \frac{n-1}{2}$. It has an asymptotics of the form

$$u \sim \sum_{N \geq 0} \rho^{\mu+N} a_N(f) + \sum_{N \geq 0} \rho^{n-1-\mu+N} b_N(f).$$

The eigenfunction u gives rise to a holomorphic family of measures on the half-plane $\Re(\lambda) > -(n-1)/2$ by

$$\langle M_u(\lambda), \varphi \rangle = \int_{H^+} \mathcal{H}_0^\lambda u\varphi db, \ \varphi \in C^\infty(S^n).$$

The support of these measures is contained in $\overline{H^+}$. Moreover, the restriction of $M_u(\lambda)$ to the open H^+ can be identified with the holomorphic family of functions $\lambda \mapsto \mathcal{H}_0^\lambda u$. But holomorphy of $\langle M_u(\lambda), \varphi \rangle$ on $\mathbb{C}$ is lost for test functions $\varphi \in C^\infty(S^n)$ the support of which intersects with S^{n-1}. However, thanks to the controlled boundary behaviour of u in terms of its asymptotics, the family $M_u(\lambda)$ admits a meromorphic continuation to $\mathbb{C}$ as a family of *distributions*. In order to study the continuation, we write

$$\langle M_u(\lambda), \varphi \rangle = \sum_{N \geq 0} \int_0^1 \rho^{\lambda+\mu+N} \left(\int_{S^{n-1}} a_N(f)(x)\varphi(\rho,x)dx \right) (1-\rho^2)^{\frac{n-2}{2}} d\rho$$

$$+ \sum_{N \geq 0} \int_0^1 \rho^{\lambda+n-1-\mu+N} \left(\int_{S^{n-1}} b_N(f)(x)\varphi(\rho,x)dx \right) (1-\rho^2)^{\frac{n-2}{2}} d\rho.$$

Using a Taylor series for $(1-\rho^2)^{\frac{n-2}{2}}$, and repeated partial integration in ρ, we construct a meromorphic continuation of $M_u(\lambda)$ with simple poles in the disjoint ladders

$$\lambda \in -\mu-1-\mathbb{N}_0, \quad \lambda \in -(n-1-\mu)-1-\mathbb{N}_0.$$

The respective residues are linear combinations of terms of the form

$$\left(\frac{\partial}{\partial\rho} \right)^j i_* a_k(f), \ \left(\frac{\partial}{\partial\rho} \right)^j i_* b_k(f) \in C^{-\infty}(\Lambda(\Gamma)).$$

But the property $\mathcal{H}_0 \in C^\infty(S^n)_1^{G^n}$ implies that the maps

$$C^\infty(S^{n-1}) \ni f \mapsto \mathrm{Res}_{-\mu-1-N}(M_u(\lambda)) \in C^{-\infty}(\Lambda(\Gamma))$$

are G^n-equivariant. Up to a renormalization, these families (in the spectral parameter μ) are the adjoint families of the G^n-equivariant operator families to be constructed.

The above arguments show how to compose tangential families and normal derivatives to equivariant families. The real value of the method, however, is its capacity to generate the residue families in a much more general setting (Section 1.6).

1.6 Q-curvature, holography and residue families

A construction of central significance in spectral geometry is the *zeta-regularized determinant* or *functional determinant* of geometric differential operators such as the Laplace-Beltrami operator on functions and the Hodge-Laplacians on differential forms. Among other things, it gave rise to the fundamental notion of analytic torsion ([207]), and was shown to have remarkable extremal properties in dimension $n = 2$ ([191]) as well as in higher dimensions ([31], [48], [37], [72], [189]). The conformal covariance of an operator has essential influence on the behaviour of its functional determinant, and the work of Branson ([31], [32]) revealed the fundamental role of Q-curvature in this context.

Q-curvature Q_n of a Riemannian manifold (M, g) of even dimension n is a scalar n^{th}-order Riemannian curvature invariant which satisfies the remarkable transformation property

$$e^{n\varphi} Q_n(e^{2\varphi} g) = Q_n(g) + (-1)^{\frac{n}{2}} P_n(g)(\varphi) \tag{1.6.1}$$

under conformal changes of the metric. This identity will be called the *fundamental identity*. Here P_n is the critical GJMS-operators of M (Section 1.3).

On a manifold of dimension n, Q_n appears together with its lower order relatives $Q_{2,n}, \ldots, Q_{n-2,n}$. However, in many respects, the quantity $Q_{n,n} = Q_n$ is the most important one. It will be called the *critical Q-curvature*.

The critical Q-curvature generalizes Gauß curvature K of a surface and the Paneitz curvature quantity

$$Q_4 = 2(\mathsf{J}^2 - |\mathsf{P}|^2) - \Delta \mathsf{J} = \frac{1}{6}\left(-3|\operatorname{Ric}|^2 + \tau^2 - \Delta\tau\right) \tag{1.6.2}$$

of a four-manifold.

For closed M, the fundamental identity (1.6.1) implies that the total Q-curvature

$$\int_{M^n} Q_n \operatorname{vol} \tag{1.6.3}$$

is a global conformal invariant. For $n = 2$, this is obvious since, by the Gauß-Bonnet formula, the integral is proportional to the Euler characteristic. In dimension 4, the conformal invariance of the integral also follows from the Gauß-Bonnet-Chern formula using the fact that the Pfaffian is a constant multiple of $\mathsf{J}^2 - |\mathsf{P}|^2 + \frac{1}{8}|\mathsf{C}|^2$.

The conformal invariance of the total Q-curvature has strong implications on the structure of Q-curvature itself. The question to describe *all* scalar local curvature invariants with a conformally invariant total integral is of significance in quantum field theory (anomalies). Deser and Schwimmer proposed a geometric classification of these quantities in [81]. After preparations in [5] and [6], Alexakis accomplished a detailed proof of that classification in the monumental work [3].

These studies are closely related to conformal anomalies of functional determinants of conformally covariant differential operators. In a series of works (starting with [45]), Branson and Ørsted explored the relation of the conformal anomaly of the functional determinant to the constant term in the heat kernel asymptotics. In such a context, Branson and Ørsted called the constant term a_n the conformal index density. The latter name is suggested by the role of analogous quantities in the heat equation approach to index theory ([107]). For closed M, the total integral of the conformal anomaly is conformally invariant. The search for universal structural results for conformal anomalies of determinants led Branson to the idea that the anomalies are deeply linked with Q_n ([31]). For more details we refer to the discussion at the end of the present section.

Now how is Q_n defined? In dimensions $n = 2$ and $n = 4$, it is given by the explicit formulas $Q_2 = K$ and (1.6.2). But in higher even dimension n, the Q-curvature quantities $Q_{2N,n}$ ($2N \leq n$) are defined only implicitly through the GJMS-operators P_{2N}: for $2N < n$, the constant term $P_{2N}(1)$ of P_{2N} on an n-manifold has the form

$$(-1)^N \left(\frac{n}{2} - N \right) Q_{2N,n}.$$

The subcritical Q-curvature $Q_{2N,n}$ is a scalar Riemannian curvature invariant of order $2N$. In order to simplify notation, we often write Q_{2N} if the dimension of the underlying space is evident. In the critical case $2N = n$, we set

$$Q_n = Q_{n,n}.$$

A comment on the nature of these definitions is in order. These definitions are *extrinsic* in the following sense. Q-curvature is derived from GJMS-operators. The GJMS-operators $P_{2N}(g)$ are induced by the powers of the Laplacian of the Fefferman-Graham ambient metric associated to g. This is an extrinsic definition since the interesting object is generated by a construction on a certain ambient space of two dimensions higher.

Now (1.3.4) and (1.3.5) show that

$$Q_{2,n} = \mathsf{J} \quad \text{and} \quad Q_{4,n} = \frac{n}{2}\mathsf{J}^2 - 2|\mathsf{P}|^2 - \Delta\mathsf{J}.$$

In the respective critical cases $n = 2$ and $n = 4$, we recover $Q_2 = \tau/2 = K$ and (1.6.2). Explicit formulas for Q_6 and Q_8 in terms of Riemannian invariants were derived in [116]. The complexity of such formulas increases exponentially with the order. It is one purpose of the present book to reveal the structure of these quantities.

In the following, we apply a version of the residue method of Section 1.5 to define the so-called *residue families*. These families will be used to study the structure of Q-curvatures. Note that the residue families, however, do not cover the construction in Section 1.5.

The relation of residue families to GJMS-operators and Q-curvature is as follows. For any metric h on M^n and any integer $N \in [1, \frac{n}{2}]$, there is a residue family of order $2N$ which, for a certain value of the family parameter, specializes to the GJMS-operator $P_{2N}(h)$. The conformal covariance of the latter operator is embedded in the conformal covariance of the corresponding residue family.

The most interesting one is the critical residue family of order n. It is related to the critical Q-curvature. The critical residue family specializes to the critical GJMS-operators P_n at $\lambda = 0$, and the linear term of its Taylor series in λ "sees" Q_n. The fundamental identity of Q_n is a direct consequence of the conformal covariance of the family.

Now let (M, h) be a compact Riemannian manifold of *even* dimension n. We view (M, h) as the conformal infinity of an associated Poincaré-Einstein metric g on $(0, \varepsilon) \times M$ in the normal form

$$g = r^{-2}(dr^2 + h_r), \ h_0 = h$$

with

$$h_r = h_0 + r^2 h_{(2)} + \cdots + r^n \left(h_{(n)} + \bar{h}_{(n)} \log r\right) + \cdots . \tag{1.6.4}$$

The metric g satisfies the Einstein condition

$$\mathrm{Ric}(g) + ng = 0 \tag{1.6.5}$$

to high order near the boundary ($r = 0$). The condition (1.6.5) has the consequence that the coefficients $h_{(2)}, \ldots, h_{(n-2)}$, $\bar{h}_{(n)}$ and the trace of $h_{(n)}$ are locally determined by h (see [122] and [96] for the details). This generalizes the case of the hyperbolic space $\mathbb{H}^{n+1}$ of sectional curvature -1 with (a multiple of) the round metric on the sphere at infinity. For odd n, the analogous series of h_r does not contain $\log r$-terms, and all coefficients are determined by $h = h_0$. For Einstein g, the series (1.6.4) will be referred to as the *Fefferman-Graham expansion*.

The following constructions will depend only on those terms in the Taylor series (1.6.4) which are determined by the leading term h_0. Moreover, for even n, the trace-free coefficient $\bar{h}_{(n)}$ of $\log r$ will not play a role. Therefore, we apply the convention that h_r denotes the finite sum

$$h_0 + r^2 h_{(2)} + \cdots + r^{n-2} h_{(n-2)} + r^n h_{(n)}$$

for some $h_{(n)}$ (with determined trace). This convention will be used throughout without further notice.

Now let $u \in \ker(\Delta_g + \mu(n-\mu))$ ($\Re(\mu) = \frac{n}{2}$ and $\mu \neq \frac{n}{2}$) be an eigenfunction of the Laplacian Δ_g with the formal asymptotics

$$u \sim \sum_{j \geq 0} r^{\mu+2j} a_{2j}(h; \mu) + \sum_{j \geq 0} r^{n-\mu+2j} b_{2j}(h; \mu), \ r \to 0 \tag{1.6.6}$$

with leading term $a_0 = f \in C^\infty(M)$. f is called the boundary value of u. We extend the family of functions $r^\lambda u$, $\lambda \in \mathbb{C}$ on $(0, \varepsilon) \times M$ through the boundary

$r = 0$ to a family of distributions on $[0, \varepsilon) \times M$. More precisely, let $M_u(\lambda)$ be defined by

$$\langle M_u(\lambda), \varphi \rangle = \int_{[0,\varepsilon) \times M} r^\lambda u \varphi \operatorname{vol}(dr^2 + h_r), \ \Re(\lambda) \gg 0, \ \varphi \in C_0^\infty([0,\varepsilon) \times M).$$

Then $M_u(\lambda)$ admits a meromorphic continuation with simple poles in the ladder $-\mu - 1 - \mathbb{N}_0$ (we ignore the analogous simple poles in the ladder $-(n - \mu) - 1 - \mathbb{N}_0$). The residues are of the form

$$\operatorname{Res}_{-\mu - 1 - N} \langle M_u(\lambda), \varphi \rangle = \int_M f \delta_N(h; \mu)(\varphi) \operatorname{vol}(h)$$

with certain families of differential operators

$$\delta_N(h; \lambda) : C^\infty([0, \varepsilon) \times M) \to C^\infty(M).$$

The coefficients $a_{2j}(h; \mu)$, $j \leq \frac{n}{2}$ are completely determined by $h_0, h_{(2)}, \dots, h_{(n-2)}$ and $\operatorname{tr}(h_{(n)})$, i.e., by h. Hence the families $\delta_N(h; \lambda)$, $2N \leq n$ are natural in h. The residue families $D_N^{\mathrm{res}}(h; \lambda)$ are certain renormalizations of $\delta_N(h; \lambda)$.

The evaluation of the above construction motivates the following definition (see Definition 6.6.2 for the general case).

Definition 1.6.1 (Residue families). *For even n and $2N \leq n$, let the family of differential operators*

$$D_{2N}^{\mathrm{res}}(h; \lambda) : C^\infty([0, \varepsilon) \times M^n) \to C^\infty(M^n)$$

be defined by

$$D_{2N}^{\mathrm{res}}(h; \lambda) = 2^{2N} N! \left[\left(-\frac{n}{2} - \lambda + 2N - 1 \right) \cdots \left(-\frac{n}{2} - \lambda + N \right) \right] \delta_{2N}(h; \lambda + n - 2N)$$

with

$$\delta_{2N}(h; \lambda)$$
$$= \sum_{j=0}^{N} \frac{1}{(2N - 2j)!} \left[T_{2j}^*(h; \lambda) \circ v_0 + \cdots + T_0^*(h; \lambda) \circ v_{2j} \right] \circ i^* \circ (\partial / \partial r)^{2N - 2j},$$

where i^* restricts functions to $r = 0$.

The rational families $T_{2j}(h; \lambda)$ are determined by the relation $T_{2j}(h; \lambda)f = a_{2j}(h; \lambda)$, i.e., the operator $T_{2j}(h; \lambda)$ maps the leading coefficient in (1.6.6) to the coefficient of $r^{\lambda + 2j}$. The renormalized families

$$P_{2j}(h; \lambda) = 2^{2j} j! \left(\frac{n}{2} - \lambda - 1 \right) \cdots \left(\frac{n}{2} - \lambda - j \right) T_{2j}(h; \lambda)$$

are holomorphic and satisfy $P_{2j}(\cdot; \lambda) = \Delta^j + LOT$. Formal adjoints are defined with respect to the metric h on M.

The *holographic coefficients* $v_{2j} \in C^\infty(M)$ are the Taylor coefficients of the volume function

$$v_r(r, \cdot) = \frac{\text{vol}(h_r)}{\text{vol}(h)} = v_0 + r^2 v_2 + \cdots + r^n v_n + \cdots, \quad v_0 = 1; \tag{1.6.7}$$

here $\text{vol}(h_r)$ denotes the Riemannian volume form of h_r. The coefficients v_{2j} $(j = 0, \ldots, \frac{n}{2})$ in (1.6.7) are given by local formulas in terms of the curvature of the metric h (and its covariant derivatives). The coefficient v_n has found much attention in recent years. It is called the *holographic anomaly* of the asymptotic volume of the Poincaré-Einstein metric g (see [119] and the discussion below).

The relation

$$P_{2N}\left(h; \frac{n}{2} - N\right) = P_{2N}$$

and the self-adjointness of P_{2N} imply that

$$D_{2N}^{\text{res}}\left(h; -\frac{n}{2} + N\right) = P_{2N}(h)i^*, \tag{1.6.8}$$

i.e., the residue families specialize to GJMS-operators for appropriate values of the family parameters. This should be compared with the formula

$$\text{Res}_{\frac{n}{2}+N}(\mathcal{S}) = -c_N P_{2N}(h)$$

(Theorem 3.2.1).

$D_n^{\text{res}}(h; \lambda)$ is called the *critical residue family*. It depends on the coefficients $h_0 = h, h_{(2)}, \ldots, h_{(n-2)}$ and the h-trace of $h_{(n)}$. Since all these terms are determined by h, the family is completely determined by h.

For odd n, residue families are defined analogously and (1.6.8) continues to hold true. By parity reasons, there is no critical residue family in that case.

It should be emphasized that the residue families are *not* families in the sense of Section 1.8 since they are defined only for *one* specific background metric $dr^2 + h_r$, which extends h into a neighbourhood. Instead, the critical residue family satisfies the transformation formula

$$e^{-(\lambda-n)\varphi} \circ D_n^{\text{res}}(\hat{h}; \lambda) = D_n^{\text{res}}(h; \lambda) \circ \kappa_* \circ \left(\frac{\kappa^*(r)}{r}\right)^\lambda, \tag{1.6.9}$$

where $\hat{h} = e^{2\varphi}h$, and κ is the diffeomorphism which relates the corresponding Poincaré-Einstein metrics

$$\kappa^*\left(r^{-2}(dr^2 + h_r)\right) = r^{-2}(dr^2 + \hat{h}_r) \tag{1.6.10}$$

([99], Theorem 2.3, [126], Section 5 and [96], Theorem 4.4). Here we apply the above convention and (1.6.10) is to be understood as an identity up to terms in

$O(r^{n-2})$ and vanishing trace $\mathrm{tr}(r^{-n+2}\cdot)$. κ is induced by the gradient flow of the function ρ with the properties

$$|d\rho|^2_{\rho^2 g} = 1, \ g = r^{-2}(dr^2 + h_r)$$

and

$$i^*(\rho^2 g) = \hat{h}.$$

Note that the Taylor-coefficients of $\kappa_* \circ (\kappa^*(r)/r)^\lambda$ up to r^n are completely determined by h.

Similar transformation formulas describe the behaviour of $D^{\mathrm{res}}_{2N}(h;\lambda)$ under conformal changes $h \mapsto e^{2\varphi}h$. The diffeomorphisms κ appear in the physics literature also under the name of PBH transformations ([147], [218]).

Now (1.6.9) would be a consequence of an identity

$$D^{\mathrm{res}}_n(h;\lambda) = D_n(dr^2 + h_r;\lambda)$$

for some conformally covariant family $D_n(g;\lambda)$ as in Section 1.8. In fact, by the naturality of $D_n(g;\lambda)$,

$$D_n(\kappa^*(dr^2 + h_r);\lambda) = \kappa^* \circ D_n(dr^2 + h_r;\lambda) \circ \kappa_*.$$

Hence

$$\begin{aligned}
D^{\mathrm{res}}_n(\hat{h};\lambda) &= D_n(dr^2 + \hat{h}_r;\lambda) \\
&= D_n\left(\left(\frac{r}{\kappa^*(r)}\right)^2 \kappa^*(dr^2 + h_r);\lambda\right) \\
&= e^{(\lambda-n)\varphi} \circ D_n(dr^2 + h_r;\lambda) \circ \kappa_* \circ \left(\frac{\kappa^*(r)}{r}\right)^\lambda \\
&= e^{(\lambda-n)\varphi} \circ D^{\mathrm{res}}_n(h;\lambda) \circ \kappa_* \circ \left(\frac{\kappa^*(r)}{r}\right)^\lambda
\end{aligned}$$

(see Remark 6.6.3). Now since $D^{\mathrm{res}}_n(h;\lambda)$ satisfies the identity

$$D^{\mathrm{res}}_n(h;0) = P^*_n(h;0)i^* = P^*_n(h)i^* = P_n(h)i^*, \tag{1.6.11}$$

it is reasonable to ask whether also

$$\dot{D}^{\mathrm{res}}_n(h;0)(1) = -(-1)^{\frac{n}{2}}Q_n(h). \tag{1.6.12}$$

In the following, we analyze the meaning of the relation (1.6.12) and confirm its validity (see Theorem 6.6.1).

In view of $\dot{P}_n(h;0)(1) = (-1)^{\frac{n}{2}}Q_n(h)$, (1.6.12) is equivalent to the formula

$$2(-1)^{\frac{n}{2}}Q_n(h) = \delta_n(h)(1) + 2^n \left(\frac{n}{2}\right)! \left(\frac{n}{2}-1\right)! \sum_{j=0}^{\frac{n}{2}-1} T^*_{2j}(h;0)(v_{n-2j}), \tag{1.6.13}$$

where
$$\delta_n(h) \overset{\text{def}}{=} \dot{P}_n(h;0) - \dot{P}_n^*(h;0)$$

(Corollary 6.6.1). In terms of the operators $P_{2j}(h;0)$, that identity reads

$$2(-1)^{\frac{n}{2}}Q_n(h) = \delta_n(h)(1) + \left(\frac{n}{2}\right)! \sum_{j=0}^{\frac{n}{2}-1} 2^{n-2j}\frac{(\frac{n}{2}-j-1)!}{j!}P_{2j}^*(h;0)(v_{n-2j}). \quad (1.6.14)$$

Note that in (1.6.13) and (1.6.14), the nature of the contribution, which is defined by $\delta_n(h) = \dot{P}_n(h;0) - \dot{P}_n^*(h;0)$ (acting on 1), differs from the remaining terms. But calculations up to $n = 6$ show that this contribution can be written also as a linear combination of the other terms. Indeed, this is a general fact (see Theorem 6.6.4).

Theorem 1.6.1. *For even n,*

$$n\delta_n(h)(1) = -2^n \left(\frac{n}{2}\right)! \left(\frac{n}{2}-1\right)! \sum_{j=0}^{\frac{n}{2}-1} 2j\mathcal{T}_{2j}^*(h;0)(v_{n-2j}).$$

(1.6.13) and Theorem 1.6.1 imply that (1.6.12) is equivalent to

$$(-1)^{\frac{n}{2}}nQ_n(h) = 2^{n-1} \left(\frac{n}{2}\right)! \left(\frac{n}{2}-1\right)! \sum_{j=0}^{\frac{n}{2}-1} (n-2j)\mathcal{T}_{2j}^*(h;0)(v_{n-2j}).$$

The latter formula, however, admits an independent proof (see [125] and Theorem 6.6.6). This proves (1.6.12).

Theorem 1.6.2 (The holographic formula). *The critical Q-curvature $Q_n(h)$ of (M^n, h) is given by the formula*

$$(-1)^{\frac{n}{2}}nQ_n(h) = 2^{n-1} \left(\frac{n}{2}\right)! \left(\frac{n}{2}-1\right)! \sum_{j=0}^{\frac{n}{2}-1} (n-2j)\mathcal{T}_{2j}^*(h;0)(v_{n-2j}). \qquad (1.6.15)$$

In terms of the operators $P_{2j}(0)$, Theorem 1.6.2 reads

$$(-1)^{\frac{n}{2}}Q_n = \left(\frac{n}{2}-1\right)! \sum_{j=0}^{\frac{n}{2}-1} 2^{n-1-2j}\frac{(\frac{n}{2}-j)!}{j!}P_{2j}^*(0)(v_{n-2j}). \qquad (1.6.16)$$

Theorem 1.6.2 relates Q_n to the coefficients v_{2j} and the asymptotics of harmonic functions. In particular, it emphasizes the significance of *all* holographic coefficients v_{2j}.

We continue with the description of some direct consequences of Theorem 1.6.2. Further consequences will be discussed in Section 1.7.

Corollary 1.6.1. *For closed M^n and $u \in \ker(P_n)$, the sum*

$$I_n(u) = \sum_{j=0}^{\frac{n}{2}-1} (n-2j) \int_{M^n} v_{n-2j} T_{2j}(0)(u) \, \mathrm{vol}$$

is conformally invariant.

Note that $\ker(P_n)$ is a conformally invariant space. Note also that the statement of Corollary 1.6.1 does not refer to Q-curvature.

For compact M, integration of (1.6.15) yields the formula

$$2 \int_M Q_n \, \mathrm{vol} = (-1)^{\frac{n}{2}} 2^n \left(\frac{n}{2}\right)! \left(\frac{n}{2}-1\right)! \int_M v_n \, \mathrm{vol}. \qquad (1.6.17)$$

This relation was proved in [128] (see also [97]). It is a relation between two *global* conformal invariants. The more precise relation (1.6.14), of course, makes sense without the compactness assumption on M.

(1.6.16) shows that $(-1)^{\frac{n}{2}} Q_n$ contains the contribution

$$-P_{n-2}^*(0)(\mathsf{J}) = -\Delta^{\frac{n}{2}-1}(\mathsf{J}) + LOT.$$

It is well known that $\Delta^{\frac{n}{2}-1}(\mathsf{J})$ contributes in this way to Q_n (see [31], Corollary 1.4 or [32], Corollary 1.5).

For a conformally flat metric h, the holographic coefficients in (1.6.16) are given by

$$v_{2j} = (-2)^{-j} \, \mathrm{tr}\left(\wedge^j(\mathsf{P})\right),$$

where the Schouten tensor P is identified with the endomorphism $\mathsf{P}^\sharp : TM \to TM$. Hence (Theorem 6.14.1)

Theorem 1.6.3. *Let (M^n, h) be conformally flat. Then*

$$Q_n = \left(\frac{n}{2}-1\right)! \sum_{j=0}^{\frac{n}{2}-1} (-1)^j 2^{\frac{n}{2}-1-j} \frac{\left(\frac{n}{2}-j\right)!}{j!} P_{2j}^*(0) \left(\mathrm{tr}\wedge^{\frac{n}{2}-j}(\mathsf{P})\right).$$

In particular,

$$Q_n = 2^{\frac{n}{2}-1}\left(\frac{n}{2}-1\right)! \left(\frac{n}{2}\right)! \, \mathrm{tr}\wedge^{\frac{n}{2}}(\mathsf{P}) + \textit{divergence terms}$$

$$= (-1)^{\frac{n}{2}} \left[(n-2)(n-4)\cdots 2\right] \mathrm{Pf}_n + \textit{divergence terms},$$

where Pf *denotes the Pfaffian form. On the round sphere the divergence terms vanish.*

The fact that for a conformally flat metric on a closed M^n the integral of Q_n is proportional to the Euler characteristic of M^n goes back to [39]. Theorem 1.6.3 is a local version of that result.

Theorem 1.6.2 (or formula (1.6.16)) can be made more explicit for $n \leq 6$. In particular, we find the formula

$$Q_6 = -8 \cdot 48 v_6 - 32 P_2^*(0)(v_4) - 2 P_4^*(0)(v_2) \tag{1.6.18}$$

(see (6.10.4)) for the critical Q-curvature in dimension 6. Here

$$v_2 = -\frac{1}{2} \operatorname{tr}(\mathsf{P}), \quad v_4 = \frac{1}{4} \operatorname{tr}(\wedge^2 \mathsf{P}) \quad \text{and} \quad v_6 = -\frac{1}{8} \operatorname{tr}(\wedge^3 \mathsf{P}) - \frac{1}{48}(\mathsf{P}, \mathcal{B}),$$

where $\mathcal{B}$ is a version of the Bach tensor. An alternative proof of (1.6.18) rests on the direct evaluation of (1.6.13) for $n = 6$.

Theorem 1.6.2 deals only with the critical Q-curvature. In dimension $n \geq 6$, analogous holographic formulas for the subcritical Q-curvatures Q_2, Q_4 and Q_6 can be summarized in the form

$$\begin{pmatrix} Q_6 \\ Q_4 \\ Q_2 \end{pmatrix} = \begin{pmatrix} -8 \cdot 48 & -32 P_2^*(\frac{n}{2}-3) & -2 P_4^*(\frac{n}{2}-3) \\ 0 & 16 & 2 P_2^*(\frac{n}{2}-2) \\ 0 & 0 & -2 \end{pmatrix} \begin{pmatrix} v_6 \\ v_4 \\ v_2 \end{pmatrix}. \tag{1.6.19}$$

For the general case we refer to Conjecture 6.9.1. The relation between subcritical Q_{2N} and subcritical $D_{2N}^{\mathrm{res}}(-\frac{n}{2}+N)(1)$ is more complicated than in the critical case. It will be discussed in Section 1.7.

We refer to the identities (1.6.14), (1.6.15), (1.6.16) and (1.6.19) as *holographic formulas* for Q-curvature. These formulas describe Q-curvature in terms of the asymptotic geometry of an associated metric in one higher dimension. A similar terminology is used in physics in connection with the AdS/CFT-correspondence. It proposes relations between quantum field theories and gravitational theories in one more dimension (see [171], [241], the reviews [172], [82], [29], [227], [146] and [11] for mathematical aspects in the case of pure gravity). Establishing a dictionary is a central part of the efforts. Such relations between theoretical concepts in different dimensions are reminiscent of the technical method of holography ([223]).

We describe how the holographic formula (1.6.16) for Q-curvature is linked with ideas around the AdS/CFT-correspondence through the *functional determinants* of conformally covariant differential operators. The relations rest on the connection of both quantities Q_n and v_n to the conformal index densities a_n of such operators. From the physical perspective, a_n and v_n are interpreted as conformal anomalies in quantum field theories and gravitation, respectively. In order to describe the relation between a_n and v_n in Figure 1.1, we recall that Witten ([241]) proposed to formulate the AdS/CFT-correspondence as identities of partition functions. In the case of pure gravity, a prototype of such a relation identifies the partition function $Z_{CFT}[h]$ of an appropriate conformal field theory on M (with the background metric h) with the quantity $\exp(-S[g])$ which is given by the value of the Einstein-Hilbert action $S[\cdot]$ on a solution g of the Einstein equation (on a manifold X with boundary M) with $[h]$ as conformal infinity. More

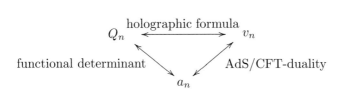

Figure 1.1: *Q-curvature and the conformal anomalies a_n and v_n*

precisely, let

$$S[g] = -\int_X (\tau_g - 2\Lambda)\,\mathrm{vol}(g),$$

and let g satisfy the Einstein equation $\mathrm{Ric}(g) + ng = 0$. Then $\tau_g = -n(n+1)$, and g is a solution of the equation

$$\mathrm{Ric}(g) - \frac{\tau}{2}g + \Lambda g = 0$$

for $\Lambda = -\frac{n(n-1)}{2}$, i.e., g is critical for the action S with the cosmological constant Λ. The value of S for such a metric is given by the divergent integral

$$2n\int_X \mathrm{vol}(g).$$

In order to renormalize the volume, *holographic renormalization* ([219], [80] and [11]) proceeds as follows. We assume that g is a global conformally compact Einstein metric on $X \setminus \partial X$ with conformal infinity $[h]$. A choice of a representing metric h on M gives rise to a defining function r such that $|dr|^2_{r^2 g} = 1$ near ∂X, and g can be written in the normal form $r^{-2}(dr^2 + h_r)$. For even n, we find an asymptotics

$$\int_{r \geq \varepsilon} \mathrm{vol}(g) = c_0 \varepsilon^{-n} + c_2 \varepsilon^{-(n-2)} + \cdots + c_{n-2}\varepsilon^{-2} - L\log\varepsilon + V + o(1) \quad (1.6.20)$$

for $\varepsilon \to 0$. The *renormalized* (or *asymptotic*) *volume* of g with respect to h is defined as the constant term $V = V(g; h)$ in this asymptotics, and we set

$$S[g; h] = 2nV(g; h). \quad (1.6.21)$$

Although $V(g; \cdot)$ is not conformally invariant, it satisfies

$$(d/dt)|_{t=0}V(e^{2t\varphi}h) = \int_M \varphi v_n(h)\,\mathrm{vol}(h) \quad (1.6.22)$$

for all $\varphi \in C^\infty(M)$, i.e., the holographic anomaly v_n describes the failure of V being conformally invariant. Moreover, the functional $L = \int_M v_n\,\mathrm{vol}$ is conformally invariant. For the proofs we refer to [119].

Now the functional $h \mapsto \exp(-S[g;h])$ is to be compared with the partition function of an appropriate conformal field theory on M. On the level of conformal anomalies this leads to a comparison of the holographic anomaly v_n with the conformal anomalies

$$(d/dt)|_{t=0} \left(\log \det D(e^{2t\varphi}h) \right) \tag{1.6.23}$$

of the determinants of certain conformally covariant differential operators D which act on the various types of fields. Here the determinant is the zeta-regularized determinant, and the basic fact is that (1.6.23) is given by a local formula. These formulas generalize a formula of Polyakov ([203]) for the determinant of the Laplacian in dimension $n = 2$.

We briefly recall the structure of such formulas. Under appropriate technical conditions, for a (power of a) conformally covariant self-adjoint differential operator D of order m with positive definite principal symbol, the generalized Polyakov formula states that

$$-(d/dt)|_{t=0} \left(\log \det D(e^{2t\varphi}h) \right) = m \int_M \varphi a_n(D(h)) \operatorname{vol}(h) \tag{1.6.24}$$

for all $\varphi \in C^\infty(M)$. The coefficient $a_n(D) \in C^\infty(M)$ is taken from the heat asymptotics

$$\operatorname{tr}(\varphi e^{-tD}) \sim \sum_{j \geq 0} t^{-\frac{n-j}{m}} \int_M \varphi a_j(D) \operatorname{vol}, \ t \to 0. \tag{1.6.25}$$

Moreover, the conformal index theorem of Branson and Ørsted states that the integral $\int_M a_n(D) \operatorname{vol}$ is conformally invariant. We call $a_n(D)$ the *conformal index density* (of D). (1.6.24) says that the function $a_n(D)$ describes the conformal anomaly of the determinant of D. For details we refer to [45], [46], [47] and also [31]. Note also that the coefficient $a_{n-m}(D) \in C^\infty(M)$ of t^{-1} in (1.6.25) is conformally invariant ([194], [107], [198], [204], [36]).

In dimension $n = 4$, Henningson and Skenderis ([141]) verified that the holographic anomaly v_4 actually *coincides* with a specific linear combination of the conformal anomalies of determinants of differential operators on the various types of fields of supersymmetric Yang-Mills theory (for the details see Section 6.15).

In order to study extremal properties of determinants, it is important to pass from the infinitesimal Polyakov formulas (1.6.24) to formulas for the quotients

$$\det(D(\hat{h}))/\det(D(h)). \tag{1.6.26}$$

Branson discovered the role of Q-curvature in that process. In fact, one expects that (if $\operatorname{vol}(\hat{h}) = \operatorname{vol}(h)$ and $\ker D$ is trivial) the quotients (1.6.26) can be written always in the *universal form*

$$-\log \frac{\det(\hat{D})}{\det(D)} = c(D) \int_M \varphi \left[Q_n \operatorname{vol} + \hat{Q}_n \widehat{\operatorname{vol}} \right] + \int_M \left[F \operatorname{vol} - \hat{F} \widehat{\operatorname{vol}} \right], \tag{1.6.27}$$

where F is a local scalar invariant, and Q_n possibly has to be modified by a local conformal invariant (if h is not conformally flat). A non-trivial kernel of D creates an additional term. Likewise, some more terms appear if the volume is not fixed. For more details we refer to [41] and its references, [43], [31], [35] and [63].

The universal Q-curvature term is generated by the following arguments. We observe that

$$\log \det D(\hat{h}) - \log \det D(h) = \int_0^1 (d/dt) \left(\log \det D(e^{2t\varphi}h)\right) dt$$

$$= -m \int_0^1 \int_M \varphi a_n(D(e^{2t\varphi}h)) \operatorname{vol}(e^{2t\varphi}h) dt$$

using (1.6.24). Now we split off a multiple of Q_n from $a_n(D)$ and find

$$\int_0^1 \int_M \varphi Q_n(e^{2t\varphi}h) e^{nt\varphi} \operatorname{vol}(h) dt = \int_0^1 \int_M \varphi \left[Q_n(h) + (-1)^{\frac{n}{2}} P_n(h)(t\varphi)\right] \operatorname{vol}(h) dt$$

$$= \int_M \varphi \left[Q_n(h) + \frac{1}{2}(-1)^{\frac{n}{2}} P_n(h)(\varphi)\right] \operatorname{vol}(h)$$

$$= \frac{1}{2} \int_M \varphi \left[Q_n(h) \operatorname{vol}(h) + Q_n(\hat{h}) \operatorname{vol}(\hat{h})\right].$$

Thus, in order to organize the terms as in (1.6.27), one needs to know that, up to a local conformal invariant, $a_n(D)$ differs from a multiple of Q_n by a quantity G so that $\int_M \varphi G(h) \operatorname{vol}(h)$ is the conformal variation of $\int_M F(h) \operatorname{vol}(h)$ for some local Riemannian invariant F. Conjecture 1 in [41] states that this is always possible. Any progress here would involve a deeper understanding of the structure of the divergence terms in $a_n(D)$ and Q_n. In that connection it also seems natural to deal with a notion of Q-curvatures which covers more than just Q_n in dimension n (see [35]). All Q-curvature quantities in the extended sense admit a conformal transformation law which is governed by a linear differential operator with leading term a power of the Laplacian. For Q_n, the holographic formula provides a description of the divergence contributions in terms of lower order holographic data: v_{2j} for $j < \frac{n}{2}$. The structure of the divergence contributions to conformal index densities $a_n(D)$, however, is much less clear. In particular, it is not known whether the lower order heat coefficients play some role here.

We close the present section with some comments on the case $n = 2$. Both v_2 and $a_2(-\Delta)$ are constant multiples of the Gauß-curvature K. Hence

$$2(V(e^{2\varphi}h) - V(h))$$

$$= 2 \int_0^1 \int_M \varphi v_2(e^{2t\varphi}h) e^{2t\varphi} \operatorname{vol}(h) dt = - \int_0^1 \int_M \varphi K(e^{2t\varphi}h) e^{2t\varphi} \operatorname{vol}(h) dt$$

$$= - \int_0^1 \int_M \varphi \left[K(h) - t\Delta_h(\varphi)\right] \operatorname{vol}(h) dt = - \int_M \varphi \left[K(h) - \frac{1}{2}\Delta_h(\varphi)\right] \operatorname{vol}(h)$$

using $e^{2\varphi} K(\hat{h}) = K(h) - \Delta_h(\varphi)$, i.e.,

$$2(V(e^{2\varphi}h) - V(h))$$
$$= -\frac{1}{2} \int_M \varphi \left[K(h)\operatorname{vol}(h) + K(\hat{h})\operatorname{vol}(\hat{h}) \right] = -\frac{1}{2} \int_M \left(2K(h)\varphi + |d\varphi|_h^2 \right) \operatorname{vol}(h).$$

On the other hand, we have the Polyakov formula

$$\log \frac{\det(-\Delta_{e^{2\varphi}h})}{\det(-\Delta_h)} = -\frac{1}{12\pi} \int_M \left(2K(h)\varphi + |d\varphi|_h^2 \right) \operatorname{vol}(h) \qquad (1.6.28)$$

(if $\operatorname{vol}(\hat{h}) = \operatorname{vol}(h)$). For S^2 with the round metric h_c, an inequality of Onofri implies that

$$\det(-\Delta_{e^{2\varphi}h_c}) \le \det(-\Delta_{h_c})$$

(if $\operatorname{vol}(\hat{h}_c) = \operatorname{vol}(h_c)$). Equality holds true exactly for those metrics which are conformally diffeomorphic to h_c. The renormalized volume is maximal for these metrics. For detailed proofs we refer to [63].

In [229], the authors studied the asymptotic volume of Kleinian manifolds $X^3 = \Gamma\backslash\mathbb{H}^3$ with compact boundary $M^2 = \Gamma\backslash\Omega(\Gamma)$, and related it to the Liouville action functional (refining [160]). Here the asymptotics of the volume of X^3 is defined by a family of equidistant (Epstein) hypersurfaces which exhaust the ends of X^3. Epstein used equidistant hypersurfaces some years before, in order to prove that for convex-cocompact Kleinian manifolds $X = \Gamma\backslash\mathbb{H}^{n+1}$ of odd dimension the coefficient of $\log \varepsilon$ in the asymptotics of the volume is proportional to the Euler-characteristic $\chi(\Gamma\backslash\Omega(\Gamma))$, $\Gamma\backslash\Omega(\Gamma) = \partial X$ (the proof is given in an appendix of [195]). That result played a key role in the identification of the multiplicities of the topological singularities of Selberg zeta functions for such Kleinian groups (see [195] and [151] for a detailed discussion). Epstein did not analyze the renormalized volume, however. For further results on the asymptotic volume of hyperbolic 3-manifolds we refer to [237], [161].

1.7   Factorization of residue families. Recursive relations

In Section 1.6, we have seen how the critical residue families $D_n^{\mathrm{res}}(h; \lambda)$ are related to the fundamental identity *and* the holographic formula for Q_n. However, it is less clear whether the fundamental identity can be derived more directly from the holographic formula. In fact, it seems that these two results actually should be regarded as two different facets of the more fundamental critical residue family. In the present section, we discuss one more of these facets: recursive formulas. The situation is summarized in Figure 1.2.

According to Theorem 1.4.2, any family $D_{2N}^c(\lambda) : C^\infty(S^n) \to C^\infty(S^{n-1})$ satisfies a system of $2N$ factorization identities. These identities are consequences

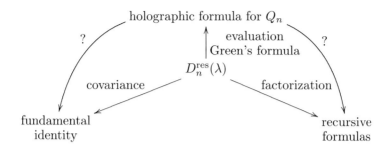

Figure 1.2: 3 facets of the critical residue family

of corresponding factorization identities for the families $\mathcal{D}_{2N}(\lambda)$ of homomorphisms of Verma modules.

The factorization identities continue to hold true for the residue families $D_{2N}^{\mathrm{res}}(h;\lambda)$ of Section 1.6, at least for conformally flat metrics. These identities give rise to *recursive relations* for

$$Q\text{-curvatures}, \quad Q\text{-polynomials} \quad \text{and} \quad \text{GJMS-operators}.$$

Even for conformally flat metrics, the complexity of explicit formulas for the critical Q-curvature Q_n, in terms of Riemannian invariants, increases exponentially with the dimension n. This makes recursive relations interesting.

We start with the formulation of the factorization identities of residue families for conformally flat metrics (Theorem 6.11.1).

Theorem 1.7.1. *Let h be conformally flat. Then the residue family $D_{2N}^{\mathrm{res}}(h;\lambda)$ factorizes for*

$$\lambda \in \left\{-\frac{n+1}{2},\ldots,-\frac{n+1}{2}+N\right\} \cup \left\{-\frac{n}{2}+N,\ldots,-\frac{n}{2}+2N\right\}$$

into products of (lower order) residue families and GJMS-operators. More precisely, for $j = 1,\ldots,N$, the identities

$$D_{2N}^{\mathrm{res}}\left(h;-\frac{n}{2}+2N-j\right) = P_{2j}(h) \circ D_{2N-2j}^{\mathrm{res}}\left(h;-\frac{n}{2}+2N-j\right)$$

and

$$D_{2N}^{\mathrm{res}}\left(h;-\frac{n+1}{2}+j\right) = D_{2N-2j}^{\mathrm{res}}\left(h;-\frac{n+1}{2}-j\right) \circ P_{2j}(dr^2+h_r)$$

hold true. Here P_{2j} are GJMS-operators with respect to the metrics h and dr^2+h_r.

In particular, we have the relation $D_{2N}^{\mathrm{res}}\left(h;-\frac{n}{2}+N\right) = P_{2N}(h)i^*$ (see (1.6.8)).

Theorem 1.7.1 offers a recursive method to determine GJMS-operators. On the one hand, $D_{2N}^{\mathrm{res}}(h;\lambda)$ satisfies $2N$ factorization identities. On the other hand, it is a polynomial of degree N, i.e., is determined by $N+1$ coefficients. Hence the right-hand sides of the factorization identities satisfy $N-1$ linear relations. In particular, for $n \geq 6$, $P_n(h)i^*$ can be written as a linear combination of the compositions

$$P_{n-2}(h) \circ D_2^{\text{res}}(h; 1), \ldots, P_2(h) \circ D_{n-2}^{\text{res}}\left(h; \frac{n}{2} - 1\right)$$

and

$$D_{n-2}^{\text{res}}\left(h; -\frac{n+3}{2}\right) \circ P_2(dr^2 + h_r), \quad D_{n-4}^{\text{res}}\left(h; -\frac{n+5}{2}\right) \circ P_4(dr^2 + h_r).$$

But the residue families $D_{2N}^{\text{res}}(h; \lambda)$ can be written recursively as linear combinations of compositions of GJMS-operators $P_2(h), \ldots, P_{2N}(h)$ *and* the Yamabe operator $P_2(dr^2 + h_r)$. This follows from the system

$$D_{2N}^{\text{res}}\left(h; -\frac{n}{2} + 2N - j\right) = P_{2j}(h) \circ D_{2N-2j}^{\text{res}}\left(h; -\frac{n}{2} + 2N - j\right), \quad j = 1, \ldots, N,$$

$$D_{2N}^{\text{res}}\left(h; -\frac{n-1}{2}\right) = D_{2N-2}^{\text{res}}\left(h; -\frac{n+3}{2}\right) \circ P_2(dr^2 + h_r)$$

of $N+1$ identities. Thus $P_n(h)i^*$ can be written as a linear combination of compositions of the lower order GJMS-operators $P_{n-2}(h), \ldots, P_2(h)$ *and* the Yamabe operator $P_2(dr^2 + h_r)$ and the Paneitz operator $P_4(dr^2 + h_r)$.

In particular, for $n = 6$ and $N = 3$ we find

$$P_6(h)i^* = \alpha P_4(h)D_2^{\text{res}}(h; 1) + \beta P_2(h)D_4^{\text{res}}(h; 2)$$
$$+ \gamma D_4^{\text{res}}(h; -9/2)P_2(dr^2 + h_r) + \delta D_2^{\text{res}}(h; -11/2)P_4(dr^2 + h_r)$$

for $(\alpha, \beta, \gamma, \delta) = \frac{1}{21}(18, -5, -4, 12)$. An evaluation of the latter formula yields the following result (Corollary 6.11.6).

Theorem 1.7.2. *For M^6 with a conformally flat metric h, the critical GJMS-operator P_6 can be written in the form*

$$P_6 = 2(P_4 P_2 + P_2 P_4) - 3P_2^3 + 2(\mathcal{L} + \mathcal{L}^*), \tag{1.7.1}$$

where the operator $\mathcal{L}$ is defined by

$$i^* \left[P_2(dr^2 + h_r), P_4(dr^2 + h_r)\right] = \mathcal{L}(h)i^*.$$

(1.7.1) actually holds true in all dimensions $n \geq 6$. For the round sphere S^n we find $\mathcal{L} = 3P_2$. For general metrics, $\mathcal{L} + \mathcal{L}^*$ is of second order. In fact, a formula for the Q-polynomial $Q_6^{\text{res}}(\lambda)$ (see Theorem 1.7.7) leads to the alternative recursive formula (1.7.13) for P_6 as the sum of the leading part $2(P_4 P_2 + P_2 P_4) - 3P_2^3$ and an explicit second-order operator.

Now we combine (1.6.12) with Theorem 1.7.1. We describe how this leads to the *recursive formula*

$$Q_n = \sum_I a_I^{(n)} P_{2I}(Q_{n-2|I|}) + (-1)^{\frac{n}{2}-1} \frac{(n-2)(n-4)\cdots 2}{(n-3)(n-5)\cdots 1} \bar{P}_2^{\frac{n}{2}-1}(\bar{Q}_2) \qquad (1.7.2)$$

for the critical Q-curvature of a conformally flat metric (Conjecture 6.11.1). In (1.7.2), the sum runs over all ordered partitions I of integers in the interval $[1, \frac{n}{2}-1]$ as sums of natural numbers. For $I = (I_1, \ldots, I_m)$, the operator P_{2I} is defined as the composition $P_{2I_1} \circ \cdots \circ P_{2I_m}$ of GJMS-operators, and we set $|I| = \sum_i I_i$. Finally, we use the notation

$$\bar{P}_2(h) = P_2(dr^2+h_r), \quad \bar{Q}_2(h) = Q_2(dr^2+h_r).$$

(1.7.2) expresses Q_n in terms of lower order Q-curvatures and lower order GJMS-operators. The right-hand side of (1.7.2) contains $2^{\frac{n}{2}-1}$ terms. In general, the coefficients $a_I^{(n)}$ are not known explicitly, but admit algorithmic descriptions ([95]).

The holographic formula (1.6.15) and the recursive relation (1.7.2) for $Q_n(h)$ use the corresponding Poincaré-Einstein metric $g = r^{-2}(dr^2+h_r)$ in fundamentally different ways: (1.6.15) involves holographic coefficients and the asymptotics of harmonic functions of the Laplacian of g, (1.7.2) contains high order powers of the Yamabe operator of dr^2+h_r.

In order to describe the mechanism which generates (1.7.2), we restrict to the special cases $n = 4$ and $n = 6$. In these cases, the rational coefficients a_I are explicit and we actually prove (1.7.2) (even for general metrics).

The following result is valid for *all* metrics (Theorem 6.11.2).

Theorem 1.7.3. *For (M^4, h), the critical residue family*

$$D_4^{\mathrm{res}}(h; \lambda) : C^\infty([0, \varepsilon) \times M) \to C^\infty(M)$$

satisfies the identities

$$D_4^{\mathrm{res}}(h; 0) = P_4(h)i^*,$$
$$D_4^{\mathrm{res}}(h; 1) = P_2(h) \circ D_2^{\mathrm{res}}(h; 1),$$
$$D_4^{\mathrm{res}}\left(h; -\frac{3}{2}\right) = D_2^{\mathrm{res}}\left(h; -\frac{7}{2}\right) \circ P_2(dr^2+h_r),$$
$$D_4^{\mathrm{res}}\left(h; -\frac{1}{2}\right) = i^* P_4(dr^2+h_r).$$

The first three identities in Theorem 1.7.3 suffice to determine the family $D_4^{\mathrm{res}}(h; \lambda)$ in terms of their right-hand sides. But $D_2^{\mathrm{res}}(h; \lambda)$ can be written as a linear combination of $P_2(h)$ and $\bar{P}_2(h)$. A combination of the resulting formula with the relation

$$Q_4 = -\dot{D}_4^{\mathrm{res}}(0)(1)$$

yields the recursive formula

$$Q_4(h) = P_2(h)Q_2(h) - 2i^* \bar{P}_2(h)\bar{Q}_2(h). \tag{1.7.3}$$

(1.7.3) is a special case of (1.7.2).

The recursive formula (1.7.3) has an interesting and important feature: although the arguments which yield (1.7.3) apply only in dimension $n = 4$, the formula holds true in *all* dimensions $n \geq 4$ (Lemma 6.11.5). In other words, this type of formula is distinguished among other formulas for Q_4, and we refer to this property as *universality*. In that sense, (1.7.1) is universal.

Next, we derive a universal formula for Q_6. The following result is valid for *all* metrics (Theorem 6.11.4).

Theorem 1.7.4. *For (M^6, h), the critical residue family $D_6^{\text{res}}(h; \lambda)$ satisfies the identities*

$$D_6^{\text{res}}(h; 0) = i^* P_6(h),$$
$$D_6^{\text{res}}(h; 1) = P_4(h) \circ D_2^{\text{res}}(h; 1),$$
$$D_6^{\text{res}}(h; 2) = P_2(h) \circ D_4^{\text{res}}(h; 2),$$
$$D_6^{\text{res}}(h; -5/2) = D_4^{\text{res}}(h; -9/2) \circ P_2(dr^2 + h_r).$$

Now we proceed similarly as for Q_4. The four identities in Theorem 1.7.4 suffice to determine the family $D_6^{\text{res}}(h; \lambda)$. Combining the resulting formula with the relation

$$Q_6(h) = \dot{D}_6^{\text{res}}(h; 0)(1),$$

we find that $Q_6(h)$ coincides with a linear combination of

$$D_4^{\text{res}}(h; -9/2) P_2(dr^2 + h_r), \quad P_2(h)D_4^{\text{res}}(h; 2) \quad \text{and} \quad P_4(h)D_2^{\text{res}}(h; 1),$$

acting on the function $u = 1$. Since the families $D_4^{\text{res}}(h; \lambda)$ and $D_2^{\text{res}}(h; \lambda)$, in turn, can be written as linear combinations of compositions of $P_4(h)$, $P_2(h)$ and $P_2(dr^2 + h_r)$, it follows that $Q_6(h)$ finally can be written as a linear combination of compositions of $P_4(h)$, $P_2(h)$ and the Yamabe operator $P_2(dr^2 + h_r)$ (acting on $u = 1$). This method yields the following special case of (1.7.2) (Theorem 6.11.5).

Theorem 1.7.5. *The critical Q-curvature of (M^6, h) satisfies*

$$Q_6 = \frac{2}{3}(P_2 Q_4 + P_4 Q_2) - \frac{5}{3}P_2^2 Q_2 + \frac{8}{3}i^* \bar{P}_2^2 \bar{Q}_2, \tag{1.7.4}$$

where all quantities are understood with respect to h.

The proof of Theorem 1.7.5 uses the observation that (1.7.3) holds true also in dimension $n = 6$ (universality). This is the way how Q_4 enters the formula for Q_6.

Among other formulas for Q_6, (1.7.4) again is distinguished by the fact that, like (1.7.3), the same formula (with the same numerical coefficients) holds true for the subcritical Q-curvatures $Q_{6,n}$ in *all* (even) dimensions $n > 6$ (Theorem 6.11.6). In other words, (1.7.4) is universal.

For general n, a proof of the recursive formula (1.7.2) along these lines would rest on the formula

$$\dot{D}_n^{\text{res}}(h;0)(1) = -(-1)^{\frac{n}{2}} Q_n(h),$$

the first half of the system of factorization identities in Theorem 1.7.1, and the identity

$$D_n^{\text{res}}\left(h; -\frac{n-1}{2}\right) = D_{n-2}^{\text{res}}\left(h; -\frac{n+3}{2}\right) \circ \bar{P}_2(h). \tag{1.7.5}$$

In addition, such a proof uses the analogous factorization identities for lower order families, and identities which relate the subcritical Q-curvatures Q_{2N}, $2N < n$ to the quantities $\bar{P}_2^N(1)$. For $n \le 6$, such formulas are provided by the universality of the respective recursive formulas. For general n, this property remains to be established.

Now we turn to the discussion of the Q-polynomials

$$Q_{2N}^{\text{res}}(h; \lambda) \overset{\text{def}}{=} -(-1)^N D_{2N}^{\text{res}}(h; \lambda)(1).$$

We start with the formulation of some results in low order cases. The proof of Theorem 1.7.5 rests (among other things) on the identity $Q_6 = \dot{D}_6^{\text{res}}(0)(1)$. For the subcritical Q-curvatures $Q_{4,n}$ and $Q_{6,n}$, there are analogous, but more complicated relations to the values

$$\dot{D}_4^{\text{res}}\left(-\frac{n}{2}+2\right)(1) \quad \text{and} \quad \dot{D}_6^{\text{res}}\left(-\frac{n}{2}+3\right)(1),$$

respectively. The following formulas hold true for all metrics and $n \ge 4$ and $n \ge 6$, respectively (Theorem 6.11.7).

Theorem 1.7.6.

$$\dot{D}_4^{\text{res}}\left(-\frac{n}{2}+2\right)(1) = -Q_{4,n} - \frac{n-4}{2}(Q_{4,n} + P_2 Q_{2,n}),$$

$$\dot{D}_6^{\text{res}}\left(-\frac{n}{2}+3\right)(1) = Q_{6,n} + \frac{n-6}{2}\left(\frac{3}{2}Q_{6,n} + P_2 Q_{4,n} - 2P_4 Q_{2,n} + \frac{3}{2}P_2^2 Q_{2,n}\right).$$

Note that these relations are a bit surprising. In fact, we recall that in the critical case the relation between $\dot{D}_n^{\text{res}}(h;0)(1)$ and $Q_n(h)$ is suggested by the formula $D_n^{\text{res}}(h;0) = P_n(h)i^*$ and the transformation formula for the quantity $\dot{D}_n^{\text{res}}(h;0)(1)$ which follows from the conformal covariance of the family. Although

$$D_{2N}^{\text{res}}\left(-\frac{n}{2}+N\right) = P_{2N}i^*$$

(see (1.6.8)), this argument does not yield information on the quantity $\dot{D}_{2N}^{\text{res}}(-\frac{n}{2}+N)(1)$. On the other hand, the subcritical Q-curvatures do not obey transformation laws, under conformal changes $h \mapsto e^{2\varphi}h$, which are linear in φ.

Theorem 1.7.6 and Theorem 1.7.4 imply the following result for the Q-polynomial $Q_6^{\mathrm{res}}(h; \lambda)$ (for all metrics h) (Theorem 6.11.8).

Theorem 1.7.7. *For even* $n \geq 6$,

$$Q_6^{\mathrm{res}}(\lambda) = \frac{1}{2}\lambda \left(\lambda + \frac{n}{2} - 4\right)\left(\lambda + \frac{n}{2} - 5\right)Q_6$$

$$+ \lambda \left(\lambda + \frac{n}{2} - 3\right)\left(\lambda + \frac{n}{2} - 4\right)P_2\left(Q_4 + \frac{3}{2}P_2 Q_2\right)$$

$$- \lambda \left(\lambda + \frac{n}{2} - 3\right)\left(\lambda + \frac{n}{2} - 5\right)P_4 Q_2.$$

The critical case $n = 6$ of the latter result is a special case of the following recursive formula for the *critical Q-polynomial* $Q_n^{\mathrm{res}}(\lambda)$ (Theorem 6.11.10).

Theorem 1.7.8. *Let* (M, h) *be a Riemannian manifold of even dimension* n *with a conformally flat metric. Assume that*

$$Q_{2j}^{\mathrm{res}}(0) = 0 \quad \text{for} \quad j = 1, \ldots, \frac{n}{2}, \tag{1.7.6}$$

and define the polynomials $\mathcal{Q}_{2j}^{\mathrm{res}}(\lambda)$ *by*

$$Q_{2j}^{\mathrm{res}}(\lambda) = \lambda \mathcal{Q}_{2j}^{\mathrm{res}}(\lambda).$$

Then

$$Q_n^{\mathrm{res}}(\lambda) = (-1)^{\frac{n}{2}-1}\lambda \prod_{k=1}^{\frac{n}{2}-1}\left(\frac{\lambda - \frac{n}{2} + k}{k}\right)Q_n$$

$$+ \lambda \sum_{j=1}^{\frac{n}{2}-1}(-1)^j \prod_{\substack{k=1 \\ k \neq j}}^{\frac{n}{2}}\left(\frac{\lambda - \frac{n}{2} + k}{k - j}\right)P_{2j}\mathcal{Q}_{n-2j}^{\mathrm{res}}\left(\frac{n}{2} - j\right). \tag{1.7.7}$$

Here all quantities are understood with respect to h.

Theorem 1.7.8 is a consequence of

$$\dot{D}_n^{\mathrm{res}}(h; 0)(1) = -(-1)^{\frac{n}{2}}Q_n(h)$$

and the first half of the system of factorization identities in Theorem 1.7.1. The condition (1.7.6) is conjectured to be always satisfied (Remark 6.11.5).

(1.7.7) relates the critical Q-polynomial $Q_n^{\mathrm{res}}(\lambda)$ to the critical Q-curvature Q_n, lower order GJMS-operators and values of lower order Q-polynomials. It motivates us to find analogs of (1.7.7) for subcritical residue families. The conjectural identities (6.11.38) provide such analogs. For their proofs it would be enough to establish appropriate generalizations of Theorem 1.7.6, i.e., relations between the subcritical Q-curvatures $Q_{2N,n}$ ($2N < n$) and the respective values $D_{2N}^{res}(-\frac{n}{2}+N)(1)$ (Theorem 6.11.11). These formulas are subcritical analogs of the identity $\dot{D}_n^{res}(0)(1) = -(-1)^{\frac{n}{2}} Q_n$.

A combination of (1.7.7) with its subcritical analogs implies a formula of the form

$$Q_n^{res}(\lambda) = \sum_I c_I(\lambda) P_{2I_1} \dots P_{2I_r} Q_{n-2|I|} \tag{1.7.8}$$

for the critical Q-polynomial in terms of Q-curvatures $Q_n, \dots, Q_2$ and GJMS-operators $P_{n-2}, \dots, P_2$. Here the sum runs over all ordered partitions I of integers in $[0, \frac{n}{2} - 1]$ as sums of non-negative integers, and the coefficients $c_I(\lambda)$ are polynomials of degree $\frac{n}{2}$. The formula

$$Q_4^{res}(\lambda) = -\lambda(\lambda-1)Q_4 - \lambda^2 P_2 Q_2$$

and Theorem 1.7.7 (for $n = 6$) are the first special cases.

The identity (1.7.8) relates the holographic data of the bulk space Poincaré-Einstein metric, which define the polynomial $Q_n^{res}(\lambda)$, to Q-curvatures and GJMS-operators of its conformal infinity. In other words, it can be regarded as a generalization of the holographic formula for Q-curvature. More precisely, by comparing the coefficients of powers of λ, the identity (1.7.7) is equivalent to a *set* of $\frac{n}{2}$ identities which relate

$$(-1)^{\frac{n}{2}} 2^{n-1} \left(\frac{n}{2}\right)! \left(\frac{n}{2}-1\right)! v_n \quad \text{and} \quad Q_n.$$

The linear coefficient is responsible for the identity $\dot{Q}_n^{res}(0) = Q_n$, and the highest power of λ yields (Corollary 6.11.5)

Corollary 1.7.1. *Under the assumption* (1.7.6),

$$\left(\frac{n}{2}-1\right)! (Q_n^{res})^{top} = (-1)^{\frac{n}{2}-1} Q_n - \sum_{j=1}^{\frac{n}{2}-1} \binom{\frac{n}{2}-1}{j-1} P_{2j} Q_{n-2j}^{res} \left(\frac{n}{2}-j\right). \tag{1.7.9}$$

Here the superscript "top" indicates the coefficient of the highest power of λ. The main difference between (1.6.15) and (1.7.9) concerns the way in which both formulas explain the divergence terms. Whereas in (1.6.15) these contributions are formulated in terms of the operators $T_{2j}^*(0)$ (which describe the asymptotics of harmonic functions of the Poincaré-Einstein metric) acting on holographic coefficients, in (1.7.9) these are covered by lower order GJMS-operators acting on lower order Q-curvatures.

Among the terms which define the Q-polynomial $Q_n^{res}(\lambda)$, the term $P_n^*(\lambda)(1)$ plays a special role. In fact, Theorem 1.6.1 and Theorem 1.6.2 (together with the identity $\dot{P}_n(0)(1) = (-1)^{\frac{n}{2}} Q_n$) imply that $\dot{P}_n^*(0)(1)$ is a linear combination of the other items. The following conjecture states how this effect extends to the full polynomial $P_n^*(\lambda)(1)$ (see Conjecture 6.11.2 and (6.11.55)).

Conjecture 1.7.1.

$$\sum_{j=0}^{\frac{n}{2}}(n+2j)\mathcal{T}_{2j}^{*}(\lambda)(v_{n-2j}) \equiv 0. \tag{1.7.10}$$

(1.7.10) can be verified for $n \leq 6$ by direct calculation using explicit formulas for the terms involved. It can be applied to evaluate the left-hand side of (1.7.9). For $n = 6$, this leads to the recursive formula

$$Q_6 = [P_2(\mathcal{Q}_4^{\mathrm{res}}(2)) + 2P_4(\mathcal{Q}_2^{\mathrm{res}}(1))] - 6[Q_4 + P_2(Q_2)]Q_2 - 2^6 3!v_6. \tag{1.7.11}$$

In Section 6.12, we will use this formula to derive the following recursive formula for the *critical* GJMS-operator P_6.

Theorem 1.7.9. *The critical GJMS-operator P_6 is given by the recursive formula*

$$P_6 u = [2(P_2 P_4 + P_4 P_2) - 3P_2^3]^0 u - 48\delta(\mathsf{P}^2 \# du) - 8\delta(\mathcal{B} \# du). \tag{1.7.12}$$

Here $[\cdot]^0$ denotes the non-constant part of the respective operator.

Among other possible formulas for the critical P_6, (1.7.12) is distinguished by the property that all subcritical cubes P_6 admit the analogous recursive presentations

$$P_6 u = [2(P_2 P_4 + P_4 P_2) - 3P_2^3]^0 u$$
$$- 48\delta(\mathsf{P}^2 \# du) - \frac{16}{n-4}\delta(\mathcal{B} \# du) - \left(\frac{n}{2}-3\right)Q_6 u, \tag{1.7.13}$$

where Q_6 is given by (1.7.11) (Corollary 6.12.2). Formula (1.7.13) shows the role of the Bach tensor $\mathcal{B}$, or rather of the conformally covariant second-order operator $-\delta(\mathcal{B} \# d) + (\mathcal{B}, \mathsf{P})$ (in dimension $n = 4$), as an obstruction to the existence of P_6 in dimension $n = 4$ ([118]).

On the round spheres S^n, the GJMS-operators are commuting intertwining operators, and (1.7.13) reduces to the non-linear relation

$$P_6 = 4P_2 P_4 - 3P_2^3 + 12P_2. \tag{1.7.14}$$

Similarly, for $n = 8$ (assuming (1.7.10)) we find the recursive formula

$$Q_8 = -[P_2(\mathcal{Q}_6^{\mathrm{res}}(3)) + 3P_4(\mathcal{Q}_4^{\mathrm{res}}(2)) + 3P_6(Q_2)]$$
$$- 12[Q_6 - P_2(\mathcal{Q}_4^{\mathrm{res}}(1)) - 2P_4(Q_2)]Q_2$$
$$- 18[Q_4 + P_2(Q_2)]^2 + 4!3!2^7 v_8 \tag{1.7.15}$$

(Theorem 6.13.1). This formula emphasizes the significance of the Q-polynomials.

Finally, Conjecture 1.7.1 implies the alternative formula

$$nQ_n^{\mathrm{res}}(\lambda) = 2^n \left(\frac{n}{2}\right)! \left[\lambda(\lambda-1)\dots\left(\lambda-\frac{n}{2}+1\right)\right] \sum_{j=0}^{\frac{n}{2}} 2j\, T_{2j}^*(\lambda)(v_{n-2j})$$

(Theorem 6.11.13) for the Q-polynomial $Q_n^{\mathrm{res}}(\lambda)$. When combined with a formula from [70], it yields a link between the *quadratic* term in the *total* critical Q-polynomial

$$\lambda \mapsto \int_M Q_n^{\mathrm{res}}(\lambda)\,\mathrm{vol}$$

of a closed manifold (M, h) and the renormalized volume of a global conformally compact Einstein metric with h being a representative of its conformal infinity (Theorem 6.11.15).

Theorem 1.7.10. *Under Conjecture* 1.7.1, *the sum*

$$(-1)^{\frac{n}{2}} \left[c_{\frac{n}{2}} \int_M Q_n^{\mathrm{res}}(\lambda)\,\mathrm{vol} + \frac{\lambda(\lambda-1)\dots(\lambda-\frac{n}{2}+1)}{(\frac{n}{2}-1)!} \int_M \mathcal{S}(n-\lambda)(1)\,\mathrm{vol} \right] + \lambda^2 V$$

vanishes up to order 2 *at* $\lambda = 0$.

Theorem 1.7.10 should be compared with the result that, for odd dimension n, the renormalized volume V is given by the integral $-\int_M \dot{\mathcal{S}}(n)(1)\,\mathrm{vol}$ ([97]). By the recursive structure of $Q_n^{\mathrm{res}}(\lambda)$, the additional contribution

$$\int_M \ddot{Q}_n^{\mathrm{res}}(0)\,\mathrm{vol}$$

can be expressed in terms of GJMS-operators acting on Q-curvatures.

The latter result allows us to determine the infinitesimal conformal variation of the quadratic coefficient of the total Q-polynomial; we recall that the linear coefficient is the conformal invariant $\int_M Q_n\,\mathrm{vol}$. We prove (see Theorem 6.11.16 and the discussion following it)

Theorem 1.7.11. *For even* n *and closed* M,

$$\frac{1}{2} \left(\int_M \ddot{Q}_n^{\mathrm{res}}(0)\,\mathrm{vol} \right)^{\bullet}[\varphi] = \int_M \varphi \left[2Q_n - (-1)^{\frac{n}{2}} c_{\frac{n}{2}}^{-1} v_n \right] \mathrm{vol}\,.$$

Note that, by Theorem 1.6.2, the conformal anomaly on the right-hand side is the divergence of a natural one-form. Theorem 1.7.11 involves the conformal variation only of local quantities. It will be derived from the transformation law of $D_n^{\mathrm{res}}(\lambda)$ *without* appealing to Conjecture 1.7.1.

We finish with some comments concerning the extension of the above results beyond the conformally flat case. The main technique in the proofs are the factorization identities. For general dimension n, and for general order, the full system of factorization identities is proved for conformally flat metrics (Theorem 1.7.1). The identities in the first half of the system actually hold true for general metrics. It follows that Theorem 1.7.8 and Corollary 1.7.1 extend to general metrics: the proofs rest only on these factorization identities and $\dot{D}_n^{res}(0)(1) = -(-1)^{\frac{n}{2}} Q_n$. For the proof of the recursive formula (1.7.2) it suffices to establish only one additional factorization identity: (1.7.5). In the cases $n = 4$ and $n = 6$, it is covered by Theorem 1.7.3 and Theorem 1.7.4. For more details we refer to the comments at the end of Section 6.11.

The observation that residue families satisfy factorization identities, which are literally the same as those for families of homomorphisms of Verma modules, being responsible for the flat case, suggests that the former are induced in some way by the latter. The holographic duality (Section 1.10) will give that idea a precise form.

1.8 Families of conformally covariant differential operators

In Section 1.6, we associated to any Riemannian manifold (Σ, h) a canonical sequence of one-parameter families of linear differential operators

$$D_N^{res}(h; \lambda) : C^\infty([0, \varepsilon) \times \Sigma) \to C^\infty(\Sigma),\ 0 \le N \le n = \dim \Sigma$$

which satisfy appropriate analogs of the transformation law (1.6.9) under conformal changes $h \mapsto e^{2\varphi}h$ of the metric, and which specialize to GJMS-operators on Σ for appropriate values of λ. We also pointed out that the transformation laws of the families $D_N^{res}(\cdot; \lambda)$ actually suggest that we ask for more general constructions which contain the residue families as a special case. More precisely, we are led to consider the following setting.

Let M be a manifold and Σ an oriented codimension one submanifold. Let $i : \Sigma \hookrightarrow M$ be the embedding. We consider one-parameter families of natural linear differential operators

$$D_N(M, \Sigma; g; \lambda) : C^\infty(M) \to C^\infty(\Sigma),\ \lambda \in \mathbb{C}$$

of order $N \ge 0$ which are conformally covariant in the sense that

$$e^{-(\lambda-N)i^*(\varphi)} \circ D_N(M, \Sigma; e^{2\varphi}g; \lambda) \circ e^{\lambda\varphi} = D_N(M, \Sigma; g; \lambda) \tag{1.8.1}$$

for all $\varphi \in C^\infty(M)$ and all $\lambda \in \mathbb{C}$.

Here a series of comments is in order. For any fixed family parameter the operators $D_N(M, \Sigma; g; \lambda)$ are supposed to be compositions of differential operators on M and Σ with the restriction map i^*. In that sense, we talk about differential operators which map functions on M to functions on a submanifold Σ. The order of a differential operator in a family may depend on the family parameter. The order of the operators for a generic parameter is called the order of the family. In (1.8.1) it is, obviously, enough to assume that the functions φ are defined in an open neighborhood of Σ. We emphasize that in (1.8.1) the order of the family is coupled in a specific way to the conformal weights: the order N coincides with the difference of the conformal weights λ and $\lambda-N$. In view of the hoped-for applications to Q-curvature, we restrict to the given version of conformal covariance. Other versions might be of interest in different contexts. Finally, we do not specify the quality of the families as functions of the family parameter. There are two reasons for that omission: we will not discuss any aspect of a possible general theory of such families, and in the specific constructions of families, which are analyzed here, that quality varies from polynomial to rational.

The problem of constructing and analyzing conformally covariant families $D_N(M, \Sigma; g; \lambda)$ should be viewed as a problem in conformal submanifold theory. In particular, the relation between conformally covariant families and Q-curvature motivates us to regard Q-curvature as a subject in conformal submanifold theory.

We note that, to a large extent, conformal submanifold theory is concerned with invariants with respect to conformal diffeomorphisms of a fixed background metric. A typical subject is the study of properties of immersions of surfaces in S^n ($n \geq 3$) which are invariant under the group of conformal diffeomorphisms of the round sphere (S^n, g_c). Sometimes that area is called Möbius geometry ([142]). One specific problem, which triggered its own large body of literature, was Willmore's conjecture concerning embeddings $i : T^2 \hookrightarrow \mathbb{R}^n$ of tori for which the functional $\mathcal{W} = \int_{T^2} |\mathcal{H}|^2 \, \text{vol}$ ($\mathcal{H}$ denotes the mean curvature vector) is minimal. Since the functional $\mathcal{W}$ is Möbius-invariant, this is a problem in conformal submanifold theory.

For a surface $i : \Sigma^2 \hookrightarrow M^3$ in a general background metric g, the appropriate generalization of $\mathcal{W}$ is the conformally invariant functional $\int_\Sigma |L_0|^2 \, \text{vol}$ (L_0 is the trace-free part of the second fundamental form). Here conformal invariance refers to arbitrary conformal changes of the metric.

It seems that the intrinsic aspect of conformal submanifold theory is still much less developed than the intrinsic theory on a given manifold. Notable exceptions, however, are Fialkow's classical work [100] and the more recent work [52] which approaches classical problems of the subject using modern technology of conformal differential geometry.

Now we describe a simple pattern by which critical conformally covariant families give rise to fundamental pairs.

A conformally covariant family will be called *critical* if its order coincides with the dimension of the target space Σ. A *fundamental pair* consists of a curvature quantity and a linear differential operator so that under conformal changes of the metric these satisfy an analog of the fundamental identity for Q-curvature.

For even $n = \dim \Sigma$ and a critical family $D_n(g; \lambda)$, assume that the operator $D_n(g; 0)$ lives on Σ, i.e., is of the form

$$D_n(g; 0) = \mathbf{P}_n(g)i^*$$

for some natural differential operator $\mathbf{P}_n$ on Σ. It is not required here that $\mathbf{P}_n(g)$ only depends on the induced metric $i^*(g)$ on Σ. If $\mathbf{P}_n(\cdot)$ only depends on $i^*(g)$, it is a conformally covariant differential operator on Σ, i.e.,

$$e^{n\varphi}\mathbf{P}_n(e^{2\varphi}g) = \mathbf{P}_n(g)$$

for all $\varphi \in C^\infty(\Sigma)$. Next, we define the quantity

$$\mathbf{Q}_n(g) \stackrel{\text{def}}{=} -(-1)^{\frac{n}{2}}\dot{D}_n(g;0)(1) \in C^\infty(\Sigma). \qquad (1.8.2)$$

It is a consequence of the conformal covariance of the family that

$$e^{n\varphi}\mathbf{Q}_n(e^{2\varphi}g) = \mathbf{Q}_n(g) + (-1)^{\frac{n}{2}}[\mathbf{P}_n(g),\varphi](1) \qquad (1.8.3)$$

for all $\varphi \in C^\infty(M)$. In fact, the operator $\dot{D}_n(\cdot;0)$ satisfies the identity

$$e^{n\varphi}\dot{D}_n(e^{2\varphi}g;0) = \dot{D}_n(g;0) + [\varphi, D_n(g;0)],$$

which for $u = 1$ implies (1.8.3). In order to simplify the formulas, we have suppressed here all pull-back operators i^*.

(1.8.3) is an analog of the fundamental identity

$$e^{n\varphi}Q_n(e^{2\varphi}g) = Q_n(g) + (-1)^{\frac{n}{2}}[P_n(g),\varphi](1) = Q_n(g) + (-1)^{\frac{n}{2}}P_n(g)(\varphi)$$

for Q_n (see (1.6.1)). Therefore, those critical families for which the fundamental pair $(\mathbf{P}_n, \mathbf{Q}_n)$ actually reproduces the pair (P_n, Q_n) deserve particular interest. However, in general, there is no reason to expect that the quantity $\mathbf{Q}_n$ only depends on the induced metric $i^*(g)$ on Σ even if $D_n(g;0)$ has this property.

The above pattern is obviously suggested by the theory of the residue families. Therefore, of particular interest are those conformally covariant families which specialize to the families

$$D_N^{\text{res}}(h;\lambda) \quad \text{for} \quad (M,\Sigma;g) = \left([0,\varepsilon) \times \Sigma; dr^2 + h_r\right).$$

The families $D_N^c(\lambda)$ provide another test case for any construction of conformally covariant families. These are actually tractor families in the sense of Section 1.9 for $(M,\Sigma) = (S^{n+1}, S^n)$. We recall that $D_n^c(0) = P_n(S^n, g_c)i^*$ (Theorem 1.4.2), i.e., the operator $D_n^c(0)$ actually lives on the submanifold S^n, and is given by the critical GJMS-operator for the round metric g_c. Moreover, the quantity $-(-1)^{\frac{n}{2}}\dot{D}_n^c(0)(1)$ equals $Q_n(S^n, g_c) = (n-1)!$ (Lemma 6.1.2).

The above derivation of (1.8.3) is an analog of a family proof of the fundamental identity for Q_n via scattering theory ([128]). In fact, near $\lambda = n$ the scattering operator $\mathcal{S}(h;\lambda) : C^\infty(\partial M) \to C^\infty(\partial M)$ has the form

$$\mathcal{S}(h;\lambda) = -\frac{c(n)P_n(h)}{\lambda - n} + \mathcal{S}_0(h;\lambda)$$

with a holomorphic family $\mathcal{S}_0(h; \lambda)$ so that $\mathcal{S}_0(h; n)(1) = c(n)(-1)^{\frac{n}{2}} Q_n(h)$. $c(n)$ is a constant which only depends on n. Hence the family

$$\Phi(h; \lambda) = \frac{n - \lambda}{c(n)} \mathcal{S}(h; \lambda)$$

is holomorphic near n, and satisfies $\Phi(h; n) = P_n(h)$ and $-(-1)^{\frac{n}{2}} \dot{\Phi}(h; n)(1) = Q_n(h)$. Moreover, under conformal changes of h the scattering operator satisfies

$$\mathcal{S}(e^{2\varphi} h; \lambda) = e^{-\lambda \varphi} \circ \mathcal{S}(h; \lambda) \circ e^{(n-\lambda)\varphi}, \quad \varphi \in C^\infty(\partial M) \qquad (1.8.4)$$

(see (3.2.6)). The fundamental identity then follows by a similar argument as above (now applied to $\Phi(\lambda)$).

However, a significant difference between both transformation rules (1.8.1) (for $N = n$) and (1.8.4) deserves to be mentioned: n is in (1.8.4) and in (1.8.1) the difference and the sum of the conformal weights, respectively.

The above arguments also work for *odd* n, and yield a pair $(\mathbf{P}_n, \mathbf{Q}_n)$ which satisfies the identity

$$e^{n\varphi} \mathbf{Q}_n(e^{2\varphi} g) = \mathbf{Q}_n(g) + [D_n(g; 0), \varphi](1)$$

for all $\varphi \in C^\infty(M)$. Here $\mathbf{Q}_n \in C^\infty(\Sigma)$ is defined as in (1.8.2) but without the sign $(-1)^{\frac{n}{2}}$. In that case, the independence of the fundamental pair $(\mathbf{P}_n, \mathbf{Q}_n)$ of the metric near Σ is no issue. In fact, for (S^{n+1}, S^n) (with odd n) the operator $D_n^c(0)$ does *not* live on S^n. Thus the fundamental pair $(\mathbf{P}_n, \mathbf{Q}_n)$ has to be regarded as being associated to the embedding. Chang and Qing ([68]) discovered such a fundamental pair (P_3, T) on a compact 4-manifold with boundary in their study of Polyakov-formulas for determinants of elliptic boundary value problems in dimension 4.

We will not address a general theory of conformally covariant families. Instead, we restrict to the discussion of one construction of such families which is of special significance: the *tractor families*. The tractor families form a sequence of conformally covariant families which are canonically associated to the given data $(M, \Sigma; g)$ of a hypersurface Σ in M and a background metric. The conformal covariance of the tractor families is a direct consequence of its definition which rests on the conformally invariant tractor calculus. This explains its name. The tractor families are rational in λ, and the above construction of fundamental pairs faces the additional problem that the critical tractor families are not holomorphic at $\lambda = 0$. For more details we refer to Section 1.9 and Section 1.10.

We close the present section with a brief discussion of a formula for a conformally covariant family of first order. Assume that dim $\Sigma \geq 2$. We introduce some more notation. We choose a unit normal vector field $N = N(g)$ on Σ. Let

$$L(X, Y) = -g(\nabla_X(Y), N) = g(\nabla_X(N), Y), \quad X, Y \in \mathcal{X}(\Sigma) \qquad (1.8.5)$$

be the second fundamental form of Σ and

$$H = \frac{\mathrm{tr}\, L}{n-1} \in C^\infty(\Sigma) \tag{1.8.6}$$

be the corresponding mean curvature. In this convention, the mean curvature of the round sphere S^{n-1} in $\mathbb{R}^n$ for $n \geq 3$ with respect to the exterior normal is 1. Now the family

$$D_1(M, \Sigma; g; \lambda) = i^* \nabla_{N(g)} - \lambda H(g) i^* : C^\infty(M) \to C^\infty(\Sigma) \tag{1.8.7}$$

is conformally covariant in the sense that

$$D_1(e^{2\varphi} g; \lambda) = e^{(\lambda-1)i^*(\varphi)} \circ D_1(g; \lambda) \circ e^{-\lambda\varphi} \tag{1.8.8}$$

for all $\varphi \in C^\infty(M)$ and all $\lambda \in \mathbb{C}$ (Theorem 6.2.1). $D_1(M, \Sigma; g; \lambda)$ is not critical since $\dim \Sigma \geq 2$. For a curve Σ in a surface M, the corresponding critical family is

$$D_1(M, \Sigma; g; \lambda) = i^* \nabla_{N(g)} - \lambda\kappa(g) i^*,$$

where $\kappa(g) = g(\nabla_{\dot{c}}(\dot{c}), N(g)) \in C^\infty(\Sigma)$ (c is a natural parametrization of Σ) is the corresponding geodesic curvature. The corresponding fundamental pair is

$$(i^* \nabla_{N(g)}, \kappa(g)).$$

For a second-order example we refer to Theorem 1.9.1. This family interpolates the Yamabe operators on M and Σ. It is critical if $\dim \Sigma = 2$.

1.9   Curved translation and tractor families

In order to find conformally covariant families in the sense of Section 1.8, direct calculations are effective only for small orders. (1.8.7) gives a first-order family. The following result yields a second-order critical family $D_2(M, \Sigma; g; \lambda)$ for an oriented surface Σ in a manifold M of dimension 3 with an arbitrary background metric g.

Theorem 1.9.1. *The natural family*

$$\begin{aligned}
D_2(M, \Sigma; g; \lambda) = {}&{-2\lambda i^* P_2(M, g) + (2\lambda+1) P_2(\Sigma, i^*(g)) i^*} \\
&+ 2\lambda(2\lambda+1) H(g) i^* \nabla_{N(g)} \\
&+ \lambda(2\lambda+1) \left(Q_2(\Sigma; i^*(g)) i^* - i^* Q_2(M; g) - \lambda H(g)^2 i^* \right)
\end{aligned}$$

is conformally covariant, i.e., it satisfies the relations

$$e^{-(\lambda-2)i^*(\varphi)} \circ D_2(M, \Sigma; e^{2\varphi} g; \lambda) \circ e^{\lambda\varphi} = D_2(M, \Sigma; g; \lambda)$$

for all $\varphi \in C^\infty(M)$ and all $\lambda \in \mathbb{C}$. Recall that $Q_2 = \mathsf{J}$.

Theorem 1.9.1 extends to higher dimensions (Theorem 6.4.1). The resulting families $D_2(M, \Sigma; g; \lambda)$ specialize, for instance, to $D_2^c(\lambda)$ (see (1.4.7)), and generalize the residue family $D_2^{\mathrm{res}}(h; \lambda)$ since

$$D_2(dr^2 + h_r; \lambda) = D_2^{\mathrm{res}}(h; \lambda),$$

where Σ is considered as the hypersurface $r^{-1}(0)$ in $M = [0, \varepsilon) \times \Sigma$ (Theorem 6.7.1). Although $D_2(M, \Sigma; g; \lambda)$ is polynomial of degree 3 in λ, under certain curvature conditions its degree can be smaller. In fact, if $H = 0$, then the degree is ≤ 2, and it is 1 iff $H = 0$ and $Q_2(M; g)$ restricts to $Q_2(\Sigma; i^*(g))$. Similarly, $D_1(M, \Sigma; g; \lambda)$ is polynomial of degree 1 for general metrics, but does not depend on λ iff $H = 0$.

In order to find higher order families, more systematic constructions are required. Such a tool is Eastwood's *curved translation principle* ([90], [20], [88]). We describe the idea of the method, and how it will be applied here. It is a curved version of Zuckerman translation of homomorphisms of Verma modules ([243]). Zuckerman translation constructs homomorphisms of Verma modules from given ones by tensoring with finite-dimensional representations. The construction translates the infinitesimal character of the Verma modules involved (whence the name). We will not go here into the details of that powerful method in any greater generality (see [236]). For our purpose, it will be enough to deal with it in a very special case. Its *curved version* mimics the construction in the framework of conformally covariant differential operators. In fact, the idea is to rephrase the construction in terms of differential intertwining operators in the conformally flat case of the sphere, and to find a conformally covariant substitute of the resulting formula. A key point is that the splitting operators, which realize the embedding of irreducibles into tensor products and projections to subquotients of tensor products, are differential operators of small order.

In order to apply curved translation, we prove that all families $\mathcal{D}_N(\lambda)$ arise by iterated translation of two extremely simple families: $\mathcal{D}_0(\lambda) = i$ and $\mathcal{D}_1(\lambda) = Y_n^- i$, where $i : \mathfrak{g}_n \hookrightarrow \mathfrak{g}_{n+1}$. $\mathcal{D}_0$ and $\mathcal{D}_1$ generate all *even* and *odd* order families, respectively, by iteration. The construction of even order families is as follows. We define families

$$\Theta_{2N}(\lambda) : \mathcal{M}_{\lambda-2N}(\mathfrak{g}_n) \to \mathcal{M}_\lambda(\mathfrak{g}_{n+1})$$

by

$$\Theta_{2N}(\lambda) = p(\lambda) \circ [\Theta_{2N-2}(\lambda-1) \otimes I_n)] \circ i(\lambda-(2N-1)), \quad \Theta_0(\lambda) = i \qquad (1.9.1)$$

using the embeddings

$$i(\lambda) : \mathcal{U}(\mathfrak{g}_n) \otimes \mathbb{C}(\lambda-1) \to \mathcal{U}(\mathfrak{g}_n) \otimes \mathbb{C}(\lambda) \otimes F_n$$

and projections

$$p(\lambda) : \mathcal{U}(\mathfrak{g}_{n+1}) \otimes \mathbb{C}(\lambda-1) \otimes F_{n+1} \to \mathcal{U}(\mathfrak{g}_{n+1}) \otimes \mathbb{C}(\lambda),$$

which induce families of homomorphisms

$$\mathcal{M}_{\lambda-1}(\mathfrak{g}_n) \to \mathcal{M}_\lambda(\mathfrak{g}_n) \otimes F_n, \qquad \mathcal{M}_{\lambda-1}(\mathfrak{g}_{n+1}) \otimes F_{n+1} \to \mathcal{M}_\lambda(\mathfrak{g}_{n+1}).$$

Here F_m denotes the finite-dimensional standard representation of G^m. In (1.9.1), $I_n : F_n \hookrightarrow F_{n+1}$ denotes the G^n-equivariant embedding which corresponds to the embedding $G^n \hookrightarrow G^{n+1}$. The rational families $i(\lambda)$ and $p(\lambda)$ are given in Lemma 6.18.2. Then the translation $\Theta_2(\lambda) : \mathcal{M}_{\lambda-2}(\mathfrak{g}_n) \to \mathcal{M}_\lambda(\mathfrak{g}_{n+1})$ of Θ_0 is induced by right multiplication with

$$(n-3+\lambda)\mathcal{D}_2(\lambda),$$

up to a rational multiple which is caused by the chosen normalizations of $i(\lambda)$ and $p(\lambda)$. Here

$$\mathcal{D}_2(\lambda) = -(n-3+2\lambda)\Delta_n^- + (n-2+2\lambda)\Delta_{n-1}^-$$

(Theorem 6.18.2). More generally, we identify all iterated translates of $\mathcal{D}_0$ in (Theorem 6.18.7)

Theorem 1.9.2. *The family* $\Theta_{2N}(\lambda) : \mathcal{M}_{\lambda-2N}(\mathfrak{g}_n) \to \mathcal{M}_\lambda(\mathfrak{g}_{n+1})$ *is induced by right multiplication with*

$$\left[\prod_{j=1}^N (n-1-(j+N)+\lambda)\right] \mathcal{D}_{2N}(\lambda), \tag{1.9.2}$$

up to a rational multiple which is caused by the chosen normalizations of $i(\lambda)$ *and* $p(\lambda)$.

The proof of Theorem 1.9.2 rests on Theorem 1.4.3. The construction actually lifts to the semi-holonomic category of Verma modules (see Section 6.17 and [92]). This has the consequence that it can be combined with the mechanisms of inductions via Cartan connections ([19]). Explicit formulas in the semi-holonomic category, however, quickly become very complicated with increasing order. The semi-holonomic lifts of $\mathcal{D}_{2N}(\lambda)$ are rational in λ with at most simple poles at the zeros of the product in (1.9.2). One can regard the existence of poles as an algebraic fact which points to the possible non-existence of certain curved analogs.

Now we write $\mathcal{D}_{2N}(\lambda)$ as an Nth Zuckerman translate, and replace each factor in the resulting formula by a conformally covariant operator. Such substitutes are given in terms of *tractor D-operators*. The simplest of these operators is given by

$$D_M(g; \lambda) : C^\infty(M) \ni u \mapsto \begin{pmatrix} \frac{1}{n-2+2\lambda}(\Delta_g + \lambda \mathsf{J}(g))u \\ du \\ 2\lambda u \end{pmatrix} \in \Gamma(\mathcal{T}^*M)$$

(Theorem 6.20.1). It replaces the projection

$$p(\lambda) : \mathcal{M}_{\lambda-1} \otimes F \to \mathcal{M}_\lambda$$

given by

$$p(\lambda) : \begin{cases} 1 \otimes 1 \otimes v_- \mapsto \frac{1}{n-2+2\lambda}(\Delta_n^- \otimes 1), \\ 1 \otimes 1 \otimes v_i \mapsto Y_i^- \otimes 1, \\ 1 \otimes 1 \otimes v_+ \mapsto 2\lambda \otimes 1, \end{cases}$$

(Lemma 6.18.2). For the sphere S^n with the round metric g_c, the conformal co-variance of $D_M(g;\lambda)$ generalizes the equivariance of the intertwining map

$$C^\infty(G, \mathbb{C}(\lambda))^P \to (C^\infty(G, \mathbb{C}(\lambda-1)) \otimes F^*)^P$$

induced by $p(\lambda)$. Analogous tractor D-operators are used on the hypersurface Σ. An additional ingredient is a substitute for the embedding $I_n : F_n \to F_{n+1}$. It is given by a conformally covariant projection operator $\Pi_\Sigma : \Gamma(\mathcal{T}^*M) \to \Gamma(\mathcal{T}^*\Sigma)$ (Theorem 6.20.6).

The formal definitions of the conformally covariant tractor families

$$D_N^T(M, \Sigma; g; \lambda) : C^\infty(M) \to C^\infty(\Sigma)$$

are given in Definition 6.21.1, Definition 6.21.2 and Definition 6.21.3. More precisely, the first two definitions describe what is meant by iterated curved translation of the embedding $i : \Sigma \hookrightarrow M$ and of (versions of) the conformally covariant normal derivative $D_1(M, \Sigma; g; \lambda)$. Then the tractor families $D_N^T(M, \Sigma; g; \lambda)$ are defined by division of the resulting compositions by certain polynomials in λ (Definition 6.21.3). This is the curved analog of the factorization in Theorem 1.9.2. In general, the tractor families have simple poles with non-vanishing residues. For a detailed discussion of the role of these poles we refer to the introduction of Chapter 6.

For $(M, \Sigma) = (S^{n+1}, S^n)$, the tractor families give rise to intertwining families. More precisely, we have $D_N^T(S^{n+1}, S^n; g_c; \lambda) = D_N^c(\lambda)$. In particular, these special cases can be used as test cases for general statements on tractor families. We recall that the value of the critical family $D_n^c(\lambda)$ at $\lambda = 0$ lives on S^n, and is given by the critical GJMS-operator $P_n(S^n, g_c)$ on the sphere. Moreover, $\dot{D}_n^c(0)(1) = -(-1)^{\frac{n}{2}} Q_n(S^n; g_c)$. This generalizes (1.6.12) and fits with the pattern formulated in Section 1.8. In contrast to the general case, the tractor families $D_N^c(\lambda)$ do not have poles.

The equivariant families $D_{2N}^c(\lambda) : C^\infty(S^{n+1}) \to C^\infty(S^n)$ can be constructed in two ways: either by induction by homomorphisms of Verma modules or by the residue method of Section 1.5. On the other hand, they are not covered by the general theory of residue families outlined in Section 1.6. This is a consequence of the observation that the respective polynomial degrees of $D_{2N}^c(\lambda)$ and $D_{2N}^{res}(S^n; h_c; \lambda)$ are $2N$ and N. The fact that $D_{2N}^c(\lambda)$ is a tractor family, which also admits a residue type construction, suggests to ask whether such constructions exist for other tractor families as well. One such case are the families $D_N^b(\lambda) : C^\infty(\mathbb{R}^{n+1}) \to C^\infty(S^n)$ (Section 5.4). These families intertwine (the non-compact model of) spherical principal series on $\mathbb{R}^{n+1}$ with spherical principal series on S^n for those conformal maps of $\mathbb{R}^{n+1} \cup \{\infty\}$ which leave the sphere S^n invariant. Like $D_N^c(\lambda)$ the families $D_N^b(\lambda)$ are induced by homomorphisms of Verma modules. We note that the polynomial degrees of the induced families $D_{2N}^c(\lambda)$ and $D_{2N}^b(\lambda)$ are $2N$ and $3N$, respectively. These different degrees reflect the properties of the second fundamental form L in both cases: (S^{n+1}, S^n) and $(\mathbb{R}^{n+1}, S^n)$ both are totally umbilic, but the mean curvatures differ. A consequence of the latter observation concerning degrees is that, in general, tractor families are not determined recursively by (analogs of) factorization identities.

If Σ is the boundary of a compact manifold M, the tractor families, for specific choices of the parameter λ, were used before by Branson and Gover ([40]) for the construction of conformally covariant elliptic self-adjoint boundary value problems. Such boundary value problems give rise to conformally covariant pseudo-differential operators on Σ (see also [111]).

The tractor families are curved versions of those semi-holonomic lifts of $\mathcal{D}_N(\lambda)$ which are constructed by iterated translation. In general, there exist semi-holonomic lifts which differ from this specific construction (Section 6.18).

The tractor families specialize to residue families (at least) in the conformally flat case. This and further aspects of the construction will be discussed in Section 1.10.

1.10 Holographic duality. Extrinsic Q-curvature. Odd order Q-curvature

Associated to a Riemannian manifold (Σ, h) (dim $\Sigma = n$) we have the sequence

$$D_N^{\mathrm{res}}(h; \lambda) : C^\infty([0, \varepsilon) \times \Sigma) \to C^\infty(\Sigma)$$

($N \leq n$ for even n) of residue families and the sequence

$$D_N^T(dr^2 + h_r; \lambda) : C^\infty([0, \varepsilon) \times \Sigma) \to C^\infty(\Sigma)$$

($N \leq n$ for even n) of tractor families. Both sequences are completely determined by the metric h. Moreover, both sequences coincide in the flat case $(\Sigma^n, h) = (\mathbb{R}^n, h_c)$. This follows from the interpretation of the residue families as homomorphisms of Verma modules which, in turn, can be decomposed as products of tractor operators for the Euclidean metrics on $\mathbb{R}^{n+1}$ and $\mathbb{R}^n$. The holographic duality concerns generalizations of that relation to the curved case (see Theorem 6.21.2).

Theorem 1.10.1 (Holographic duality). *For any conformally flat Riemannian manifold (Σ, h),*

$$D_N^{\mathrm{res}}(h; \lambda) = D_N^T(dr^2 + h_r; \lambda), \ N \in \mathbb{N}.$$

Theorem 1.10.1 deals with residue families of any order. The odd order families are defined similarly as the even order ones (Definition 1.6.1). We compare the inputs of both sides of the duality. The definition of the residue family $D_N^{\mathrm{res}}(\lambda)$ rests on the holographic coefficients v_{2j} and a sequence of differential operators which describe the asymptotics of an eigenfunction of the Laplacian for the Poincaré-Einstein metric of h. On the other hand, the tractor family $D_N^T(dr^2+h_r; \lambda)$ is defined in terms of the tractor connections of the conformal compactification dr^2+h_r of the Poincaré-Einstein metric of h and of the metric h. The relation of both constructions gives tractor formulas for the eigenfunctions. Note that, in contrast to general metrics h, the families in Theorem 1.10.1 exist for all orders N, since, for conformally flat h, the Taylor series of h_r breaks off at r^4.

The holographic duality can be regarded either as a formula for the complicated product on the right-hand side, or as the statement that the left-hand side is the specialization of a conformally covariant family to the metric dr^2+h_r. Theorem 1.10.1, in particular, implies that the rational tractor families are polynomial in λ. We expect that the duality extends to the pseudo-Riemannian case.

The following consequences of Theorem 1.10.1 are of special importance.

Corollary 1.10.1. *For a conformally flat metric h,*

$$D_n^{\mathrm{res}}(h; 0) = D_n^T(dr^2+h_r; 0),$$
$$D_n^{\mathrm{res}}(h; \lambda)(1) = D_n^T(dr^2+h_r; \lambda)(1).$$

Corollary 1.10.1 covers the critical cases of even and odd order. For *even n*, it implies the relation

$$D_n^T(dr^2+h_r; 0) = D_n^{\mathrm{res}}(h; 0) = P_n(h)i^* \tag{1.10.1}$$

(by (1.6.11)) and the relation

$$Q_n^{\mathrm{res}}(h; \lambda) = -(-1)^{\frac{n}{2}} Q_n^T(dr^2+h_r; \lambda) \tag{1.10.2}$$

of the Q-polynomials

$$Q_n^{\mathrm{res}}(h; \lambda) = -(-1)^{\frac{n}{2}} D_n^{\mathrm{res}}(h; \lambda)(1) \quad \text{and} \quad Q_n^T(g; \lambda) \overset{\mathrm{def}}{=} D_n^T(g; \lambda)(1).$$

Since $Q_n^{\mathrm{res}}(h; 0) = 0$ and $\dot{Q}_n^{\mathrm{res}}(h; 0) = Q_n(h)$ (see (1.6.12)), Theorem 1.10.1 says that

$$\dot{Q}_n^T(dr^2+h_r; 0) = -(-1)^{\frac{n}{2}} Q_n(h).$$

This is a tractor formula for the critical Q-curvature $Q_n(h)$. (1.10.1) is a tractor formula for the critical GJMS-operator $P_n(h)$ (for a conformally flat metric h).

It is a challenge to extend the holographic duality to metrics which are not conformally flat. We prove analogs of Corollary 1.10.1 for second-order families (in the non-critical case $n \geq 3$), for order 3 families (in the case $n \geq 3$) and for fourth-order families (in the critical case $n = 4$) ((6.22.5), Theorem 6.25.4, Theorem 6.22.3 and Theorem 6.22.4).

Theorem 1.10.1 implies that for conformally flat metrics the tractor families $D_N^T(dr^2 + h_r; \lambda)$ satisfy systems of factorization identities: these are consequences of corresponding identities for the residue families (Theorem 1.7.1). It would be interesting to determine to which extent, and in which form, the full theory of factorization identities generalizes to tractor families.

The conformally covariant critical tractor family $D_n^T(M, \Sigma; g; \lambda)$ ($\dim \Sigma = n$) is defined for *any* hypersurface $i : \Sigma \hookrightarrow M$. $D_n^T(M, \Sigma; g; \lambda)$ is rational in λ with a possible simple pole at $\lambda = 0$. However, for even n, it is conjectured to be regular at $\lambda = 0$ (as in Theorem 1.10.1). Then the pair

$$\left(D_n^T(M, \Sigma; g; 0), \dot{D}_n^T(M, \Sigma; g; 0)(1) \right)$$

is well defined. In the situation of Corollary 1.10.1, this pair is given by

$$\left(P_n(h)i^*, -(-1)^{\frac{n}{2}} Q_n(h) \right).$$

In particular, it only depends on the metric $h = i^*(g)$, $g = dr^2 + h_r$. In general, the situation is more complicated, and the behaviour of the critical tractor family near $\lambda = 0$ is described in the following conjecture.

Conjecture 1.10.1 (Decomposition). *The operator*

$$P_n^T(\cdot, \Sigma; g) = D_n^T(\cdot, \Sigma; g; 0)$$

admits a decomposition

$$P_n^T(\cdot, \Sigma; g) = \mathbf{P}_n(\Sigma, i^*(g))i^* + P_n^e(\cdot, \Sigma; g)$$

into conformally covariant differential operators

$$\mathbf{P}_n : C^\infty(\Sigma) \to C^\infty(\Sigma), \quad P_n^e : C^\infty(M) \to C^\infty(\Sigma),$$

where $\mathbf{P}_n$ has leading term $\Delta^{\frac{n}{2}}$. The operator P_n^e depends on the metric in a small neighborhood of Σ, but lives on Σ. Similarly,

$$Q_n^T(\cdot, \Sigma; g) = \dot{D}_n^T(\cdot, \Sigma; g; 0)(1)$$

admits a decomposition

$$-(-1)^{\frac{n}{2}} Q_n^T(\cdot, \Sigma; g) = \mathbf{Q}_n(\Sigma, i^*(g)) + Q_n^e(\cdot, \Sigma; g).$$

Moreover, the pairs $(\mathbf{P}_n, \mathbf{Q}_n)$ and (P_n^e, Q_n^e) satisfy the respective fundamental identities

$$e^{n\varphi} \mathbf{Q}_n(\hat{g}) = \mathbf{Q}_n(g) - \mathbf{P}_n(g)(\varphi)$$

on Σ and

$$e^{n\varphi} Q_n^e(\hat{g}) = Q_n^e(g) - P_n^e(g)(\varphi).$$

Q_n^e will be called the *extrinsic Q-curvature*.

For $n = 4$, we confirm Conjecture 1.10.1 by proving the following result (see Section 6.22).

Theorem 1.10.2. *For $(M^5, \Sigma^4; g)$, the operator $P_4^T(g; 0)$ and the function $Q_4^T(g)$ decompose as*

$$D_4^T(g; 0) = P_4(i^*(g))i^* + P_4^e(g)i^*$$

and

$$-Q_4^T(g) = Q_4(i^*(g)) + Q_4^e(g),$$

respectively. Here (P_4, Q_4) is the Paneitz pair on Σ for the metric $i^*(g)$. The extrinsic pair (P_4^e, Q_4^e) is given by the formulas

$$P_4^e = -4\delta(\mathcal{J} \# d)$$

and

$$Q_4^e = 4\delta(\delta \mathcal{J}) - 2\Delta(\operatorname{tr} \mathcal{J}),$$

where $\mathcal{J}$ is the conformally invariant symmetric bilinear form

$$\mathcal{J} = i^*(\mathsf{P}_M) - \mathsf{P}_\Sigma + HL - \frac{1}{2}H^2 g \tag{1.10.3}$$

with trace

$$\operatorname{tr}(\mathcal{J}) = \frac{1}{6}|L_0|^2.$$

Here L_0 is the trace-free part of the second fundamental form L. The pair (P_4^e, Q_4^e) satisfies the fundamental identity

$$e^{4\varphi}\hat{Q}_4^e = Q_4^e + P_4^e(\varphi).$$

Moreover, for the conformal compactification $g = dr^2 + h_r$ of the Poincaré-Einstein metric extending the metric h on Σ, the extrinsic pair (P_4^e, Q_4^e) vanishes since the invariant $\mathcal{J}$ vanishes.

Theorem 1.10.2 implies that for any metric g, which satisfies the conformally invariant condition

$$\mathcal{J}(g) = 0,$$

the tractor family $D_4^T(\cdot, \Sigma^4; g; \lambda)$ yields the pair (P_4, Q_4) of Σ with the induced metric. In a conformally flat background g, the invariant $\mathcal{J}$ can be written in the form

$$\frac{1}{n-2}\left(L_0^2 - \frac{1}{2(n-1)}|L_0^2|g\right).$$

(Lemma 6.23.3). Thus $\mathcal{J}$ relates to L_0^2 in the same way as the Schouten tensor relates to the Ricci tensor.

An obvious problem is to extend that picture to general dimensions by determining a conformally invariant condition on g which forces the extrinsic pair (P_n^e, Q_n^e) to vanish. In the case (S^5, S^4) with the round metric, we easily find $\mathcal{J} = 0$, i.e., the extrinsic pair vanishes.

The tractor construction of P_4 requires a choice of some embedding $i : \Sigma \to M$ with a vanishing $\mathcal{J}$-tensor and a formal calculation. One should compare this with the construction of P_4 using Poincaré-Einstein metrics. It requires us to find the Poincaré-Einstein extension of the given metric h on Σ (i.e., to determine $h_{(2)}$ and $\operatorname{tr} h_{(4)}$ in terms of h) followed by a calculation of the first two terms in the asymptotics of eigenfunctions.

For *odd* n, the critical tractor family $D_n^T(M, \Sigma; g; \lambda)$ has a simple pole at $\lambda = 0$. Its residue is a conformally covariant differential operator. If it vanishes, then the fundamental pair

$$\left(D_n^T(M, \Sigma; g; 0), \dot{D}_n^T(M, \Sigma; g; 0)(1) \right) \tag{1.10.4}$$

is well defined and satisfies a fundamental identity. The first component is a conformally covariant odd order differential operator $C^\infty(M) \to C^\infty(\Sigma)$. The second component

$$Q_n^T(M, \Sigma; g) = \dot{D}_n^T(M, \Sigma; g; 0)(1) \tag{1.10.5}$$

defines a notion of *odd order* Q-curvature. In contrast to even n, the pair (1.10.4) is not an invariant of the metric $i^*(g)$, but depends on the embedding. It is the analog of the extrinsic Q-curvature defined above for even n. For odd n, there is no intrinsic Q-curvature.

In the case $n = 3$, the critical tractor family $D_3^T(M, \Sigma; g; \lambda)$ is regular at $\lambda = 0$ if L_0 vanishes, and the resulting pair

$$P_3^T(g) = D_3^T(g; 0) \quad \text{and} \quad Q_3^T(g) = \dot{D}_3^T(0)(1)$$

is related to a pair which was discovered by Chang and Qing ([68]) in connection with Polyakov-formulas for boundary value problems on 4-manifolds. In particular, the curvature quantity Q_3^T is the T-curvature of [68]. T-curvature generalizes geodesic curvature of the boundary of a surface. It gives rise to interesting geometric flows on 4-manifolds with boundary ([186]) which supplement Q-curvature flows ([49]).

The rational tractor families $D_N^T(M, \Sigma; g; \lambda)$ give rise to

$$Q_N^T(M, \Sigma; g; \lambda) = D_N^T(M, \Sigma; g; \lambda)(1) \in C^\infty(\Sigma).$$

These are analogs of the Q-polynomials $Q_N^{\text{res}}(\Sigma; h; \lambda) \in C^\infty(\Sigma)$. For a conformally flat h, we have (Corollary 1.10.1)

$$D_n^T(dr^2 + h_r; \lambda)(1) = D_n^{\text{res}}(h; \lambda)(1),$$

i.e., in this case the critical tractor Q-polynomial coincides with the critical Q-polynomial $Q_n^{\mathrm{res}}(h; \lambda)$. The main problems here are to determine to what extent the recursive structure of the polynomials $Q_{2N}^{\mathrm{res}}(h; \lambda)$ (and the families $D_{2N}^{\mathrm{res}}(h; \lambda)$) (discussed in Section 1.7) continues to exist for the tractor Q-polynomials (it is not yet known that these are polynomials), and which information on the embedding $(\Sigma \hookrightarrow M, g)$ is encoded in $Q_n^T(M, \Sigma; g; \lambda)$ and the lower order analogs.

For the tractor families $D_N^c(\lambda) = D_N^T(S^{n+1}, S^n; g_c; \lambda)$, the corresponding Q-polynomials are determined in Lemma 5.2.11 and Lemma 5.2.14. In particular, for even n,

$$Q_n^T(S^{n+1}, S^n; g_c; \lambda) = (-1)^{\frac{n}{2}} \lambda(\lambda-1) \dots (\lambda-n+1).$$

Theorem 1.10.2 emphasizes the submanifold perspective towards Q-curvature in an example. We close the present section with a description of the main features of that perspective. The critical Q-curvature is viewed as a quantity which naturally appears in (certain) critical families

$$D_n^T(M, \Sigma; g; \lambda)$$

of conformally covariant differential operators. These families live in a neighborhood of $\Sigma \hookrightarrow X$. In that sense, Q-curvature emerges from an extrinsic point of view. The tractor families, in general, only lead to a Q-like curvature, and the actual relation to *the* Q-curvature remains to be determined. The approach shares the extrinsic point of view (in one form or another) with other approaches. We recall that the Fefferman-Graham ambient metric construction ([99]) is the basis of the construction of the GJMS-operators P_n ([124]). The ambient metric construction is a specific embedding of a given conformal manifold into an ambient space with a Ricci flat (ambient) metric. The powers of the Laplacian of the ambient metric induce the GJMS-operators on the original manifold (Section 3.1). Here we adopt a different point of view. Although we work with an ambient framework as well, we do not pose an apriori condition on the extension of the metric. Instead, we choose a specific conformally covariant construction for *all* metrics which pull back to a given one under the embedding. Separating contributions which depend on the extension from those which do not depend on it, finally leads to intrinsic results. For specific background metrics (as those, for example, which are related to the ambient metric or rather the associated Poincaré-Einstein metric), the extrinsic contributions vanish (if Σ has even dimension).

1.11 Review of the contents

In Chapter 2, we introduce basic notation and summarize background material. We describe Lie groups and their actions which are related to various models of hyperbolic spaces (Section 2.1). Of central significance are suitable models for the spherical principal series representations. These are usually defined as induced representations from parabolic subgroups. We relate these definitions to geometric definitions which are adapted to their role in conformal geometry (Section 2.3). These issues are closely related to various versions of Helgason-Poisson transformations. A construction of Nayatani ([185]) provides us with an intrinsically defined canonical invariant metric on the complement of limit sets of Kleinian groups (Section 2.4). In Section 2.5, we fix our conventions in Riemannian geometry, and prove the basic facts concerning the conformal transformation properties of various curvature quantities.

The construction of conformally covariant powers of Laplacian (GJMS-operators) is reviewed in Chapter 3. Here we describe results of Graham et al. ([124]), Branson ([32]) and Graham-Zworski ([128]).

In Chapter 4, we discuss in detail the example of the fourth-order Paneitz operator and the related Q-curvature. In particular, we prove the conformal transformation law of Q_4.

The main algebraic result of the book is Theorem 1.4.3. It yields the families $\mathcal{D}_N(\lambda) \in \mathcal{U}(\mathfrak{n}_{n+1}^-)$. Its proof splits into two parts according to the parity of the order N (Theorem 5.1.1 for even N and Theorem 5.1.3 for odd N). We characterize these families in terms of Gegenbauer polynomials (Theorem 5.1.2, Theorem 5.1.4). The families $\mathcal{D}_N(\lambda)$ induce families of equivariant differential operators $D_N^c(\lambda) : C^\infty(S^n) \to C^\infty(S^{n-1})$ and $D_N^{nc}(\lambda) : C^\infty(\mathbb{R}^n) \to C^\infty(\mathbb{R}^{n-1})$. The induction and the mutual relations between these families are discussed in Section 5.2. The explicit formula for $D_N^{nc}(\lambda)$ directly follows from the corresponding formula for $\mathcal{D}_N(\lambda)$. We prove Theorem 1.4.1 by using the recursive method which rests on the system of factorization identities.

The fact that the families $\mathcal{D}_N(\lambda)$ can be read off from the asymptotics of eigenfunctions of the Laplacian for the hyperbolic metric on the upper half-space is proved in Theorem 5.2.5 and Theorem 5.2.6. In Section 5.1.3, we prove that Theorem 5.1.1 and Theorem 5.1.3 find their natural interpretations as statements about homomorphism of Verma modules.

In Section 5.3, we derive explicit formulas for the intertwining families $D_N^c(\lambda)$ of order ≤ 4. In Section 5.4, we discuss a further series of examples. We use the relation between the residue method and geometric induction in order to find explicit formulas for low order families $D_N^b(\lambda) : C^\infty(\mathbb{R}^n) \to C^\infty(S^{n-1})$ which are induced by $\mathcal{D}_N(\lambda)$. We use the asymptotics of eigenfunctions of the Laplacian of the Poincaré metric on the ball $\mathbb{B}^n$ in order to derive explicit formulas for the families of order at most 3. In that case, the polynomial degree of $D_{2N}^b(\lambda)$ is $3N$. The detailed presentation serves also as a preparation for the later discussion of curved analogs in Chapter 6. As an application, we prove that the operator $D_3^b(0) : C^\infty(\overline{\mathbb{B}^4}) \to C^\infty(S^3)$ is proportional to the conformally covariant Chang-Qing operator P_3 ([68]) specialized to the present conformally flat situation. The general case is treated in Section 6.26.

In Section 5.5, we describe the role of the families $D_N^c(\lambda)$ in the theory of automorphic distributions and Selberg zeta functions.

In the extensive Chapter 6, we leave the framework of group-induced families and conformally flat situations. For a detailed description of the contents of Chapter 6 we refer to the introduction of that chapter. Here we only review the main points.

In Section 6.2, we discuss a first-order family. In Section 6.3, we prove Theorem 1.9.1. In Section 6.4, we generalize that result to codimension one oriented submanifolds Σ of manifolds M of dimension $n \geq 4$ and arbitrary background metrics (Theorem 6.4.1). In Section 6.24, the same families will be constructed using tractor calculus.

Section 6.5 and Section 6.6 are devoted to a discussion of the residue method in the curved case. Here we construct the residue families and discuss the holographic formulas (1.6.14), (1.6.15) and (1.6.16) for the critical Q-curvature.

In Section 6.7, we prove that the second-order residue family, actually coincides with the family discussed in Section 6.4, specialized to the background metric given by the conformal compactification of a Poincaré-Einstein metric. In the same spirit, we identify in Section 6.8 the value at $\lambda = 0$ of the order 3 residue family with the corresponding Chang-Qing operator $P_3(dr^2 + h_r)$ (which will be discussed in detail in Section 6.26).

Section 6.9 and Section 6.10 contain alternative direct proofs of the holographic formulas for $n \leq 6$. These results provide additional cross-checks of the holographic formulas. For that purpose, we determine the first three terms in the Fefferman-Graham expansion (Theorem 6.9.1). These are used to derive explicit formulas for the holographic coefficients v_2, v_4 and v_6 (Theorem 6.9.2). Section 6.10 is devoted to a detailed discussion of Q-curvature Q_6 in all dimensions $n \geq 6$.

In Section 6.11, we discuss the recursive structures outlined in Section 1.7. The central tools are the factorization identities for residue families (Theorem 1.7.1). In particular, we describe the algorithm which leads to the recursive formula (1.7.2) for the critical Q-curvature, prove Theorem 1.7.3 and Theorem 1.7.4, derive the explicit formula for Q_6 directly from the factorizations, describe the recursive structure of the Q-polynomials, prove the alternative recursive formulas (1.7.11) for Q_6, and relate its quadratic coefficient to the renormalized volume of Poincaré-Einstein metrics.

In the following two sections, we discuss two applications of some results in Section 6.11. In Section 6.12, we derive the recursive formula (1.7.12) for P_6 by infinitesimal conformal variation of (1.7.11) and observe its universality. In Section 6.13, we derive the recursive formula (1.7.15) for Q_8.

In Section 6.14, we determine the holographic coefficients v_{2j} for conformally flat metrics. We describe the consequences of (1.6.16) in the conformally flat case. In particular, we detect the Euler-form in the Q-curvature. We also show that for conformally flat metrics the explicit formula for Q_6 was derived in [12]. Finally, we emphasize similarities between holographic renormalization of the Einstein-Hilbert action and Weyl's tube formula.

In Section 6.15, we describe the Henningson-Skenderis test of the AdS/CFT duality. It interprets the holographic conformal anomaly v_4 as a linear combination of quantum conformal anomalies. Here the field content of the $SU(N)$ super Yang-Mills multiplet suggest the correct choice of operators. The calculations rest on explicit formulas for heat-equation coefficients.

In Section 6.16, we demonstrate that the holographic formula implies Gover's formula for the Q-curvature of an Einstein metric.

In Section 6.17 and Section 6.18, we recall basic ideas on geometric induction using Verma modules and Cartan connections. That approach cannot be used to construct the critical GJMS-operators. The obstruction comes from the non-existence of a certain lift of a homomorphism of Verma-modules to a homomorphism of semi-holonomic Verma modules.

In Section 6.19, we prepare the translation from the Lie algebra framework to the curved framework.

Some tools from conformal tractor calculus are developed in Section 6.20. The discussion in these sections is self-contained, and can be read as an introduction to tractor calculus. For more background material on tractor calculus see [17], [88], [54], [109] and the references in [41].

In Section 6.21, we formulate the principles around the construction of the tractor families $D_N^T(\cdot, \Sigma; g; \lambda)$. In particular, we introduce the notion of extrinsic Q-curvature of a codimension one embedding, formulate the holographic duality, and define an odd order version of Q-curvature.

In Section 6.22, we analyze the constructions of Section 6.21 in special cases, and supply evidence for the conjectures. In particular, we determine the extrinsic pair (P_4^e, Q_4^e) (Theorem 1.10.2).

In Section 6.24, we recover the order 2 family constructed in Section 6.4 by the general tractor construction.

In Section 6.25 and Section 6.26, we prove that the order 3 tractor family contains the Chang-Qing pair (P_3, T) (i.e., an order 3 operator together with a curvature invariant) if $L_0 = 0$.

1.12   Some further perspectives

The residue method of the present work applies Poincaré-Einstein metrics in a neighborhood of the submanifold $\Sigma \hookrightarrow M$. Another application of Poincaré-Einstein metrics is suggested by the following observations. The case $(M, \Sigma) = (S^n, S^{n-1})$ with the canonical metrics g_c on spheres can be viewed as the boundary situation associated to an isometric embedding $\mathbb{H}^n \hookrightarrow \mathbb{H}^{n+1}$ of hyperbolic spaces. Thus $(S^{n-1}, [g_c])$ is the conformal infinity of the submanifold $\mathbb{H}^n$ of $\mathbb{H}^{n+1}$ with conformal infinity $(S^n, [g_c])$. A natural problem is to construct the families $D_N^c(S^n, S^{n-1}; \lambda)$ from that point of view using asymptotic analysis of eigenfunctions of Laplacians on hyperbolic spaces. The appropriate mathematical tool here, of course, is the theory of Poisson transformations ([140]). These group-equivariant transforms can be used to detect the counterpart in the spectral theory on hyperbolic spaces of the family theory on the boundary. Here Gegenbauer polynomials play a central role, and this connection actually can be viewed as the reason for their occurrence in the boundary theory ([150]). The above situation is a typical special case of those considered in connection with the AdS/CFT correspondence for submanifold observables. Here the expectation value of a k-dimensional submanifold Σ of M (a k-brane) is calculated using a path integral with a brane wrapped on a $k + 1$-dimensional submanifold with boundary Σ of an Einstein space with M as boundary. Note that the renormalized volume of a surface observable leads to the Willmore functional ([127], [7]). That perspective suggests that we construct conformally covariant families by using submanifolds of Poincaré-Einstein spaces associated to the conformal infinity $(M, \Sigma; [g])$. A closely related problem is to find ambient metric constructions.

It seems natural to extend the theory to submanifolds Σ of *higher* codimension. A typical case again is that of a standard embedding $S^m \hookrightarrow S^n$ of spheres. The theory of the Selberg zeta function for Kleinian groups with limit set S^m again suggests the existence of G^{m+1}-equivariant families of differential intertwining operators $C^\infty(S^n) \to C^\infty(S^m)$. An extension of the residue method would construct such families from an asymptotic analysis of generalized eigenfunctions of the Laplacian on the complement Ω of S^m in S^n. The first obvious question is which metric is to be used on Ω. For the application to zeta functions and related automorphic distributions, that metric must be invariant under the conformal group of the complement. This group is isomorphic to $SO(1, m+1) \times SO(n-m)$, and a canonical choice of such a metric in the conformal class of g_c is the Nayatani metric ([162], [185]). It provides an isometry

$$S^n \setminus S^m \to \mathbb{H}^{m+1} \times S^{n-m-1}$$

with the direct product of an Einstein metric with negative scalar curvature and an Einstein metric with constant positive curvature. In the case $m = n - 1$ discussed in the present work, the second factor degenerates to $S^0 = \{\pm\}$, i.e., Ω is isometric to two copies of hyperbolic space. The generalized eigenfunctions of the Laplacian of that metric are determined by functions (or hyperfunctions) on the geodesic boundary S^m of $\mathbb{H}^{m+1}$ and eigenfunctions of the Laplacian on the second factor. Spherical harmonics then serve as an additional parameter (besides the spectral parameter) for the families $C^\infty(S^n) \to C^\infty(S^m)$.

The product space $\mathbb{H}^5 \times S^5$ (or rather $AdS_5 \times S^5$) is the framework of a basic special case of the AdS/CFT correspondence which proposes a duality between gravity on the ten-dimensional $AdS_5 \times S^5$ and conformal field theory on the boundary S^4 (or rather the compactified Minkowski space M^4) of $\mathbb{H}^5$ (or rather AdS_5). The metric on the product is the direct product of the canonical metrics (see [172] and [241]), i.e., it coincides with the Nayatani metric.

As an example of a conformally covariant family in a higher codimension situation, we note the following generalization of Theorem 1.9.1. For any isometric embedding $\Sigma^m \hookrightarrow M^n$ $(m < n)$ in the background metric g, the family

$$
\begin{aligned}
D_2(M, \Sigma; g; \lambda) = & -(m-2+2\lambda)i^* P_2(M, g) + (n-2+2\lambda)P_2(\Sigma, g)i^* \\
& + (n-2+2\lambda)(m-2+2\lambda)i^* \nabla_{\mathcal{H}} \\
& + 2\left(\lambda + \frac{n-2}{2}\right)\left(\lambda + \frac{m-2}{2}\right)\left(Q_2(\Sigma; g) - Q_2(M; g) - \lambda |\mathcal{H}|^2\right)i^*
\end{aligned}
$$

is conformally covariant. Here $\mathcal{H}$ denotes the mean curvature vector. In the critical case $m = 2$, we find

$$
D_2(M; \Sigma; g; 0) = (n-2)P_2(\Sigma; g) \quad \text{and} \quad \dot{D}_2(M, \Sigma; g; 0)(1) = (n-2)Q_2(\Sigma; g).
$$

The main constructions (residue families and tractor families) extend to differential forms, and yield conformally covariant families of local operators on p-forms. In particular, the residue families yield Q-curvature operators

$$
Q_{n-p}^{(p)}(h) : \ker(d \,|\, \Omega^p(M)) \to \Omega^p(M), \quad Q_n^{(0)} = Q_n
$$

on closed forms which satisfy fundamental identities, and are given by holographic formulas. The details and the relation to Q-curvature operators as constructed in [42] using ambient metric and tractor constructions will be discussed elsewhere. More generally, versions for other vector bundles are of interest, too. Such families should combine with Bernstein-Gelfand-Gelfand-sequences on M and Σ ([60]).

Although the present work only deals with the Riemannian case, it is clear that most of the arguments will generalize to the broader setting of metrics with non-trivial signature. In particular, the asymptotic behaviour of eigenfunctions, the relation to Verma module theory ([16]) and the tractor constructions generalize appropriately. In fact, Q-curvature is defined, and the holographic formula actually is valid in the general case ([125]).

In another direction, the conformal setting can be enlarged to the setting of parabolic geometry ([54], [109], [222]). In particular, this would cover CR-manifolds ([98], [215], [214], [143], [144]).

The holographic formulas for Q-curvature offer natural ways to organize the Riemannian invariants which contribute. Explicit formulae are important for applications to problems in geometric analysis such as generalized Polyakov-formulas, Q-curvature prescription problems, Q-curvature flows, geometric relevance of Q-curvature etc. ([63], [64], [66], [71], [74], [84], [49]). It would be interesting to see applications in such directions beyond 4 dimensions.

Finally, we note that it is actually not too surprising that, starting from Selberg zeta functions, we came close to ideas in the AdS/CFT-correspondence and holography. In fact, the idea of studying the Selberg zeta function of a discrete subgroup Γ of $SO(1, n)^\circ$ in terms of the conformal action of Γ on the boundary of hyperbolic space can be viewed as a version of a holography principle. For other discussions of relations between Kleinian groups and holography we refer to [154] and [160]. Such relations for Schottky groups in hyperbolic 3-space are naturally connected also with arithmetic geometry ([174]). Patterson's conjecture concerning the divisor of Selberg zeta functions for Kleinian groups is another natural part of the picture ([50], [190], [151]). As described in Section 1.1, any Kleinian group Γ induces a hyperbolic quotient space $X = \Gamma \backslash \mathbb{H}^n$ (Kleinian manifold), as well as a conformal dynamics on S^{n-1} and a continuous dynamics (geodesic flow of X). The interplay between these three pictures has been the source of many exciting developments ([79], [151] and the references therein). In the arithmetical special case $PSL(2, \mathbb{Z})$, it led to the work [61], [167].

Chapter 2

Spaces, Actions, Representations and Curvature

The present chapter contains background material. We fix conventions on the structure of the Lie group $SO(1, n)$ and its Lie-algebra. Starting from the light-cone model, we derive formulas for its actions on S^{n-1} and $\mathbb{R}^{n-1} \cup \{\infty\}$ using stereographic projection. These actions induce two models of the spherical principal series representations. We define these in a form which emphasizes the conformal nature of the geometric actions with respect to canonical metrics in each model. Moreover, we prove the equivariance of the corresponding Poisson transformations. These intertwine the principal series representations with eigenspace representations for the Laplacian on hyperbolic space. Finally, we derive the standard transformation laws for the Riemannian curvature, Ricci curvature and scalar curvature with respect to conformal changes of the metric.

2.1 Lie groups, Lie algebras, spaces and actions

Let $O(1, n)$ be the Lie group of all linear automorphisms of $(\mathbb{R}^{1,n}, q)$, where q is the quadratic form $q(x, y) = -x_0 y_0 + x_1 y_1 + \cdots + x_n y_n$. Let $SO(1, n) = \{T \in O(1, n) \mid \det T = 1\}$, and let $G^n = SO(1, n)^\circ$ be the connected component of the identity in $SO(1, n)$. G^n is a real simple Lie group of real rank one. Let $\mathfrak{g}_n$ be the Lie algebra of G^n.

We write elements in $G^n = SO(1, n)^\circ$ as matrices of the block forms

$$\begin{pmatrix} 1 \times 1 & 1 \times 1 & 1 \times n-1 \\ 1 \times 1 & 1 \times 1 & 1 \times n-1 \\ n-1 \times 1 & n-1 \times 1 & n-1 \times n-1 \end{pmatrix}, \quad \begin{pmatrix} 1 \times 1 & 1 \times n \\ n \times 1 & n \times n \end{pmatrix}.$$

$K^n = SO(n)$ embeds into G^n by $T \hookrightarrow \begin{pmatrix} 1 & 0 \\ 0 & T \end{pmatrix}$ as a maximal compact subgroup.

Let A be the group of the matrices

$$a_t = \begin{pmatrix} \cosh t & \sinh t & 0 \\ \sinh t & \cosh t & 0 \\ 0 & 0 & 1 \end{pmatrix}, \; t \in \mathbb{R}.$$

Its Lie algebra $\mathfrak{a}$ is generated by

$$H_0 \overset{\text{def}}{=} d/dt|_{t=0}(a_t) = \begin{pmatrix} 0 & 1 & 0 \\ 1 & 0 & 0 \\ 0 & 0 & 0 \end{pmatrix}.$$

Let M^n be the centralizer of $\mathfrak{a}$ in K^n. Then $M^n \subset K^n$ is the group of all matrices of the form

$$\begin{pmatrix} 1 & 0 \\ 0 & T \end{pmatrix}, \; T \in SO(n-1).$$

The matrices $\begin{pmatrix} 0 & 0 \\ 0 & M_{ij} \end{pmatrix}$, $i, j = 1, \ldots, n-1$, $i < j$ such that

$$(M_{ij})_{rs} = \delta_{ir}\delta_{js} - \delta_{is}\delta_{jr}$$

form a basis of the Lie algebra $\mathfrak{m}_n$ of M^n. The restricted root system $\Delta(\mathfrak{g}, \mathfrak{a})$ consists of two elements $\pm \alpha_0$. We fix a positive root α_0 by the condition $\alpha_0(H_0) = 1$. Let $\mathfrak{a}^+ = \{X \in \mathfrak{a} \,|\, \alpha_0(X) > 0\}$ be the positive Weyl chamber. The Weyl group $W = W(\mathfrak{g}, \mathfrak{a})$ consists of two elements. Let w be the non-trivial one.

$\mathfrak{g}_n$ decomposes as

$$\mathfrak{g}_n = \mathfrak{n}_n^+ \oplus \mathfrak{m}_n \oplus \mathfrak{a} \oplus \mathfrak{n}_n^-,$$

where $\mathfrak{n}_n^{\pm} = \{Y \in \mathfrak{g}_n \,|\, [X, Y] = \pm \alpha_0(X)Y, \; X \in \mathfrak{a}\}$. The Lie algebras $\mathfrak{n}_n^+$ and $\mathfrak{n}_n^-$ consist of the respective matrices

$$\begin{pmatrix} 0 & 0 & Y^+ \\ 0 & 0 & Y^+ \\ (Y^+)^t & -(Y^+)^t & 0 \end{pmatrix}, \quad \begin{pmatrix} 0 & 0 & Y^- \\ 0 & 0 & -Y^- \\ (Y^-)^t & (Y^-)^t & 0 \end{pmatrix}$$

where $Y^{\pm} = (y_1^{\pm}, \ldots, y_{n-1}^{\pm}) \in \mathbb{R}^{n-1}$. $\mathfrak{n}_n^{\pm}$ are the Lie algebras of the abelian Lie groups N_n^+ and N_n^-. We choose

$$Y_j^+ = \begin{pmatrix} 0 & 0 & e_j \\ 0 & 0 & e_j \\ e_j^t & -e_j^t & 0 \end{pmatrix}, \quad Y_j^- = \begin{pmatrix} 0 & 0 & e_j \\ 0 & 0 & -e_j \\ e_j^t & e_j^t & 0 \end{pmatrix}, \; j = 1, \ldots, n-1 \quad (2.1.1)$$

as *standard bases* of $\mathfrak{n}_n^{\pm}$. Here $e_j \in \mathbb{R}^{n-1}$ has the components $(e_j)_k = \delta_{jk}$. In these terms, the following *commutator relations* hold true:

$$\begin{aligned} [Y_i^+, Y_j^-] &= 2\delta_{ij} H_0 + 2M_{ij}, \\ [M_{ij}, Y_r^{\pm}] &= \delta_{jr} Y_i^{\pm} - \delta_{ir} Y_j^{\pm}, \\ [H_0, Y_j^{\pm}] &= \pm Y_j^{\pm}, \\ [H_0, M_{ij}] &= 0. \end{aligned} \qquad (2.1.2)$$

The abelian groups $(N^n)^\pm$ are realized as

$$(N^n)^\pm = \left\{ n_s^\pm = \begin{pmatrix} 1 + |s|^2/2 & \mp|s|^2/2 & s \\ \pm|s|^2/2 & 1 - |s|^2/2 & \pm s \\ s^t & \mp s^t & 1_{(n-1,n-1)} \end{pmatrix} \right\},$$

where $s = (s_1, \ldots, s_{n-1}) \in \mathbb{R}^{n-1}$ and $|s| = \sum_{j=1}^{n-1} s_j^2$.

Now let

$$C(-1) = \left\{ \xi = (\xi_0, \xi') \in \mathbb{R}^{1,n} \,|\, |\xi'|^2 - \xi_0^2 = -1, \, \xi_0 > 0 \right\}.$$

$G^n = SO(1,n)^\circ$ operates transitively on $C(-1)$, and as homogeneous spaces $C(-1) \simeq SO(1,n)^\circ/SO(n)$, where $SO(n)$ is the isotropy group of $(1,0)$. $SO(1,n)^\circ$ also operates transitively on the upper light cone

$$C(0) = \left\{ \xi = (\xi_0, \xi') \in \mathbb{R}^{1,n} \,|\, |\xi'|^2 - \xi_0^2 = 0, \, \xi_0 > 0 \right\}.$$

Since the action $G^n \times C(0) \to C(0)$ commutes with the action $\xi \mapsto \lambda\xi$, $\lambda > 0$ on $C(0)$, it induces an action on the sphere

$$S^{n-1} \simeq \left\{ (1, \xi') \in \mathbb{R}^{1,n} \,|\, |\xi'| = 1 \right\} \subset C(0)$$

via the composition

$$\mathbb{R}^n \supset S^{n-1} \ni \xi' \mapsto (1, \xi') \mapsto g(1, \xi') \mapsto \left(1, \frac{g(1,\xi')'}{g(1,\xi')_0} \right) \mapsto \frac{g(1,\xi')'}{g(1,\xi')_0} \in S^{n-1}. \tag{2.1.3}$$

As homogeneous spaces, $S^{n-1} \simeq G^n/P^n$, where P^n is the parabolic subgroup which fixes the ray $\mathbb{R}^+(1,1,0,\ldots,0)$. The G^n-equivariant identification $S^{n-1} \simeq G^n/P^n$ with $P = MAN^+$ shows that the geometric G-action on S^{n-1} can be identified with the G-action

$$g : kM \mapsto \kappa(g^{-1}k)M$$

on $K^n/M^n = SO(n)/SO(n-1)$ which is induced by the Iwasawa decomposition $G = KAN^+$. Here $\kappa(g) \exp H(g) n^+(g)$ is the Iwasawa decomposition of g.

For $g = \begin{pmatrix} d & c \\ b^t & A \end{pmatrix}$ (with arrows b, c of length n and a scalar d), we have $g(1, x) = (d + (c, x), b + xA^t)$, i.e., g induces the map

$$x \mapsto \frac{xA^t + b}{(c, x) + d} \tag{2.1.4}$$

on $S^{n-1} \subset \mathbb{R}^n$.

The restriction of the Lorentzian metric $-d\zeta_0^2 + \sum_{i=1}^n d\zeta_i^2$ on $\mathbb{R}^{1,n}$ to S^{n-1} coincides with the round metric g_c induced by the embedding $S^{n-1} \hookrightarrow \mathbb{R}^n$. Moreover, (2.1.3) implies that the maps (2.1.4) are *conformal* with respect to the metric g_c:

$$g_*(g_c) = \frac{1}{(d - (b,x))^2} g_c, \quad g = \begin{pmatrix} d & c \\ b^t & A \end{pmatrix}. \tag{2.1.5}$$

The action (2.1.4) of G^n on S^{n-1} extends the *isometric* action on the unit ball $\mathbb{B}^n$ in $\mathbb{R}^n$ with the metric

$$\frac{4}{(1 - |x|^2)^2} \sum_{i=1}^n dx_i^2. \tag{2.1.6}$$

In order to see that, we use the fact that

$$C(-1) \ni (\xi_0, \xi') \mapsto \frac{\xi'}{1 + \xi_0} \in \mathbb{B}^n$$

is an isometry ([206], Section 4.5). Its inverse is given by

$$\mathbb{B}^n \ni x \mapsto \left(\frac{1 + |x|^2}{1 - |x|^2}, \frac{2x}{1 - |x|^2} \right) \in C(-1).$$

These formulas show that $\begin{pmatrix} d & c \\ b^t & A \end{pmatrix}$ operates on $\mathbb{B}^n$ by

$$x \mapsto \frac{2x A^t + b(1 + |x|^2)}{1 - |x|^2 + 2(c,x) + d(1 + |x|^2)}. \tag{2.1.7}$$

For $x \in S^{n-1} = \partial \mathbb{B}^n$, the latter formula reproduces (2.1.4).

Remark 2.1.1. *The above conventions differ from those in part of the conformal geometry literature ([17], [88], [109]). In fact, several authors define $SO(1,n)$ by the quadratic form $2x_0 x_n + x_1^2 + \cdots + x_{n-1}^2$. In that situation it is convenient to write elements in the block form*

$$\begin{pmatrix} 1 \times 1 & 1 \times n{-}1 & 1 \times 1 \\ n{-}1 \times 1 & n{-}1 \times n{-}1 & n{-}1 \times 1 \\ 1 \times 1 & 1 \times n{-}1 & 1 \times 1 \end{pmatrix}.$$

Then M and A are the groups of matrices

$$\begin{pmatrix} 1 & 0 & 0 \\ 0 & T & 0 \\ 0 & 0 & 1 \end{pmatrix}, \ T \in SO(n{-}1), \quad \begin{pmatrix} e^t & 0 & 0 \\ 0 & 1 & 0 \\ 0 & 0 & e^{-t} \end{pmatrix}, \ t \in \mathbb{R}.$$

In terms of

$$H_0 = \begin{pmatrix} 1 & 0 & 0 \\ 0 & 0 & 0 \\ 0 & 0 & -1 \end{pmatrix}, \quad Y_j^+ = \begin{pmatrix} 0 & e_j & 0 \\ 0 & 0 & -e_j^t \\ 0 & 0 & 0 \end{pmatrix}, \quad Y_j^- = \begin{pmatrix} 0 & 0 & 0 \\ e_j^t & 0 & 0 \\ 0 & -e_j & 0 \end{pmatrix}$$

the commutator relations of that model are

$$\left[Y_i^+, Y_j^- \right] = \delta_{ij} H_0 + M_{ij},$$
$$\left[M_{ij}, Y_r^\pm \right] = \delta_{jr} Y_i^\pm - \delta_{ir} Y_j^\pm,$$
$$\left[H_0, Y_j^\pm \right] = \pm Y_j^\pm,$$
$$\left[H_0, M_{ij} \right] = 0.$$

The choice of the flat model will later be reflected, for instance, in the explicit form of the conformally invariant tractor calculus (Chapter 6).

2.2 Stereographic projection

Let i denote the mappings $\mathbb{R}^{n-1} \ni x' \mapsto (x',0) \in \mathbb{R}^n$ and $\mathbb{R}^n \ni x \mapsto (x,0) \in \mathbb{R}^{n+1}$. The second map induces a map $i : S^{n-1} \to S^n$ of the standard spheres. Then $S^{n-1} \subset S^n$ is the intersection of S^n with the hyperplane $x_{n+1} = 0$, i.e., $\mathcal{H}_0(x) = x_{n+1}$ is a defining function. $\mathcal{H}_0$ will be called the *height*.

Let

$$\kappa' : \mathbb{R}^{n-1} \to S^{n-1} \setminus \{\mathcal{N}\}, \quad \kappa : \mathbb{R}^n \to S^n \setminus \{\mathcal{N}\}$$

be the stereographic projections with respect to the pole $\mathcal{N} = (1,0,\ldots,0) \in S^{n-1} \subset S^n$. Then

$$\kappa' = \kappa'_{\mathcal{N}} : x' = (x_1, \ldots, x_{n-1}) \mapsto \left(\frac{|x'|^2 - 1}{|x'|^2 + 1}, \frac{2x'}{1 + |x'|^2} \right) \in S^{n-1}$$

and

$$\kappa = \kappa_{\mathcal{N}} : x = (x_1, \ldots, x_n) \mapsto \left(\frac{|x|^2 - 1}{|x|^2 + 1}, \frac{2x}{1 + |x|^2} \right) \in S^n.$$

The diagram

$$
\begin{array}{ccc}
S^{n-1} & \xrightarrow{\;i\;} & S^n \\
\big\uparrow{\scriptstyle \kappa'} & & \big\uparrow{\scriptstyle \kappa} \\
\mathbb{R}^{n-1} & \xrightarrow{\;i\;} & \mathbb{R}^n
\end{array}
$$

commutes. We have $\kappa_{\mathcal{N}}(\infty) = \mathcal{N}$ and $\kappa_{\mathcal{N}}(0) = \mathcal{S} = -\mathcal{N}$. The inverse of $\kappa_{\mathcal{N}}$ is given by

$$\kappa_{\mathcal{N}}^{-1}(x_1, \ldots, x_{n+1}) = \frac{1}{1 - x_1}(x_2, \ldots, x_{n+1}).$$

We use κ to transport the round metric g_c on S^n to a metric on $\mathbb{R}^n$. The identity

$$\kappa_* \left(\frac{\partial}{\partial x_i} \right) = \frac{4x_i}{(1+|x|^2)^2} \frac{\partial}{\partial x_1} + \sum_{j=2}^{n+1} \frac{2(1+|x|^2)\delta_{ij-1} - 4x_i x_{j-1}}{(1+|x|^2)^2} \frac{\partial}{\partial x_j}, \quad i = 1, \ldots, n$$

gives

$$\kappa^*(g_c) \left(\frac{\partial}{\partial x_i}, \frac{\partial}{\partial x_j} \right) = \frac{4\delta_{ij}}{(1+|x|^2)^2}, \quad i, j = 1, \ldots, n, \tag{2.2.1}$$

i.e., κ is an isometry if we use on $\mathbb{R}^n$ the *spherical* metric

$$g_s = \frac{4}{(1+|x|^2)^2} \sum_{i=1}^{n} dx_i^2. \tag{2.2.2}$$

The pull-back $\kappa^*(\mathcal{H}_0)$ of the height function $\mathcal{H}_0 \in C^\infty(S^n)$ will be denoted also by $\mathcal{H}_0$. Then

$$\mathcal{H}_0(x) = \frac{2x_n}{1+|x|^2} \in C^\infty(\mathbb{R}^n)$$

and

$$d\mathcal{H}_0(x) = -\sum_{j=1}^{n-1} \frac{4x_j x_n}{(1+|x|^2)^2} dx_j + \frac{2(1+|x|^2 - 2x_n^2)}{(1+|x|^2)^2} dx_n,$$

and the gradient of $\mathcal{H}_0$ with respect to the metric g_s is given by

$$\mathrm{grad}_{g_s}(\mathcal{H}_0) = -\sum_{j=1}^{n-1} x_j x_n \frac{\partial}{\partial x_j} + \frac{1}{2}(1+|x|^2 - 2x_n^2) \frac{\partial}{\partial x_n}. \tag{2.2.3}$$

In particular,

$$i^* \mathrm{grad}_{g_s}(\mathcal{H}_0) = \frac{1}{2}(1+|x|^2) \frac{\partial}{\partial x_n} \Big|_{x_n=0}. \tag{2.2.4}$$

It follows that

$$\kappa^*(\mathcal{H}_0^{-2} g_c) = \left(\frac{1+|x|^2}{2x_n} \right)^2 \frac{4}{(1+|x|^2)^2} \sum_{i=1}^{n} dx_i^2 = x_n^{-2} \sum_{i=1}^{n} dx_i^2,$$

i.e., κ realizes an isometry of the upper hemisphere H^+ with the metric $\mathcal{H}_0^{-2} g_c$ and the hyperbolic upper half-space $\mathbb{H}^n = \{(x', x_n) \in \mathbb{R}^n \mid x_n > 0\}$ with the metric $x_n^{-2} \sum_{i=1}^{n} dx_i^2$.

We use κ to transport the action of $SO(1, n+1)$ on S^n to an action on $\mathbb{R}^n \cup \{\infty\}$. For $g \in SO(n+1) \subset SO(1, n+1)$ of the form

$$g = \begin{pmatrix} 1 & 0 & 0 \\ 0 & d & c \\ 0 & b^t & T \end{pmatrix},$$

we find

$$x \mapsto \frac{2xT^t + b(|x|^2 - 1)}{1 + |x|^2 - 2(c, x) + d(1 - |x|^2)}. \tag{2.2.5}$$

In particular, if $b = c = 0$, $d = 1$ and $T \in SO(n)$, the action is given by $x \mapsto xT^t$, i.e., T operates by standard rotation. Next, a calculation shows that

$$\kappa^{-1}(a_t \kappa(x)) = e^t x, \quad \kappa^{-1}(n_s^+ \kappa(x)) = x + s. \tag{2.2.6}$$

Similar formulas hold true for κ'.

The action of

$$\begin{pmatrix} SO(1, n) & 0 \\ 0 & 1 \end{pmatrix} \subset SO(1, n+1)$$

on S^n obviously preserves the hemispheres $H^\pm$ and the hypersphere $S^{n-1} = \partial H^\pm$. Hence under the action on $\mathbb{R}^n \cup \{\infty\}$ the subgroup $G^n = \begin{pmatrix} SO(1, n) & 0 \\ 0 & 1 \end{pmatrix} \subset G^{n+1}$ preserves the half-spaces $\mathbb{R}_\pm^n$ and their boundary $\mathbb{R}^{n-1} \cup \{\infty\}$. Moreover, the action preserves the hyperbolic metric $x_n^{-2} \sum_{i=1}^n dx_i^2$ on $\mathbb{R}_n^+$. In fact, for

$$g = \begin{pmatrix} 1 & 0 & 0 \\ 0 & T & 0 \\ 0 & 0 & 1 \end{pmatrix} \in SO(1, n+1), \ T \in SO(n)$$

this is clear since g preserves $\mathcal{H}_0 \in C^\infty(S^n)$ and the round metric g_c on H^+. On the other hand, (2.2.6) shows that $n^+(s) \in (N^n)^+$ and $a_t \in A$ preserve the metric $x_n^{-2} \sum_{i=1}^n dx_i^2$. This proves the claim using the Iwasawa decomposition $G^n = K^n A(N^n)^+$.

(2.2.6) also implies that stereographic projection can be interpreted naturally in terms of the Bruhat decomposition. We recall that

$$G_{\mathcal{S}}^n = (P^n)^- = M^n A(N^n)^- \quad \text{and} \quad G_{\mathcal{N}}^n = (P^n)^+ = M^n A(N^n)^+.$$

We consider the stereographic projections

$$\kappa_{\mathcal{S}}' : \mathbb{R}^{n-1} \to S^{n-1} \setminus \{\mathcal{S}\} \quad \text{and} \quad \kappa_{\mathcal{N}}' : \mathbb{R}^{n-1} \to S^{n-1} \setminus \{\mathcal{N}\}$$

with respect to the poles $\mathcal{S} = (-1, 0, \ldots, 0) = -\mathcal{N}$ and $\mathcal{N}$. Then

$$\kappa_{\mathcal{S}}'(x') = \left(\frac{1 - |x'|^2}{1 + |x'|^2}, \frac{2x'}{1 + |x'|^2} \right) \in S^{n-1} \setminus \{\mathcal{S}\}.$$

Note that $\kappa_{\mathcal{S}}'(x') = -\kappa_{\mathcal{N}}'(-x')$. Now

$$G = N^- P^+ \cup w P^+ = N^+ P^- \cup w P^-,$$

where $w \in SO(n)$ is a rotation so that $N^+ = wN^-w^{-1}$ and we have $w(\mathcal{S}) = \mathcal{N}$. These decompositions of G are the Bruhat decompositions. It follows that the sphere S^{n-1} decomposes as

$$S^{n-1} = G/G_{\mathcal{N}} = (N^n)^-(\mathcal{N}) \cup w\mathcal{N} = (N^n)^-(\mathcal{N}) \cup \mathcal{S} \qquad (2.2.7)$$

and

$$S^{n-1} = G/G_{\mathcal{S}} = (N^n)^+(\mathcal{S}) \cup w\mathcal{S} = (N^n)^+(\mathcal{S}) \cup \mathcal{N}. \qquad (2.2.8)$$

Lemma 2.2.1. $n_s^-(\mathcal{N}) = \kappa'_{\mathcal{S}}(s)$ and $n_s^+(\mathcal{S}) = \kappa'_{\mathcal{N}}(s)$.

Proof. (2.2.6) implies

$$(\kappa'_{\mathcal{N}})^{-1}(n_s^+ \kappa'_{\mathcal{N}}(0)) = s.$$

Now $\kappa'_{\mathcal{N}}(0) = \mathcal{S}$ implies the second assertion. The first follows from this by conjugation with w. $\square$

Finally, the stereographic projection

$$\kappa'_{\mathcal{S}} : \mathbb{R}^{n-1} \cup \{\infty\} \to S^{n-1} \setminus \mathcal{S}, \quad \mathcal{S} = (-1, 0, \ldots, 0)$$

is related to the restriction of the action of $\sigma_0 = \begin{pmatrix} 1 & 0 \\ 0 & T_0 \end{pmatrix} \in G^{n+1}$, where

$$T_0 = \begin{pmatrix} 0 & 0 & \ldots & 0 & 1 \\ -1 & 0 & \ldots & 0 & 0 \\ 0 & -1 & \ldots & 0 & 0 \\ & & \ldots & & \\ 0 & 0 & \ldots & -1 & 0 \end{pmatrix} \in SO(n+1). \qquad (2.2.9)$$

In fact, (2.2.5) yields

$$\sigma_0 : \mathbb{R}^n \ni x = (x_1, \ldots, x_n)$$

$$\mapsto \left(\frac{1 - |x|^2}{1 + |x|^2 - 2x_n}, \frac{-2x_1}{1 + |x|^2 - 2x_n}, \ldots, \frac{-2x_{n-1}}{1 + |x|^2 - 2x_n} \right) \in \mathbb{R}^n.$$

The map $x \mapsto \sigma_0(\tau(x))$, where $\tau(x', x_n) = (x', -x_n)$ realizes an isometry $\mathbb{R}^n_+ \to \mathbb{B}^n$ for the respective hyperbolic metrics

$$g_c = \frac{1}{x_n^2} \sum_{i=1}^n dx_i^2, \quad g_c = \frac{4}{(1 - |x|^2)^2} \sum_{i=1}^n dx_i^2.$$

In fact, $|\sigma_0(x', x_n)|^2 < 1$ iff $0 < -x_n(|x'|^2 + (x_n - 1)^2)$. Hence

$$(\sigma_0)_*(\text{vol}(\mathbb{R}^n_-, g_c)) = \text{vol}(\mathbb{B}^n, g_c),$$

i.e., we obtain the identity

$$(\sigma_0)_*(x_n) = \frac{1 - |x|^2}{2} \left(\frac{(\sigma_0)_*(dx_1 \ldots dx_n)}{dx_1 \ldots dx_n} \right)^{1/n} \tag{2.2.10}$$

which will be useful in Section 5.4.

For $x_n = 0$, it restricts to

$$\mathbb{R}^{n-1} \ni x' \mapsto - \left(\frac{|x'|^2 - 1}{|x'|^2 + 1}, \frac{2x'}{|x'|^2 + 1} \right) = -\kappa'_{\mathcal{N}}(x') = \kappa'_{\mathcal{S}}(-x') \in S^{n-1}.$$

2.3 Poisson transformations and spherical principal series

It will be convenient to fix the following convention concerning the notation of *standard metrics* on Euclidean spaces $\mathbb{R}^m$, on the spheres S^m and on the models $\mathbb{H}^m$, $\mathbb{B}^m$ of hyperbolic space. In all cases, the respective flat, curvature 1 and curvature -1 metric will be denoted by g_c (c for canonical). The meaning of g_c will be always clear from the context. Thus

- $g_c = \sum_{i=1}^m dx_i^2$ for the Euclidean space $\mathbb{R}^m$,
- $g_c = i^* \sum_{i=1}^m dx_i^2$ for the sphere $i : S^{m-1} \hookrightarrow \mathbb{R}^m$,
- $g_c = x_m^{-2} \sum_{i=1}^m dx_i^2$ for the upper half-space $\mathbb{H}^m$,
- $g_c = 4(1 - |x|^2)^{-2} \sum_{i=1}^m dx_i^2$ for the unit ball $\mathbb{B}^m$.

First, we consider the ball-model $\mathbb{B}^m$ with boundary S^{m-1}. G^m operates by isometries on $\mathbb{B}^m$ and on S^{m-1} by conformal diffeomorphisms of g_c.

Theorem 2.3.1. *The Poisson kernel*

$$P(x, b) \stackrel{\text{def}}{=} \frac{1 - |x|^2}{|x - b|^2}, \quad x \in \mathbb{B}^m, \ b \in S^{m-1} \tag{2.3.1}$$

of the ball $\mathbb{B}^m$ satisfies the identities

$$P(g \cdot \mathcal{O}, b) = \left(\frac{g_*(\text{vol}(g_c))}{\text{vol}(g_c)} \right)^{\frac{1}{m-1}} = e^{\langle g \cdot \mathcal{O}, b \rangle} \tag{2.3.2}$$

for $g \in G^m$. Here $\text{vol}(g_c) = \text{vol}(S^{m-1}, g_c)$ is the volume form of g_c on S^{m-1} and $\langle x, b \rangle$ for $x \in \mathbb{B}^m$, $b \in S^{m-1}$ is the (signed) distance of the unique horosphere $\xi_b(x)$ through x with normal b from the center $\mathcal{O}$ of the ball.

For a proof see [140].

Note that the first equality in (2.3.2) follows from the version

$$g_*(g_c) = P(g \cdot \mathcal{O}, \cdot)^2 g_c, \ g \in G^m \tag{2.3.3}$$

of the conformality of the action with respect to g_c. In fact, using (2.1.7) we find
$g \cdot \mathcal{O} = b/(1 + d)$ for $g = \begin{pmatrix} d & c \\ b^t & A \end{pmatrix}$, and a calculation (using $-d^2 + |b|^2 = -1$)
shows that
$$P(g \cdot \mathcal{O}, y) = \frac{1}{(d - (b, y))}, \quad y \in S^{m-1}.$$

Thus (2.3.3) coincides with (2.1.5).

For $\lambda \in \mathbb{C}$, we define a G^m-module structure on $C^\infty(S^{m-1})$ by

$$\pi_\lambda^c(g)u \overset{\text{def}}{=} \left(\frac{g_*(\text{vol}(g_c))}{\text{vol}(g_c)} \right)^{-\frac{\lambda}{m-1}} g_*(u). \tag{2.3.4}$$

The representation π_λ^c is called the *compact model* of the spherical principal series representation of G^m. The space $C^\infty(S^{m-1})$ equipped with the G^m-action π_λ^c will be denoted by $C^\infty(S^{m-1})_\lambda$. By the second identity in (2.3.2), the definition is equivalent to

$$\pi_\lambda^c(g)u(b) = e^{-\lambda\langle g \cdot \mathcal{O}, b\rangle} u(g^{-1}(b)). \tag{2.3.5}$$

The latter definition, in turn, can be rephrased in group-theoretical terms. We identify
$$S^{m-1} \simeq SO(m)/SO(m-1) = K^m/M^m \simeq G^m/P^m,$$

where P^m is the parabolic subgroup which fixes $(1, 0, \dots, 0) \in S^{m-1}$. It has the Langlands decomposition
$$P^m = M^m A(N^m)^+.$$

Then (2.3.5) can be identified with the representation

$$\pi_\lambda^c(g)u(k) = e^{\lambda H(g^{-1}k)} u(\kappa(g^{-1}k)) \tag{2.3.6}$$

on $C^\infty(K)^M$, where $g = \kappa(g) \exp H(g) n^+(g) \in KAN^+$ is the Iwasawa decomposition. Here we identify $H(g) \in \mathfrak{a}$ with the corresponding multiple of H_0.

The definitions (2.3.4), (2.3.5) and (2.3.6) emphasize different aspects of the representation π_λ^c. The version (2.3.4) emphasizes the conformal action of G^m on (S^{m-1}, g_c). (2.3.5) uses the relation to the hyperbolic space and (2.3.6) emphasizes that it can be regarded as an induced representation.

The proof of the following result demonstrates the convenience of definition (2.3.4).

Corollary 2.3.1. *For the Poisson transformation*

$$C^\infty(S^{m-1}) \ni \omega(b) \mapsto \mathcal{P}_\lambda(\omega)(x) \overset{\text{def}}{=} \int_{S^{m-1}} P(x, b)^\lambda \omega(b) \, \text{vol}(g_c) \tag{2.3.7}$$

we have

$$\mathcal{P}_\lambda \circ \pi_{\lambda-(m-1)}^c(g) = g_* \circ \mathcal{P}_\lambda.$$

Proof. Let vol $= \text{vol}(g_c)$. (2.3.2) implies

$$
\mathcal{P}_\lambda(\omega)(g'g \cdot \mathcal{O}) = \int_{S^{m-1}} \left(\frac{(g'g)_*(\text{vol})}{\text{vol}} \right)^{\frac{\lambda}{m-1}} \omega \, \text{vol}
$$

$$
= \int_{S^{m-1}} \left(\frac{g_*(\text{vol})}{\text{vol}} \right)^{\frac{\lambda}{m-1}} \left(\frac{g'^*(\text{vol})}{\text{vol}} \right)^{-\frac{\lambda}{m-1}} g'^*(\omega \, \text{vol})
$$

$$
= \int_{S^{m-1}} \left(\frac{g_*(\text{vol})}{\text{vol}} \right)^{\frac{\lambda}{m-1}} \left(\frac{g'^*(\text{vol})}{\text{vol}} \right)^{1-\frac{\lambda}{m-1}} g'^*(\omega) \, \text{vol}
$$

using

$$
g'^* \left(\frac{(g'g)_*(\text{vol})}{\text{vol}} \right) = \frac{g_*(\text{vol})}{g'^*(\text{vol})} = \frac{g_*(\text{vol})}{\text{vol}} \left(\frac{g'^*(\text{vol})}{\text{vol}} \right)^{-1}.
$$

Hence

$$
\mathcal{P}_\lambda(\omega)(g'g \cdot \mathcal{O}) = \mathcal{P}_\lambda(\pi^c_{\lambda-(m-1)}(g'^{-1})\omega)(g \cdot \mathcal{O}).
$$

The proof is complete. □

Now we consider the case of a totally umbilic hypersphere Σ in S^m. Such spheres play a central role in classical conformal submanifold theory, where one associates to any hypersurface of the sphere the *mean curvature sphere congruence* or *central sphere congruence* consisting of those totally umbilic subspheres which are tangential of first order and have the same mean curvatures as the hypersurface at the points of contact ([142]).

It is a classical result that all totally umbilic hypersurfaces are of the form $\sigma(S^{m-1})$ for some $\sigma \in G^{m+1}$. Σ inherits a Riemannian metric g_c from the sphere and we define

$$
\pi_\lambda(\Sigma, g_c)(g) = \left(\frac{g_*(\text{vol}(\Sigma, g_c))}{\text{vol}(\Sigma, g_c)} \right)^{-\frac{\lambda}{m-1}} g_* \tag{2.3.8}
$$

for all $g \in G^{m+1}$ which preserve Σ (vol$(\Sigma, g_c) \in \Omega^{m-1}(\Sigma)$ denotes the Riemannian volume form). The spheres $\sigma(S^{m-1})$ is totally umbilic since $S^{m-1} \hookrightarrow S^m$ is and σ is a conformal diffeomorphism which preserves the trace-free part of the second fundamental form L.

For $\Sigma = \sigma(S^{m-1})$, the representations $\pi_\lambda(\Sigma, g_c)$ and $\pi_\lambda(S^{m-1}, g_c)$ are related by the obvious identity

$$
\left(\frac{\sigma_*(\text{vol}(S^{m-1}, g_c))}{\text{vol}(\Sigma, g_c)} \right)^{-\frac{\lambda}{m-1}} \circ \sigma_* \circ \pi_\lambda(\Sigma, g_c)(g)
$$

$$
= \pi_\lambda(S^{m-1}, g_c)(\sigma g \sigma^{-1}) \circ \left(\frac{\sigma_*(\text{vol}(S^{m-1}, g_c))}{\text{vol}(\Sigma, g_c)} \right)^{-\frac{\lambda}{m-1}} \circ \sigma_*.
$$

In addition to the compact model, we also need the *non-compact* model of the principal series. It is induced by the action of G^{m+1} on $N^- \simeq \mathbb{R}^m$ which is induced by the Bruhat decomposition. We begin with the discussion of the following analog of Theorem 2.3.1 for the upper half-space.

Theorem 2.3.2. *The Poisson kernel*

$$P(x, \zeta) \overset{\text{def}}{=} \frac{x_m}{|(x', x_m) - (\zeta, 0)|^2}, \qquad x = (x', x_m) \in \mathbb{R}_+^m, \ \zeta \in \mathbb{R}^{m-1} \tag{2.3.9}$$

of the upper half-space satisfies the identity

$$P(g \cdot \mathcal{O}, \zeta) = \left(\frac{g_* \left(\frac{d\zeta}{(1+|\zeta|^2)^{m-1}} \right)}{d\zeta} \right)^{\frac{1}{m-1}}, \ g \in G^m, \ \zeta \in \mathbb{R}^{m-1}, \tag{2.3.10}$$

i.e.,

$$P(g \cdot \mathcal{O}, \zeta)(1 + |\zeta|^2) = \left(\frac{g_* \left(\frac{d\zeta}{(1+|\zeta|^2)^{m-1}} \right)}{\frac{d\zeta}{(1+|\zeta|^2)^{m-1}}} \right)^{\frac{1}{m-1}} = e^{\langle g \cdot \mathcal{O}, \zeta \rangle}, \tag{2.3.11}$$

where $\mathcal{O} = (0, 0, \dots, 1)$ and $d\zeta$ is the (translation invariant) Riemannian volume form for the Euclidean metric on $\mathbb{R}^{m-1}$.

Proof. By the Iwasawa decomposition $G = KAN^+$, it suffices to prove (2.3.10) for g in the subgroups $K \simeq SO(m)$, $A \simeq \mathbb{R}$ and $N^+ \simeq \mathbb{R}^{m-1}$. We use the fact that a_t operates on $\mathbb{R}^m$ by $x \mapsto e^t x$ (see (2.2.6)). Now

$$(a_t)_* \left(\frac{d\zeta}{(1 + |\zeta|^2)^{m-1}} \right) = \frac{e^{-(m-1)t} d\zeta}{(1 + e^{-2t}|\zeta|^2)^{m-1}},$$

i.e.,

$$\left(\frac{(a_t)_* \left(\frac{d\zeta}{(1+|\zeta|^2)^{m-1}} \right)}{d\zeta} \right)^{\frac{1}{m-1}} = \frac{e^{-t}}{1 + e^{-2t}|\zeta|^2} = \frac{e^t}{e^{2t} + |\zeta|^2}.$$

On the other hand, we have

$$P(a_t \cdot \mathcal{O}, \zeta) = \frac{e^t}{|(0, e^t) - (\zeta, 0)|^2} = \frac{e^t}{|\zeta|^2 + e^{2t}}.$$

Next, we use the fact that N^+ operates by translations on $\mathbb{R}^m$ (see (2.2.6)), i.e.,

$$(x', x_m) \mapsto (x' + y', x_m) \quad \text{and} \quad \zeta \mapsto \zeta + y'.$$

The latter operation pushes forward the volume form $\frac{d\zeta}{(1+|\zeta|^2)^{m-1}}$ into

$$\frac{d\zeta}{(1 + |\zeta - y'|^2)^{m-1}}.$$

On the other hand, for the Poisson kernel we get

$$P((y', 1), \zeta) = \frac{1}{|(y', 1) - (\zeta, 0)|^2}.$$

Finally, for $g \in SO(m)$ the assertion is equivalent to

$$\frac{d\zeta}{(1+|\zeta|^2)^{m-1}} = g_* \left(\frac{d\zeta}{(1+|\zeta|^2)^{m-1}} \right), \tag{2.3.12}$$

i.e., the $SO(m)$-invariance of the volume form

$$\frac{d\zeta}{(1+|\zeta|^2)^{m-1}}.$$

But this is a consequence of (2.2.2). $\qquad \square$

Note that (2.3.9) implies

$$\lim_{\zeta \to \infty} P(x,\zeta)(1+|\zeta|^2) = x_m,$$

which is consistent with (2.3.11) in view of $x_m = \langle x, \infty \rangle$.

Since G^{m+1} operates on (S^m, g_c) by conformal diffeomorphisms, the action (2.2.5) consists of conformal diffeomorphisms of the metric $g_s = \kappa^*(g_c)$. Hence it acts by conformal diffeomorphisms of the Euclidean metric on $\mathbb{R}^m$. The following definition emphasizes the conformality of this action. Let

$$\pi_\lambda^{nc}(g) \overset{\text{def}}{=} \left(\frac{g_*(\operatorname{vol}(g_c))}{\operatorname{vol}(g_c)} \right)^{-\frac{\lambda}{m}} g_*, \tag{2.3.13}$$

where $\operatorname{vol}(g_c)$ denotes the Riemannian volume of the Euclidean metric g_c on $\mathbb{R}^m$. (2.3.13) is well defined on the representation space

$$V_\lambda(\mathbb{R}^m) \overset{\text{def}}{=} \left\{ u \in C^\infty(\mathbb{R}^m) \,|\, |x|^{2\lambda} u \left(\frac{x}{|x|^2} \right) \in C^\infty(\mathbb{R}^m) \right\}. \tag{2.3.14}$$

Note that the function

$$x \mapsto |x|^{2\lambda} u \left(\frac{x}{|x|^2} \right)$$

is C^∞ only for $x \neq 0$. The condition in (2.3.14) requires that it extends through 0 as a C^∞ function. An equivalent condition is

$$\pi_\lambda^{nc}(w)u \in C^\infty(\mathbb{R}^m) \text{ for } w = \left(\begin{pmatrix} -1 & 0 \\ 0 & -1 \end{pmatrix} \quad 0 \\ 0 \qquad 1_{(m-1,m-1)} \right) \in SO(m+1).$$

Note that $w(\mathcal{N}) = \mathcal{S}$. In order to prove the equivalence, we observe that by (2.2.5)

$$w(x) = \frac{(-x_1, x_2, \dots, x_n)}{|x|^2}.$$

Hence, by the $SO(m+1)$-invariance of g_s,

$$\frac{w_*(g_c)}{g_c} = \frac{w_*(g_s)}{g_s}\left(\frac{w_*(1+|x|^2)}{1+|x|^2}\right)^2 = \frac{1}{|x|^4}.$$

Thus

$$\left(\frac{w_*(\mathrm{vol}(g_c))}{\mathrm{vol}(g_c)}\right)^{-\frac{\lambda}{n}} u(w(x)) = |x|^{2\lambda} u\left(\frac{(-x_1, x_2, \dots, x_n)}{|x|^2}\right).$$

This implies the equivalence.

Corollary 2.3.2. *For the Poisson transformation*

$$\omega \mapsto \mathcal{P}_\lambda(\omega)(x) \stackrel{\mathrm{def}}{=} \int_{\mathbb{R}^{m-1}} P(x,\zeta)^\lambda \omega(\zeta)d\zeta \qquad (2.3.15)$$

we have

$$\mathcal{P}_\lambda \circ \pi^{nc}_{\lambda-(m-1)}(g) = g_* \circ \mathcal{P}_\lambda, \ g \in G^m.$$

The Poisson transformation is well defined on the space $V_{\lambda-(m-1)}(\mathbb{R}^{m-1})$.

The following proof again demonstrates the convenience of the definition (2.3.13).

Proof. Similarly as in the proof of Corollary 2.3.1, we calculate

$$\mathcal{P}_\lambda(\omega)(g'g \cdot \mathcal{O}) = \int_{\mathbb{R}^{m-1}} P(g'g \cdot \mathcal{O}, \zeta)^\lambda \omega(\zeta)d\zeta$$

$$= \int_{\mathbb{R}^{m-1}} \left(\frac{(g'g)_*(\mathrm{vol}(g_s))}{\mathrm{vol}(g_s)}\right)^{\frac{\lambda}{m-1}} \Phi^{-\lambda}_{m-1}\omega \, \mathrm{vol}(g_c)$$

$$= \int_{\mathbb{R}^{m-1}} \left(\frac{g_*(\mathrm{vol}(g_s))}{\mathrm{vol}(g_s)}\right)^{\frac{\lambda}{m-1}} \left(\frac{g'^*(\mathrm{vol}(g_s))}{\mathrm{vol}(g_s)}\right)^{-\frac{\lambda}{m-1}} g'^*(\Phi^{-\lambda}_{m-1}\omega \, \mathrm{vol}(g_c))$$

$$= \int_{\mathbb{R}^{m-1}} \left(\frac{g_*(\mathrm{vol}(g_s))}{\mathrm{vol}(g_s)}\right)^{\frac{\lambda}{m-1}} \left(\frac{g'^*(\mathrm{vol}(g_c))}{\mathrm{vol}(g_c)}\right)^{1-\frac{\lambda}{m-1}}$$

$$\times \left(\frac{g'^*(\Phi_{m-1})}{\Phi_{m-1}}\right)^\lambda g'^*(\Phi^{-\lambda}_{m-1})g'^*(\omega) \, \mathrm{vol}(g_c)$$

$$= \int_{\mathbb{R}^{m-1}} \left(\frac{g_*(\mathrm{vol}(g_s))}{\mathrm{vol}(g_s)}\right)^{\frac{\lambda}{m-1}} \pi^{nc}_{\lambda-(m-1)}(g'^{-1})(\omega)\Phi^{-\lambda}_{m-1} \, \mathrm{vol}(g_c)$$

$$= \int_{\mathbb{R}^{m-1}} P(g \cdot \mathcal{O}, \zeta)^\lambda \pi^{nc}_{\lambda-(m-1)}(g'^{-1})(\omega) \, \mathrm{vol}(g_c)$$

using $g_s = 4\Phi^{-2}_n g_c$. $\qquad\square$

In the proof of Corollary 2.3.2, we have used that $g_s = 4\Phi^{-2}_n g_c$ for $\Phi_n(x) = 1+|x|^2$ implies

$$\frac{g_*(\mathrm{vol}(g_s))}{\mathrm{vol}(g_s)} = \frac{g_*(\Phi^{-n}_n \, \mathrm{vol}(g_c))}{\Phi^{-n}_n \, \mathrm{vol}(g_c)}.$$

Hence

$$\left(\frac{g_*(\mathrm{vol}(g_s))}{\mathrm{vol}(g_s)}\right)^{-\frac{\lambda}{n}} = \left(\frac{g_*(\Phi_n)}{\Phi_n}\right)^{\lambda}\left(\frac{g_*(\mathrm{vol}(g_c))}{\mathrm{vol}(g_c)}\right)^{-\frac{\lambda}{n}}.$$

Therefore, the obvious relation

$$\kappa_* \circ \pi_\lambda(\mathbb{R}^n, g_s)(g) = \pi_\lambda(S^n, g_c)(g) \circ \kappa_* \tag{2.3.16}$$

can be written in the form

$$\kappa_* \circ \left(\left(\frac{g_*(\Phi_n)}{\Phi_n}\right)^{\lambda} \circ \pi_\lambda(\mathbb{R}^n, g_c)(g)\right) = \pi_\lambda(S^n, g_c)(g) \circ \kappa_*,$$

i.e.,

$$(\kappa_* \circ \Phi_n^{-\lambda}) \circ \pi_\lambda(\mathbb{R}^n, g_c)(g) = \pi_\lambda(S^n, g_c)(g) \circ (\kappa_* \circ \Phi_n^{-\lambda}). \tag{2.3.17}$$

Similarly as for S^n, we can also consider principal series representations on (generalized) hyperspheres Σ in $\mathbb{R}^n$. These can be written in the form $\Sigma = \sigma(\mathbb{R}^{n-1})$ for some $\sigma \in G^{n+1}$ (acting on $\mathbb{R}^n$). Then we define

$$\pi_\lambda(\Sigma, g_c)(g) = \left(\frac{g_*(\mathrm{vol}(\Sigma, g_c))}{\mathrm{vol}(\Sigma, g_c)}\right)^{-\frac{\lambda}{n-1}} \circ g_*, \tag{2.3.18}$$

for all $g \in G^{n+1}$ which preserve Σ. Here g_c is the restriction of the Euclidean metric on $\mathbb{R}^n$, and $\mathrm{vol}(\Sigma, g_c) \in \Omega^{n-1}(\Sigma)$ denotes the associated Riemannian volume form.

The group $\sigma_0 G^n \sigma_0^{-1}$ operates on $\Sigma_0 = S^{n-1} = \partial\mathbb{B}^n$ since G^n leaves the subspace $\mathbb{R}^{n-1} = \ker(x_n) \subset \mathbb{R}^n$ invariant. As a special case of (2.3.18), we obtain the principal series representation

$$\pi_\lambda(\Sigma_0, g_c)(g) = \left(\frac{g_*(\mathrm{vol}(\Sigma_0, g_c))}{\mathrm{vol}(\Sigma_0, g_c)}\right)^{-\frac{\lambda}{n-1}} g_*$$

for $g \in \sigma_0 G^n \sigma_0^{-1}$ and the induced metric g_c on Σ_0. It is related to $\pi_\lambda(\mathbb{R}^{n-1}, g_c)$ by the obvious relation

$$\left(\frac{(\sigma_0)_*(\mathrm{vol}(\mathbb{R}^{n-1}, g_c))}{\mathrm{vol}(\Sigma_0, g_c)}\right)^{-\frac{\lambda}{n-1}} \circ (\sigma_0)_* \circ \pi_\lambda(\mathbb{R}^{n-1}, g_c)(g)$$

$$= \pi_\lambda(\Sigma_0, g_c)(\sigma_0 g \sigma_0^{-1}) \circ \left(\frac{(\sigma_0)_*(\mathrm{vol}(\mathbb{R}^{n-1}, g_c))}{\mathrm{vol}(\Sigma_0, g_c)}\right)^{-\frac{\lambda}{n-1}} \circ (\sigma_0)_*$$

for $g \in G^n$, i.e.,

$$(\sigma_0)_* \left(\frac{\mathrm{vol}(\mathbb{R}^{n-1}, g_c))}{\sigma_0^*(\mathrm{vol}(\Sigma_0, g_c))}\right)^{-\frac{\lambda}{n-1}} \circ \pi_\lambda(\mathbb{R}^{n-1}, g_c)(g)$$

$$= \pi_\lambda(\Sigma_0, g_c)(\sigma_0 g \sigma_0^{-1}) \circ (\sigma_0)_* \circ \left(\frac{\mathrm{vol}(\mathbb{R}^{n-1}, g_c))}{\sigma_0^*(\mathrm{vol}(\Sigma_0, g_c))}\right)^{-\frac{\lambda}{n-1}}. \tag{2.3.19}$$

We illustrate these matters in the case $n = 2$. We use complex notation. Then $\mathbb{B}^2 = \mathbb{D} = \{z \in \mathbb{C} \,|\, |z| < 1\}$ is the unit disk with the hyperbolic metric

$$\frac{4}{(1-|z|^2)^2}(dx^2 + dy^2).$$

$G^3 = SO(1,3) = SL(2,\mathbb{C})$ operates on $\hat{\mathbb{C}}$ by

$$\begin{pmatrix} a & b \\ c & d \end{pmatrix} : z \mapsto \frac{az+b}{cz+d}.$$

The subgroup $G^2 = SO(1,2) = SL(2,\mathbb{R})$ preserves $\hat{\mathbb{R}} = \mathbb{R} \cup \{\infty\}$ and the half-planes

$$\mathbb{H}^+ = \{z \in \mathbb{C} \,|\, y > 0\}, \quad \mathbb{H}^- = \{z \in \mathbb{C} \,|\, y < 0\}.$$

Now

$$\sigma_0 = \frac{1}{\sqrt{2}} \begin{pmatrix} 1 & -i \\ -i & 1 \end{pmatrix} \in SL(2,\mathbb{C}),$$

i.e., $z \mapsto (z-i)/(-iz+1)$ maps

$$\mathbb{H}^+ \to \mathbb{D}, \ \hat{\mathbb{R}} \to S^1 = \partial\mathbb{D}$$

so that $0 \mapsto -i$, $\infty \mapsto i$ and $i \mapsto 0$. It is an isometry if we use the metric $y^{-2}(dx^2 + dy^2)$ on $\mathbb{H}^+$. The restriction $\sigma_0 : \mathbb{R} \to S^1$ relates to the stereographic projection $\kappa_S : \mathbb{R} \to S^1$ by $\kappa_S = i\sigma_0$. In fact,

$$\sigma_0(x) = \frac{x-i}{-ix+1} = \frac{2x}{1+x^2} - i\frac{1-x^2}{1+x^2} \quad \text{and} \quad \kappa_S(x) = \frac{1-x^2}{1+x^2} + i\frac{2x}{1+x^2}.$$

The group

$$SU(1,1) = \sigma_0 SL(2,\mathbb{R})\sigma_0^{-1} \subset SL(2,\mathbb{C})$$

preserves S^1 and $\mathbb{D}$. Now the non-compact models of the principal series of the groups $SL(2,\mathbb{C})$ and $SL(2,\mathbb{R})$ are defined by

$$\pi_\lambda^{nc}(g) = \pi_\lambda(\mathbb{C}, g_c)(g) = \left(\frac{g_*(d\zeta \wedge d\bar{\zeta})}{d\zeta \wedge d\bar{\zeta}}\right)^{-\frac{\lambda}{2}} g_*, \ g \in SL(2,\mathbb{C})$$

and

$$\pi_\lambda^{nc}(g) = \pi_\lambda(\mathbb{R}, g_c)(g) = \left(\frac{g_*(dx)}{dx}\right)^{-\lambda} g_*, \ g \in SL(2,\mathbb{R}).$$

Lemma 2.3.1.

$$\pi_\lambda^{nc}(g)u(\zeta) = |-c\zeta+a|^{2\lambda} u\left(\frac{d\zeta-b}{-c\zeta+a}\right)$$

for $g = \begin{pmatrix} a & b \\ c & d \end{pmatrix} \in SL(2,\mathbb{C})$, $\zeta \in \mathbb{C}$. *Similarly,*

$$\pi_\lambda^{nc}(g)u(x) = |-cx+a|^{2\lambda}u\left(\frac{dx-b}{-cx+a}\right)$$

for $g = \begin{pmatrix} a & b \\ c & d \end{pmatrix} \in SL(2,\mathbb{R})$, $x \in \mathbb{R}$.

Proof. By the holomorphy of the action of g on $\hat{\mathbb{C}}$,

$$g_*(d\zeta \wedge d\bar{\zeta}) = \left|\frac{\partial g^{-1}(\zeta)}{\partial \zeta}\right|^2 d\zeta \wedge d\bar{\zeta}.$$

But using

$$\left|\frac{\partial g^{-1}(\zeta)}{\partial \zeta}\right|^2 = \frac{1}{|-c\zeta+a|^4},$$

we get

$$\left(\frac{g_*(d\zeta \wedge d\bar{\zeta})}{d\zeta \wedge d\bar{\zeta}}\right)^{-\frac{\lambda}{2}} = |-c\zeta+a|^{2\lambda},$$

and the proof is complete for $SL(2,\mathbb{C})$. We omit the analogous argument for $SL(2,\mathbb{R})$. $\qquad\square$

The action of $SU(1,1)$ on $\Sigma_0 = S^1 = \partial\mathbb{D}$ gives rise to the representation

$$\pi_\lambda(\Sigma_0, g_c)(g) = (g_*(d\theta)/d\theta)^{-\lambda} g_*.$$

Lemma 2.3.2.

$$\pi_\lambda(\Sigma_0, g_c)(g)u(\zeta) = |-\bar{\beta}\zeta+\alpha|^{2\lambda}u\left(\frac{\alpha\zeta-\beta}{-\bar{\beta}\zeta+\alpha}\right)$$

for $g = \begin{pmatrix} \alpha & \beta \\ \bar{\beta} & \bar{\alpha} \end{pmatrix}$ *and* $\zeta \in S^1$.

Proof. Let $\zeta = e^{i\theta}$. In order to calculate $g_*(d\theta)/d\theta$, we note that $d\zeta/\zeta = id\theta$ and hence

$$\frac{g_*(d\theta)}{d\theta} = \frac{g_*(d\zeta)}{d\zeta}\frac{\zeta}{g_*(\zeta)}.$$

But since

$$g_*(d\zeta) = d(g^{-1}\zeta) = d\left(\frac{\bar{\alpha}\zeta - \beta}{-\bar{\beta}\zeta + \alpha}\right) = \frac{1}{(-\bar{\beta}\zeta + \alpha)^2}d\zeta$$

we get

$$\frac{g_*(d\theta)}{d\theta}$$

$$= \frac{1}{(-\bar{\beta}\zeta+\alpha)^2} \frac{\zeta}{\frac{\bar{\alpha}\zeta-\beta}{-\beta\zeta+\alpha}} = \frac{\zeta}{(-\bar{\zeta}+\alpha)(\bar{\alpha}\zeta-\beta)} = \frac{1}{(-\bar{\zeta}+\alpha)(\bar{\alpha}-\beta\bar{\zeta})} = \frac{1}{|-\bar{\beta}\zeta+\alpha|^2}$$

and the proof is complete. $\qquad\qquad\qquad\qquad\qquad\qquad\qquad\qquad\qquad\qquad\qquad\square$

The relation between $\pi_\lambda(\Sigma_0, g_c)$ and $\pi_\lambda(\mathbb{R}, g_c)$ is given by

$$\left((\sigma_0)_* \Phi_1^{-\lambda}\right) \circ \pi_\lambda(\mathbb{R}, g_c)(g) = \pi_\lambda(\Sigma_0, g_c)(\sigma_0 g \sigma_0^{-1}) \circ \left((\sigma_0)_* \Phi_1^{-\lambda}\right)$$

for $g \in SL(2, \mathbb{R})$ (see (2.3.19)) since

$$\Phi_1(x) = (1+x^2)/2 = \mathrm{vol}(\mathbb{R}^1, g_c)/\sigma_0^*(\mathrm{vol}(\sigma_0, g_c))$$

(see (2.2.1)). That identity can be proved directly. In fact, it is equivalent to

$$\pi_\lambda(\mathbb{R}, g_c)(g) \circ \left(\Phi_1^\lambda \sigma_0^*\right) = \left(\Phi_1^\lambda \sigma_0^*\right) \circ \pi_\lambda(\Sigma_0, g_c)(\sigma_0 g \sigma_0^{-1}). \qquad (2.3.20)$$

We calculate the left-hand side. We have

$$\pi_\lambda^{nc}(g)\Phi_1^\lambda \sigma_0^* u(x) = \pi_\lambda^{nc}(g)\left(2^{-\lambda}|-ix+1|^{2\lambda}\sigma_0^*(u)\right)(x)$$

$$= 2^{-\lambda}|-cx+a|^{2\lambda}\left|-i(\frac{dx-b}{-cx+a})+1\right|^{2\lambda} g_* \sigma_0^*(u)(x)$$

$$= 2^{-\lambda}|(a+ib)-x(c+id)|^{2\lambda} g_* \sigma_0^*(u)(x)$$

for $g = \begin{pmatrix} a & b \\ c & d \end{pmatrix} \in SL(2, \mathbb{R})$. On the other hand, using

$$\sigma_0 g \sigma_0^{-1} = \begin{pmatrix} \alpha & \beta \\ \bar{\beta} & \bar{\alpha} \end{pmatrix} = \frac{1}{2}\begin{pmatrix} a+d+i(b-c) & b+c+i(a-d) \\ b+c-i(a-d) & a+d-i(b-c) \end{pmatrix}$$

we obtain

$$\Phi_1^\lambda \sigma_0^* \pi_\lambda(\Sigma_0, g_c)(\sigma_0 g \sigma_0^{-1}) = 2^{-\lambda}|-ix+1|^{2\lambda}\left|-\bar{\beta}\kappa^{-1}(x)+\alpha\right|^{2\lambda}(g\sigma_0^{-1})_* u(x)$$

$$= 2^{-\lambda}|-ix+1|^{2\lambda}\left|-\bar{\beta}\frac{x-i}{-ix+1}+\alpha\right|^{2\lambda}(g\sigma_0^{-1})_* u(x)$$

$$= 2^{-\lambda}|(\alpha+i\bar{\beta})-ix(\alpha-i\bar{\beta})|^{2\lambda} g_* \sigma_0^* u(x).$$

Now the explicit formulas for α and β imply $\alpha + i\bar{\beta} = a + ib$ and $\alpha - i\bar{\beta} = c + id$. The proof of (2.3.20) is complete.

2.4 The Nayatani metric

We briefly describe the construction of a canonical metric on the complement of a round sphere $S^m \subset S^n$ viewed as the limit set of a Kleinian group $\Gamma \subset G^{n+1} = SO(n+1,1)^\circ$. We first recall that there is a canonical Γ-invariant measure μ_{PS} which is supported on the limit set $\Lambda(\Gamma) = S^m \subset S^n$ (Patterson-Sullivan measure) ([188], [178]). We consider μ_{PS} as a distribution in $C^{-\infty}(S^n)$ with support in $\Lambda(\Gamma)$, i.e., $\mu_{PS} \in C^{-\infty}(\Lambda(\Gamma))$, and describe its invariance in terms of the (distributional) spherical principal series representation π_λ^c of G^{n+1}. In these terms,

$$\pi_{m-n}^c(\gamma)\mu_{PS} = \mu_{PS}, \ \gamma \in \Gamma. \tag{2.4.1}$$

Since the limit set is a smooth submanifold, μ_{PS} can be written explicitly as

$$\mu_{PS}(\varphi) = \int_{S^m} i^*(\varphi)\,\mathrm{vol}(S^m, g_c) \Big/ \int_{S^m} \mathrm{vol}(S^m, g_c), \ \varphi \in C^\infty(S^n),$$

where volumes are Riemannian volume form with respect to the standard metric g_c. μ_{PS} is actually not only Γ-invariant (as for more general Γ) but G^{m+1}-invariant, where $G^{m+1} \subset G^{n+1}$ is the subgroup which leaves the subsphere invariant. We prove this fact in order to illustrate the formalism used in the present work. By definition,

$$\left\langle \pi_{m-n}^c(g)\mu_{PS}, \varphi \right\rangle = \left\langle \mu_{PS}, \pi_{-m}^c(g^{-1})\varphi \right\rangle, \ g \in G^{m+1}$$

and

$$\pi_{-m}^c(g^{-1})\varphi = \left(\frac{g^*(\mathrm{vol}_n)}{\mathrm{vol}_n} \right)^{\frac{m}{n}} g^*(\varphi) = \frac{g^*(\mathrm{vol}_m)}{\mathrm{vol}_m} g^*(\varphi)$$

on S^m. Thus

$$\left\langle \pi_{m-n}^c(g)\mu_{PS}, \varphi \right\rangle = \int_{S^m} g^*(i^*(\varphi))g^*(\mathrm{vol}_m) \Big/ \int_{S^m} \mathrm{vol}_m = \left\langle \mu_{PS}, \varphi \right\rangle.$$

Nayatani [185] observed that the measure μ_{PS} induces a *canonical* Γ-invariant metric on the proper set $\Omega(\Gamma)$ as follows. Here the induced metric is actually G^{m+1}-invariant. In order to construct the metric, we use the intertwining operator

$$I_\lambda : C^\infty(S^n)_\lambda \to C^\infty(S^n)_{-\lambda-n}, \quad I_\lambda(\varphi)(x) \stackrel{\mathrm{def}}{=} \int_{S^n} \frac{\varphi(y)}{|x-y|^{2\lambda+2n}}\,\mathrm{vol}_n(y) \tag{2.4.2}$$

for the spherical principal series. It satisfies the intertwining relation

$$I_\lambda \circ \pi_\lambda^c(g) = \pi_{-\lambda-n}^c(g) \circ I_\lambda \tag{2.4.3}$$

for $g \in G^{n+1}$ which extends to distributions. Now define

$$\Phi \stackrel{\mathrm{def}}{=} I_{m-n}(\mu_{PS}) \in C^{-\infty}(S^n)_{-m}^{G^{m+1}}.$$

The restriction $\mathrm{res}_{\Omega(\Gamma)}(\Phi)$ of Φ to $\Omega(\Gamma) = S^n \setminus S^m$ is C^∞ since I_λ does not increase singular support. More explicitly, we have

$$\Phi(x) = \left\langle \mu_{PS}(y), \frac{1}{|x-y|^{2m}} \right\rangle$$

for $x \in \Omega(\Gamma)$. The integral is well defined since x is in the complement of the support of μ_{PS}. Moreover, $\mathrm{res}_{\Omega(\Gamma)}(\Phi) > 0$. Hence

$$0 < \Phi_0 = (\mathrm{res}_{\Omega(\Gamma)}(\Phi))^{\frac{1}{m}} \in C^\infty(\Omega(\Gamma))_{-1}^{G^n}. \tag{2.4.4}$$

Now we define the G^n-invariant Nayatani metric on $\Omega(\Gamma)$ by

$$g_N = \Phi_0^2 g_c. \tag{2.4.5}$$

In the case of the equatorial subsphere $S^{n-1} \subset S^n$, the $\pi_{-1}^c(G^n)$-invariant function Φ_0 has also a simple *extrinsic* description. On the upper hemisphere $H^+ \subset S^n$ the function Φ_0^{-1} is a *constant* multiple of the restriction $\mathcal{H}_0$ of the height function x_{n+1} to S^n. This follows from the simple

Lemma 2.4.1.
$$\left\langle \mu_{PS}(y), \frac{1}{|x-y|^{2(n-1)}} \right\rangle = c \left(\frac{1}{|\mathcal{H}_0(x)|} \right)^{n-1}$$

for $x \in \Omega(\Gamma)$.

Thus the metric

$$\frac{1}{\mathcal{H}_0^2} g_c$$

on H^+ is a constant multiple of the Nayatani metric. For $m < n-1$, the corresponding Nayatani metric on $S^n \setminus S^m$ plays a natural role in an analogous theory of equivariant families $C^\infty(S^n) \to C^\infty(S^m)$.

2.5   Riemannian curvature and conformal change

In the present section we derive the basic identities which describe the behaviour of the Levi-Civita connection ∇, the Riemann curvature tensor R, the Ricci curvature tensor Ric and the scalar curvature τ with respect to conformal changes of the metric. In this connection, it is natural to introduce the rescaled scalar curvature J and the Schouten tensor P. In addition, we associate a symmetric bilinear form G to a hypersurface.

We recall first how the Riemann curvature tensor of (M^n, g) transforms under $g \mapsto \hat{g}$. Let ∇ and $\hat{\nabla}$ denote the corresponding Levi-Civita connections. Then ([24], [85]) the Koszul formula

$$2g(\nabla_X Y, Z) = X\left(g(Y,Z)\right) + Y\left(g(X,Z)\right) - Z\left(g(X,Y)\right)$$
$$+ g\left([X,Y], Z\right) + g\left([Z,X], Y\right) + g\left([Z,Y], X\right)$$

implies
$$\hat{\nabla}_X Y = \nabla_X Y + X(\varphi)Y + Y(\varphi)X - g(X,Y)\operatorname{grad}\varphi \qquad (2.5.1)$$

for all vector fields $X, Y \in \mathcal{X}(M)$. For the curvature

$$R(X,Y)Z = \nabla_X \nabla_Y Z - \nabla_Y \nabla_X Z - \nabla_{[X,Y]}Z, \quad R(X,Y,Z,W) = g(R(X,Y)Z,W),$$

it follows that

$$\begin{aligned}
\hat{R}(X,Y)Z = {} & R(X,Y)Z + g(\nabla_X(\operatorname{grad}\varphi), Z)Y - g(\nabla_Y(\operatorname{grad}\varphi), Z)X \\
& + \nabla_Y(\operatorname{grad}\varphi)g(X,Z) - \nabla_X(\operatorname{grad}\varphi)g(Y,Z) \\
& + \left(Y(\varphi)Z(\varphi) - g(Y,Z)|\operatorname{grad}\varphi|^2\right)X \\
& - \left(X(\varphi)Z(\varphi) - g(X,Z)|\operatorname{grad}\varphi|^2\right)Y \\
& + (X(\varphi)g(Y,Z) - Y(\varphi)g(X,Z))\operatorname{grad}\varphi,
\end{aligned}$$

i.e.,

$$\begin{aligned}
e^{-2\varphi}\hat{R}(X,Y,Z,W) = {} & e^{-2\varphi}\hat{g}(\hat{R}(X,Y)Z, W) = R(X,Y,Z,W) \\
& + \Xi(X,Z)g(Y,W) - \Xi(Y,Z)g(X,W) + \Xi(Y,W)g(X,Z) - \Xi(X,W)g(Y,Z)
\end{aligned}$$
$$(2.5.2)$$

for

$$\Xi(X,Y) = \Xi_{(g,\varphi)}(X,Y) \overset{\text{def}}{=} g(\nabla_X(\operatorname{grad}\varphi), Y) - X(\varphi)Y(\varphi) + \frac{1}{2}|\operatorname{grad}\varphi|^2 g(X,Y).$$

Here
$$\operatorname{Hess}(\varphi)(X,Y) = g(\nabla_X(\operatorname{grad}\varphi), Y) = \langle \nabla_X(d\varphi), Y \rangle \qquad (2.5.3)$$

is the covariant *Hessian* of φ. For the Ricci tensor

$$\operatorname{Ric}(X,Y) = \sum_i R(X, e_i, e_i, Y) = \sum_i g(R(X, e_i)e_i, Y),$$

we obtain

$$\begin{aligned}
\widehat{\operatorname{Ric}}(X,Y) = {} & \operatorname{Ric}(X,Y) \\
& + \sum_i \Xi(X, e_i)g(e_i, Y) + \Xi(e_i, Y)g(X, e_i) - \Xi(e_i, e_i)g(X,Y) - \Xi(X,Y)g(e_i, e_i)
\end{aligned}$$

i.e.,

$$\begin{aligned}
\widehat{\operatorname{Ric}}(X,Y) = {} & \operatorname{Ric}(X,Y) - (n-2)\Xi(X,Y) - g(X,Y)\sum_i \Xi(e_i, e_i) \\
= {} & \operatorname{Ric}(X,Y) - (n-2)g(\nabla_X \operatorname{grad}\varphi, Y) + g(X,Y)\Delta\varphi \\
& - (n-2)|\operatorname{grad}\varphi|^2 g(X,Y) + (n-2)X(\varphi)Y(\varphi) \qquad (2.5.4)
\end{aligned}$$

using the identity

$$\sum_i g(\nabla_{e_i}(\operatorname{grad}\varphi), e_i) = -\Delta\varphi. \tag{2.5.5}$$

(2.5.4) yields for the scalar curvature $\tau = \sum_i \operatorname{Ric}(e_i, e_i)$ the transformation rule

$$\hat{\tau} = e^{-2\varphi}\left(\tau + (2n-2)\Delta\varphi - (n-1)(n-2)|\operatorname{grad}\varphi|^2\right), \tag{2.5.6}$$

i.e.,

$$\hat{\mathsf{J}} = e^{-2\varphi}\left(\mathsf{J} + \Delta\varphi - \frac{n-2}{2}|\operatorname{grad}\varphi|^2\right) \tag{2.5.7}$$

for

$$\mathsf{J} = \frac{\tau}{2(n-1)}. \tag{2.5.8}$$

These results imply the transformation formula

$$\hat{\mathsf{P}}(X,Y) = \mathsf{P}(X,Y) - \frac{1}{2}|\operatorname{grad}\varphi|^2 g(X,Y) - g(\nabla_X(\operatorname{grad}\varphi), Y) + X(\varphi)Y(\varphi)$$
$$= \mathsf{P}(X,Y) - \Xi(X,Y) \tag{2.5.9}$$

for the *Schouten* tensor

$$\mathsf{P} = \frac{1}{n-2}\left(\operatorname{Ric} - \mathsf{J}g\right). \tag{2.5.10}$$

(2.5.1) implies that the Hessian (2.5.3) satisfies the conformal transformation law

$$\widehat{\operatorname{Hess}(u)}(X,Y) = \operatorname{Hess}(u)(X,Y)$$
$$\quad - \langle du, X\rangle\langle d\varphi, Y\rangle - \langle d\varphi, X\rangle\langle du, Y\rangle + (du, d\varphi)g(X,Y). \tag{2.5.11}$$

We write (2.5.2) also in the form

$$e^{-2\varphi}\hat{R} = R + \Xi \owedge g \tag{2.5.12}$$

in terms of the Kulkarni-Nomizu product

$$(\Xi \owedge g)(X,Y,Z,W)$$
$$= \Xi(X,Z)g(Y,W) - \Xi(Y,Z)g(X,W) + \Xi(Y,W)g(X,Z) - \Xi(X,W)g(Y,Z)$$

(see [24], Definition 1.110). Now for the *Weyl tensor* C being defined by

$$R = \mathsf{C} - \mathsf{P} \owedge g \tag{2.5.13}$$

we obtain

$$e^{-2\varphi}\hat{\mathsf{C}} = e^{-2\varphi}\left(\hat{R} + \hat{\mathsf{P}} \owedge \hat{g}\right) = R + \Xi \owedge g + (\mathsf{P} - \Xi)\owedge g = R + \mathsf{P} \owedge g = \mathsf{C}. \tag{2.5.14}$$

In particular, we have

$$e^{4\varphi}|\hat{\mathsf{C}}|^2 = |\mathsf{C}|^2.$$ (2.5.15)

A general reference for the basic formulas (2.5.12), (2.5.4) and (2.5.7) is [24], Theorem 1.159. Note that in [24] the curvature tensor has opposite sign, but Ricci curvature and scalar curvature coincide with ours, of course. The (somewhat) unusual sign in (2.5.13) is due to the conventions. A contraction of (2.5.13) in the second and third argument yields Ric on both sides since C is trace-free.

Finally, we fix some notational conventions. For a symmetric bilinear form B on a vector space V with a scalar product g, we define its g-trace as follows. B gives rise to a linear map

$$B^\sharp : V \to V$$

by $g(B^\sharp(X), Y) = B(X, Y)$ for $X, Y \in V$. In particular, $g^\sharp = \mathrm{id}$. We set

$$\mathrm{tr}_g(B) = \mathrm{tr}(B^\sharp).$$ (2.5.16)

Then for an arbitrary basis $e_1, \ldots, e_n$ of V,

$$\mathrm{tr}_g(B) = g^{rs} B_{rs},$$

where $B_{rs} = B(e_r, e_s) = g(B^\sharp(e_r), e_s)$ and $g_{rs} = g(e_r, e_s)$. Here we use the Einstein sum convention, and raise and lower indices as usual using g. In particular, for a g-orthonormal basis,

$$\mathrm{tr}_g(B) = \sum_i g(B^\sharp(e_i), e_i) = \sum_i B(e_i, e_i).$$

Note that

$$\mathrm{tr}((B^\sharp)^2) = \mathrm{tr}((g^{-1}B)^2).$$ (2.5.17)

In fact, we calculate

$$\mathrm{tr}((B^\sharp)^2) = g^{rs} g((B^\sharp)^2(e_r), e_s) = g^{rs} B_r^t g(B^\sharp(e_t), e_s)$$
$$= g^{rs} B_r^t B_{ts} = (g^{rs} B_{rq})(g^{qt} B_{ts}) = \mathrm{tr}((g^{-1}B)^2).$$

In order to simplify formulas, we shall often write $\mathrm{tr}(B)$ instead of $\mathrm{tr}_g(B)$, if the metric is clear from the context.

The components of the curvature tensor R will be denoted by R^l_{ijk}. Then

$$R_{ijkl} = \sum_r g_{rl} R^r_{ijk},$$

$$\mathrm{Ric}_{ij} = \sum_{r,s} R_{irsj} g^{rs} = \sum_{r,s} R_{rijs} g^{rs} = \sum_r R^r_{rij}$$

and

$$\tau = \mathrm{tr}_g(\mathrm{Ric}) = \mathrm{tr}(\mathrm{Ric}^\sharp) = \sum_{i,j} g^{ij} \, \mathrm{Ric}_{ij} \, .$$

For a hypersurface $\Sigma \subset M$ with the unit normal vector field N, we shall write ∇_N, or just N, for the corresponding normal derivative. We define the symmetric bilinear form

$$G(X, Y) = R(X, N, N, Y) = g(R(X, N)N, Y) \qquad (2.5.18)$$

on $T(\Sigma)$ and let $T = \mathrm{Ric} - G$. Then

$$\mathrm{tr}(G) = \mathrm{Ric}(N, N) \quad \text{and} \quad \mathrm{tr}(T) = \tau_M - 2\,\mathrm{Ric}(N, N).$$

We shall also use the notation

$$F = \mathrm{Ric}(N, N) = \mathrm{Ric}_N . \qquad (2.5.19)$$

Chapter 3

Conformally Covariant Powers of the Laplacian, Q-curvature and Scattering Theory

In the present chapter, we review the GJMS-construction of conformally covariant powers of the Laplacian, describe Branson's definition of Q-curvature, and discuss the relation of both to geometric scattering theory on conformally compact Einstein manifolds.

3.1 GJMS-operators and Q-curvature

In the present section, we review the construction of conformally covariant powers of the Laplacian which are known as GJMS-operators. These generalize the *conformal Laplacian* or *Yamabe operator*

$$P_2 = Y : g \mapsto \Delta_g - \frac{n-2}{4(n-1)}\tau_g$$

on a manifold of dimension n. P_2 is of conformal weight $(\frac{n}{2} - 1, \frac{n}{2} + 1)$, i.e., it satisfies the transformation formula

$$P_2(e^{2\varphi}g) = e^{-(\frac{n}{2}+1)\varphi} \circ P_2(g) \circ e^{(\frac{n}{2}-1)\varphi}$$

for all $\varphi \in C^\infty(M)$. Here we use the convention that $-\Delta_g$ is non-negative. A theorem of Graham, Jenne, Mason and Sparling ([124]) yields analogous conformally covariant modifications of powers of the Laplacian by lower order terms on any manifold M. More precisely,

Theorem 3.1.1 ([124]). *Let $n \geq 3$ and let*

$$\begin{cases} 1 \leq N \leq \frac{n}{2} & \text{if } n \text{ is even,} \\ N \geq 1 & \text{if } n \text{ is odd.} \end{cases}$$

Then on any manifold M of dimension n there exists a natural conformally co-variant differential operator P_{2N} of the form

$$\Delta^N + LOT$$

and of conformal weight $(\frac{n}{2} - N, \frac{n}{2} + N)$. In other words, $P_{2N}(g)$ differs from the power Δ_g^N of the Laplacian by lower order terms and satisfies the transformation formula

$$P_{2N}(e^{2\varphi}g) = e^{-(\frac{n}{2}+N)\varphi} \circ P_{2N}(g) \circ e^{(\frac{n}{2}-N)\varphi} \tag{3.1.1}$$

for all $\varphi \in C^\infty(M)$. On $\mathbb{R}^n$ with the Euclidean metric g_c the operator $P_{2N}(g_c)$ coincides with $\Delta_{\mathbb{R}^n}^N$.

Note that there is no uniqueness assertion in Theorem 3.1.1. The GJMS-operator P_{2N} in Theorem 3.1.1 provides *one* specific construction of an operator with the listed properties. On the other hand, extending a result of Graham ([118]), Gover and Hirachi ([112]) proved that the statement is sharp in the sense that for general metrics the condition on the order (for even dimension n) cannot be relaxed. Theorem 3.1.1 holds true for pseudo-Riemannian manifolds.

In the following, we write $P_{2N,n}$ if there is reason to emphasize both the dimension of the underlying manifold and the order of the GJMS-operator. For even dimension n, the operator $P_n = P_{n,n}$ will be called the *critical* GJMS-operator. This terminology makes it natural to refer to the operators $P_{2N,n}$, $2N < n$ (for even n) as to the subcritical GJMS-operators.

We briefly describe the main ideas of the proof ([124]) of Theorem 3.1.1. It rests on the Fefferman-Graham ambient metric introduced in [99]. For a Riemannian manifold (M, g), we form the space

$$\mathcal{G}_M = \{(x, tg_x) \,|\, t > 0\} \subset S^2(T^*M).$$

$\mathcal{G}_M$ is a ray bundle on M. Its sections are the metrics which are conformally equivalent to g, i.e., $\mathcal{G}_M$ is an invariant of the conformal class c of g. $\mathcal{G}_M$ admits a canonical $\mathbb{R}^+$-action

$$\delta_s : (x, g_x) \mapsto (x, s^2 g_x)$$

which enables us to define spaces of homogeneous functions

$$\mathcal{E}_M(\lambda) = \{u \in C^\infty(\mathcal{G}_M) \,|\, \delta_s^*(u) = s^\lambda u, \; s > 0\}$$

for any $\lambda \in \mathbb{C}$. Elements of $\mathcal{E}_M(\lambda)$ can be identified with sections of the associated line bundle on M which is defined by the $\mathbb{R}^+$- character s^λ. A choice of a metric in the conformal class c induces trivializations of these line bundles and isomorphisms $\mathcal{E}_M(\lambda) \simeq C^\infty(M)$.

The ambient manifold $\tilde{\mathcal{G}}_M$ is $\mathcal{G}_M \times (-1, 1)$. Let $\iota : \mathcal{G}_M \to \tilde{\mathcal{G}}_M$, $\iota : p \mapsto (p, 0)$ be the canonical embedding. The action δ_s extends naturally to $\tilde{\mathcal{G}}_M$. The metric g induces a tautological symmetric bilinear form $\mathbf{g}$ on the ray bundle $\mathcal{G}_M$ which satisfies $\delta_s^*(\mathbf{g}) = s^2 \mathbf{g}$.

The Fefferman-Graham ambient metric $\tilde{g}$ is a (formal pseudo-Riemannian) metric on $\tilde{\mathcal{G}}_M$ along the hypersurface $\mathcal{G}_M$ (i.e., it exists in a sufficiently small neighbourhood) so that

(i) $\iota^*(\tilde{g}) = \mathbf{g}$, i.e., it extends $\mathbf{g}$,

(ii) $\delta_s^*(\tilde{g}) = s^2 \tilde{g}$, $s > 0$

and

(iii) $\mathrm{Ric}(\tilde{g}) = 0$ along $\mathcal{G}_M$.

Theorem 2.1 in [99] states that if $n = \dim M$ is odd then, up to $\mathbb{R}^+$-equivariant diffeomorphisms which fix $\mathcal{G}_M$, there is a *unique* formal power series solution of $(i), (ii), (iii)$. For even n the condition (iii), in general, cannot be satisfied. However, up to $\mathbb{R}^+$-equivariant diffeomorphisms which fix $\mathcal{G}_M$, and up to terms which vanish of order $\frac{n}{2}$, there is a unique formal power series solution $\tilde{g}$ so that $\mathrm{Ric}(\tilde{g})$ vanishes of order $\frac{n}{2} - 2$ along $\mathcal{G}_M$ and the tangential component of $\mathrm{Ric}(\tilde{g})$ vanishes of order $\frac{n}{2} - 1$ along $\mathcal{G}_M$. For full details see [96].

Now for $N \geq 1$ let $\Delta_{\tilde{g}}^N$ be a power of the Laplacian of an ambient metric $\tilde{g}$. It induces operators

$$\Delta_{\tilde{g}}^N : \tilde{\mathcal{E}}_M(\lambda) \to \tilde{\mathcal{E}}_M(\lambda - 2N)$$

on homogeneous functions on $\tilde{\mathcal{G}}_M$, and for a specific value of the weight λ it descends to a differential operator on homogeneous functions on $\mathcal{G}_M$. More precisely, if N is subject to the conditions in Theorem 3.1.1, then it induces a differential operator

$$\mathcal{E}_M\left(-\frac{n}{2} + N\right) \to \mathcal{E}_M\left(-\frac{n}{2} - N\right)$$

by the composition

$$u \mapsto \iota^* \Delta_{\tilde{g}}^N(\tilde{u}),$$

where $\tilde{u}$ denotes an arbitrary homogeneous extension of u. Here the crucial observation is that for $\lambda = -\frac{n}{2} + N$, the composition does not depend on the choice of the extension of u. By the uniqueness of $\tilde{g}$, up to diffeomorphisms which fix the boundary, the resulting operator is an invariant of the conformal class c of g. It corresponds to the conformally covariant differential operator $P_{2N}(g)$ on $C^\infty(M)$. The condition $N \leq \frac{n}{2}$ (for even n) arises from the obstructions to the existence of $\tilde{g}$ satisfying the condition (iii) to high order.

By the convention $-\Delta_g \geq 0$, the operator $P_{2N}(g)$ differs from the operator $P_{2N}(g)$ in [124] by the sign $(-1)^N$.

The operators $P_{2N}(g)$ are formally self-adjoint ([128]). Branson ([32]) proved that the operator $P_{2N}(g)$ is of the form

$$P_{2N} = P_{2N}^0 + (-1)^N \left(\frac{n}{2} - N\right) Q_{2N}, \qquad (3.1.2)$$

where $Q_{2N,n} = Q_{2N}$ is a local *scalar* invariant and P^0_{2N} annihilates constants. Due to the sign $(-1)^N$ in (3.1.2), this definition of $Q_{2N}(g)$ coincides with that in [32]. (3.1.2) implies that

$$P_{2N}(g)(1) = \left(\frac{n}{2} - N\right)(-1)^N Q_{2N}(g). \tag{3.1.3}$$

(3.1.1) applied to $u = 1$ gives

$$e^{(\frac{n}{2}+N)\varphi} P_{2N}(e^{2\varphi}g)(1) = P_{2N}(g)(e^{(\frac{n}{2}-N)\varphi}). \tag{3.1.4}$$

But now (3.1.2) and $P^0_{2N}(g)(1) = 0$ yield

$$\left(\frac{n}{2} - N\right) e^{(\frac{n}{2}+N)\varphi} Q_{2N}(e^{2\varphi}g)$$
$$= (-1)^N P^0_{2N}(g)(e^{(\frac{n}{2}-N)\varphi} - 1) + \left(\frac{n}{2} - N\right) e^{(\frac{n}{2}-N)\varphi} Q_{2N}(g). \tag{3.1.5}$$

We divide by $\frac{n}{2} - N$ and set $N = \frac{n}{2}$ (for even n). This yields

$$e^{n\varphi} Q_n(e^{2\varphi}g) = (-1)^{\frac{n}{2}} P^0_n(g)(\varphi) + Q_n(g) = (-1)^{\frac{n}{2}} P_n(g)(\varphi) + Q_n(g)$$

since P_n (for even n) has no constant term. The identity

$$e^{n\varphi} Q_n(e^{2\varphi}g) = (-1)^{\frac{n}{2}} P_n(g)(\varphi) + Q_n(g) \tag{3.1.6}$$

will be called the *fundamental identity*. The latter *formal* argument is due to Branson and Ørsted ([48], [31], [32]), and known as analytic continuation in dimension.

Definition 3.1.1 (Q-curvatures). *Let $n \geq 2$ and N be as in Theorem 3.1.1. Then for a Riemannian manifold (M, g) of dimension n, the quantities $Q_{2N}(g) = Q_{2N,n}(g)$ will be called the Q-curvatures of (M, g). For manifolds of even dimension n, we distinguish between the critical Q-curvature $Q_n(g) = Q_{n,n}(g)$ and the subcritical Q-curvatures $Q_{2N}(g) = Q_{2N,n}(g)$, $2N < n$.*

For $2N < n$, (3.1.5) gives

$$\left(\frac{n}{2} - N\right) Q_{2N}(e^{2t\varphi}g)$$
$$= e^{-(\frac{n}{2}+N)t\varphi}(-1)^N P^0_{2N}(g)(e^{(\frac{n}{2}-N)t\varphi} - 1) + \left(\frac{n}{2} - N\right) e^{-2Nt\varphi} Q_{2N}(g).$$

Differentiating at $t = 0$ yields

$$\left(\frac{n}{2} - N\right) Q^\bullet_{2N}(g)[\varphi] = (-1)^N \left(\frac{n}{2} - N\right) P^0_{2N}(g)(\varphi) - \left(\frac{n}{2} - N\right) 2N\varphi Q_{2N}(g),$$

i.e.,

$$Q^\bullet_{2N}(g)[\varphi] = (-1)^N P^0_{2N}(g)(\varphi) - 2N\varphi Q_{2N}(g). \tag{3.1.7}$$

Here

$$Q^{\bullet}_{2N}(g)[\varphi] = (d/dt)|_{t=0}(Q_{2N}(e^{2t\varphi}g))$$

denotes the infinitesimal conformal variation of Q_{2N}.

(3.1.6) yields the analogous result for $2N = n$. But whereas the infinitesimal conformal transformation law for the critical and the subcritical Q-curvatures are analogous, the subcritical Q-curvatures do *not* satisfy a conformal transformation law which is governed by a linear differential operator.

The relation $\mathrm{vol}(e^{2\varphi}g) = e^{n\varphi}\,\mathrm{vol}(g)$ and (3.1.6) imply that for a closed manifold M^n the *total Q-curvature*

$$\int_M Q_n(g)\,\mathrm{vol}(g)$$

is conformally invariant. For the proof we use that $P_n(g)$ is self-adjoint and annihilates constants. Partial integration yields

$$\int_{M^n} Q_n(e^{2\varphi}g)\,\mathrm{vol}(e^{2\varphi}g) = \int_{M^n} Q_n(e^{2\varphi}g)e^{n\varphi}\,\mathrm{vol}(g)$$

$$= (-1)^{\frac{n}{2}}\int_{M^n} P_n(g)(\varphi)\,\mathrm{vol}(g) + \int_{M^n} Q_n(g)\,\mathrm{vol}(g).$$

But the first integral vanishes. Thus

$$\int_{M^n} Q_n(e^{2\varphi}g)\,\mathrm{vol}(e^{2\varphi}g) = \int_{M^n} Q_n(g)\,\mathrm{vol}(g). \tag{3.1.8}$$

The fundamental identity (3.1.6) generalizes the Gauß curvature prescription equation

$$e^{2\varphi}K(e^{2\varphi}g) = -\Delta_g(\varphi) + K(g)$$

on a surface M^2. For $N = 1$ and $n \geq 3$, (3.1.4) states

$$-\frac{n-2}{4(n-1)}\tau(e^{2\varphi}g)e^{(\frac{n}{2}+1)\varphi} = \left(\Delta_g - \frac{n-2}{4(n-1)}\tau(g)\right)(e^{(\frac{n}{2}-1)\varphi}),$$

i.e.,

$$\left(-\Delta_g + \frac{n-2}{4(n-1)}\tau(g)\right)(u) = \frac{n-2}{4(n-1)}\tau(e^{2\varphi}g)u^{\frac{n+2}{n-2}} \tag{3.1.9}$$

for $u = e^{(\frac{n}{2}-1)\varphi}$. The latter identity is known as the *Yamabe equation*.

3.2 Scattering theory

A major development ([128]) was the identification of the GJMS-operators $P_{2N}(g)$ and the Q-curvature $Q_n(h)$ on a closed manifold M^n in the *scattering operator*

$$S(h;\lambda) : C^\infty(M) \to C^\infty(M)$$

of the Laplacian for an associated Poincaré-Einstein metric via the formulas

$$\text{Res}_{\frac{n}{2}+N}(\mathcal{S}(h;\lambda)) = -c_N P_{2N}(h)$$

and

$$\mathcal{S}(h;n)(1) = (-1)^{\frac{n}{2}} c_{\frac{n}{2}} Q_n(h)$$

(Theorem 3.2.1 and Theorem 3.2.2). The first formula generalizes the relation between the operators P_{2N} on the round sphere (S^n, g_c) and the residues of the standard Knapp-Stein intertwining operators for the spherical principal series viewed
as the scattering operator of the free hyperbolic space with boundary S^n at infinity.

The scattering operator is a rather complicated object even in simple cases
and Q_n is only a very small piece of information coded in that family of pseudo-
differential operators. A more direct formula for $Q_n(h)$ is

$$(-1)^{\frac{n}{2}} \dot{P}_n(h;0)(1),$$

where $P_n(\lambda)$ is a certain family of differential operators which arises by formally
solving the eigenequation $-\Delta_g u = \lambda(n-\lambda)u$ (see (3.2.20)). That result is also due
to [128]. Eliminating all superfluous constructions Fefferman and Graham ([97])
later found a simpler characterization of Q_n in term of the Poincaré metric. That
result will play an important role in Section 6.6.

Let M^n be the boundary of a compact manifold X^{n+1} (if M does not bound
any X one can take $X = [0, \varepsilon] \times M$ with $\partial X = M \sqcup M$). We consider metrics of
the form

$$g = \rho^{-2}\overline{g}, \tag{3.2.1}$$

where $\rho \in C^\infty(X)$ is a defining function of ∂X in X, and $\overline{g}$ is a smooth metric on X.
g is called a conformally compact metric on X and $\overline{g}$ a conformal compactification.
If we assume, in addition, that $|d\rho|_{\overline{g}} = 1$ on ∂X, then g is asymptotically hyperbolic,
i.e., the sectional curvatures of g approach -1 at ∂X. A conformally compact g
has many conformal compactifications and induces a conformal class $[i^*(\rho^2 g)]$,
$i : \partial X \hookrightarrow X$. This class is called the conformal infinity of g.

Conversely, for a given conformal class $[h]$ of metrics on ∂X, one can associate
a distinguished class of conformally compact metrics on X with $[h]$ as conformal
infinity. These are known as Poincaré metrics since they generalize the correspondence between the round metric on the sphere S^n and the constant curvature
Poincaré metric on the unit ball with S^n as its sphere at infinity. Poincaré metrics
correspond to ambient metrics ([99], [96]).

Let g_{AH} be an asymptotically hyperbolic metric on X with conformal infinity
$[h]$. We choose a representing metric h on ∂X. Then there exists a unique $\rho \in$
$C^\infty(X)$ near ∂X so that

$$|d\rho|_{\rho^2 g_{AH}} = 1 \quad \text{near } \partial X$$

and $i^*(\rho^2 g_{AH})$ coincides with h on $T(\partial X)$. The gradient flow of ρ for the conformal compactification, $\rho^2 g_{AH}$, defines a diffeomorphism

$$\alpha : [0, \varepsilon) \times \partial X \to X$$

onto a collar neighbourhood of ∂X. Then $\alpha^*(\rho) = r$ and

$$\alpha^*(g_{AH}) = \frac{1}{r^2}(dr^2 + h_r), \quad h_0 = h$$

with a curve h_r of metrics on ∂X. For more details we refer to [119].

Now we pose additional conditions on the metric. We require that the metric on $(0, \varepsilon) \times \partial X$ is Einstein,

$$\mathrm{Ric}(g) + ng = 0 \qquad (3.2.2)$$

($n = \dim \partial X$), i.e., $\tau(g) = -n(n+1)$, i.e., the scalar curvature of g coincides with that of the real hyperbolic space of dimension n. The condition (3.2.2) corresponds to the Ricci-flatness of the related ambient metric ([99], [96]).

Evaluating (3.2.2) for a formal power series in r leads to the following consequences. If n is *odd*, then there is a formal Taylor series

$$h_r \sim h_{(0)} + r^2 h_{(2)} + \cdots + r^{n-1} h_{(n-1)} + r^n h_{(n)} + \cdots, \qquad (3.2.3)$$

where all powers up to $n-1$ are even. Here the coefficients $h_{(2)}, \ldots, h_{(n-1)}$ are locally determined by the leading coefficient (conformal infinity) $h_{(0)} = h$. Moreover, $h_{(n)}$ has vanishing h-trace. h_r is formally *unique* if assumed to be *even* in r (in that case $h_{(n)} = 0$). For *even* n, the Taylor-type series of h_r is of the form

$$h_r \sim h_{(0)} + r^2 h_{(2)} + \cdots + r^{n-2} h_{(n-2)} + r^n \left(h_{(n)} + \log r \bar{h}_{(n)} \right) + \cdots, \qquad (3.2.4)$$

where all powers up to n are even. Here the coefficients $h_{(2)}, \ldots, h_{(n-2)}, \bar{h}_{(n)}$ and the h-trace of $h_{(n)}$ are locally determined by the conformal infinity $h_{(0)} = h$. Moreover, $\bar{h}_{(n)}$ has vanishing h-trace. The logarithmic terms obstruct the existence of a formal smooth solution. Note also that the condition

$$\mathrm{Ric}(g) + ng = O(\rho^{n-2})$$

uniquely determines the coefficients $h_{(2)}, \ldots, h_{(n-2)}$, and the additional condition

$$\mathrm{tr}(\mathrm{Ric}(g) + ng) = O(\rho^{n+2})$$

can be satisfied and uniquely determines the h-trace of $h_{(n)}$. See [99], [119], [128], [122].

Now a (true) metric on X is called a *Poincaré-Einstein metric* associated to $[h]$ if

$$\mathrm{Ric}(g) + ng = \begin{cases} O(\rho^\infty) & \text{if } n \text{ is odd,} \\ O(\rho^{n-2}) & \text{if } n \text{ is even,} \end{cases}$$

and induces the given class $[h]$. Moreover, it is required to be even in ρ if n is odd, and satisfies the vanishing trace condition

$$\operatorname{tr}(\operatorname{Ric}(g) + ng) = O(\rho^{n+2})$$

if n is even (see [128] and the references therein).

Geometric scattering theory ([181]) on X with an asymptotically hyperbolic metric g defines a scattering operator

$$\mathcal{S}(h; \lambda) : C^\infty(\partial X) \to C^\infty(\partial X).$$

For λ so that $\Re(\lambda) = \frac{n}{2}$, the scattering operator is determined by the property

$$\mathcal{S}(h; \lambda)f = g,$$

where near the boundary

$$u = \rho^{n-\lambda} F + \rho^\lambda G, \quad F, G \in C^\infty(X) \tag{3.2.5}$$

describes the boundary asymptotic behaviour of an eigenfunction $-\Delta_g u = \lambda(n-\lambda)u$ of the Laplacian of g, and $f = F|_{\partial X}$, $g = G|_{\partial X}$. Varying h in its conformal class, results in a conjugate scattering operator. More precisely,

$$\mathcal{S}(e^{2\varphi}h; \lambda) = e^{-\lambda\varphi} \circ \mathcal{S}(h; \lambda) \circ e^{(n-\lambda)\varphi}. \tag{3.2.6}$$

Theorem 3.2.1 ([128]). *Let g be a Poincaré metric associated to h. Then $\mathcal{S}(h; \lambda)$ has a meromorphic continuation to $\mathbb{C}$. Let $\lambda = \frac{n}{2} + N$, where*

$$N \in \begin{cases} 1, 2, \ldots, \frac{n}{2} & \text{if } n \text{ is even,} \\ \mathbb{N} & \text{if } n \text{ is odd,} \end{cases}$$

and assume that $\lambda(n-\lambda)$ is not an L^2-eigenvalue of $-\Delta_g$. Then $\mathcal{S}(h; \lambda)$ has a simple pole at λ with residue

$$\operatorname{Res}_{\frac{n}{2}+N}(\mathcal{S}(h; \lambda)) = -c_N P_{2N}(h), \tag{3.2.7}$$

where

$$c_N = \frac{1}{2^{2N} N! (N-1)!}.$$

We recall that our choice of the positive Laplacian is opposite to that in [128]. Let n be even. In view of $P_n(h)(1) = 0$, the formula (3.2.7) implies that $\mathcal{S}(h; \lambda)(1)$ is regular at $\lambda = n$. Graham and Zworski proved

Theorem 3.2.2 ([128]). *In the situation of Theorem 3.2.1,*

$$\mathcal{S}(h; n)(1) = (-1)^{\frac{n}{2}} c_{\frac{n}{2}} Q_n(h) \tag{3.2.8}$$

for even n.

Therefore, at $\lambda = n$ the Laurent series of the scattering operator has the form

$$\mathcal{S}(h; \lambda) = -c_{\frac{n}{2}} \frac{P_n(h)}{\lambda - n} + \mathcal{S}_0(h; \lambda) \qquad (3.2.9)$$

such that $\mathcal{S}_0(h; n)(1) = c_{\frac{n}{2}} Q_n(h)$. This identity can be used to give the following *family proof* of (3.1.6). (3.2.6) applied to the function 1 gives

$$\mathcal{S}(h; \lambda)(e^{(n-\lambda)\varphi}) = e^{\lambda \varphi} \mathcal{S}(e^{2\varphi} h; \lambda) 1.$$

Now we use (3.2.8) and (3.2.9) to calculate the limit $\lambda \to n$. We obtain

$$c_{\frac{n}{2}} \lim_{\lambda \to n} P_n(h) \left(\frac{e^{(n-\lambda)\varphi}}{n - \lambda} \right) + (-1)^{\frac{n}{2}} c_{\frac{n}{2}} Q_n(h) = e^{n\varphi} (-1)^{\frac{n}{2}} c_{\frac{n}{2}} Q_n(e^{2\varphi} h),$$

i.e., using $P_n(h)(1) = 0$, we get

$$e^{n\varphi} Q_n(e^{2\varphi} h) = (-1)^{\frac{n}{2}} \lim_{\lambda \to n} P_n(h) \left(\frac{e^{(n-\lambda)\varphi} - 1}{n - \lambda} \right) + Q_n(h)$$
$$= (-1)^{\frac{n}{2}} P_n(h)(\varphi) + Q_n(h).$$

The proof of (3.1.6) is complete.

The pole of the scattering operator $\mathcal{S}(h; \lambda)$ at $\lambda = \frac{n}{2} + N$, $N \in \mathbb{N}_0$ corresponds to a specific asymptotic behaviour of the eigenfunctions for the corresponding eigenvalue. For these values of λ, a genuine eigenfunction u is of the form

$$u = \rho^{n-\lambda} F + \rho^\lambda \log \rho G \qquad (3.2.10)$$

for $F, G \in C^\infty(X)$, i.e., has a logarithmic singularity, and for $\lambda = \frac{n}{2} + N$, we have the relation

$$G|_{\partial X} = 2 \operatorname{Res}_{\frac{n}{2}+N}(\mathcal{S}(\lambda) f), \quad f = F|_{\partial X}.$$

The occurrence of a logarithmic singularity for a genuine eigenfunction is related to the simple poles of the rational coefficients in the formal solutions

$$\sum_{j \geq 0} a_j(\lambda) \rho^{\lambda+j} \quad \text{and} \quad \sum_{j \geq 0} b_j(\lambda) \rho^{n-\lambda+j}$$

of the equation $-\Delta_g u = \lambda(n - \lambda) u$. In fact, the set $\frac{n}{2} + \frac{1}{2}\mathbb{Z}$ is the set of all λ for which the two ladders $\lambda + \mathbb{N}_0$ and $n - \lambda + \mathbb{N}_0$ intersect non-trivially. At the pole $\frac{n}{2} + N$ both sums contribute to the asymptotics of a genuine eigenfunction and induce a logarithmic singularity.

It is instructive to describe the details in the case of the hyperbolic upper half-space $\mathbb{H}^{n+1}$. The sums

$$\sum_{j \geq 0} a_{2j}(x'; \lambda) x_{n+1}^{\lambda+2j} \quad \text{and} \quad \sum_{j \geq 0} b_{2j}(x'; \lambda) x_{n+1}^{n-\lambda+2j}$$

are only *formal* eigenfunctions. Any *genuine* eigenfunction u is the *Helgason-Poisson* transformation of a hyperfunction vector of the corresponding principal series on the boundary ([152]). For the present discussion, we consider an eigenfunction with boundary values in $C_0^\infty(\mathbb{R}^n)$, i.e., let

$$u(x) = \int_{\mathbb{R}^n} P(x,\zeta)^\lambda \omega(\zeta) d\zeta, \quad \omega \in C_0^\infty(\mathbb{R}^n),$$

where

$$P(x,\zeta) = \frac{x_{n+1}}{|x-\zeta|^2}, \quad x \in \mathbb{H}^{n+1}, \zeta \in \mathbb{R}^n$$

is the Poisson kernel of the upper half-space. Then $-\Delta_{\mathbb{H}^{n+1}} u = \lambda(n-\lambda)u$ and at least for $\lambda \in \frac{n}{2} + i\mathbb{R}$, $\lambda \neq \frac{n}{2}$ we have the asymptotics

$$u(x', x_{n+1}) \sim \sum_{j \geq 0} a_{2j}(x',\lambda) x_{n+1}^{\lambda+2j} + \sum_{j \geq 0} b_{2j}(x',\lambda) x_{n+1}^{n-\lambda+2j} \qquad (3.2.11)$$

for $x_{n+1} \to 0$ with the leading terms

$$a_0(\lambda) = I_\lambda(\omega) \quad \text{and} \quad b_0(\lambda) = \pi^{\frac{n}{2}} \frac{\Gamma(\lambda-\frac{n}{2})}{\Gamma(\lambda)} \omega \stackrel{\text{def}}{=} c(\lambda)\omega \qquad (3.2.12)$$

(see [195]). Here I_λ denotes the meromorphic continuation of the integral operator with the kernel

$$\frac{1}{|x-y|^{2\lambda}}$$

and $c(\lambda)$ is the spherical Harish-Chandra c-function of $\mathbb{H}^{n+1}$. I_λ defines an intertwining operator for the (non-compact model of the) spherical principal series representations. It satisfies the intertwining relation

$$I_\lambda \circ \pi_{\lambda-n}^{nc}(g) = \pi_{-\lambda}^{nc}(g) \circ I_\lambda.$$

The analog of I_λ in the compact model appears in (2.4.2). (3.2.12) shows that $S(\lambda)b_0(\lambda) = a_0(\lambda)$, i.e.,

$$S(\lambda) = \frac{I_\lambda}{c(\lambda)}.$$

(5.2.52) shows that the coefficients $a_{2j}(x',\lambda)$ are given by

$$a_{2j}(x';\lambda) = \frac{1}{2^j j!} \frac{1}{\{(n-2j-2\lambda)\cdots(n-2-2\lambda)\}^\#} \Delta_n^j a_0(x';\lambda)$$

and

$$b_{2j}(x';\lambda) = \frac{1}{2^j j!} \frac{1}{\{(-n-2j+2\lambda)\cdots(-n-2+2\lambda)\}^\#} \Delta_n^j b_0(x';\lambda),$$

i.e.,

$$u \sim \sum_{j \geq 0} \frac{1}{2^j j!} \frac{1}{\{(n-2j-2\lambda)\cdots(n-2-2\lambda)\}^{\#}} \Delta_n^j I_\lambda(\omega) x_{n+1}^{\lambda+2j}$$
$$+ c(\lambda) \sum_{j \geq 0} \frac{1}{2^j j!} \frac{1}{\{(-n-2j+2\lambda)\cdots(-n-2+2\lambda)\}^{\#}} \Delta_n^j(\omega) x_{n+1}^{n-\lambda+2j}. \quad (3.2.13)$$

(3.2.13) holds true if $\Re(\lambda) > \frac{n}{2}$, $\lambda \notin \frac{n}{2} = \mathbb{N}_0$; then I_λ and $c(\lambda)$ have no poles.

Now we take a closer look at the asymptotics of u for $\lambda \in \frac{n}{2} + \mathbb{N}$. Let $\lambda = \frac{n}{2} + N$. In (3.2.13) both sums have a simple pole at $\lambda = \frac{n}{2} + N$ with respective residues

$$\sum_{j \geq 0} \frac{1}{2^j j!} \frac{1}{\{(-2j-2N)\cdots(-2-2N)\}^{\#}} \Delta_n^j \operatorname{Res}_{\frac{n}{2}+N}(I_\lambda)(\omega) x_{n+1}^{\frac{n}{2}+N+2j}$$

and

$$\sum_{j \geq 0} \frac{1}{2^j j!} \operatorname{Res}_{\frac{n}{2}+N} \left(\frac{1}{\{(-n-2j+2\lambda)\cdots(-n-2+2\lambda)\}^{\#}} \right) \Delta_n^j(\omega) x_{n+1}^{\frac{n}{2}-N+2j}.$$

The latter sum contributes only for $j \geq N$. The coefficient of $x_{n+1}^{\frac{n}{2}+N+2j}$ is

$$\frac{1}{2^j j!} \frac{1}{\{(-2j-2N)\cdots(-2-2N)\}^{\#}} \mu_n \Delta_n^{j+N}(\omega)$$
$$+ c\left(\frac{n}{2}+N\right) \frac{1}{2^{j+N}(j+N)!}$$
$$\times \operatorname{Res}_{\frac{n}{2}+N} \left(\frac{1}{\{(-n-2j-2N+2\lambda)\cdots(-n-2+2\lambda)\}^{\#}} \right) \Delta_n^{j+N}(\omega)$$

using

$$\operatorname{Res}_{\frac{n}{2}+N}(I_\lambda) = \mu_N \Delta_n^N. \quad (3.2.14)$$

We claim that these coefficients vanish, i.e., there are no poles on the right-hand side of (3.2.13) at $\lambda \in \frac{n}{2} + \mathbb{N}$. The assertion is equivalent to the identity

$$\mu_N = c\left(\frac{n}{2}+N\right) \frac{1}{2^{2N}} \frac{1}{N!(N-1)!} = \pi^{\frac{n}{2}} \frac{1}{\Gamma(\frac{n}{2})} \frac{1}{\{n \cdots(n-2+2N)\}^{\#}} \frac{1}{2^N n!}. \quad (3.2.15)$$

Now it is well known ([105]) that $|x|^{-\lambda}$ is a meromorphic family of distributions on $\mathbb{R}^n$ with simple poles at $\lambda \in n + 2\mathbb{N}_0$ and residue

$$\operatorname{vol}(S^{n-1}) \frac{1}{2^N N! n(n+2)\cdots(n-2+2N)} \Delta_n^N \delta_0$$

at $\lambda = n+2N$. It follows that I_λ has a meromorphic continuation to $\mathbb{C}$ with simple poles at $\lambda \in \frac{n}{2} + \mathbb{N}_0$ and residues

$$\frac{1}{2} \frac{\text{vol}(S^{n-1})}{2^N N! \{n\cdots(n-2+2N)\}^\#} \Delta_n^N = \frac{\pi^{\frac{n}{2}}}{\Gamma(\frac{n}{2})2^N N! \{n\cdots(n-2+2N)\}^\#} \Delta_n^N,$$

i.e., we obtain the formula

$$\mu_N = \frac{\pi^{\frac{n}{2}}}{\Gamma(\frac{n}{2})2^N N! \{n\cdots(n-2+2N)\}^\#}.$$

This proves (3.2.15), i.e., the vanishing of the residues.

It follows that at $\lambda = \frac{n}{2} + N$ the asymptotics (3.2.13) takes the form

$$\sum_{j\geq 0} \frac{1}{2^j j!} \frac{1}{\{(-2j-2N)\cdots(-2-2N)\}^\#} \Delta_n^{j+N}(\omega) \mu_N x_{n+1}^{\frac{n}{2}+N+2j} \log x_{n+1}$$

$$- c\left(\frac{n}{2}+N\right) \sum_{j\geq 0} \frac{1}{2^j j!} \text{Res}_{\frac{n}{2}+N}\left(\frac{1}{\{(-n-2j+2\lambda)\cdots(-n-2+2\lambda)\}^\#}\right)$$

$$\times x_{n+1}^{\frac{n}{2}-N+2j} \log x_{n+1} \Delta_n^j(\omega) + F x_{n+1}^{\frac{n}{2}-N}$$

with $F|_{x_{n+1}} = c(\frac{n}{2}+N)\omega$. The second sum contributes only for $j \geq N$. Moreover, the coefficient of $x_{n+1}^{\frac{n}{2}+N} \log x_{n+1}$ is

$$\mu_N \Delta_n^N(\omega)$$

$$- c\left(\frac{n}{2}+N\right) \frac{1}{2^N N!} \text{Res}_{\frac{n}{2}+N}\left(\frac{1}{\{(-n-2N+2\lambda)\cdots(-n-2+2\lambda)\}^\#}\right) \Delta_n^N(\omega).$$

But we have seen above that

$$\mu_N + c\left(\frac{n}{2}+N\right) \frac{1}{2^N N!} \text{Res}_{\frac{n}{2}+N}\left(\frac{1}{\{(-n-2N+2\lambda)\cdots(-n-2+2\lambda)\}^\#}\right) = 0.$$

Hence the coefficient of $x_{n+1}^{\frac{n}{2}+N} \log x_{n+1}$ is

$$2\mu_N \Delta_n^N(\omega) = 2 \text{Res}_{\frac{n}{2}+N}(I_\lambda)(\omega).$$

It follows that the asymptotics of u at $\lambda = \frac{n}{2} + N$ takes the form

$$u \sim x_{n+1}^{n-\lambda} F + \left(x_{n+1}^\lambda \log x_{n+1}\right) G, \quad x_{n+1} \to 0$$

with

$$F|_{x_{n+1}=0} = c\left(\frac{n}{2}+N\right)\omega \quad \text{and} \quad G|_{x_{n+1}=0} = 2 \text{Res}_{\frac{n}{2}+N}(I_\lambda)(\omega).$$

This corresponds to the asymptotics (3.2.10).

It is remarkable that analogous results hold true for general Poincaré-Einstein metrics ([128]). Since the residues of the scattering operator yield the GJMS-operators, it follows that these can be seen in the coefficients of the log-term in the asymptotics of *genuine* eigenfunctions.

In contrast to the above perspective, in the present work it is more important that the relevant conformally covariant differential operators can be seen in the asymptotics of formal eigenfunctions. In fact, in the case of eigenfunctions on $\mathbb{H}^{n+1}$ in the formal asymptotics

$$\sum_{j \geq 0} b_{2j}(x'; \lambda) x_{n+1}^{n-\lambda+2j}$$

the coefficient of $x_{n+1}^{n-\lambda+2N}$ is given by

$$b_{2N}(\lambda) = c(\lambda) \frac{1}{2^N N!} \frac{1}{\{(-n+2\lambda-2N) \cdots (-n-2+2\lambda)\}^{\#}} \Delta_n^N(\omega).$$

It has a simple pole at $\lambda = \frac{n}{2} + N$ with residue given by a multiple of the GJMS-operator Δ_n^N.

That picture extends to the case of a Poincaré metric, too. In fact, Graham and Zworski ([128]) gave the following construction of GJMS-operators P_{2N} in terms of Poincaré metrics. Let M^n be a closed manifold with metric h. We consider an associated Poincaré metric

$$g = \frac{1}{r^2}(dr^2 + h_r), \quad h_0 = h$$

on $(0, \varepsilon) \times M$ and construct a *formal* solution u of the equation $-\Delta_g u = \mu(n-\mu)u$ by using a power series ansatz $u \sim \sum_{j \geq 0} r^{\mu+j} a_j$. For odd n, all coefficients a_{odd} vanish and

$$a_{2j}(h; \mu) = \frac{1}{2^{2j} j!} \frac{\Gamma(\frac{n}{2} - \mu - j)}{\Gamma(\frac{n}{2} - \mu)} P_{2j}(h; \mu)(a_0(h))$$

$$= \frac{1}{2^{2j} j! \left(\frac{n}{2} - \mu - 1\right) \cdots \left(\frac{n}{2} - \mu - j\right)} P_{2j}(h; \mu)(a_0(h)) \qquad (3.2.16)$$

for certain *polynomial families* of natural differential operators $P_{2j}(h; \mu)$ of order $2j$ on M so that $P_{2j}(h; \mu) = \Delta_h^j + LOT$. Then

$$P_{2j}\left(h; \frac{n}{2} - j\right) = P_{2j}(h), \quad j \geq 1. \qquad (3.2.17)$$

For even n, the situation is analogous. In particular, we have

$$a_n(h; \mu) = \frac{1}{2^{2n} \left(\frac{n}{2}\right)! \left(\frac{n}{2} - \mu - 1\right) \cdots (-\mu)} P_n(h; \mu)(a_0(h))$$

and the formula

$$P_n(h) = P_n(h; 0) \tag{3.2.18}$$

for the critical GJMS-operator on M^n. The family $P_{2j}(h; \mu)$ satisfies the identity

$$P_{2j}(h; \mu)(1) = \mu Q_{2j}(h; \mu), \quad j = 1, \dots, \frac{n}{2}$$

for certain polynomial families of scalar Riemannian invariants $Q_{2j}(h; \mu)$. In particular, the operators $P_{2j}(h; 0)$ annihilate constants. Moreover, using (3.1.3) and (3.2.17), it follows that

$$Q_{2j}(h) = (-1)^j Q_{2j}\left(h; \frac{n}{2} - j\right) \tag{3.2.19}$$

and

$$Q_n(h) = (-1)^{\frac{n}{2}} \dot{P}_n(h; 0)(1). \tag{3.2.20}$$

For the details we refer to [128].

On the round sphere (S^n, g_c) the notion of a natural conformally covariant differential operator and the notion of a differential intertwining operator are closely related. We briefly describe that relation. On a general manifold M the naturality of D implies

$$D(M; f_*(g)) = f_* \circ D(M; g) \circ f^* \tag{3.2.21}$$

for a diffeomorphism $f : M \to M$. Now for (S^n, g_c) we have $\tau(S^n, g_c) = n(n-1)$. The group $G^{n+1} = SO(1, n+1)^\circ$ operates by conformal diffeomorphims on S^n, i.e.,

$$g_*(g_c) = e^{2\Phi_g} g_c, \quad g \in G^{n+1}, \quad g_* = (g^{-1})^* \tag{3.2.22}$$

for certain smooth functions $0 < \Phi_g \in C^\infty(S^n)$. It follows that for a conformally covariant operator D of weight (a, b) on S^n,

$$e^{-b\Phi_g} \circ D(S^n; g_c) \circ e^{a\Phi_g} = D(S^n; e^{2\Phi_g} g_c) = D(S^n; g_*(g_c)) = g_* \circ D(S^n; g_c) \circ g^* \tag{3.2.23}$$

i.e.,

$$D(S^n; g_c) \circ e^{a\Phi_g} g_* = e^{b\Phi_g} g_* \circ D(S^n; g_c). \tag{3.2.24}$$

Now (3.2.22) yields $g_*(\text{vol}(g_c)) = e^{n\Phi_g} \text{vol}(g_c)$ for the Riemannian volume form and the spherical principal series representation can be written in the form

$$\pi_\lambda^c(g)u = e^{-\lambda\Phi_g} g_*(u), \quad u \in C^\infty(S^n).$$

Thus the intertwining relation (3.2.24) reads

$$D(S^n; g_c) \circ \pi_{-a}^c(g) = \pi_{-b}^c(g) \circ D(S^n; g_c), \quad g \in G^{n+1}.$$

For a generic parameter a, the spherical principal series π_a^c is irreducible. Moreover, for generic parameters $\pi_a^c \simeq \pi_b^c$ iff $a + b = -n$, where the equivalence is realized by constant multiples of the standard Knapp-Stein intertwining operator I_λ (see (2.4.2)). The meromorphic family I_λ of pseudo-differential operators has *differential operator* residues at $\lambda \in -\frac{n}{2} + \mathbb{N}_0$. More precisely, up to a constant multiple, the residue of I_λ at $\lambda = -\frac{n}{2} + N$ has leading part $\Delta_{S^n}^N$ and is given by ([32], [23])

$$\prod_{j=1}^{N}\left(\Delta_{S^n} - \left(\frac{n-1}{2}\right)^2 + \left(j-\frac{1}{2}\right)^2\right) = \prod_{j=\frac{n}{2}}^{\frac{n}{2}+N-1}\left(\Delta_{S^n} - j(n-1-j)\right). \qquad (3.2.25)$$

Formula (3.2.25) is a special case of an analogous product formula for Einstein spaces.

Theorem 3.2.3 ([110]). *Let (M^n, g) be Einstein and n even. Then*

$$P_{2N}(M, g) = \prod_{j=\frac{n}{2}}^{\frac{n}{2}+N-1}\left(\Delta_g - \frac{j(n-1-j)}{n(n-1)}\tau_g\right) \qquad (3.2.26)$$

and

$$Q_n(g) = (n-1)!\left(\frac{\tau_g}{n(n-1)}\right)^{\frac{n}{2}}. \qquad (3.2.27)$$

For more information around Theorem 3.2.3 (including a proof) we refer to Section 6.16. Note that Theorem 3.2.3 yields the product formula

$$\prod_{j=\frac{n}{2}}^{\frac{n}{2}+N-1}\left(\Delta + j(n-1-j)\right) \qquad (3.2.28)$$

for the GJMS-operators of the compact hyperbolic manifolds $M^n = \Gamma\backslash\mathbb{H}^n$ of even dimension $n \geq 4$. We close the present section with an outline of a derivation of formula (3.2.28) using scattering theory. For this purpose, we realize $M^n = \Gamma\backslash\mathbb{H}^n$ as one boundary component of a hyperbolic cylinder

$$X^{n+1} = (0, \pi) \times M^n, \quad \partial X^{n+1} = M^n \sqcup M^n$$

and apply an explicit formula for the scattering operator. X is defined as the quotient $X = \Gamma\backslash\mathbb{H}^{n+1}$, where the cocompact torsion-free $\Gamma \subset G^n = SO(1, n)^\circ$ is viewed as a subgroup of G^{n+1} acting on $\mathbb{H}^{n+1}$. The map

$$\kappa : \mathbb{H}^n \times (0, \pi) \ni (x', x_n, \theta) \mapsto (x', x_n \cos\theta, x_n \sin\theta) \in \mathbb{H}^{n+1}$$

satisfies

$$\kappa\left(\mathbb{H}^n, \frac{\pi}{2}\right) = \mathbb{H}^{n+1} \cap \{x_n = 0\} \simeq \mathbb{H}^n$$

and pulls back the hyperbolic metric $g_{\mathbb{H}^{n+1}}$ on $\mathbb{H}^{n+1}$ to the metric

$$g = (\sin\theta)^{-2}(d\theta^2 + g_{\mathbb{H}^n})$$

on the cylinder $(0, \pi) \times \mathbb{H}^n$. The conformal compactification $(x_n \sin \theta)^2 g$ restricts to $g_{\mathbb{H}^n}$ on $\kappa(0) \cup \kappa(\pi)$. $G^n \subset G^{n+1}$ operates trivially on $(0, \pi)$, and for a cocompact $\Gamma \subset G^n$, we have

$$X^{n+1} \simeq (0, \pi) \times M^n, \quad \partial X^{n+1} = \kappa(M^n, 0) \cup \kappa(M^n, \pi),$$

where $M^n = \Gamma \backslash \mathbb{H}^n$ and $X^{n+1} = \Gamma \backslash \mathbb{H}^{n+1}$. The scattering operator

$$\mathcal{S}(\lambda) : C^\infty(\partial X) \to C^\infty(\partial X)$$

can be identified with an operator on $C^\infty(M) \oplus C^\infty(M)$. The following result provides an explicit formula for the latter operator.

Theorem 3.2.4. $\mathcal{S}(\lambda)$ *decomposes into the direct sum of the operators*

$$\mathcal{S}(\lambda, \mu) =$$
$$\frac{1}{\pi} 2^{n-2\lambda} \frac{\Gamma(\frac{n}{2} - \lambda)}{\Gamma(\lambda - \frac{n}{2})} \Gamma(\lambda - \mu) \Gamma(\lambda - (n-1-\mu)) \begin{pmatrix} \sin \pi(\frac{n}{2} - \mu) & \sin \pi(\frac{n}{2} - \lambda) \\ \sin \pi(\frac{n}{2} - \lambda) & \sin \pi(\frac{n}{2} - \mu) \end{pmatrix} \quad (3.2.29)$$

acting on the space $\mathcal{E}(\mu) \oplus \mathcal{E}(\mu)$, *where* $\mathcal{E}(\mu) = \ker(\Delta + \mu(n-1-\mu)) \subset C^\infty(M)$.

For a brief outline of the proof of Theorem 3.2.4 we refer to [195]. Full details are given in [150]. A direct calculation shows that (3.2.29) implies the functional equation

$$\mathcal{S}(\lambda, \mu) \mathcal{S}(n - \lambda, \mu) = \mathrm{id}.$$

We analyze the poles of $\mathcal{S}(\lambda)$. Let n be even and $1 \leq N \leq \frac{n}{2}$. (3.2.29) is invariant under the substitution $\mu \mapsto n-1-\mu$ and we choose μ in $(\frac{n-1}{2}, n-1] \cup \frac{n-1}{2} + i\mathbb{R}$; this is no restriction of generality since $-\Delta \geq 0$. If $\mu \notin \mathbb{R}$, then $\mathcal{S}(\lambda, \mu)$ has a simple pole at $\lambda = \frac{n}{2} + N$. Furthermore, if μ is real, but $\mu \notin \mathbb{Z}$, then $\Gamma(\lambda - \mu)$ is regular at $\lambda = \frac{n}{2} + N$, i.e., $\mathcal{S}(\lambda, \mu)$ has a simple pole at $\lambda = \frac{n}{2} + N$. We determine its residue in these cases. In view of $\Gamma(z)\Gamma(1-z) = \frac{\pi}{\sin(\pi z)}$, we find

$$\mathrm{Res}_{\lambda = \frac{n}{2} + N}(\mathcal{S}(\cdot, \mu))$$
$$= -\frac{1}{\pi} 2^{-2N} \frac{(-1)^N}{N!} \frac{1}{\Gamma(N)} \Gamma\left(\frac{n}{2} - \mu + N\right) \Gamma\left(-\left(\frac{n}{2} - \mu\right) + 1 + N\right) \sin \pi \left(\frac{n}{2} - \mu\right) \mathrm{id}$$
$$= -\frac{1}{N!(N-1)! 2^{2N}} \prod_{j=\frac{n}{2}}^{\frac{n}{2}+N-1} (j - \mu)(n-1-j-\mu) \, \mathrm{id}.$$

Hence the residue is given by

$$-\frac{1}{N!(N-1)! 2^{2N}} \prod_{j=\frac{n}{2}}^{\frac{n}{2}+N-1} (-\mu(n-1-\mu) + j(n-1-j)) \, \mathrm{id}$$
$$= -\frac{1}{N!(N-1)! 2^{2N}} \prod_{j=\frac{n}{2}}^{\frac{n}{2}+N-1} (\Delta + j(n-1-j)).$$

If $\mu \in \mathbb{Z}$, then the matrix in (3.2.29) has the form

$$\begin{pmatrix} 0 & \sin \pi(\frac{n}{2} - \lambda) \\ \sin \pi(\frac{n}{2} - \lambda) & 0 \end{pmatrix}$$

and both Γ-factors $\Gamma(\frac{n}{2} - \lambda)$ and $\Gamma(\lambda - \mu)$ contribute a simple pole at $\lambda = \frac{n}{2} + N$ iff $\frac{n}{2} + N - \mu \in -\mathbb{N}_0$, i.e., iff

$$\mu \in \left\{ \frac{n}{2} + N, \ldots, n-1 \right\}. \tag{3.2.30}$$

If (3.2.30) is not satisfied, i.e., iff

$$\mu \in \left\{ \frac{n}{2}, \ldots, \frac{n}{2} + N - 1 \right\},$$

then $\mathcal{S}(\lambda, \mu)$ has no pole at $\lambda = \frac{n}{2} + N$. In fact, the latter values of μ correspond to the eigenspaces of Δ in the kernel of the residue. In the cases (3.2.30), $\mathcal{S}(\lambda, \mu)$ has a simple pole at $\lambda = \frac{n}{2} + N$ with a residue being proportional to

$$\begin{pmatrix} 0 & 1 \\ 1 & 0 \end{pmatrix}.$$

In particular, the residue is *not* local. Now the cases (3.2.30) are characterized by the existence of L^2-eigenfunctions of the Laplacian on the infinite-volume cylinder X. This illustrates the role of the technical condition in Theorem 3.2.1 that $\lambda(n-\lambda)$ is not an L^2-eigenvalue of $-\Delta_X$.

In the present situation, it is actually possible to give explicit formulas for the L^2-eigenfunctions which have to be ruled out. Generalized eigenfunctions of $-\Delta_X$ can be constructed by integration of the Eisenstein kernel [1]

$$E(x, \zeta; \lambda) = \sum_{\gamma \in \Gamma} P(\gamma \cdot x, \zeta)^\lambda, \ x \in \mathbb{H}^{n+1}, \ \zeta_n \neq 0, \ \Re(\lambda) > n$$

against automorphic functions in $C^\infty(M) \oplus C^\infty(M)$. For $\omega \in \ker(\Delta_M + \mu(n-1-\mu))$, we find the generalized eigenfunction

$$\pi^{\frac{n}{2}} 2^{n-2\lambda} (\sin \theta)^\lambda \frac{\Gamma(\lambda - \mu)\Gamma(\lambda - (n-1-\mu))}{\Gamma(\lambda)\Gamma(\lambda - \frac{n}{2} + 1)}$$

$$\times \, {}_2F_1\left(\lambda - \mu, \lambda - (n-1-\mu); \lambda - \frac{n}{2} + 1; \frac{1 + \cos \theta}{2}\right) \omega(x', x_n) \tag{3.2.31}$$

in $\ker(\Delta_X + \lambda(n-\lambda))$. (3.2.31) is holomorphic in $\Re(\lambda) > n$ and admits a meromorphic continuation in λ to $\mathbb{C}$. It has a pole at $\lambda = \frac{n}{2} + N$ if μ satisfies (3.2.30).

---

[1] This is an Eisenstein series of a convex-cocompact Kleinian group. For the details we refer to [173], [200], [51] and the last chapter of [151].

In that case, the residue of (3.2.31) is an eigenfunction for the eigenvalue $\lambda(n-\lambda)$, $\lambda = \frac{n}{2} + N$. It is a constant multiple of the function

$$(\sin\theta)^{\frac{n}{2}+N} {}_2F_1\left(-M, 2N+M+1; N+1; \frac{1+\cos\theta}{2}\right) w(x', x_n),$$

where we write $\frac{n}{2} + N - \mu = -M$. Here the hypergeometric function degenerates to a polynomial. More precisely,

$$ {}_2F_1\left(-M, 2N+M+1; N+1; \frac{1-r}{2}\right) \sim C_M^{N+\frac{1}{2}}(r),$$

where $C_M^\alpha(r)$ is a Gegenbauer polynomial of degree M (see [21], 3.15, formula (3)). The latter formula implies that the eigenfunction is in $L^2(X)$ with respect to the hyperbolic volume $(\sin\theta)^{-n} \operatorname{vol}(\mathbb{H}^n) \operatorname{vol}(0,\pi)$.

Now the relation

$$\operatorname{Res}_{\lambda=\frac{n}{2}+N}(\mathcal{S}) = -c_N P_{2N}$$

(see (3.2.7)) implies that the GJMS-operators P_{2N} on $\Gamma\backslash\mathbb{H}^n$ are given by (3.2.28).

Theorem 3.2.4 also implies that

$$\mathcal{S}(\lambda;0) = \frac{1}{\pi} 2^{n-2\lambda} \frac{\Gamma(\frac{n}{2}-\lambda)}{\Gamma(\lambda-\frac{n}{2})}\Gamma(\lambda)\Gamma(\lambda-(n-1))\sin\pi\left(\frac{n}{2}-\lambda\right)\begin{pmatrix}0 & 1\\ 1 & 0\end{pmatrix}.$$

Hence

$$\mathcal{S}(\lambda)\begin{pmatrix}1\\1\end{pmatrix} = 2^{n-2\lambda}\frac{\Gamma(\lambda)\Gamma(\lambda-(n-1))}{\Gamma(\lambda-\frac{n}{2})}\frac{1}{\Gamma(1-(\frac{n}{2}-\lambda))}\begin{pmatrix}1\\1\end{pmatrix}$$

so that for $\lambda \to n$ we obtain

$$\mathcal{S}(n)\begin{pmatrix}1\\1\end{pmatrix} = 2^{-n}\frac{\Gamma(n)}{\Gamma(\frac{n}{2})\Gamma(\frac{n}{2}+1)}\begin{pmatrix}1\\1\end{pmatrix} = c_{\frac{n}{2}}(n-1)!\begin{pmatrix}1\\1\end{pmatrix}.$$

Now (3.2.8) yields the formula

$$Q_n = (-1)^{\frac{n}{2}}(n-1)!$$

for the Q-curvature of the hyperbolic space $\Gamma\backslash\mathbb{H}^n$. The latter formula is a special case of (3.2.27).

Chapter 4

Paneitz Operator and Paneitz Curvature

In the present chapter we derive basic facts concerning the Q-curvature Q_4 in dimension 4. In particular, we prove its conformal transformation law, i.e., the *fundamental identity*, which involves the Paneitz operator P_4.

The early history of the pair (P_4, Q_4) is as follows. The operator P_4 was discovered in the middle of the 1980s independently by Eastwood-Singer ([91]), Paneitz ([192]) and Riegert ([208]) almost at the same time.

Paneitz worked in general dimension $n \geq 3$ and discovered the conformal covariance of the operator $P_{4,n}$. Although the original paper of Paneitz remained unpublished until recently ([192]), it had a major influence. In [30], Branson supplied detailed proofs of the conformal covariance of the operator $P_{4,n}$. The works [91] and [208] contain this result in dimension $n = 4$.

Although the curvature quantity $Q_{4,n}$ in dimension $n \geq 3$ appeared in [192] in the constant term of $P_{4,n}$, Paneitz did not discuss the transformation properties of $Q_4 = Q_{4,4}$ under conformal changes of the metric. This aspect is important in Riegert's independent work [208]. Riegert discovered that the transformation law of a certain linear combination of Q_4 and $|C|^2$, under conformal changes of the metric, is governed by a *linear* differential operator: P_4. Later work ([12]) along these lines led to a formula for Q_6 for conformally flat metrics (see the discussion at the end of Section 6.14).

In [30], Branson introduced the notation Q for $Q_{4,n}$, and emphasized that this quantity (for $n \neq 1, 2, 4$) gives rise to a fourth-order analog of the Yamabe problem. In [48], Branson and Ørsted applied analytic continuation in dimension to derive the conformal transformation property of Q_4, i.e., the fundamental identity. The seminal work [31] embedded the material into the general context of GJMS-operators and introduced the hierarchy of Q-curvatures (as in Definition 3.1.1).

For detailed information on the more recent role of Q_4 in geometric analysis on four-manifolds we refer to [34], [73], [41], [43], [63], [64], [83], [170], [177] and the bibliographies of these references. Analogous problems in higher dimension are discussed, for instance, in [49], [93], [187], [205].

4.1 P_4, Q_4 and their transformation properties

We introduce some notation. For a Riemannian manifold (M, g) of dimension n, let $*_g$ be the Hodge star operator on forms. Then the operators

$$\delta_g : \Omega^p(M) \to \Omega^{p-1}(M), \quad \delta_g = (-1)^{np+n+1} *_g d*_g$$

are the formal adjoints of the exterior differentials on forms with respect to the Hodge scalar product

$$\int_M \omega \wedge *_g \eta$$

([239]). Let $\Delta = \delta d$ denote the non-negative Laplacian (this sign convention is opposite to Chapter 3).

A symmetric bilinear form T induces on $\Omega^1(M)$ the linear operator

$$T\# : \omega \mapsto (X \mapsto T(X, \omega^\#)).$$

Here $\omega^\# \in \mathcal{X}(M)$ denotes the vector field dual to ω, i.e., $g(\omega^\#, Y) = \langle \omega, Y \rangle$ for all $Y \in \mathcal{X}(M)$. In other words, T acts by the dual of the linear operator $T^\#$ on vector fields which corresponds to T by $g(T^\#(X), Y) = T(X, Y)$. Similarly, for the vector field X, let $X^\flat \in \Omega^1(M)$ be defined by $\langle X^\flat, Y \rangle = g(X, Y)$ for all $Y \in \mathcal{X}(M)$. Note that $g\#$ acts as the identity on $\Omega^1(M)$.

The critical Q-curvature of a Riemannian manifold of dimension $n = 4$ is defined by

$$Q_4 = 2J^2 - 2|P|^2 + \Delta J = \frac{1}{6}\tau^2 - \frac{1}{2}|\operatorname{Ric}|^2 + \frac{1}{6}\Delta\tau. \tag{4.1.1}$$

We shall refer to this fourth-order curvature invariant also as to the *Paneitz curvature* or *Paneitz quantity* in order to distinguish it from Q-curvatures of orders $\neq 4$.

The Paneitz curvature Q_4 has the following conformal transformation law.

Theorem 4.1.1 (The fundamental identity for Q_4). *For all $\varphi \in C^\infty(M)$,*

$$e^{4\varphi}Q_4(e^{2\varphi}g) = Q_4(g) + P_4(g)(\varphi), \tag{4.1.2}$$

where the Paneitz operator P_4 is defined by

$$P_4 = \Delta^2 + \delta(2Jg - 4P)\#d = \Delta^2 + \delta\left(\frac{2}{3}\tau g - 2\operatorname{Ric}\right)\#d. \tag{4.1.3}$$

Although the individual terms in (4.1.1) transform with up to fourth-order powers of φ under conformal transformations, all non-linearities cancel in the given specific linear combination. This property distinguishes the quantity Q_4 among other fourth-order scalar Riemannian curvature invariants.

The fourth-order Paneitz-operator P_4 is conformally covariant of weight $(0,4)$, i.e.,

Theorem 4.1.2. *For all $\varphi \in C^\infty(M)$,*

$$e^{4\varphi} \circ P_4(e^{2\varphi}g) = P_4(g). \tag{4.1.4}$$

Proof. The assertion is a consequence of Theorem 4.1.1. In fact, we find

$$e^{4\varphi} P_4(e^{2\varphi}g)u = e^{4\varphi}\left(e^{4u}Q_4(e^{2(u+\varphi)}g) - Q_4(e^{2\varphi}g)\right) \qquad \text{(by (4.1.2))}$$
$$= e^{4(\varphi+u)}Q_4(e^{2(u+\varphi)}g) - e^{4\varphi}Q_4(e^{2\varphi}g)$$
$$= Q_4(g) + P_4(g)(u+\varphi) - [Q_4(g) + P_4(g)\varphi] \qquad \text{(by (4.1.2))}$$
$$= P_4(g)u.$$

The proof is complete. ∎

Alternatively, (4.1.2) follows from the generalization

$$e^{-(\frac{n}{2}+2)\varphi} \circ P_{4,n}(e^{2\varphi}g) = P_{4,n}(g) \circ e^{(\frac{n}{2}-2)\varphi} \tag{4.1.5}$$

of (4.1.4) for the Paneitz-operator

$$P_{4,n} = \Delta^2 + \delta\left((n-2)\mathsf{J}g - 4\mathsf{P}\right)\#d + \frac{n-4}{2}Q_{4,n} \tag{4.1.6}$$

on Riemannian manifolds of dimension $n \geq 3$ (see Section 3.1). Here the constant term of $P_{4,n}$ is given by the fourth-order curvature invariant

$$Q_{4,n} = \frac{n}{2}\mathsf{J}^2 - 2|\mathsf{P}|^2 + \Delta\mathsf{J}. \tag{4.1.7}$$

$Q_{4,n}$ specializes to Q_4 for $n = 4$.

In terms of Ricci and scalar curvature, $P_{4,n}$ and $Q_{4,n}$ can be written in the less illuminating form

$$\Delta^2 + \delta\left(a_n \tau g + b_n \operatorname{Ric}\right)\#d + \frac{n-4}{2}Q_{4,n}$$

with

$$a_n = \frac{(n-2)^2 + 4}{2(n-1)(n-2)}, \quad b_n = -\frac{4}{n-2},$$

and

$$Q_{4,n} = c_n|\operatorname{Ric}|^2 + d_n\tau^2 + \frac{1}{2(n-1)}\Delta\tau$$

with

$$c_n = -\frac{2}{(n-2)^2}, \quad d_n = \frac{n^3 - 4n^2 + 16n - 16}{8(n-1)^2(n-2)^2}.$$

4.2   The fundamental identity for the Paneitz curvature

In the present section we give a direct proof of Theorem 4.1.1. The proof will be interwoven with an informal description of how the basic principle of equivalence between conformal covariance and its infinitesimal version acts in the given special case. Moreover, we use the instance to establish some important facts for later reference.

We continue to use $\hat{\ }$ to indicate the conformal change $g \mapsto e^{2\varphi}g$. In particular, $\hat{\delta}$ and $\hat{\Delta}$ denote the respective operators for the metric $e^{2\varphi}g$. We need the following simple result.

Lemma 4.2.1. *The operators*

$$e^{a\varphi} \circ d \circ e^{-a\varphi} : \Omega^p(M) \to \Omega^{p+1}(M), \quad e^{(a+2)\varphi} \circ \hat{\delta} \circ e^{-a\varphi} : \Omega^p(M) \to \Omega^{p-1}(M)$$

coincide with

$$\omega \mapsto d\omega - ad\varphi \wedge \omega, \quad \omega \mapsto \delta\omega - (n-2p-a)i_{\mathrm{grad}\,\varphi}(\omega)$$

for all $a \in \mathbb{C}$, respectively. Hence the operator

$$e^{(a+2)\varphi} \circ \hat{\Delta} \circ e^{-a\varphi} : C^\infty(M) \to C^\infty(M), \quad \Delta = \delta d$$

coincides with

$$\omega \mapsto \Delta\omega - a\delta(d\varphi \wedge \omega) - (n-2-a)i_{\mathrm{grad}\,\varphi}(d\omega) + a(n-2-a)|d\varphi|^2\omega.$$

Proof. The assertion for d is obvious. Note that $\hat{*} = e^{(n-2p)\varphi}* : \Omega^p \to \Omega^{n-p}$. Hence

$$(-1)^{np+n+1}\hat{\delta}\omega = \hat{*}d\hat{*}\omega$$

$$= e^{(n-2(n-p+1))\varphi} * d * e^{(n-2p)\varphi}\omega$$

$$= e^{-2\varphi}\left(*d*\omega + (n-2p)*d\varphi \wedge *\omega\right)$$

$$= e^{-2\varphi}\left(*d*\omega + (n-2p)(-1)^{n-p}i_{\mathrm{grad}(\varphi)}(*^2\omega)\right)$$

$$= e^{-2\varphi}\left(*d*\omega + (-1)^{np+n}(n-2p)i_{\mathrm{grad}(\varphi)}(\omega)\right)$$

using $*^2 = (-1)^{np+p}$ on Ω^p. Hence

$$e^{2\varphi}\hat{\delta}\omega = \delta\omega - (n-2p)i_{\mathrm{grad}(\varphi)}(\omega).$$

This proves the assertion on δ for $a = 0$. The proof for general a is analogous. In order to prove the assertion for the Laplacian, we write

$$e^{(a+2)\varphi}\hat{\Delta}(e^{-a\varphi}\omega) = \left(e^{(a+2)\varphi} \circ \hat{\delta} \circ e^{-a\varphi}\right)\left(e^{a\varphi}d(e^{-a\varphi}\omega)\right)$$

and apply the identities for d and δ. The proof is complete. $\square$

Now $\delta = - * d*$ in dimension $n = 4$. We get

$$\widehat{\Delta}\widehat{\mathsf{J}} = \widehat{\delta}d\widehat{\mathsf{J}}$$

$$= e^{-4\varphi}\delta e^{2\varphi}d\widehat{\mathsf{J}} \qquad \text{(by Lemma 4.2.1)}$$

$$= e^{-4\varphi}\delta d(\mathsf{J} + \Delta\varphi - |d\varphi|^2) - 2e^{-4\varphi}\delta(\mathsf{J} + \Delta\varphi - |d\varphi|^2)d\varphi$$

$$= e^{-4\varphi}\big(\Delta\mathsf{J} + \Delta^2\varphi - 2\mathsf{J}\Delta\varphi + 2(d\mathsf{J}, d\varphi)$$

$$- \Delta|d\varphi|^2 - 2(\Delta\varphi)^2 + 2(d\Delta\varphi, d\varphi) + 2|d\varphi|^2\Delta\varphi - 2(d|d\varphi|^2, d\varphi)\big)$$

and

$$e^{4\varphi}\widehat{\mathsf{J}}^2 = \mathsf{J}^2 + 2\mathsf{J}\Delta\varphi + (\Delta\varphi)^2 - 2\mathsf{J}|d\varphi|^2 - 2|d\varphi|^2\Delta\varphi + |d\varphi|^4$$

by (2.5.7). In view of (2.5.9), the squared norm $|\widehat{\mathsf{P}}|^2$ is given by the product of $e^{-4\varphi}$ and the squared norm of

$$\mathsf{P} - \operatorname{Hess}(\varphi) + d\varphi \otimes d\varphi - \frac{1}{2}|d\varphi|^2 g.$$

A calculation yields

$$e^{4\varphi}\widehat{Q}_4 = Q_4 + \Big[\Delta^2\varphi + 2\mathsf{J}\Delta\varphi + 2(d\mathsf{J}, d\varphi) + 4\sum_{i,j}\mathsf{P}^{ij}\operatorname{Hess}_{ij}(\varphi)\Big]$$

$$+ \Big[-\Delta|d\varphi|^2 + 2(d\Delta\varphi, d\varphi) - 2\mathsf{J}|d\varphi|^2 - 2|\operatorname{Hess}(\varphi)|^2 - 4\sum_{i,j}\mathsf{P}^{ij}\varphi_i\varphi_j\Big] \tag{4.2.1}$$

$$+ \Big[-2(d|d\varphi|^2, d\varphi) + 4\sum_{i,j}\operatorname{Hess}^{ij}(\varphi)\varphi_i\varphi_j\Big]. \tag{4.2.2}$$

We rewrite this result in the form

Lemma 4.2.2.

$$e^{4\varphi}\widehat{Q}_4 = Q_4 + P_4(\varphi) + \frac{1}{2}\left(4\varphi P_4 + P_4^\bullet[\varphi]\right)(\varphi), \tag{4.2.3}$$

where

$$P_4^\bullet(g)[\varphi] = (d/dt)|_{t=0}\left(P_4(e^{2t\varphi}g)\right). \tag{4.2.4}$$

Now the relation (4.2.3) and the *infinitesimal conformal covariance*

$$4\varphi \circ P_4 + P_4^\bullet[\varphi] = 0 \tag{4.2.5}$$

of P_4 imply the fundamental identity (4.1.2).

Note that the conditions (4.1.4) (conformal covariance) and (4.2.5) (infinitesimal conformal covariance) are *equivalent*. In fact, (4.2.5) implies

$$e^{4\varphi} \circ P_4(e^{2\varphi}g) - P_4(g) = \int_0^1 (d/dt) \left(e^{4t\varphi} \circ P_4(e^{2t\varphi}g)\right) dt$$

$$= \int_0^1 (d/ds)|_0 \left(e^{4(t+s)\varphi} \circ P_4(e^{2(t+s)\varphi}g)\right) dt$$

$$= \int_0^1 e^{4t\varphi} \circ \left[4\varphi \circ P_4(e^{2t\varphi}g) + P_4^{\bullet}(e^{2t\varphi}g)[\varphi]\right] dt = 0.$$

Thus it remains to prove (4.2.3) and (4.2.5). For the proof of (4.2.3) we have to verify that

$$P_4(u) = \Delta^2 u + 2\mathsf{J}\Delta u + 2(d\mathsf{J}, du) + 4(\mathsf{P}, \mathrm{Hess}(u)), \qquad (4.2.6)$$

and that (4.2.1) coincides with

$$\frac{1}{2}(d/dt)|_{t=0} \left(e^{4t\varphi}P_4(e^{2t\varphi}g)\right)(\varphi) = \frac{1}{2}\left(4\varphi P_4(g) + P_4^{\bullet}(g)[\varphi]\right)(\varphi); \qquad (4.2.7)$$

the vanishing of (4.2.2) is obvious by

$$(d(d\varphi, d\varphi), d\varphi) = \nabla_{\mathrm{grad}\,\varphi}(d\varphi, d\varphi) = 2\left(\nabla_{\mathrm{grad}\,\varphi}d\varphi, d\varphi\right).$$

We note that this proof of the fundamental identity is an explicit version of the following argument showing that (4.1.2) follows from its infinitesimal version

$$4\varphi Q_4 + Q_4^{\bullet}[\varphi] = P_4(\varphi), \qquad (4.2.8)$$

where

$$Q_4^{\bullet}(g)[\varphi] = (d/dt)|_{t=0}(Q_4(e^{2t\varphi}g)),$$

and the conformal covariance (4.1.4).

Lemma 4.2.3. *(4.2.8) and (4.1.4) for all metrics in the conformal class of g imply* (4.1.2) *for g.*

Proof. We calculate

$$e^{4\varphi}Q_4(e^{2\varphi}g) - Q_4(g) = \int_0^1 (d/dt) \left(e^{4t\varphi}Q_4(e^{2t\varphi}g)\right) dt$$

$$= \int_0^1 (d/ds)|_0 \left(e^{4(t+s)\varphi}Q_4(e^{2(t+s)\varphi}g)\right) dt$$

$$= \int_0^1 e^{4t\varphi} \left(4\varphi Q_4(e^{2t\varphi}g) + Q_4^{\bullet}(e^{2t\varphi}g)[\varphi]\right) dt$$

$$= \int_0^1 e^{4t\varphi} P_4(e^{2t\varphi}g)\varphi dt \qquad \text{(by (4.2.8))}$$

$$= P_4(g)\varphi. \qquad \text{(by (4.1.4))}$$

The proof is complete.   □

In particular, (4.2.6) (being equivalent to (4.2.8)) and (4.2.5) (being equivalent to (4.1.4)) imply (4.1.2).

In order to prove (4.2.5), we calculate

$$(d/dt)|_{t=0} \left(e^{4t\varphi} P_4(e^{2t\varphi}g)u \right). \tag{4.2.9}$$

That calculation can be performed by using one of the equivalent formulas (4.1.3) or (4.2.6) for the operator P_4. It will be convenient to use the latter one. For the evaluation of (4.2.9) we apply Lemma 4.2.1 for $p = 0$ and $a = \frac{n}{2} - 1$. We find

$$e^{(\frac{n}{2}+1)\varphi} \circ \hat{\Delta} \circ e^{-(\frac{n}{2}-1)\varphi} = \Delta - \left(\frac{n}{2}-1\right) \left[\delta(\cdot d\varphi) + i_{\mathrm{grad}(\varphi)}d\right] + \left(\frac{n}{2}-1\right)^2 |d\varphi|^2$$

$$= \Delta - \left(\frac{n}{2}-1\right)\Delta(\varphi) + \left(\frac{n}{2}-1\right)^2 |d\varphi|^2.$$

Using

$$e^{2\varphi}\hat{J} = J + \Delta(\varphi) - \left(\frac{n}{2}-1\right)|d\varphi|^2,$$

we conclude that

$$e^{(\frac{n}{2}+1)\varphi} \circ \left(\hat{\Delta} + \left(\frac{n}{2}-1\right) \hat{J} \right) \circ e^{-(\frac{n}{2}-1)\varphi} = \Delta + \left(\frac{n}{2}-1\right) J.$$

This proves the conformal covariance of the Yamabe operator P_2. In view of

$$e^{4\varphi} P_4(e^{2\varphi}g)u = e^{4\varphi} \left(\hat{\Delta}^2 u + 2\hat{J}\hat{\Delta}u + 2\hat{g}(d\hat{J}, du) + 4\hat{g}(\hat{P}, \widehat{\mathrm{Hess}}(u)) \right)$$

(see (4.2.6)), Lemma 4.2.1, the factorization $e^{4\varphi}\hat{\Delta}^2 = \left(e^{4\varphi}\hat{\Delta}e^{-2\varphi} \right) \left(e^{2\varphi}\hat{\Delta} \right)$, (2.5.7) and (2.5.9) show that (4.2.9) equals

$$- 2\delta(d i_{\mathrm{grad}\,\varphi} + d\varphi \wedge \delta)du$$
$$+ 2(-2\mathrm{J}i_{\mathrm{grad}\,\varphi}du + \Delta\varphi\Delta u) + 2\sum_{r,s} g^{rs}((\Delta\varphi)_r - 2\mathrm{J}\varphi_r)u_s$$
$$+ 4\sum_{i,j} \mathsf{P}^{ij} \left(-\varphi_i u_j - \varphi_j u_i + (du, d\varphi)g_{ij}\right) - 4(\mathrm{Hess}(\varphi), \mathrm{Hess}(u))$$
$$= -2\Delta(du, d\varphi) + 2(d\Delta u, d\varphi) + 2(du, d\Delta\varphi) - 4\mathrm{J}(du, d\varphi)$$
$$- 4(\mathrm{Hess}(\varphi), \mathrm{Hess}(u)) - 4\sum_{i,j} \mathsf{P}^{ij} \left(\varphi_i u_j + \varphi_j u_i\right). \tag{4.2.10}$$

For $u = \varphi$, (4.2.10) simplifies to

$$-2\Delta|d\varphi|^2 + 4(d\Delta\varphi, d\varphi) - 4\mathrm{J}|d\varphi|^2 - 4|\mathrm{Hess}(\varphi)|^2 - 8\sum_{i,j} \mathsf{P}^{ij}\varphi_i\varphi_j,$$

and this coincides with (4.2.1), up to the factor 2. Thus (4.2.1) coincides with (4.2.7). But Lemma 4.2.4 shows that (4.2.10) vanishes. This completes the proof of (4.2.5).

Lemma 4.2.4.

$$\Delta(d\varphi, d\psi) = (d\Delta\varphi, d\psi) - 2(\text{Hess}(\varphi), \text{Hess}(\psi)) + (d\varphi, d\Delta\psi) - 2(\text{Ric}, d\varphi \otimes d\psi)$$

for $\varphi, \psi \in C^\infty(M)$.

Proof. For $\alpha, \beta \in \Omega^1(M)$ we have

$$\Delta(\alpha, \beta) = -(\Delta^B \alpha, \beta) - 2(\nabla\alpha, \nabla\beta) - (\alpha, \Delta^B \beta)$$

with the Bochner Laplacian Δ^B. But the Bochner-Weitzenböck formula $\Delta^B = -\Delta + \text{Ric}$ ([24], 1.155) implies

$$\Delta(d\varphi, d\psi) = (\Delta d\varphi, d\psi) - 2(\text{Hess}(\varphi), \text{Hess}(\psi)) - 2(\text{Ric}, d\varphi \otimes d\psi) + (d\varphi, \Delta d\psi).$$

This proves the assertion.   □

It only remains to verify that P_4, defined by (4.1.3), can be written in the form (4.2.6). That observation is the special case $n = 4$ of the fact that $P_{4,n}$ (see (4.1.6)) coincides with the operator

$$\Delta^2 u + (n-2)\mathsf{J}\Delta u + (6-n)(d\mathsf{J}, du) + 4(\mathsf{P}, \text{Hess}(u)) + \frac{n-4}{2}Q_{4,n}u. \qquad (4.2.11)$$

That assertion is obviously equivalent to

Lemma 4.2.5.
$$-\delta(\mathsf{P}\#du) = (d\mathsf{J}, du) + (\mathsf{P}, \text{Hess}(u)). \qquad (4.2.12)$$

This result follows from

Lemma 4.2.6. $\delta(\mathsf{P}\#du) = (\delta\mathsf{P}, du) - (\mathsf{P}, \text{Hess}(u))$, *where the divergence* δ *acts on a symmetric bilinear form* h *by* $\delta h(X) = -\sum_i \nabla_i(h)(e_i, X)$ *using an orthonormal frame.*

and the relation $-\delta\mathsf{P} = d\mathsf{J}$, which is a consequence of the second Bianchi identity. For the convenience of later reference and for the sake of completeness, these consequences are discussed in the following lemma.

Lemma 4.2.7 (Bianchi identities).

$$\nabla_Z(R)(U, V, X, Y) + \nabla_Y(R)(U, V, Z, X) + \nabla_X(R)(U, V, Y, Z) = 0 \qquad (4.2.13)$$

and

$$\text{div}_1(C)(X, Y, Z) = \sum_i \nabla_i(C)(e_i, X, Y, Z)$$

$$= (n-3)\left(\nabla_Z(\mathsf{P})(X, Y) - \nabla_Y(\mathsf{P})(X, Z)\right), \qquad (4.2.14)$$

$$-\delta\mathsf{P} = d\mathsf{J}. \qquad (4.2.15)$$

The tensor on the right-hand side of (4.2.14) is called the Cotton tensor.

Proof of Lemma 4.2.6. We observe that for $\alpha \in \Omega^1(M)$,

$$\nabla_X(\mathsf{P}\#\alpha)(Y) = X\left((\mathsf{P}\#\alpha)(Y)\right) - (\mathsf{P}\#\alpha)(\nabla_X(Y))$$
$$= X(\mathsf{P}(Y, \alpha^\sharp)) - \mathsf{P}(\nabla_X(Y), \alpha^\sharp)$$
$$= \nabla_X(\mathsf{P})(Y, \alpha^\sharp) + \mathsf{P}(Y, \nabla_X(\alpha^\sharp)).$$

Hence

$$\delta(\mathsf{P}\#du) = -\sum_i \nabla_i(\mathsf{P}\#du)(e_i)$$
$$= -\sum_i \nabla_i(\mathsf{P})(e_i, (du)^\sharp) - \sum_i \mathsf{P}(e_i, \nabla_i((du)^\sharp))$$
$$= \langle \delta(\mathsf{P}), (du)^\sharp \rangle - (\mathsf{P}, \mathrm{Hess}(u))$$
$$= (\delta(\mathsf{P}), du) - (\mathsf{P}, \mathrm{Hess}(u)).$$

This completes the proof. □

Proof of Lemma 4.2.7. The first identity is the differential Bianchi identity. Since both sides of (4.2.14) are linear in X, Y and Z, it suffices to prove the assertion for a geodesic frame e_i at m. We contract (4.2.13) in the arguments U and X. Then

$$\nabla_k(R)(e_s, e_j, e_s, e_i) + \nabla_i(R)(e_s, e_j, e_k, e_s) + \nabla_s(R)(e_s, s_j, e_i, e_k) = 0.$$

Now using that C is trace-free and $R = \mathsf{C} - \mathsf{P} \oslash g$, we find

$$\sum_s \nabla_s(\mathsf{C})(e_s, e_j, e_i, e_k) = \sum_s \Big\{ e_k(\mathsf{P} \oslash g)(e_s, e_j, e_s, e_i)$$
$$+ e_i(\mathsf{P} \oslash g)(e_s, e_j, e_k, e_s) + e_s(\mathsf{P} \oslash g)(e_s, e_j, e_i, e_k) \Big\}.$$

The sums on the right-hand side simplify to

$$(n-3)\left(e_k(\mathsf{P}_{ji}) - e_i(\mathsf{P}_{jk})\right) + \sum_s [e_s(\mathsf{P}_{is})\delta_{jk} - e_s(\mathsf{P}_{ks})\delta_{ji}] + e_k(\mathsf{J})\delta_{ji} - e_i(\mathsf{J})\delta_{jk}.$$

Since, by (4.2.15), $\sum_s e_s(\mathsf{P}_{is}) = \mathsf{J}_i$, we find

$$\sum_s \nabla_s(\mathsf{C})(e_s, e_j, e_i, e_k) = (n-3)\left(\nabla_k(\mathsf{P})(e_j, e_i) - \nabla_i(\mathsf{P})(e_j, e_k)\right),$$

i.e.,

$$\sum_i \nabla_i(\mathsf{C})(e_i, X, Y, Z) = (n-3)\left(\nabla_Z(\mathsf{P})(X, Y) - \nabla_Y(\mathsf{P})(X, Z)\right).$$

In order to prove (4.2.15), we first contract (4.2.13) in U and X. Then

$$-\nabla_k(\mathrm{Ric})_{ij} + \nabla_i(\mathrm{Ric})_{jk} + \sum_s \nabla_s(R)(e_s, e_j, e_i, e_k) = 0.$$

A second contraction yields

$$-\nabla_k(\tau) + \sum_i \nabla_i(\mathrm{Ric})_{ik} + \sum_s \nabla_s(\mathrm{Ric})_{sk} = 0,$$

i.e.,

$$2\sum_i \nabla_i(\mathrm{Ric})_{ik} = \nabla_k(\tau).$$

The latter identity implies $\sum_i \nabla_i(\mathsf{P})_{ik} = \nabla_k(\mathsf{J})$. In fact,

$$\sum_i \nabla_i(\mathsf{P})_{ik} = \frac{1}{n-2} \sum_i (\nabla_i(\mathrm{Ric})_{ik} - \nabla_i(\mathsf{J})\delta_{ik})$$

$$= \frac{1}{n-2}\left(\frac{1}{2}\nabla_k(\tau) - \nabla_k(\mathsf{J})\right)$$

$$= \frac{1}{n-2}\left(\frac{1}{2} - \frac{1}{2(n-1)}\right)\nabla_k(\tau)$$

$$= \frac{1}{2(n-1)}\nabla_k(\tau) = \nabla_k(\mathsf{J}).$$

The proof is complete. $\square$

     An alternative proof of Theorem 4.1.1 will be given in Chapter 6. In Section 6.6 (see Remark 6.6.5), we shall derive the formula (4.1.6) from the asymptotics of eigenfunctions of the Laplacian of an associated Poincaré-Einstein metric. For $n = 4$, this result will later imply that the residue family $D_4^{\mathrm{res}}(h; \lambda)$ specializes to $P_4(h)i^*$ at $\lambda = 0$. The conformal covariance of $D_4^{\mathrm{res}}(h; \lambda)$ then yields the conformal covariance of $P_4(h)$. Moreover, we will prove that the quantity $\dot{D}_4^{\mathrm{res}}(h; 0)(1)$ coincides with $-Q_4(h)$. From this perspective, the fundamental identity for $Q_4(h)$ is just a consequence of the conformal covariance of the family $D_4^{\mathrm{res}}(h; \lambda)$. Note that the latter argument does not require going to dimensions $\neq 4$. Section 6.22 will contain a further proof of the fundamental identity for Q_4. It generalizes the latter argument but replaces the residue family $D_4^{\mathrm{res}}(h; \lambda)$ by the tractor family $D_4^T(g; \lambda)$.

4.3 Q_4 and v_4

The discussion in Section 4.2 clearly shows that the fundamental identity for

$$Q_4 = 2(\mathsf{J}^2 - |\mathsf{P}|^2) + \Delta\mathsf{J} \tag{4.3.1}$$

is a consequence of a subtle interplay between the quantities $v_4 = \mathsf{J}^2 - |\mathsf{P}|^2$ and $\Delta\mathsf{J}$. The behaviour of v_4 in a conformal class is *not* governed by a linear operator.

Its infinitesimal conformal transformation law is

$$v_4^\bullet[\varphi] + 4v_4\varphi = 2\mathsf{J}\Delta\varphi + 2(\mathsf{P}, \text{Hess}(\varphi))$$
$$= 3\mathsf{J}\Delta\varphi + (\text{Ric}, \text{Hess}(\varphi)) = (S, \text{Hess}(\varphi)), \qquad (4.3.2)$$

where $S = \text{Ric} - \frac{\tau}{2}g$ is the Einstein tensor on M^4.

In addition to its relation to Q_4, the quantity $\mathsf{J}^2 - |\mathsf{P}|^2$ appears naturally in a number of other contexts.

The multiple $\frac{1}{8}(\mathsf{J}^2 - |\mathsf{P}|^2)$ is the holographic anomaly of the renormalized volume of Poincaré-Einstein metrics in dimension 5 ([141], [119]). In this connection, the relation (4.3.2) appears, e.g., in [147]. The same relation plays an important role in the study ([66], [63]) of the fully non-linear partial differential equation for φ so that $v_4(e^{2\varphi}g) = f \in C^\infty(M)$.

The quantity $\mathsf{J}^2 - |\mathsf{P}|^2$ is also of interest in dimensions $\neq 4$. In fact, for a closed manifold M^n of dimension $n \geq 3$, we notice that the functional

$$\mathcal{F}_2 = \int_M (\mathsf{J}^2 - |\mathsf{P}|^2)\,\text{vol}$$

satisfies

$$(d/dt)|_{t=0}(\mathcal{F}_2(e^{2t\varphi}g))$$
$$= (n-4)\int_M \varphi(\mathsf{J}^2 - |\mathsf{P}|^2)\,\text{vol}(g) + 2\int_M (\mathsf{J}\Delta\varphi + (\text{Hess}(\varphi), \mathsf{P}))\,\text{vol}$$
$$= (n-4)\int_M \varphi(\mathsf{J}^2 - |\mathsf{P}|^2)\,\text{vol}(g) - \int_M \delta(\mathsf{P}\#d\varphi)\,\text{vol}$$
$$= (n-4)\int_M \varphi v_4\,\text{vol}$$

(Lemma 4.2.5). It follows that $\mathcal{F}_2$ is conformally invariant for $n = 4$, and that for $n > 4$ the critical points of the functional

$$\left(\int_M (\mathsf{J}^2 - |\mathsf{P}|^2)\,\text{vol}\right) \Big/ \left(\int_M \text{vol}\right)^{\frac{n-4}{n}}$$

on a conformal class are solutions of the equation $\mathsf{J}^2 - |\mathsf{P}|^2 = \text{constant}$. This observation generalizes the description of the equation $\mathsf{J} = \text{constant}$ as the Euler-Lagrange equation of the Yamabe functional ([165]). More generally, Viaclovsky ([235]) discovered that for $2k < n$ ($k \geq 3$) and $C = 0$, the critical points of the functional

$$\mathcal{F}_k = \int_M \sigma_k(\mathsf{P})\,\text{vol}, \qquad \sigma_k(\mathsf{P}) = \text{tr}(\wedge^k \mathsf{P})$$

are metrics with constant σ_k. In Lemma 6.14.1, we will see that the symmetric polynomials $\sigma_k(\mathsf{P})$ of the eigenvalues of the Schouten tensor are related to the holographic coefficients v_{2k} of conformally flat metrics. For general metrics, the equations $v_{2k} = $ constant are the Euler-Lagrange equations of the respective functionals

$$\left(\int_M v_{2k}\, \text{vol} \right) \Big/ \left(\int_M \text{vol} \right)^{\frac{n-2k}{n}}$$

([65], [117]).

Finally, we emphasize the relation between the quantity v_4 and the Gauß-Bonnet integrand of M^4. The observation is that the Euler form of M is proportional to

$$8\mathsf{J}^2 - 8|\mathsf{P}|^2 + |\mathsf{C}|^2$$

(see (6.14.15)). Hence for closed M the integral

$$\int_M (\mathsf{J}^2 - |\mathsf{P}|^2)\, \text{vol}$$

depends only on the conformal class of g. This re-proves the conformal invariance of $\mathcal{F}_2$. The latter integral is a multiple of the total Q-curvature $\int_M Q_4\, \text{vol}$.

The theory of residue families in Chapter 6 will recognize the relations between P_4, Q_4 and v_4 as special cases of general facts. In particular, the holographic formula will give the two summands $2(\mathsf{J}^2 - |\mathsf{P}|^2)$ and $\Delta\mathsf{J}$ in (4.3.1) a natural interpretation.

Chapter 5

Intertwining Families

In the present chapter, we construct two sequences of polynomial families $D_N^c(\lambda)$: $C^\infty(S^n) \to C^\infty(S^{n-1})$ and $D_N^{nc}(\lambda) : C^\infty(\mathbb{R}^n) \to C^\infty(\mathbb{R}^{n-1})$ of differential intertwining operators for spherical principal series representations. These families are induced by families $\mathcal{D}_N(\lambda)$ of homomorphisms of Verma modules. We show how $D_N^{nc}(\lambda)$ naturally arises from the asymptotics of eigenfunctions of the Laplacian of the hyperbolic metric on the upper half-space. This is the simplest special case of the construction of residue families in Section 6.6. In Chapter 6, the coincidence of two constructions of different nature for $D_N^{nc}(\lambda)$ will be interpreted as the simplest case of a relation between residue families and tractor families (holographic duality). The induction and the mutual relations between the families are discussed in Section 5.2. We prove that all families $\mathcal{D}_N(\lambda)$ satisfy a system of factorization identities. These relations imply corresponding identities for the induced families of differential operators. In turn, these give rise to a recursive algorithm which allows determination of explicit formulas for the families $D_N^c(\lambda)$. In Section 6.11, analogs of these factorization identities for residue families will shed light on the recursive structure of Q-curvatures and $GJMS$-operators.

5.1 The algebraic theory

In this section, we construct the families $\mathcal{D}_N(\lambda)$, describe them as homomorphisms of generalized Verma modules, and relate them to the asymptotics of eigenfunctions of the hyperbolic Laplacian on the upper half-space.

5.1.1 Even order families $\mathcal{D}_{2N}(\lambda)$

We introduce the notation

$$\Delta_{n-1}^- = \sum_{j=1}^{n-1}(Y_j^-)^2 \in \mathcal{U}(\mathfrak{n}_n^-) \subset \mathcal{U}(\mathfrak{g}_n), \quad \Delta_n^- = \sum_{j=1}^{n}(Y_j^-)^2 \in \mathcal{U}(\mathfrak{n}_{n+1}^-) \subset \mathcal{U}(\mathfrak{g}_{n+1}).$$

Theorem 5.1.1. *For any non-negative integer N and any $\lambda \in \mathbb{C}$, the element*

$$\mathcal{D}_{2N} = \sum_{j=0}^{N} a_j (\Delta_{n-1}^-)^j (Y_n^-)^{2N-2j} \in \mathcal{U}(\mathfrak{n}_{n+1}^-)$$

satisfies

$$[Y_i^+, \mathcal{D}_{2N}] \in \mathcal{U}(\mathfrak{n}_{n+1}^-)(\mathfrak{m}_{n+1} \oplus \mathbb{C}(H_0 - \lambda)), \ i = 1, \ldots, n-1 \qquad (5.1.1)$$

iff the coefficients a_j are subject to the recursive relations

$$(N-j+1)(2N-2j+1)a_{j-1} + j(n-1+2\lambda-4N+2j)a_j = 0, \ j = 1, \ldots, N. \quad (5.1.2)$$

Since $(N-j+1)(2N-2j+1) \neq 0$ for all $j = 0, \ldots, N$, the recursive relations (5.1.2) imply that the coefficients a_j are determined by a_N. In particular, for all λ the element $\mathcal{D}_{2N} \in \mathcal{U}(\mathfrak{n}_{n+1}^-)$ is *uniquely determined* by (5.1.1), up to a multiple. More precisely, we have

Corollary 5.1.1. *For $0 \leq j \leq N-1$, we have*

$$a_j(\lambda) = \frac{N!}{j!(2N-2j)!}(-2)^{N-j} \prod_{k=j}^{N-1} (2\lambda-4N+2k+n+1)a_N(\lambda).$$

Let $\mathcal{D}_{2N}^0(\lambda) \in \mathcal{U}(\mathfrak{n}_{n+1}^-)$ be the unique element satisfying (5.1.1) which is normalized by the condition $a_N(\lambda) = 1$, i.e., $\mathcal{D}_{2N}^0(\lambda)$ is given by

$$\mathcal{D}_{2N}^0(\lambda) = a_0(\lambda)(Y_n^-)^{2N} + a_1(\lambda)\Delta_{n-1}^-(Y_n^-)^{2N-2} + \cdots + (\Delta_{n-1}^-)^N \qquad (5.1.3)$$

with the polynomial coefficients

$$a_j(\lambda) = \frac{N!}{j!(2N-2j)!}(-2)^{N-j} \prod_{k=j}^{N-1} (2\lambda-4N+2k+n+1) \qquad (5.1.4)$$

for $j = 0, \ldots, N-1$. We shall write $a_j^{(N)}(\lambda)$ if it is appropriate to emphasize the order of the corresponding operator.

The coefficients $a_j^{(N)}(\lambda)$ are coefficients of a *Gegenbauer polynomial* of degree $2N$. The Gegenbauer polynomial $C_n^\lambda(r)$ is defined by the power series

$$\frac{1}{(1 - 2rx + x^2)^\lambda} = \sum_{n \geq 0} C_n^\lambda(r)x^n. \qquad (5.1.5)$$

(5.1.5) implies

$$C_n^\lambda(-r) = (-1)^n C_n^\lambda(r). \qquad (5.1.6)$$

On the other hand, we have

$$C_n^\lambda(r) = \frac{\Gamma(2\lambda+n)}{\Gamma(2\lambda)} \frac{1}{n!} F\left(-n, 2\lambda+n; \lambda+\frac{1}{2}; \frac{1-r}{2}\right), \qquad (5.1.7)$$

i.e., $C_n^\lambda(r)$ is a hypergeometric polynomial. For details see [21], Section 3.15.

Theorem 5.1.2.

$$a_0^{(N)}(\lambda)x^{2N} - a_1^{(N)}(\lambda)x^{2N-2} \pm \cdots + (-1)^N x^0 = \frac{N!}{(-\lambda-\frac{n-1}{2})_N} C_{2N}^{-\lambda-\frac{n-1}{2}}(x).$$

Proof. We recall the explicit formula ([21], 3.15, formula (9))

$$C_{2N}^\lambda(x) = \sum_{j=0}^N \frac{(-1)^j(\lambda)_{2N-j}}{j!(2N-2j)!}(2x)^{2N-2j},$$

where $(\lambda)_m = \lambda(\lambda+1)\cdots(\lambda+m-1)$. Now, by (5.1.4), we have

$$a_j^{(N)}(\lambda) = \frac{N!}{j!(2N-2j)!}(-2)^{N-j}2^{N-j}\prod_{k=j}^{N-1}\left(\lambda-2N+k+\frac{n+1}{2}\right),$$

i.e.,

$$a_j^{(N)}(-\lambda) = \frac{N!}{j!(2N-2j)!}2^{2N-2j}\left(\lambda-\frac{n-1}{2}+2N-j-1\right)\cdots\left(\lambda-\frac{n-1}{2}+N\right).$$

Hence we obtain

$$a_0(-\lambda)x^{2N} - a_1(-\lambda)x^{2N-2} \pm \cdots + (-1)^N x^0$$

$$= \frac{N!}{(\lambda-\frac{n-1}{2})_N}\sum_{j=0}^N(-1)^j\frac{(\lambda-\frac{n-1}{2})_{2N-j}}{j!(2N-2j)!}(2x)^{2N-2j}$$

$$= \frac{N!}{(\lambda-\frac{n-1}{2})_N}C_{2N}^{\lambda-\frac{n-1}{2}}(x).$$

The proof is complete. $\qquad\square$

Corollary 5.1.2. *For all $N \in \mathbb{N}_0$,*

$$\mathcal{D}_{2N}^0\left(-\frac{n}{2}+N\right) = (\Delta_n^-)^N \quad and \quad \mathcal{D}_{2N}^0\left(-\frac{n-1}{2}+N\right) = (\Delta_{n-1}^-)^N.$$

Proof. The coefficients $a_r = a_r\left(-\frac{n}{2}+N\right)$, $0 \le r \le N$ of $\mathcal{D}_{2N}^0(-\frac{n}{2}+N)$ are characterized by

$$(N-r+1)(2N-2r+1)a_{r-1} + r(-1-2N+2r)a_r = 0, \quad a_N = 1,$$

i.e.,

$$a_{N-1} + N(-1)a_N = 0,$$
$$6a_{N-2} + (N-1)(-3)a_{N-1} = 0,$$
$$\cdots$$
$$(N-1)(2N-3)a_1 + 2(-2N+3)a_2 = 0,$$
$$N(2N-1)a_0 + (-2N+1)a_1 = 0.$$

Hence

$$a_0 = \frac{1}{N}a_1 = \frac{2!}{N(N-1)}a_2 = \cdots = \frac{N!}{N!}a_N = a_N = 1,$$

i.e., $a_r = \binom{N}{r}$. Therefore,

$$\mathcal{D}_{2N}^0 \left(-\frac{n}{2}+N\right) = \sum_{j=0}^{N} \binom{N}{j} (\Delta_{n-1}^-)^j (Y_n^-)^{2N-2j} = (\Delta_{n-1}^- + (Y_n^-)^2)^N = (\Delta_n^-)^N.$$

For the proof of the second assertion we note that the coefficients

$$a_r = a_r \left(-\frac{n-1}{2}+N\right), \quad 0 \le r \le N$$

are characterized by

$$(N-r+1)(2N-2r+1)a_{r-1} + r(-2N+2r)a_r = 0, \quad a_N = 1,$$

i.e.,

$$a_{N-1} = 0,$$
$$6a_{N-2} + (N-1)(-2)a_{N-1} = 0,$$
$$\cdots$$
$$N(2N-1)a_0 + (-2N+2)a_1 = 0.$$

Hence $a_{N-1} = \cdots = a_0 = 0$, $a_N = 1$.   □

We split the proof of Theorem 5.1.1 into a series of partial results on commutators.

Lemma 5.1.1. $\left[H_0, (Y_r^-)^j\right] = -j(Y_r^-)^j$ for $1 \le r \le n$ and $j \ge 1$.

Proof. The assertion is obvious for $j = 1$. We argue by induction and calculate

$$\begin{aligned}
H_0(Y_r^-)^{j+1} &= \left[H_0, (Y_r^-)^j\right] Y_r^- + (Y_r^-)^j H_0 Y_r^- \\
&= -j(Y_r^-)^{j+1} + (Y_r^-)^j \left[H_0, Y_r^-\right] + (Y_r^-)^{j+1} H_0 \quad \text{(by assumption)} \\
&= -(j+1)(Y_r^-)^{j+1} + (Y_r^-)^{j+1} H_0.
\end{aligned}$$

The proof is complete.   □

Lemma 5.1.2. $\left[M_{1r}, (Y_r^-)^{2j}\right] = 2jY_1^-(Y_r^-)^{2j-1}$ for $1 \le r \le n$ and $j \ge 1$.

Proof. The calculation

$$\begin{aligned}
M_{1r}(Y_r^-)^2 &= \left[M_{1r}, Y_r^-\right] Y_r^- + Y_r^- M_{1r} Y_r^- \\
&= Y_1^- Y_r^- + Y_r^- \left[M_{1r}, Y_r^-\right] + (Y_r^-)^2 M_{1r} \\
&= Y_1^- Y_r^- + Y_r^- Y_1^- + (Y_r^-)^2 M_{1r} \quad\quad\quad\quad\quad (5.1.8)
\end{aligned}$$

proves the assertion for $j = 1$. Now in sense of an induction we have

$$M_{1r}(Y_r^-)^{2(j+1)} = \left[M_{1r}, (Y_r^-)^{2j}\right](Y_r^-)^2 + (Y_r^-)^{2j}M_{1r}(Y_r^-)^2$$
$$= 2jY_1^-(Y_r^-)^{2j+1} + (Y_r^-)^{2j}\left[M_{1r}, (Y_r^-)^2\right] + (Y_r^-)^{2(j+1)}M_{1r}$$

by assumption and thus

$$M_{1r}(Y_r^-)^{2(j+1)} = 2(j+1)Y_r^-(Y_r^-)^{2(j+1)} + (Y_r^-)^{2(j+1)}M_{1r}$$

by (5.1.8). The proof is complete. □

Lemma 5.1.3.

$$\left[Y_1^+, (Y_r^-)^{2N}\right] = \begin{cases} -2N(2N-1)(Y_1^-)^{2N-1} + 4N(Y_1^-)^{2N-1}H_0, & r = 1, \\ 2N(2N-1)Y_1^-(Y_r^-)^{2N-2} + 4N(Y_r^-)^{2N-1}M_{1r}, & 2 \le r \le n. \end{cases}$$

Proof. We first prove the assertion for $r = 1$. We find

$$Y_1^+(Y_1^-)^2 = \left[Y_1^+, Y_1^-\right]Y_1^- + Y_1^-Y_1^+Y_1^-$$
$$= 2H_0Y_1^- + Y_1^-\left[Y_1^+, Y_1^-\right] + (Y_1^-)^2Y_1^+$$
$$= 2\left[H_0, Y_1^-\right] + 2Y_1^-H_0 + 2Y_1^-H_0 + (Y_1^-)^2Y_1^+$$
$$= 4Y_1^-H_0 - 2Y_1^- + (Y_1^-)^2Y_1^+,$$

i.e.,

$$\left[Y_1^+, (Y_1^-)^2\right] = -2Y_1^- + 4Y_1^-H_0. \tag{5.1.9}$$

Now we proceed by induction. We calculate

$$Y_1^+(Y_1^-)^{2N+2} = \left[Y_1^+, (Y_1^-)^{2N}\right](Y_1^-)^2 + (Y_1^-)^{2N}Y_1^+(Y_1^-)^2$$
$$= \left[Y_1^+, (Y_1^-)^{2N}\right](Y_1^-)^2 + (Y_1^-)^{2N}\left[Y_1^+, (Y_1^-)^2\right] + (Y_1^-)^{2N+2}Y_1^+$$
$$= -2N(2N-1)(Y_1^-)^{2N+1} + 4N(Y_1^-)^{2N-1}H_0(Y_1^-)^2$$
$$+ 4(Y_1^-)^{2N+1}H_0 - 2(Y_1^-)^{2N+1} + (Y_1^-)^{2N+2}Y_1^+,$$

by assumption and by (5.1.9), i.e., we get

$$\left[Y_1^+, (Y_1^-)^{2N+2}\right] = (-2N(2N-1) - 2 - 8N)(Y_1^-)^{2N+1} + (4N+4)(Y_1^-)^{2N+1}H_0$$
$$= -2(N+1)(2N+1)(Y_1^-)^{2N+1} + 4(N+1)(Y_1^-)^{2N+1}H_0$$

using Lemma 5.1.1. This completes the proof of the claim for $r = 1$. For $2 \le r \le n$ and $N = 1$, the assertion follows from

$$Y_1^+(Y_r^-)^2 = \left[Y_1^+, Y_r^-\right]Y_r^- + Y_r^-Y_1^+Y_r^-$$
$$= 2M_{1r}Y_r^- + Y_r^-\left[Y_1^+, Y_r^-\right] + (Y_r^-)^2Y_1^+$$
$$= 2\left[M_{1r}, Y_r^-\right] + 2Y_r^-M_{1r} + 2Y_r^-M_{1r} + (Y_r^-)^2Y_1^+$$
$$= 2Y_1^- + 4Y_r^-M_{1r} + (Y_r^-)^2Y_1^+. \tag{5.1.10}$$

We proceed by induction. We obtain

$$
\begin{aligned}
Y_1^+(Y_r^-)^{2N+2} &= \left[Y_1^+,(Y_r^-)^{2N}\right](Y_r^-)^2 + (Y_r^-)^{2N}Y_1^+(Y_r^-)^2 \\
&= \left[Y_1^+,(Y_r^-)^{2N}\right](Y_r^-)^2 + (Y_r^-)^{2N}\left[Y_1^+,(Y_r^-)^2\right] + (Y_r^-)^{2N+2}Y_1^+ \\
&= 2N(2N-1)Y_1^-(Y_r^-)^{2N} + 4N(Y_r^-)^{2N-1}M_{1r}(Y_r^-)^2 \\
&\quad + (Y_r^-)^{2N}(2Y_1^- + 4Y_r^-M_{1r}) + (Y_r^-)^{2N+2}Y_1^+
\end{aligned}
$$

by using the assumption and (5.1.10). Now $\left[M_{1r},(Y_r^-)^2\right] = 2Y_1^-Y_r^-$ (Lemma 5.1.2) implies

$$
\begin{aligned}
&\left[Y_1^+,(Y_r^-)^{2N+2}\right] \\
&= (2N(2N-1)+2)Y_1^-(Y_r^-)^{2N} + 8NY_1^-(Y_r^-)^{2N} + 4(N+1)(Y_r^-)^{2N+1}M_{1r} \\
&\qquad = 2(N+1)(2N+1)Y_1^-(Y_r^-)^{2N} + 4(N+1)(Y_r^-)^{2N+1}M_{1r}.
\end{aligned}
$$

The proof is complete. $\square$

Lemma 5.1.4. *For $j \geq 1$,*

$$
\left[Y_1^+,(\Delta_{n-1}^-)^j\right] - 2j(n-1-2j)Y_1^-(\Delta_{n-1}^-)^{j-1} - 4jY_1^-(\Delta_{n-1}^-)^{j-1}H_0 \in \mathcal{U}(\mathfrak{n}_n^-)\mathfrak{m}_n.
$$

Proof. We write $(\Delta_{n-1}^-)^j$ in the form

$$
(\Delta_{n-1}^-)^j = \sum_{|a|=j} \frac{j!}{a!}(Y_1^-)^{2a_1}\cdots(Y_{n-1}^-)^{2a_{n-1}}, \quad a = (a_1,\ldots,a_{n-1}),
$$

where $|a| = a_1 + \cdots + a_{n-1}$ and apply Lemma 5.1.3 successively. The first step yields

$$
Y_1^+(\Delta_{n-1}^-)^j = \sum_{|a|=j} \frac{j!}{a!}\left(T_1(a) + (Y_1^-)^{2a_1}Y_1^+(Y_2^-)^{2a_2}\cdots(Y_{n-1}^-)^{2a_{n-1}}\right),
$$

where

$$
\begin{aligned}
&T_1(a) \\
&\quad = \left(-2a_1(2a_1-1)(Y_1^-)^{2a_1-1} + 4a_1(Y_1^-)^{2a_1-1}H_0\right)(Y_2^-)^{2a_2}\cdots(Y_{n-1}^-)^{2a_{n-1}}.
\end{aligned}
\tag{5.1.11}
$$

In the next step, we use Lemma 5.1.3 for $r = 2$. We get

$$
\begin{aligned}
&Y_1^+(\Delta_{n-1}^-)^j \\
&\quad = \sum_{|a|=j} \frac{j!}{a!}\left(T_1(a) + T_2(a) + (Y_1^-)^{2a_1}(Y_2^-)^{2a_2}Y_1^+(Y_3^-)^{2a_3}\cdots(Y_{n-1}^-)^{2a_{n-1}}\right),
\end{aligned}
$$

where

$$T_2(a) = 2a_2(2a_2-1)(Y_1^-)^{2a_1+1}(Y_2^-)^{2a_2-1}(Y_3^-)^{2a_3}\cdots(Y_{n-1}^-)^{2a_{n-1}}$$
$$+ 4a_2(Y_1^-)^{2a_1}(Y_2^-)^{2a_2-1}M_{12}(Y_3^-)^{2a_3}\cdots(Y_{n-1}^-)^{2a_{n-1}}.$$

The continuation of this method finally gives the formula

$$[Y_1^+,(\Delta_{n-1}^-)^j] = \sum_{|a|=j}\frac{j!}{a!}(T_1(a)+T_2(a)+\cdots+T_{n-1}(a))$$

with $T_1(a)$ given by (5.1.11) and

$$T_k(a) = 2a_k(2a_k-1)(Y_1^-)^{2a_1+1}(Y_2^-)^{2a_2}\cdots(Y_k^-)^{2a_k-2}\cdots(Y_{n-1}^-)^{2a_{n-1}}$$
$$+ 4a_k(Y_1^-)^{2a_1}\cdots(Y_k^-)^{2a_k-1}M_{1k}\cdots(Y_{n-1}^-)^{2a_{n-1}},$$

$k=2,\ldots,n-1$. Here we use that $\mathfrak{n}^-$ is abelian to get the powers of Y_1^- left. It follows that

$$[Y_1^+,(\Delta_{n-1}^-)^j] = -\sum_{|a|=j}\frac{j!}{a!}2a_1(2a_1-1)Y_1^-(Y_1^-)^{2a_1-2}(Y_2^-)^{2a_2}\cdots(Y_{n-1}^-)^{2a_{n-1}}$$

$$+\sum_{|a|=j}\frac{j!}{a!}2a_2(2a_2-1)Y_1^-(Y_1^-)^{2a_1}(Y_2^-)^{2a_2-2}\cdots(Y_{n-1}^-)^{2a_{n-1}}$$

$$+\cdots+$$

$$+\sum_{|a|=j}\frac{j!}{a!}2a_{n-1}(2a_{n-1}-1)Y_1^-(Y_1^-)^{2a_1}(Y_2^-)^{2a_2}\cdots(Y_{n-1}^-)^{2a_{n-1}-2}$$

$$+2\sum_{|a|=j}\frac{j!}{a!}2a_1(Y_1^-)^{2a_1-1}H_0(Y_2^-)^{2a_2}\cdots(Y_{n-1}^-)^{2a_{n-1}}$$

$$+2\sum_{|a|=j}\frac{j!}{a!}2a_2(Y_1^-)^{2a_1}(Y_2^-)^{2a_2-1}M_{12}\cdots(Y_{n-1}^-)^{2a_{n-1}}$$

$$+\cdots+$$

$$+2\sum_{|a|=j}\frac{j!}{a!}2a_{n-1}(Y_1^-)^{2a_1}\cdots(Y_{n-1}^-)^{2a_{n-1}-1}M_{1n-1}. \tag{5.1.12}$$

Next, we sum up the first $n-1$ sums in (5.1.12). Since $\mathfrak{n}^-$ is abelian, it is enough to apply the following result.

Lemma 5.1.5.

$$\left(-\frac{\partial^2}{\partial y_1^2}+\frac{\partial^2}{\partial y_2^2}+\cdots+\frac{\partial^2}{\partial y_{n-1}^2}\right)(|y|^{2j}) = 2j(n-5+2j)|y|^{2(j-1)}-8j(j-1)y_1^2|y|^{2(j-2)},$$

where $|y|^2 = y_1^2+\cdots+y_{n-1}^2$.

Proof. We use the identities

$$\frac{\partial^2}{\partial y_1^2}\left(|y|^{2j}\right) = 2j\frac{\partial}{\partial y_1}\left(y_1|y|^{2j-2}\right) = 2j|y|^{2j-2} + 4j(j-1)y_1^2|y|^{2j-4} \qquad (5.1.13)$$

and

$$\Delta_{\mathbb{R}^{n-1}}\left(|y|^{2j}\right) = 2j\sum_{r=1}^{n-1}\frac{\partial}{\partial y_r}(y_r|y|^{2j-2})$$
$$= 2j(2j-2)|y|^{2j-2} + 2j(n-1)|y|^{2j-2} = 2j(n-3+2j)|y|^{2j-2} \quad (5.1.14)$$

to calculate

$$\left(-\frac{\partial^2}{\partial y_1^2} + \frac{\partial^2}{\partial y_2^2} + \cdots + \frac{\partial^2}{\partial y_{n-1}^2}\right)(|y|^{2j}) = -2\frac{\partial^2}{\partial y_1^2}\left(|y|^{2j}\right) + \Delta\left(|y|^{2j}\right)$$
$$= -4j|y|^{2j-2} - 8j(j-1)y_1^2|y|^{2j-4} + 2j(n-3+2j)|y|^{2j-2}$$
$$= 2j(n-5+2j)|y|^{2j-2} - 8j(j-1)y_1^2|y|^{2j-4}.$$

The proof is complete. $\qquad\qquad\qquad\qquad\qquad\qquad\qquad\qquad\qquad\qquad\qquad\square$

Now Lemma 5.1.5 implies that the first $n-1$ sums in (5.1.12) can be simplified to

$$2j(n-5+2j)Y_1^-(\Delta_{n-1}^-)^{j-1} - 8j(j-1)(Y_1^-)^3(\Delta_{n-1}^-)^{j-2}. \qquad (5.1.15)$$

Next, successive application of Lemma 5.1.1 gives

$$2\sum_{|a|=j}\frac{j!}{a!}2a_1(Y_1^-)^{2a_1-1}H_0(Y_2^-)^{2a_2}\cdots(Y_{n-1}^-)^{2a_{n-1}} \qquad (5.1.16)$$

$$= 2\sum_{|a|=j}\frac{j!}{a!}2a_1(Y_1^-)^{2a_1-1}(Y_2^-)^{2a_2}\cdots(Y_{n-1}^-)^{2a_{n-1}}H_0$$

$$- 2\sum_{|a|=j}\frac{j!}{a!}2a_12a_2(Y_1^-)^{2a_1-1}(Y_2^-)^{2a_2}\cdots(Y_{n-1}^-)^{2a_{n-1}}$$

$$- \cdots -$$

$$- 2\sum_{|a|=j}\frac{j!}{a!}2a_12a_{n-1}(Y_1^-)^{2a_1-1}(Y_2^-)^{2a_2}\cdots(Y_{n-1}^-)^{2a_{n-1}}.$$

In order to simplify the sums in (5.1.16), we apply the identities

$$\frac{\partial}{\partial y_1}\left(|y|^{2j}\right) = 2jy_1|y|^{2(j-1)} \qquad (5.1.17)$$

$$\frac{\partial}{\partial y_1}\left(\sum_{r=2}^{n-1}y_r\frac{\partial}{\partial y_r}\left(|y|^{2j}\right)\right) = 4j(j-1)\left(y_1|y|^{2(j-1)} - y_1^3|y|^{2(j-2)}\right). \qquad (5.1.18)$$

For the proof of (5.1.18) we calculate

$$\frac{\partial}{\partial y_1}\left(\sum_{r=2}^{n-1} y_r \frac{\partial}{\partial y_r}\left(|y|^{2j}\right)\right) = \frac{\partial}{\partial y_1}\left(\sum_{r=1}^{n-1} y_r \frac{\partial}{\partial y_r}\left(|y|^{2j}\right)\right) - \frac{\partial}{\partial y_1}\left(y_1 \frac{\partial}{\partial y_1}\left(|y|^{2j}\right)\right)$$

$$= 2j \frac{\partial}{\partial y_1}\left(|y|^{2j}\right) - y_1 \frac{\partial^2}{\partial y_1^2}\left(|y|^{2j}\right) - \frac{\partial}{\partial y_1}\left(|y|^{2j}\right).$$

The latter sum simplifies to

$$(2j-1)2jy_1|y|^{2(j-1)} - 2jy_1|y|^{2(j-1)} - 4j(j-1)y_1^3|y|^{2(j-2)}$$

$$= 4j(j-1)\left(y_1|y|^{2(j-1)} - y_1^3|y|^{2(j-1)}\right)$$

using (5.1.13) and (5.1.17). Now (5.1.16) and (5.1.18) imply

$$2 \sum_{|a|=j} \frac{j!}{a!} 2a_1 (Y_1^-)^{2a_1-1} H_0 (Y_2^-)^{2a_2} \cdots (Y_{n-1}^-)^{2a_{n-1}} \tag{5.1.19}$$

$$= 4j Y_1^- (\Delta_{n-1}^-)^{j-1} H_0 - 8j(j-1) Y_1^- (\Delta_{n-1}^-)^{j-1} + 8j(j-1)(Y_1^-)^3 (\Delta_{n-1}^-)^{j-2}.$$

Finally, we notice that $[M_{1r}, Y_s^-] = 0$ for $r < s$. Hence the last $n - 2$ sums in (5.1.12) define an element in $\mathcal{U}(\mathfrak{n}_n^-)\mathfrak{m}_n$. Thus we see that by (5.1.15) and (5.1.19)

$$\left[Y_1^+, (\Delta_{n-1}^-)^j\right] = 2j(2j+n-5) Y_1^- (\Delta_{n-1}^-)^{j-1}$$
$$- 8j(j-1)(Y_1^-)^3 (\Delta_{n-1}^-)^{j-2} + 4j Y_1^- (\Delta_{n-1}^-)^{j-1} H_0$$
$$- 8j(j-1) Y_1^- (\Delta_{n-1}^-)^{j-1} + 8j(j-1)(Y_1^-)^3 (\Delta_{n-1}^-)^{j-2} \quad (\mathrm{mod}\ \mathcal{U}(\mathfrak{n}_n^-)\mathfrak{m}_m),$$

i.e.,

$$\left[Y_1^+, (\Delta_{n-1}^-)^j\right]$$
$$= 2j(n-1-2j) Y_1^- (\Delta_{n-1}^-)^{j-1} + 4j Y_1^- (\Delta_{n-1}^-)^{j-1} H_0 \quad (\mathrm{mod}\ \mathcal{U}(\mathfrak{n}_n^-)\mathfrak{m}_m).$$

The proof is complete. $\qquad\square$

Now since M^n fixes Y_n^- Lemma 5.1.4 implies

$$Y_1^+ (\Delta_{n-1}^-)^j (Y_n^-)^{2k} = \left[Y_1^+, (\Delta_{n-1}^-)^j\right](Y_n^-)^{2k} + (\Delta_{n-1}^-)^j Y_1^+ (Y_n^-)^{2k}$$
$$= 2j(n-1-2j) Y_1^- (\Delta_{n-1}^-)^{j-1} (Y_n^-)^{2k} + 4j Y_1^- (\Delta_{n-1}^-)^{j-1} H_0 (Y_n^-)^{2k}$$
$$+ (\Delta_{n-1}^-)^j Y_1^+ (Y_n^-)^{2k} \quad (\mathrm{mod}\ \mathcal{U}(\mathfrak{n}_{n+1}^-)\mathfrak{m}_n).$$

Hence using Lemma 5.1.1 and Lemma 5.1.3 (for $r = n$), we obtain

$$
\begin{aligned}
\left[Y_1^+, (\Delta_{n-1}^-)^j (Y_n^-)^{2k}\right] &= 2j(n-1-2j)Y_1^-(\Delta_{n-1}^-)^{j-1}(Y_n^-)^{2k} \\
&\quad + 4jY_1^-(\Delta_{n-1}^-)^{j-1}(Y_n^-)^{2k}H_0 - 8jkY_1^-(\Delta_{n-1}^-)^{j-1}(Y_n^-)^{2k} \\
&\quad + 2k(2k-1)Y_1^-(\Delta_{n-1}^-)^j(Y_n^-)^{2k-2} \\
&\quad + 4k(\Delta_{n-1}^-)^j(Y_n^-)^{2k-1}M_{1n} \quad (\mathrm{mod}\ \mathcal{U}(\mathfrak{n}_{n+1}^-)\mathfrak{m}_n) \\
&= \{2j(n-1-2j)+4j\lambda-8jk\}\,Y_1^-(\Delta_{n-1}^-)^{j-1}(Y_n^-)^{2k} \\
&\quad + 2k(2k-1)Y_1^-(\Delta_{n-1}^-)^j(Y_n^-)^{2k-2} \qquad\qquad (5.1.20)
\end{aligned}
$$

$(\mathrm{mod}\ \mathcal{U}(\mathfrak{n}_{n+1}^-)(\mathfrak{m}_{n+1}\oplus\mathbb{C}(H_0-\lambda)))$.

Now we are ready to complete the

Proof of Theorem 5.1.1. We use (5.1.20) to determine the commutator $\left[Y_1^+,\mathcal{D}_{2N}\right]$, where

$$
\mathcal{D}_{2N} = \sum_{j=0}^{N} a_j (\Delta_{n-1}^-)^j (Y_n^-)^{2N-2j}.
$$

We obtain

$$
\left[Y_1^+,\mathcal{D}_{2N}\right] = \sum_{j=0}^{N} a_j\Big\{ (2j(n-1-2j)+4j\lambda-8j(N-j))\,Y_1^-(\Delta_{n-1}^-)^{j-1}(Y_n^-)^{2N-2j} \\
+2(N-j)(2N-2j-1)Y_1^-(\Delta_{n-1}^-)^j(Y_n^-)^{2N-2j-2}\Big\}
$$

up to an element in $\mathcal{U}(\mathfrak{n}_{n+1}^-)(\mathfrak{m}_{n+1}\oplus\mathbb{C}(H_0-\lambda))$. It follows that

$$
\left[Y_1^+,\mathcal{D}_{2N}\right] \equiv 0 \quad (\mathrm{mod}\ \mathcal{U}(\mathfrak{n}_{n+1}^-)(\mathfrak{m}_{n+1}\oplus\mathbb{C}(H_0-\lambda)))
$$

is equivalent to the set of conditions

$$
\{2r(n-1-2r)+4r\lambda-8r(N-r)\}\,a_r + 2(N-r+1)(2N-2r+1)a_{r-1} = 0
$$

for $r = 1,\ldots,N$. The latter conditions are equivalent to those given in Theorem 5.1.1. Now for each $2 \leq j \leq n-1$ there exists $m_j \in M^n = SO(n-1)$ such that $\mathrm{Ad}(m_j)Y_1^+ = Y_j^+$. Then

$$
\begin{aligned}
\left[Y_j^+,\mathcal{D}_{2N}\right] &= \left[\mathrm{Ad}(m_j)Y_1^+,\mathcal{D}_{2N}\right] \\
&= \mathrm{Ad}(m_j)\left[Y_1^+,\mathrm{Ad}(m_j^{-1})\mathcal{D}_{2N}\right] = \mathrm{Ad}(m_j)\left[Y_1^+,\mathcal{D}_{2N}\right]
\end{aligned}
$$

since Δ_{n-1}^- and Y_n^- are fixed under $\mathrm{Ad}(M^n)$. It follows that the set of conditions

$$
\left[Y_j^+,\mathcal{D}_{2N}\right] \equiv 0 \quad (\mathrm{mod}\ \mathcal{U}(\mathfrak{n}_{n+1}^-)(\mathfrak{m}_{n+1}\oplus\mathbb{C}(H_0-\lambda)))
$$

for $1 \leq j \leq n-1$, is equivalent to the condition for $j = 1$. The proof is complete. $\qquad\square$

5.1.2 Odd order families $\mathcal{D}_{2N+1}(\lambda)$

We start with the following analog of Theorem 5.1.1 for families in $\mathcal{U}(\mathfrak{n}_{n+1}^-)$ of *odd* order. It is the algebraic core of the theory of odd order families.

Theorem 5.1.3. *For any non-negative integer N and any $\lambda \in \mathbb{C}$, the element*

$$\mathcal{D}_{2N+1} = \sum_{j=0}^{N} b_j (\Delta_{n-1}^-)^j (Y_n^-)^{2N-2j+1} \in \mathcal{U}(\mathfrak{n}_{n+1}^-)$$

satisfies

$$\left[Y_i^+, \mathcal{D}_{2N+1} \right] \in \mathcal{U}(\mathfrak{n}_{n+1}^-)(\mathfrak{m}_{n+1} \oplus \mathbb{C}(H_0 - \lambda)), \ i = 1, \ldots, n-1 \qquad (5.1.21)$$

iff the coefficients b_j are subject to the recursive relations

$$(N - j + 1)(2N - 2j + 3) b_{j-1} + j (n - 3 + 2\lambda - 4N + 2j) b_j = 0 \qquad (5.1.22)$$

for $j = 1, \ldots, N$.

Proof. The proof is similar to that of Theorem 5.1.1. Lemma 5.1.4 gives

$$Y_1^+ (\Delta_{n-1}^-)^j (Y_n^-)^{2k+1} = \left[Y_1^+, (\Delta_{n-1}^-)^j \right] (Y_n^-)^{2k+1} + (\Delta_{n-1}^-)^j Y_1^+ (Y_n^-)^{2k+1}$$
$$= 2j(n-1-2j)Y_1^- (\Delta_{n-1}^-)^{j-1}(Y_n^-)^{2k+1} + 4jY_1^- (\Delta_{n-1}^-)^{j-1} H_0 (Y_n^-)^{2k+1}$$
$$+ (\Delta_{n-1}^-)^j Y_1^+ (Y_n^-)^{2k+1} \pmod{\mathcal{U}(\mathfrak{n}_{n+1}^-)\mathfrak{m}_n}.$$

Hence by Lemma 5.1.1 we get

$$\left[Y_1^+, (\Delta_{n-1}^-)^j (Y_n^-)^{2k+1} \right] = 2j(n-1-2j)Y_1^- (\Delta_{n-1}^-)^{j-1}(Y_n^-)^{2k+1}$$
$$+ 4jY_1^- (\Delta_{n-1}^-)^{j-1}(Y_n^-)^{2k+1} H_0 - 4j(2k+1)Y_1^- (\Delta_{n-1}^-)^{j-1}(Y_n^-)^{2k+1}$$
$$+ (\Delta_{n-1}^-)^j \left[Y_1^+, (Y_n^-)^{2k+1} \right] \pmod{\mathcal{U}(\mathfrak{n}_{n+1}^-)(\mathfrak{m}_{n+1})}.$$

Now we use Lemma 5.1.3 in order to determine the commutator $\left[Y_1^+, (Y_n^-)^{2k+1} \right]$. We have

$$\left[Y_1^+, (Y_n^-)^{2N+1} \right] = \left[Y_1^+, (Y_n^-)^{2N} \right] Y_n^- + (Y_n^-)^{2N} Y_1^+ Y_n^- - (Y_n^-)^{2N+1} Y_1^+$$
$$= 2N(2N-1)Y_1^- (Y_n^-)^{2N-1} + 4N(Y_n^-)^{2N-1} M_{1n} Y_n^- + (Y_n^-)^{2N} \left[Y_1^+, Y_n^- \right]$$
$$= 2N(2N-1)Y_1^- (Y_n^-)^{2N-1} + 4N(Y_n^-)^{2N-1} Y_1^- + (4N+2)(Y_n^-)^{2N} M_{1n}.$$

Hence

$$\left[Y_1^+, (\Delta_{n-1}^-)^j (Y_n^-)^{2k+1} \right] = (2j(n-1-2j) - 4j(2k+1)) Y_1^- (\Delta_{n-1}^-)^{j-1}(Y_n^-)^{2k+1}$$
$$+ 4jY_1^- (\Delta_{n-1}^-)^{j-1}(Y_n^-)^{2k+1} H_0$$
$$+ (2k(2k-1)+4k)(\Delta_{n-1}^-)^j Y_1^- (Y_n^-)^{2k-1} \pmod{\mathcal{U}(\mathfrak{n}_{n+1}^-)\mathfrak{m}_{n+1}},$$

i.e.,

$$\left[Y_1^+, (\Delta_{n-1}^-)^j (Y_n^-)^{2k+1}\right]$$
$$= \{2j(n-1-2j) - 4j(2k+1) + 4j\lambda\} Y_1^- (\Delta_{n-1}^-)^{j-1} (Y_n^-)^{2k+1}$$
$$+ 2k(2k+1)(\Delta_{n-1}^-)^j Y_1^- (Y_n^-)^{2k-1} \quad (\mathrm{mod}\ \mathcal{U}(\mathfrak{n}_{n+1}^-)(\mathfrak{m}_{n+1} \oplus \mathbb{C}(H_0 - \lambda))),$$

i.e., (5.1.20) remains true if we replace $2k$ on both sides by $2k+1$.

Now for $\mathcal{D}_{2N+1} = \sum_{j=0}^N b_j (\Delta_{n-1}^-)^j (Y_n^-)^{2N-2j+1} \in \mathcal{U}(\mathfrak{n}_{n+1}^-)$, it follows that

$$\left[Y_1^+, \mathcal{D}_{2N+1}\right]$$
$$= \sum_{j=0}^N b_j \Big\{ 2j\,(n-1-2j-2(2N-2j+1)+2\lambda)\,Y_1^- (\Delta_{n-1}^-)^{j-1} (Y_n^-)^{2N-2j+1}$$
$$+ 2(N-j)(2N-2j+1)Y_1^- (\Delta_{n-1}^-)^j (Y_n^-)^{2N-2j-1} \Big\}$$

up to an element in $\mathcal{U}(\mathfrak{n}_{n+1}^-)(\mathfrak{m}_{n+1} \oplus \mathbb{C}(H_0 - \lambda))$. Hence, for $i = 1$, the condition in Theorem 5.1.3 is equivalent to the set of conditions

$$\{2r(n-1-2r) + 4r\lambda - 8r(N-r) - 4r\}\,b_r + 2(N-r+1)(2N-2r+3)b_{r-1} = 0$$

for $r = 1, \ldots, N$. These conditions for the coefficients are equivalent to those formulated in the theorem. As for $\mathcal{D}_{2N}$, it follows that the remaining conditions for $i = 2, \ldots, n-1$ do not imply additional conditions. The proof is complete. $\qquad \square$

Corollary 5.1.3. *For* $0 \le j \le N-1$,

$$b_j(\lambda) = \frac{N!}{j!(2N-2j+1)!}(-2)^{N-j} q \prod_{k=j}^{N-1} (2\lambda - 4N + 2k + n - 1) b_N(\lambda).$$

Let $\mathcal{D}_{2N+1}^0(\lambda) \in \mathcal{U}(\mathfrak{n}_{n+1}^-)$ be the unique element satisfying (5.1.22) which is normalized by the condition $b_N(\lambda) = 1$, i.e., $\mathcal{D}_{2N+1}^0(\lambda)$ is given by

$$\mathcal{D}_{2N+1}^0(\lambda) = b_0(\lambda)(Y_n^-)^{2N+1} + b_1(\lambda)\Delta_{n-1}^-(Y_n^-)^{2N-1} + \cdots + (\Delta_{n-1}^-)^N Y_n^-, \quad (5.1.23)$$

with the polynomial coefficients

$$b_j(\lambda) = \frac{N!}{j!(2N-2j+1)!}(-2)^{N-j} \prod_{k=j}^{N-1} (2\lambda - 4N + 2k + n - 1) \qquad (5.1.24)$$

for $j = 0, \ldots, N-1$. Note that $\mathcal{D}_1^0(\lambda) = Y_n^-$. Instead of $b_j(\lambda)$ we shall also write $b_j^{(N)}(\lambda)$ if it is appropriate to emphasize the order of the operator.

The following theorem extends Theorem 5.1.2 to the odd order case.

Theorem 5.1.4.

$$b_0^{(N)}(\lambda)x^{2N+1} - b_1^{(N)}(\lambda)x^{2N-1} \pm \cdots + (-1)^N x = \frac{N!}{2\left(-\lambda-\frac{n-1}{2}\right)_{N+1}} C_{2N+1}^{-\lambda-\frac{n-1}{2}}(x).$$

Proof. We recall the explicit formula

$$C_{2N+1}^\lambda(x) = \sum_{j=0}^N \frac{(-1)^j(\lambda)_{2N+1-j}}{j!(2N+1-2j)!}(2x)^{2N+1-2j}.$$

Now, by (5.1.24), we have

$$b_j^{(N)}(\lambda) = \frac{N!}{j!(2N+1-2j)!}(-2)^{N-j}2^{N-j}\prod_{k=j}^{N-1}\left(\lambda-2N+k+\frac{n-1}{2}\right),$$

i.e.,

$$b_j^{(N)}(-\lambda) = \frac{N!}{j!(2N+1-2j)!}2^{2N-2j}\left(\lambda-\frac{n-1}{2}+2N-j\right)\cdots\left(\lambda-\frac{n-1}{2}+N+1\right)$$

$$= \frac{N!}{j!(2N+1-2j)!}2^{2N-2j}\frac{\left(\lambda-\frac{n-1}{2}\right)_{2N+1-j}}{\left(\lambda-\frac{n-1}{2}\right)_{N+1}}.$$

Hence

$$b_0^{(N)}(\lambda)x^{2N+1} - b_1^{(N)}(\lambda)x^{2N-2} \pm \cdots + (-1)^N x^1$$

$$= \frac{1}{2}\frac{N!}{\left(\lambda-\frac{n-1}{2}\right)_{N+1}}\sum_{j=0}^N(-1)^j\frac{\left(\lambda-\frac{n-1}{2}\right)_{2N+1-j}}{j!(2N+1-2j)!}(2x)^{2N+1-2j}$$

$$= \frac{1}{2}\frac{N!}{\left(\lambda-\frac{n-1}{2}\right)_{N+1}}C_{2N+1}^{\lambda-\frac{n-1}{2}}(x).$$

The proof is complete. □

5.1.3 $\mathcal{D}_N(\lambda)$ as homomorphism of Verma modules

We prove that both families $\mathcal{D}_{2N}(\lambda)$ (Theorem 5.1.1) and $\mathcal{D}_{2N+1}(\lambda)$ (Theorem 5.1.3) are naturally interpreted as families of homomorphisms of generalized Verma modules. We abuse notation by writing $\mathcal{D}_{2N}(\lambda)$ for the normalized family $\mathcal{D}_{2N}^0(\lambda)$ with polynomial coefficients. The same convention applies to the odd order families. Let

$$\mathfrak{p}_m = \mathfrak{m}_m \oplus \mathfrak{a} \oplus \mathfrak{n}_m^+ \subset \mathfrak{g}_m$$

be a parabolic subalgebra. $\mathfrak{p}_m$ is the Lie algebra of the minimal parabolic subgroup $P^m = M^m A(N^m)^+ \subset G^m$. For $\lambda \in \mathbb{C}$, we consider the character

$$\xi_\lambda : A \ni a_t \mapsto e^{\lambda t} \in Aut(\mathbb{C}).$$

ξ_λ gives rise to a character of P^m by $\xi_\lambda(man^+) = \xi_\lambda(a)$. Let ξ_λ also denote the corresponding character of $\mathfrak{p}_m$. Let

$$I_\lambda(\mathfrak{g}_m) \subset \mathcal{U}(\mathfrak{g}_m) \otimes \mathbb{C}(\lambda)$$

be the left $\mathcal{U}(\mathfrak{g}_m)$-ideal which is generated by the elements

$$X \otimes 1 - 1 \otimes \xi_\lambda(X)1 \in \mathcal{U}(\mathfrak{g}_m) \otimes \mathbb{C}(\lambda), \ X \in \mathfrak{p}_m.$$

We consider the left $\mathcal{U}(\mathfrak{g}_m)$-module

$$\mathcal{M}_\lambda(\mathfrak{g}_m) = (\mathcal{U}(\mathfrak{g}_m) \otimes \mathbb{C}(\lambda))/I_\lambda(\mathfrak{g}_m).$$

$\mathcal{M}_\lambda(\mathfrak{g}_m)$ is a *generalized Verma module* for $\mathfrak{g}_m$.

Theorem 5.1.5. *Let $\mathcal{D}_N(\lambda) \in \mathcal{U}(\mathfrak{n}_{n+1}^-)$ be as in Theorem 5.1.1 (even N) and Theorem 5.1.3 (odd N). Then for all $\lambda \in \mathbb{C}$ the map*

$$\mathcal{U}(\mathfrak{g}_n) \otimes \mathbb{C}(\lambda - N) \ni T \otimes 1 \mapsto i(T)\mathcal{D}_N(\lambda) \otimes 1 \in \mathcal{U}(\mathfrak{g}_{n+1}) \otimes \mathbb{C}(\lambda)$$

induces a homomorphism

$$\mathcal{M}_{\lambda-N}(\mathfrak{g}_n) \to \mathcal{M}_\lambda(\mathfrak{g}_{n+1})$$

of $\mathcal{U}(\mathfrak{g}_n)$-modules. Here $i: \mathfrak{g}_n \hookrightarrow \mathfrak{g}_{n+1}$ is the canonical inclusion.

Proof. It is enough to prove that $\mathcal{D}_N(\lambda)$ induces a map $I_{\lambda-N}(\mathfrak{g}_n) \to I_\lambda(\mathfrak{g}_{n+1})$. For $X = H_0 \in \mathfrak{p}_n$, we have

$$
\begin{aligned}
I_{\lambda-N}(\mathfrak{g}_n) \ni (H_0 \otimes 1 - 1 \otimes \xi_{\lambda-N}&(H_0)1) = (H_0 \otimes 1 - 1 \otimes (\lambda - N)) \\
\xrightarrow{\mathcal{D}_N(\lambda)} H_0 \mathcal{D}_N&(\lambda) \otimes 1 - \mathcal{D}_N(\lambda) \otimes (\lambda - N) \\
&= (\mathcal{D}_N(\lambda)H_0 - N\mathcal{D}_N(\lambda)) \otimes 1 - \mathcal{D}_N(\lambda) \otimes (\lambda - N) \\
&= \mathcal{D}_N(\lambda)(H_0 \otimes 1 - 1 \otimes \lambda) \\
&= \mathcal{D}_N(\lambda)(H_0 \otimes 1 - 1 \otimes \xi_\lambda(H_0)1) \in I_\lambda(\mathfrak{g}_{n+1})
\end{aligned}
$$

using $[H_0, \mathcal{D}_N(\lambda)] = -N\mathcal{D}_N(\lambda)$. Next, for $X \in \mathfrak{m}_n$, we obtain

$$I_{\lambda-N}(\mathfrak{g}_n) \ni (X \otimes 1 - 1 \otimes \xi_{\lambda-N}(X)1) = X \otimes 1$$
$$\xrightarrow{\mathcal{D}_N(\lambda)} X\mathcal{D}_N(\lambda) \otimes 1 = \mathcal{D}_N(\lambda)X \otimes 1 = \mathcal{D}_N(\lambda)(X \otimes 1 - 1 \otimes \xi_\lambda(X)1) \in I_\lambda(\mathfrak{g}_{n+1}).$$

Here we have used that $[\mathfrak{m}_n, \mathcal{D}_N(\lambda)] = 0$ which is a consequence of the definition of $\mathcal{D}_N(\lambda)$ and the commutator relations $[X, Y_n^-] = 0$ and $[X, \Delta_{n-1}^-] = 0$ for $X \in \mathfrak{m}_n$. Finally, for $X \in \mathfrak{n}_n^+$, i.e., for $X = Y_j^+$, $j = 1, \ldots, n-1$ we find that $\mathcal{D}_N(\lambda)$ maps

$$I_{\lambda-N}(\mathfrak{g}_n) \ni (X \otimes 1 - 1 \otimes \xi_{\lambda-N}(X)1) = X \otimes 1$$

into

$$XD_N(\lambda) \otimes 1 = D_N(\lambda)X \otimes 1 + \mathcal{R}_N(\lambda)X \otimes 1$$
$$= D_N(\lambda)\,(X \otimes 1 - 1 \otimes \xi_\lambda(X)1) + \mathcal{R}_N(\lambda)X \otimes 1,$$

where

$$\mathcal{R}_N(\lambda)X \otimes 1 \in \mathcal{U}(\mathfrak{n}_{n+1}^-)(\mathfrak{m}_{n+1} \otimes 1) \oplus \mathcal{U}(\mathfrak{n}_{n+1}^-)(H_0 \otimes 1 - 1 \otimes \xi_\lambda(H_0)1) \in I_\lambda(\mathfrak{g}_{n+1})$$

according to Theorem 5.1.1 and Theorem 5.1.3. We notice that there is no analog of the latter calculation for $X = Y_n^+$. The proof is complete. $\qquad\square$

5.2 Induced families

5.2.1 Induction

In the present section, we use the polynomial families $\mathcal{D}_{2N}^0(\lambda) \in \mathcal{U}(\mathfrak{n}_{n+1}^-)$ to induce G^n-equivariant families of differential operators. First of all, we have a model on induced representation spaces. Restriction of that picture to K and N^- yields the respective compact and non-compact models. In addition, we consider restrictions to totally umbilic subspheres Σ with a conformal group isomorphic to G^n.

For $\lambda \in \mathbb{C}$, let ξ_λ be the (non-unitary) A-character $\xi_\lambda(a_t) = e^{\lambda t}$ acting on the space $\mathbb{C}(\lambda)$. Let

$$\operatorname{Ind}_{P^m}^{G^m}(\xi_\lambda) = C^\infty(G^m, \mathbb{C}(\lambda))^{P^m}$$
$$= \{u \in C^\infty(G^m) \mid u(gman) = \xi_\lambda(a)u(g),\ man \in P^m\}.$$

G^m operates by left translation on the space $\operatorname{Ind}_{P^m}^{G^m}(\xi_\lambda)$. Note that the latter definition slightly differs from the conventions in representation theory ([156]) in that the above space usually would be considered to be induced by the character ξ_λ^{-1}.

The following result is a consequence of Theorem 5.1.1.

Theorem 5.2.1. *For any non-negative integer N, the family $\mathcal{D}_N^0(\lambda)$ induces a polynomial family of G^{n+1}-equivariant differential operators*

$$D_N^0(\lambda) : \operatorname{Ind}_{P^{n+1}}^{G^{n+1}}(\xi_\lambda) \to \operatorname{Ind}_{P^n}^{G^{n+1}}(\xi_{\lambda-N}). \tag{5.2.1}$$

Hence the composition

$$D_N(\lambda) = i^* \circ D_N^0(\lambda)$$

of $D_N^0(\lambda)$ with the G^n-equivariant restriction map $i^* : C^\infty(G^{n+1}) \to C^\infty(G^n)$ defines a polynomial family of G^n-equivariant operators

$$D_N(\lambda) : \operatorname{Ind}_{P^{n+1}}^{G^{n+1}}(\xi_\lambda) \to \operatorname{Ind}_{P^n}^{G^n}(\xi_{\lambda-N}). \tag{5.2.2}$$

Proof. We let $\mathcal{U}(\mathfrak{g})$ operate from the right on $C^\infty(G)$, i.e., the action of $\mathcal{U}(\mathfrak{g})$ extends the representation

$$R(X) : u(g) \mapsto (d/dt)|_{t=0}(u(g\exp(tX))), \ X \in \mathfrak{g}$$

of $\mathfrak{g}$. Now let $\mathcal{D}_N^0(\lambda)$ operate on $u \in \operatorname{Ind}_{P^{n+1}}^{G^{n+1}}(\xi_\lambda) \subset C^\infty(G)$. We have to prove that

$$\mathcal{D}_N^0(\lambda)u(gma_tn) = e^{(\lambda-N)t}\mathcal{D}_N^0(\lambda)u(g) \tag{5.2.3}$$

for $ma_tn \in P^n = M^n A N^n$, $g \in G^{n+1}$. Now $\operatorname{Ad}(m)\Delta_{n-1}^- = \Delta_{n-1}^-$ and $\operatorname{Ad}(m)Y_n^- = Y_n^-$ for $m \in M^n$ immediately imply

$$\mathcal{D}_N^0(\lambda)u(gm) = \mathcal{D}_N^0(\lambda)u(g), \ m \in M^n.$$

Next, we observe that for $1 \le j \le n-1$ and $u \in \operatorname{Ind}_{P^{n+1}}^{G^{n+1}}(\xi_\lambda)$,

$$\begin{aligned}
Y_j^+\mathcal{D}_N^0(\lambda)u(g) &= \left[Y_j^+, \mathcal{D}_N^0(\lambda)\right]u(g) + \mathcal{D}_N^0(\lambda)Y_j^+u(g) \\
&= \left[Y_j^+, \mathcal{D}_N^0(\lambda)\right]u(g) \qquad \text{(by right } N^n\text{-invariance of } u) \\
&= 0 \qquad \text{(by Theorem 5.1.1 or Theorem 5.1.3).}
\end{aligned}$$

Hence $\mathcal{D}_N^0(\lambda)u(gn) = \mathcal{D}_N^0(\lambda)u(g)$, $n \in N^n$. Finally, we note that

$$\begin{aligned}
H_0\mathcal{D}_N^0(\lambda)u(g) &= \left[H_0, \mathcal{D}_N^0(\lambda)\right]u(g) + \mathcal{D}_N^0(\lambda)H_0u(g) \\
&= -N\mathcal{D}_N^0(\lambda)u(g) + \lambda\mathcal{D}_N^0(\lambda)u(g) \qquad \text{(by Lemma 5.1.1)} \\
&= (\lambda-N)\mathcal{D}_N^0(\lambda)u(g),
\end{aligned}$$

i.e., $\mathcal{D}_N^0(\lambda)u(ga) = \xi_{\lambda-N}(a)\mathcal{D}_N^0(\lambda)u(g)$. The proof of (5.2.3) is complete. The remaining assertions are obvious. $\square$

Restriction from $G = KAN^+$ to K defines a G^m-isomorphism

$$C^\infty(G^m, \mathbb{C}(\lambda))^{P^m} \to C^\infty(K^m)_\lambda^{M^m},$$

where G^m acts on $C^\infty(K^m)_\lambda^{M^m}$ by

$$\varphi(k) \mapsto \varphi(\kappa(g^{-1}k))e^{\lambda H(g^{-1}k)} = \varphi(\kappa(g^{-1}k))e^{-\lambda\langle g\cdot\mathcal{O},kM\rangle}.$$

We compose this isomorphism with $C^\infty(K^m)^{M^m} \simeq C^\infty(K^m/M^m) \simeq C^\infty(S^{m-1})$ and obtain an isomorphism

$$\alpha : \operatorname{Ind}_{P^m}^{G^m}(\xi_\lambda) \to C^\infty(S^{m-1})_\lambda \tag{5.2.4}$$

of G^m-modules. Hence $D_N(\lambda)$ induces a G^n-equivariant operator

$$D_N(S^n, S^{n-1}; \lambda) : C^\infty(S^n)_\lambda \to C^\infty(S^{n-1})_{\lambda-N}. \tag{5.2.5}$$

The renormalized operator

$$D_N^c(\lambda) = 2^{-N} D_N(S^n, S^{n-1}; \lambda) \qquad (5.2.6)$$

is called the *compact model* of $D_N(\lambda)$. The renormalization in (5.2.6) is introduced in order to recover the GJMS-operators for the round metric of S^n as values of the families $D_{2N}^c(\lambda)$ at $\lambda = -\frac{n}{2}+N$ (compare Lemma 5.2.2 with Lemma 5.2.8).

Similarly, restriction from G to $N^- \subset N^- M A N^+ = N^- P^+ \subset G$ induces a map

$$\beta : \mathrm{Ind}_{Pm}^{Gm}(\xi_\lambda) \to C^\infty(N^-) \simeq C^\infty(\mathbb{R}^{m-1}) \qquad (5.2.7)$$

and a *non-compact model*

$$D_N^{nc}(\lambda) : C^\infty(\mathbb{R}^n) \to C^\infty(\mathbb{R}^{n-1}) \qquad (5.2.8)$$

of $D_N(\lambda)$. By Lemma 2.2.1, the G^m-action on $\mathbb{R}^{m-1}$ which is induced by transport from S^{m-1} using κ_S coincides with the G^m-action which is induced by restriction to $N^- \simeq \mathbb{R}^{m-1}$. The following result is obvious.

Lemma 5.2.1.

$$D_{2N}^{nc}(\lambda) = \sum_{j=0}^{N} a_j(\lambda) \Delta_{\mathbb{R}^{n-1}}^j i^* \left(\partial/\partial x_n \right)^{2N-2j},$$

where $i : \mathbb{R}^{n-1} \ni x' \mapsto (x',0) \in \mathbb{R}^n$, $\Delta_{\mathbb{R}^{n-1}} = \sum_{j=1}^{n-1} \partial^2/\partial x_i^2$, *and the coefficients satisfy* (5.1.2) *and* $a_N(\lambda) = 1$. *Similarly,*

$$D_{2N+1}^{nc}(\lambda) = \sum_{j=0}^{N} b_j(\lambda) \Delta_{\mathbb{R}^{n-1}}^j i^* \left(\partial/\partial x_n \right)^{2N+1-2j},$$

where the coefficients satisfy (5.1.22) *and* $b_N(\lambda) = 1$.

Now we establish an explicit relation between the induced picture on the one hand, and the compact and the non-compact picture on the other hand.

Lemma 5.2.2. $\Phi_m^\lambda \circ \kappa_S^* \circ \alpha = \beta$ *on* $\mathrm{Ind}_{Pm}^{Gm}(\xi_\lambda)$, *where* $\Phi_m(x) = 1 + |x|^2$, $x \in \mathbb{R}^m$.

Proof. Let $n_s^- = k_s a_s n_s^+ \in K A N^+$ be the Iwasawa decomposition of n_s^-. Then $n_s^- \mathcal{N} = k_s \mathcal{N} \in S^{m-1}$ and for $u \in \mathrm{Ind}_P^G(\xi_\lambda)$ we get

$$\beta(u)(n_s^-) = u(n_s^-) = u(k_s a_s n_s^+)$$
$$= \xi_\lambda(a_s) u(k_s) = \xi_\lambda(a_s) \alpha(u)(k_s \mathcal{N}) = \xi_\lambda(a_s) \alpha(u)(n_s^- \mathcal{N}).$$

Hence Lemma 2.2.1 gives

$$\beta(u)(n_s^-) = \xi_\lambda(a_s) \alpha(u)(\kappa_S(s)) = \xi_\lambda(a_s) \left(\kappa_S^* \circ \alpha \right)(u)(s),$$

and it only remains to prove that

$$\xi_\lambda(a_s) = (1+|s|^2)^\lambda, \ s \in \mathbb{R}^{m-1}. \tag{5.2.9}$$

In order to prove (5.2.9), we note that, by M-invariance, it is clearly enough to verify it in the case $n = 2$, i.e., $G = SO(1,2)^\circ$. Then the poles are $\mathcal{N} = (1,0)$ and $\mathcal{S} = (-1,0)$ in $\partial \mathbb{B}^2 = S^1 \subset \mathbb{R}^2 \simeq \mathbb{C}$. The element

$$n_s^- = \begin{pmatrix} 1 + s^2/2 & s^2/2 & s \\ -s^2/2 & 1 - s^2/2 & -s \\ s & s & 1 \end{pmatrix}, \quad s \in \mathbb{R}$$

fixes $\mathcal{S}$, and for its action on $\mathbb{B}^2$ we have

$$n_s^-(\mathcal{O}) = \left(\frac{-s^2}{4 + s^2}, \frac{2s}{4 + s^2} \right) = \frac{-s}{s + 2i} \in \mathbb{B}^2.$$

Now

$$\xi_\lambda(a_s) = e^{-\lambda \langle a_s^{-1}(\mathcal{O}), 1 \rangle}.$$

Since $(n_s^-)^{-1}(\mathcal{O}) = (n_s^+)^{-1}(a_s)^{-1}(\mathcal{O})$, we get

$$\langle (n_s^-)^{-1}(\mathcal{O}), 1 \rangle = \langle (a_s)^{-1}(\mathcal{O}), 1 \rangle,$$

i.e.,

$$\xi_\lambda(a_s) = e^{-\lambda \langle (n_s^-)^{-1}(\mathcal{O}), 1 \rangle}.$$

Now

$$(n_s^-)^{-1}(\mathcal{O}) = n_{-s}^-(\mathcal{O}) = \frac{s}{-s + 2i} \in \mathbb{B}^2$$

is on the circle $|z - a| = (1-a)^2$, where $a = 1/s^2 + 2$. It intersects the real axis in $-1 < x = -s^2/s^2 + 2 \leq 0$. Now the hyperbolic distance of x and $\mathcal{O}$ is

$$-\log \frac{1 + |x|}{1 - |x|} = -\log(1 + s^2)$$

($\mathcal{O}$ is inside the horosphere through x with normal 1). Hence

$$\xi_\lambda(a_s) = e^{-\lambda \langle (n_s^-)^{-1}(\mathcal{O}), 1 \rangle} = (1 + s^2)^\lambda$$

and the proof is complete. $\square$

Lemma 5.2.2 and (2.3.17) show that β transports the induced representation $\mathrm{Ind}_{Pm}^{G^m}(\xi_\lambda)$ to the representation π_λ^{nc}. Moreover, Lemma 5.2.2 implies the relation

$$\left(\Psi_{n-1}^{\lambda - N} \circ \kappa_\mathcal{S}^* \right) \circ D_N^c(\lambda) = D_N^{nc}(\lambda) \circ \left(\Psi_n^\lambda \circ \kappa_\mathcal{S}^* \right), \quad \Psi_m = \Phi_m/2 \tag{5.2.10}$$

between the models $D_N^c(\lambda)$ and $D_N^{nc}(\lambda)$. In fact, using

$$\beta^{-1} \circ (\Phi_m^\lambda \circ \kappa_S^*) \circ \alpha = \text{id}$$

for $m = n$ and $m = n-1$, we get

$$\beta^{-1} \circ (\Phi_{n-1}^{\lambda-N} \circ \kappa_S^*) \circ \alpha \circ D_N(\lambda) = D_N(\lambda) \circ \beta^{-1} \circ (\Phi_{n-1}^\lambda \circ \kappa_S^*) \circ \alpha,$$

i.e.,

$$\Phi_{n-1}^{\lambda-N} \circ \kappa_S^* \circ (\alpha \circ D_N(\lambda) \circ \alpha^{-1}) = \beta \circ D_N(\lambda) \circ \beta^{-1} \circ (\Phi_{n-1}^\lambda \circ \kappa_S^*).$$

(5.2.10) then follows from the definitions

$$2^N D_N^c(\lambda) = \alpha \circ D_N(\lambda) \circ \alpha^{-1} \quad \text{and} \quad D_N^{nc}(\lambda) = \beta \circ D_N(\lambda) \circ \beta^{-1}.$$

(5.2.10) implies

Lemma 5.2.3.

$$\kappa_S^* \circ D_N^c(\lambda) \circ (\kappa_S)_* = \Psi_{n-1}^{-\lambda+N} \circ D_N^{nc}(\lambda) \circ \Psi_n^\lambda. \tag{5.2.11}$$

In particular, the equivariance

$$\pi_{\lambda-N}^{nc}(g) \circ D_N^{nc}(\lambda) = D_N^{nc}(\lambda) \circ \pi_\lambda^{nc}(g), \ g \in G^n$$

of the non-compact model is a consequence of the equivariance of the compact model and (2.3.17).

Finally, we induce operators from S^n to totally umbilic hypersurfaces Σ as in Section 2.3. As before, let $G^n \subset G^{n+1}$ so that, if G^{n+1} is considered as the conformal group of (S^n, g_c), then G^n is the subgroup which leaves the subsphere $S^{n-1} = S^n \cap \{x_{n+1} = 0\}$ invariant. We have seen above that the family $\mathcal{D}_N(\lambda) \in \mathcal{U}(\mathfrak{n}_{n+1}^-)$ induces the G^{n+1}-equivariant family

$$D_N^0(\lambda) : C^\infty(G^{n+1}, \mathbb{C}(\lambda))^{P^{n+1}} \to C^\infty(G^{n+1}, \mathbb{C}(\lambda-N))^{P^n}$$

by differentiation from the right. Its composition with the restriction i^* to $G^n \subset G^{n+1}$ then defines the left G^n-equivariant family $D_N(\lambda)$ (see Theorem 5.2.1). Using respective restrictions to $K^{n+1} \subset G^{n+1}$ and $K^n \subset G^n$, it can be identified with the G^n-equivariant family

$$D_N(S^n, S^{n-1}; \lambda) : C^\infty(S^n)_\lambda \to C^\infty(S^{n-1})_{\lambda-N}$$

and we set $D_N^c(S^n; \lambda) = 2^{-N} D_N(S^n, S^{n-1}; \lambda)$. Then

$$D_N^c(\lambda) \circ \pi_\lambda^c(S^n)(g) = \pi_{\lambda-N}^c(S^{n-1})(g) \circ D_N^c(\lambda), \ g \in G^n.$$

This is a consequence of the fact that the restriction to K^m defines the G^m-module isomorphism α. Now the families $\mathcal{D}_N(\lambda)$ (N even) also induces equivariant families

$$D_N(S^n, \Sigma; \lambda) : C^\infty(S^n) \to C^\infty(\Sigma)$$

for all totally umbilic hypersurfaces Σ of S^n. We describe the corresponding induction mechanism and the equivariance of such families. We write $\Sigma = \sigma(S^{n-1}) \subset S^n$ for some $\sigma \in G^{n+1}$. The group $\sigma G^n \sigma^{-1}$ leaves Σ invariant. We consider the submanifold $G^n[\sigma] = \sigma G^n \subset G^{n+1}$. The left-action of $\sigma G^n \sigma^{-1}$ on G^{n+1} restricts to a left-action of this group on the manifold $G^n[\sigma]$:

$$L_{\sigma g_0 \sigma^{-1}} : G^n[\sigma] \to G^n[\sigma], \ \sigma g \mapsto \sigma g_0 g.$$

It induces representations

$$L_{\sigma g_0 \sigma^{-1}} : C^\infty(G^n[\sigma]) \to C^\infty(G^n[\sigma]), \ u(\sigma g) \mapsto u(\sigma g_0^{-1} g).$$

Then the composition of $D_N^0(\lambda)$ with the restriction i_σ^* to $G^n[\sigma]$ is left $\sigma G^n \sigma^{-1}$-equivariant. The map

$$\beta_\sigma : C^\infty(G^n[\sigma], \mathbb{C}(\lambda))^{P^n} \to C^\infty(\sigma(S^{n-1})) = C^\infty(\Sigma)$$
$$u(\sigma g) \mapsto (\sigma(kM) \mapsto u(\sigma k))$$

is an isomorphism. Hence we can use the commutative diagram

$$\begin{array}{ccc}
C^\infty(G^{n+1}, \mathbb{C}(\lambda))^{P^{n+1}} & \xrightarrow{\ i_\sigma^* \circ D_N^0(\lambda)\ } & C^\infty(G^n[\sigma], \mathbb{C}(\lambda - N))^{P^n} \\
\Big\downarrow{\alpha} & & \Big\downarrow{\alpha_\sigma} \\
C^\infty(S^n) & \xrightarrow{\ D_N(S^n, \Sigma; \lambda)\ } & C^\infty(\Sigma)
\end{array}$$

to define the family $D_N(S^n, \Sigma; \lambda)$. The following result describes its equivariance.

Lemma 5.2.4.

$$\pi_{\lambda-N}(\Sigma, g_c)(g) \circ D_N(S^n, \Sigma; \lambda) = D_N(S^n, \Sigma; \lambda) \circ \pi_\lambda(S^n, g_c)(g)$$

for all $g \in \sigma G^n \sigma^{-1} \subset G^{n+1}$, i.e., for all $g \in G^{n+1}$ which leave Σ invariant.

Proof. We recall that

$$\pi_\lambda(\Sigma, g_c)(g) = \left(\frac{g_*(\mathrm{vol}(\Sigma, g_c))}{\mathrm{vol}(\Sigma, g_c)} \right)^{-\frac{\lambda}{n-1}} \circ g_*,$$

on functions on Σ (see (2.3.18)). Here $g \in G^{n+1}$ preserves Σ and $\mathrm{vol}(\Sigma, g_c)$ is the Riemannian volume form on Σ for the restriction of the round metric. It suffices to prove that the maps α and α_σ satisfy the intertwining identities $\alpha \circ L_{g_0} = \pi_\lambda(S^n, g_c)(g_0) \circ \alpha$ and

$$\alpha_\sigma \circ L_{\sigma g_0 \sigma^{-1}} = \pi_\lambda(\Sigma, g_c)(\sigma g_0 \sigma^{-1}) \circ \alpha_\sigma, \ g_0 \in G^n.$$

The equivariance of α has been established above. We prove the second assertion by identifying $\pi_\lambda(\Sigma, g_c)(\sigma g_0 \sigma^{-1})$ as the bottom line in the commutative diagram

$$
\begin{array}{ccc}
C^\infty(G^n[\sigma], \mathbb{C}(\lambda))^{P^n} & \xrightarrow{\ L_{\sigma g \sigma^{-1}}\ } & C^\infty(G^n[\sigma], \mathbb{C}(\lambda))^{P^n} \\
\Big\downarrow \alpha_\sigma & & \Big\downarrow \alpha_\sigma \\
C^\infty(\Sigma) & \xrightarrow{\quad\quad} & C^\infty(\Sigma).
\end{array}
\tag{5.2.12}
$$

Let $u \in C^\infty(G^n[\sigma], \mathbb{C}(\lambda))^{P^n}$. Then $\alpha_\sigma(u)$ is the function

$$
\Sigma \ni \sigma(kM) \mapsto u(\sigma k) = u(\kappa(\sigma k))e^{\lambda H(\sigma k)}.
$$

On the other hand, we get $L_{\sigma g \sigma^{-1}}(u)(\sigma g) = u(\sigma g_0^{-1} g)$. Hence the composition $\alpha_\sigma L_{\sigma g \sigma^{-1}}(u)$ is the function

$$
\sigma(kM) \mapsto u(\sigma g_0^{-1} k) = u(\kappa(\sigma g_0^{-1} k))e^{\lambda H(\sigma g_0^{-1} k)}.
$$

It follows that the bottom line in (5.2.12) is given by the operator

$$
\begin{aligned}
e^{\lambda H(\sigma g_0^{-1} k) - \lambda H(\sigma k)}(\sigma g_0 \sigma^{-1})_* &= e^{-\lambda \langle g_0 \sigma^{-1} \cdot \mathcal{O}, b \rangle + \lambda \langle \sigma^{-1} \cdot \mathcal{O}, b \rangle}(\sigma g_0 \sigma^{-1})_* \\
&= e^{-\lambda \langle \sigma g_0 \sigma^{-1} \cdot \mathcal{O}, \sigma(b) \rangle}(\sigma g_0 \sigma^{-1})_* \\
&= \Phi_{\sigma g_0 \sigma^{-1}}^{-\lambda}(\sigma(b))(\sigma g_0 \sigma^{-1})_*, \ b = kM,
\end{aligned}
$$

where $\Phi_g(b) = e^{\langle g \cdot \mathcal{O}, b \rangle}$. Here we have used the identity

$$
\langle g_2 g_1 \cdot \mathcal{O}, b \rangle = \langle g_2 \cdot \mathcal{O}, b \rangle + \langle g_1 \cdot \mathcal{O}, g_2^{-1}(b) \rangle
\tag{5.2.13}
$$

and its consequence $\langle g \cdot \mathcal{O}, b \rangle = -\langle g^{-1} \cdot \mathcal{O}, g^{-1}(b) \rangle$. (5.2.13) is equivalent to the cocycle identity

$$
\Phi_{g_2 g_1} = \Phi_{g_2} \cdot (g_2)_*(\Phi_{g_1}).
\tag{5.2.14}
$$

Now it only remains to interpret the exponential term as

$$
\left(\frac{g_*(\mathrm{vol}(\Sigma, g_c))}{\mathrm{vol}(\Sigma, g_c)} \right)^{-\frac{\lambda}{n-1}}
$$

at $\sigma(b)$ for $g = \sigma g_0 \sigma^{-1}$, i.e., we have to prove

$$
\frac{(\sigma g \sigma^{-1})_*(\mathrm{vol}(\Sigma, g_c))}{\mathrm{vol}(\Sigma, g_c)} = \Phi_{\sigma g_0 \sigma^{-1}}^{n-1}(\sigma(b)).
\tag{5.2.15}
$$

We first note that $\sigma^*(g_c|_\Sigma) = \sigma^*(g_c)|_{S^{n-1}} = (\Phi_{\sigma^{-1}}^2 g_c)|_{S^{n-1}}$. Hence

$$
g_c|_\Sigma = \sigma_*(\Phi_{\sigma^{-1}}^2|_{S^{n-1}})\sigma_*(g_c|_{S^{n-1}}) = (\Phi_\sigma^{-2})|_{S^{n-1}}\sigma_*(g_c|_{S^{n-1}}),
$$

and therefore
$$\text{vol}(\Sigma, g_c) = \sigma_*(\text{vol}(S^{n-1}, g_c))/(\Phi_\sigma^{n-1})|_{S^{n-1}}.$$

Now we calculate

$$\begin{aligned}
(\sigma g_0 \sigma^{-1})_*(\text{vol}(\Sigma, g_c)) &= \sigma_*(g_0)_* \sigma^*(\text{vol}(\Sigma, g_c)) \\
&= \sigma_*(g_0)_* \left(\text{vol}(S^{n-1}, g_c)\Phi_{\sigma^{-1}}^{n-1}\right) \\
&= \sigma_* \left(\Phi_{g_0}^{n-1}\text{vol}(S^{n-1}, g_c)(g_0)_*(\Phi_{\sigma^{-1}}^{n-1})\right) \\
&= \sigma_*(\Phi_{g_0}^{n-1})\sigma_*(\text{vol}(S^{n-1}, g_c))(\sigma g_0)_*(\Phi_{\sigma^{-1}}^{n-1}) \\
&= \sigma_*(\Phi_{g_0}^{n-1})(\sigma g_0)_*(\Phi_{\sigma^{-1}}^{n-1})\Phi_\sigma^{n-1}\text{vol}(\Sigma, g_c) \\
&= \left(\frac{\Phi_{\sigma g_0}}{\Phi_\sigma}\right)^{n-1}\left(\frac{\Phi_{\sigma g_0 \sigma^{-1}}}{\Phi_{\sigma g_0}}\right)^{n-1}\Phi_\sigma^{n-1}\text{vol}(\Sigma, g_c) \\
&= \Phi_{\sigma g_0 \sigma^{-1}}^{n-1}\text{vol}(\Sigma, g_c)
\end{aligned}$$

using (5.2.14). The proof is complete. $\square$

The definitions immediately imply that the families

$$D_N(S^n, S^{n-1}; \lambda) \quad \text{and} \quad D_N(S^n, \Sigma; \lambda)$$

are conjugate. More precisely, we have

Lemma 5.2.5. *Let* $\Sigma = \sigma(S^{n-1})$. *Then*

$$\left(\frac{\sigma_*(\text{vol}(S^{n-1}, g_c))}{\text{vol}(\Sigma, g_c)}\right)^{-\frac{\lambda-N}{n-1}} \circ \sigma_* \circ D_N(S^n, S^{n-1}; \lambda) = D_N(S^n, \Sigma; \lambda) \circ \pi_\lambda(S^n)(\sigma). \tag{5.2.16}$$

We omit the proof. Lemma 5.2.5 can be used to derive the equivariance of the family $D_N(S^n, \Sigma; \lambda)$ (see Lemma 5.2.4) from the equivariance of the family $D_N(S^n, S^{n-1}; \lambda)$. Lemma 5.2.5 can also be used as an equivalent alternative definition of $D_N(S^n, \Sigma; \lambda)$. In fact, $D_N(S^n, \Sigma; \lambda)$ is well defined by (5.2.16) since by the G^n-equivariance of the family $D_N(S^n, S^{n-1}; \lambda)$ (see (5.2.5)) it only depends on the class of σ in G^{n+1}/G^n.

As noticed above, using respective restrictions to $(N^{n+1})^- \subset G^{n+1}$ and $(N^n)^- \subset G^n$, the G^n-equivariant family $D_N(\lambda)$ induces a G^n-equivariant family

$$D_N^{nc}(\lambda) = D_N(\mathbb{R}^n, \mathbb{R}^{n-1}; \lambda) : C^\infty(\mathbb{R}^n) \to C^\infty(\mathbb{R}^{n-1}),$$

i.e., we have

$$D_N^{nc}(\lambda) \circ \pi_\lambda^{nc}(\mathbb{R}^n)(g) = \pi_{\lambda-N}^{nc}(\mathbb{R}^{n-1})(g) \circ D_N^{nc}(\lambda), \ g \in G^n.$$

Moreover, for a (generalized) sphere $\Sigma = \sigma(\mathbb{R}^{n-1}) \subset \mathbb{R}^n$, $\sigma \in G^{n+1}$ we can use $D_N(\lambda)$ to induce an equivariant family $D_N(\mathbb{R}^n, \Sigma; \lambda) : C^\infty(\mathbb{R}^n) \to C^\infty(\Sigma)$. We have no reason to describe that construction in more detail for general Σ. However, for later purposes, we formulate the basic properties of such families in the special case $\Sigma = S^{n-1} = \partial \mathbb{B}^n \subset \mathbb{R}^n$.

Lemma 5.2.6.

$$\pi_{\lambda - N}(S^{n-1}, g_c)(g) \circ D_N(\mathbb{R}^n, S^{n-1}; \lambda) = D_N(\mathbb{R}^n, S^{n-1}; \lambda) \circ \pi_\lambda(\mathbb{R}^n, g_c)(g)$$

for all $g \in G^{n+1}$ which leave $S^{n-1} \subset \mathbb{R}^n$ invariant.

Of course, in Lemma 5.2.6 we use on the left-hand side the action on $S^{n-1} \subset \mathbb{R}^n$ which is induced by that on $\mathbb{R}^n$.

The families $D_N(\mathbb{R}^n, \mathbb{R}^{n-1}; \lambda)$ and $D_N(\mathbb{R}^n, S^{n-1}; \lambda)$ are conjugate. More precisely,

Lemma 5.2.7. *Let $S^{n-1} = \partial \mathbb{B}^n \subset \mathbb{R}^n$ and let $\sigma_0 : \mathbb{R}^{n-1} \to S^{n-1}$ be as in (2.2.9). Then*

$$\left(\frac{(\sigma_0)_* (\mathrm{vol}(\mathbb{R}^{n-1}, g_c))}{\mathrm{vol}(S^{n-1}, g_c)} \right)^{-\frac{\lambda - N}{n-1}} \circ (\sigma_0)_* \circ D_N(\mathbb{R}^n, \mathbb{R}^{n-1}; \lambda)$$

$$= D_N(\mathbb{R}^n, S^{n-1}; \lambda) \circ \pi_\lambda(\mathbb{R}^n, g_c)(\sigma_0).$$

In Section 5.4, the last result will be used to find explicit formulas for the induced families $D_N(\mathbb{R}^n, S^{n-1}; \lambda)$ for $N \leq 3$. In particular, it will be shown that the Chang-Qing boundary operator P_3 on $\mathbb{B}^4$ is an induced one.

5.2.2 Even order families: $D_{2N}^{nc}(\lambda)$ and $D_{2N}^c(\lambda)$

The main results of the present section are the factorization identities for the compact and the non-compact models of $\mathcal{D}_N(\lambda)$.

Theorem 5.2.2 (Factorization). *For $j = 1, \ldots, N$, the operators $D_{2N}^{nc}(-\frac{n}{2} + j)$ and $D_{2N}^{nc}(2N - j - \frac{n-1}{2})$ factorize as maps*

$$C^\infty(\mathbb{R}^n)_{-\frac{n}{2} + j} \to C^\infty(\mathbb{R}^n)_{-\frac{n}{2} - j} \to C^\infty(\mathbb{R}^{n-1})_{-\frac{n}{2} + j - 2N},$$

$$C^\infty(\mathbb{R}^n)_{-\frac{n-1}{2} + 2N - j} \to C^\infty(\mathbb{R}^{n-1})_{-\frac{n-1}{2} + j} \to C^\infty(\mathbb{R}^{n-1})_{-\frac{n-1}{2} - j}$$

in the form

$$D_{2N}^{nc}\left(-\frac{n}{2} + j\right) = D_{2N-2j}^{nc}\left(-\frac{n}{2} - j\right) \circ P_{2j}(\mathbb{R}^n), \tag{5.2.17}$$

$$D_{2N}^{nc}\left(2N - j - \frac{n-1}{2}\right) = P_{2j}(\mathbb{R}^{n-1}) \circ D_{2N-2j}^{nc}\left(2N - j - \frac{n-1}{2}\right). \tag{5.2.18}$$

We note that for $j = N$, the factorizations (5.2.17) and (5.2.18) state that

$$D_{2N}^{nc}\left(-\frac{n}{2} + N\right) = i^* \circ (\Delta_{\mathbb{R}^n})^N \quad \text{and} \quad D_{2N}^{nc}\left(-\frac{n-1}{2} + N\right) = (\Delta_{\mathbb{R}^{n-1}})^N \circ i^*.$$

Proof. It is enough to prove the factorizations

$$\mathcal{D}_{2N}^0\left(-\frac{n}{2}+j\right) = \mathcal{D}_{2N-2j}^0\left(-\frac{n}{2}-j\right)(\Delta_n^-)^j,$$

$$\mathcal{D}_{2N}^0\left(2N-j-\frac{n-1}{2}\right) = (\Delta_{n-1}^-)^j\mathcal{D}_{2N-2j}^0\left(2N-j-\frac{n-1}{2}\right)$$

in $\mathcal{U}(\mathfrak{n}_{n+1}^-)$. For both sets of identities we use different types of arguments. We begin with the proof of the first set of identities. We first notice that the right-hand side is of the form

$$\left(\sum_{r=0}^{N-j} a_r(\Delta_{n-1}^-)^r(Y_n^-)^{2N-2j-2r}\right)\sum_{s=0}^{j}\binom{j}{s}(\Delta_{n-1}^-)^s(Y_n^-)^{2(j-s)}$$

$$=\sum_{r=0}^{N-j}\sum_{s=0}^{j} a_r\binom{j}{s}(\Delta_{n-1}^-)^{r+s}(Y_n^-)^{2N-2r-2s}$$

$$=\sum_{k=0}^{N} s_k(\Delta_{n-1}^-)^k(Y_n^-)^{2N-2k}$$

and the coefficient of $(\Delta_{n-1}^-)^N$ is equal to 1. In order to apply the *uniqueness* consequence of Theorem 5.1.1, it is enough to verify that the product

$$T_{(j,N)} = \mathcal{D}_{2N-2j}^0\left(-\frac{n}{2}-j\right)(\Delta_n^-)^j \in \mathcal{U}(\mathfrak{n}_{n+1}^-)$$

satisfies the same commutator relations as $\mathcal{D}_{2N}^0\left(-\frac{n}{2}+j\right)$, i.e.,

$$[Y_i^+, T_{(j,N)}] \in \mathcal{U}(\mathfrak{n}_{n+1}^-)\left(\mathfrak{m}_{n+1} \oplus \mathbb{C}\left(H_0+\left(\frac{n}{2}-j\right)\right)\right). \tag{5.2.19}$$

Now we write

$$[Y_i^+, T_{(j,N)}] = \left[Y_i^+, \mathcal{D}_{2N-2j}^0\left(-\frac{n}{2}-j\right)\right](\Delta_n^-)^j + \mathcal{D}_{2N-2j}^0\left(-\frac{n}{2}-j\right)[Y_i^+, (\Delta_n^-)^j]$$

$$\in \mathcal{U}(\mathfrak{n}_{n+1}^-)\left(\mathfrak{m}_{n+1} \oplus \mathbb{C}\left(H_0+\left(\frac{n}{2}+j\right)\right)\right)(\Delta_n^-)^j$$

$$+ \mathcal{U}(\mathfrak{n}_{n+1}^-)\left(\mathfrak{m}_{n+1} \oplus \mathbb{C}\left(H_0+\left(\frac{n}{2}-j\right)\right)\right)$$

using $(\Delta_n^-)^j = \mathcal{D}_{2j}^0(-\frac{n}{2}+j)$ (Corollary 5.1.2). But M^{n+1} leaves Δ_n^- invariant and $[H_0, (\Delta_n^-)^j] = -2j(\Delta_n^-)^j$ (Lemma 5.1.1). Hence

$$\mathcal{U}(\mathfrak{n}_{n+1}^-)\left(\mathfrak{m}_{n+1} \oplus \mathbb{C}\left(H_0+\left(\frac{n}{2}+j\right)\right)\right)(\Delta_n^-)^j$$

$$\subseteq \mathcal{U}(\mathfrak{n}_{n+1}^-)\left(\mathfrak{m}_{n+1} \oplus \mathbb{C}\left(H_0+\left(\frac{n}{2}-j\right)\right)\right).$$

This proves (5.2.19), i.e., $T_{(j,N)}$ and $\mathcal{D}_{2N}^0(-\frac{n}{2}+j)$ satisfy the same commutator relations *and* are normalized both by the condition that the coefficient of Δ_{n-1}^- is 1. Theorem 5.1.1 implies that these two elements coincide.

A proof of the second set of identities along the above lines does *not* work since M^{n+1} does *not* leave $\mathcal{D}_{2N}^0(2N-j-\frac{n-1}{2})$ invariant. However, here the assertion follows from Corollary 5.1.1. In view of

$$\mathcal{D}_{2N}^0\left(2N-j-\frac{n-1}{2}\right) = \sum_{r=0}^{N} a_r^{(N)}\left(2N-j-\frac{n-1}{2}\right)(\Delta_{n-1}^-)^r(Y_n^-)^{2N-2r}$$

and

$$(\Delta_{n-1}^-)^j \mathcal{D}_{2N-2j}^0\left(2N-j-\frac{n-1}{2}\right)$$

$$= \sum_{r=0}^{N-j} a_r^{(N-j)}\left(2N-j-\frac{n-1}{2}\right)(\Delta_{n-1}^-)^{r+j}(Y_n^-)^{2N-2j-2r}$$

$$= \sum_{r=j}^{N} a_{r-j}^{(N-j)}\left(2N-j-\frac{n-1}{2}\right)(\Delta_{n-1}^-)^r(Y_n^-)^{2N-2r},$$

the assertion is equivalent to

$$a_r^{(N)}\left(2N-j-\frac{n-1}{2}\right) = 0 \quad \text{for } r = 0,\ldots,j-1 \tag{5.2.20}$$

and

$$a_r^{(N)}\left(2N-j-\frac{n-1}{2}\right) = a_{r-j}^{(N-j)}\left(2N-j-\frac{n-1}{2}\right) \quad \text{for } r = j,\ldots,N. \tag{5.2.21}$$

Now Corollary 5.1.1 gives

$$a_r^{(N)}\left(2N-j-\frac{n-1}{2}\right) = \frac{N!}{r!(2N-2r)!}(-2)^{N-r}\prod_{k=r}^{N-1}(-2j+2k+2) \tag{5.2.22}$$

and thus (5.2.20) is obvious. For the proof of (5.2.21), we have to compare (5.2.22) with

$$a_{r-j}^{(N-j)}\left(2N-j-\frac{n-1}{2}\right) = \frac{(N-j)!}{(r-j)!(2N-2r)!}(-2)^{N-r}\prod_{k=r-j}^{N-j-1}(2j+2k+2) \tag{5.2.23}$$

(Corollary 5.1.1). But (5.2.22) simplifies to

$$\frac{N!}{r!(2N-2r)!}(-2)^{N-r}2^{N-r}\frac{(N-j)!}{(r-j)!},$$

whereas (5.2.23) coincides with

$$\frac{(N-j)!}{(r-j)!(2N-2r)!}(-2)^{N-r}2^{N-r}\frac{N!}{r!}.$$

The proof is complete. Notice that the direct type of arguments used in the proof of the second set of identities do *not* give an analogous proof of the first set since these lead to a set of non-trivial identities for binomial coefficients. □

Now we prove analogous factorizations for the compact models.

Theorem 5.2.3 (Factorization). *For $j = 1, \ldots, N$, the operators $D_{2N}^c(-\frac{n}{2}+j)$ and $D_{2N}^c(2N-j-\frac{n-1}{2})$ factorize as equivariant maps*

$$C^\infty(S^n)_{-\frac{n}{2}+j} \to C^\infty(S^n)_{-\frac{n}{2}-j} \to C^\infty(S^{n-1})_{-\frac{n}{2}+j-2N},$$
$$C^\infty(S^n)_{-\frac{n-1}{2}+2N-j} \to C^\infty(S^{n-1})_{-\frac{n-1}{2}+j} \to C^\infty(S^{n-1})_{-\frac{n-1}{2}-j}$$

in the form

$$D_{2N}^c\left(-\frac{n}{2}+j\right) = D_{2N-2j}^c\left(-\frac{n}{2}-j\right) \circ P_{2j}(S^n), \qquad (5.2.24)$$

$$D_{2N}^c\left(2N-j-\frac{n-1}{2}\right) = P_{2j}(S^{n-1}) \circ D_{2N-2j}^c\left(2N-j-\frac{n-1}{2}\right). \qquad (5.2.25)$$

Proof. We prove the first set of identities. The identity

$$D_{2N}^{nc}\left(-\frac{n}{2}+j\right) = D_{2N-2j}^{nc}\left(-\frac{n}{2}-j\right) \circ P_{2j}(\mathbb{R}^n, g_c)$$

(see (5.2.17)) and Lemma 5.2.3 imply

$$\kappa_S^* \circ D_{2N}^c\left(-\frac{n}{2}+j\right) \circ (\kappa_S)_*$$
$$= \Psi_{n-1}^{\frac{n}{2}-j-2N} \circ D_{2N}^{nc}\left(-\frac{n}{2}+j\right) \circ \Psi_n^{-\frac{n}{2}+j}$$
$$= \left(\Psi_{n-1}^{\frac{n}{2}-j-2N} \circ D_{2N-2j}^{nc}\left(-\frac{n}{2}-j\right) \circ \Psi_n^{-\frac{n}{2}-j}\right) \circ \left(\Psi_n^{\frac{n}{2}+j} \circ P_{2j}(\mathbb{R}^n, g_c) \circ \Psi_n^{-\frac{n}{2}+j}\right)$$
$$= \left(\kappa_S^* \circ D_{2N-2j}^c\left(-\frac{n}{2}-j\right) \circ (\kappa_S)_*\right) \circ P_{2j}(\mathbb{R}^n, g_s) \quad \text{(by Lemma 5.2.8)}$$
$$= \kappa_S^* \circ D_{2N-2j}^c\left(-\frac{n}{2}-j\right) \circ ((\kappa_S)_* \circ P_{2j}(\mathbb{R}^n, g_s))$$
$$= \kappa_S^* \circ D_{2N-2j}^c\left(-\frac{n}{2}-j\right) \circ P_{2j}(S^n, g_c) \circ (\kappa_S)_* \quad \text{(by (2.2.1))}.$$

Now Lemma 5.2.3 implies the first set of identities. The proof of the second set is similar. □

Note that for $e^{\varphi(x)} = 2/(1+|x|^2)$, the identity (3.1.1) implies

Lemma 5.2.8.

$$\left(\frac{1+|x|^2}{2}\right)^{N+\frac{n}{2}} \circ P_{2N}(\mathbb{R}^n, g_c) \circ \left(\frac{1+|x|^2}{2}\right)^{N-\frac{n}{2}} = P_{2N}(\mathbb{R}^n, g_s),$$

where g_s is given by (2.2.2).

Now we use Theorem 5.2.3 to define a *recursive method* which can be used to deduce explicit formulas for $D_{2N}^c(\lambda) : C^\infty(S^n) \to C^\infty(S^{n-1})$ in terms of $P_{2M}(S^N)$ and $P_{2M}(S^{n-1})$ for $1 \leq M \leq N-1$. For that purpose, we prove that $\lambda \mapsto D_{2N}^c(\lambda)$ is a polynomial operator family of degree $2N$ with a scalar leading coefficient, i.e.,

$$D_{2N}^c(\lambda) = A_{2N} i^* \lambda^{2N} + A_{2N-1}\lambda^{2N-1} + \cdots + A_1\lambda + A_0 \tag{5.2.26}$$

for some $A_{2N} \in \mathbb{C}$. Then the $2N$ *unknown* operator-valued coefficients A_j, $j = 0, \ldots, 2N-1$ can be determined by solving the system

$$D_{2N}^c\left(-\frac{n}{2}+j\right) = D_{2N-2j}^c\left(-\frac{n}{2}-j\right) \circ P_{2j}(S^n),$$

$$D_{2N}^c\left(2N-j-\frac{n-1}{2}\right) = P_{2j}(S^{n-1}) \circ D_{2N-2j}^c\left(2N-j-\frac{n-1}{2}\right)$$

for $j = 1, \ldots, N$ using the *known* families $D_{2M}^c(\lambda)$, $0 \leq M \leq N-1$. Note that the determinant of the system is the Vandermonde determinant

$$\det \begin{pmatrix} \varepsilon_1^{2N-1} & \cdots & \varepsilon_1^0 \\ \vdots & & \vdots \\ \varepsilon_{2N}^{2N-1} & \cdots & \varepsilon_{2N}^0 \end{pmatrix} = \prod_{j>i}(\varepsilon_j - \varepsilon_i) \neq 0$$

for $\varepsilon_j = -\frac{n}{2} + j$ and $\varepsilon_{j+N} = 2N - j - \frac{n-1}{2}$ for $j = 1, \ldots, N$. Hence the recursive method is well defined.

Lemma 5.2.9. $\lambda \mapsto D_{2N}^c(\lambda)$ *is a polynomial of degree $2N$. The leading coefficient is scalar. More precisely,*

$$A_{2N} = (-1)^N.$$

Proof. The proof rests on the formula

$$\kappa_S^* \circ D_{2N}^c(\lambda) \circ (\kappa_S)_* = \Psi_{n-1}^{-\lambda+2N} \circ D_{2N}^{nc}(\lambda) \circ \Psi_n^\lambda \tag{5.2.27}$$

(Lemma 5.2.3) and the following result.

Lemma 5.2.10. *Let $a, b \in \mathbb{N}_0$. Then*

$$\lambda \mapsto \Psi_{n-1}^{-\lambda+2(a+b)} \circ \Delta_{\mathbb{R}^{n-1}}^b \circ i^* \circ \left(\frac{\partial^2}{\partial x_n^2}\right)^a \circ \Psi_n^\lambda$$

is an operator-valued polynomial of degree $2b + a$ with leading term

$$2^{-2a} \frac{(2a)!}{a!} |x'|^{2b} (1+|x'|^2)^a \lambda^{2b+a} i^*.$$

Proof. We write the operator polynomial as the composition

$$\left(\Psi_{n-1}^{-\lambda+2(a+b)} \circ \Delta_{\mathbb{R}^{n-1}}^b \circ \Psi_{n-1}^{\lambda-2a} \right) \circ \left(\Psi_{n-1}^{-\lambda+2a} \circ i^* \circ \left(\frac{\partial^2}{\partial x_n^2} \right)^a \circ \Psi_n^\lambda \right). \qquad (5.2.28)$$

Now the second factor

$$\lambda \mapsto \Psi_{n-1}^{-\lambda+2a} \circ i^* \circ \left(\frac{\partial^2}{\partial x_n^2} \right)^a \circ \Psi_n^\lambda$$

is a polynomial of degree a the leading coefficient of which is the operator

$$2^{-2a} \frac{(2a)!}{a!} (1+|x'|^2)^a i^* = 2^{-a} \frac{(2a)!}{a!} \Psi_{n-1}^a(x') i^*.$$

In fact, this follows from

$$c^{-\lambda+2a} \left(\frac{d^2}{dx^2} \right)^a \Big|_{x=0} \left((c+x^2)^\lambda \right) = c^{2a} \left(\frac{d^2}{dx^2} \right)^a \Big|_{x=0} \left(\left(1 + \left(\frac{x}{\sqrt{c}} \right)^2 \right)^\lambda \right)$$

$$= c^{2a} \left(\frac{d^2}{dx^2} \right)^a \Big|_{x=0} \sum_{n \geq 0} \binom{\lambda}{n} \left(\frac{x}{\sqrt{c}} \right)^{2n}$$

$$= c^a (2a)! \binom{\lambda}{a}.$$

Now it is easy to see that the leading term of the polynomial $\lambda \mapsto \Psi_{n-1}^{-\lambda+2(a+b)} \circ \Delta_{\mathbb{R}^{n-1}}^b \circ \Psi_{n-1}^{\lambda-a}$ is the operator

$$|x'|^{2b} \Psi_{n-1}^a(x') \lambda^{2b}.$$

It follows that the composition (5.2.28) is a polynomial of degree $2b + a$ with a leading term as claimed. □

Now Lemma 5.2.10 and the definition

$$D_{2N}^{nc}(\lambda) = \sum_{j=0}^N a_j^{(N)}(\lambda) \Delta_{\mathbb{R}^{n-1}}^j \circ i^* \circ \left(\frac{\partial^2}{\partial x_n^2} \right)^{N-j}$$

with polynomial coefficients $a_j^{(N)}(\lambda)$ of degree $N-j$ (see (5.1.4)) imply that the leading term of $D_{2N}^c(\lambda)$ is given by the composition of the sum

$$\sum_{j=0}^{N} \left[2^{2N-2j}(-1)^{N-j} \frac{N!}{j!(2N-2j)!} \right]$$

$$\times 2^{-2(N-j)} \frac{(2N-2j)!}{(N-j)!} |x'|^{2j} (1+|x'|^2)^{N-j} \lambda^{N-j} \lambda^{2j+N-j}$$

$$= \sum_{j=0}^{N} \binom{N}{j} (-1)^{N-j} |x'|^{2j} (1+|x'|^2)^{N-j} \lambda^{2N}$$

$$= (|x'|^2 - (1+|x'|^2))^N \lambda^{2N} = (-1)^N \lambda^{2N}$$

with i^*. This completes the proof of Lemma 5.2.9.   □

We use these results for the

Proof of Theorem 1.4.1. In fact, Lemma 5.2.9 shows that $D_{2N}^c(\lambda)$ is of the form

$$(-1)^N \lambda^{2N} i^* + A_{2N-1} \lambda^{2N-1} + \cdots + A_1 \lambda + A_0.$$

By the above recursive method, the $2N$ coefficients $A_{2N-1}, \ldots, A_0$ are determined by $2N$ factorization identities. More precisely, they are given as compositions of the operators $P_2(S^m), \ldots, P_{2N}(S^m)$ for $m = n$ and $m = n-1$ and the values

$$D_{2N-2j}^c \left(-\frac{n}{2} - j \right), \quad D_{2N-2j}^c \left(2N - j - \frac{n-1}{2} \right)$$

for $j = 1, \ldots, N$ of families of lower order. Now the factorization identities applied to the latter families imply formulas for these in terms of $P_2(S^m), \ldots, P_{2N-2}(S^m)$ for $m = n$ and $m = n-1$ and certain values of the families $D_{\leq 2N-4}^c(\lambda)$. A continuation of the procedure proves the assertion in Theorem 1.4.1 concerning the structure of the coefficients A_j.

We determine the normal order of $D_{2N}^c(\lambda)$. The behaviour of the normal order can be read off from the non-compact model $D_{2N}^{nc}(\lambda)$. The explicit formula for $\mathcal{D}_{2N}^0(\lambda)$ (see (5.1.3)) shows that the normal order of $D_{2N}^{nc}(\lambda)$ degenerates, i.e., is smaller than the order $2N$ of the family, iff

$$\prod_{k=0}^{N-1} (2\lambda - 4N + 2k + n + 1) = 0,$$

i.e., iff

$$\lambda \in \left\{ -\frac{n-1}{2} + 2N - 1, \ldots, -\frac{n-1}{2} + N \right\}.$$

More precisely, for $\lambda = -\frac{n-1}{2} + N + M$ with $0 \leq M \leq N - 1$, precisely the coefficients $a_0(\lambda), \ldots, a_{N-M-1}(\lambda)$ vanish, i.e., the normal order is $2N - 2(N-M) =$

$2M$. In the most extreme case $M = 0$, the normal order is 0, i.e., for $\lambda = -\frac{n-1}{2} + N$ the family is tangential. In fact, this operator coincides with $P_{2N}(S^{n-1})i^*$. This completes the proof of Theorem 1.4.1/(i). $\square$

The $SO(n)$-equivariance of $D^c_{2N}(\lambda) : C^\infty(S^n) \to C^\infty(S^{n-1})$ implies that the function $D^c_{2N}(\lambda)(1)$ is constant on S^{n-1}, i.e.,

$$Q_{2N}(\lambda) = D^c_{2N}(\lambda)(1) \tag{5.2.29}$$

for a polynomial $Q_{2N}(\lambda)$ of degree $2N$. We determine an explicit formula for this polynomial. It is a special case of the tractor Q-polynomial which generalizes Q-curvature (Definition 6.21.5). The identity (5.2.27) implies

$$Q_{2N}(\lambda) = D^c_{2N}(\lambda)(1) = \Psi_{n-1}^{-\lambda+2N} D^{nc}_{2N}(\lambda)(\Psi_n^\lambda). \tag{5.2.30}$$

We use this identity to prove that

$$Q_{2N}(2N) = (-1)^N (2N)!. \tag{5.2.31}$$

In order to verify (5.2.31), it is enough to determine the value of

$$D^{nc}_{2N}(2N)(\Psi_n^{2N})(x') \quad \text{at} \quad x' = 0$$

(since it does not depend on x'), i.e., to calculate the sum

$$2^{-2N} \sum_{j=0}^N a_j^{(N)}(2N) \Delta^j_{\mathbb{R}^{n-1}} \left(\frac{\partial^2}{\partial x_n^2} \right)^{N-j} \left((1 + |x'|^2 + x_n^2)^{2N} \right)$$

at $(x', x_n) = 0$. We obtain

$$2^{-2N} \sum_{j=0}^N a_j^{(N)}(2N) \sum_{r=0}^{2N} \binom{2N}{r}$$
$$\times \Delta^j_{\mathbb{R}^{n-1}} \left(|x'|^{2r} \right) \big|_{x'=0} \left(\frac{\partial^2}{\partial x_n^2} \right)^{N-j} \left((1 + x_n^2)^{2N-2r} \right) \big|_{x_n=0}. \tag{5.2.32}$$

But

$$\Delta^k_{\mathbb{R}^{n-1}} \left(|x'|^{2r} \right) \big|_{x'=0} = \delta_{rk} r! 2^r \left\{ (n-3+2r) \cdots (n-1) \right\}^{\#}$$

(see (5.1.14)) and

$$\left(\frac{\partial^2}{\partial x_n^2} \right)^k \left((1 + x_n^2)^M \right) \big|_{x_n=0} = \binom{M}{k} (2k)!.$$

Hence (5.2.32) reads

$$\sum_{j=0}^{N} a_j^{(N)}(2N) \binom{2N}{j} 2^{j-2N} j! \binom{2N-j}{N-j} (2N-2j)! \{(n-3+2j)\cdots(n-1)\}^{\#}$$

$$= \sum_{j=0}^{N} a_j^{(N)}(2N) 2^{j-2N} \frac{(2N)!(2N-2j)!}{N!(N-j)!} \{(n-3+2j)\cdots(n-1)\}^{\#}.$$

Now using Corollary 5.1.1, it follows that the last expression coincides with

$$(-1)^N 2^{-N}(2N)! \sum_{j=0}^{N} \frac{(-1)^j}{j!(N-j)!} \frac{\{(n-1+2N)\cdots(n-1)\}^{\#}}{n-1+2j}.$$

Therefore, for the proof of (5.2.31) it only remains to verify that the latter sum is equal to 2^N. However, this follows from the identity

$$\sum_{j=0}^{N} \frac{(-1)^j}{j!(N-j)!} \frac{\{x\cdots(x+2N)\}^{\#}}{x+2j} = 2^N. \tag{5.2.33}$$

In order to prove (5.2.33), we notice that the left-hand side is a polynomial $q_N(x)$ of degree N in x. But

$$q_N(-2r) = \frac{(-1)^r}{r!(N-r)!} \{(-2r)\cdots(-2)\}^{\#} \{2\cdots(-2r+2N)\}^{\#} = 2^N$$

for $0 \le r \le N$. Hence $q_N(x) = 2^N$ and the proof of (5.2.31) is complete.

In addition to the identity (5.2.31) we have a series of recursive relations for the polynomials $Q_{2N}(\lambda)$. These follow from the identities (5.2.24) and (5.2.25). We get

$$Q_{2N}\left(-\frac{n}{2}+j\right) = Q_{2N-2j}\left(-\frac{n}{2}-j\right) P_{2j}(S^n)(1), \tag{5.2.34}$$

$$Q_{2N}\left(2N-j-\frac{n-1}{2}\right) = P_{2j}(S^{n-1})(1)Q_{2N-2j}\left(2N-j-\frac{n-1}{2}\right), \tag{5.2.35}$$

for $j = 1, \ldots, N$. Note that

$$P_{2j}(S^n)(1) = \prod_{r=\frac{n}{2}}^{\frac{n}{2}+j-1} (-r)(n-1-r) \quad \text{(for all } n\text{)}. \tag{5.2.36}$$

The following lemma gives the unique polynomials which satisfy (5.2.31), (5.2.34) and (5.2.35).

Lemma 5.2.11. $Q_{2N}(\lambda) = (-1)^N \lambda(\lambda-1)\ldots(\lambda-2N+1)$.

Proof. (5.2.31) is obviously satisfied. In order to prove that the given polynomial satisfies the identities (5.2.34) and (5.2.35) for $j = 1, \ldots, N$, we use induction over N. Moreover, we use the formula

$$P_{2N}(S^n)(1) = \prod_{r=\frac{n}{2}}^{\frac{n}{2}+N-1} (-r(n-1-r)) = (-1)^N \left(-\frac{n}{2}-N+1\right) \cdots \left(-\frac{n}{2}+N\right).$$

Assuming that $Q_{2M}(\lambda) = (-1)^M (\lambda-1) \cdots (\lambda-2M+1)$ for $1 \leq M \leq N-1$, we obtain for the right-hand side of (5.2.34)

$$\left((-1)^{N-j} \left(-\frac{n}{2}-j\right) \left(-\frac{n}{2}-j-1\right) \cdots \left(-\frac{n}{2}+j-2N+1\right)\right)$$
$$\times \left((-1)^j \left(-\frac{n}{2}+j\right) \cdots \left(-\frac{n}{2}-j+1\right)\right) = (-1)^N \left(-\frac{n}{2}+j\right) \cdots \left(-\frac{n}{2}+j-2N+1\right)$$
$$= Q_{2N} \left(-\frac{n}{2}+j\right).$$

This proves (5.2.34). The proof of (5.2.35) is analogous. $\qquad\square$

5.2.3   Odd order families: $D_{2N+1}^{nc}(\lambda)$ and $D_{2N+1}^{c}(\lambda)$

In the present section, we extend the results of Section 5.2.2 to the odd order families $D_{2N+1}^{nc}(\lambda)$ and $D_{2N+1}^{c}(\lambda)$ which are induced by the families $\mathcal{D}_{2N+1}^{0}(\lambda)$.

Theorem 5.2.4 (Factorization). *For $j = 1, \ldots, N$, the operators $D_{2N+1}^{c}(-\frac{n}{2}+j)$ and $D_{2N+1}^{c}(2N+1-j-\frac{n-1}{2})$ factorize as equivariant maps*

$$C^\infty(S^n)_{-\frac{n}{2}+j} \to C^\infty(S^n)_{-\frac{n}{2}-j} \to C^\infty(S^{n-1})_{-\frac{n}{2}+j-2N-1},$$
$$C^\infty(S^n)_{-\frac{n-1}{2}+2N+1-j} \to C^\infty(S^{n-1})_{-\frac{n-1}{2}+j} \to C^\infty(S^{n-1})_{-\frac{n-1}{2}-j}$$

in the form

$$D_{2N+1}^{c} \left(-\frac{n}{2}+j\right) = D_{2N-2j+1}^{c} \left(-\frac{n}{2}-j\right) \circ P_{2j}(S^n),$$
$$D_{2N+1}^{c} \left(2N+1-j-\frac{n-1}{2}\right) = P_{2j}(S^{n-1}) \circ D_{2N+1-2j}^{c} \left(2N+1-j-\frac{n-1}{2}\right).$$

There are analogous factorizations for the non-compact models $D_{2N+1}^{nc}(\lambda)$.

Proof. For the proof of the factorization identities for $D_{2N+1}^{nc}(\lambda)$, it suffices to prove the $2N$ identities

$$\mathcal{D}_{2N+1}^{0} \left(-\frac{n}{2}+j\right) = \mathcal{D}_{2N+1-2j}^{0} \left(-\frac{n}{2}-j\right) (\Delta_n^-)^j, \qquad (5.2.37)$$
$$\mathcal{D}_{2N+1}^{0} \left(2N+1-j-\frac{n-1}{2}\right) = (\Delta_{n-1}^-)^j \mathcal{D}_{2N+1-2j}^{0} \left(2N+1-j-\frac{n-1}{2}\right) \qquad (5.2.38)$$

in $\mathcal{U}(\mathfrak{n}_{n+1}^-)$. For the proof of the first set of identities we note that the right-hand sides of (5.2.37) is of the form

$$\left(\sum_{r=0}^{N-j} b_r (\Delta_{n-1}^-)^r (Y_n^-)^{2N+1-2j-2r}\right) (\Delta_{n-1}^- + (Y_n^-)^2) = \sum_{s=0}^{N} d_s (\Delta_{n-1}^-)^s (Y_n^-)^{2N+1-2s}$$

and that the coefficient of $(\Delta_{n-1}^-)^N$ in the product is 1. $\mathcal{D}_{2N+1}^0(-\frac{n}{2}+j)$ has the same properties. Moreover, the product on the right-hand side of (5.2.37) satisfies the same commutator relations as $\mathcal{D}_{2N+1}^0(-\frac{n}{2}+j)$. This can be seen as in the proof of Theorem 5.2.3; we recall that the main point of the argument is that $\mathfrak{m}_{n+1}$ commutes with Δ_n^-. Thus the asserted identity follows from Theorem 5.1.3. For the proof of (5.2.38) we have to compare

$$\sum_{r=1}^{N} b_r^{(N)} \left(2N+1-j-\frac{n-1}{2}\right) (\Delta_{n-1}^-)^r (Y_n^-)^{2N+1-2r}$$

with

$$\sum_{r=1}^{N-j} b_r^{(N-j)} \left(2N+1-j-\frac{n-1}{2}\right) (\Delta_{n-1}^-)^{r+j} (Y_n^-)^{2N+1-2j-2r}.$$

In view of Corollary 5.1.3, that means to compare

$$\sum_{r=0}^{N} \frac{N!}{r!(2N-2r+1)!}(-2)^{N-r} \prod_{k=r}^{N-1} (2k-2j+2) \qquad (5.2.39)$$

with

$$\sum_{r=0}^{N-j} \frac{(N-j)!}{r!(2N-2j-2r+1)!}(-2)^{N-j-r} \prod_{k=r}^{N-j-1} (2k+2j+2). \qquad (5.2.40)$$

Now (5.2.39) coincides with

$$\sum_{r=0}^{N-j} \frac{N!}{(r+j)!(2N-2j-2r+1)!}(-2)^{N-r-j} \prod_{k=r}^{N-1-j} (2k+2).$$

But since

$$\frac{(N-j)!}{r!} \prod_{k=r}^{N-j-1} (2k+2j+2) = \frac{N!}{(r+j)!} \prod_{k=r}^{N-1-j} (2k+2),$$

the proof of (5.2.38) is complete. The assertions for the compact model follow as in the proof of Theorem 5.2.3. $\square$

As for the even order operator families, we now define a *recursive* method which allows us to determine the family $D_{2N+1}^c(\lambda)$ by using the families $D_{2M+1}^c(\lambda)$ for $0 \leq M \leq N-1$. The method rests on the $2N$ factorization identities for the family $D_{2N+1}^c(\lambda)$ (Theorem 5.2.4) and the following result (compare with (5.2.26) and Lemma 5.2.9).

Lemma 5.2.12. *The operator families* $D^c_{2N+1}(\lambda)$ *are polynomials of degree $2N$ of the form*

$$B_{2N}\lambda^{2N} + B_{2N-1}\lambda^{2N-1} + \cdots + B_1\lambda + B_0 \qquad (5.2.41)$$

with $B_{2N} = (-1)^N D^c_1 = (-1)^N i^* \operatorname{grad}(\mathcal{H}_0)$.

Proof. The proof rests on the identity

$$D^c_{2N+1}(\lambda) = \Psi^{-\lambda+2N+1}_{n-1} \circ D^{nc}_{2N+1}(\lambda) \circ \Psi^\lambda_n \qquad (5.2.42)$$

(Lemma 5.2.3) and the following result.

Lemma 5.2.13. *For $a, b \in \mathbb{N}_0$,*

$$\lambda \mapsto \Psi^{-\lambda+2(a+b)+1}_{n-1} \circ \Delta^b_{\mathbb{R}^{n-1}} \circ i^* \circ \left(\frac{\partial}{\partial x_n}\right)^{2a+1} \circ \Psi^\lambda_n$$

is an operator-valued polynomial of degree $2b+a$ with leading term

$$2^{-2a-1}\frac{(2a+1)!}{a!}|x'|^{2b}(1+|x'|^2)^{a+1}\left(i^*\frac{\partial}{\partial x_n}\right)\lambda^{2b+a}.$$

Proof. We write the polynomial as the composition

$$\left(\Psi^{-\lambda+2(a+b)+1}_{n-1} \circ \Delta^b_{\mathbb{R}^{n-1}} \circ \Psi^{\lambda-2a}_{n-1}\right) \circ \left(\Psi^{-\lambda+2a}_{n-1} \circ i^* \circ \left(\frac{\partial}{\partial x_n}\right)^{2a+1} \circ \Psi^\lambda_n\right).$$

Similarly as in the proof of Lemma 5.2.10, it can be shown that the second factor is a polynomial with leading term

$$2^{-2a}\frac{(2a+1)!}{a!}(1+|x'|^2)^a\left(i^*\frac{\partial}{\partial x_n}\right)\lambda^a.$$

From here the assertion follows easily. $\qquad\square$

Now Lemma 5.2.13 and the definition

$$D^{nc}_{2N+1}(\lambda) = \sum_{j=0}^{N} b^{(N)}_j(\lambda)\Delta^j_{\mathbb{R}^{n-1}} \circ i^* \circ \left(\frac{\partial}{\partial x_n}\right)^{2N+1-2j}$$

with polynomial coefficients $b^{(N)}_j(\lambda)$ of degree $N-j$ (see (5.1.24)) imply that the leading term of $D^c_{2N+1}(\lambda)$ is given by the composition of the sum

$$\sum_{j=0}^{N}\left[2^{2N-2j}(-1)^{N-j}\frac{N!}{j!(2N-2j+1)!}\right]$$

$$\times\, 2^{-2(N-j)-1}\frac{(2N-2j+1)!}{(N-j)!}|x'|^{2j}(1+|x'|^2)^{N-j+1}\lambda^{N-j}\lambda^{2j+(N-j)}$$

with $i^* \partial/\partial x_n$. Since the latter sum simplifies to

$$\frac{1}{2}\sum_{j=0}^{N}(-1)^{N-j}\binom{N}{j}|x'|^{2j}(1+|x'|^2)^{N-j+1}\lambda^{2N} = (-1)^N\frac{1}{2}(1+|x'|^2)\lambda^{2N},$$

it is enough to notice that

$$i^*\operatorname{grad}(\mathcal{H}_0) = \frac{1}{2}(1+|x'|^2)\frac{\partial}{\partial x_n}\Big|_{x_n=0}$$

(see (2.2.4)) to complete the proof of Lemma 5.2.12. $\square$

We complete the

Proof of Theorem 1.4.1. The proof of Theorem 1.4.1/(ii) is analogous to that of part (i) on page 145. Lemma 5.2.12 shows that $D^c_{2N+1}(\lambda)$ is of the form

$$(-1)^N i^*\nabla_N\lambda^{2N} + B_{2N-1}\lambda^{2N-1} + \cdots + B_1\lambda + B_0.$$

By the recursive method, the $2N$ coefficients $B_{2N-1},\ldots,B_0$ are determined by $2N$ factorization identities (Theorem 5.2.4). More precisely, these are given as compositions of the operators $P_2(S^m),\ldots,P_{2N}(S^m)$ for $m = n$ and $m = n-1$ and the values

$$D^c_{2N+1-2j}\left(-\frac{n}{2}-j\right),\qquad D^c_{2N+1-2j}\left(2N+1-j-\frac{n-1}{2}\right)$$

for $j = 1,\ldots,N$ of lower order families. This proves the assertion in Theorem 1.4.1/(ii) concerning the structure of the coefficients B_j.

We determine the normal order of $D^c_{2N+1}(\lambda)$. In order to calculate the normal order, we use the non-compact model $D^{nc}_{2N+1}(\lambda)$. By the explicit formula for $D^0_{2N+1}(\lambda)$ (see (5.1.23)), the normal order of $D^{nc}_{2N+1}(\lambda)$ degenerates iff

$$\prod_{k=0}^{N-1}(2\lambda-4N+2k+n-1) = 0,$$

i.e., iff

$$\lambda \in \left\{-\frac{n-1}{2}+2N,\ldots,-\frac{n-1}{2}+N+1\right\}.$$

More precisely, for $\lambda = -\frac{n-1}{2}+N+M$ with $1 \le M \le N$ precisely the coefficients $b_0(\lambda),\ldots,b_{N-M}(\lambda)$ vanish, i.e., the normal order is $2N+1-2(N-M+1) = 2M-1$. In the most extreme case $M = 1$, the normal order of $D^c_{2N+1}(-\frac{n-3}{2}+N)$ is 1. In fact, this operator equals $P_{2N}(S^{n-1})D^c_1$. This completes the proof of Theorem 1.4.1/(ii). $\square$

The K^n-equivariance of $D^c_{2N+1}(\lambda) : C^\infty(S^n) \to C^\infty(S^{n-1})$ and the K^n-invariance of $\mathcal{H}_0$ imply that $D^c_{2N+1}(\lambda)(\mathcal{H}_0) \in C^\infty(S^{n-1})$ is $SO(n)$-invariant, i.e., a constant function. We define

$$Q_{2N+1}(\lambda) = D^c_{2N+1}(\lambda)(\mathcal{H}_0). \tag{5.2.43}$$

Note that since $D^c_1(\mathcal{H}_0) = 1$ the leading coefficient of $Q_{2N+1}(\lambda)$ coincides with the coefficient of D^c_1 in B_{2N}. We continue with the determination of the degree $2N$ polynomial $Q_{2N+1}(\lambda)$.

Lemma 5.2.14. $Q_{2N+1}(\lambda) = (-1)^N(\lambda-1)\cdots(\lambda-2N)$.

In order to prove Lemma 5.2.14, we use induction over N. The idea is to find $2N+1$ conditions for the polynomial family $Q_{2N+1}(\lambda)$ which uniquely determine it using the families $Q_{2M+1}(\lambda)$ for $0 \le M \le N-1$. The following lemma provides the first condition.

Lemma 5.2.15. $Q_{2N+1}(2N+1) = (-1)^N(2N)!$.

Proof. We evaluate the right-hand side of the identity

$$Q_{2N+1}(2N+1) = D^c_{2N+1}(2N+1)(\mathcal{H}_0) \qquad \text{(by definition)}$$
$$= 2^{-2N} D^{nc}_{2N+1}(2N+1)((1+|x'|^2+x_n^2)^{2N} x_n)$$

at $x' = 0$. The result coincides with

$$2^{-2N} \sum_{j=0}^{N} b_j^{(N)}(2N+1)\Delta^j_{\mathbb{R}^{n-1}}\left(\frac{\partial}{\partial x_n}\right)^{2N+1-2j} \left((1+|x'|^2+x_n^2)^{2N} x_n\right)\big|_{x'=x_n=0}$$

i.e.,

$$2^{-2N} \sum_{j=0}^{N} b_j^{(N)}(2N+1)$$
$$\times \sum_{r=0}^{2N} \binom{2N}{r} \Delta^j_{\mathbb{R}^{n-1}}\left(|x'|^{2r}\right)\big|_{x'=0} \left(\frac{\partial}{\partial x_n}\right)^{2N+1-2j} \left((1+x_n^2)^{2N-r} x_n\right)\big|_{x_n=0}.$$

But using

$$\Delta^k_{\mathbb{R}^{n-1}}\left(|x'|^{2r}\right)\big|_{x'=0} = \delta_{rk} r! 2^r \left\{(n-3+2r)\cdots(n-1)\right\}^\#$$

and

$$\left(\frac{\partial}{\partial x_n}\right)^{2k+1} \left((1+x_n^2)^M x_n\right)\big|_{x_n=0} = \binom{M}{k}(2k+1)!,$$

we obtain the following formula for $Q_{2N+1}(2N+1)$.

$$2^{-2N} \sum_{j=0}^{N} b_j^{(N)}(2N+1)$$

$$\times \sum_{r=0}^{2N} \binom{2N}{r} \delta_{rj} j! 2^r \{(n-3+2r)\cdots(n-1)\}^{\#} \binom{2N-r}{N-j}(2N-2j+1)!$$

$$= 2^{-2N} \sum_{j=0}^{N} b_j^{(N)}(2N+1) 2^{j-2N} \frac{(2N)!(2N-2j+1)!}{N!(N-j)!} \{(n-3+2r)\cdots(n-1)\}^{\#}.$$

Now we apply Corollary 5.1.3 and (5.2.33). It follows that $Q_{2N+1}(2N+1)$ is given by

$$(-1)^N 2^{-N}(2N)!$$

$$\times \sum_{j=0}^{N} \frac{(-1)^j}{j!(N-j)!} \{\{(n-1)\cdots(n-3+2j)\}^{\#}\{(n+1+2j)\cdots(n-1+2N)\}^{\#}$$

$$= (-1)^N 2^{-N}(2N)! \sum_{j=0}^{N} \frac{(-1)^j}{j!(N-j)!} \frac{\{(n-1)\cdots(n-1+2N)\}^{\#}}{n-1+2j} = (-1)^N(2N)!.$$

The proof is complete. $\square$

Now we apply the identities in Theorem 5.2.4 to the function $\mathcal{H}_0$. We obtain

$$Q_{2N+1}\left(-\frac{n}{2}+j\right) = D^c_{2N+1-2j}\left(-\frac{n}{2}-j\right) P_{2j}(S^n)(\mathcal{H}_0), \qquad (5.2.44)$$

$$Q_{2N+1}\left(2N-j-\frac{n-1}{2}\right) = Q_{2N+1-2j}\left(2N+1-j-\frac{n-1}{2}\right) P_{2j}(S^{n-1})(1) \quad (5.2.45)$$

for $j = 1, \ldots, N$. $\mathcal{H}_0$ is the restriction of the degree 1 harmonic polynomial x_n on $\mathbb{R}^{n+1}$ to S^n. It follows that $\mathcal{H}_0$ is an eigenfunction of Δ_{S^n}:

$$\Delta_{S^n}(\mathcal{H}_0) = -n\mathcal{H}_0. \qquad (5.2.46)$$

Hence

$$P_{2j}(S^n)(\mathcal{H}_0) = \prod_{r=\frac{n}{2}}^{\frac{n}{2}+j-1} (-n-r(n-1-r)) = \prod_{r=\frac{n}{2}}^{\frac{n}{2}+j-1} (r-n)(r+1).$$

Together with (5.2.36) we see that (5.2.44) and (5.2.45) are equivalent to

$$Q_{2N+1}\left(-\frac{n}{2}+j\right) = Q_{2N+1-2j}\left(-\frac{n}{2}-j\right) \prod_{r=\frac{n}{2}}^{\frac{n}{2}+j-1} (r-n)(r+1) \qquad (5.2.47)$$

and

$$Q_{2N+1}\left(2N-j-\frac{n-1}{2}\right) = Q_{2N+1-2j}\left(2N+1-j-\frac{n-1}{2}\right) \prod_{r=\frac{n-1}{2}}^{\frac{n-1}{2}+j-1} (-r)(n-1-r) \tag{5.2.48}$$

for $j = 1, \ldots, N$.

Now we are able to complete the proof of Lemma 5.2.14. We use induction over N to prove

$$Q_{2N+1}(2N+1) = (-1)^N (2N)!, \tag{5.2.49}$$

(5.2.47) and (5.2.48). (5.2.49) is obvious. For the right-hand side of (5.2.47) we obtain

$$(-1)^{N-j}\left\{\left(-\frac{n}{2}-j-1\right)\cdots\left(-\frac{n}{2}+j-2N\right)\right\}$$
$$\times\left\{\left(-\frac{n}{2}\right)\cdots\left(-\frac{n}{2}+j-1\right)\left(\frac{n}{2}+1\right)\cdots\left(\frac{n}{2}+j\right)\right\}$$
$$= (-1)^N\left\{\left(-\frac{n}{2}+j-1\right)\cdots\left(-\frac{n}{2}\right)\right\}$$
$$\times\left\{\left(-\frac{n}{2}-1\right)\cdots\left(-\frac{n}{2}-j\right)\right\}\left\{\left(-\frac{n}{2}-j-1\right)\cdots\left(-\frac{n}{2}+j-2N\right)\right\}$$
$$= (-1)^N\left(-\frac{n}{2}+j-1\right)\cdots\left(-\frac{n}{2}+j-2N\right) = Q_{2N+1}\left(-\frac{n}{2}+j\right).$$

The proof of (5.2.48) is analogous. The proof is complete.   □

5.2.4   Eigenfunctions of $\Delta_{\mathbb{H}^n}$ and the families $D_N^{nc}(\lambda)$

In the present section, we relate the families $\mathcal{D}_N(\lambda)$ to the asymptotics of eigenfunctions of the Laplacian on the upper half-space with respect to the hyperbolic metric of curvature -1. That relation leads to an alternative proof of a system of factorization identities for the families $D_{2N}^{nc}(\lambda)$. We also show how these families appear in the residues of a family of distributions, and that the latter observation implies their equivariance. In Section 6.6, these results will be generalized within a much wider framework.

Let

$$u(x', x_n) \sim \sum_{j \geq 0} a_j(x') x_n^{\lambda+j}, \quad x_n > 0, \ a_{odd} \equiv 0 \tag{5.2.50}$$

be a formal solution of the eigenequation $-\Delta_{\mathbb{H}^n} u = \lambda(n-1-\lambda)u$, where

$$\Delta_{\mathbb{H}^n} = x_n^2 \sum_{i=1}^n \frac{\partial^2}{\partial x_i^2} - (n-2)x_n\frac{\partial}{\partial x_n} = x_n^2 \Delta_{\mathbb{R}^{n-1}} + x_n^2\frac{\partial^2}{\partial x_n^2} - (n-2)x_n\frac{\partial}{\partial x_n}.$$

The coefficients a_j are determined recursively by the first coefficient a_0. More precisely, the maps

$$T_{2j}(\lambda) : a_0(\cdot) \mapsto a_{2j}(\lambda, \cdot)$$

are differential operators (with rational coefficients in λ) of order $2j$ on $\mathbb{R}^{n-1}$ with $T_0(\lambda) = \mathrm{id}$. We use these data to define for any $N \in \mathbb{N}_0$ the family

$$\mathcal{S}_{2N}(\lambda) : C^\infty(\mathbb{R}^n) \to C^\infty(\mathbb{R}^{n-1})$$

by

$$\mathcal{S}_{2N}(\lambda) = \sum_{j=0}^{N} \frac{1}{(2N-2j)!} T_{2j}(\lambda) i^* \left(\partial/\partial x_n\right)^{2N-2j}, \qquad (5.2.51)$$

where $i : \mathbb{R}^{n-1} \hookrightarrow \mathbb{R}^n$, $x' \mapsto (x', 0)$.

On the other hand, we have seen in Section 5.2.1 that $\mathcal{D}_{2N}^0(\lambda)$ induces a family $D_{2N}^{nc}(\lambda) : C^\infty(\mathbb{R}^n) \to C^\infty(\mathbb{R}^{n-1})$. The following result establishes a relation between both families.

Theorem 5.2.5. *The families $\mathcal{S}_{2N}(\lambda+n-1-2N)$ and $D_{2N}^{nc}(\lambda)$ are proportional by a rational function in λ.*

Proof. The series in (5.2.50) corresponds to an eigenfunction iff

$$\sum_{j \geq 0} x_n^2 \left(x_n^{\lambda+j} \Delta_{\mathbb{R}^{n-1}} a_j + x_n^{\lambda+j-2}(\lambda+j)(\lambda+j-1)a_j\right) - (n-2)(\lambda+j)x_n^{\lambda+j} a_j$$

$$= -\lambda(n-1-\lambda) \sum_{j \geq 0} x_n^{\lambda+j} a_j.$$

Comparing the coefficients of $x_n^{\lambda+j}$ we obtain

$$((\lambda+j)(\lambda+j-1) - (\lambda+j)(n-2)) a_j + \Delta_{\mathbb{R}^{n-1}} a_{j-2} = -\lambda(n-1-\lambda)a_j.$$

The latter relations are equivalent to

$$(\lambda+j)(\lambda+j-(n-1))a_j + \Delta_{\mathbb{R}^{n-1}} a_{j-2} = -\lambda(n-1-\lambda)a_j,$$

i.e.,

$$\Delta_{\mathbb{R}^{n-1}} a_{j-2} = j(n-1-2\lambda-j)a_j, \ j \geq 2. \qquad (5.2.52)$$

It follows that we can write $\mathcal{S}_{2N}(\lambda)$ in the form

$$A_0(\lambda)\frac{1}{(2N)!}i^* \left(\frac{\partial^2}{\partial x_n^2}\right)^N + A_2(\lambda)\frac{1}{(2N-2)!}\Delta_{\mathbb{R}^{n-1}} i^* \left(\frac{\partial^2}{\partial x_n^2}\right)^{N-1}$$

$$+ \cdots + A_{2N}(\lambda)\Delta_{\mathbb{R}^{n-1}}^N i^*$$

with coefficients $A_{2j}(\lambda)$ which are determined recursively by

$$2j(n-1-2\lambda-2j)A_{2j}(\lambda) = A_{2j-2}(\lambda), \ A_0(\lambda) = 1. \qquad (5.2.53)$$

Equivalently, we can write $\mathcal{S}_{2N}(\lambda)$ as

$$B_0(\lambda)i^* \left(\frac{\partial^2}{\partial x_n^2}\right)^N + B_2(\lambda)\Delta_{\mathbb{R}^{n-1}}i^* \left(\frac{\partial^2}{\partial x_n^2}\right)^{N-1} + \cdots + B_{2N}(\lambda)\Delta_{\mathbb{R}^{n-1}}^N i^*$$

with coefficients $B_{2j}(\lambda)$ which are determined recursively by

$$\frac{2j(n-1-2\lambda-2j)}{(2N-2j+2)(2N-2j+1)}B_{2j}(\lambda) = B_{2j-2}(\lambda) \quad \text{and} \quad B_0(\lambda) = \frac{1}{(2N)!}. \quad (5.2.54)$$

According to Lemma 5.2.1, the family $D_{2N}^{nc}(\lambda)$ has the form

$$a_0(\lambda)i^* \left(\frac{\partial^2}{\partial x_n^2}\right)^N + a_1(\lambda)\Delta_{\mathbb{R}^{n-1}}i^* \left(\frac{\partial^2}{\partial x_n^2}\right)^{N-1} + \cdots + a_N(\lambda)\Delta_{\mathbb{R}^{n-1}}^N i^*,$$

where the coefficients $a_j(\lambda)$ are determined recursively by

$$a_{j-1}(\lambda) = -\frac{j(n-1+2\lambda-4N+2j)}{(N-j+1)(2N-2j+1)}a_j(\lambda) \quad \text{and} \quad a_N(\lambda) = 1. \quad (5.2.55)$$

But (5.2.54) implies

$$B_{2j}(\lambda+n-1-2N)\frac{j(-(n-1)-2\lambda+4N-2j)}{(N-j+1)(2N-2j+1)} = B_{2j-2}(\lambda+n-1-2N).$$

Comparing these relations with (5.2.55), it follows that the quotients

$$\frac{B_{2j}(\lambda+n-1-2N)}{a_j(\lambda)} \quad (5.2.56)$$

do not depend on j. The proof is complete. $\square$

It is easy to determine the proportionality coefficient in Theorem 5.2.5. In fact, it is enough to find the quotient (5.2.56) for $j = N$. Then $a_N(\lambda) = 1$ and we calculate, using (5.2.54),

$$B_{2N}(\lambda) = \frac{(2N)!}{\{2N\cdots2\}\#}\frac{1}{\{(n-1-2\lambda-2N)\cdots(n-1-2\lambda-2)\}\#}B_0(\lambda)$$

$$= \frac{1}{2^N N!\{(n-1-2\lambda-2N)\cdots(n-3-2\lambda)\}\#}.$$

Here $\#$ indicates that each second term in the corresponding product is deleted. Hence

$$\mathcal{S}_{2N}(\lambda+n-1-2N)$$

$$= \frac{1}{2^N N!\{(-(n-1)-2\lambda+2N)\cdots(-(n-1)-2\lambda+4N-2)\}\#}D_{2N}^{nc}(\lambda), \quad (5.2.57)$$

i.e., $D_{2N}^{nc}(\lambda)$ is of the form

$$\frac{2^N N!}{(2N)!}\{(-(n-1)-2\lambda+2N)\cdots(-(n-1)-2\lambda+4N-2)\}^{\#}i^*\left(\frac{\partial^2}{\partial x_n^2}\right)^N + \cdots + \Delta_{\mathbb{R}^{n-1}}^N i^*.$$

Theorem 5.2.5 extends to *odd order* families. For any $N \in \mathbb{N}_0$, we define the family

$$\mathcal{S}_{2N+1}(\lambda) : C^\infty(\mathbb{R}^n) \to C^\infty(\mathbb{R}^{n-1})$$

by

$$\mathcal{S}_{2N+1}(\lambda) = \sum_{j=0}^{N} \frac{1}{(2N+1-2j)!} T_{2j}(\lambda) i^* (\partial/\partial x_n)^{2N+1-2j}. \tag{5.2.58}$$

The following result is the analog of Theorem 5.2.5.

Theorem 5.2.6. *The families* $\mathcal{S}_{2N+1}(\lambda+n-1-(2N+1))$ *and* $D_{2N+1}^{nc}(\lambda)$ *are proportional by a rational function in* λ.

Proof. By the arguments in the proof of Theorem 5.2.5, we can write $\mathcal{S}_{2N+1}(\lambda)$ in the form

$$A_0(\lambda)\frac{1}{(2N+1)!}i^*\left(\frac{\partial}{\partial x_n}\right)^{2N+1}$$

$$+ A_2(\lambda)\frac{1}{(2N-1)!}\Delta_{\mathbb{R}^{n-1}}i^*\left(\frac{\partial}{\partial x_n}\right)^{2N-1} + \cdots + A_{2N}(\lambda)\Delta_{\mathbb{R}^{n-1}}^N i^*\left(\frac{\partial}{\partial x_n}\right)$$

with coefficients $A_{2j}(\lambda)$, $j = 0, \ldots, N$ which are determined recursively by (5.2.53). Equivalently, we can write $\mathcal{S}_{2N+1}(\lambda)$ as

$$C_0(\lambda)i^*\left(\frac{\partial}{\partial x_n}\right)^{2N+1} + C_2(\lambda)\Delta_{\mathbb{R}^{n-1}}i^*\left(\frac{\partial}{\partial x_n}\right)^{2N-1} + \cdots + C_{2N}(\lambda)\Delta_{\mathbb{R}^{n-1}}^N i^*\left(\frac{\partial}{\partial x_n}\right)$$

with coefficients $C_{2j}(\lambda)$ which are determined recursively by

$$\frac{2j(n-1-2\lambda-2j)}{(2N-2j+3)(2N-2j+2)}C_{2j}(\lambda) = C_{2j-2}(\lambda) \quad \text{and} \quad C_0(\lambda) = \frac{1}{(2N+1)!}. \tag{5.2.59}$$

Comparing (5.2.59) with (5.1.22) gives the desired result. $\qquad\square$

In order to determine the proportionality factor in Theorem 5.2.6, it is enough to find the quotient

$$\frac{C_{2N}(\lambda+(n-1)-(2N+1))}{b_N(\lambda)}$$

with $b_N(\lambda) = 1$. Now (5.2.59) gives

$$C_{2N}(\lambda) = \frac{(2N+1)!}{2^N N!\{(n-1-2\lambda-2N)\cdots(n-3-2\lambda)\}^{\#}}C_0(\lambda).$$

Hence

$$\mathcal{S}_{2N+1}(\lambda+(n-1)-(2N+1))$$
$$= \frac{1}{2^N N!\{(-(n-1)-2\lambda+2(N+1))\cdots(-(n-1)-2\lambda+4N)\}^\#} D_{2N+1}^{nc}(\lambda),$$
$$(5.2.60)$$

i.e., $D_{2N+1}^{nc}(\lambda)$ is of the form

$$\frac{2^N N!}{(2N+1)!}\{(-(n-1)-2\lambda+2(N+1))\cdots(-(n-1)-2\lambda+4N)\}^\#$$
$$\times i^*\,(\partial/\partial x_n)^{2N+1} + \cdots + \Delta_{\mathbb{R}^{n-1}}^N i^*\,(\partial/\partial x_n)\,.$$

The interpretation of $D_N^{nc}(\lambda)$ in terms of eigenfunctions of the Laplacian on the hyperbolic upper half-space allows us to give a conceptually alternative proof of one half of the system of factorization identities for these families. The following result restates the corresponding part of Theorem 5.2.2. For convenience of later use we shift the dimension by 1.

Theorem 5.2.7. *The families $D_{2N}^{nc}(\lambda) : C^\infty(\mathbb{R}^{n+1}) \to C^\infty(\mathbb{R}^n)$ factorize in the form*

$$D_{2N}^{nc}\left(2N-k-\frac{n}{2}\right) = \Delta_{\mathbb{R}^n}^k \circ D_{2N-2k}^{nc}\left(2N-k-\frac{n}{2}\right)$$

for $k = 0,\ldots,N$.

Proof. We recall that

$$D_{2N}^{nc}(\lambda) = \sum_{j=0}^N a_j^{(N)}(\lambda)\Delta_{\mathbb{R}^n}^j i^*\,(\partial/\partial x_{n+1})^{2N-2j}$$

(Lemma 5.2.1), where

$$a_j^{(N)}(\lambda) = \frac{N!}{j!(2N-2j)!}(-2)^{N-j}\prod_{k=j}^{N-1}(2\lambda-4N+2k+n+2).$$

Since

$$a_0^{(N)}\left(2N-k-\frac{n}{2}\right) = \cdots = a_{k-1}^{(N)}\left(2N-k-\frac{n}{2}\right) = 0,$$

it is enough to prove that

$$a_j^{(N)}\left(2N-k-\frac{n}{2}\right) = a_{j-k}^{(N-k)}\left(2N-k-\frac{n}{2}\right)\Delta_{\mathbb{R}^n}^k$$

for all $j = k,\ldots,N$. By Theorem 5.2.5, this is equivalent to the proportionality of the operators

$$\mathrm{Res}_{\frac{n}{2}-k}(T_{2j}(\cdot)) \quad \text{and} \quad T_{2j-2k}\left(\frac{n}{2}+k\right)\Delta_{\mathbb{R}^n}^k$$

for all $j = k, \ldots, N$. But for $\lambda = \frac{n}{2} + k$, the two ladders in the asymptotics

$$u \sim \sum_{j \geq 0} T_{2j}(\lambda)(a_0)x_{n+1}^{\lambda+2j} + \sum_{j \geq 0} T_{2j}(n-\lambda)(b_0)x_{n+1}^{n-\lambda+2j}, \quad x_{n+1} \to 0$$

overlap, and we find the relation

$$T_{2j-2k}\left(\frac{n}{2}+k\right)\operatorname{Res}_{\frac{n}{2}+k}(\mathcal{S}(\cdot)) = \operatorname{Res}_{\frac{n}{2}-k}(T_{2j}(\cdot))$$

for all $j \geq k$, where $a_0 = \mathcal{S}(\lambda)b_0$ with $\mathcal{S}(\lambda) = I_\lambda/c(\lambda)$ (see Section 3.2). Here the residue $\operatorname{Res}_{\frac{n}{2}+k}(\mathcal{S})$ of the scattering operator is a multiple of $\Delta_{\mathbb{R}^n}^k = P_{2k}(\mathbb{R}^n)$. This completes the proof. $\qquad\square$

We close the present section with the outline of an alternative argument for the equivariance of the families $\mathcal{S}_N(\lambda)$ which are defined by (5.2.51) and (5.2.58). We first prove that, up to a meromorphic coefficient, $\mathcal{S}_N(\mu)$ coincides with the operator $\mathcal{R}_N(\mu)$ given by

$$\operatorname{Res}_{\lambda=-\mu-1-N}\left(\int_{\mathbb{R}^n_+} x_n^\lambda u\varphi dx\right) = \int_{\mathbb{R}^{n-1}} \mathcal{R}_N(\mu)(\varphi)\omega d\zeta,$$

where the eigenfunction u is the Poisson transformation (see Section 2.3)

$$u(x) = \mathcal{P}_{n-1-\mu}(\omega)(x) = \int_{\mathbb{R}^{n-1}} P(x,\zeta)^{n-1-\mu}\omega(\zeta)d\zeta$$

of $\omega \in C_0^\infty(\mathbb{R}^{n-1})$. Then the equivariance of $\mathcal{S}_N(\lambda)$ will be a direct consequence. In order to determine the operator $\mathcal{R}_N(\mu)$, we use repeated partial integration. Assume that $\Re(\mu) = \frac{n-1}{2}$, $\mu \neq \frac{n-1}{2}$. Then for $\Re(\lambda) > -\frac{n-1}{2}$,

$$\int_{\mathbb{R}^n_+} x_n^\lambda u\varphi dx = \int_0^\infty \int_{\mathbb{R}^{n-1}} x_n^\lambda u(x',x_n)\varphi(x',x_n)dx'dx_n$$

$$= \int_0^\infty \int_{\mathbb{R}^{n-1}} \sum_{j \geq 0} x_n^{\lambda+\mu+2j} T_{2j}(\mu)(a_0)(x')\varphi(x',x_n)dx'dx_n$$

up to an analogous integral with powers $x_n^{\lambda+n-1-\mu+2j}$; in the following we ignore the second integral. It is enough to establish the meromorphic continuation of the integrals

$$\int_0^\infty x_n^{\lambda+\mu+2j}\varphi(x',x_n)dx_n, \quad \Re(\lambda) > -\frac{n-1}{2}$$

in λ. The latter integral is holomorphic on $\Re(\lambda) > -\Re(\mu)-1-2j$. We integrate $M \geq 1$ times by parts and find

$$\int_0^\infty x_n^{\lambda+\mu+2j}\varphi(x',x_n)dx'dx_n$$

$$= \frac{(-1)^M}{(\lambda+\mu+2j+1)\cdots(\lambda+\mu+2j+M)} \int_0^\infty x_n^{\lambda+\mu+2j+M}\varphi^{(M)}(x',x_n)dx'dx_n,$$

where $\varphi^{(M)}$ denotes the Mth derivative in the variable x_n. Hence the integral has a meromorphic continuation to $\Re(\lambda) > -\Re(\mu) - 2j - M - 1$ with simple poles at $\lambda = -\mu - 2j - 1, \ldots, \lambda = -\mu - 2j - M$ and

$$\mathrm{Res}_{\lambda = -\mu - 2j - M} \left(\int_0^\infty \int_{\mathbb{R}^{n-1}} x_n^{\lambda + \mu + 2j} T_{2j}(\mu)(a_0)(x')\varphi(x', x_n) dx' dx_n \right)$$
$$= \frac{1}{(M-1)!} \int_{\mathbb{R}^{n-1}} T_{2j}(\mu)(a_0)(x')\varphi^{(M-1)}(x', 0) dx'.$$

Hence we find the residues

$$\mathrm{Res}_{\lambda = -\mu - 1 - L} \left(\int_{\mathbb{R}^n_+} x_n^\lambda u\varphi dx \right) = \sum_{j=0}^{\left[\frac{L}{2}\right]} \frac{1}{(L-2j)!} \int_{\mathbb{R}^{n-1}} T_{2j}(\mu)(a_0)\varphi^{(L-2j)}(x', 0) dx'$$

for $L \geq 0$. In other words,

$$\mathcal{R}_L(\mu)(\varphi)(x') = c(n-1-\mu) \sum_{j=0}^{\left[\frac{L}{2}\right]} \frac{1}{(L-2j)!} T_{2j}^*(\mu) \left(\frac{\partial}{\partial x_n} \right)^{L-2j} (\varphi)(x', 0).$$

Now (5.2.51) and (5.2.58) follow from the self-adjointness of T_{2j}^*. Here we have used that the leading coefficient a_0 and ω are related by

$$a_0 = c(n-1-\mu)\omega, \quad c(\lambda) = \pi^{\frac{n-1}{2}} \frac{\Gamma(\lambda - \frac{n-1}{2})}{\Gamma(\lambda)}$$

since

$$\int_{\mathbb{R}^{n-1}} \frac{x_n^\lambda}{(|x' - \zeta|^2 + x_n^2)^\lambda} d\zeta = x_n^{n-1-\lambda} c(\lambda).$$

Now the equivariance of the family $\mathcal{R}_N(\mu)$ follows from the equivariance of the Poisson transformation. The G^n-equivariance of $\mathcal{P}_\lambda$ states that

$$\mathcal{P}_\lambda \circ \pi_{\lambda-(n-1)}^{nc}(g) = g_* \circ \mathcal{P}_\lambda$$

(Corollary 2.3.2). Now observe that $x_n \in C^\infty(\mathbb{R}^n)_1^{G^n}$. In fact,

$$\pi_1^{nc}(g)(x_n) = \left(\frac{g_*(\mathrm{vol}(g_c))}{\mathrm{vol}(g_c)} \right)^{-\frac{1}{n}} g_*(x_n) = \left(\frac{g_*(\mathrm{vol}(x_n^{-2}g_c))}{\mathrm{vol}(x_n^{-2}g_c)} \right)^{-\frac{1}{n}} x_n = x_n.$$

Hence

$$C_0^\infty(\mathbb{R}^{n-1}) \ni \omega \mapsto x_n^\lambda \mathcal{P}_{n-1-\mu}(\omega) \in C^{-\infty}(\mathbb{R}^n) \tag{5.2.61}$$

intertwines $\pi_{-\mu}^{nc}$ on functions on $\mathbb{R}^{n-1}$ with π_λ^{nc} on distributions on $\mathbb{R}^n$. In particular, the residue of the meromorphic continuation of (5.2.61) at $\lambda = -\mu - 1 - N$ intertwines $\pi_{-\mu}^{nc}$ and $\pi_{-\mu-1-N}^{nc}$, i.e.,

$$\mathcal{R}_N^*(\mu) \left(\pi_{-\mu}^{nc}(g)\omega \right) = \pi_{-\mu-1-N}^{nc}(g)\mathcal{R}_N^*(\mu)\omega, \tag{5.2.62}$$

where

$$\langle \mathcal{R}_N^*(\mu)\omega, \varphi \rangle = \langle \omega, \mathcal{R}_N(\mu)\varphi \rangle = \int_{\mathbb{R}^{n-1}} \omega \mathcal{R}_N(\mu)\varphi d\zeta.$$

Therefore, (5.2.62) yields

$$\int_{\mathbb{R}^{n-1}} \omega \pi_{\mu-(n-1)}^{nc}(g^{-1})\mathcal{R}_N(\mu)(\varphi)d\zeta = \int_{\mathbb{R}^n} \omega \mathcal{R}_N(\mu)\left(\pi_{\mu-(n-1)+N}^{nc}(g^{-1})\varphi\right)d\zeta.$$

This proves the equivariance

$$\mathcal{R}_N(\lambda+n-1-N) \circ \pi_\lambda^{nc}(g) = \pi_{\lambda-N}^{nc}(g) \circ \mathcal{R}_N(\lambda+n-1-N), \ g \in G^n. \quad (5.2.63)$$

5.3 Some low order examples

Here we derive explicit formulas for the intertwining families $D_N^c(\lambda) : C^\infty(S^n) \to C^\infty(S^{n-1})$ of order $N \le 4$ in terms of GJMS-operators on S^n and S^{n-1}. These results serve as illustration and as convenient future reference.

The first-order family is induced by $\mathcal{D}_1^0(\lambda) = Y_n^-$, and we find $D_1^c(\lambda) = i^* \operatorname{grad}(\mathcal{H}_0)$. The second-order family is induced by

$$\mathcal{D}_2^0(\lambda) = -(2\lambda+n-3)(Y_n^-)^2 + \Delta_{n-1}^- \in \mathcal{U}(\mathfrak{n}_{n+1}^-). \quad (5.3.1)$$

Lemma 5.3.1.

$$D_2^c(\lambda) = -\lambda^2 i^* + \lambda\left(2P_2(S^{n-1})i^* - 2i^*P_2(S^n) - \left(n-\frac{5}{2}\right)i^*\right)$$
$$+ (n-2)P_2(S^{n-1})i^* - (n-3)i^*P_2(S^n) - \frac{(n-2)(n-3)}{4}i^*,$$

i.e.,

$$D_2^c(\lambda) = -(2\lambda+n-3)i^*P_2(S^n) + (2\lambda+n-2)P_2(S^{n-1})i^*$$
$$- \left(\lambda+\frac{n-2}{2}\right)\left(\lambda+\frac{n-3}{2}\right)i^*. \quad (5.3.2)$$

These formulas are equivalent to

$$D_2^c(\lambda) = -(2\lambda+n-3)i^*\Delta_{S^n} + (2\lambda+n-2)\Delta_{S^{n-1}}i^* - \lambda(\lambda-1)i^*$$
$$= -(2\lambda+n-3)i^*(\operatorname{grad}\mathcal{H}_0)^2 + (\Delta_{S^{n-1}} - \lambda(\lambda-1))\, i^*. \quad (5.3.3)$$

Proof. We determine $D_2^c(\lambda) = -\lambda^2 i^* + a\lambda + b$ (recall that by Lemma 5.2.9 the leading term of $D_2^c(\lambda)$ is $-i^*$) so that the factorization identities

$$D_2^c\left(-\frac{n}{2}+1\right) = i^*P_2(S^n), \quad D_2^c\left(-\frac{n-3}{2}\right) = P_2(S^{n-1})i^* \quad (5.3.4)$$

are satisfied. This yields the first two formulas for $D_2^c(\lambda)$. For the proof of the remaining two formulas we use the identity

$$i^*(\text{grad}(\mathcal{H}_0))^2 = i^*\Delta_{S^n} - \Delta_{S^{n-1}}i^* \qquad (5.3.5)$$

and the formulas

$$P_2(S^n) = \Delta_{S^n} - \frac{n}{2}\left(\frac{n}{2}-1\right), \quad P_2(S^{n-1}) = \Delta_{S^{n-1}} - \frac{n-1}{2}\left(\frac{n-1}{2}-1\right).$$

For the proof of (5.3.5) we use the coordinates

$$\kappa : S^{n-1} \times (0,1) \ni (y,\rho) \mapsto (\sqrt{1-\rho^2}y, \rho) \in S^n$$

on S^n. Then

$$\kappa^*(g_{S^n}) = (1-\rho^2)g_{S^{n-1}} + \frac{1}{1-\rho^2}d\rho^2$$

and the Laplacian on $S^{n-1} \times (0,1)$ is given by

$$\Delta_{S^{n-1}\times(0,1)} = (1-\rho^2)\frac{\partial^2}{\partial\rho^2} - n\rho\frac{\partial}{\partial\rho} + \frac{1}{1-\rho^2}\Delta_{S^{n-1}}. \qquad (5.3.6)$$

Finally, we have the following formula for the gradient of the defining function $\rho = \kappa^*(\mathcal{H}_0)$ with respect to the metric $\kappa^*(g_{S^n})$.

$$\text{grad}(\rho) = (1-\rho^2)\frac{\partial}{\partial\rho}. \qquad (5.3.7)$$

In these terms, (5.3.5) reads

$$i^*\left((1-\rho^2)\frac{\partial^2}{\partial\rho^2} - n\rho\frac{\partial}{\partial\rho} + \frac{1}{1-\rho^2}\Delta_{S^{n-1}}\right) - \Delta_{S^{n-1}}i^* = i^*(\text{grad}(\rho))^2,$$

where now $i : y \mapsto (y,0)$. But the latter identity is equivalent to

$$i^*\frac{\partial^2}{\partial\rho^2} = i^*(\text{grad}(\rho))^2, \qquad (5.3.8)$$

which is a direct consequence of (5.3.7). The proof is complete. $\qquad\square$

The proof of Lemma 5.3.1 rests on the relations (5.3.4). Next, we illustrate how the residue method (see Section 1.5) works in that special case. The calculation of the residue of the pole of $M_u(\lambda)$ at $\lambda = -\mu - 3$ gives the distribution

$$\frac{1}{2!}\frac{\partial^2}{\partial\rho^2}i_*a_0(f) + i_*\left(a_2(f) - \frac{n-2}{2}a_0(f)\right)$$

(we ignore the contributions of the coefficients b_N). Here $a_2(f)$ is determined by $a_0(f)$ via

$$2(2\mu-(n-3))a_2(f) = \mu(\mu+1)a_0(f) - \Delta_{S^{n-1}}a_0(f)$$

(see (1.5.1)), i.e.,

$$a_2(f) = \frac{\mu(\mu+1)}{2(2\mu-(n-3))}a_0(f) - \frac{1}{2(2\mu-(n-3))}\Delta_{S^{n-1}}a_0(f).$$

Hence we arrive at the operator

$$\frac{1}{2}i^*\frac{\partial^2}{\partial\rho^2} + \frac{\mu(\mu+1)}{2(2\mu-(n-3))}i^* - \frac{1}{2(2\mu-(n-3))}\Delta_{S^{n-1}}i^* - \frac{n-2}{2}i^*$$

(compare with the definition of $\mathcal{S}_2(\mu)$). In order to find $D_2^c(\mu)$, we have to replace μ by $\mu+n-3$ and to renormalize. We obtain

$$(2\mu+n-3)i^*\frac{\partial^2}{\partial\rho^2} + [(\mu+n-3)(\mu+n-2) - (n-2)(2\mu+n-3)]\,i^* - \Delta_{S^{n-1}}i^*$$

$$= (2\mu+n-3)i^*\frac{\partial^2}{\partial\rho^2} - [\Delta_{S^{n-1}} - \mu(\mu-1)]\,i^*.$$

Now (5.3.8) shows that this operator coincides with $-D_2^c(\mu)$.

We continue with the discussion of the third-order family. It is induced by

$$\mathcal{D}_3^0(\lambda) = -\frac{1}{3}(2\lambda+n-5)(Y_n^-)^3 + \Delta_{n-1}^- Y_n^- \in \mathcal{U}(\mathfrak{n}_{n+1}^-).$$

Now $D_3^c(\lambda)$ satisfies the factorization identities

$$D_3^c\left(-\frac{n}{2}+1\right) = D_1^c P_2(S^n), \quad D_3^c\left(-\frac{n-5}{2}\right) = P_2(S^{n-1})D_1^c.$$

Similarly as for $D_2^c(\lambda)$, we apply these two identities to find an explicit formula for $D_3^c(\lambda)$ by using that it has the form $D_3^c(\lambda) = -\lambda^2 D_1^c + \lambda A + B$ with coefficients A and B to be determined (Lemma 5.2.12). We find

Lemma 5.3.2.

$$D_3^c(\lambda) = -\lambda^2 D_1^c + \lambda\left(\frac{2}{3}\left(P_2(S^{n-1})D_1^c - D_1^c P_2(S^n)\right) - \left(n-\frac{7}{2}\right)D_1^c\right)$$

$$+ \left(-\frac{n-5}{3}D_1^c P_2(S^n) + \frac{n-2}{3}P_2(S^{n-1})D_1^c - \frac{(n-2)(n-5)}{4}D_1^c\right),$$

i.e.,

$$D_3^c(\lambda) = -\frac{1}{3}(2\lambda+n-5)D_1^c P_2(S^n) + \frac{1}{3}(2\lambda+n-2)P_2(S^{n-1})D_1^c$$

$$- \left(\lambda+\frac{n-5}{2}\right)\left(\lambda+\frac{n-2}{2}\right)D_1^c. \quad (5.3.9)$$

Similarly, we get

Lemma 5.3.3.

$$D_4^c(\lambda) = (2\lambda+n-5)(2\lambda+n-7)i^* \left(\frac{1}{3}P_4(S^n) + \frac{1}{2}(2\lambda+n-4)P_2(S^n) \right)$$

$$+ (2\lambda+n-2)(2\lambda+n-4) \left(\frac{1}{3}P_4(S^{n-1}) - \frac{1}{2}(2\lambda+n-5)P_2(S^{n-1}) \right) i^*$$

$$- \frac{2}{3}(2\lambda+n-4)(2\lambda+n-5)P_2(S^{n-1})i^* P_2(S^n)$$

$$+ \frac{1}{16}(2\lambda+n-2)(2\lambda+n-4)(2\lambda+n-5)(2\lambda+n-7)i^*.$$

The zeros of the four monomials $(2\lambda + n - 2)$, $(2\lambda + n - 4)$, $(2\lambda + n - 5)$, $(2\lambda + n - 7)$ coincide with the special values of λ for which the family satisfies a factorization identity.

Proof. The family is of the form $\lambda^4 i^* + a_3\lambda^2 + a_2\lambda^2 + a_1\lambda + a_0$ and satisfies the characterizing identities

$$D_4^c \left(-\frac{n}{2}+2 \right) = i^* P_4(S^n), \quad D_4^c \left(-\frac{n}{2}+1 \right) = D_2^c \left(-\frac{n}{2}-1 \right) P_2(S^n)$$

and

$$D_4^c \left(2-\frac{n-1}{2} \right) = P_4(S^{n-1})i^*, \quad D_4^c \left(3-\frac{n-1}{2} \right) = P_2(S^{n-1})D_2^c \left(3-\frac{n-1}{2} \right).$$

It suffices to prove that the given family satisfies these identities. The respective first two identities are obvious. We prove the factorization identity for $\lambda = -\frac{n}{2}+1$. The definition of $D_4^c(\lambda)$ and the explicit formulas (see (3.2.25))

$$P_4(S^n) = P_2(S^n) \left(\Delta_{S^n} - \left(\frac{n}{2}+1 \right) \left(\frac{n}{2}-2 \right) \right), \quad P_2(S^n) = \Delta_{S^n} - \frac{n}{2} \left(\frac{n}{2}-1 \right)$$

imply

$$D_4^c \left(-\frac{n}{2}+1 \right) = 5P_4(S^n) - 15P_2(S^n) - 4P_2(S^{n-1})P_2(S^n)$$

$$= \left(5 \left(\Delta_{S^n} - \left(\frac{n}{2}+1 \right) \left(\frac{n}{2}-2 \right) \right) - 4P_2(S^{n-1}) - 15 \right) P_2(S^n)$$

$$= \left(5P_2(S^n) - 4P_2(S^{n-1}) - 5 \right) P_2(S^n)$$

$$= D_2^c \left(-\frac{n}{2}-1 \right) P_2(S^n);$$

here we omit i^*. For the latter identity we have used Lemma 5.3.1. The proof of the fourth identity is symmetric to the latter argument. In fact, we have

$$D_4^c\left(3 - \frac{n-1}{2}\right) = 5P_4(S^{n-1}) - 15P_2(S^{n-1}) - 4P_2(S^{n-1})P_2(S^n)$$

$$= P_2(S^{n-1})\left(5\left(\Delta_{S^{n-1}} - \left(\frac{n+1}{2}\right)\left(\frac{n-5}{2}\right)\right) - 4P_2(S^n) - 15\right)$$

$$= P_2(S^{n-1})\left(-4P_2(S^n) + 5P_2(S^{n-1}) - 5\right)$$

$$= P_2(S^{n-1})D_2^c\left(3 - \frac{n-1}{2}\right).$$

The proof is complete. $\qquad\qquad\qquad\qquad\qquad\qquad\qquad\qquad\qquad\qquad\qquad\square$

5.4 Families for $(\mathbb{R}^n, S^{n-1})$

In the present section, we take a closer look at the intertwining families

$$D_N^b(\lambda) = D_N(\mathbb{R}^n, S^{n-1}; \lambda) : C^\infty(\mathbb{R}^n) \to C^\infty(S^{n-1})$$

which are induced by $\mathcal{D}_N(\lambda)$. The fact that they are induced by homomorphisms of Verma modules implies that they are special cases of tractor families (Section 6.21). Therefore, they serve as an illustration of the general theory of tractor families. In particular, the examples $D_N^b(\lambda)$ demonstrate that, in general, tractor families are *not* determined by the factorization identities they satisfy. In Section 5.4.1, we describe the induced families $D_N^b(\lambda)$ in terms of the asymptotics of eigenfunctions of the Poincaré metric on the unit ball. From the perspective of Chapter 6 that relation is interpreted as follows. The residue construction in Section 5.4.1 is a special case of the general construction in Section 6.5. The families constructed here are conjugate to the residue families $D_N^{\mathrm{res}}(\lambda)$ (defined in Section 6.6) for the round metric on S^{n-1}. Both arise from two different conformal compactifications of the Poincaré metric on the unit ball (see also the discussion at the end of Section 6.5 and Example 6.6.2). The relation between two types of constructions (via induction and via Poincaré metric) is a prototype of general results concerning the relation between residue families and tractor families (Section 6.21). Here we use that relation to determine $D_N^b(\lambda)$ for $N \leq 3$. Last not least, the family $D_3^b(\lambda)$ (for $n = 4$) deserves special interest. In Section 5.4.3, we apply an explicit formula for that family to relate its value at $\lambda = 0$ to an operator of Chang and Qing.

5.4.1 The families $D_N^b(\lambda)$

The families $D_N^b(\lambda) : C^\infty(\mathbb{R}^n) \to C^\infty(S^{n-1})$ are equivariant with respect to principal series representations of the subgroup $G_0^n = \sigma_0 G^n \sigma_0^{-1} \subset G^{n+1}$ (acting on $\mathbb{R}^n$) which leaves the sphere S^{n-1} invariant. More precisely,

$$\pi_{\lambda-N}(S^{n-1}, g_c)(g) \circ D_N^b(\lambda) = D_N^b(\lambda) \circ \pi_\lambda(\mathbb{R}^n, g_c)(g)$$

for all $g \in G_0^n$ (see Lemma 5.2.6). In order to derive explicit formulas for the families $D_N^b(\lambda)$, we identify them as residues. For that purpose, we regard the families as boundary operators $C^\infty(\overline{\mathbb{B}^n}) \to C^\infty(S^{n-1})$ on the unit ball $\mathbb{B}^n = \{x \in \mathbb{R}^n \,|\, |x| < 1\}$. We consider $\mathbb{B}^n$ as a hyperbolic manifold with the metric

$$\frac{4}{(1-|x|^2)^2}(dx_1^2 + \cdots + dx_n^2).$$

Let $\Delta_{\mathbb{B}^n}$ be the associated Laplacian and $u \in C^\infty(\mathbb{B}^n)$ an eigenfunction

$$-\Delta_{\mathbb{B}^n} u = \mu(n-1-\mu)u,$$

where $\Re(\mu) = (n-1)/2$. We associate to u a family of measures on $\mathbb{R}^n$ by

$$\langle M_u(\lambda), \varphi \rangle = \int_{\mathbb{B}^n} (1-|x|^2)^\lambda u(x)\varphi(x)dx = \int_{\mathbb{R}^n} (1-|x|^2)^\lambda u(x)\varphi(x)dx$$

for test functions $\varphi \in C_0^\infty(\mathbb{R}^n)$. In the second integral, u denotes the extension of u by 0 in the complement of $\mathbb{B}^n$. The integral converges if $\Re(\lambda) > -\frac{n-1}{2}$. We construct a meromorphic continuation of the family $M_u(\lambda)$. For test functions φ with compact support in $\mathbb{B}^n$, the family $\langle M_u(\lambda), \varphi \rangle$ is holomorphic on $\mathbb{C}$. Thus in order to see the poles it is enough to choose φ with support in a neighborhood of the boundary. The existence of the continuation follows from the existence of an asymptotic expansion of the eigenfunction u. We assume that

$$u(x) = \int_{S^{n-1}} e^{(n-1-\mu)\langle x,b \rangle} \omega(b)db$$

for some $\omega \in C^\infty(S^{n-1})$; here the distance $\langle x, b \rangle$ is defined using the metric of curvature -1 ([140]). In order to write the asymptotic expansion of u, we use polar coordinates $(0,1) \times S^{n-1} \to \mathbb{B}^n$. In these coordinates, the metric is given by

$$\frac{4}{(1-r^2)^2}(dr^2 + r^2 g_c)$$

with the round metric g_c on S^{n-1}, and for the hyperbolic Laplacian we have the formula

$$4\Delta_{\mathbb{B}^n} = (1-r^2)^2 \left(\frac{\partial^2}{\partial r^2} + \frac{n-1}{r}\frac{\partial}{\partial r} + \frac{1}{r^2}\Delta_{S^{n-1}} \right) + 2(n-2)(1-r^2)r\frac{\partial}{\partial r}.$$

Now we write the asymptotics of u in the form

$$u(r,b) \sim \sum_{N \geq 0} (1-r^2)^{\mu+N} a_N(\mu,b) + \sum_{N \geq 0} (1-r^2)^{n-1-\mu+N} b_N(\mu,b). \qquad (5.4.1)$$

u is a spherical function if ω is constant. In that case, it can be identified with a hypergeometric function in the argument $|x|^2$. $a_0(\mu,b)$ coincides with $\omega(b)$, up to a coefficient which only depends on μ (c-function).

A routine calculation shows that the coefficients $a_N(\mu, b)$, $N \geq 1$ are determined by $a_0(b)$ via the *recursion* formula

$$4(k+1)(n-2-2\mu-k)a_{k+1}$$
$$= 2(\mu+k)(n-2-2(\mu+k))a_k + \Delta_{S^{n-1}}(a_{k-1} + \cdots + a_0). \quad (5.4.2)$$

A similar relation holds true for $b_N(\mu, b)$, $N \geq 1$.

Now let $N \in \mathbb{N}_0$. We construct an equivariant differential operator of order N in terms of the operators $T_j(\mu) : a_0(\mu, b) \mapsto a_j(\mu, b)$, $j = 0, \ldots, N$. We first rewrite the integral

$$\int_{\mathbb{B}^n} (1-|x|^2)^\lambda u(x)\varphi(x)dx$$

in polar coordinates as

$$\int_0^1 \int_{S^{n-1}} (1-r^2)^\lambda u(r, b)\varphi(r, b)r^{n-1}drdb.$$

In view of (5.4.1), we are led to study the continuation of the function

$$\lambda \mapsto \int_0^1 \int_{S^{n-1}} (1-r^2)^\lambda \left(\sum_{N\geq 0} (1-r^2)^{\mu+N} a_N(\mu, b) \right) r^{n-1}\varphi(r, b)drdb.$$

For convenience we change to the coordinates $(\rho, b) \in (0, 1) \times S^{n-1}$, $\rho = 1 - r^2$. Let ψ be defined by $\varphi(r, b) = \psi(1 - r^2, b)$. Then we rewrite the integral as the sum of

$$\frac{1}{2}\int_0^1 \int_{S^{n-1}} \sum_{N\geq 0} \rho^{\lambda+\mu+N} a_N(\mu, b)(1-\rho)^{\frac{n-2}{2}}\psi(\rho, b)d\rho db$$

$$= \frac{1}{2}\int_0^1 \int_{S^{n-1}} \sum_{N\geq 0} \rho^{\lambda+\mu+N} a_N(\mu, b) \left(\sum_{m\geq 0} \frac{(-\frac{n-2}{2})_m}{m!}\rho^m \right) \psi(\rho, b)d\rho db$$

$$= \frac{1}{2}\int_0^1 \int_{S^{n-1}} \sum_{N\geq 0} \rho^{\lambda+\mu+N} \left(\sum_{j=0}^N \frac{(-\frac{n-2}{2})_{N-j}}{(N-j)!}a_j(\mu, b) \right) \psi(\rho, b)d\rho db$$

$$= \frac{1}{2}\int_0^1 \int_{S^{n-1}} \sum_{N\geq 0} \rho^{\lambda+\mu+N} \left(\sum_{j=0}^N \frac{(-\frac{n-2}{2})_{N-j}}{(N-j)!}T_j(\mu)(a_0(\mu, b)) \right) \psi(\rho, b)d\rho db,$$

and a similar term using $b_0(\mu, b)$. We introduce the abbreviation

$$A_N(\mu) = \sum_{j=0}^N \frac{(-\frac{n-2}{2})_{N-j}}{(N-j)!}T_j(\mu).$$

In the integrals

$$\int_0^1 \int_{S^{n-1}} \rho^{\lambda+\mu+N} A_N(\mu)(a_0)\psi(\rho,b)d\rho db$$

we use $M \geq 1$ partial integrations in the variable ρ to get

$$\frac{1}{\lambda+\mu+N+1} \int_{S^{n-1}} A_N(\mu)(a_0) \left[\rho^{\lambda+\mu+N+1}\psi(\rho,b)\right]_0^1 db$$

$$-\frac{1}{(\lambda+\mu+N+1)(\lambda+\mu+N+2)} \int_{S^{n-1}} A_N(\mu)(a_0) \left[\rho^{\lambda+\mu+N+2}\psi'(\rho,b)\right]_0^1 db$$

$$\pm \ldots$$

$$+\frac{(-1)^{M-1}}{(\lambda+\mu+N+1)\cdots(\lambda+\mu+N+M)} \int_{S^{n-1}} A_N(\mu)(a_0) \left[\rho^{\lambda+\mu+N+M}\psi^{(M-1)}(\rho,b)\right]_0^1 db$$

$$+\frac{(-1)^M}{(\lambda+\mu+N+1)\cdots(\lambda+\mu+N+M)} \int_0^1 \int_{S^{n-1}} \rho^{\lambda+\mu+N+M} A_N(\mu)(a_0)\psi^{(M)}(\rho,b)d\rho db$$

$$=\frac{(-1)^M}{(\lambda+\mu+N+1)\cdots(\lambda+\mu+N+M)} \int_0^1 \int_{S^{n-1}} \rho^{\lambda+\mu+N+M} A_N(\mu)(a_0)\psi^{(M)}(\rho,b)d\rho db$$

for $\Re(\lambda) > -\mu-1-N$. We use this formula for meromorphic continuation to $\Re(\lambda) > -\mu-1-N-M$ with simple poles at

$$\lambda = -\mu-1-N, \ldots, -\mu-N-M.$$

The residue at $\lambda = -\mu-N-M$ is given by

$$\frac{-1}{(M-1)!} \int_0^1 \int_{S^{n-1}} A_N(\mu)(a_0)\psi^{(M)}(\rho,b)d\rho db$$

$$=\frac{1}{(M-1)!} \int_{S^{n-1}} A_N(\mu)(a_0)\psi^{(M-1)}(0,b)db.$$

It follows that $\langle M_u(\lambda), \varphi \rangle$ has a meromorphic continuation to $\mathbb{C}$ with simple poles in the ladder $-\mu-1-\mathbb{N}_0$. More precisely, only the integrals which involve $A_0, \ldots, A_P$ contribute to the residue of the pole in $\lambda = -\mu-1-P, P \geq 0$ and

$$\mathrm{Res}_{-\mu-1-P} \left(\langle M_u(\lambda), \varphi \rangle\right)$$

$$=\frac{1}{2} \left\{ \int_{S^{n-1}} A_P(\mu)(a_0)(b)\psi(0,b)db + \cdots + \frac{1}{P!} \int_{S^{n-1}} A_0(\mu)(a_0)(b)\psi^{(P)}(0,b)db \right\}.$$

But since the operators A_j are self-adjoint we obtain

$$\mathrm{Res}_{-\mu-1-N} \left(\langle M_u(\lambda), \varphi \rangle\right) = \sum_{j=0}^N \frac{1}{(N-j)!} \int_{S^{n-1}} a_0(\mu,b) A_j(\mu)\psi^{(N-j)}(0,b)db.$$

In order to formulate the result in a compact form, we introduce an operator family $\mathcal{R}_N(\mu)$ by

$$\psi(\rho, b)$$

$$\mapsto \sum_{j=0}^{N} \frac{1}{(N-j)!} \left(\frac{(-\frac{n-2}{2})_j}{j!} \mathcal{T}_0(\mu) + \cdots + \frac{(-\frac{n-2}{2})_1}{1!} \mathcal{T}_{j-1}(\mu) + \mathcal{T}_j(\mu) \right) i^* \frac{\partial^{N-j}}{\partial \rho^{N-j}} \psi,$$

$$(5.4.3)$$

where i^* restricts functions in ρ to $\rho = 0$. We regard $\mathcal{R}_N(\mu)$ also as an operator family $\mathcal{R}_N(\mu) : C^\infty(\overline{\mathbb{B}^n}) \to C^\infty(S^{n-1})$. The above result can be formulated in the following form.

Lemma 5.4.1. *Let u be the Poisson-transform*

$$\int_{S^{n-1}} e^{(n-1-\mu)\langle x, b \rangle} \omega(b) db$$

of $\omega \in C^\infty(S^{n-1})$. Assume that $\Re(\mu) = (n-1)/2$ and $\mu \neq (n-1)/2$. Then u is an eigenfunction

$$-\Delta_{\mathbb{B}^n} u = \mu(n-1-\mu) u.$$

The family $\langle M_u(\lambda), \varphi \rangle$ has simple poles at $\lambda = -\mu - 1 - N$ for all $N \in \mathbb{N}_0$. The residues are

$$\mathrm{Res}_{-\mu-1-N}\left(\langle M_u(\lambda), \varphi \rangle \right) = \int_{S^{n-1}} a_0(\mu, b) \mathcal{R}_N(\mu)(\psi)(b) db,$$

where a_0 is one of the leading terms in the asymptotics of u.

Next, we relate the family $\mathcal{R}_N(\mu)$ to the induced family $D_N(\mathbb{R}^n, S^{n-1}; \mu)$. The method is as follows. We use σ_0 to establish a connection between the distribution $M_u(\lambda)$ with support in $\overline{\mathbb{B}^n}$ and a corresponding distribution with support in $\overline{\mathbb{R}^n_+}$. The residues of the latter one are induced families. The main identity for that argument is given in the following result.

Lemma 5.4.2.

$$\int_{\mathbb{B}^n} \left(\frac{1 - |x|^2}{2} \right)^\lambda u(x) \varphi(x) dx = \int_{\mathbb{R}^n_+} x_n^\lambda \sigma_0^*(u) \mathcal{I}_{\lambda+n}(\varphi) dx, \qquad (5.4.4)$$

where

$$\mathcal{I}_\lambda = \left(\frac{\sigma_0^*(\mathrm{vol}(\mathbb{R}^n, g_c))}{\mathrm{vol}(\mathbb{R}^n, g_c)} \right)^{\lambda/n} \sigma_0^*, \quad dx = \mathrm{vol}(\mathbb{R}^n, g_c).$$

Proof. By the identity (2.2.10), the left-hand side equals

$$\int_{\mathbb{B}^n} (\sigma_0)_*(x_n^\lambda) \left(\frac{dx}{(\sigma_0)_*(dx)} \right)^{\lambda/n} u\varphi dx = \int_{\mathbb{R}^n_+} x_n^\lambda \left(\frac{\sigma_0^*(dx)}{dx} \right)^{\lambda/n} \sigma_0^*(u\varphi)\sigma_0^*(dx)$$

$$= \int_{\mathbb{R}^n_+} x_n^\lambda \sigma_0^*(u) \left(\frac{\sigma_0^*(dx)}{dx} \right)^{\lambda/n+1} \sigma_0^*(\varphi)dx.$$

The proof is complete.   □

Now we calculate residues at $\lambda = -\mu-1-N$ on both sides of (5.4.4).

$$2^{\mu+N} \int_{S^{n-1}} a_0(\mu, b)\mathcal{R}_N(\mu)(\varphi)(b)db$$

$$= 2^\mu \int_{\mathbb{R}^{n-1}} \sigma_0^*(a_0) \left(\frac{\sigma_0^*(\mathrm{vol}(S^{n-1}, g_c))}{\mathrm{vol}(\mathbb{R}^{n-1}, g_c)} \right)^{\frac{\mu}{n-1}} \mathcal{S}_N(\mu)\mathcal{I}_{-\mu-N+n-1}(\varphi)dx. \quad (5.4.5)$$

Here we use the fact that the residue at $\lambda = -\mu-1-N$ of the family

$$\lambda \mapsto \int_{\mathbb{R}^n_+} x_n^\lambda u(x)\varphi(x)dx, \quad u \in \ker(\Delta_{\mathbb{R}^n} + \mu(n-1-\mu)), \ \varphi \in C_c^\infty(\mathbb{R}^n)$$

is given by

$$\int_{\mathbb{R}^{n-1}} a_0(\mu, x')\mathcal{S}_N(\mu)(\varphi)(x')dx',$$

where $a_0(\mu, x') \in C^\infty(\mathbb{R}^{n-1})$ is its leading term in the asymptotics $u(x', x_n) \sim x_n^\mu a_0(\mu, x') + \cdots$ for $x_n \to 0$. The proof of that result is analogous to the above one for $M_u(\lambda)$. We omit the details. Moreover, we have used that an asymptotics

$$u(x) \sim (1-|x|^2)^\mu a_0(\mu, b) + \cdots \quad \text{for } |x| \to 1$$

implies an asymptotics

$$\sigma_0^*(u) \sim x_n^\mu 2^\mu \sigma_0^*(a_0) \left(\frac{\sigma_0^*(\mathrm{vol}(S^{n-1}, g_c))}{\mathrm{vol}(\mathbb{R}^{n-1}, g_c)} \right)^{\frac{\mu}{n-1}} + \cdots \quad \text{for } x_n \to 0.$$

In fact, (2.2.10) implies

$$\frac{\sigma_0^*\left((1-|x|^2)^\mu\right)}{x_n^\mu} = 2^\mu \left(\frac{\sigma_0^*(dx)}{dx} \right)^{\frac{\mu}{n}} = 2^\mu \left(\frac{\sigma_0^*(\mathrm{vol}(S^{n-1}, g_c))}{\mathrm{vol}(\mathbb{R}^{n-1}, g_c)} \right)^{\frac{\mu}{n-1}}.$$

Now the integral on the right-hand side of (5.4.5) can be rewritten as

$$\int_{S^{n-1}} a_0(\mu, b) \left(\frac{db}{(\sigma_0)_*(dx)} \right)^{\frac{\mu}{n-1}} (\sigma_0)_* (\mathcal{S}_N(\mu)\mathcal{I}_{-\mu-N+n-1}(\varphi)) \frac{(\sigma_0)_*(dx)}{db} db$$

$$= \int_{S^{n-1}} a_0(\mu, b) \left(\frac{db}{(\sigma_0)_*(dx)} \right)^{\frac{\mu-(n-1)}{n-1}} (\sigma_0)_* (\mathcal{S}_N(\mu)\mathcal{I}_{-\mu-N+n-1}(\varphi)) \, db.$$

Hence

$$2^N \left(\frac{\sigma_0^*(db)}{dx} \right)^{\frac{-\mu+n-1}{n-1}} \circ \sigma_0^* \circ \mathcal{R}_N(\mu) = \mathcal{S}_N(\mu) \circ \left(\frac{\sigma_0^*(dx)}{dx} \right)^{\frac{-\mu+n-1-N}{n}} \circ \sigma_0^*.$$

Now we shift μ into $\mu + n - 1 - N$. Then the last identity states

$$2^N \left(\frac{\sigma_0^*(db)}{dx} \right)^{-\frac{\mu-N}{n-1}} \circ \sigma_0^* \circ \mathcal{R}_N(\mu+n-1-N) = \mathcal{S}_N(\mu+n-1+N) \circ \left(\frac{\sigma_0^*(dx)}{dx} \right)^{-\frac{\mu}{n}} \circ \sigma_0^*,$$

which, in turn, is equivalent to

$$\left(\frac{db}{(\sigma_0)_*(dx)} \right)^{\frac{\mu-N}{n-1}} \circ (\sigma_0)_* \circ \mathcal{S}_N(\mu+n-1+N)$$

$$= 2^N \mathcal{R}_N(\mu+n-1-N) \circ \left(\frac{dx}{(\sigma_0)_*(dx)} \right)^{\frac{\mu}{n}} \circ (\sigma_0)_*, \quad (5.4.6)$$

i.e.,

$$\left(\frac{(\sigma_0)_*(\mathrm{vol}(\mathbb{R}^{n-1}, g_c))}{\mathrm{vol}(S^{n-1}, g_c)} \right)^{-\frac{\mu-N}{n-1}} \circ (\sigma_0)_* \circ \mathcal{S}_N(\mu+n-1+N)$$

$$= 2^N \mathcal{R}_N(\mu+n-1-N) \circ \pi_\mu(\mathbb{R}^n, g_c)(\sigma_0).$$

Now Theorem 5.2.5 and Theorem 5.2.6 identify $\mathcal{S}_N(\lambda+n-1+N)$ with $D_N^{nc}(\lambda)$ (up to a normalizing coefficient which only depends on λ). Thus, by Lemma 5.2.7, we can identify $\mathcal{R}_N(\lambda+n-1-N)$ with the induced family $D_N(\mathbb{R}^n, S^{n-1}; \lambda)$ (up to a polynomial normalizing coefficient). The following theorem summarizes these results.

Theorem 5.4.1. *The family $\mathcal{R}_N(\lambda+n-1-N)$ coincides with the induced family $D_N^b(\lambda) = D_N(\mathbb{R}^n, S^{n-1}; \lambda)$, up to a rational coefficient in λ. More precisely,*

$$D_N(\mathbb{R}^n, S^{n-1}; \lambda) = 2^N k_N(\lambda) \mathcal{R}_N(\lambda+n-1-N), \quad (5.4.7)$$

where

$$k_N(\lambda) \mathcal{S}_N(\lambda+n-1-N) = D_N(\mathbb{R}^n, \mathbb{R}^{n-1}; \lambda).$$

Here

$$k_{2N}(\lambda) = 2^N N! \{(-(n-1)-2\lambda+2N) \cdots (-(n-1)-2\lambda+4N-2)\}^\#$$

and

$$k_{2N+1}(\lambda) = 2^N N! \{(-(n-1)-2\lambda+2N+2) \cdots (-(n-1)-2\lambda+4N)\}^\#.$$

For the formula for $k_N(\lambda)$ we refer to (5.2.57) and (5.2.60).

5.4.2 $D_1^b(\lambda)$, $D_2^b(\lambda)$ and $D_3^b(\lambda)$

In the present section, we determine explicit formulas for the induced families $D_N^b(\lambda)$ for $N \leq 3$. For $N = 3$, the result allows us to recognize the Chang-Qing operator P_3 as a member of the family $D_3^b(\lambda)$ in dimension 4. The method is to apply Theorem 5.4.1 and to determine the families $\mathcal{R}_N(\lambda)$ of order $N \leq 3$ via a calculation of the relevant residues of the extension of the family $M_u(\lambda)$. Explicit formulas for $D_N^b(\lambda)$, at least in principle, can be derived also *directly* from the identity in Lemma 5.2.7. The method used here, however, is easier to handle.

For $k = 0$, (5.4.2) shows that $\mathcal{T}_1(\lambda)(a_0(\lambda)) = a_1(\lambda) = \frac{1}{2}\lambda a_0(\lambda)$. Therefore, $\mathcal{R}_1(\lambda)$ is given by

$$\mathcal{R}_1(\lambda)(\psi(\rho, b)) = \frac{\partial \psi}{\partial \rho}(0, b) + \left(-\left(\frac{n-2}{2}\right) + \frac{\lambda}{2}\right)\psi(0, b).$$

In terms of (r, b)-coordinates, we obtain

$$\mathcal{R}_1(\lambda) = -\frac{1}{2}i^*\frac{\partial}{\partial r} - \frac{n-2-\lambda}{2}i^*, \tag{5.4.8}$$

where i^* restricts functions in r to $r = 1$. Now, by Theorem 5.4.1,

$$D_1^b(\lambda) = 2\mathcal{R}_1(\lambda + n - 2) = -i^*\frac{\partial}{\partial r} + \lambda i^*. \tag{5.4.9}$$

Thus $D_1^b(\lambda)$ coincides with $D_1(\mathbb{B}^n, S^{n-1}; \lambda)$ (Theorem 6.2.1) for the inner normal field $-\partial/\partial r$; we recall that in our convention the mean curvature for the inner normal is -1. In particular,

$$D_1^b\left(-\frac{n}{2}+1\right) : C^\infty(\overline{\mathbb{B}^n})_{-\frac{n}{2}+1} \to C^\infty(S^{n-1})_{-\frac{n}{2}}$$

is the boundary operator

$$-i^*\frac{\partial}{\partial r} - \frac{n-2}{2}i^* = -i^*\frac{\partial}{\partial r} - \frac{n-2}{2}Hi^*,$$

i.e., $D_1^b(-\frac{n}{2}+1)$ is a *Robin boundary condition* in the *conformal Neumann problem*

$$\left(P_2(\mathbb{B}^n), D_1^b\left(-\frac{n}{2}+1\right)\right) : C^\infty(\overline{\mathbb{B}^n})_{-\frac{n}{2}+1} \to C^\infty(\overline{\mathbb{B}^n})_{-\frac{n}{2}-1} \oplus C^\infty(S^{n-1})_{-\frac{n}{2}}.$$

We continue with the discussion of the second-order family. Theorem 5.4.1 yields the first assertion of the following result.

Lemma 5.4.3. $D_2^b(\lambda) = -(2\lambda+n-3)8\mathcal{R}_2(\lambda+n-3)$ *is induced by* $\mathcal{D}_2^0(\lambda)$ *(see (5.3.1))*. *Moreover,*

$$D_2^b(\lambda) = -(2\lambda+n-3)i^*\frac{\partial^2}{\partial r^2}$$

$$+ (2\lambda+n-3)(2\lambda-1)i^*\frac{\partial}{\partial r} + (\Delta_{S^{n-1}} + \lambda(1-\lambda)(2\lambda+n-2))i^*. \tag{5.4.10}$$

Proof. It only remains to establish the given formula for $D_2^b(\lambda)$. (5.4.3) shows that $\mathcal{R}_2(\lambda)$ is given by

$$\frac{1}{2}i^*\frac{\partial^2}{\partial\rho^2} - \frac{n-2-\lambda}{2}i^*\frac{\partial}{\partial\rho} + \left[\frac{(-\frac{n}{2}+1)(-\frac{n}{2}+2)}{2!}T_0(\lambda) + \frac{(-\frac{n}{2}+1)}{1!}T_1(\lambda) + T_2(\lambda)\right]i^*.$$

But $T_0(\lambda) = 1$, $T_1(\lambda) = \frac{\lambda}{2}$ and

$$T_2(\lambda) = \frac{1}{8(n-3-2\lambda)}\left(\Delta_{S^{n-1}} + \lambda(\lambda+1)(n-4-2\lambda)\right) \tag{5.4.11}$$

by (5.4.2). Now the result follows by a direct calculation. $\qquad\square$

(5.4.10) re-proves the factorization identities

$$D_2^b\left(-\frac{n}{2}+\frac{3}{2}\right) = \left(\Delta_{S^{n-1}} - \left(\frac{n-3}{2}\right)\left(\frac{n-1}{2}\right)\right)i^* = P_2(S^{n-1})i^*$$

and

$$D_2^b\left(-\frac{n}{2}+1\right) = i^*\left(\frac{\partial^2}{\partial r^2} + \frac{n-1}{r}\frac{\partial}{\partial r} + \Delta_{S^{n-1}}\right) = i^*\Delta_{\mathbb{R}^n} = i^*P_2(\mathbb{R}^n).$$

In terms of Laplacians $\Delta_{\mathbb{R}^n}$, $\Delta_{S^{n-1}}$ and $\partial/\partial r$, we find

$$D_2^b(\lambda) = -(2\lambda+n-3)i^*\Delta_{\mathbb{R}^n}$$
$$+ (2\lambda+n-2)\left(\Delta_{S^{n-1}} + \lambda(1-\lambda)\right)i^* + (2\lambda+n-3)(2\lambda+n-2)i^*\frac{\partial}{\partial r}. \tag{5.4.12}$$

In Section 6.3, we shall recognize the family $D_2^b(\lambda)$ as a special case of a general construction of a conformally covariant second-order family. Notice that $\partial/\partial r$ is the (exterior) unit normal geodesic vector field.

We continue with the discussion of $D_3^b(\lambda)$. Theorem 5.4.1 implies the first part of the following result.

Lemma 5.4.4. $D_3^b(\lambda) = -16(2\lambda+n-5)\mathcal{R}_3(\lambda+n-4)$ *is induced by* $\mathcal{D}_3^0(\lambda)$. *Moreover,*

$$-D_3^b(\lambda) = -\frac{1}{3}(2\lambda+n-5)i^*\frac{\partial^3}{\partial r^3} + (2\lambda+n-5)(\lambda-1)i^*\frac{\partial^2}{\partial r^2}$$
$$- (\lambda-1)\left((\lambda-1)(2\lambda+n-5) + (\lambda-2)\right)i^*\frac{\partial}{\partial r} + \Delta_{S^{n-1}}i^*\frac{\partial}{\partial r}$$
$$- \lambda\Delta_{S^{n-1}}i^* + \frac{1}{3}(2\lambda+n-2)\lambda(\lambda-1)(\lambda-2)i^*.$$

$D_3^b(\lambda)$ is a polynomial family of degree 4 with leading term $\frac{2}{3}\lambda^4$. Since there are only two factorization identities at disposal, these do not suffice to determine the family.

Proof. By definition, $\mathcal{R}_3(\lambda)$ is given by the sum

$$\frac{1}{3!}i^*\frac{\partial^3}{\partial\rho^3} + \left(\frac{(-\frac{n-2}{2})_1}{1!}T_0(\lambda) + T_1(\lambda)\right)\frac{1}{2!}i^*\frac{\partial^2}{\partial\rho^2}$$

$$+ \left(\frac{(-\frac{n-2}{2})_2}{2!}T_0(\lambda) + \frac{(-\frac{n-2}{2})_1}{1!}T_1(\lambda) + T_2(\lambda)\right)i^*\frac{\partial}{\partial\rho}$$

$$+ \left(\frac{(-\frac{n-2}{2})_3}{3!}T_0(\lambda) + \frac{(-\frac{n-2}{2})_2}{2!}T_1(\lambda) + \frac{(-\frac{n-2}{2})_1}{1!}T_2(\lambda) + T_3(\lambda)\right)i^*,$$

where $T_0(\lambda) = 1$, $T_1(\lambda) = \frac{\lambda}{2}$, $T_2(\lambda)$ is given by (5.4.11), and

$$T_3(\lambda) = \frac{1}{12(n-4-2\lambda)}\left[\Delta_{S^{n-1}} + \frac{\lambda}{2}\Delta_{S^{n-1}}\right.$$

$$\left. + \frac{2(\lambda+2)(n-6-2\lambda)}{8(n-3-2\lambda)}\left(\Delta_{S^{n-1}} + \lambda(\lambda+1)(n-4-2\lambda)\right)\right].$$

The latter formula for $T_3(\lambda)$ can be simplified to

$$T_3(\lambda) = \frac{\lambda+2}{16(n-3-2\lambda)}\left(\Delta_{S^{n-1}} + \frac{1}{3}\lambda(\lambda+1)(n-6-2\lambda)\right).$$

Hence (by some calculation)

$$\mathcal{R}_3(\lambda) = \frac{1}{6}i^*\frac{\partial^3}{\partial\rho^3} - \frac{n-2-\lambda}{4}i^*\frac{\partial^2}{\partial\rho^2}$$

$$+ \frac{1}{8(n-3-2\lambda)}\left[\Delta_{S^{n-1}} + (n-4-2\lambda)(\lambda-n+3)(\lambda-n+2)\right]i^*\frac{\partial}{\partial\rho}$$

$$+ \left[\frac{(-\frac{n}{2}+1)(-\frac{n}{2}+2)(-\frac{n}{2}+3)}{6} + \frac{(-\frac{n}{2}+1)(-\frac{n}{2}+2)}{2}\frac{\lambda}{2}\right.$$

$$- \frac{n-2}{2}\frac{1}{8(n-3-2\lambda)}\left(\Delta_{S^{n-1}} + \lambda(\lambda+1)(n-4-2\lambda)\right)$$

$$\left. + \frac{\lambda+2}{16(n-3-2\lambda)}\Delta_{S^{n-1}} + \frac{1}{48}\frac{\lambda(\lambda+1)(\lambda+2)(n-6-2\lambda)}{(n-3-2\lambda)}\right]i^*.$$

Another calculation yields the formula

$$\frac{1}{6}i^*\frac{\partial^3}{\partial\rho^3} - \frac{2-\lambda}{4}i^*\frac{\partial^2}{\partial\rho^2} - \frac{1}{8(n-5+2\lambda)}\left[\Delta_{S^{n-1}} - (\lambda-1)(\lambda-2)(n-4+2\lambda)\right]i^*\frac{\partial}{\partial\rho}$$

$$- \frac{1}{8(n-5+2\lambda)}\frac{\lambda}{2}\Delta_{S^{n-1}}i^* + \frac{1}{48(n-5+2\lambda)}(n-2+2\lambda)\lambda(\lambda-1)(\lambda-2)i^*$$

for $\mathcal{R}_3(\lambda+n-4)$. Multiplying the latter formula with $16(n-5+2\lambda)$, gives

$$\frac{8}{3}(2\lambda+n-5)i^*\frac{\partial^3}{\partial\rho^3} - 4(2-\lambda)(2\lambda+n-5)i^*\frac{\partial^2}{\partial\rho^2} + 2(\lambda-1)(\lambda-2)(2\lambda+n-4)i^*\frac{\partial}{\partial\rho}$$
$$- 2\Delta_{S^{n-1}}i^*\frac{\partial}{\partial\rho} - \lambda\Delta_{S^{n-1}}i^* + \frac{1}{3}(2\lambda+n-2)\lambda(\lambda-1)(\lambda-2)i^*.$$

Finally, we use the identities

$$2\frac{\partial\psi}{\partial\rho}(1,b) = -\frac{\partial\varphi}{\partial r}(0,b),$$

$$4\frac{\partial^2\psi}{\partial\rho^2}(1,b) = \frac{\partial^2\varphi}{\partial r^2}(0,b) - \frac{\partial\varphi}{\partial r}(0,b),$$

$$8\frac{\partial^3\psi}{\partial\rho^3}(1,b) = -\frac{\partial^3\varphi}{\partial r^3}(0,b) + 3\frac{\partial^2\varphi}{\partial r^2}(0,b) - 3\frac{\partial\varphi}{\partial r}(0,b)$$

to return to the coordinates (r,b). $\qquad\square$

Lemma 5.4.4 re-proves the factorization identities

$$D_3^b\left(-\frac{n-5}{2}\right) = P_2(S^{n-1})D_1^b\left(-\frac{n-5}{2}\right)$$

and

$$D_3^b\left(-\frac{n}{2}+1\right) = D_1^b\left(-\frac{n}{2}-1\right)P_2(\mathbb{R}^n).$$

The following equivalent formula for $D_3^b(\lambda)$ in terms of $\Delta_{\mathbb{R}^n}$, $\Delta_{S^{n-1}}$ and $\partial/\partial r$ will be useful in Section 5.4.3.

Lemma 5.4.5.

$$-D_3^b(\lambda) = -\frac{1}{3}(2\lambda+n-5)\left\{i^*\frac{\partial}{\partial r}\Delta_{\mathbb{R}^n} - (3\lambda+n-4)i^*\Delta_{\mathbb{R}^n}\right\}$$
$$+ \frac{1}{3}(2\lambda+n-2)\left\{\Delta_{S^{n-1}}i^*\frac{\partial}{\partial r} - (3\lambda+n-5)\Delta_{S^{n-1}}i^*\right.$$
$$\left. - [(3\lambda+n-4)\lambda+(n-3)(n-4)]i^*\frac{\partial}{\partial r} + \lambda(\lambda-1)(\lambda-2)i^*\right\}.$$

In turn, the formula for $D_3^b(\lambda)$ becomes somewhat more beautiful if the family is expressed in terms of Yamabe operators on $\mathbb{R}^n$ and S^{n-1}, and $D_1^b(\lambda) = -\partial/\partial r + \lambda$. One should compare the following result with (5.3.9).

Lemma 5.4.6.

$$- D_3^b(\lambda) = \frac{1}{3}(2\lambda+n-5)D_1^b(\lambda-2)P_2(\mathbb{R}^n) - \frac{1}{3}(2\lambda+n-2)P_2(S^{n-1})D_1^b(\lambda)$$

$$+ \left(\lambda+\frac{n-5}{2}\right)\left(\lambda+\frac{n-2}{2}\right)(2\lambda+n-3)D_1^b(\lambda)$$

$$+ \frac{1}{3}(2\lambda+n-5)(2\lambda+n-2)\left[P_2(\mathbb{R}^n)-P_2(S^{n-1})-\left(\lambda+\frac{n-1}{2}\right)\left(\lambda+\frac{n-3}{2}\right)\right].$$

5.4.3 $D_3^b(0)$ for $n=4$ and (P_3,T) for $(\mathbb{B}^4, S^3)$

The constant term of the family $D_3^b(\lambda)$ and the polynomial $Q_3(\lambda) = D_3^b(\lambda)(1)$ are quantities which are of special interest. Lemma 5.4.5 implies

$$D_3^b(0) = \frac{n-5}{3}\frac{\partial}{\partial r}\Delta_{\mathbb{R}^n} - \frac{n-2}{3}\Delta_{S^{n-1}}\frac{\partial}{\partial r} - \frac{(n-4)(n-5)}{3}\Delta_{\mathbb{R}^n}$$

$$+ \frac{(n-2)(n-5)}{3}\Delta_{S^{n-1}} + \frac{(n-2)(n-3)(n-4)}{3}\frac{\partial}{\partial r} \quad (5.4.13)$$

(here we have omitted i^* in order to simplify the statement of the formula). Moreover, we have

$$Q_3^b(\lambda) \overset{\text{def}}{=} D_3^b(\lambda)(1) = -\frac{1}{3}(2\lambda+n-2)\lambda(\lambda-1)(\lambda-2),$$

and therefore $\dot{Q}_3^b(0) = -\frac{2}{3}(n-2)$. In particular, for $n=4$, we find

$$D_3^b(0) = -\frac{1}{3}i^*\frac{\partial}{\partial r}\Delta_{\mathbb{R}^4} - \frac{2}{3}\left(\Delta_{S^3}i^*\frac{\partial}{\partial r} + \Delta_{S^3}i^*\right) \quad (5.4.14)$$

and

$$Q_3^b(\lambda) \overset{\text{def}}{=} D_3^b(\lambda)(1) = -\frac{2}{3}(\lambda+1)\lambda(\lambda-1)(\lambda-2), \quad \dot{Q}_3^b(0) = -\frac{4}{3}. \quad (5.4.15)$$

For a compact four-manifold M with boundary, Chang and Qing ([68]) discovered a conformally covariant operator $P_3(M,\partial M) : C^\infty(M) \to C^\infty(\partial M)$ such that

$$P_3(M,\partial M; e^{2\varphi}g) = e^{-3\varphi}P_3(M,\partial M; g),$$

and introduced an associated curvature invariant $T \in C^\infty(\partial M)$. The formulas for P_3 and T are rather complicated for general metrics. We refer to Section 6.26 for more details on (P_3, T). In the special case of a *vanishing* second fundamental form L, these are given by

$$P_3(M,\partial M; g) = \frac{1}{2}i^* N_g \Delta_{(M,g)} + \Delta_{(\Sigma,g)}i^* N_g + \left(\text{tr}(G) - \frac{1}{3}\tau_M(g)\right)i^* N_g \quad (5.4.16)$$

and

$$T(M, \partial M; g) = -\frac{1}{12} N_g \tau_M(g), \tag{5.4.17}$$

where normal derivatives are taken with respect to the *inner* normal N_g and

$$G_i^j = R_{iNN}^j \tag{5.4.18}$$

(see (2.5.18)). We emphasize that in [68] the sign convention for the Laplacian is opposite to ours. The pair (P_3, T) satisfies the fundamental identity

$$e^{3\varphi} T(e^{2\varphi} g) = T(g) + P_3(g)(\varphi).$$

In the cases (S^4, S^3) and $(\mathbb{B}^4, S^3)$, the operator P_3 and the curvature T are given in the following lemma.

Lemma 5.4.7. *In the models (S^4, S^3) and $(\mathbb{B}^4, S^3)$, we have the explicit formulas*

$$P_3(S^4, S^3; g_c) = \frac{1}{2} i^* \operatorname{grad}(\mathcal{H}_0) \Delta_{S^4} + \Delta_{S^3} i^* \operatorname{grad}(\mathcal{H}_0) - i^* \operatorname{grad}(\mathcal{H}_0), \quad T = 0$$

and

$$P_3(\mathbb{B}^4, S^3; g_c) = -\frac{1}{2} i^* \frac{\partial}{\partial r} \Delta_{\mathbb{R}^4} - \Delta_{S^3} i^* \frac{\partial}{\partial r} - \Delta_{S^3} i^*, \quad T = 2.$$

Proof. For the proof of the first part we use (5.4.16) and (5.4.17). The vector field $\operatorname{grad}(\mathcal{H}_0)$ has length 1 (on S^3) and points into the *interior* of the upper hemisphere. The second fundamental form L vanishes since S^3 is a totally geodesic submanifold. Hence $H = 0$. Moreover, $\operatorname{tr}(G) = 3$ and $\tau(S^4) = 12$. The vanishing of T follows from the constance of curvature of S^4 and the vanishing of L. This proves the assertions in the first case. In the second case, we use the formulas for P_3 and T as given in Theorem 6.26.1. Now L is the identity, i.e., $H = 3$. Moreover, $\operatorname{tr}(G) = 0$ and $\tau(\mathbb{B}^4) = 0$. On S^3 the vector field $\partial/\partial r$ has length 1 and points into the *exterior* of the ball. This proves the formula for P_3 in the second case. Finally, in order to find T, we calculate

$$T = 3H^3 - \frac{1}{3} \operatorname{tr}(L^3) = 3 - 1 = 2.$$

The proof is complete. The formula for the ball is also stated in [68]. $\square$

Corollary 5.4.1.

$$D_3^c(S^4, S^3; 0) = \frac{2}{3} P_3(S^4, S^3), \quad D_3^b(\mathbb{B}^4, S^3; 0) = \frac{2}{3} P_3(\mathbb{B}^4, S^3).$$

Proof. By (5.3.9), we have

$$D_3^c(0) = \frac{1}{3} D_1^c \Delta_{S^4} + \frac{2}{3} \Delta_{S^3} D_1^c - \frac{2}{3} D_1^c$$

on S^4. The formula in the first part of Lemma 5.4.7 and $D_1^c = i^* \mathrm{grad}(\mathcal{H}_0)$ prove
the first identity. In the case of the ball $\mathbb{B}^4$, we have, by Lemma 5.4.5,

$$D_3^b(0) = -\frac{1}{3}i^*\frac{\partial}{\partial r}\Delta_{\mathbb{R}^4} - \frac{2}{3}(\Delta_{S^3}i^*\frac{\partial}{\partial r} + \Delta_{S^3}i^*).$$

Comparing the latter formula with the second formula of Lemma 5.4.7 completes
the proof.   □

5.5   Automorphic distributions

In the present section, we return to the theory of Selberg zeta functions for Kleinian
groups as described in Section 1.2. We formulate a result from [150] on the structure of certain spaces of automorphic distributions on S^n. It rests on the theory
of the families $D_N^c(\lambda) : C^\infty(S^n) \to C^\infty(S^{n-1})$.

The general situation is as follows. We consider a discrete subgroup Γ of
$G^{n+1} = SO(1, n+1)^\circ$ acting on the hyperbolic space

$$\mathbb{H}^{n+1} = \{x \in \mathbb{R}^{n+1} \,|\, |x| < 1\}$$

and its geodesic boundary S^n. We assume that Γ is torsion-free and convex-
cocompact, i.e., Γ acts without fixed points on $\mathbb{H}^{n+1}$ and with a compact quotient on the convex hull of the limit set $\Lambda(\Gamma) \subset S^n$. Let $\Omega(\Gamma) \subset S^n$ be the
complement of $\Lambda(\Gamma)$. Then the quotient $\Gamma\backslash\Omega(\Gamma)$ is a compact smooth manifold
which can be viewed as the boundary of a compactification of the quotient space
$X^{n+1} = \Gamma\backslash\mathbb{H}^{n+1}$. Associated to this situation is the Selberg zeta function

$$Z_\Gamma(s) = \prod_{\text{p.p.o. } c} \prod_{N \geq 0} \det(\mathrm{id} - S^N((P_c^{n-1})^-)e^{-s|c|}), \ \Re(s) > \delta(\Gamma)$$

of the geodesic flow of X^{n+1}. Here the first product runs over its prime periodic
orbits. These can be identified with the closed oriented geodesics in X^{n+1}. But, in
turn, closed oriented geodesics in X^{n+1} are in bijection with Γ-conjugacy classes
of loxodromic elements, i.e., with Γ-orbits of attracting fixed points of elements
of Γ. The set of fixed points is dense in $\Lambda(\Gamma)$. Therefore, one can regard Z_Γ as
being naturally associated to the fixed points of the action of Γ on S^n. Then the
natural question is to describe the divisor of its meromorphic continuation to $\mathbb{C}$
in terms of analysis on $\Lambda(\Gamma)$. An ingredient of such a theory are theorems which
relate the multiplicities of zeros and poles of Z_Γ to the dimensions of spaces of
Γ-automorphic distributions which are supported on $\Lambda(\Gamma)$. This motivates us to
describe the spaces

$$C^{-\infty}(\Lambda(\Gamma))_\lambda^\Gamma, \ \lambda \in \mathbb{C}$$

of Γ-automorphic distributions supported on the limit set. The general problem
will not be addressed here since we will be interested in that problem only for

a very special choice of Γ. In fact, we assume that $\Gamma \subset G^{n+1}$ is a cocompact subgroup of the subgroup $G^n \subset G^{n+1}$ which leaves the equatorially embedded sphere $S^{n-1} \subset S^n$ invariant. Then the limit set of Γ acting on $\mathbb{H}^{n+1}$ is $\Lambda(\Gamma) = S^{n-1} \subset S^n$. Let $X^n = \Gamma \backslash \mathbb{H}^n$.

Theorem 5.5.1. *Let $\Gamma \subset G^n \subset G^{n+1}$ be as above. Then if $\lambda \notin \mathbb{R}$ the space*

$$C^{-\infty}(\Lambda(\Gamma))_\lambda^\Gamma$$

of automorphic distributions is non-trivial iff $\lambda \in -\mu-1-\mathbb{N}_0$, where $\mu(n-1-\mu)$ is an eigenvalue of the Laplacian on the compact hyperbolic manifold X^n. Here any eigenvalue gives rise to two values for μ. More precisely, the space

$$C^{-\infty}(\Lambda(\Gamma))_{-\mu-1-N}^\Gamma, \ N \in \mathbb{N}_0$$

has finite dimension, and is spanned by the distributions

$$\delta_N^c(\mu+N-(n-1))^*(\omega),$$

where for each eigenfunction u of the Laplacian on X for the eigenvalue $\mu(n-1-\mu)$, there is a canonically associated distribution $\omega \in C^{-\infty}(S^{n-1})_{-\mu}^\Gamma$.

The distributions ω in Theorem 5.5.1 are the boundary distributions of (automorphic) eigenfunctions in the sense of the theory of Poisson transformations.

Note that $\lambda \notin \mathbb{R}$ implies $\mu \notin \mathbb{R}$. Hence $\mu \in \frac{n-1}{2} + i\mathbb{R}$. For $\lambda \in \mathbb{R}$ (in particular, for integers λ), the situation is more complicated due to possible topological contributions.

We briefly describe the main idea of the proof. The first step is to prove that the relevant distributions arise in the range of the residues of an *extension operator*

$$\mathrm{ext}_\lambda : C^\infty(\Omega(\Gamma))_\lambda^\Gamma \to C^{-\infty}(S^n)_\lambda^\Gamma$$

which extends Γ-automorphic functions on $\Omega(\Gamma)$ through the limit set to automorphic distributions on the sphere S^n. The second step is to prove that this construction is exhaustive.

The construction of the family ext_λ is very close to the residue method in the present work. It is also related to scattering theory of the hyperbolic manifold X^{n+1}. For details see [50] and [151].

In order to define ext_λ, we first use $\mathcal{H}_0 \in C^\infty(S^n)_1^{G^n}$ (see Section 2.2) to identify

$$C^\infty(\Gamma \backslash \Omega(\Gamma)) \simeq C^\infty(\Omega(\Gamma))_0^\Gamma \ni u \mapsto |\mathcal{H}_0|^\lambda u \in C^\infty(\Omega(\Gamma))_\lambda^\Gamma.$$

Now for $\Re(\lambda) > -1$ and $u \in C^\infty(\Gamma \backslash \Omega(\Gamma))$ define

$$\langle \mathrm{ext}_\lambda(|\mathcal{H}_0|^\lambda u), \varphi \rangle = \int_{\Gamma \backslash \Omega(\Gamma)} |\mathcal{H}_0|^\lambda u \pi_{*,-\lambda-n}^\Gamma(\varphi) db, \ \varphi \in C^\infty(S^n)$$

with the average

$$\pi_{*,\lambda}^{\Gamma}(\varphi) = \sum_{\gamma \in \Gamma} \pi_{\lambda}^{c}(\gamma)(\varphi).$$

The restriction on λ is imposed in order to guarantee the convergence of the series $\pi_{*,-\lambda-n}^{\Gamma}(\varphi)$. No further condition is needed since $\Gamma \backslash \Omega(\Gamma)$ is compact. But

$$\pi_{*,\lambda}^{\Gamma} : C^{\infty}(S^n) \to C^{\infty}(\Omega(\Gamma))_{\lambda}^{\Gamma}$$

is well defined if $\Re(\lambda) < -\delta(\Gamma) = -(n-1)$.

Now we observe that for $\Re(\lambda) > -1$, we have

$$\mathrm{ext}_\lambda(|\mathcal{H}_0|^\lambda u) = M_u(\lambda).$$

In fact, using a fundamental domain F for the action of Γ on $\Omega(\Gamma)$, we calculate

$$\langle \mathrm{ext}_\lambda(|\mathcal{H}_0|^\lambda u), \varphi \rangle = \int_F (|\mathcal{H}_0|^\lambda u) \sum_{\gamma \in \Gamma} \left(\frac{\gamma_*(db)}{db} \right)^{\frac{\lambda}{n}+1} \gamma_*(\varphi) db$$

$$= \sum_{\gamma \in \Gamma} \int_{\gamma^{-1}(F)} \gamma^*(|\mathcal{H}_0|^\lambda u) \left(\frac{db}{\gamma^*(db)} \right)^{\frac{\lambda}{n}+1} \varphi \frac{\gamma^*(db)}{db} db$$

$$= \sum_{\gamma \in \Gamma} \int_{\gamma^{-1}(F)} \gamma^*(|\mathcal{H}_0|)^\lambda u \left(\frac{\gamma^*(db)}{db} \right)^{-\frac{\lambda}{n}} \varphi db$$

$$= \sum_{\gamma \in \Gamma} \int_{\gamma^{-1}(F)} |\mathcal{H}_0|^\lambda u \varphi db$$

$$= \int_{\Omega(\Gamma)} |\mathcal{H}_0|^\lambda u \varphi db = \langle M_u(\lambda), \varphi \rangle. \tag{5.5.1}$$

Since u is bounded (recall that Γ acts cocompactly on $\Omega(\Gamma)$), the latter formula again proves that $\mathrm{ext}_\lambda(|\mathcal{H}_0|^\lambda u)$ is well defined for $\Re(\lambda) > -1$.

By construction, $\mathrm{ext}_\lambda(|\mathcal{H}_0|^\lambda u) \in C^{-\infty}(S^n)_\lambda^\Gamma$. We are interested in a meromorphic continuation of the extension operator. The importance of such a continuation is due to the following simple but crucial observation. Since the composition $\mathrm{res}_{\Omega(\Gamma)} \circ \mathrm{ext}_\lambda$ of ext_λ with the restriction of distributions to the open set $\Omega(\Gamma)$ is the identity on $C^{\infty}(\Omega(\Gamma))_\lambda^\Gamma$, it follows that at a simple pole of ext_λ we have

$$\mathrm{res}_{\Omega(\Gamma)} \circ \mathrm{rg}\,\mathrm{Res}_\lambda(\mathrm{ext}) = 0,$$

i.e.,

$$\mathrm{rg}\,\mathrm{Res}_\lambda(\mathrm{ext}) \subset C^{-\infty}(\Lambda(\Gamma))_\lambda^\Gamma.$$

The theory for general *convex-cocompact* Γ provides a meromorphic continuation of ext_λ via its relation to the scattering operator of X^{n+1} ([50], [151]). In the present special case, the situation, however, is much easier. In fact, we use the relation (5.5.1) and a direct *formal* analysis of $M_u(\lambda)$ to make the existence of the meromorphic continuation plausible and to derive formulas for the residues. The problem for general $u \in C^\infty(\Gamma\backslash\Omega(\Gamma))$ can be decomposed into the easier problems for eigenfunctions of the Laplacian on $\Omega(\Gamma)$ for the hyperbolic metric $g_N = |\mathcal{H}_0|^{-2} g_c$. Since Γ operates with a compact quotient on $\Omega(\Gamma)$, the spectral decomposition of the Laplacian implies the desired result for general u.

Now let $u \in C^\infty(\Gamma\backslash\Omega(\Gamma)) \simeq C^\infty(\Gamma\backslash\Omega^+(\Gamma)) \oplus C^\infty(\Gamma\backslash\Omega^-(\Gamma)) \simeq C^\infty(\Gamma\backslash\mathbb{H}^n) \oplus C^\infty(\Gamma\backslash\mathbb{H}^n)$ be given by an eigenfunction for the Laplacian of the constant curvature metric g_N for the eigenvalue $\mu(n-1-\mu)$ on $\Omega^+(\Gamma)$. We set $u = 0$ on $\Omega^-(\Gamma)$. As before, we use a formal asymptotics of u near $\ker(\mathcal{H}_0) = \Lambda(\Gamma) = S^{n-1}$ to argue that $\langle M_u(\lambda), \varphi \rangle$ has a meromorphic continuation to $\mathbb{C}$ with simple poles in the ladder $-\mu-1-\mathbb{N}_0$ and residues of the form

$$\mathrm{Res}_{\lambda=-\mu-1-N}\left(\langle M_u(\lambda), \varphi\rangle\right) = \langle \omega, \delta_N(\mu+N-(n-1))(\varphi)\rangle_{S^{n-1}} \qquad (5.5.2)$$

with certain G^n-equivariant families

$$\delta_N^c(\lambda) : C^\infty(S^n)_\lambda \to C^\infty(S^{n-1})_{\lambda-N}.$$

Here we have to use generalized asymptotic expansions of eigenfunctions with distributional coefficients in $C^{-\infty}(S^{n-1})$ ([233]). The distribution

$$\omega \in C^{-\infty}(S^{n-1})_{-\mu}^\Gamma$$

is the leading term a_0 in the generalized asymptotics

$$u \sim \sum_{j\geq 0} \mathcal{H}_0^{\mu+j} a_j(\mu) + \sum_{j\geq 0} \mathcal{H}_0^{n-1-\mu+j} b_j(\mu).$$

It is straightforward to verify the asserted Γ-invariance of $\omega = a_0$. Now (5.5.2) gives the result

$$\mathrm{Res}_{\lambda=-\mu-1-N}\left(\langle M_u(\lambda), \varphi\rangle\right) = \langle \delta_N^c(\mu+N-(n-1))^*(\omega), \varphi\rangle_{S^n},$$

i.e.,

$$\mathrm{Res}_{\lambda=-\mu-1-N}\left(M_u(\lambda)\right) = \delta_N^c(\mu+N-(n-1))^*(\omega)$$

(see also Lemma 6.5.1). Here we define the adjoint family $\delta_N^c(\lambda)^*$ by

$$\langle \delta_N^c(\lambda)^*(\omega), \varphi\rangle_{S^n} = \langle \omega, \delta_N^c(\lambda)\varphi\rangle_{S^{n-1}}, \quad \varphi \in C^\infty(S^n).$$

Then $\delta_N^c(\lambda)^*$ defines a G^n-equivariant operator

$$C^{-\infty}(S^{n-1})_{-\lambda+N-(n-1)} \to C^{-\infty}(S^n)_{-\lambda-n}.$$

In particular, we see that

$$\delta_N^c(\mu+N-(n-1))^*(\omega) \in C^{-\infty}(S^n)_{-\mu-1-N}^\Gamma$$

with support in the limit set S^{n-1}. In [150], the above arguments are replaced by arguments using analysis on $\mathbb{H}^{n+1}$.

Chapter 6

Conformally Covariant Families

Let M be a smooth oriented manifold and $i : \Sigma \hookrightarrow M$ an oriented codimension one submanifold (hypersurface). In this chapter, we shall discuss conformally covariant families of differential operators $C^\infty(M) \rightarrow C^\infty(\Sigma)$ in the presence of general background metrics on M. The model cases $(\mathbb{R}^{n+1}, \mathbb{R}^n)$ with the Euclidean metric on $\mathbb{R}^{n+1}$, (S^{n+1}, S^n) with the round metric on S^{n+1} and $(\mathbb{B}^{n+1}, S^n)$ with the Euclidean metric on $\mathbb{B}^{n+1}$ (Chapter 5) are now viewed as special cases in a much wider framework.

The three model cases already indicate some aspects of the influence of curvature on the structure of such families. The first model is Riemannian flat, and the resulting families are not influenced by curvature, i.e., they are algebraic objects. The families in the second model reflect the non-trivial curvature of M and Σ in terms of higher polynomial degree. In contrast to these two models, the third model deals with a hypersurface with a non-vanishing second fundamental form. This leads to a further increase of the polynomial degree of the associated families. In the present chapter, we substantially refine these observations.

We start with a review of the contents.

We shall call a family *critical* if its order coincides with the dimension of the target manifold Σ. In Section 6.1, we associate to any critical conformally covariant family $D_n(M, \Sigma; g, \lambda)$ the fundamental pair

$$(P_n(M, \Sigma; g), Q_n(M, \Sigma; g)) = \left(D_n(M, \Sigma; g, 0), \dot{D}_n(M, \Sigma; g, 0)(1) \right).$$

We show that the conformal covariance of the family implies that the associated fundamental pair satisfies an identity which resembles the fundamental identity for the *critical Q-curvature* Q_n of (Σ^n, g). We consider some low order special cases, and detect the GJMS-operator $P_n(S^n)$ and $Q_n(S^n; g_c)$ in the family $D_n^c(S^{n+1}, S^n; \lambda)$.

Of course, this leaves open the existence of critical conformally covariant families. The remainder of the chapter will be devoted to the discussion of two major constructions of such families and their consequences on Q-curvature.

In Section 6.2, we discuss the first-order family $D_1(M, \Sigma; g; \lambda)$.

Section 6.3 and Section 6.4 are devoted to a discussion of an elementary construction of second-order families for an oriented codimension one submanifold Σ^n and a general background metric g. Section 6.3 deals with the *critical* case $n = 2$ and Section 6.4 describes the generalization to $n \geq 3$.

The residue method generates *equivariant* families from the asymptotics of eigenfunctions in the model cases $\mathbb{R}^n \hookrightarrow \mathbb{R}^{n+1}$, $S^n \hookrightarrow S^{n+1}$ and $S^n \hookrightarrow \mathbb{R}^{n+1}$. In Section 6.5, we introduce a version of that method if (Σ, h) is the boundary of $[0, \varepsilon) \times \Sigma$ with the metric $dr^2 + h_r$ so that $r^{-2}(dr^2 + h_r)$ is a Poincaré-Einstein extension of h. In particular, we introduce the critical *residue family* $D_n^{res}(h; \lambda)$. It satisfies $D_n^{res}(h; 0) = P_n(\Sigma; h)i^*$. An evaluation of the relation

$$-(-1)^{\frac{n}{2}} \dot{D}_n^{res}(h; 0)(1) = Q_n(h) \tag{6.0.1}$$

leads to the formula

$$2(-1)^{\frac{n}{2}} Q_n = \left(\dot{P}_n(0) - \dot{P}_n^*(0) \right)(1) + 2^n \left(\frac{n}{2} \right)! \sum_{j=0}^{\frac{n}{2}-1} \frac{(\frac{n}{2}-j-1)!}{2^{2j} j!} P_{2j}^*(0)(v_{n-2j})$$

for the Q-curvature (Theorem 6.6.2). It expresses $Q_n(h)$ in terms of the *holographic coefficients* v_{2j}, which describe the asymptotics of the volume form of the Poincaré-Einstein metric, and the structure of *harmonic* functions for the corresponding Laplacian, up to the divergence term

$$\delta_n(1) = \left(\dot{P}_n(0) - \dot{P}_n^*(0) \right)(1).$$

The result that the divergence term $\delta_n(1)$ also can be written in terms of the operators $P_{2j}^*(0)$ and the coefficients v_{2j} leads to the more explicit version

$$Q_n = (-1)^{\frac{n}{2}} 2^{n-1} \left(\frac{n}{2} - 1 \right)! \sum_{j=0}^{\frac{n}{2}-1} \frac{(\frac{n}{2}-j)!}{2^{2j} j!} P_{2j}^*(0)(v_{n-2j}) \tag{6.0.2}$$

of the holographic formula (Theorem 6.6.6). The latter formula has an independent proof. This fact can be seen as a proof of the relation (6.0.1) used to derive it. The details are discussed in Section 6.6.

(6.0.2) expresses Q-curvature $Q_n(h)$ of (Σ, h) in dimension n in terms of the asymptotic geometry of an associated Poincaré-Einstein metric $r^{-2}(dr^2 + h_r)$ in $n + 1$ dimensions. Thus from a physical perspective, (6.0.2) says that $Q_n(h)$ can be viewed as part of an n-dimensional hologram of geometry in dimension $n+1$. This motivates the terminology.

For closed Σ, the holographic formula (6.0.2) reproduces the result of Graham and Zworski ([128]) that the total Q-curvature is proportional to the integrated holographic anomaly. Moreover, it is compatible with a formula of Branson describing the contribution of $\Delta^{\frac{n}{2}-1}(\mathsf{J})$ to Q_n, and it implies Gover's formula ([110])

for the Q-curvature of a (Riemannian) Einstein metric. These issues are discussed in Section 6.6 and Section 6.16. Along the way, we derive formulas for the Paneitz operator $P_{4,n}$ and the curvature function $Q_{4,n}$ (from Chapter 4) using formal asymptotic series for eigenfunctions for the Poincaré metric (see Remark 6.6.5). Moreover, we introduce the Q-polynomial $Q_n^{\mathrm{res}}(\lambda)$.

In Section 6.7, we prove that in the framework of Section 6.6 the families of Section 6.4 coincide with the second-order residue families. In Section 6.8, we analyze the critical order 3 residue families.

In order to make the holographic formulas more explicit, formulas for the holographic coefficients v_{2j} are required. In Section 6.9, we derive such formulas for the first few of them: v_2, v_4 and v_6. These formulas have been stated in one form or another in the literature (usually without proofs). Here we give proofs which rest on explicit formulas for the first three terms in the Fefferman-Graham expansion (Theorem 6.9.1). In the formula for v_6, a version of the Bach-tensor in dimension $n > 4$ contributes. It is caused by the contribution of the Bach tensor to the term $h_{(4)}$ in the Fefferman-Graham expansion for the Poincaré metric. The latter result is naturally reflected also in the curvature tensor of the Fefferman-Graham ambient metric ([99], [96]). The results are used to derive explicit formulas for the first two terms in the asymptotics of eigenfunctions of the Laplacian for the Poincaré-Einstein metric. Finally, we formulate an extension of the holographic formula to all *subcritical* Q-curvatures (Conjecture 6.9.1).

In Section 6.10, we give an alternative proof of the holographic formulas for the critical Q-curvature Q_6. Here we derive an explicit formula Q_6 from the asymptotics of eigenfunctions of the Laplacian for the Poincaré-Einstein metric. The method works in all dimensions $n \geq 6$ and yields formulas for the subcritical $Q_{6,n}$ ($n > 6$) as well. We prove that the resulting formula for $Q_{6,n}$ coincides with the one derived in [116]. Finally, these formulas are used to prove the holographic formula

$$Q_{6,n} = -8 \cdot 48 v_6 - 32 P_2^* \left(\frac{n}{2} - 3 \right)(v_4) - 2 P_4^* \left(\frac{n}{2} - 3 \right)(v_2)$$

for the subcritical $Q_{6,n}$ (Theorem 6.10.4). The latter result supports Conjecture 6.9.1.

The results in Section 6.11 have central significance. They are consequences of the fact that, for a *conformally flat* metric h, the residue family $D_{2N}^{\mathrm{res}}(h; \lambda)$ ($2N \leq n$) satisfies a system of $2N$ factorization identities (Theorem 6.11.1). These identities are curved analogs of those for the family $\mathcal{D}_{2N}(\lambda)$ of homomorphisms of Verma modules (Theorem 5.2.2). Partial results support the conjecture that the factorizations hold true for general metrics.

The factorization identities give rise to *recursive relations* for Q-curvature, Q-polynomials and GJMS-operators. We describe some typical results.

The fact that the number of factorization identities for $D_{2N}^{\mathrm{res}}(\lambda)$ exceeds its degree as a polynomial implies recursive relations which allow us to express the critical GJMS-operator P_n in terms of lower order GJMS-operators. In particular, we find a formula for $P_6(h)$ in terms of $P_4(h)$, $P_2(h)$ and

$$\bar{P}_4(h) = P_4(dr^2 + h_r) \quad \text{and} \quad \bar{P}_2(h) = P_2(dr^2 + h_r)$$

(Corollary 6.11.6).

One part of the discussion of recursive formulas for Q-curvatures is concerned with the recursive representation

$$Q_n = \sum_{|I| \leq \frac{n}{2}-1} a_I^{(n)} P_{2I}(Q_{n-2|I|}) + (-1)^{\frac{n}{2}-1} \frac{(n-2)!!}{(n-3)!!} i^* \bar{P}_2^{\frac{n}{2}-1}(\bar{Q}_2) \tag{6.0.3}$$

of the critical Q-curvature in terms of lower order Q-curvature and lower order GJMS-operators (Conjecture 6.11.1). For $n = 4$ and $n = 6$, it takes the respective explicit forms

$$Q_4 = P_2(Q_2) - 2i^* \bar{P}_2(\bar{Q}_2) \tag{6.0.4}$$

and

$$Q_6 = \frac{2}{3} P_2(Q_4) + \left[-\frac{5}{3} P_2^2 + \frac{2}{3} P_4 \right](Q_2) + \frac{8}{3} i^* \bar{P}_2^2(\bar{Q}_2) \tag{6.0.5}$$

(Theorem 6.11.5). The latter formula is equivalent to the explicit formula for Q_6 in Theorem 6.10.3 (for $n = 6$). It holds true for *all* metrics.

For conformally flat h, the existence of the presentation (6.0.3) for the critical Q-curvature is a consequence of (6.0.1) combined with the factorization identities for residue families and analogs of (6.0.3) for *all* lower order, i.e., subcritical, Q-curvatures. Here a crucial role is played by the distinguished feature of the presentation (6.0.3) to hold true also for subcritical Q-curvatures. We refer to this property as *universality*. The conjectural status of (6.0.3) is due to the conjectural universality. For an alternative description of the coefficients a_I we refer to [95].

Furthermore, the factorization identities imply a recursive formula for the critical Q-polynomial

$$Q_n^{\text{res}}(\lambda) = -(-1)^{\frac{n}{2}} D_n^{\text{res}}(\lambda)(1)$$

in terms of Q_n, lower order GJMS-operators and lower order Q-polynomials (Theorem 6.11.10). Then $Q_n^{\text{res}}(0) = 0$, and Q_n is its linear coefficient. In addition, Q_n contributes to *all* coefficients of the polynomial $Q_n^{\text{res}}(\lambda)$. More precisely, the holographic formula $\dot{Q}_n^{\text{res}}(0) = Q_n$ is only one identity among $\frac{n}{2}$ similar ones which relate Q_n and v_n. In particular, for the critical Q_6 we find

$$Q_6 = [P_2(Q_4^{\text{res}}(2)) + 2P_4(Q_2)] - 6[Q_4 + P_2(Q_2)]Q_2 - 2^6 3! v_6, \tag{6.0.6}$$

where

$$Q_4^{\text{res}}(\lambda) = -\lambda Q_4 - (\lambda+1) P_2(Q_2).$$

(6.0.6) is both holographic (in view of the term v_6) and recursive. In Section 6.13, we derive an analog of (6.0.6) for Q_8.

In Section 6.12, we apply the relation (6.0.6) to derive the following recursive formula for the critical GJMS-operator P_6:

$$P_6 u = [2(P_2 P_4 + P_4 P_2) - 3P_2^3]^0 u - 48\delta(\mathsf{P}^2 \# du) - 8\delta(\mathcal{B} \# du) \qquad (6.0.7)$$

(Corollary 6.12.1). A similar formula yields the subcritical P_6 (Corollary 6.12.2). The astonishing simplicity of these formulas suggests that we ask for their analogs for higher order GJMS-operators. In Section 6.13, we speculate on such a universal recursive formula for the GJMS-operators P_8 which generalizes the formula

$$P_8 = 6P_2 P_6 - 32P_2^2 P_4 + 9P_4^2 + 18P_2^4 + 3!4!P_2$$

on all round spheres S^n. For an alternative but not explicit construction of a conformally covariant cube of the Laplacian see [56].

Finally, in Section 6.11 we discuss the meaning of the quadratic coefficient in $Q_n^{\mathrm{res}}(\lambda)$ for the renormalized volume of Poincaré-Einstein metrics (Theorem 6.11.15 and Theorem 6.11.16). A part of the picture remains conjectural (see Conjecture 6.11.2).

For conformally flat metrics, explicit formulas for the holographic coefficients v_{2j} (Lemma 6.14.1) lead to a more explicit version of the holographic formula (6.0.2) (Theorem 6.14.1). It shows that the Euler form naturally splits off as the non-divergence part. This result suggests splitting off, for general metrics, the contributions

$$\Delta^j \operatorname{tr}(\wedge^{\frac{n}{2}-j}(\mathsf{P})), \; j = 0, \dots, \frac{n}{2} - 1$$

in Q_n. The study of variational problems defined by the traces $\operatorname{tr}(\wedge^j(\mathsf{P}))$ was initiated in [235], and has led to an extensive literature around the σ_k-Yamabe problem (see [135], [134], [217]).

By the classification of Deser and Schwimmer (conjectured in [81] and established by Alexakis in [5], [6] and [3]), it is natural to relate v_n in dimension n to the Pfaffian. This can be seen as an analog of the fact that the leading coefficient in Weyl's tube formula for a submanifold of $\mathbb{R}^n$ is given by the integral of the Pfaffian. In Section 6.14, we emphasize this analogy between the holographic coefficients v_{2j} and the coefficients in Weyl's formula for the volume of tubes.

In Section 6.15, we discuss the Henningson-Skenderis test of the AdS/CFT duality which relates the holographic conformal anomaly v_4 to a linear combination of quantum conformal anomalies. The calculations rest on explicit formulas for heat-equation coefficients (conformal index densities).

In Section 6.16, we discuss the meaning of the holographic formula for Q-curvature for Einstein metrics h. In that case, it is possible to calculate the Q-curvature explicitly. In fact, Gover ([110]) proved that

$$Q_n(h) = (n-1)! \left(\frac{\tau(h)}{n(n-1)} \right)^{\frac{n}{2}}.$$

This generalizes the formula $Q_n(S^n, g_c) = (n-1)!$. Here we prove that the latter formula is a direct consequence of the holographic formula. The argument rests on an explicit formula for the Poincaré-Einstein metric which also enables us to derive a formula ([110]) for the critical GJMS-operator for the Einstein metric. In the present special case, the relation to Weyl tube invariants is made explicit.

In the remaining sections, we shall use a version of Eastwood's method of curved translation ([19], [88], [90], [20]) in order to construct curved analogs of $\mathcal{D}_N(\lambda)$ in full generality. As a preparation, we briefly recall in Section 6.17 the relation between conformally covariant differential operators and homomorphisms of generalized Verma modules. This leads to a lifting problem (to semi-holonomic Verma modules) which has a negative solution within the framework of critical GJMS-operators. We formulate an analogous problem for families. In Section 6.18, we solve the lifting problem for all families $\mathcal{D}_N(\lambda)$ using a version of Zuckerman translation ([243]). Here the key observation is that $\mathcal{D}_N(\lambda)$ can be obtained by *iterated translation*. In particular, all families $\mathcal{D}_N(\lambda)$ can be constructed from the embedding $\mathcal{D}_0(\lambda) = i$ and $\mathcal{D}_1(\lambda) = Y_n^- i$. Now translation of homomorphisms of Verma modules has an analog for homomorphisms of semi-holonomic Verma modules. This yields the desired lifts. Moreover, we work out explicit formulas for the lifts of families of orders 2, 3 and 4. The resulting formulas in the semi-holonomic category quickly become quite complicated with increasing order. In addition, we prove the existence of other lifts.

The identification of $\mathcal{D}_N(\lambda)$ as an iterated Zuckerman translation is the key to the construction of the *tractor family*. The idea is to replace the embeddings and projections by tractor D-operators. In Section 6.19, we prepare the translation from Lie theory to the curved framework. The necessary tools from the conformal tractor calculus are developed in Section 6.20. The central objects are the conformally invariant tractor connection and the conformally covariant tractor D-operator. The discussion in these sections is self-contained, and can be read as an introduction to tractor calculus.

After these preparations, in Section 6.21, we define the tractor families

$$D_N^T(M, \Sigma; g; \lambda) : C^\infty(M) \to C^\infty(\Sigma),$$

and formulate their relations to GJMS-operators and Q-curvature. To some extent, the material here is conjectural. The outlined perspective will be explicated in the following sections by a list of particular results.

We summarize the main points. First of all, the quality of the situation depends on the parity of the dimension n of the target manifold Σ.

For odd n, the family $D_{2N}^T(\cdot, \Sigma; g; \lambda)$ is regular at $\lambda = -\frac{n}{2} + N$. Its value at $\lambda = -\frac{n}{2} + N$ is a conformally covariant differential operator on Σ which only depends on $i^*(g)$. The critical family $D_n^T(\cdot, \Sigma; g; \lambda)$ has a simple pole at $\lambda = 0$. Its residue is a conformally covariant differential operator. For a vanishing residue, the value of the family at $\lambda = 0$ defines a conformally covariant operator. The value on the constant function 1 of its derivative at $\lambda = 0$ defines a notion of *odd* *order* Q-curvature which satisfies a fundamental identity. It involves the value of the family at $\lambda = 0$. That notion of odd order Q-curvature is an invariant of the embedding. The obstructing residue is expected to vanish if the trace-free part L_0 of L vanishes. This is the case if $g = dr^2 + h_r$ on $(0, \varepsilon) \times \Sigma^n$ so that $r^{-2}(dr^2 + h_r)$ is a Poincaré metric with conformal infinity $h_0 = h$. For another notion of odd order Q-curvature we refer to [97].

For even n, the behaviour of tractor families depends on the order.

The critical family $D_n^T(\cdot, \Sigma^n; g; \lambda)$ has a simple pole at $\lambda = 0$. However, its residue is expected to vanish. More generally, the family $D_{2N}^T(\cdot, \Sigma; g; \lambda)$ has a simple pole at $\lambda = -\frac{n}{2} + N$ if $N > \frac{n}{2}$ (supercritical case). The non-vanishing residue can be viewed as an obstruction to the existence of a conformally invariant power Δ^N of the Laplacian on the manifold Σ^n.

In the subcritical case $N < \frac{n}{2}$, the family $D_{2N}^T(\cdot, \Sigma; g; \lambda)$ is regular at $\lambda = -\frac{n}{2} + N$. Its value only depends on $i^*(g)$ and is given by a conformally covariant power of the Laplacian.

In the critical case $2N = n$, its value at $\lambda = 0$ is more complicated and depends on the embedding (Conjecture 6.21.3). The critical family $D_n^T(\cdot, \Sigma^n; g; \lambda)$ induces the function

$$Q_n^T(\cdot, \Sigma^n; g; \lambda) = D_n^T(\cdot, \Sigma^n; g; \lambda)(1).$$

For even n, we expect that $Q_n^T(\cdot, \Sigma^n; g; 0) = 0$ and

$$\dot{Q}_n^T(\cdot, \Sigma^n; g; 0) = -(-1)^{\frac{n}{2}} \left(\mathbf{Q}_n(i^*(g)) + Q_n^e(g) \right),$$

where $\mathbf{Q}_n(i^*(g))$ is a version of Q-curvature of $(\Sigma, i^*(g))$. The additional term Q_n^e depends on the embedding, and we shall call it the *extrinsic Q-curvature* of the embedding. It satisfies a fundamental identity which involves a conformally covariant operator P_n^e. Analogous results are expected in the supercritical case (if the obstructions vanish).

For $(M, \Sigma) = (\mathbb{R}^{n+1}, \mathbb{R}^n)$ with the Euclidean metric g_c, the residue families and the tractor families both specialize to $D_{2N}^{nc}(\lambda)$. Such relations continue to hold true for certain curved cases. In particular, we prove the *holographic duality*

$$D_{2N}^{res}(h; \lambda) = D_{2N}^T(dr^2 + h_r; \lambda)$$

for conformally flat h (Theorem 6.21.2). We expect that the duality holds true for more general metrics h (possibly in a modified form).

In Section 6.22, we collect results along the lines of the perspectives developed in Section 6.21. In particular, we prove that for $n \neq 4$, the value $D_4^T(\cdot, \Sigma^n; g; -\frac{n}{2} + 2)$ does not depend on the metric off Σ and, more precisely, is given by the Paneitz operator $P_4(i^*(g))i^*$ (Theorem 6.22.2). In the critical case $n = 4$, we prove that $D_4^T(\cdot; \Sigma^4; g; \lambda)$ is regular at $\lambda = 0$, determine its value at $\lambda = 0$, and study the quantity $\dot{Q}_4^T(\cdot, \Sigma^4; g; 0)$. We prove that the intrinsic Paneitz pair (P_4, Q_4) for (Σ^4, g) naturally appears in these data *together* with an additional extrinsic pair (P_4^e, Q_4^e)

for which explicit formula are derived. It turns out that the extrinsic pair is characterized completely by the conformally invariant symmetric bilinear form

$$\mathcal{J} \overset{\mathrm{def}}{=} (i^*(\mathsf{P}_M) - \mathsf{P}_\Sigma) + HL - \frac{1}{2}H^2 g$$

on Σ. $\mathcal{J}$ is conformally invariant in all dimensions $n \geq 3$ and can be generalized to submanifolds of higher codimension. The trace of the invariant $\mathcal{J}$ is proportional to the Chen-Willmore quantity $H^2 - \tau_e \simeq |L_0|^2$. It is related to the conformally invariant fundamental forms of Fialkow ([100]). In Section 6.23, we describe that relation, and prove that, for a hypersurface Σ^n, it can be written in the form

$$\mathcal{J} = \frac{1}{n-2} \left(L_0^2 - \frac{1}{2(n-1)}|L_0|^2 g + c_{23}(\mathsf{C}) \right),$$

where $c_{23}(\mathsf{C})_{ij} = \mathsf{C}_{iNNj}$. It follows that, for a hypersurface $\Sigma^n \hookrightarrow S^{n+1}$ in the Möbius space (S^{n+1}, g_c), the invariant $\mathcal{J}$ vanishes iff Σ is totally umbilic ($L_0 = 0$). More generally, for a conformally flat background metric on M^5, the extrinsic pair (P_4^e, Q_4^e) vanishes iff Σ is totally umbilic.

$\mathcal{J}$ and the extrinsic pair vanish for a background metric which is given by a Poincaré-Einstein metric which extends a given metric on Σ (in these cases all three individual terms in the definition of $\mathcal{J}$ vanish, or equivalently, Σ is totally umbilic and C vanishes). This emphasizes the special role of the latter metrics. We prove directly that for such a background metric g the tractor Q-polynomial $Q_4^T(\cdot, \Sigma^4; g; \lambda)$ actually coincides with the Q-polynomial introduced in Section 6.6.

In Section 6.24, we prove that the second-order family $D_2(M, \Sigma; g; \lambda)$ constructed in Section 6.4 actually *is* a composition of conformally covariant operators in terms of tractors. Moreover, we find a generalization of the second-order family which acts on standard tractors instead of functions. This family appears in the proofs in Section 6.21.

In Section 6.25 and Section 6.26, we prove that the order 3 family naturally yields the Chang-Qing pair (P_3, T) (i.e., an order 3 operator together with a curvature invariant) if $L_0 = 0$. In this case, there is only an extrinsic fundamental pair.

6.1   Fundamental pairs and critical families

Here we describe a pattern by which any critical conformally covariant family $D_n(M, \Sigma; g; \lambda)$ of differential operators gives rise to a conformally covariant linear differential operator and a curvature quantity with a conformal transformation law which is governed by the corresponding linear operator. Such pairs are called *fundamental pairs*.

Let M be a manifold of dimension $n+1$ and let $i : \Sigma \hookrightarrow M$ be a hypersurface. For any Riemannian metric g, we fix a unit normal vector field $N(g)$ on Σ. Let

$$D_n(M, \Sigma; g; \lambda) : C^\infty(M) \to C^\infty(\Sigma)$$

be a holomorphic family of natural differential operators which is conformally covariant in the sense that

$$e^{-(\lambda-n)i^*(\varphi)} \circ D_n(M, \Sigma; e^{2\varphi}g; \lambda) \circ e^{\lambda\varphi} = D_n(M, \Sigma; g; \lambda) \qquad (6.1.1)$$

for all $\varphi \in C^\infty(M)$ and all $\lambda \in \mathbb{C}$. For the following it will be enough to assume that the family is holomorphic at $\lambda = 0$. We define the *fundamental pair*

$$(P_n(M, \Sigma; g), Q_n(M, \Sigma; g)) = \left(D_n(M, \Sigma; g, 0), \dot{D}_n(M, \Sigma; g, 0)(1)\right) \qquad (6.1.2)$$

consisting of a differential operator $C^\infty(M) \to C^\infty(\Sigma)$ and a scalar function in $C^\infty(\Sigma)$. The following simple but important observation is a direct consequence of the conformal covariance of the family.

Lemma 6.1.1. *For all $\varphi \in C^\infty(M)$,*

$$e^{ni^*(\varphi)} P_n(M, \Sigma; e^{2\varphi}g) = P_n(M, \Sigma; g)$$

and

$$e^{ni^*(\varphi)} Q_n(M, \Sigma; e^{2\varphi}g) = Q_n(M, \Sigma; g) - [P_n(M, \Sigma; g), \varphi](1).$$

Proof. The first assertion is obvious. (6.1.1) is equivalent to

$$D_n(M, \Sigma; e^{2\varphi}g; \lambda) = e^{(\lambda-n)\varphi} \circ D_n(M, \Sigma; g; \lambda) \circ e^{-\lambda\varphi}.$$

Differentiating the latter identity with respect to λ at $\lambda = 0$ we get

$$\dot{D}_n(e^{2\varphi}g; 0)(u) = \varphi e^{-n\varphi} D_n(g; 0)(u) + e^{-n\varphi} \dot{D}_n(g; 0)(u) - e^{-n\varphi} D_n(g; 0)(\varphi u).$$

Hence, for $u = 1$, we obtain

$$\dot{D}_n(e^{2\varphi}g; 0)(1) = e^{-n\varphi} \dot{D}_n(g; 0)(1) - e^{-n\varphi} [D_n(g; 0), \varphi](1),$$

i.e.,

$$e^{n\varphi} Q_n(M, \Sigma; e^{2\varphi}g) = Q_n(M, \Sigma; g) - [P_n(M, \Sigma; g), \varphi](1).$$

The proof is complete. $\qquad\qquad\qquad\qquad\qquad\qquad\qquad\qquad\qquad\qquad\quad$ $\square$

Moreover, if $P_n(M, \Sigma; g)$ annihilates constants, i.e., if $P_n(M, \Sigma; g)(1) = 0$, then

$$e^{ni^*(\varphi)} Q_n(M, \Sigma; e^{2\varphi}g) = Q_n(M, \Sigma; g) - P_n(M, \Sigma; g)(\varphi) \qquad (6.1.3)$$

for all $\varphi \in C^\infty(M)$. For even n, (6.1.3) resembles the fundamental identity

$$e^{n\varphi} Q_n(\Sigma, e^{2\varphi}g) = Q_n(\Sigma; g) + (-1)^{\frac{n}{2}} P_n(\Sigma; g)(\varphi), \quad \varphi \in C^\infty(\Sigma) \qquad (6.1.4)$$

for Q-curvature on Σ (see (3.1.6)). Although $Q_n(M, \Sigma; g; \lambda)$ and $Q_n(\Sigma; g)$ are curvature quantities on Σ, it is important to emphasize that (6.1.3) and (6.1.4) deal with different situations: in (6.1.4) the metric g and the functions φ are objects on Σ, but in (6.1.3) g and φ live in a neighbourhood of Σ in M. However, if $P_n(M, \Sigma; g) = D_n(M, \Sigma; g; 0)$ lives on Σ, i.e., if it is of the form $\mathbf{P}_n(g)i^*$ for a natural differential operator $\mathbf{P}_n$ on Σ, then (6.1.3) and (6.1.4) can be expected to be related. This case was discussed in Section 1.8 (for even n).

Now we discuss three examples of fundamental identities from the point of view of the above pattern.

Example 6.1.1. *For $n = 1$, let*

$$D_1(g; \lambda) = i^* \nabla_{N(g)} - \lambda \kappa(g)i^* : C^\infty(M) \to C^\infty(\Sigma),$$

where $\kappa(g)$ is the geodesic curvature of Σ in the surface M which corresponds to the choice of $N(g)$. Then $D_1(g; \lambda)$ is conformally covariant, i.e.,

$$e^{-(\lambda-1)\varphi} \circ D_1(e^{2\varphi}g; \lambda) \circ e^{\lambda\varphi} = D_1(g; \lambda), \ \varphi \in C^\infty(M).$$

We differentiate the identity with respect to λ at $\lambda = 0$ and find the well-known relation

$$e^\varphi \kappa(e^{2\varphi}g) = \kappa(g) + i^* \nabla_{N(g)}(\varphi) \tag{6.1.5}$$

on Σ. Note that $D_1(g; 0) = i^*$ does not live on Σ. For the details we refer to Section 6.2.

Example 6.1.2. *Let $n = 2$ and assume that $D_2(M, \Sigma; g; \lambda) : C^\infty(M) \to C^\infty(\Sigma)$ is a holomorphic family which satisfies the identity*

$$e^{-(\lambda-2)\varphi} \circ D_2(e^{2\varphi}g; \lambda) \circ e^{\lambda\varphi} = D_2(g; \lambda)$$

for all $\lambda \in \mathbb{C}$ (near $\lambda = 0$) and all $\varphi \in C^\infty(M)$. Differentiating the identity at $\lambda = 0$ yields the equation

$$e^{2\varphi} \dot{D}_2(e^{2\varphi}g; 0)(1) = \dot{D}_2(g; 0)(1) - D_2(g; 0)(\varphi), \ \varphi \in C^\infty(M) \tag{6.1.6}$$

if $D_2(g; 0)(1) = 0$. The latter identity is to be compared with the relation

$$e^{2\varphi} K(e^{2\varphi}g) = K(g) - \Delta_g(\varphi), \ \varphi \in C^\infty(\Sigma) \tag{6.1.7}$$

for the Gauß curvature $K = \mathsf{J} = \frac{\tau}{2}$ (see (2.5.7)).

Theorem 6.3.1 *yields a conformally covariant second-order family. For this family,*

$$D_2(g; 0) = P_2(i^*(g))i^* = \Delta_{i^*(g)} i^*,$$

i.e., $D_2(g; 0)$ lives in Σ. Moreover, the quantity $\dot{D}_2(g; 0)(1)$ coincides with $K(g)$. Thus the fundamental identity (6.1.6) coincides with (6.1.7).

Special cases are $D_2^c(S^3, S^2; \lambda)$ and $D_2^b(\mathbb{B}^3, S^2; \lambda)$. On S^3 we have $D_2^c(0) = P_2(S^2)i^* = \Delta_{S^2}i^*$ and $\dot{D}_2^c(0)(1) = 1$ (Lemma 5.3.1). The Gauß curvature of S^2 is 1. On $\mathbb{B}^3$ we have seen that $D_2^b(0) = \Delta_{S^2}i^*$ and $\dot{D}_2(0)(1) = 1$ ((5.4.12)).

Example 6.1.3. *Let $n = 3$ and assume that $D_3(M, \Sigma; g; \lambda) : C^\infty(M) \to C^\infty(\Sigma)$ is a family which satisfies the identity*

$$e^{-(\lambda-3)\varphi} \circ D_3(e^{2\varphi} g; \lambda) \circ e^{\lambda\varphi} = D_3(g; \lambda)$$

for all $\lambda \in \mathbb{C}$ (near $\lambda = 0$) and all $\varphi \in C^\infty(M)$. Differentiating the identity at $\lambda = 0$ yields the relation

$$e^{3\varphi} \dot{D}_3(e^{2\varphi} g; 0)(1) = \dot{D}_3(g; 0)(1) - D_3(g; 0)(\varphi), \quad \varphi \in C^\infty(M)$$

if $D_3(g; 0)(1) = 0$. We compare the latter identity with the relation

$$e^{3\varphi} T(e^{2\varphi} g) = T(g) + P_3(g)(\varphi), \quad \varphi \in C^\infty(M) \tag{6.1.8}$$

for the operator P_3 and the curvature function T introduced in [62], [68], [73] (see Section 6.26). In Theorem 6.26.2 we will establish the proportionality of both pairs

$$(D_3(g; 0), \dot{D}_3(g; 0)(1)) \quad and \quad (-P_3(g), T(g)) \tag{6.1.9}$$

for $D_3(g; \lambda)$ given by the tractor family $D_3^T(g; \lambda)$, if the trace-free part L_0 of L vanishes. $P_3(g)$ does not live on Σ, i.e., the fundamental pair is associated to the embedding $i : \Sigma \hookrightarrow M$.

This covers the special cases $D_3^c(S^4, S^3; \lambda)$ and $D_3^b(\mathbb{B}^4, S^3; \lambda)$. In fact, on S^4 we have

- $D_3^c(S^4, S^3; 0) = \frac{2}{3} P_3(S^4, S^3)$ *(Corollary 5.4.1)*,
- $\dot{D}_3^c(S^4, S^3; 0)(1) = 0$ *(by (5.3.9)) and*
- $T = 0$ *(Lemma 5.4.7)*.

On $\mathbb{B}^4$ we have

- $D_3^b(\mathbb{B}^4, S^3; 0) = \frac{2}{3} P_3^b(\mathbb{B}^4, S^3)$ *(Corollary 5.4.1)*,
- $\dot{D}_3^b(\mathbb{B}^4, S^3; 0)(1) = -\frac{4}{3}$ *(by (5.4.15)) and*
- $T = 2$ *(by Lemma 5.4.7)*.

In particular, the proportionality coefficient is $-3/2$.

Finally, we show that the GJMS-operator $P_n(S^n)$ and the Q-curvature $Q_n(S^n)$ of (S^n, g_c) can be read off from the family $D_n^c(S^{n+1}, S^n; \lambda)$ by forming the corresponding fundamental pair. We consider the family

$$D_{2N}^c(\lambda) = D_{2N}^c(S^{n+1}, S^n; \lambda) : C^\infty(S^{n+1}) \to C^\infty(S^n).$$

By Lemma 5.2.11, we have

$$D_{2N}^c(\lambda)(1) = Q_{2N}(\lambda) = (-1)^N \lambda(\lambda-1) \cdots (\lambda-2N+1) \tag{6.1.10}$$

for any n, and by (5.2.25) (for $j = N$), we know that

$$D_{2N}^c \left(-\frac{n}{2}+N\right) = P_{2N}(S^n)i^*.$$

These two results imply

$$P_{2N}(S^n)(1) = P_{2N}(S^n)(i^*(1)) = D_{2N}^c\left(-\frac{n}{2}+N\right)(1)$$
$$= \left(\frac{n}{2}-N\right)(-1)^N\left\{\left(\frac{n}{2}-N+1\right)\cdots\left(\frac{n}{2}+N-1\right)\right\}.$$

On the other hand, by definition of $Q_{2N}(S^n;g_c)$,

$$P_{2N}(S^n)(1) = \left(\frac{n}{2}-N\right)(-1)^N Q_{2N}(S^n;g_c),$$

i.e.,

$$Q_{2N}(S^n;g_c) = \left(\frac{n}{2}-N+1\right)\cdots\left(\frac{n}{2}+N-1\right).$$

In particular, we obtain

Lemma 6.1.2. *For even n, $D_n^c(S^{n+1},S^n;0) = P_n(S^n)i^*$ and*

$$\dot{D}_n^c(S^{n+1},S^n;0)(1) = (-1)^{\frac{n}{2}-1}Q_n(S^n;g_c) = (-1)^{\frac{n}{2}-1}(n-1)!.$$

Proof. By (6.1.10),

$$D_n^c(S^{n+1},S^n;\lambda)(1) = Q_n(\lambda) = (-1)^{\frac{n}{2}}\lambda(\lambda-1)\cdots(\lambda-n+1).$$

Hence

$$\dot{D}_n^c(S^{n+1},S^n;0)(1) = (-1)^{\frac{n}{2}}(n-1)!(-1)^{n-1} = (-1)^{\frac{n}{2}-1}(n-1)!.$$

Using $Q_n(S^n;g_c) = (n-1)!$, the proof is complete. $\square$

For odd order families, we have $D_{2N+1}^c(\lambda)(1) = 0$.

6.2   The family $D_1(g;\lambda)$

For any hypersurface $i : \Sigma \hookrightarrow M$, there is an obvious conformally covariant family of first order (in the sense of Section 1.8).

Theorem 6.2.1. *For a hypersurface $\Sigma \hookrightarrow M^n$ ($n \geq 3$) and a Riemannian metric g on M, let*

$$D_1(M,\Sigma;g;\lambda) = i^*\nabla_{N(g)} - \lambda H(g)i^*,$$

where $N(g)$ is a unit normal vector field and $H(g)$ is the corresponding mean-curvature (see (1.8.6)). Then $D_1(M,\Sigma;g;\lambda)$ is conformally covariant, i.e.,

$$e^{-(\lambda-1)i^*(\varphi)} \circ D_1(M,\Sigma;e^{2\varphi}g;\lambda) \circ e^{\lambda\varphi} = D_1(M,\Sigma;g;\lambda) \qquad (6.2.1)$$

for all $\lambda \in \mathbb{C}$ and all $\varphi \in C^\infty(M)$.

Proof. (6.2.1) is equivalent to

$$e^{-(\lambda-1)i^*(\varphi)} \left[i^* \nabla_{N(e^{2\varphi}g)}(u) - \lambda H(e^{2\varphi}g)i^*u \right]$$
$$= e^{-\lambda\varphi} \left[-\lambda i^* \nabla_{N(g)}(\varphi)u + i^* \nabla_{N(g)}(u) - \lambda H(g)i^*u \right]$$

for all $u \in C^\infty(M)$. Using $N(e^{2\varphi}g) = e^{-\varphi}N(g)$, the latter identity, in turn, is equivalent to

$$H(e^{2\varphi}g) = e^{-\varphi} \left(H(g) + \nabla_{N(g)}(\varphi) \right). \tag{6.2.2}$$

In order to prove (6.2.2), we recall that

$$\hat{\nabla}_X Y - \nabla_X Y = \langle d\varphi, X \rangle Y + \langle d\varphi, Y \rangle X - g(X,Y)\,\mathrm{grad}(\varphi)$$

(see (2.5.1)). The convention $L(X,Y) = -g(\nabla_X(Y), N)$ implies

$$e^{-\varphi}\hat{L}(X,Y) - L(X,Y) = g(X,Y)g(\mathrm{grad}(\varphi), N).$$

Hence

$$e^{-\varphi} \mathrm{tr}_g(\hat{L}) - \mathrm{tr}_g(L) = (n-1)\nabla_{N(g)}(\varphi),$$

i.e., $e^\varphi H(\hat{g}) - H(g) = \nabla_{N(g)}(\varphi)$. This proves (6.2.2). The proof is complete. □

Remark 6.2.1. *For a curve Σ in a surface M, an analogous critical family is given by*

$$D_1(M, \Sigma; g; \lambda) = i^* \nabla_{N(g)} - \lambda \kappa(g)i^*.$$

Here

$$\kappa(g) = g(\nabla_{\dot{c}}(\dot{c}), N(g)) \in C^\infty(\Sigma)$$

(c is a natural parametrization of Σ) is the corresponding geodesic curvature. In that case, the conformal covariance of the family is equivalent to

$$e^\varphi \kappa(e^{2\varphi}g) = \kappa(g) + \nabla_{N(g)}(\varphi). \tag{6.2.3}$$

In the following, the identity (6.2.2), i.e.,

$$e^\varphi H(e^{2\varphi}g) = H(g) + \nabla_{N(g)}(\varphi) \tag{6.2.4}$$

will be called the fundamental identity for H.

6.3 $D_2(g; \lambda)$ for a surface in a 3-manifold

Let $i : \Sigma^2 \hookrightarrow M^3$ be an oriented surface in a 3-manifold M. In the present section, we construct a second-order polynomial family $D_2(M, \Sigma; g; \lambda) : C^\infty(M) \to C^\infty(\Sigma)$ of differential operators so that

$$e^{-(\lambda-2)\varphi} \circ D_2(M, \Sigma; e^{2\varphi}g; \lambda) \circ e^{\lambda\varphi} = D_2(M, \Sigma; g; \lambda) \tag{6.3.1}$$

for all $\lambda \in \mathbb{C}$ and all $\varphi \in C^\infty(M)$, and

$$D_2(M, \Sigma; g; 0) = P_2(\Sigma; g)i^*, \quad \dot{D}_2(M, \Sigma; g; 0)(1) = Q_2(\Sigma; g). \qquad (6.3.2)$$

This is a critical case: the order of the family coincides with the dimension of the target surface Σ. In Section 6.24, we will use tractor calculus for an alternative construction of its non-critical version (Theorem 6.4.1).

Theorem 6.3.1. *Let N_g be a unit geodesic normal vector field in a neighborhood of Σ. Then the natural family*

$$D_2(M, \Sigma; g; \lambda) = -2\lambda^3 H_g^2 i^* + \lambda^2 \left(4H_g i^* N_g + c_g i^*\right)$$
$$+ \lambda \left(-2i^* N_g^2 - 2H_g i^* N_g + K_g i^*\right) + \Delta_{(\Sigma, g)} i^*, \qquad (6.3.3)$$

with $c_g = 2K_g - H_g^2 - \frac{1}{2}\tau_g$, satisfies (6.3.1) and (6.3.2).

In Theorem 6.3.1 it is important to choose the *geodesic* extension of the unit normal vector field on Σ to an open neighborhood. The choice of the extension influences the second normal derivative in the coefficient of λ. The following result yields an equivalent formula in terms of Yamabe operators of (M, g) and (Σ, g).

Theorem 6.3.2. *Let N_g be a unit normal vector field on Σ. Then*

$$D_2(M, \Sigma; g; \lambda) = -2\lambda i^* P_2(M, g) + (2\lambda + 1) P_2(\Sigma, g) i^*$$
$$+ 2\lambda(2\lambda + 1) H_g i^* N_g$$
$$+ \lambda(2\lambda + 1) \left(Q_2(\Sigma; g) - Q_2(M; g) - \lambda H_g^2\right) i^*. \qquad (6.3.4)$$

An advantage of Theorem 6.3.2 is that it does not require a choice of the normal field in a neighbourhood of Σ. The second-order normal derivatives are covered by the Yamabe operator on M.

Although the sign of the mean curvature depends on the choice of a normal vector field, the family $D_2(M, \Sigma; g; \lambda)$ does not depend on such a choice.

If the mean curvature H vanishes, the polynomial degree of the family is at most 2. The coefficient of λ^2 is a combination of curvature contributions of both metrics and mean curvature. The family is of degree 1 iff $H = 0$ and $Q_2(M; g)$ restricts to $Q_2(\Sigma; g)$.

Example 6.3.1. *For the standard sphere $\Sigma = S^2 \subset \mathbb{R}^3$, we have $H_c = 1$, $K_c = 1$ and $\partial/\partial n_c = \partial/\partial r$. Therefore, (6.3.3) specializes to $D_2^b(\mathbb{R}^3, S^2; \lambda)$ (Lemma 5.4.3). Moreover, we recognize (5.4.12) for $D_2^b(\mathbb{R}^3, S^2; \lambda)$ as a special case of (6.3.4) (for $g = g_c$).*

Example 6.3.2. *(6.3.3) and (6.3.4) reproduce the family $D_2^c(\lambda): C^\infty(S^3) \to C^\infty(S^2)$. In fact, in this case $H = 0$, $K = 1$ and $\tau(S^3) = 6$. Hence (6.3.3) yields the operator*

$$-\lambda^2 + \lambda \left(-2\frac{\partial^2}{\partial n^2} + 1\right) + \Delta_{S^2} = -2\lambda \frac{\partial^2}{\partial n^2} + (\Delta_{S^2} - \lambda(\lambda - 1))$$

(see (5.3.3)), where the normal derivative is the Lie derivative with respect to a unit length geodesic normal vector field. On the other hand, (6.3.4) gives

$$-2\lambda i^* P_2(S^3) + (2\lambda+1) P_2(S^2) i^* + \lambda(2\lambda+1)\left(1 - \frac{6}{4}\right) i^*$$

(see Lemma 5.3.1).

The polynomial $Q_2(M, \Sigma; g; \lambda) = D_2(M, \Sigma; g; \lambda)(1)$ will be called the Q-polynomial. Analogous polynomials will be associated to higher order families in Section 6.6. Theorem 6.3.1 implies

Corollary 6.3.1.

$$Q_2(M, \Sigma; g; \lambda) = \lambda \left[\left(-2H_g^2\right) \lambda^2 + \left(-H_g^2 + 2Q_2(\Sigma; g) - 2Q_2(M; g)\right) \lambda + Q_2(\Sigma; g) \right].$$

The Q-polynomial $Q_2(M, \Sigma; g; \lambda)$ has no constant term and the coefficient of λ is $Q_2(\Sigma; g)$. Moreover, the coefficient of λ^2 admits the following interpretation. We write $-c_g$ in the form

$$\left(H_g^2 - K_g + \overline{K}_g\right) - K_g + \left(\frac{\tau_g}{2} - \overline{K}_g\right), \tag{6.3.5}$$

where $\overline{K}_g$ denotes the sectional curvature of the planes $T\Sigma \subset TM$ with respect to the metric g on M. Now the volume form

$$\left(H_g^2 - K_g + \overline{K}_g\right) \mathrm{vol}(g) \in \Omega^2(\Sigma) \tag{6.3.6}$$

is conformally invariant. In fact, by the Gauß equation, the difference $K_g - \overline{K}_g$ coincides with the product of the principal curvatures and thus $H_g^2 - K_g + \overline{K}_g$ is $\frac{1}{4}$ times the squared difference of the principal curvatures. That observation implies the conformal invariance of (6.3.6). Integration of (6.3.6) (for compact Σ) yields

$$\int_\Sigma \left(H_g^2 + \overline{K}_g\right) \mathrm{vol}(g) - \int_\Sigma K_g \,\mathrm{vol}(g) = W(\Sigma, g) - 2\pi\chi(\Sigma),$$

where W denotes the Willmore functional. The integral of the second term in (6.3.5) is conformally invariant, too. The third term in (6.3.5) gives rise to the form

$$\left(\frac{\tau_g}{2} - \overline{K}_g\right) \mathrm{vol}(g) \in \Omega^2(\Sigma).$$

In view of $\frac{\tau_g}{2} = \overline{K}_g + \mathrm{tr}(G_g)$, where G denotes the tensor $G_i^j = R_{iNN}^j$ (see (2.5.18)), the latter volume form can also be written in the form $\mathrm{tr}(G_g) \,\mathrm{vol}(g) \in \Omega^2(\Sigma)$, i.e., we have proved the conformal invariance of the integral

$$\int_\Sigma \left[\frac{1}{2}\ddot{Q}_2(g; 0) + \mathrm{tr}(G_g)\right] \mathrm{vol}(g).$$

We continue with the proof of Theorem 6.3.1. As a preparation, we recall the following well-known result.

Lemma 6.3.1. *For* $(M, \Sigma; g)$ *and geodesic normals,*

$$i^* \Delta_{(M,g)} = i^* \frac{\partial^2}{\partial n_g^2} + 2H_g i^* \frac{\partial}{\partial n_g} + \Delta_{(\Sigma,g)} i^*.$$

Now we derive a formula for the conformal change of second-order geodesic normal derivatives $N_g^2 = \partial^2 / \partial n_g^2$. The following results are valid in all dimensions.

Lemma 6.3.2. *For* $\hat{g} = e^{2\varphi} g$, $N_{\hat{g}} = e^{-\varphi} N_g$ *and*

$$N_{\hat{g}}^2 = e^{-2\varphi} \left(N_g^2 - \frac{\partial \varphi}{\partial n_g} N_g + \text{grad}_g^\Sigma (\varphi) \right)$$

as identities of differential operators $C^\infty(M) \to C^\infty(\Sigma)$.

Proof. The first formula is obvious. For the proof of the second assertion we use

$$\Delta_{(M,\hat{g})} = e^{-2\varphi} \Delta_{(M,g)} + e^{-2\varphi} \text{grad}_g^M (\varphi)$$

(see Lemma 4.2.1) and apply Lemma 6.3.1 to both sides. Then

$$N_{\hat{g}}^2 + 2H_{\hat{g}} N_{\hat{g}} + \Delta_{(\Sigma,\hat{g})} = e^{-2\varphi} \left(N_g^2 + 2H_g N_g + \Delta_{(\Sigma,g)} + \text{grad}_g^M (\varphi) \right).$$

The assertion follows from

$$H_{\hat{g}} N_{\hat{g}} = e^{-2\varphi} \left(H_g + N_g(\varphi) \right) N_g. \tag{6.3.7}$$

The proof is complete. $\square$

Using (6.3.3), we obtain

$$e^{(\lambda-2)\varphi} D_2(M, \Sigma; g; \lambda)(e^{-\lambda\varphi} u)$$
$$= \lambda^3 e^{-2\varphi} \left\{ -2H_g^2 u \right\}$$
$$+ \lambda^2 e^{-2\varphi} \left\{ 4H_g \left(-\lambda \frac{\partial \varphi}{\partial n_g} u + \frac{\partial u}{\partial n_g} \right) + c_g u \right\}$$
$$+ \lambda e^{-2\varphi} \left\{ 2\lambda \frac{\partial^2 \varphi}{\partial n_g^2} u + 4\lambda \frac{\partial \varphi}{\partial n_g} \frac{\partial u}{\partial n_g} - 2 \frac{\partial^2 u}{\partial n_g^2} - 2\lambda^2 \left(\frac{\partial \varphi}{\partial n_g} \right)^2 u \right.$$
$$\left. + 2\lambda H_g \frac{\partial \varphi}{\partial n_g} u - 2H_g \frac{\partial u}{\partial n_g} + K_g u \right\}$$
$$+ e^{(\lambda-2)\varphi} \Delta_{(\Sigma,g)} (e^{-\lambda\varphi} u)$$

$$= -2\lambda^3 e^{-2\varphi} \left\{ H_g^2 + 2H_g \frac{\partial\varphi}{\partial n_g} + \left(\frac{\partial\varphi}{\partial n_g}\right)^2 \right\} u$$

$$+ \lambda^2 e^{-2\varphi} \left\{ 2\frac{\partial^2\varphi}{\partial n_g^2} u + 4\frac{\partial\varphi}{\partial n_g}\frac{\partial u}{\partial n_g} + 4H_g\frac{\partial u}{\partial n_g} + 2H_g\frac{\partial\varphi}{\partial n_g}u + c_g u \right\}$$

$$+ \lambda e^{-2\varphi} \left\{ -2\frac{\partial^2 u}{\partial n_g^2} - 2H_g\frac{\partial u}{\partial n_g} + K_g u \right\}$$

$$+ e^{(\lambda-2)\varphi}\Delta_{(\Sigma,g)}(e^{-\lambda\varphi}u).$$

In order to evaluate the term which involves the Laplacian on Σ, we apply

Lemma 6.3.3.

$$\Delta_{(\Sigma,g)}(e^{\lambda\varphi}u) = \Delta_{(\Sigma,g)}(e^{\lambda\varphi})u + 2\lambda e^{\lambda\varphi}\left(\mathrm{grad}_g^\Sigma(\varphi), \mathrm{grad}_g^\Sigma(u)\right)_g + e^{\lambda\varphi}\Delta_{(\Sigma,g)}(u)$$

and

$$\Delta_{(\Sigma,g)}(e^{\lambda\varphi}) = \lambda e^{\lambda\varphi}\Delta_{(\Sigma,g)}(\varphi) + \lambda^2 e^{\lambda\varphi}\left|\mathrm{grad}_g^\Sigma(\varphi)\right|_g^2.$$

and find

$$e^{(\lambda-2)\varphi}\Delta_{(\Sigma,g)}(e^{-\lambda\varphi}u) = -\lambda e^{-2\varphi}\Delta_{(\Sigma,g)}(\varphi)u + \lambda^2 e^{-2\varphi}\left|\mathrm{grad}_g^\Sigma(\varphi)\right|_g^2 u$$

$$- 2\lambda e^{-2\varphi}\left(\mathrm{grad}_g^\Sigma(\varphi), \mathrm{grad}_g^\Sigma(u)\right)_g + e^{-2\varphi}\Delta_{(\Sigma,g)}(u).$$

Thus we obtain the identity

$$e^{(\lambda-2)\varphi}D_2(M,\Sigma; g; \lambda)(e^{-\lambda\varphi}u) \tag{6.3.8}$$

$$= -2\lambda^3 e^{-2\varphi}\left\{ H_g^2 + 2H_g\frac{\partial\varphi}{\partial n_g} + \left(\frac{\partial\varphi}{\partial n_g}\right)^2 \right\} u$$

$$+ \lambda^2 e^{-2\varphi}\left\{ 2\frac{\partial^2\varphi}{\partial n_g^2}u + 4\frac{\partial\varphi}{\partial n_g}\frac{\partial u}{\partial n_g} + 4H_g\frac{\partial u}{\partial n_g} + 2H_g\frac{\partial\varphi}{\partial n_g}u + \left|\mathrm{grad}_g^\Sigma(\varphi)\right|_g^2 u + c_g u \right\}$$

$$+ \lambda e^{-2\varphi}\left\{ -2\frac{\partial^2 u}{\partial n_g^2} - 2H_g\frac{\partial u}{\partial n_g} - \Delta_{(\Sigma,g)}(\varphi)u - 2\left(\mathrm{grad}_g^\Sigma(\varphi), \mathrm{grad}_g^\Sigma(u)\right)_g + K_g u \right\}$$

$$+ e^{-2\varphi}\Delta_{(\Sigma,g)}(u).$$

We claim that the coefficients of the powers of λ in (6.3.8) coincide with the coefficients of the corresponding powers in $D_2(M, \Sigma; \hat{g}; \lambda)$. First of all, the identity $e^{-2\varphi}\Delta_{(\Sigma,g)} = \Delta_{(\Sigma,\hat{g})}$ on the surface Σ establishes the assertion for the constant term. The linear term of $D_2(\hat{g}; \lambda)$ is given by

$$-2\frac{\partial^2}{\partial n_{\hat{g}}^2} - 2H_g\frac{\partial}{\partial n_{\hat{g}}} + K_{\hat{g}}.$$

Now (6.3.7) and

$$K_{\hat{g}} = e^{-2\varphi}(K_g - \Delta_{(\Sigma,g)}(\varphi)), \qquad (6.3.9)$$

together with Lemma 6.3.2, show that the linear term equals

$$-2e^{-2\varphi}\frac{\partial^2}{\partial n_g^2} - 2e^{-2\varphi}H_g\frac{\partial}{\partial n_g} - 2e^{-2\varphi}\operatorname{grad}_g^{\Sigma}(\varphi) + e^{-2\varphi}(K_g - \Delta_{(\Sigma,g)}(\varphi)).$$

The last sum coincides with the coefficient of λ in (6.3.8). (6.2.4) also shows that the coefficients of λ^3 coincide. Thus it only remains to identify the coefficients of λ^2. In other words, it only remains to prove that

$$e^{-2\varphi}\left\{2\frac{\partial^2\varphi}{\partial n_g^2} + \left|\operatorname{grad}_g^{\Sigma}(\varphi)\right|_g^2 + 2H_g\frac{\partial\varphi}{\partial n_g} + 4\frac{\partial\varphi}{\partial n_g}\frac{\partial}{\partial n_g} + 4H_g\frac{\partial}{\partial n_g} + c_g\right\}$$
$$= 4H_{\hat{g}}\frac{\partial}{\partial n_{\hat{g}}} + c(\hat{g})$$

on Σ. (6.3.7) shows that the latter identity is equivalent to

$$e^{-2\varphi}\left\{2\frac{\partial^2\varphi}{\partial n_g^2} + 2H_g\frac{\partial\varphi}{\partial n_g} + \left|\operatorname{grad}_g^{\Sigma}(\varphi)\right|_g^2 + 2K_g - H_g^2 - \frac{1}{2}\tau_g\right\}$$
$$= 2K_{\hat{g}} - H_{\hat{g}}^2 - \frac{1}{2}\tau_{\hat{g}} \quad (6.3.10)$$

on Σ. But (6.2.4), (6.3.7), (6.3.9) and (2.5.6) imply

$$2K_{\hat{g}} - H_{\hat{g}}^2 - \frac{1}{2}\tau_{\hat{g}} = e^{-2\varphi}\left\{2K_g - 2\Delta_{(\Sigma,g)}(\varphi) - \left(H_g + \frac{\partial\varphi}{\partial n_g}\right)^2\right\}$$
$$+ e^{-2\varphi}\left\{2\Delta_{(M,g)}(\varphi) + \left|\operatorname{grad}_g^{M}(\varphi)\right|_g^2\right\} - \frac{1}{2}e^{-2\varphi}\tau_g$$
$$= e^{-2\varphi}\left\{2K_g - H_g^2 - \frac{1}{2}\tau_g - 2H_g\frac{\partial\varphi}{\partial n_g} + \left|\operatorname{grad}_g^{M}(\varphi)\right|_g^2\right.$$
$$\left. - \left(\frac{\partial\varphi}{\partial n_g}\right)^2 + 2\Delta_{(M,g)}(\varphi) - 2\Delta_{(\Sigma,g)}(\varphi)\right\}$$
$$= e^{-2\varphi}\left\{2K_g - H_g^2 - \frac{1}{2}\tau_g + \left|\operatorname{grad}_g^{\Sigma}(\varphi)\right|_g^2 + 2\frac{\partial^2\varphi}{\partial n_g^2} + 2H_g\frac{\partial\varphi}{\partial n_g}\right\}$$

using Lemma 6.3.1. This proves (6.3.10) and the proof of Theorem 6.3.1 is complete.   □

Theorem 6.3.2 is a direct consequence of Theorem 6.3.1 and Lemma 6.3.1. We omit the details.

6.4 Second-order families. General case

In the present section, we generalize Theorem 6.3.2 to oriented codimension one submanifolds Σ^n of an arbitrary manifold M^{n+1} of dimension $n+1$ ($n > 2$) with a background metric g. The arguments are extensions of those in Section 6.3.

Theorem 6.4.1. *The family*

$$D_2(M, \Sigma; g; \lambda) = -(2\lambda+n-2)i^* P_2(M, g) + (2\lambda+n-1)P_2(\Sigma, g)i^*$$
$$+ (2\lambda+n-1)(2\lambda+n-2)H_g i^* N_g$$
$$+ 2\left(\lambda + \frac{n-1}{2}\right)\left(\lambda + \frac{n-2}{2}\right)\left(Q_2(\Sigma; g) - Q_2(M; g) - \lambda H_g^2\right)i^* \quad (6.4.1)$$

is conformally covariant, i.e.,

$$e^{-(\lambda-2)\varphi} \circ D_2(M, \Sigma; e^{2\varphi}g; \lambda) \circ e^{\lambda\varphi} = D_2(M, \Sigma; g; \lambda) \qquad (6.4.2)$$

for all $\varphi \in C^\infty(M)$.

The family $\lambda \mapsto D_2(g; \lambda)$ is of degree 3 and satisfies the factorization relations

$$D_2\left(g; -\frac{n}{2}+1\right) = P_2(\Sigma, g)i^* \quad \text{and} \quad D_2\left(g; -\frac{n-1}{2}\right) = i^* P_2(M, g).$$

Example 6.4.1. *Let $(M, \Sigma) = (S^{n+1}, S^n)$ with the round metric $g = g_c$. Then (6.4.1) specializes to*

$$D_2(S^{n+1}, S^n; g_c; \lambda) = -(2\lambda+n-2)i^* P_2(S^{n+1}, g_c)$$
$$+ (2\lambda+n-1)P_2(S^n, g_c)i^* - \left(\lambda + \frac{n-1}{2}\right)\left(\lambda + \frac{n-2}{2}\right)i^*$$

using $H_{g_c} = 0$ and $Q_2(S^m; g_c) = \frac{m}{2}$, i.e., the family coincides with $D_2^c(S^{n+1}, S^n; \lambda)$ (see (5.3.2)).

Example 6.4.2. *Let $(M, \Sigma) = (\mathbb{R}^{n+1}, S^n)$ with the Euclidean metric $g = g_c$. Then (6.4.1) specializes to*

$$D_2(\mathbb{R}^{n+1}, S^n; g_c; \lambda) = -(2\lambda+n-2)i^* \Delta_{\mathbb{R}^{n+1}} + (2\lambda+n-1)\left(\Delta_{S^n} - \frac{n}{2}\left(\frac{n}{2}-1\right)\right)i^*$$
$$+ (2\lambda+n-1)(2\lambda+n-2)i^*\partial/\partial r$$
$$+ 2\left(\lambda + \frac{n-1}{2}\right)\left(\lambda + \frac{n-2}{2}\right)\left(\frac{n}{2} - \lambda\right)i^*$$
$$= -(2\lambda+n-2)i^* \Delta_{\mathbb{R}^{n+1}} + (2\lambda+n-1)\left(\Delta_{S^n} + \lambda(1-\lambda)\right)i^*$$
$$+ (2\lambda+n-1)(2\lambda+n-2)i^*\partial/\partial r$$

using $H = 1$ for the exterior normal derivative $\partial/\partial r$. Hence the family coincides with $D_2^b(\mathbb{R}^{n+1}, S^n; g_c; \lambda)$ (see (5.4.12)).

In Section 6.24, the families $D_2(M, \Sigma; g; \lambda)$ will be recognized as tractor families. These are curved versions of families which are induced by homomorphisms of Verma modules. In particular, the latter observation that $D_2(M, \Sigma; g; \lambda)$ specializes to the induced families $D_2^c(\lambda)$ and $D_2^b(\lambda)$ finds its natural explanation.

In order to prove Theorem 6.4.1, we have to verify four identities for the four Taylor coefficients of both sides of (6.4.2) at $\lambda = -\frac{n}{2} + 1$. Here we restrict to a detailed proof for the coefficients of the powers 0, 1 and 3. The remaining identity can be proved similarly.

An alternative proof will be given in Section 6.24. It relates the family $D_2(M, \Sigma; g; \lambda)$ to a construction in terms of the conformally invariant tractor calculus. More precisely, $D_2(M, \Sigma; g; \lambda)$ can be written as the composition of certain universal conformally covariant families (tractor D-operators).

The fact that the coefficients of $(\lambda + \frac{n}{2} - 1)^0$ on both sides of (6.4.2) coincide is equivalent to

$$e^{-(\frac{n}{2}+1)\varphi} \circ D_2\left(M, \Sigma; \hat{g}; -\frac{n}{2}+1\right) \circ e^{(-\frac{n}{2}+1)\varphi} = D_2\left(M, \Sigma; g; -\frac{n}{2}+1\right).$$

But this follows from $D_2(M, \Sigma; g; -\frac{n}{2}+1) = P_2(\Sigma, g)i^*$.

Next, we find that on the left-hand side of (6.4.2) the coefficient of $(\lambda + \frac{n}{2} - 1)^3$ is given by

$$\left[-2H_{\hat{g}}^2 + 4H_{\hat{g}}\frac{\partial\varphi}{\partial n_{\hat{g}}} - 2\left(\frac{\partial\varphi}{\partial n_{\hat{g}}}\right)^2\right]e^{2\varphi} = -2\left(H_{\hat{g}} - \frac{\partial\varphi}{\partial n_{\hat{g}}}\right)^2 e^{2\varphi}.$$

In view of

$$H_g = e^\varphi H_{\hat{g}} - \frac{\partial\varphi}{\partial n_g} = e^\varphi\left(H_{\hat{g}} - \frac{\partial\varphi}{\partial n_{\hat{g}}}\right),$$

this coincides with the corresponding coefficient $-2H_g^2$ of the right-hand side of (6.4.2).

Next, we differentiate (6.4.2) at $\lambda = -\frac{n}{2} + 1$ and find

$$[P_2(\Sigma, g), \varphi] + e^{(\frac{n}{2}+1)\varphi} \circ \dot{D}_2\left(\hat{g}; -\frac{n}{2}+1\right) \circ e^{(-\frac{n}{2}+1)\varphi} = \dot{D}_2\left(g; -\frac{n}{2}+1\right). \quad (6.4.3)$$

On the other hand, (6.4.1) yields

$$\dot{D}_2\left(g; -\frac{n}{2}+1\right) = -2\Delta_{(M,g)} + 2\Delta_{(\Sigma,g)}$$
$$+ 2H_g\frac{\partial}{\partial n_g} + \left((n-2)Q_2(M; g) - (n-3)Q_2(\Sigma; g) + \left(\frac{n}{2}-1\right)H_g^2\right). \quad (6.4.4)$$

We prove that (6.4.3) is satisfied by the latter operator. In the first step we verify (6.4.3) for $u = 1$. Let

$$q_2(g) = \dot{D}_2\left(g; -\frac{n}{2}+1\right)(1) \in C^\infty(\Sigma).$$

Then

$$q_2(g) = (n-2)Q_2(M;g) - (n-3)Q_2(\Sigma;g) + \left(\frac{n}{2}-1\right)H_g^2 \in C^\infty(\Sigma). \tag{6.4.5}$$

In these terms, the assertion is

$$q_2(g) = \Delta_{(\Sigma,g)}(\varphi) + e^{\left(\frac{n}{2}+1\right)\varphi}\left[-2\Delta_{(M,\hat{g})} + 2\Delta_{(\Sigma,\hat{g})} + 2H_{\hat{g}}\frac{\partial}{\partial n_{\hat{g}}} + q_2(\hat{g})\right]\left(e^{\left(-\frac{n}{2}+1\right)\varphi}\right),$$

i.e.,

$$q_2(g) = \Delta_{(\Sigma,g)}(\varphi) - 2e^{\left(\frac{n}{2}+1\right)\varphi}\left[\frac{\partial^2}{\partial n_{\hat{g}}^2} + (n-1)H_{\hat{g}}\frac{\partial}{\partial n_{\hat{g}}}\right]\left(e^{\left(-\frac{n}{2}+1\right)\varphi}\right) + q_2(\hat{g})e^{2\varphi}$$

using the appropriate analog of Lemma 6.3.1 (with geodesic normals). We apply Lemma 6.3.2 in order to make the latter formula more explicit. It follows that the differential operator on its right-hand side equals

$$\Delta_{(\Sigma,g)}(\varphi)$$
$$- 2e^{\left(\frac{n}{2}-1\right)\varphi}\left[\frac{\partial^2}{\partial n_g^2} + (n-1)H_g\frac{\partial}{\partial n_g} + (n-2)\frac{\partial\varphi}{\partial n_g}\frac{\partial}{\partial n_g} + \mathrm{grad}_g^\Sigma(\varphi)\right]\left(e^{\left(-\frac{n}{2}+1\right)\varphi}\right),$$

i.e.,

$$\Delta_{(\Sigma,g)}(\varphi) - 2e^{\left(\frac{n}{2}-1\right)\varphi}\frac{\partial^2}{\partial n_g^2}\left(e^{\left(-\frac{n}{2}+1\right)\varphi}\right)$$

$$+ (n-1)(n-2)H_g\frac{\partial\varphi}{\partial n_g} + (n-2)^2\left(\frac{\partial\varphi}{\partial n_g}\right)^2 + (n-2)\left|\mathrm{grad}_g^\Sigma(\varphi)\right|_g^2$$

$$= \Delta_{(\Sigma,g)}(\varphi) + (n-2)\frac{\partial^2\varphi}{\partial n_g^2} + \frac{(n-2)^2}{2}\left(\frac{\partial\varphi}{\partial n_g}\right)^2$$

$$+ (n-1)(n-2)H_g\frac{\partial\varphi}{\partial n_g} + (n-2)\left|\mathrm{grad}_g^\Sigma(\varphi)\right|_g^2.$$

Therefore, the claim is

$$q_2(g) = q_2(\hat{g})e^{2\varphi} + \left[\Delta_{(\Sigma,g)}(\varphi) + (n-2)\frac{\partial^2\varphi}{\partial n_g^2} + \frac{(n-2)^2}{2}\left(\frac{\partial\varphi}{\partial n_g}\right)^2\right.$$

$$\left. + (n-1)(n-2)H_g\frac{\partial\varphi}{\partial n_g} + (n-2)\left|\mathrm{grad}_g^\Sigma(\varphi)\right|_g^2\right]. \tag{6.4.6}$$

But the latter identity (6.4.6) is a consequence of Yamabe's equations (see (3.1.9)) for the scalar curvatures of M and Σ, and the fundamental identity for the mean curvature. In fact, we recall that, by Yamabe's equation,

$$\frac{n-2}{4(n-1)}\tau_X(\hat{g}) = \frac{n-2}{4(n-1)}e^{-2\varphi}\tau_X(g) - e^{-\frac{n+2}{2}\varphi}\Delta_{(X,g)}\left(e^{\frac{n-2}{2}\varphi}\right)$$

on an n-manifold X. We write this in the form

$$(n-2)Q_2(X;\hat{g}) = (n-2)Q_2(X;g)e^{-2\varphi} - 2e^{-\frac{n+2}{2}\varphi}\Delta_{(X,g)}(e^{\frac{n-2}{2}\varphi}).$$

In particular, we have (for $n > 2$)

$$(n-2)Q_2(M;\hat{g}) = (n-2)Q_2(M;g)e^{-2\varphi} - 2\frac{n-2}{n-1}e^{-\frac{n+3}{2}\varphi}\Delta_{(M,g)}(e^{\frac{n-1}{2}\varphi}) \quad (6.4.7)$$

and

$$(n-3)Q_2(\Sigma;\hat{g}) = (n-3)Q_2(\Sigma;g)e^{-2\varphi} - 2\frac{n-3}{n-2}e^{-\frac{n+2}{2}\varphi}\Delta_{(\Sigma,g)}(e^{\frac{n-2}{2}\varphi}). \quad (6.4.8)$$

Now, by Lemma 6.3.3, (6.4.7) is equivalent to

$$(n-2)Q_2(M;\hat{g})e^{2\varphi}$$
$$= (n-2)Q_2(M;g) - (n-2)\Delta_{(M,g)}(\varphi) - \frac{(n-1)(n-2)}{2}\left|\text{grad}_g^M(\varphi)\right|_g^2.$$

Similarly, (6.4.8) is equivalent to

$$(n-3)Q_2(\Sigma;\hat{g})e^{2\varphi}$$
$$= (n-3)Q_2(\Sigma;g) - (n-3)\Delta_{(\Sigma,g)}(\varphi) - \frac{(n-2)(n-3)}{2}\left|\text{grad}_g^\Sigma(\varphi)\right|_g^2.$$

Hence we get

$$q_2(\hat{g})e^{2\varphi} = q_2(g) - (n-2)\Delta_{(M,g)}(\varphi) - \frac{(n-1)(n-2)}{2}\left|\text{grad}_g^M(\varphi)\right|_g^2$$
$$+ (n-3)\Delta_{(\Sigma,g)}(\varphi) + \frac{(n-2)(n-3)}{2}\left|\text{grad}_g^\Sigma(\varphi)\right|_g^2$$
$$+ \left(\frac{n}{2}-1\right)\left(\left(H_g + \frac{\partial\varphi}{\partial n_g}\right)^2 - H_g^2\right).$$

Now using

$$\Delta_{(M,g)} = \frac{\partial^2\varphi}{\partial n_g^2} + nH_g\frac{\partial\varphi}{\partial n_g} + \Delta_{(\Sigma,g)}, \quad (6.4.9)$$

we rewrite the above formula for the difference $q_2(\hat{g})e^{2\varphi} - q_2(g)$ as

$$- \Delta_{(\Sigma,g)}(\varphi) - n(n-2)H_g\frac{\partial\varphi}{\partial n_g} - (n-2)\frac{\partial^2\varphi}{\partial n_g^2}$$
$$- \frac{(n-1)(n-2)}{2}\left|\text{grad}_g^M(\varphi)\right|_g^2 + \frac{(n-2)(n-3)}{2}\left|\text{grad}_g^\Sigma(\varphi)\right|_g^2$$
$$+ (n-2)H_g\frac{\partial\varphi}{\partial n_g} + \frac{n-2}{2}\left(\frac{\partial\varphi}{\partial n_g}\right)^2$$

$$= -\Delta_{(\Sigma,g)}(\varphi) - (n-2)\frac{\partial^2 \varphi}{\partial n_g^2} - (n-1)(n-2)H_g\frac{\partial \varphi}{\partial n_g}$$

$$- \frac{(n-2)^2}{2}\left(\frac{\partial \varphi}{\partial n_g}\right)^2 - (n-2)\left|\text{grad}_g^\Sigma(\varphi)\right|_g^2$$

using

$$\left|\text{grad}_g^M(\varphi)\right|_g^2 = \left|\text{grad}_g^\Sigma(\varphi)\right|_g^2 + \left(\frac{\partial \varphi}{\partial n_g}\right)^2.$$

The latter sum coincides with the sum in (6.4.6), i.e., (6.4.6) is a consequence of the Yamabe equations and the fundamental identity for H_g.

Now having established (6.4.3) for $u = 1$, it is easy to prove (6.4.3) for all $u \in C^\infty(M)$. In view of (6.4.4) and (6.4.9), we have to prove that

$$q_2(g) - 2\frac{\partial^2}{\partial n_g^2} - 2(n-1)H_g\frac{\partial}{\partial n_g}$$

$$= e^{(\frac{n}{2}+1)\varphi} \circ \left[-2\frac{\partial^2}{\partial n_{\hat{g}}^2} - 2(n-1)H_{\hat{g}}\frac{\partial}{\partial n_{\hat{g}}} + q_2(\hat{g})\right] \circ e^{(-\frac{n}{2}+1)\varphi} + \left[\Delta_{(\Sigma,g)}, \varphi\right].$$

$$(6.4.10)$$

By the above calculations, the right-hand side of (6.4.10) coincides with

$$q_2(\hat{g})e^{2\varphi}$$

$$+ \left\{\Delta_{(\Sigma,g)}(\varphi) + (n-2)\frac{\partial^2 \varphi}{\partial n_g^2} + \frac{(n-2)^2}{2}\left(\frac{\partial \varphi}{\partial n_g}\right)^2\right.$$

$$\left. + (n-1)(n-2)H_g\frac{\partial \varphi}{\partial n_g} + (n-2)\left|\text{grad}_g^\Sigma(\varphi)\right|_g^2\right\}$$

$$+ e^{2\varphi}\left[-2\frac{\partial^2}{\partial n_{\hat{g}}^2} + 2(n-2)\frac{\partial \varphi}{\partial n_{\hat{g}}}\frac{\partial}{\partial n_{\hat{g}}} - 2(n-1)H_{\hat{g}}\frac{\partial}{\partial n_{\hat{g}}}\right] + 2\,\text{grad}_g^\Sigma(\varphi).$$

By (6.4.6), the first two terms sum up to $q_2(g)$. Therefore, it only remains to prove that

$$\frac{\partial^2}{\partial n_g^2} + (n-1)H_g\frac{\partial}{\partial n_g}$$

$$= e^{2\varphi}\left[\frac{\partial^2}{\partial n_{\hat{g}}^2} - (n-2)\frac{\partial \varphi}{\partial n_{\hat{g}}}\frac{\partial}{\partial n_{\hat{g}}} + (n-1)H_{\hat{g}}\frac{\partial}{\partial n_{\hat{g}}}\right] - \text{grad}_g^\Sigma(\varphi).$$

Now it suffices to apply Lemma 6.3.2 and (6.3.7). This proves (6.4.10).

We close the present section with some comments. We consider the Q-polynomial $Q_2(g; \lambda)$ more closely. Theorem 6.4.1 implies the formula

$$Q_2\left(g; -\frac{n}{2}+1\right) = (n-2)Q_2(M;g) - (n-3)Q_2(\Sigma;g) + \left(\frac{n}{2}-1\right)H_g^2.$$

Hence for $n > 2$ the function $Q_2(g; -\frac{n}{2}+1)$ is composed of Q-curvatures and (squared) mean curvature. However, in the critical case $n = 2$, $Q_2(g; 0)$ is completely determined only by $Q_2(\Sigma; g)$. It is immediate from the definition that

$$\frac{1}{2}\ddot{Q}_2\left(g; -\frac{n}{2}+1\right) = (n-3)H_g^2 + 2\left(Q_2(\Sigma; g) - Q_2(M; g)\right).$$

Thus we can identify the Q-polynomial.

Lemma 6.4.1. *The Q-polynomial $Q_2(M, \Sigma; g; \lambda)$ is given by*

$$- 2H_g^2 \left(\lambda + \frac{n}{2} - 1\right)^3$$
$$+ \left[(n-3)H_g^2 + 2Q_2(\Sigma; g) - 2Q_2(M; g)\right]\left(\lambda + \frac{n}{2} - 1\right)^2$$
$$+ \left[\left(\frac{n}{2} - 1\right)H_g^2 + (n-2)Q_2(M; g) - (n-3)Q_2(\Sigma; g)\right]\left(\lambda + \frac{n}{2} - 1\right)$$
$$- \left(\frac{n}{2} - 1\right)Q_2(\Sigma).$$

This result extends Corollary 6.3.1.

We notice that (6.4.1) immediately yields $D_2(M, \Sigma; g; 0)(1) = 0$, i.e., $D_2(g; 0)$ annihilates constants. Hence it is also natural to write the Q-polynomial as a Taylor polynomial at $\lambda = 0$ (as in Corollary 6.3.1). We find

$$Q_2(M, \Sigma; g; \lambda) = -2\lambda^3 H_g^2 + \lambda^2 \left[2Q_2(\Sigma; g) - 2Q_2(M; g) - (2n-3)H_g^2\right]$$
$$+ \lambda\left[(n-1)Q_2(\Sigma; g) - (n-2)Q_2(M; g) - \frac{(n-1)(n-2)}{2}H_g^2\right]. \quad (6.4.11)$$

Example 6.4.3. *In the case of the trivial embedding $M^{n+1} = \mathbb{R} \times \Sigma^n$ with the metric $g_M = dr^2 + g_\Sigma$,*

$$Q_2(M, \Sigma; g_M; \lambda) = \frac{2\lambda(\lambda+n-1)}{n}Q_2(\Sigma; g_\Sigma).$$

In particular, for $n = 2$, we find

$$Q_2(M, \Sigma; g_M; \lambda) = \lambda(\lambda+1)Q_2(\Sigma; g_\Sigma).$$

Thus $Q_2(g; \lambda)$ is a quadratic polynomial which is determined by $Q_2(g; \Sigma)$. Of course, if $H = 0$ and $i^*Q_2(M; g) = Q_2(\Sigma; g)$ then the degree of the polynomial is only 1. In Section 6.11, we shall consider a class of embeddings with this property.

Proof. We have $H = 0$, $\tau_M = \tau_\Sigma$ and $Q_2(g_M) = \frac{\tau_M}{2n}$, $Q_2(g_\Sigma) = \frac{\tau_\Sigma}{2(n-1)}$. Hence (6.4.11) yields

$$Q_2(M, \Sigma; g_M; \lambda) = \lambda^2 \left(\frac{1}{n-1} - \frac{1}{n}\right)\tau + \lambda\left(\frac{1}{2} - \frac{n-2}{2n}\right)\tau = \frac{1}{n}\lambda(\lambda+n-1)\frac{\tau}{n-1}.$$

This implies the assertion. $\square$

Although the quantity

$$\frac{(n-1)(n-2)}{2} H_g^2 + (n-2)Q_2(M; g) - (n-1)Q_2(\Sigma; g)$$

is *not* conformally invariant, it is related to a conformal invariant. We recall a result of Chen. Let

$$\tau_e(g) \overset{\text{def}}{=} \frac{2}{n(n-1)} \sum_{i<j} \lambda_i \lambda_j,$$

where λ_i are the principal curvatures. τ_e is called the *extrinsic scalar curvature*. Chen ([76]) observed that

$$\left(H_g^2 - \tau_e(g)\right) g \qquad (6.4.12)$$

is a conformal invariant. In fact, $H^2 - \tau_e$ is proportional to the squared norm $|L_0|^2$ of the trace-free part L_0 of L (see Lemma 6.22.4). The conformal invariance of the symmetric bilinear form (6.4.12) has been observed long before. In particular, it has been the basis of Fialkow's profound study of conformal invariants of submanifolds ([100]). For more details see Section 6.23.

For $n = 2$, (6.4.12) reduces to $(H_g^2 - K_g + \overline{K}_g)g$ used in Section 6.3.

Lemma 6.4.2.

$$\frac{(n-1)(n-2)}{2} H_g^2 + (n-2)Q_2(M; g) - (n-1)Q_2(\Sigma; g)$$

$$= \frac{(n-1)(n-2)}{2} \left(H_g^2 - \tau_e(g)\right) - \frac{1}{n} \left(\tau_\Sigma(g) + (n-2)\operatorname{tr}(G_g)\right),$$

where $G_i^j = R_{iNN}^j$. *The trace* $\operatorname{tr}(G) = \sum_i G_i^i = \sum_i \operatorname{Ric}_{NN}$ *is the Ricci curvature in the direction of the unit normal.*

Proof. We have

$$H_g^2 + \frac{2Q_2(M; g)}{n-1} - \frac{2Q_2(\Sigma; g)}{n-2} = \left(H_g^2 + \frac{\tau_M(g)}{n(n-1)}\right) - \frac{\tau_\Sigma(g)}{(n-1)(n-2)}.$$

But the Gauß equation

$$R_M(X, Y, Z, W) = R_\Sigma(X, Y, Z, W) - L(X, W)L(Y, Z) + L(X, Z)L(Y, W)$$

for $X, Y, Z, W \in \mathcal{X}(\Sigma)$ implies

$$\tau_M = \tau_\Sigma + 2\operatorname{tr}(G) - \left(\sum_i \lambda_i\right)^2 + \sum_i \lambda_i^2 = \tau_\Sigma + 2\operatorname{tr}(G) - n(n-1)\tau_e. \qquad (6.4.13)$$

Hence the above sum can be written as

$$\left(H_g^2 - \tau_e(g)\right) + \frac{2}{n(n-1)} \operatorname{tr}(G_g) - \frac{2}{n(n-1)(n-2)} \tau_\Sigma(g).$$

This proves the assertion. $\qquad \square$

It follows that the difference

$$\dot{Q}_2(M, \Sigma; g; 0)g - \frac{1}{n}\left[\tau_\Sigma(g) - (n-2)\operatorname{tr}(G_g)\right]g$$

is conformally invariant and coincides with a multiple of $(H_g^2 - \tau_e(g))g$.

6.5   Families and the asymptotics of eigenfunctions

We start with some motivating observations in the conformally flat case (S^n, S^{n-1}) with the canonical metric g_c. Let $u \in C^\infty(H^+)$ be an eigenfunction

$$-\Delta_N u = \mu(n-1-\mu)u, \quad \Re(\mu) = (n-1)/2$$

of the Laplacian Δ_N of the metric $\mathcal{H}_0^{-2}g_c$ with boundary value $f \in C^\infty(S^{n-1})$ so that

$$u \sim \sum_{N \geq 0} \mathcal{H}_0^{\mu+N} a_N(f) + \sum_{N \geq 0} \mathcal{H}_0^{n-1-\mu+N} b_N(f)$$

with $a_0(f) = f$. The eigenfunction u gives rise to the holomorphic family

$$\langle M_u(\lambda), \varphi \rangle = \int_{H^+} \mathcal{H}_0^\lambda u\varphi \operatorname{vol}(S^n, g_c), \quad \varphi \in C^\infty(S^n)$$

of distributions on the half-plane $\Re(\lambda) > -\frac{n-1}{2}$.

Lemma 6.5.1. *Let the operator family*

$$\delta_N^c(\mu) : C^\infty(S^n) \to C^\infty(S^{n-1})$$

be defined by the residue formula

$$\operatorname{Res}_{\lambda=-\mu-1-N}\left(\langle M_u(\lambda), \varphi \rangle\right) = \int_{S^{n-1}} f\delta_N^c(\mu+N-(n-1))(\varphi)\operatorname{vol}(S^{n-1}, g_c). \quad (6.5.1)$$

Then

$$\delta_N^c(\mu) \circ \pi_\mu^c(g) = \pi_{\mu-N}^c(g) \circ \delta_N^c(\mu), \quad g \in G^n. \quad (6.5.2)$$

Proof. The G^n-equivariance of $\delta_N^c(\mu)$ is a consequence of $\mathcal{H}_0 \in C^\infty(S^n)_1^{G^n}$. In fact,

$$\pi_1^c(g)(\mathcal{H}_0) = g_*(\mathcal{H}_0)\left(\frac{g_*(\operatorname{vol}(S^n, g_c))}{\operatorname{vol}(S^n, g_c)}\right)^{-\frac{1}{n}} = \mathcal{H}_0 \quad (6.5.3)$$

implies

$$\langle M_u(\lambda), \varphi \rangle = \int_{H^+} g_*(\mathcal{H}_0)^\lambda g_*(u)g_*(\varphi)\left(\frac{g_*(\operatorname{vol}(S^n, g_c))}{\operatorname{vol}(S^n, g_c)}\right)\operatorname{vol}(S^n, g_c)$$

$$= \int_{H^+} \mathcal{H}_0^\lambda g_*(u)\left(\frac{g_*(\operatorname{vol}(S^n, g_c))}{\operatorname{vol}(S^n, g_c)}\right)^{\frac{\lambda}{n}+1} g_*(\varphi)\operatorname{vol}(S^n, g_c)$$

$$= \int_{H^+} \mathcal{H}_0^\lambda g_*(u)\pi_{-\lambda-n}^c(g)(\varphi)\operatorname{vol}(S^n, g_c). \quad (6.5.4)$$

Now (6.5.3) and the asymptotics $u \sim \mathcal{H}_0^\mu f + \cdots$ yield $g_*(u) \sim \mathcal{H}_0^\mu \pi_{-\mu}^c(g) f + \cdots$. We apply (6.5.1) to (6.5.4) and obtain the identity

$$\int_{S^{n-1}} f \delta_N^c(\mu + N - (n-1))(\varphi) \operatorname{vol}(S^{n-1}, g_c)$$

$$= \int_{S^{n-1}} \pi_{-\mu}^c(g)(f) \delta_N^c(\mu + N - (n-1)) \left(\pi_{\mu + N - (n-1)}^c(g)(\varphi) \right) \operatorname{vol}(S^{n-1}, g_c)$$

$$= \int_{S^{n-1}} f \pi_{\mu - (n-1)}^c(g^{-1}) \delta_N^c(\mu + N - (n-1)) \pi_{\mu + N - (n-1)}^c(g)(\varphi) \operatorname{vol}(S^{n-1}, g_c).$$

The proof is complete. $\qquad\square$

We refer also to (5.2.63) for the analogous results in the non-compact model. (6.5.2) can be rewritten as

$$g_* \circ \delta_N^c(g) \circ g^* = \Phi_g^{\mu - N} \circ \delta_N^c(\mu) \circ \Phi_g^{-\mu}, \quad g \in G^n \qquad (6.5.5)$$

for Φ_g so that $g_*(g_c) = \Phi_g^2 g_c$. Now we consider $\delta_N^c(\mu)$ as being associated to the metric g_c and reflect this in the notation $\delta_N^c(\mu) = \delta_N^c(g_c; \mu)$. More generally, (6.5.5) suggests to regard $g_* \circ \delta_N^c(g) \circ g^*$ as being associated to the metric $g_*(g_c) = \Phi_g^2 g_c$. Therefore, we write also $\delta_N^c(g_*(g_c); \mu)$ for this composition. Now the residual interpretation of $\delta_N^c(\mu)$ (Lemma 6.5.1) suggests the following analogous residual interpretation of $\delta_N^c(g_*(g_c); \mu)$.

Lemma 6.5.2. *For u as in Lemma 6.5.1 and $g \in G^n$, let*

$$\langle M_u^g(\lambda), \varphi \rangle = \int_{H^+} \mathcal{H}_g^\lambda u \varphi \operatorname{vol}(S^n, g_*(g_c)),$$

where $\mathcal{H}_g = g_*(\mathcal{H}_0)$. Then

$$\operatorname{Res}_{\lambda = -\mu - 1 - N} \left(\langle M_u^g(\lambda), \varphi \rangle \right) = \int_{S^{n-1}} \left(\frac{g_*(\operatorname{vol}(S^{n-1}, g_c))}{\operatorname{vol}(S^{n-1}, g_c)} \right)^{-\frac{\mu}{n-1}} f$$

$$\times \delta_N^c(g_*(g_c); \mu + N - (n-1))(\varphi) \operatorname{vol}(S^{n-1}, g_*(g_c)).$$

Proof. In view of the relation

$$\langle M_u^g(\lambda), \varphi \rangle = \int_{H^+} \mathcal{H}_0^\lambda g^*(u) g^*(\varphi) \operatorname{vol}(S^n, g_c) = \langle M_{g^*(u)}(\lambda), g^*(\varphi) \rangle,$$

(6.5.1) yields for the residues

$$\operatorname{Res}_{\lambda = -\mu - 1 - N} \left(\langle M_u^g(\lambda), \varphi \rangle \right)$$

$$= \int_{S^{n-1}} \pi_{-\mu}^c(g^{-1}) f \delta_N^c(\mu + N - (n-1)) g^*(\varphi) \operatorname{vol}(S^{n-1}, g_c)$$

$$= \int_{S^{n-1}} \left(\frac{g^*(\text{vol}(S^{n-1}, g_c))}{\text{vol}(S^{n-1}, g_c)} \right)^{\frac{\mu}{n-1}} g^*(f) \delta_N^c(\mu + N - (n-1)) g^*(\varphi) \, \text{vol}(S^{n-1}, g_c)$$

$$= \int_{S^{n-1}} \left(\frac{\text{vol}(S^{n-1}, g_c)}{g_*(\text{vol}(S^{n-1}, g_c))} \right)^{\frac{\mu}{n-1}} f \delta_N^c(g_*(g_c); \mu + N - (n-1)) \varphi \, \text{vol}(S^{n-1}, g_*(g_c)).$$

The proof is complete. □

Notice that

$$\left(\frac{g_*(\text{vol}(S^{n-1}, g_c))}{\text{vol}(S^{n-1}, g_c)} \right)^{-\frac{\mu}{n-1}} f$$

is the coefficient of $\mathcal{H}_g^\mu$ in the asymptotics of u. Thus Lemma 6.5.2 states that the family $\delta_N^c(g_*(g_c); \mu)$ naturally arises from the asymptotics of u in powers of the function $\mathcal{H}_g$.

Now let M be a compact manifold with boundary $\Sigma^n = \partial M$. Assume that on M an asymptotically hyperbolic metric g_{AH} is given. Let Δ_{AH} be the corresponding Laplacian. Then to an eigenfunction u,

$$-\Delta_{AH} u = \mu(n - \mu)u, \quad \Re(\mu) = \frac{n}{2}$$

and a conformal compactification $\bar{g}_{AH}(\rho) \overset{\text{def}}{=} \rho^2 g_{AH}$, $\rho \in C^\infty(M)$ we associate the family

$$\langle M_u(\lambda; \rho), \varphi \rangle \overset{\text{def}}{=} \int_M \rho^\lambda u \varphi \, \text{vol}(\bar{g}_{AH}(\rho)), \quad \varphi \in C^\infty(M), \, \Re(\lambda) > -\frac{n}{2} - 1$$

of distributions on M. The existence of a formal asymptotic expansion

$$u \sim \sum_{j \geq 0} \rho^{\mu + j} a_j(\mu) + \sum_{j \geq 0} \rho^{n - \mu + j} b_j(\mu), \quad \rho \to 0$$

with $a_j, b_j \in C^\infty(\partial M)$ implies the existence of a meromorphic continuation of $M_u(\lambda; \rho)$ to $\mathbb{C}$ with simple poles in the ladders

$$-\mu - 1 - \mathbb{N}_0, \quad -(n - \mu) - 1 - \mathbb{N}_0.$$

Its residue at $\lambda = -\mu - 1 - N$ has the form

$$\varphi \mapsto \int_\Sigma a_0 \delta_N(\mu + N - n; \bar{g}_{AH}(\rho))(\varphi) \, \text{vol}(\Sigma, \rho), \tag{6.5.6}$$

where $\text{vol}(\Sigma, \rho)$ is the Riemannian volume of the pull-back of $\bar{g}_{AH}(\rho)$ to Σ and

$$\delta_N(\cdot; \bar{g}_{AH}(\rho)) : C^\infty(M) \to C^\infty(\Sigma)$$

is a family of differential operators. These families are *conformally covariant* with respect to a change of the conformal compactification. More precisely, let

$$\bar{g}_{AH}(\hat{\rho}) = \hat{\rho}^2 g_{AH}$$

be another conformal compactification of g_{AH}. Then for $\Re(\lambda) >> 0$,

$$\langle M_u(\lambda;\hat\rho),\varphi\rangle \overset{\text{def}}{=} \int_M \hat\rho^\lambda u\varphi\,\text{vol}(\bar{g}_{AH}(\hat\rho))$$

$$= \int_M \rho^\lambda u\varphi e^{\lambda\Phi}e^{(n+1)\Phi}\,\text{vol}(\bar{g}_{AH}(\rho))$$

$$= \langle M_u(\lambda;\rho),e^{(\lambda+n+1)\Phi}\varphi\rangle \qquad (6.5.7)$$

if $\hat\rho = e^\Phi\rho$. We use the asymptotic expansions of u in powers of ρ and $\hat\rho$ to construct the families

$$\delta_N(\mu+N-n;\bar{g}_{AH}(\rho)) \quad\text{and}\quad \delta_N(\mu+N-n;\bar{g}_{AH}(\hat\rho)).$$

(6.5.7) implies that these satisfy the relation

$$\int_{\partial M} \hat{a}_0\delta_N(\mu+N-n;\bar{g}_{AH}(\hat\rho))(\varphi)\,\text{vol}(\Sigma,\hat\rho)$$

$$= \int_{\partial M} a_0\delta_N(\mu+N-n;\bar{g}_{AH}(\rho))(e^{(-\mu+n-N)\Phi}\varphi)\,\text{vol}(\Sigma,\rho).$$

But $\hat{a}_0\hat\rho^\mu = a_0\rho^\mu$, i.e., $\hat{a}_0 = e^{-\mu\Phi}a_0$ on ∂M. Hence

$$e^{(-\mu+n)\Phi}\circ\delta_N(\mu+N-n;\bar{g}_{AH}(\hat\rho))\circ e^{(\mu-n+N)\Phi} = \delta_N(\mu+N-n;\bar{g}_{AH}(\rho)),$$

i.e.,

$$e^{-(\lambda-N)\Phi}\circ\delta_N(\lambda;e^{2\Phi}\bar{g}_{AH}(\rho))\circ e^{\lambda\Phi} = \delta_N(\lambda;\bar{g}_{AH}(\rho)). \qquad (6.5.8)$$

(6.5.8) is the claimed conformal covariance.

Note that the resulting families only depend on a *finite* part of the formal asymptotics of u near ∂M. In the above discussion, we also suppressed a discussion of the continuation in λ. In fact, the families are meromorphic and polynomial families arise by renormalization.

Now we regard the families $\delta_N(\lambda;\bar{g}_{AH}(\rho))$ as being associated to the conformal class on Σ which is defined by the conformal compactifications of g_{AH}. That perspective is reasonable if the resulting families depend naturally on that class. In Section 6.6, we shall deal with such a case in detail.

Details of the residue method in the cases of the unit ball $\mathbb{B}^n$ and the hemisphere H^+ were given in Section 5.4.1 and Section 5.3. In these two cases, we worked with conformal compactifications with a large conformal group and we emphasized the equivariance property of the families with respect to the associated group representations.

Finally, we illustrate the residue method for a conformal compactification of the ball $\mathbb{B}^n$ which is natural from the point of view of the geometry of the Poincaré-Einstein metric. Let

$$g_E = \frac{4}{(1-|x|^2)^2}(dx_1^2 + \cdots + dx_n^2)$$

be the Poincaré-Einstein metric on $\mathbb{B}^n$. In contrast to the conformal compactification

$$\bar{g}_E(\rho) = dx_1^2 + \cdots + dx_n^2$$

given by $\rho(x) = \frac{1-|x|^2}{2}$ (see Section 5.4.1), we consider now the compactification
defined by

$$\hat{\rho}(x) = \frac{1-|x|}{1+|x|}.$$

Then

$$\bar{g}_E(\hat{\rho}) = \frac{4}{(1+|x|)^4}(dx_1^2 + \cdots + dx_n^2).$$

We consider a formal asymptotic development

$$u \sim \sum_{j \geq 0} \left(\frac{1-|x|}{1+|x|}\right)^{\mu+j} \hat{a}_j$$

of an eigenfunction $-\Delta_{g_E} u = \mu(n-1-\mu)u$. Since

$$\frac{1-|x|}{1+|x|} = \exp-d(0,x),$$

the latter form of the asymptotics is the typical form of radial asymptotics used in
harmonic analysis ([140], [233]). In order to determine the asymptotics, we write
Δ_{g_E} in terms of polar coordinates $(r,\cdot) \in (0,1) \times S^{n-1}$ and use the substitution
$s = \frac{1-r}{1+r}$. In the coordinates $(s,\cdot) \in (0,1) \times S^{n-1}$, we have

$$g_E = \frac{1}{s^2}\left(ds^2 + \left(\frac{1-s^2}{2}\right)^2 g_{S^{n-1}}\right), \quad \bar{g}_E(\hat{\rho}) = ds^2 + \left(\frac{1-s^2}{2}\right)^2 g_{S^{n-1}}, \quad (6.5.9)$$

and the corresponding Laplacian reads

$$s^2\frac{\partial^2}{\partial s^2} + \left(n - \frac{2(n-1)}{1-s^2}\right)s\frac{\partial}{\partial s} + \frac{4s^2}{(1-s^2)^2}\Delta_{S^{n-1}}. \qquad (6.5.10)$$

The power series ansatz $s^\mu \sum_{j \geq 0} s^j \hat{a}_j$ yields

$$\sum_{j \geq 0}(\mu+j)(\mu+j-1)s^j\hat{a}_j + \left(n - \frac{2(n-1)}{1-s^2}\right)\sum_{j \geq 0}(\mu+j)s^j\hat{a}_j$$

$$+ \frac{4s^2}{(1-s^2)^2}\sum_{j \geq 0}s^j\Delta_{S^{n-1}}\hat{a}_j = -\mu(n-1-\mu)\sum_{j \geq 0}s^j\hat{a}_j.$$

We write $(1-s^2)^{-1} = \sum_{n \geq 0} s^{2n}$ and obtain a recursive relation for the coefficients
$\hat{a}_j$ saying that

$$N(n-1-2\mu-N)\hat{a}_N$$

is a linear combination of second-order differential operators with polynomial coefficients in μ acting on $\hat{a}_{N-2}, \hat{a}_{N-4}, \ldots$. Thus the leading coefficient $\hat{a}_0$ is free, $\hat{a}_1 = 0$, and all other coefficients are determined by $\hat{a}_0$. In particular, we have

$$-2(n-3-2\mu)\hat{a}_2 = 2(n-1)\mu\hat{a}_0 - 4\Delta_{S^{n-1}}\hat{a}_0,$$

i.e.,

$$\hat{a}_2 = \frac{1}{n-3-2\mu} \left(2\Delta_{S^{n-1}}\hat{a}_0 - \mu(n-1)\hat{a}_0 \right).$$

We use these results to determine $\hat{\delta}_j(\lambda) = \delta_j(\lambda; \bar{g}_E(\hat{\rho}))$ $(j = 1, 2)$. We rewrite the integral

$$\int_{\mathbb{B}^n} \hat{\rho}^\lambda u\varphi \operatorname{vol}(\bar{g}_E(\hat{\rho}))$$

in terms of geodesic normal coordinates based on S^{n-1}, i.e., $\hat{a} : (0,1) \times S^{n-1} \to \mathbb{B}^n$, $(t, x) \mapsto \hat{\exp}_x(tN)$, where $\hat{\exp}$ denotes the exponential map for $\bar{g}_E(\hat{\rho})$ and N is a unit normal vector field. A simple calculation shows that

$$\hat{a}(t, x) = \frac{1-t}{1+t}x.$$

Hence $\hat{a}^*(\hat{\rho})(t, x) = t$,

$$\hat{a}^* \operatorname{vol}(\bar{g}_E(\hat{\rho})) = \operatorname{vol}(\hat{a}^*(\bar{g}_E(\hat{\rho}))) = \left(\frac{1-t^2}{2}\right)^{n-1} dt \operatorname{vol}(S^{n-1}, g_c).$$

Notice that geodesic normal coordinates coincide with the coordinates defined by the gradient flow of $\hat{\rho}$ with respect to the metric $\bar{g}_E(\hat{\rho})$.

Now the integral can be written as

$$\frac{1}{2^{n-1}} \int_0^1 \int_{S^{n-1}} t^\lambda u\varphi(1-t^2)^{n-1} dt db.$$

The asymptotics $u(t, \cdot) \sim t^\mu \hat{a}_0 + t^{\mu+2}\hat{a}_2 + \cdots$ and

$$v(t) = \left(\frac{1-t^2}{2}\right)^{n-1} = v_0 + v_2 t^2 + \cdots$$

imply

$$\hat{\delta}_1(\mu+2-n)f = v_0 \frac{\partial f}{\partial t}$$

and

$$\hat{\delta}_2(\mu+3-n)f = v_0 \frac{1}{2!} \frac{\partial^2 f}{\partial t^2} + \left(v_2 + \frac{1}{n-3-2\mu}(2\Delta_{S^{n-1}} - (n-1)\mu) \right)$$

using a calculation similar to that in Section 5.4.1 (here restrictions to $t = 0$ are omitted in order to simplify the formulation). Thus

$$2^{n-1}\hat{\delta}_1(\lambda) = \frac{\partial}{\partial t}$$

and

$$-2^n(2\lambda+n-3)\hat{\delta}_2(\lambda) = -(2\lambda+n-3)\frac{\partial^2}{\partial t^2} + 4\Delta_{S^{n-1}} + 2\lambda(n-1). \qquad (6.5.11)$$

We note that for $n = 3$, the family (6.5.11) reads

$$\lambda\left(-2\frac{\partial^2}{\partial t^2} + \lambda\right) + 4\Delta_{S^2}.$$

This coincides with the natural family $D_2(\mathbb{R}^3, S^2; \bar{g}_E(\hat{\rho}); \lambda)$ (Theorem 6.3.1). In fact,

$$\bar{g}_E(\hat{\rho})|_{S^{n-1}} = \frac{1}{4}g_c,$$

$K(S^2, \frac{1}{4}g_c) = 4$, and it is easy to prove that $H = 0$ and $\tau = 16$ on S^2.

6.6   Residue families and holographic formulas for Q-curvature

We start with a summary of the content of the present section. Here we apply the method of Section 6.5 to the special case of the Poincaré-Einstein metric $g_E = r^{-2}(dr^2 + h_r)$, $h_0 = h$ on $(0, \varepsilon) \times M$ being associated to a given metric h on M (Section 3.2). In particular, the conformal compactification $\bar{g}_E = r^2 g_E$ of g_E gives rise to the *critical residue family*

$$D_n^{\mathrm{res}}(h; \lambda) : C^\infty([0, \varepsilon) \times M) \to C^\infty(M). \qquad (6.6.1)$$

The family $D_n^{\mathrm{res}}(h; \lambda)$ is completely determined by the metric h and satisfies

$$D_n^{\mathrm{res}}(h; 0) = P_n(h)i^*.$$

More generally, we introduce residue families $D_N^{\mathrm{res}}(h; \lambda)$ of order $N \leq n$ for even n, and any order for odd n, and show how they transform under conformal changes of the metric. In particular, that law suggests that residue families are specializations of conformally covariant families to $g = \bar{g}_E$.

The main result of the present section is

Theorem 6.6.1. *Let n be even. Then*

$$\dot{D}_n^{\mathrm{res}}(h; 0)(1) = -(-1)^{\frac{n}{2}}Q_n(h). \qquad (6.6.2)$$

In order to prove Theorem 6.6.1, we evaluate the left-hand side of (6.6.2) (Theorem 6.6.2 and Corollary 6.6.1). Combined with an additional identity (Theorem 6.6.4), it follows that the assertion (6.6.2) is equivalent to a formula which will be called the *holographic formula* for Q_n (Theorem 6.6.6). The latter result has an independent proof which in turn proves (6.6.2). In the remaining part of the section, we make the holographic formulas explicit for $n \leq 6$.

Let (M^n, h) (n even) be a Riemannian manifold. Let $g_E = r^{-2}(dr^2 + h_r)$ with

$$h_r = h_{(0)} + r^2 h_{(2)} + \cdots + r^n h_{(n)} + r^n \log r \bar{h}_{(n)} + \cdots$$

be the Poincaré-Einstein metric on $(0, \varepsilon) \times M^n$ which is associated to $h = h_{(0)} = h_0$. Here $\bar{h}_{(n)}$ has vanishing h-trace. The coefficients $h_{(2)}, \ldots, h_{(n-2)}, \bar{h}_{(n)}$ and the trace of $h_{(n)}$ are locally determined by $h_{(0)} = h_0 = h$.

Now the ansatz

$$u \sim \sum_{j \geq 0} r^{\mu + 2j} a_{2j}(h; \mu)$$

for a formal approximate solution of $-\Delta_{g_E} u = \mu(n - \mu)u$ determines a sequence of differential operators $T_j(h; \mu)$ so that

$$T_j(h; \mu) a_0 = a_j(h; \mu).$$

Since h_r is even in r (up to order n), we have $T_j(\mu) = 0$ for odd $j \leq n$. Moreover,

$$T_{2j}(h; \mu) = \frac{1}{2^{2j} j!} \frac{\Gamma(\frac{n}{2} - \mu - j)}{\Gamma(\frac{n}{2} - \mu)} P_{2j}(h; \mu)$$

$$= \frac{1}{2^{2j} j!} \frac{1}{(\frac{n}{2} - \mu - 1) \ldots (\frac{n}{2} - \mu - j)} P_{2j}(h; \mu) \qquad (6.6.3)$$

(see (3.2.16)), where $P_{2j}(h; \mu) = \Delta_h^j + LOT$ and $P_n(h; 0) = P_n(h)$. $P_{2j}(h; \mu)$ is a polynomial family of differential operators of order $2j$ on M.

Finally, let

$$v(r, \cdot) = \frac{\text{vol}(h_r)}{\text{vol}(h)} = v_0 + r^2 v_2 + \cdots + r^n v_n + \cdots, \quad v_0 = 1.$$

Note that there is no term $r^n \log r$ in the formal Taylor series of $v(r, b)$. In fact, that coefficient coincides with the coefficient of $r^n \log r$ in $\det(\text{id} + (r^n \log r)h^{-1}\bar{h}_{(n)})$ which is given by $\text{tr}(h^{-1}\bar{h}_{(n)}) = 0$.

Definition 6.6.1 (The critical residue family). *For even n, let*

$$D_n^{\text{res}}(h; \lambda) = 2^n \left(\frac{n}{2}\right)! \left[\left(\frac{n}{2} - \lambda - 1\right) \cdots (-\lambda)\right] \delta_n(h; \lambda)$$

with

$$\delta_n(h; \lambda) = \sum_{j=0}^{n} \frac{1}{(n-j)!} \left[T_j^*(h; \lambda) \circ v_0 + \cdots + T_0^*(h; \lambda) \circ v_j\right] \circ i^* \circ (\partial/\partial r)^{n-j}.$$

*Here i^* restricts functions to $r = 0$. The family $D_n^{\mathrm{res}}(h; \lambda) : C^\infty([0, \varepsilon) \times M^n) \to$
$C^\infty(M^n)$ is called the critical residue family.*

Since $\mathcal{T}_{odd} = 0$ and $v_{odd} = 0$, we can also write

$$\delta_n(h; \lambda) = \sum_{j=0}^{\frac{n}{2}} \frac{1}{(n-2j)!} \left[\sum_{k=0}^{j} \mathcal{T}_{2k}^*(h; \lambda) \circ v_{n-2k} \right] \circ i^* \circ (\partial/\partial r)^{n-2j}. \qquad (6.6.4)$$

Theorem 6.6.2. *(6.6.2) is equivalent to*

$$-(-1)^{\frac{n}{2}} Q_n(h) = \dot{P}_n^*(h; 0)(1) - \left(\frac{n}{2}\right)! \sum_{j=0}^{\frac{n}{2}-1} 2^{n-2j} \frac{\left(\frac{n}{2}-j-1\right)!}{j!} P_{2j}^*(h; 0)(v_{n-2j})$$

$$= \dot{P}_n^*(h; 0)(1) - 2n P_{n-2}^*(h; 0)(v_2) - \cdots - 2^n \left(\frac{n}{2}\right)! \left(\frac{n}{2}-1\right)! v_n. \qquad (6.6.5)$$

Proof. The family $\delta_n(h; \mu) : C^\infty([0, \varepsilon) \times M) \to C^\infty(M)$ arises in the residue
formula

$$\mathrm{Res}_{\lambda=-\mu-1-n}(\langle M_u(\lambda; r), \varphi \rangle) = \int_M f \delta_n(h; \mu)(\varphi) \, \mathrm{vol}(h),$$

where $a_0 = f$ is the leading term in the asymptotics of u (see (6.5.6)) and the test
function φ has compact support in $[0, \varepsilon) \times M$. In fact, repeated partial integration
in the integral

$$\langle M_u(\lambda; r), \varphi \rangle = \int_{(0,\varepsilon) \times M} r^\lambda u \varphi \, \mathrm{vol}(\bar{g}_E) = \int_0^\varepsilon \int_M r^\lambda (uv)(r, b) \varphi(r, b) dr \, \mathrm{vol}(h)$$

(see Section 5.4.1 for more details in a special case) yields the formula

$$\mathrm{Res}_{\lambda=-\mu-1-n}(\langle M_u(\lambda; r), \varphi \rangle)$$

$$= \sum_{j=0}^{n} \frac{1}{(n-j)!} \int_M [\mathcal{T}_j(\mu)(f)v_0 + \cdots + \mathcal{T}_0(\mu)(f)v_j] \frac{\partial^{n-j}\varphi}{\partial r^{n-j}}(0, b) \, \mathrm{vol}(h)$$

$$= \sum_{j=0}^{n} \frac{1}{(n-j)!} \int_M f \left[\mathcal{T}_j^*(\mu) \circ v_0 + \cdots + \mathcal{T}_0^*(\mu) \circ v_j\right] \left(\left(\frac{\partial}{\partial r}\right)^{n-j}(\varphi)(0, b)\right) \mathrm{vol}(h).$$

Hence

$$\delta_n(h; \mu)$$

$$= \sum_{j=0}^{n} \frac{1}{(n-j)!} \left[\mathcal{T}_j^*(h; \mu) \circ v_0 + \cdots + \mathcal{T}_0^*(h; \mu) \circ v_j\right] \circ i^* \circ (\partial/\partial r)^{n-j}. \qquad (6.6.6)$$

It follows that the renormalization

$$D_n^{\mathrm{res}}(h; \mu) = 2^n \left(\frac{n}{2}\right)! \left[\left(\frac{n}{2}-\mu-1\right)\ldots(-\mu)\right] \delta_n(h; \mu) \qquad (6.6.7)$$

is a family of the form

$$\sum_{j=0}^{\frac{n}{2}} \Delta_h^j \circ i^* \circ (\partial/\partial r)^{n-2j} \frac{1}{(n-2j)!} \frac{2^n}{2^{2j}} \frac{(\frac{n}{2})!}{j!} \frac{(\frac{n}{2}-\mu-1)\dots(-\mu)}{(\frac{n}{2}-\mu-1)\dots(\frac{n}{2}-\mu-j)} + LOT$$

$$= \sum_{j=0}^{\frac{n}{2}} a_j^{(n)}(\mu) \left[\Delta_h^j \circ i^* \circ (\partial/\partial r)^{n-2j} \right] + LOT,$$

where

$$a_j^{(n)}(\mu) = \frac{(\frac{n}{2})!}{j!(n-2j)!} 2^{n-2j} \prod_{k=j}^{\frac{n}{2}-1} \left(\frac{n}{2}-1-k-\mu \right), \quad a_{\frac{n}{2}}^{(n)}(\mu) = 1$$

(see (5.1.4)). In particular, we have

$$D_n^{\mathrm{res}}(h;0) = P_n(h;0)i^* = P_n(h)i^*. \tag{6.6.8}$$

Since the families $\mathcal{T}_{n-2}(\mu), \dots, \mathcal{T}_0(\mu)$ are regular at $\mu = 0$, (6.6.6) implies

$$\dot{D}_n^{\mathrm{res}}(h;0)(1) = \dot{P}_n^*(h;0)(1) - \left(\frac{n}{2} \right)! \sum_{j=0}^{\frac{n}{2}-1} 2^{n-2j} \frac{(\frac{n}{2}-j-1)!}{j!} P_{2j}^*(h;0)(v_{n-2j}).$$

Hence (6.6.2) is equivalent to (6.6.5). The proof is complete. $\qquad\square$

Remark 6.6.1. *(6.6.2) is equivalent to*

$$-(-1)^{\frac{n}{2}} Q_n(h) = \dot{P}_n^*(h;0)(1) - 2^n \left(\frac{n}{2} \right)! \left(\frac{n}{2}-1 \right)! \sum_{j=0}^{\frac{n}{2}-1} \mathcal{T}_{2j}^*(h;0)(v_{n-2j}). \tag{6.6.9}$$

More generally, we introduce residue families of order $\leq n$.

Definition 6.6.2 (Residue families. General case). *For even n and $N \leq n$, let*

$$D_N^{\mathrm{res}}(h;\lambda) : C^\infty([0,\varepsilon) \times M^n) \to C^\infty(M^n)$$

be defined as the product of

$$\begin{cases} 2^{2M} M! \left[(-\frac{n}{2}-\lambda+2M-1) \cdots (-\frac{n}{2}-\lambda+M) \right] & \text{if } N = 2M, \\ 2^{2M} M! \left[(-\frac{n}{2}-\lambda+2M) \cdots (-\frac{n}{2}-\lambda+M+1) \right] & \text{if } N = 2M+1 \end{cases} \tag{6.6.10}$$

with

$$\delta_N(h;\lambda+n-N),$$

where

$$\delta_N(h;\lambda) = \sum_{j=0}^{N} \frac{1}{(N-j)!} \left[\mathcal{T}_j^*(h;\lambda) \circ v_0 + \cdots + \mathcal{T}_0^*(h;\lambda) \circ v_j \right] \circ i^* \circ (\partial/\partial r)^{N-j}.$$

The families $D_N^{\mathrm{res}}(h;\lambda)$ with $N < n$ will be called subcritical *residue families.*

Note that $\delta_N(h; \lambda)$ satisfies

$$\mathrm{Res}_{\lambda=-\mu-1-N}(\langle M_u(\lambda; r), \varphi \rangle) = \int_M f \delta_N(h; \mu)(\varphi)\, \mathrm{vol}(h) \qquad (6.6.11)$$

and we have

$$D_{2N}^{\mathrm{res}}\left(h; -\frac{n}{2}+N\right) = P_{2N}(h)i^*,$$

$$D_{2N+1}^{\mathrm{res}}\left(h; -\frac{n}{2}+N+1\right) = P_{2N}(h)i^* \frac{\partial}{\partial r}. \qquad (6.6.12)$$

The latter two identities are examples of factorization identities (see Remark 6.11.1).

Remark 6.6.2. *For even n, residue families $D_{2N}^{\mathrm{res}}(h; \lambda)$ are well defined for general h only if $2N \leq n$. In fact, the definition of $D_{2N}^{\mathrm{res}}(h; \lambda)$ involves the Taylor coefficients $h_{(0)} = h, h_{(2)}, \ldots, h_{(2N)}$ of h_r, where $h_{(2N)}$ enters only through its trace. These terms are completely determined by h. The coefficient $\bar{h}_{(n)}$ obstructs the construction of residue families of orders exceeding n. For odd n, residue families are defined analogously for all orders since the construction of the Poincaré-Einstein metric is not obstructed. We omit the details.*

The residue families $D_{2N}^{\mathrm{res}}(h; \lambda)$ give rise to Q-polynomials.

Definition 6.6.3 (Q-polynomials). *For a Riemannian manifold (M, h) of even dimension n, the polynomials*

$$Q_{2N}^{\mathrm{res}}(h; \lambda) = -(-1)^N D_{2N}^{\mathrm{res}}(h; \lambda)(1),\ 1 \leq N \leq \frac{n}{2}$$

are called the Q-polynomials of (M, h).

Sometimes we shall distinguish between the *critical Q-polynomial* $Q_n^{\mathrm{res}}(h; \lambda)$ and the *subcritical Q-polynomials* $D_{2N}^{\mathrm{res}}(h; \lambda)$ ($N < \frac{n}{2}$).

These polynomials are called Q-polynomials since, by Theorem 6.6.1, the linear term of the critical Q-polynomial yields Q-curvature. In Section 6.11, we will further analyze the relation between these polynomials and Q-curvature.

In the critical case, (6.6.7) implies the relation

$$D_n^{\mathrm{res}}(\lambda)(1) = 2^n \left(\frac{n}{2}\right)! \left(\frac{n}{2}-\lambda-1\right)\ldots(-\lambda)\delta_n(\lambda)(1)$$

$$= 2^n \left(\frac{n}{2}\right)! \left(\frac{n}{2}-\lambda-1\right)\ldots(-\lambda)\left[\mathcal{T}_n^*(\lambda)(v_0) + \cdots + \mathcal{T}_0^*(\lambda)(v_n)\right]$$

$$= \left(\frac{n}{2}\right)! 2^n \sum_{j=0}^{\frac{n}{2}} \frac{1}{2^{2j} j!} \left(\frac{n}{2}-\lambda-j-1\right)\ldots(-\lambda)P_{2j}^*(\lambda)(v_{n-2j}). \qquad (6.6.13)$$

Since $P_{2j}(\lambda)$ has polynomial degree j (in λ), it follows that $Q_n^{\mathrm{res}}(h; \lambda)$ has degree $\frac{n}{2}$ (in λ).

More generally, (6.6.10) yields

$$D_{2N}^{\text{res}}(\lambda)(1) = 2^{2N} N! \left[\left(-\frac{n}{2} - \lambda + 2N - 1 \right) \dots \left(-\frac{n}{2} - \lambda + N \right) \right]$$
$$\times \left[\mathcal{T}_{2N}^*(\lambda + n - 2N)(v_0) + \dots + \mathcal{T}_0^*(\lambda + n - 2N)(v_{2N}) \right]$$

and it follows that $Q_{2N}^{\text{res}}(h; \lambda)$ has degree N (in λ).

Example 6.6.1. *The discussion in Section 5.2.4 shows that for $M = \mathbb{R}^n$ with the flat metric h_c, the residue family $D_{2N}^{\text{res}}(h_c; \lambda)$ coincides with $D_{2N}^{\text{nc}}(\lambda) : C^\infty(\mathbb{R}^{n+1}) \to C^\infty(\mathbb{R}^n)$.*

Example 6.6.2. *For the sphere $M = S^n$ with the round metric $\frac{1}{4} h_c$, the corresponding Poincaré-Einstein metric is*

$$g_E = r^{-2} \left(dr^2 + (1 - r^2)^2 \frac{1}{4} h_c \right).$$

g_E coincides with the usual hyperbolic metric

$$\frac{4}{(1 - |x|^2)^2} (dx_1^2 + \dots + dx_{n+1}^2)$$

on the unit ball $\mathbb{B}^{n+1}$ (see (6.5.9)). The conformal compactification

$$dr^2 + (1 - r^2)^2 \frac{1}{4} h_c,$$

i.e.,

$$\left(\frac{1 - |x|}{1 + |x|} \right)^2 g_E,$$

yields the residue families $D_N^{\text{res}}(\frac{1}{4} h_c; \lambda)$. On the other hand in Section 5.4 an analogous residue constructions with respect to the conformal compactification

$$(1 - |x|^2) g_E = 4(dx_1^2 + \dots + dx_{n+1}^2)$$

led to the families $D_N^b(\lambda)$. The arguments in Section 6.5 show that the resulting families are conjugate, i.e.,

$$e^{-(\lambda - N)\Phi} \circ D_N^b(\lambda) \circ e^{\lambda \Phi} = D_N^{\text{res}} \left(\frac{1}{4} h_c; \lambda \right), \qquad (6.6.14)$$

where

$$\left(\frac{1 - |x|}{1 + |x|} \right) e^\Phi = 1 - |x|^2,$$

i.e., $e^\Phi = (1 + |x|)^2$. Thus (6.6.14) reads

$$2^{2N} D_N^b(\lambda) \circ \left(\frac{1 + |x|}{2} \right)^{2\lambda} = D_N^{\text{res}} \left(\frac{1}{4} h_c; \lambda \right).$$

Moreover, as for $(M^n, h) = (\mathbb{R}^n, h_c)$, the renormalization (6.6.10) has the effect that

$$D^{\text{res}}_{2N}\left(h; -\frac{n}{2}+N\right) = P_{2N}(h)i^* \qquad (6.6.15)$$

extending (6.6.8). In fact, it suffices to observe that (6.6.3) yields

$$2^{2N}N!\left[\left(-\frac{n}{2}-\lambda+2N-1\right)\cdots\left(-\frac{n}{2}-\lambda+N\right)\right]T_{2N}(\lambda+n-2N) = P_{2N}(\lambda+n-2N)$$

and $P_{2N}(\frac{n}{2}-N) = P_{2N}$ (by (3.2.17)). (6.6.15) is an extreme case of the factorization identities (see Section 6.11).

Next, we discuss the behaviour of residue families under *conformal changes* of the metric. The Poincaré-Einstein metrics of h and $\hat{h} = e^{2\varphi}h$ are related by

$$\kappa^*\left(r^{-2}(dr^2 + h_r)\right) = r^{-2}(dr^2 + \hat{h}_r)$$

using a diffeomorphism κ which fixes the boundary $r = 0$ ([99], Theorem 2.3); for the present discussion we can ignore that all data are only determined to a certain order. Hence

$$\kappa^*(dr^2 + h_r) = \left(\frac{\kappa^*(r)}{r}\right)^2 (dr^2 + \hat{h}_r).$$

The latter identity implies

$$\lim_{r\to 0}\left(\frac{\kappa^*(r)}{r}\right) = e^{-\varphi}. \qquad (6.6.16)$$

Now the arguments of Section 6.5 yield (for all $N \leq n$ if n is even)

Theorem 6.6.3.

$$D^{\text{res}}_N(\hat{h}; \lambda) = e^{(\lambda-N)\varphi} \circ D^{\text{res}}_N(h; \lambda) \circ \kappa_* \circ \left(\frac{\kappa^*(r)}{r}\right)^{\lambda}. \qquad (6.6.17)$$

Proof. Let $g = r^{-2}(dr^2 + h_r)$ be the Poincaré-Einstein metric on $(0, \varepsilon) \times M$ associated to h and let $u \in \ker(\Delta_g + \mu(n-\mu))$ be an eigenfunction with smooth leading term f. Let $\psi \in C^\infty_0(X)$, $X = [0, \varepsilon) \times M$. We calculate the residue

$$\text{Res}_{\lambda=-\mu-1-N}\left(\int_X r^\lambda \kappa^*(u)\psi \,\text{vol}(r^2\kappa^*(g))\right)$$

in two ways. On the one hand, it is given by

$$\int_M e^{-\mu\varphi} f\delta_N(\hat{h}; \mu)(\psi)\,\text{vol}(\hat{h}) = \int_M e^{-(\mu-n)\varphi} f\delta_N(\hat{h}; \mu)(\psi)\,\text{vol}(h).$$

In fact, by $\Delta_{\kappa^*(g)} \circ \kappa^* = \kappa^* \circ \Delta_g$, the function $\kappa^*(u)$ is an eigenfunction of $\Delta_{\kappa^*(g)}$. Since κ restricts to the identity on $r = 0$, its leading term is

$$i^*\left(\frac{\kappa^*(r)}{r}\right)^\mu f = e^{-\mu\varphi}f$$

(using (6.6.16)). On the other hand, using $dr^2 + \hat{h}_r = r^2 \kappa^*(r^{-2}(dr^2 + h_r))$, we find

$$\mathrm{Res}_{\lambda = -\mu-1-N}\left(\int_X r^{\lambda+n+1}\kappa^*(u)\psi\kappa^*(r)^{-n-1}\kappa^*(\mathrm{vol}(dr^2 + h_r))\right)$$

$$= \mathrm{Res}_{\lambda = -\mu-1-N}\left(\int_X \kappa_*(r)^{\lambda+n+1} u\kappa_*(\psi)r^{-n-1}\mathrm{vol}(dr^2 + h_r)\right)$$

$$= \mathrm{Res}_{\lambda = -\mu-1-N}\left(\int_X r^\lambda \left(\frac{\kappa_*(r)}{r}\right)^{\lambda+n+1} u\kappa_*(\psi)\,\mathrm{vol}(dr^2 + h_r)\right)$$

$$= \int_M f\delta_N(h;\mu)\left(\left(\frac{\kappa_*(r)}{r}\right)^{-\mu+n-N}\kappa_*(\psi)\right)\mathrm{vol}(h).$$

Since f is arbitrary, we conclude

$$e^{-(\mu-n)\varphi} \circ \delta_N(\hat{h};\mu) = \delta_N(h;\mu) \circ \left(\frac{\kappa_*(r)}{r}\right)^{-\mu+n-N} \circ \kappa_*$$

$$= \delta_N(h;\mu) \circ \kappa_* \circ \left(\frac{\kappa^*(r)}{r}\right)^{\mu-n+N}.$$

For $\mu = \lambda + n - N$, the latter formula reads

$$e^{-(\lambda-N)\varphi} \circ \delta_N(\hat{h};\lambda+n-N) = \delta_N(h;\lambda+n-N) \circ \kappa * \circ \left(\frac{\kappa^*(r)}{r}\right)^\lambda.$$

(6.6.10) implies the assertion. □

In particular, for the critical residue family, we find the transformation formula

$$D_n^{\mathrm{res}}(\hat{h};\lambda) = e^{(\lambda-n)\varphi} \circ D_n^{\mathrm{res}}(h;\lambda) \circ \kappa_* \circ \left(\frac{\kappa^*(r)}{r}\right)^\lambda. \qquad (6.6.18)$$

Remark 6.6.3. *Theorem 6.6.3 suggests that we ask for the construction of natural families $D_N(X, M; g; \lambda) : C^\infty(X) \to C^\infty(M)$ with the properties*

$$e^{-(\lambda-N)i^*(\varphi)} \circ D_N(X, M; e^{2\varphi}g; \lambda) \circ e^{\lambda\varphi} = D_N(X, M; g; \lambda)$$

for all $\varphi \in C^\infty(X)$ and all $\lambda \in \mathbb{C}$ (conformal covariance), and

$$D_N(dr^2 + h_r; \lambda) = D_N^{\mathrm{res}}(h;\lambda).$$

The transformation formula (6.6.17) would be a direct consequence of such a spe-

cialization result. In fact, we find

$$D_N^{\text{res}}(\hat{h}; \lambda) = D_N(dr^2 + \hat{h}_r; \lambda)$$

$$= D_N\left(\left(\frac{\kappa^*(r)}{r}\right)^{-2}\kappa^*(dr^2 + h_r); \lambda\right)$$

$$= i^*\left(\frac{\kappa^*(r)}{r}\right)^{-(\lambda-N)} \circ D_N\left(\kappa^*(dr^2 + h_r); \lambda\right) \circ \left(\frac{\kappa^*(r)}{r}\right)^{\lambda}$$

$$= e^{(\lambda-N)\varphi} \circ D_N^{\text{res}}(dr^2 + h_r; \lambda) \circ \kappa_* \circ \left(\frac{\kappa^*(r)}{r}\right)^{\lambda}.$$

Remark 6.6.4. (6.6.18) *can be used for a proof of the fundamental identity. In fact, differentiating* (6.6.18) *at* $\lambda = 0$ *yields for* $u = 1$

$$e^{n\varphi}\dot{D}_n^{\text{res}}(\hat{h}; 0)(1)$$

$$= \varphi D_n^{\text{res}}(h; 0)(1) + \dot{D}_n^{\text{res}}(h; 0)(1) + D_n^{\text{res}}(h; 0)\left(\kappa_* \log(\kappa^*(r)/r)\right).$$

Now $D_n^{\text{res}}(h; 0) = P_n(h)i^*$, *i.e.,* $D_n^{\text{res}}(h; 0)$ *is tangential. Hence we find*

$$e^{n\varphi}\dot{D}_n^{\text{res}}(\hat{h}; 0)(1) = \dot{D}_n^{\text{res}}(h; 0)(1) - P_n(h)(\varphi),$$

i.e.,

$$e^{n\varphi}Q_n(\hat{h}) = Q_n(h) + (-1)^{\frac{n}{2}}P_n(h)(\varphi)$$

using (6.6.2). *It is actually this observation which suggests that we expect the validity of* (6.6.2).

The transformation formulas (6.6.17) can be regarded as an effective and compact formulation of the behaviour of the Fefferman-Graham coefficients $h_{(2i)}$ and the holographic coefficients v_{2i} with respect to conformal changes of the boundary metric. The (infinitesimal) conformal variations of $h_{(2i)}$ and v_{2i} have been used in [147], [211] and [212] to study structural properties of anomalies. In these references, the corresponding diffeomorphisms of bulk space appear under the name of PBH diffeomorphisms (see also [80] and [218]).

Using the relation $\dot{P}_n(0)(1) = (-1)^{\frac{n}{2}}Q_n$, we can write the formulas (6.6.9) and (6.6.5) also in the following more symmetric form.

Corollary 6.6.1. (6.6.2) *is equivalent to the identity*

$$2(-1)^{\frac{n}{2}}Q_n(h) = (\dot{P}_n(h; 0) - \dot{P}_n^*(h; 0))(1)$$

$$+ 2^n\left(\frac{n}{2}\right)!\left(\frac{n}{2}-1\right)!\sum_{j=0}^{\frac{n}{2}-1}\mathcal{T}_{2j}^*(h; 0)(v_{n-2j}) \quad (6.6.19)$$

or, equivalently, to

$$2(-1)^{\frac{n}{2}} Q_n(h) = (\dot{P}_n(h;0) - \dot{P}_n^*(h;0))(1)$$

$$+ \left(\frac{n}{2}\right)! \sum_{j=0}^{\frac{n}{2}-1} 2^{n-2j} \frac{(\frac{n}{2}-j-1)!}{j!} P_{2j}^*(h;0)(v_{n-2j}). \quad (6.6.20)$$

In contrast to $P_n(0)$, the operator $\dot{P}_n(0)$ is *not* self-adjoint, in general. In order to evaluate the latter formulas further, it remains to describe the difference

$$\delta_n(h;0)(1) \overset{\text{def}}{=} (\dot{P}_n(h;0) - \dot{P}_n^*(h;0))(1) \in C^\infty(M^n). \quad (6.6.21)$$

Note that if h is Einstein then $\dot{P}_n(h;0)$ is self-adjoint (see Section 6.16). Moreover, direct calculations show that $\delta_2(h;0)$ vanishes and $\delta_4(h;0)(u) = 2(\Delta(\mathsf{J}u) - \mathsf{J}\Delta(u))$. In particular, we find

$$\delta_4(h;0)(1) = 2\Delta\mathsf{J}$$

(see Example 6.6.4). Similarly, for $n = 6$, we shall prove in Section 6.10 that $\delta_6(h;0)(1)$ coincides with a linear combination of the quantities $P_{2j}^*(h;0)(v_{6-2j})$, $j = 1, 2$. In fact, this is a general result which leads to an explicit representation of Q_n for all n (see the discussion on page 26). The following result provides the desired formula for $\delta_n(h;0)(1)$. [1]

Theorem 6.6.4. *Let $n \geq 2$ be even. Then*

$$n\delta_n(h;0)(1) = -2^n \left(\frac{n}{2}\right)! \left(\frac{n}{2}-1\right)! \sum_{j=0}^{\frac{n}{2}-1} 2j T_{2j}^*(h;0)(v_{n-2j}).$$

Proof. We consider the coefficient of $\varepsilon^{2\lambda}$ in the asymptotics of the integral

$$\int_{\delta > r > \varepsilon} (u\Delta v - v\Delta u) \operatorname{vol}(g_E), \quad (6.6.22)$$

where

$$u = \sum_{j=0}^{\frac{n}{2}} r^{\lambda+2j} T_{2j}(h;\lambda)(f) \quad \text{and} \quad v = \sum_{j=0}^{\frac{n}{2}} r^{\lambda+2j} T_{2j}(h;\lambda)(1)$$

are approximate eigenfunctions of Δ_{g_E} for λ near 0. Here we choose $f \in C_c^\infty$. On the one hand, the coefficient of $\varepsilon^{2\lambda}$ vanishes in the asymptotic expansion of

---

[1] Theorem 6.6.4 has been found by R. Graham during the Winter School "Geometry and Physics" at Srni in January 2007. At this conference I gave a lecture discussing formula (1.6.14). Partial results on $\delta_n(h)(1)$ for $n \leq 6$ suggested that we expect a formula for Q_n for general n without the contribution $\delta_n(1)$, i.e., a formula only in terms of $P_{2j}^*(0)$ and v_{2j}.

(6.6.22). On the other hand, by Green's formula, the coefficient coincides with the coefficient of $\varepsilon^{n+2\lambda}$ in the asymptotic expansion of

$$\int_{r=\varepsilon} \left(ur\frac{\partial v}{\partial r} - vr\frac{\partial u}{\partial r} \right) \mathrm{vol}(h_r).$$

The evaluation of that coefficient yields the identity for $\lambda \to 0$. For full details we refer to [125].   □

We combine Theorem 6.6.4 with Corollary 6.6.1 and find

Theorem 6.6.5. (6.6.2) *is equivalent to*

$$(-1)^{\frac{n}{2}} 2nQ_n(h) = 2^n \left(\frac{n}{2}\right)! \left(\frac{n}{2}-1\right)! \sum_{j=0}^{\frac{n}{2}-1} (n-2j)T_{2j}^*(h;0)(v_{n-2j}),$$

i.e.,

$$Q_n(h) = (-1)^{\frac{n}{2}} \left(\frac{n}{2}-1\right)! \sum_{j=0}^{\frac{n}{2}-1} 2^{n-1-2j}\frac{\left(\frac{n}{2}-j\right)!}{j!} P_{2j}^*(h;0)(v_{n-2j}).$$

Proof. Corollary 6.6.1 and Theorem 6.6.4 prove the equivalence of (6.6.2) and

$$2(-1)^{\frac{n}{2}}Q_n(h) = \frac{1}{n}2^n \left(\frac{n}{2}\right)! \left(\frac{n}{2}-1\right)! \sum_{j=0}^{\frac{n}{2}-1} (-2j)T_{2j}^*(h;0)(v_{n-2j})$$

$$+ 2^n \left(\frac{n}{2}\right)! \left(\frac{n}{2}-1\right)! \sum_{j=0}^{\frac{n}{2}-1} T_{2j}^*(h;0)(v_{n-2j})$$

$$= 2^n \left(\frac{n}{2}\right)! \left(\frac{n}{2}-1\right)!\frac{1}{n} \sum_{j=0}^{\frac{n}{2}-1} (n-2j)T_{2j}^*(h;0)(v_{n-2j}).$$

This proves the first formula. The latter identity is equivalent to

$$2(-1)^{\frac{n}{2}}Q_n(h)$$

$$= 2^n \left(\frac{n}{2}\right)! \left(\frac{n}{2}-1\right)!\frac{1}{n} \sum_{j=0}^{\frac{n}{2}-1} (n-2j)\frac{1}{2^{2j}j!}\frac{1}{\left(\frac{n}{2}-1\right)\cdots\left(\frac{n}{2}-j\right)} P_{2j}^*(h;0)(v_{n-2j}).$$

This implies the second assertion.   □

Now the identities in Theorem 6.6.5 admit an independent proof ([125]).

Theorem 6.6.6 (The holographic formula). *For even* $n \geq 2$,

$$(-1)^{\frac{n}{2}} 2n Q_n(h) = 2^n \left(\frac{n}{2}\right)! \left(\frac{n}{2}-1\right)! \sum_{j=0}^{\frac{n}{2}-1} (n-2j) \mathcal{T}_{2j}^*(h;0)(v_{n-2j}), \qquad (6.6.23)$$

i.e.,

$$Q_n(h) = (-1)^{\frac{n}{2}} \left(\frac{n}{2}-1\right)! \sum_{j=0}^{\frac{n}{2}-1} 2^{n-1-2j} \frac{\left(\frac{n}{2}-j\right)!}{j!} P_{2j}^*(h;0)(v_{n-2j}). \qquad (6.6.24)$$

Theorem 6.6.6, in particular, completes the proof of Theorem 6.6.1. The proof of Theorem 6.6.6 ([125]) will be outlined below.

As a consequence of (6.6.24), we give a simple proof of the following result.

Corollary 6.6.2 ([128]).

$$2 \int_{M^n} Q_n \,\mathrm{vol} = (-1)^{\frac{n}{2}} 2^n \left(\frac{n}{2}\right)! \left(\frac{n}{2}-1\right)! \int_{M^n} v_n \,\mathrm{vol}.$$

Proof. (6.6.24) implies

$$\int_{M^n} Q_n \,\mathrm{vol} = (-1)^{\frac{n}{2}} 2^{n-1} \left(\frac{n}{2}-1\right)! \left(\frac{n}{2}\right)! \int_M v_n \,\mathrm{vol}$$

$$+ (-1)^{\frac{n}{2}} \left(\frac{n}{2}-1\right)! \sum_{j=1}^{\frac{n}{2}-1} 2^{n-1-2j} \frac{\left(\frac{n}{2}-j\right)!}{j!} \int_{M^n} P_{2j}^*(0)(v_{n-2j}) \,\mathrm{vol}.$$

But

$$\int_{M^n} P_{2j}^*(0)(v_{n-2j}) \,\mathrm{vol} = \int_{M^n} v_{n-2j} P_{2j}(0)(1) \,\mathrm{vol} = 0$$

since $P_{2j}(0)$ annihilates constants. The proof is complete. $\qquad \square$

The quantity

$$L = \int_{M^n} v_n \,\mathrm{vol}$$

appears as the coefficient of the log-term in the asymptotics of the g_E-volume of the set $\{r > \varepsilon\}$ for $\varepsilon \to 0$,

$$\int_\varepsilon^\delta r^{-n-1} \int_{M^n} v(r,b) \,\mathrm{vol}\, dr$$

$$\sim c_0 \varepsilon^{-n} + c_2 \varepsilon^{-(n-2)} + \cdots + c_{n-2} \varepsilon^{-2} - L \log \varepsilon + \cdots \qquad (6.6.25)$$

with

$$c_{2j} = \frac{1}{n-2j} \int_{M^n} v_{2j} \, \text{vol} \tag{6.6.26}$$

for $j = 0, \ldots, \frac{n}{2}-1$. Therefore, Corollary 6.6.2 can be restated as

$$2 \int_{M^n} Q_n \, \text{vol} = (-1)^{\frac{n}{2}} 2^n \left(\frac{n}{2}\right)! \left(\frac{n}{2}-1\right)! L. \tag{6.6.27}$$

In this form the result appears in [128].

Theorem 6.6.6 can be used to determine the contribution of $\Delta^{\frac{n}{2}-1}(J)$ to Q_n. We recall that $Q_4 = 2(J^2 - |P|^2) - \Delta J$. In all dimensions $n \geq 4$, Q_n contains an analogous contribution $\Delta^{\frac{n}{2}-1}(J)$. By Theorem 6.6.6, all contributions of the form $\Delta^{\frac{n}{2}-1}(\tau)$ come from $P^*_{n-2}(0)(v_2)$, i.e., are given by

$$(-1)^{\frac{n}{2}} 2\Delta^{\frac{n}{2}-1}(v_2) = (-1)^{\frac{n}{2}-1} \Delta^{\frac{n}{2}-1}(J)$$

using $v_2 = -\frac{1}{2}J$. The fact that this *is* the correct contribution is a result of Branson (see [31], Corollary 1.4 or [32], Corollary 1.5).

Corollary 6.6.3. *For closed M^n (n even) and $u \in \ker P_n$, the sum*

$$I_n(u) = \sum_{j=0}^{\frac{n}{2}-1} (n-2j) \int_{M^n} v_{n-2j} T_{2j}(0)(u) \, \text{vol}$$

is conformally invariant.

Proof. For $u \in \ker(P_n)$, we find that $\int_M u Q_n \, \text{vol}$ is a conformal invariant. But Theorem 6.6.6 implies that

$$\sum_{j=0}^{\frac{n}{2}-1} (n-2j) \int_{M^n} u T^*_{2j}(0)(v_{n-2j}) \, \text{vol} = \sum_{j=0}^{\frac{n}{2}-1} (n-2j) \int_{M^n} v_{n-2j} T_{2j}(0)(u) \, \text{vol}$$

is conformally invariant. $\square$

It should be noted that Corollary 6.6.3 actually does *not* refer to Q-curvature. For $u = 1$, it reduces to the conformal invariance of L.

The following comments on the nature of the various terms in the holographic formula (6.6.24) might be useful. The two main ingredients are the holographic coefficients $1 = v_0, v_2, \ldots, v_n$ which describe the asymptotics of the volume of g_E *and* the solution of the (approximate) Dirichlet problem for the Laplacian of g_E. We explain the details. The following discussion is formal. h is fixed and suppressed. We consider a genuine eigenfunction

$$-\Delta_{g_E} u = \mu(n-\mu)u \tag{6.6.28}$$

of the form

$$u = \sum_{j\geq 0} r^{n-\mu+2j} a_{2j}(n-\mu) + \sum_{j\geq 0} r^{\mu+2j} b_{2j}(n-\mu). \qquad (6.6.29)$$

Then (by definition of $\mathcal{S}$) $b_0(n-\mu) = \mathcal{S}(\mu)a_0(n-\mu)$ and (see (6.6.3))

$$a_{2j}(\mu) = \frac{1}{2^{2j}j!\left(\frac{n}{2}-\mu-1\right)\cdots\left(\frac{n}{2}-\mu-j\right)} P_{2j}(\mu)a_0(\mu).$$

In the special case $\mu = n$, the function u is *harmonic* and (6.6.29) has the form

$$u = a_0(0) + r^2 a_2(0) + \cdots + r^{n-2}a_{n-2}(0) + HOT$$

with

$$a_{2j}(0) = \frac{1}{2^{2j}j!\left(\frac{n}{2}-1\right)\cdots\left(\frac{n}{2}-j\right)} P_{2j}(0)a_0(0), \quad j = 0,\ldots,\frac{n}{2} - 1.$$

Thus the operators $P_{2j}(0)$ determine the first $\frac{n}{2}$ terms in the asymptotics of harmonic functions. The operator $a_n(\mu)$ has a simple pole at $\mu = 0$. For the boundary function $a_0(0) = 1$, its residue vanishes and we have

$$a_n(0) = -\frac{1}{2^n\left(\frac{n}{2}\right)!\left(\frac{n}{2}-1\right)!}\dot{P}_n(0)(1) = -c_{\frac{n}{2}}(-1)^{\frac{n}{2}}Q_n. \qquad (6.6.30)$$

As a preparation for the proof of Theorem 6.6.6, we recall some results of [97] and [128]. For the family $u(\mu) = \mathcal{P}(\mu)(1)$ (here $\mathcal{P}$ denotes the Poisson transformation of [128]) of eigenfunctions, we differentiate (6.6.29) and obtain

$$\dot{u}(n) = (-\log r - Ar^n \log r + B) + (Cr^n \log r + Dr^n)$$

with coefficients A, B, C and D which are even in r and satisfy

$$A|_{r=0} = a_n(0), \quad B|_{r=0} = 0, \quad C|_{r=0} = b_0(0).$$

Here we have used that $a_0(0) = 1$ and $P_{2j}(0)(1) = 0$ for $j = 1,\ldots,\frac{n}{2} - 1$. Hence

$$\dot{u}(n) = -\log r + \alpha r^n \log r + \beta$$

with even α, β so that

$$\alpha|_{r=0} = -a_n(0) + b_0(0), \quad \beta|_{r=0} = 0.$$

But

$$b_0(0) = \mathcal{S}(n)(1) = (-1)^{\frac{n}{2}}c_{\frac{n}{2}}Q_n$$

and (6.6.30) yield

$$\alpha|_{r=0} = 2(-1)^{\frac{n}{2}}Q_n.$$

Differentiation of the eigenequation (6.6.28) at $\mu = n$ shows that

$$-\Delta_{g_E} \dot{u}(n) = -nu(n) = -n\mathcal{P}(n)(1) = -n.$$

In other words, $v = \dot{u}(n)$ satisfies $\Delta_{g_E} v = n$ and has the asymptotics

$$v = -\log r + \alpha r^n \log r + \beta \tag{6.6.31}$$

with

$$\alpha|_{r=0} = 2(-1)^{\frac{n}{2}} c_{\frac{n}{2}} Q_n, \quad \beta|_{r=0} = 0. \tag{6.6.32}$$

Fefferman and Graham ([97]) proved that $v \pmod{O(r^n)}$ actually is uniquely determined by the requirements $\Delta_{g_E} v = n$ and (6.6.31) with smooth α and β so that $\beta|_{r=0} = 0$. $\beta \pmod{O(r^n)}$ and $\alpha|_{r=0}$ are locally determined by h. Moreover, $\alpha|_{r=0}$ yields Q_n. An ambient metric version of that result is proved in [98].

Next, we recall the ideas of the proof of the identity

$$2(-1)^{\frac{n}{2}} c_{\frac{n}{2}} \int_{M^n} Q_n \, \text{vol} = \int_{M^n} v_n \, \text{vol} \tag{6.6.33}$$

(Corollary 6.6.2) in [128] and [97], and outline an independent *proof of Theorem 6.6.6* ([125]). In particular, this proves the relation (6.6.2) which has been used here as the starting point to derive the explicit holographic formula for Q_n. The proof in [128] rests on the regularization of the integral

$$\int_X \left(|du(\lambda)|^2 - \lambda(n-\lambda)u^2(\lambda) \right) \text{vol} \tag{6.6.34}$$

for $u(\lambda) = \mathcal{P}(\lambda)(1)$ (λ near n). We compare the limiting behaviour for $\lambda \to n$ of the finite parts in two different calculations of the integral $\int_{r>\varepsilon}$. On the one hand, a version of Green's formula implies that the finite part is given by

$$-n \int_M \mathcal{S}(\lambda)(1) \, \text{vol}.$$

By Theorem 3.2.2, the latter integral tends to $-n(-1)^{\frac{n}{2}} c_{\frac{n}{2}} \int_M Q_n \, \text{vol}$ for $\lambda \to n$. On the other hand, an evaluation of (6.6.34) using the asymptotics of $u(\lambda)$ for $r \to 0$ yields $-n\frac{1}{2} \int_M v_n \, \text{vol}$. This proves (6.6.33).

The proof in [97] rests on the consideration of the coefficient of $\log \varepsilon$ in the asymptotics of the integral

$$\int_{\delta > r > \varepsilon} (\Delta v) \, \text{vol}(g_E), \tag{6.6.35}$$

where v is as in (6.6.31). On the one hand, Green's formula yields $-n \int_M \alpha|_{r=0} \text{vol} = -2(-1)^{\frac{n}{2}} c_{\frac{n}{2}} \int_M Q_n \text{vol}$ (using (6.6.32)). On the other hand, using $\Delta v = n$, we find

$n \int_{r>\varepsilon}$ vol and, by the definition of $v(r)$, the coefficient of $\log \varepsilon$ is $-n \int_M v_n$ vol. This proves (6.6.33).

A generalization of the latter argument can be used to prove Theorem 6.6.6 (see [125] for the details). The idea is to determine the coefficient of $\log \varepsilon$ in the asymptotics for $\varepsilon \to 0$ of the integral

$$\int_{\delta > r > \varepsilon} (u \Delta v - v \Delta u) \operatorname{vol}(g_E), \tag{6.6.36}$$

where v is as in (6.6.31) and

$$u = \sum_{j=0}^{\frac{n}{2}-1} (\mathcal{T}_{2j}(0)f) \, r^{2j}, \ f \in C_c^\infty(M).$$

On the one hand, using $\Delta u = O(r^n)$ we find

$$-n \sum_{j=0}^{\frac{n}{2}-1} \int_M v_{n-2j} \, (\mathcal{T}_{2j}(0)f) \operatorname{vol} = -n \sum_{j=0}^{\frac{n}{2}-1} \int_M f \mathcal{T}_{2j}^*(0)(v_{n-2j}) \operatorname{vol}.$$

On the other hand, by Green's formula, it suffices to determine the coefficient of $\log \varepsilon$ in

$$-\varepsilon^{1-n} \int_{r=\varepsilon} \left(u \frac{\partial v}{\partial r} - v \frac{\partial u}{\partial r} \right) \operatorname{vol}(h_r).$$

We find

$$-\int_M \left[fn\alpha + \sum_{j=0}^{\frac{n}{2}-1} 2j(\mathcal{T}_{2j}(0)f)v_{n-2j} \right] \operatorname{vol}.$$

Using (6.6.32), this proves Theorem 6.6.6.

In order to illustrate the holographic formulas, we work out these formulas in dimensions $n = 2$, $n = 4$ and $n = 6$.

Example 6.6.3. *For $n = 2$, we have $Q_2 = K$. In that case, (6.6.5) states that*

$$Q_2 = \dot{P}_2^*(0)(1) - 4v_2.$$

But

$$v_2 = -\frac{1}{2}\mathsf{J} = -\frac{\tau}{4} = -\frac{K}{2}$$

([119]) and $P_2(\mu) = \Delta - \mu K$. It is clearly better to use the more symmetric form (6.6.20) which reads

$$2Q_2 = -4v_2$$

since $P_2(\mu)$ is self-adjoint. Finally, Theorem 6.6.6 states directly $Q_2 = -2v_2$.

Example 6.6.4. *For $n = 4$, (6.6.5) states that*

$$Q_4 = -\dot{P}_4^*(0)(1) + 8P_2^*(0)(v_2) + 32v_4.$$

We confirm the latter formula by explicit calculations using Theorem 6.9.2. First, we take a closer look at the contributions of ΔJ on both sides. For the left-hand side we know that

$$Q_4 = 2\left(J^2 - |P|^2\right) - \Delta J$$

(see (4.1.1)), where $J = \mathrm{tr}(P)$ and $P = \frac{1}{2}(\mathrm{Ric} - Jg)$ (see (2.5.10)). On the right-hand side we use the self-adjoint operator $P_2(\lambda) = \Delta - \lambda J$ and $v_2 = -\frac{1}{2}\mathrm{tr}(P) = -\frac{1}{2}J$. Thus the middle term contributes $-4\Delta J$. v_4 is given by

$$v_4 = \frac{1}{8}(J^2 - |P|^2). \tag{6.6.37}$$

Finally, by (6.6.39), we have

$$\dot{P}_4^*(0)(1) = -3\Delta J + 2(J^2 - |P|^2).$$

We also recall that $\dot{P}_4(0)(1) = Q_4$ (see (6.6.38)). Again, it is clearly more efficient to use the more symmetric form (6.6.20)

$$2Q_4 = \left(\dot{P}_4(0) - \dot{P}_4^*(0)\right)(1) + 8P_2^*(0)(v_2) + 32v_4$$

which reads

$$2Q_4 = 2\Delta J + 8\Delta\left(-\frac{1}{2}J\right) + 4(J^2 - |P|^2)$$

$$= -2\Delta J + 4(J^2 - |P|^2)$$

using $(\dot{P}_4(0) - \dot{P}_4^*(0))(1) = 2\Delta J$ (see (6.6.40)). The latter formula is a special case of Theorem 6.6.4 which states

$$4(\dot{P}_4(0) - \dot{P}_4^*(0))(1) = -2^6 T_2^*(0)(v_2).$$

Since $T_2(\lambda) = \frac{1}{4(1-\lambda)}(\Delta - \lambda J)$ (see the proof of Lemma 6.6.1), we recover the above formula. Finally, Theorem 6.6.6 directly yields

$$Q_4 = 8\left(2v_4 + \frac{1}{4}\Delta(v_2)\right),$$

i.e., $Q_4 = 2(J^2 - |P|^2) - \Delta J$.

The following result establishes the missing facts.

Lemma 6.6.1. *Let $n = 4$. Then $P_2(\lambda) = \Delta - \lambda J$ and*

$$P_4(\lambda) = (\Delta - (\lambda+2)J)(\Delta - \lambda J) + 2\lambda(\lambda-1)|P|^2 + 4(\lambda-1)\delta(P\#d) + 2(\lambda-1)(dJ, d).$$

Proof. Theorem 6.9.4 implies that the ansatz $u \sim \sum_{j\geq 0} r^{\lambda+2j} a_{2j}(\lambda)$ for u so that $-\Delta_{g_E} u = \lambda(n-\lambda)u$ yields (see (6.6.3))

$$a_2(\lambda) = \mathcal{T}_2(\lambda)a_0 = \frac{1}{4(1-\lambda)}P_2(\lambda)a_0,$$

$$a_4(\lambda) = \mathcal{T}_4(\lambda)a_0 = -\frac{1}{4(1-\lambda)}\frac{1}{8\lambda}P_4(\lambda)a_0$$

with $P_2(\lambda) = \Delta - \lambda J$ and

$$a_4(\lambda) = -\frac{1}{8\lambda}\left((\Delta-(\lambda+2)J)a_2 - \frac{1}{2}\lambda|P|^2 a_0 - \delta(P\#da_0) - \frac{1}{2}(dJ, da_0)\right)$$

$$= -\frac{1}{8\lambda 4(1-\lambda)}\left[(\Delta-(\lambda+2)J)(\Delta-\lambda J)a_0 + 2\lambda(\lambda-1)|P|^2 a_0\right.$$

$$\left. + 4(\lambda-1)\delta(P\#da_0) + 2(\lambda-1)(dJ, da_0)\right].$$

The proof is complete.   □

In particular, we find

$$P_4(0)u = \Delta^2 u - 2J\Delta u - 4\delta(P\#du) - 2(dJ, du) = P_4 u$$

by (4.1.3); note that in the present context the Laplacian Δ is defined as $-\delta d$. In other words, the Paneitz operator is constructed by the asymptotics of eigenfunctions of the associated Poincaré-Einstein metric. This is a special case of the results of Graham and Zworski (Section 3.2, (3.2.18)). Moreover, Lemma 6.6.1 implies

$$\dot{P}_4(0)u = -\Delta(Ju) - J\Delta u + 2J^2 u - 2|P|^2 u + 4\delta(P\#du) + 2(dJ, du)$$

and

$$\dot{P}_4^*(0)u = -J\Delta u - \Delta(Ju) + 2J^2 u - 2|P|^2 u + 4\delta(P\#du) + 2\delta(udJ).$$

For $u = 1$, we get
$$\dot{P}_4(0)(1) = -\Delta J + 2J^2 - 2|P|^2 = Q_4 \tag{6.6.38}$$

(by (4.1.1)) and
$$\dot{P}_4^*(0)(1) = -3\Delta J + 2J^2 - 2|P|^2. \tag{6.6.39}$$

Note that

$$\delta_4 u = \left(\dot{P}_4(0) - \dot{P}_4^*(0)\right)u$$

$$= 2(dJ, du) - 2\delta(udJ) = 4(dJ, du) + 2u\Delta J = 2(\Delta(Ju) - J\Delta(u)), \tag{6.6.40}$$

i.e., the operator $\delta_4 = \dot{P}_4(0) - \dot{P}_4^*(0)$ is only of first order.

Example 6.6.5. *We consider* Q_6 *on* M^6. *The holographic formula* (6.6.5) *says that*

$$2Q_6 = -\delta_6(1) - (48 \cdot 16)v_6 - (16 \cdot 6)P_2^*(0)(v_4) - 12P_4^*(0)(v_2)$$

(for the definition of δ_6 *we refer to* (6.6.21)). *In Section 6.9 and Section 6.10, we will discuss explicit formulas for* v_6 *and* $\delta_6(1)$. *Here we focus on the remaining two terms. As in Example 6.6.3 we have* $P_2(\lambda) = \Delta - \lambda J$ *(in all dimensions). In fact,* $P_2(\frac{n}{2}-1) = P_2$ *(see* (3.2.17)). *For the formula*

$$v_4 = \frac{1}{8}\left(J^2 - |P|^2\right)$$

we refer to Theorem 6.9.2. Thus the third term in the above formula for Q_6 *yields*

$$12\Delta(|P|^2 - J^2).$$

Next, we determine the family $P_4(\lambda)$ *which is given by the relations*

$$a_2(\lambda) = T_2(\lambda)a_0 = \frac{1}{4(2-\lambda)}P_2(\lambda)a_0,$$

$$a_4(\lambda) = T_4(\lambda)a_0 = \frac{1}{32(2-\lambda)(1-\lambda)}P_4(\lambda)a_0$$

(compare with the proof of Lemma 6.6.1 dealing with the case $n = 4$). *Now Theorem 6.9.4 yields*

$$a_4(\lambda) = \frac{1}{8(1-\lambda)}\left((\Delta-(\lambda+2)J)a_2 - \frac{1}{2}\lambda|P|^2a_0 - \delta(P\#da_0) - \frac{1}{2}(dJ, da_0)\right)$$

$$= \frac{1}{32(1-\lambda)(2-\lambda)}\Big[(\Delta-(\lambda+2)J)(\Delta-\lambda J)a_0 + 2\lambda(\lambda-2)|P|^2a_0$$

$$+ 4(\lambda-2)\delta(P\#da_0) + 2(\lambda-2)(dJ, da_0)\Big],$$

i.e.,

$$P_4(\lambda)u = (\Delta-(\lambda+2)J)(\Delta-\lambda J)u$$
$$+ 2\lambda(\lambda-2)|P|^2u + 4(\lambda-2)\delta(P\#du) + 2(\lambda-2)(dJ, du) \quad (6.6.41)$$

on M^6. *In order to check the latter formula, we use the criterion* $P_4(1) = P_{4,6}$ *(by* (3.2.17) *for* $j = 2$ *and* $n = 6$). *We find*

$$P_4(1)u = \Delta^2u - \Delta(Ju) - 3J\Delta u + 3J^2u - 2|P|^2u - 4\delta(P\#du) - 2(dJ, du).$$

It is easy to prove that this operator actually coincides with the Paneitz operator

$$P_{4,6}u = \Delta^2u + \delta\left(4J - 4P\#\right)du + \left(3J^2 - 2|P|^2 - \Delta J\right)u$$

(see (4.1.6) and (4.1.7)). Now (6.6.41) yields

$$P_4(0) = (\Delta - 2\mathsf{J})\Delta - 8\delta(\mathsf{P}\#d) - 4(d\mathsf{J}, d)$$

and hence

$$P_4^*(0) = \Delta(\Delta - 2\mathsf{J}) - 8\delta(\mathsf{P}\#d) - 4\delta(\cdot d\mathsf{J}).$$

It follows that

$$P_4^*(0)(\mathsf{J}) = \Delta^2\mathsf{J} - 2\Delta(\mathsf{J}^2) - 8\delta(\mathsf{P}\#d\mathsf{J}) - 4\delta(\mathsf{J}d\mathsf{J})$$
$$= \Delta^2\mathsf{J} - 8\delta(\mathsf{P}\#d\mathsf{J}). \tag{6.6.42}$$

Thus using $v_2 = -\frac{1}{2}\mathsf{J}$, we find the formula

$$Q_6 = -\frac{1}{2}\delta_6(1) - (8 \cdot 48)v_6 + 6\Delta\left(|\mathsf{P}|^2 - \mathsf{J}^2\right) + 3\left(\Delta^2\mathsf{J} - 8\delta(\mathsf{P}\#d\mathsf{J})\right). \tag{6.6.43}$$

In order to obtain a fully explicit formula for Q_6, it remains to work out the contributions $\delta_6(1)$ and v_6. As mentioned above, this will be done in Section 6.9 and Section 6.10. Moreover, we will confirm the resulting formula for Q_6 by an independent calculation of $\dot{P}_6(0)(1)$. Here we only note that the operator δ_6 contains the contribution $-4\Delta\delta(ud\mathsf{J})$ which finally yields the contribution $\Delta^2\mathsf{J}$ to Q_6 (recall that the convention here is $\Delta = -\delta d$).

Remark 6.6.5. *The analogous formula for $a_4(\lambda)$ for general dimension n reproduces the Paneitz operator $P_{4,n}$ (see (4.1.6)). In fact, we find*

$$a_4(\lambda) = \frac{1}{32(\frac{n}{2} - \lambda - 1)(\frac{n}{2} - \lambda - 2)}\left[(\Delta - (\lambda+2)\mathsf{J})(\Delta - \lambda\mathsf{J})a_0 + 2\lambda\left(\lambda - \frac{n}{2} + 1\right)|\mathsf{P}|^2 a_0\right.$$
$$\left. + 4\left(\lambda - \frac{n}{2} + 1\right)\delta(\mathsf{P}\#da_0) + 2\left(\lambda - \frac{n}{2} + 1\right)(d\mathsf{J}, da_0)\right],$$

i.e.,

$$P_4\left(\frac{n}{2} - 2\right)u = \left(\Delta - \frac{n}{2}\mathsf{J}\right)\left(\Delta - \left(\frac{n}{2} - 2\right)\mathsf{J}\right)u - (n-4)|\mathsf{P}|^2 u - 4\delta(\mathsf{P}\#du) - 2(d\mathsf{J}, du).$$

It is easy to verify that the latter operator coincides with

$$P_{4,n} = \Delta^2 + \delta\left((n-2)\mathsf{J} - 4\mathsf{P}\#\right)d + \frac{n-4}{2}\left(\frac{n}{2}\mathsf{J}^2 - 2|\mathsf{P}|^2 - \Delta\mathsf{J}\right).$$

Remark 6.6.6. *The details of the relation between the Paneitz operator $P_{4,n}$ and the square of the Laplacian of the ambient metric can be found in [149].*

Finally, we determine the critical Q-polynomial $Q_n^{\text{res}}(M^n; h; \lambda)$ in two special cases.

Example 6.6.6. *Let $n = 2$. Then (by (6.6.13))*

$$Q_2^{\mathrm{res}}(M^2; h; \lambda) = -4\lambda P_0^*(\lambda)(v_2) + P_2^*(\lambda)(v_0)$$
$$= -4\lambda v_2 + (\Delta - \lambda\mathsf{J})(1)$$
$$= 2\lambda\mathsf{J} - \lambda\mathsf{J}$$
$$= \lambda Q_2(h).$$

Example 6.6.7. *Let $n = 4$. We apply the holographic formula (6.6.13) in order to find*

$$-Q_4^{\mathrm{res}}(\lambda) = 32(1-\lambda)(-\lambda)P_0^*(v_4) + 8(-\lambda)P_2^*(\lambda)(v_2) + P_4^*(\lambda)(v_0).$$

Now Lemma 6.6.1 implies

$$P_4^*(\lambda)(1) = -(\lambda+2)(\Delta - \lambda\mathsf{J})\mathsf{J} + 2\lambda(\lambda-1)|\mathsf{P}|^2 - 2(\lambda-1)\Delta\mathsf{J}$$
$$= -3\lambda\Delta\mathsf{J} + \lambda(\lambda+2)\mathsf{J}^2 + 2\lambda(\lambda-1)|\mathsf{P}|^2.$$

Hence

$$-Q_4^{\mathrm{res}}(\lambda) = \lambda\left[32(\lambda-1)v_4 + 4(\Delta-\lambda\mathsf{J})\mathsf{J} - 3\Delta\mathsf{J} + (\lambda+2)\mathsf{J}^2 + 2(\lambda-1)|\mathsf{P}|^2\right]$$
$$= \lambda\left[4(\lambda-1)(\mathsf{J}^2 - |\mathsf{P}|^2) + \Delta\mathsf{J} - 3\lambda\mathsf{J}^2 + 2\mathsf{J}^2 + 2(\lambda-1)|\mathsf{P}|^2\right]$$
$$= \lambda\left[\lambda(\mathsf{J}^2 - 2|\mathsf{P}|^2) + (-2\mathsf{J}^2 + 2|\mathsf{P}|^2 + \Delta\mathsf{J})\right],$$

i.e.,

$$Q_4^{\mathrm{res}}(\lambda) = \lambda^2(2|\mathsf{P}|^2 - \mathsf{J}^2) + \lambda Q_4. \tag{6.6.44}$$

The latter result can be written also in the form

$$Q_4^{\mathrm{res}}(\lambda) = -\lambda(\lambda-1)Q_4 - \lambda^2 P_2 Q_2$$

since

$$Q_4 + P_2 Q_2 = \mathsf{J}^2 - 2|\mathsf{P}|^2.$$

For analogous results on critical Q-polynomials in higher dimensions we refer to Theorem 6.11.8.

Remark 6.6.7. *For* odd *n, we find*

$$Q_n^{\mathrm{res}}(h; \lambda) = D_n^{\mathrm{res}}(h; \lambda)(1) = 0$$

since

$$\delta_n(h; \lambda)(1) = \mathcal{T}_n^*(\lambda)(v_0) + \mathcal{T}_{n-1}^*(\lambda)(v_1) + \cdots + \mathcal{T}_0^*(\lambda)(v_n)$$

and $\mathcal{T}_{odd} = 0$ and $v_{odd} = 0$ (including v_n in view of the vanishing trace of $h_{(n)}$). In particular, $\dot{Q}_n^{\mathrm{res}}(h; 0) = 0$. The latter observation rests on the fact that h_r is even in r (which implies that the boundary $r = 0$ has vanishing second fundamental form (by (6.16.10))). The situation is different if the second fundamental form L is non-trivial. As an example, we note that the calculations in Section 5.4 yield families $D_N(\mathbb{B}^{n+1}, S^n; dr^2 + r^2 g_{S^n}; \lambda)$ (the sphere corresponds to $r = 1$). In that case, L and $\dot{D}_n(dr^2 + h_r^2; 0)(1)$ do not vanish and the calculations in Section 5.4 yield a holographic formula for this quantity.

6.7 $D_2(g;\lambda)$ as a residue family

For $M^{n+1} = [0,\varepsilon) \times \Sigma^n$ and a Poincaré metric $r^{-2}(dr^2 + h_r)$ of h, the family

$$D_2(dr^2 + h_r; \lambda) : C^\infty(M) \to C^\infty(\Sigma)$$

defined in Section 6.4 (see (6.4.1)) is related to the residue family $D_2^{\text{res}}(h;\lambda)$ defined in Section 6.6. Here we prove that $D_2^{\text{res}}(h;\lambda)$ arises by specialization of $D_2(M,\Sigma;g;\lambda)$ to the background metric $g = dr^2 + h_r$. In Section 6.24, the family $D_2(M,\Sigma;g;\lambda)$ will be recognized as a special tractor family. From that perspective, Theorem 6.7.1 appears as a version of the holographic duality (Section 6.21).

We recall that the family $D_2^{\text{res}}(h;\lambda)$ arises as follows. We consider the residue at $\lambda = -\mu - 3$ of the family of distributions

$$C_0^\infty(M) \ni \varphi \mapsto \int_0^\varepsilon \int_\Sigma r^\lambda u\varphi \, \text{vol}(dr^2 + h_r)$$

which is defined by a (formal) eigenfunction $u(r,b) \sim \sum_{j\geq 0} r^{\mu+2j} a_j(\mu;b)$ of the Laplacian of $r^{-2}(dr^2 + h_r)$ for the eigenvalue $\mu(n-\mu)$. Partial integration yields the formula

$$\int_\Sigma a_0 \delta_2(h;\mu)(\varphi) \, \text{vol}(h), \quad h = h_0,$$

where

$$\delta_2(h;\mu) = \frac{1}{2} v_0 \circ i^* \frac{\partial^2}{\partial r^2} + [T_2^*(\mu) \circ v_0 + T_0^*(\mu) \circ v_2] i^*$$

(Definition 6.6.2). Now using $v_0 = 1$, $h_{(2)} = -P(h)$, $v_2 = \frac{1}{2}\text{tr}(h_{(2)}) = -\frac{1}{2}J(h)$ (Theorem 6.9.1, Theorem 6.9.2) and

$$a_2(\mu) = T_2(\mu)a_0, \quad T_2(\mu) = \frac{1}{2(n-2-2\mu)}(\Delta_h - \mu J(h)),$$

we find

$$\delta_2(h;\mu+n-2) = \frac{1}{2}i^*\frac{\partial^2}{\partial r^2} - \left[\frac{1}{2(n-2+2\mu)}(\Delta_h - (\mu+n-2)J(h)) + \frac{1}{2}J(h)\right]i^*$$

$$= \frac{1}{2(n-2+2\mu)}\left\{(n-2+2\mu)i^*\frac{\partial^2}{\partial r^2} - (\Delta_h + \mu J(h))i^*\right\}.$$

Hence

$$D_2^{\text{res}}(h;\lambda) = -2(n-2+2\lambda)\delta_2(h;\lambda+n-2) = -(n-2+2\lambda)i^*\frac{\partial^2}{\partial r^2} + (\Delta_h + \lambda J(h))i^*.$$

On the other hand, the relation $J(h) = i^*J(dr^2 + h_r)$ (Lemma 6.11.1) implies

$$D_2(M,\Sigma;dr^2 + h_r;\lambda)$$
$$= -(n-2+2\lambda)i^*(\Delta_{dr^2+h_r} + \lambda J(dr^2+h_r)) + (n-1+2\lambda)(\Delta_h + \lambda J(h))i^*,$$

i.e.,

$$D_2(M, \Sigma; dr^2 + h_r; \lambda) = -(n-2+2\lambda)i^* \Delta_{dr^2+h_r} + (n-1+2\lambda)\Delta_h i^* + \lambda J(h) i^*$$

$$= -(n-2+2\lambda)i^* \frac{\partial^2}{\partial r^2} + (\Delta_h + \lambda J(h)) \, i^* \qquad (6.7.1)$$

in view of $H = 0$ and (6.4.9). In fact, in the present case the general formula

$$\tau(\hat{g}) = e^{-2\varphi}\tau(g) - 2n\Delta_{\hat{g}}(\varphi) + n(n-1)|d\varphi|_{\hat{g}}^2$$

yields

$$\tau(dr^2 + h_r) = 2n\,\mathrm{tr}(\mathsf{P}(h)) + o(r) = 2n\mathsf{J}(h) + o(r).$$

Thus we have proved

Theorem 6.7.1. $D_2^{\mathrm{res}}(h; \lambda) = D_2(dr^2 + h_r; \lambda).$

Note that the above discussion also proves the factorization identities

$$D_2^{\mathrm{res}}\left(h; -\frac{n-2}{2}\right) = P_2(h)i^* \quad \text{and} \quad D_2^{\mathrm{res}}\left(h; -\frac{n-1}{2}\right) = i^* P_2(dr^2 + h_r). \quad (6.7.2)$$

In Section 6.11, we shall prove analogous factorizations for higher order residue families (for conformally flat metrics).

6.8 $D_3^{\mathrm{res}}(h; \lambda)$

For a 3-dimensional Riemannian manifold (M, h), we analyze the critical order 3 residue family

$$D_3^{\mathrm{res}}(h; \lambda) \stackrel{\mathrm{def}}{=} 2(1-2\lambda)\delta_3(h; \lambda) : C^\infty([0, \varepsilon) \times M) \to C^\infty(M).$$

The following result identifies the value of that family for $\lambda = 0$ with the Chang-Qing operator $P_3(dr^2 + h_r) : C^\infty([0, \varepsilon) \times M) \to C^\infty(M)$ (see Section 6.26).

Lemma 6.8.1.
$$P_3(dr^2 + h_r) = 3\delta_3(h; 0) = \frac{3}{2}D_3^{\mathrm{res}}(h; 0).$$

Proof. We determine an explicit formula for the family $\delta_3(h; \lambda)$. Let $h = h_0$. (6.6.6) gives

$$\delta_3(h; \lambda) = \frac{1}{6}i^* \frac{\partial^3}{\partial r^3} + (\mathcal{T}_2^*(h; \lambda) + v_2)i^* \frac{\partial}{\partial r}.$$

Now $v_2 = -\frac{1}{2}\mathsf{J}(h)$ (by Theorem 6.9.2) and

$$\mathcal{T}_2(\lambda) = \frac{1}{2(1-2\lambda)}(\Delta - \lambda\mathsf{J})$$

by (6.6.3). Hence for the renormalization

$$D_3^{\mathrm{res}}(h;\lambda) = 2(1-2\lambda)\delta_3(h;\lambda)$$

we find

$$D_3^{\mathrm{res}}(h;\lambda) = \frac{1}{3}(1-2\lambda)i^* \frac{\partial^3}{\partial r^3} + (\Delta_h + (\lambda-1)\mathsf{J}(h)))\, i^* \frac{\partial}{\partial r}. \qquad (6.8.1)$$

In particular,

$$D_3^{\mathrm{res}}(h;0) = \frac{1}{3}i^* \frac{\partial^3}{\partial r^3} + (\Delta_h - \mathsf{J}(h))\, i^* \frac{\partial}{\partial r}.$$

On the other hand,

$$P_3(dr^2+h_r) = \frac{1}{2}\frac{\partial}{\partial r}\Delta_{dr^2+h_r} + \Delta_h \frac{\partial}{\partial r} + \left(\mathrm{tr}(G) - 2\mathsf{J}(dr^2+h_r)\right)\frac{\partial}{\partial r}$$

by (6.26.2) using $H = 0$; here we suppress the restriction i^* to $r = 0$. Now

$$\Delta_{dr^2+h_r} = \frac{\partial^2}{\partial r^2} + \frac{1}{2}\mathrm{tr}(h_r^{-1}\dot{h}_r)\frac{\partial}{\partial r} + \Delta_{h_r}.$$

Thus using $\dot{h}_0 = 0$ and $h_0 = h$, we find

$$P_3(dr^2+h_r)$$
$$= \frac{1}{2}\left(\frac{\partial^3}{\partial r^3} + \frac{1}{2}\mathrm{tr}(h^{-1}\ddot{h}_0)\frac{\partial}{\partial r} + \Delta_h \frac{\partial}{\partial r}\right) + \Delta_h \frac{\partial}{\partial r} + \left(\mathrm{tr}(G) - 2\mathsf{J}(dr^2+h_r)\right)\frac{\partial}{\partial r}.$$

In view of Theorem 6.9.1, the latter formula simplifies to

$$\frac{1}{2}\frac{\partial^3}{\partial r^3} + \frac{3}{2}\Delta_h \frac{\partial}{\partial r} + \left(\mathrm{tr}(G) - 2\mathsf{J}(dr^2+h_r) - \frac{1}{2}\mathsf{J}(h)\right)\frac{\partial}{\partial r}.$$

Now (6.8.2) implies $\tau(dr^2+h_r) = \tau(h) + 2\,\mathrm{tr}(G) = \frac{3}{2}\tau(h)$, i.e., $\mathsf{J}(dr^2+h_r) = \mathsf{J}(h)$ for $r = 0$. Hence

$$P_3(dr^2+h_r) = \frac{1}{2}\frac{\partial^3}{\partial r^3} + \frac{3}{2}\Delta_h \frac{\partial}{\partial r} - \frac{3}{2}\mathsf{J}(h)\frac{\partial}{\partial r}.$$

This proves the assertion. It remains to verify the identity

$$\mathrm{tr}(G) = \sum_i R_{iNN}^i = \mathsf{J}(h). \qquad (6.8.2)$$

We use the formulas

$$R_{iNN}^i = \sum_s \Gamma_{NN}^s \Gamma_{is}^i - \sum_s \Gamma_{iN}^s \Gamma_{Ns}^i + (\Gamma_{NN}^i)_i - (\Gamma_{iN}^i)_N$$

and

$$\Gamma_{ij}^s = \frac{1}{2} \sum_k ((g_{jk})_i + (g_{ki})_j - (g_{ij})_k) g^{ks}. \tag{6.8.3}$$

In the present setting, we find $\Gamma_{NN}^i = 0$ and

$$\Gamma_{iN}^j = \frac{1}{2} \sum_k \frac{\partial}{\partial r}((h_r)_{ik})h_r^{jk}.$$

Hence using $\dot{h}_0 = 0$, we find

$$R_{iNN}^i = -(\Gamma_{iN}^i)_N = -\frac{1}{2} \sum_k \frac{\partial}{\partial r}\left(\frac{\partial}{\partial r}((h_r)_{ik})h_r^{ik}\right)\Big|_{r=0} = \sum_k \mathsf{P}_{ik}h^{ik},$$

i.e.,

$$\mathrm{tr}(G) = \sum_i R_{iNN}^i = \mathrm{tr}(\mathsf{P}(h)) = \mathsf{J}(h).$$

This proves (6.8.2) and completes the proof.   □

Remark 6.8.1. *For the analogous family*

$$D_3^{\mathrm{res}}(h; \lambda) = -2(n-4+2\lambda)\delta_3(h; \lambda+n-3) : C^\infty([0, \varepsilon) \times M^n) \to C^\infty(M^n)$$

in higher dimension, we find

$$\delta_3(h; \lambda) = i^* \frac{1}{6}\frac{\partial^3}{\partial r^3} + (T_2^*(h; \lambda) + v_2)\, i^* \frac{\partial}{\partial r}$$

with

$$T_2(\lambda) = \frac{1}{2(n-2-2\lambda)}(\Delta_h - \lambda\mathsf{J}(h)), \quad v_2 = -\frac{1}{2}\mathsf{J}(h).$$

Hence

$$D_3^{\mathrm{res}}(h; \lambda) = -\frac{1}{3}(n-4+2\lambda)i^* \frac{\partial^3}{\partial r^3} + (\Delta_h+(\lambda-1)\mathsf{J}(h))\, i^* \frac{\partial}{\partial r}. \tag{6.8.4}$$

Remark 6.8.2. *Using formula (6.8.4), we prove the factorization identities*

$$D_3^{\mathrm{res}}\left(h; -\frac{n-1}{2}\right) = i^* \frac{\partial}{\partial r}P_2(dr^2+h_r), \quad D_3^{\mathrm{res}}\left(h; -\frac{n}{2}+2\right) = P_2(h)i^* \frac{\partial}{\partial r}.$$

In fact, by (6.8.4) we have

$$D_3^{\mathrm{res}}\left(h; -\frac{n-1}{2}\right) = i^* \frac{\partial^3}{\partial r^3} + \left(\Delta_h - \frac{n+1}{2}\mathsf{J}(h)\right) i^* \frac{\partial}{\partial r}.$$

On the other hand, the product on the right-hand side of the first identity is

$$i^* \frac{\partial}{\partial r} \left(\frac{\partial^2}{\partial r^2} + \frac{1}{2} \mathrm{tr}(h_r^{-1} \dot{h}_r) \frac{\partial}{\partial r} + \Delta_{h_r} - \frac{n-1}{2} \mathsf{J}(dr^2 + h_r) \right)$$

$$= i^* \frac{\partial^3}{\partial r^3} - \frac{1}{2} \mathrm{tr}(\mathsf{P}(h)) i^* \frac{\partial}{\partial r} + \Delta_h i^* \frac{\partial}{\partial r} - \frac{n-1}{2} \mathsf{J}(h) i^* \frac{\partial}{\partial r}.$$

This proves the first identity. The second factorization is obvious. In Section 6.11, we shall prove analogous factorizations for higher order residue families (for conformally flat metrics).

Remark 6.8.3. *In Section 6.25, we shall use the curved translation principle in order to construct an order 3 family $D_3^T(g; \lambda)$ for a general background metric. By Lemma 6.25.3, this family satisfies analogs of both factorization identities which hold true for $\mathcal{D}_3(\lambda)$. The identities in Remark 6.8.2 are special cases. The holographic duality*

$$D_3^{\mathrm{res}}(h; \lambda) = D_3^T(dr^2 + h_r; \lambda)$$

for order 3 families will be proved in Theorem 6.25.4.

6.9 The holographic coefficients v_2, v_4 and v_6

In the present section, we give detailed proofs of a number of results which can be found (usually without proofs) at various places and in varying form in the literature. More precisely, we determine the first two terms in the Fefferman-Graham expansion, derive the holographic coefficients v_{2j}, $j = 1, 2, 3$, and calculate the first two terms in the asymptotics of eigenfunctions of the Laplacian for the Poincaré-Einstein metric. Finally, we formulate an extension of the holographic formula to subcritical Q-curvatures (Conjecture 6.9.1).

Theorem 6.9.1. *The terms $h_{(2)}$ and $h_{(4)}$ in the formal Taylor series*

$$h_r = h_{(0)} + r^2 h_{(2)} + r^4 h_{(4)} + \cdots$$

of the Poincaré-Einstein metric $g = r^{-2}(dr^2 + h_r)$ are given by

$$h_{(2)} = -\mathsf{P} \tag{6.9.1}$$

(if $n \geq 4$), where P denotes the Schouten tensor for $h_{(0)}$, and

$$h_{(4)ij} = \frac{1}{4(n-4)} \left(-\mathcal{B}_{ij} + (n-4) \mathsf{P}_i^k \mathsf{P}_{kj} \right) \tag{6.9.2}$$

(if $n \geq 6$), where

$$\mathcal{B}_{ij} \overset{\mathrm{def}}{=} \Delta(\mathsf{P})_{ij} - \nabla^k \nabla_j(\mathsf{P})_{ik} + \mathsf{P}^{kl} \mathsf{C}_{kijl}. \tag{6.9.3}$$

$\mathcal{B}$ will be called the Bach tensor. $\Delta = \nabla^k \nabla_k$ is the Bochner-Laplacian on symmetric tensors.

Remark 6.9.1. *The fact that the Bach tensor $\mathcal{B}$ appears in Theorem 6.9.1 is directly linked with the fact that the curvature tensor $\tilde{R}$ of the ambient metric $\tilde{g}$ which corresponds to the Poincaré-metric g satisfies*

$$\tilde{R}_{mijm} = \frac{t^2}{n-4}\mathcal{B}_{ij}, \; n \neq 4$$

on the ray bundle $\mathcal{G} = \{(x, th_x) \,|\, t > 0\}$ of the metric h (Proposition 3.2/(iv) in [99]). In fact, the latter result follows by combining a calculation of the curvature tensor $\tilde{R}$ with Theorem 6.9.1. The tensor $\tilde{R}_{mijm}$ describes the normal curvature of the embedding of $\mathcal{G}$ in the ambient space. In dimension $n = 4$ the Bach tensor yields the Fefferman-Graham obstruction.

Note that Lemma 4.2.7 implies that

$$\operatorname{div}_4 \operatorname{div}_1(\mathsf{C})_{ij} = (n-3)\left(\Delta(\mathsf{P})_{ij} - \nabla^k \nabla_j(\mathsf{P})_{ik}\right).$$

Hence

$$\mathcal{B}_{ij} = \frac{1}{n-3}\nabla^l\nabla^k(\mathsf{C})_{kijl} + \frac{1}{n-2}\operatorname{Ric}^{kl}\mathsf{C}_{kijl}. \tag{6.9.4}$$

For $n = 4$, the latter formula yields the familiar Bach tensor

$$\mathcal{B}_{ij} = \nabla^l\nabla^k(\mathsf{C})_{kijl} + \frac{1}{2}\operatorname{Ric}^{kl}\mathsf{C}_{kijl}$$

from general relativity which arises as the first variation of the conformally invariant functional (see (2.5.15))

$$\int_{M^4} |\mathsf{C}|^2 \operatorname{vol}.$$

A direct calculation shows that

$$e^{2\varphi}\hat{\mathcal{B}}_{ij} = \mathcal{B}_{ij} - (n-4)(\mathcal{C}_{ikj} + \mathcal{C}_{jki})\varphi^k + (n-4)\mathsf{C}_{kijl}\varphi^k\varphi^l, \tag{6.9.5}$$

where

$$\mathcal{C}(X, Y, Z) = \nabla_X(\mathsf{P})(Y, Z) - \nabla_Y(\mathsf{P})(X, Z) \tag{6.9.6}$$

is the Cotton tensor (see (4.2.14)) ([168], Section 5). Hence, for $n = 4$, the Bach tensor is conformally invariant:

$$e^{2\varphi}\hat{\mathcal{B}} = \mathcal{B}. \tag{6.9.7}$$

It plays a central role in the characterization of conformally Einstein spaces ([158], [159], [168], [115]). By (6.9.4), the higher-dimensional Bach tensor $\mathcal{B}$ vanishes if $\mathsf{C} = 0$, i.e., in the conformally flat case.

Thus Theorem 6.9.1 implies that for *conformally flat* $h = h_{(0)}$ the Fefferman-Graham expansion is of the form

$$h_r = 1 - r^2\mathsf{P} + r^4\frac{1}{4}\mathsf{P}^2 + \cdots = \left(1 - \frac{r^2}{2}\mathsf{P}\right)^2 + HOT, \quad \mathsf{P} = \mathsf{P}(h), \tag{6.9.8}$$

where we identify P with $\mathsf{P}^\sharp$. In fact, there are no higher order terms (see (6.14.2)).

Theorem 6.9.2. *The holographic coefficients v_2, v_4 and v_6 are given by the formulas*

$$v_2 = -\frac{1}{2}\mathsf{J} \quad \text{if } n \geq 2,$$

$$v_4 = \frac{1}{8}\left(\mathsf{J}^2 - |\mathsf{P}|^2\right) = \frac{1}{4}\operatorname{tr}(\wedge^2\mathsf{P}) \quad \text{if } n \geq 4,$$

$$v_6 = -\frac{1}{8}\operatorname{tr}\left(\wedge^3\mathsf{P}\right) - \frac{1}{24(n-4)}(\mathcal{B},\mathsf{P}) \quad \text{if } n \geq 6.$$

Now we combine the holographic formula (6.6.43) for Q_6 with Theorem 6.9.2 and find

Corollary 6.9.1. *On M^6 we have*

$$Q_6 = -\frac{1}{2}\delta_6(1) + 48\operatorname{tr}(\wedge^3\mathsf{P}) + 8(\mathcal{B},\mathsf{P}) + 6\Delta\left(|\mathsf{P}|^2 - \mathsf{J}^2\right) + 3\left(\Delta^2\mathsf{J} - 8\delta(\mathsf{P}\#d\mathsf{J})\right).$$

In order to compare the latter formula for Q_6 with the formula for Q_6 given in [116], it remains to work out the contribution $\delta_6(1)$. This will be done in Section 6.10 and can be considered as a test of the holographic formula.

In terms of the coordinates $r^2 = \rho$, the metric g reads

$$g = \frac{d\rho^2}{4\rho^2} + \frac{1}{\rho}h_\rho, \quad h_\rho = h_{(0)} + \rho h_{(2)} + \rho^2 h_{(4)} + \cdots. \tag{6.9.9}$$

Then the condition $\operatorname{Ric}(g) + ng = 0$ implies the system

$$-\rho\left[2\ddot{h} - 2\dot{h}h^{-1}\dot{h} + \operatorname{tr}(h^{-1}\dot{h})\dot{h}\right] + (n-2)\dot{h} + \operatorname{tr}(h^{-1}\dot{h})h + \operatorname{Ric}(h) = 0, \tag{6.9.10}$$

$$\operatorname{tr}(h^{-1}\ddot{h}) - \frac{1}{2}\operatorname{tr}(h^{-1}\dot{h}h^{-1}\dot{h}) = 0 \tag{6.9.11}$$

and the conditions

$$\nabla^k\dot{h}_{ik} - \nabla_i\operatorname{tr}(h^{-1}\dot{h}) = 0 \tag{6.9.12}$$

for all i. (6.9.10) and (6.9.11) express the respective vanishing of the tangential and the normal components of $\operatorname{Ric}(g) + ng$. Here dots denote derivatives with respect to ρ.

Proof of Theorem 6.9.1. For $\rho = 0$, condition (6.9.10) yields

$$(n-2)h_{(2)} + \operatorname{tr}(h_{(0)}^{-1}h_{(2)})h_{(0)} + \operatorname{Ric}(h_{(0)}) = 0.$$

Hence

$$(n-2)\operatorname{tr}(h_{(0)}^{-1}h_{(2)}) + n\operatorname{tr}(h_{(0)}^{-1}h_{(2)}) + \tau(h_{(0)}) = 0,$$

i.e.,

$$\operatorname{tr}(h_{(2)}) = -\frac{1}{2(n-1)}\tau(h_{(0)}) = -\mathsf{J}(h_{(0)}),$$

where the trace is with respect to $h_{(0)}$. Therefore,

$$h_{(2)} = -\frac{1}{n-2}\left(\mathrm{Ric}(h_{(0)}) - \mathsf{J}(h_{(0)})h_{(0)}\right) = -\mathsf{P}(h_{(0)}).$$

This proves the first assertion of Theorem 6.9.1.

Now (6.9.11) yields at $\rho = 0$,

$$2\,\mathrm{tr}(h_{(0)}^{-1}h_{(4)}) - \frac{1}{2}\,\mathrm{tr}((h_{(0)}^{-1}h_{(2)})^2) = 0,$$

i.e.,

$$\mathrm{tr}(h_{(4)}) = \frac{1}{4}\,\mathrm{tr}(\mathsf{P}^2) \qquad\qquad (6.9.13)$$

using $h_{(2)} = -\mathsf{P}$ and (2.5.17). We differentiate (6.9.10) once with respect to ρ and obtain

$$(n-4)\ddot{h} + 2\dot{h}h^{-1}\dot{h} - \mathrm{tr}(h^{-1}\dot{h})\dot{h} + \dot{\mathrm{Ric}}(h) + \mathrm{tr}(\dot{h}^{-1}\dot{h})h + \mathrm{tr}(h^{-1}\ddot{h})h + \mathrm{tr}(h^{-1}\dot{h})\dot{h} = 0.$$

For $\rho = 0$, we get

$$2(n-4)h_{(4)} + 2\mathsf{P}h^{-1}\mathsf{P} - \mathrm{tr}(\mathsf{P}^2)h + \frac{1}{2}\,\mathrm{tr}(\mathsf{P}^2)h + (d/d\rho)|_0(\mathrm{Ric}(h_\rho)) = 0$$

using $\dot{h^{-1}} = -h^{-1}\dot{h}h^{-1}$ and (6.9.13). Here we have simplified the notation by setting $h_{(0)} = h$. This convention will be in force in what follows. Hence

$$2(n-4)h_{(4)} - \frac{1}{2}\,\mathrm{tr}(\mathsf{P}^2)h + 2\mathsf{P}h^{-1}\mathsf{P} + (d/d\rho)|_0(\mathrm{Ric}(h_\rho)) = 0,$$

i.e., we obtain the intermediate formula

$$h_{(4)ij} = \frac{1}{n-4}\left(\frac{1}{4}|\mathsf{P}|^2 h_{ij} - \mathsf{P}_i^k\mathsf{P}_{kj} - \frac{1}{2}(d/d\rho)|_0(\mathrm{Ric}(h_\rho))_{ij}\right). \qquad (6.9.14)$$

In order to evaluate the latter formula, we use the following formula for the first variation of the Ricci tensor.

$$\dot{\mathrm{Ric}}(g)[h]_{ij} = (d/dt)|_{t=0}\,\mathrm{Ric}_{ij}(g+th)$$
$$= -\frac{1}{2}\left\{\Delta_L(h)_{ij} - [\nabla_i(\delta_g(h))_j + \nabla_j(\delta_g(h))_i] + \mathrm{Hess}_g(\mathrm{tr}_g(h))_{ij}\right\},$$

where $\Delta_L : \Gamma(S^2T^*M) \to \Gamma(S^2T^*M)$ is the Lichnerowicz-Laplacian on symmetric bilinear forms. Δ_L is given by the formula

$$\Delta_L(h)_{ij} = \nabla^k\nabla_k(h)_{ij} - \left\{\mathrm{Ric}_{ik}(g)h_j^k + \mathrm{Ric}_{jk}(g)h_i^k\right\} + 2R_{rij}{}^s h_s^r.$$

For details we refer to [24], Theorem 1.174 and [77] (Chapter 3 and Appendix A). In the present application, we consider the variation $h - t\mathsf{P}$. Hence

$$(d/d\rho)|_0(\mathrm{Ric}(h_\rho))_{ij} = \dot{\mathrm{Ric}}(h)[-\mathsf{P}]_{ij}$$

$$= \frac{1}{2}\{\Delta_{L,h}(\mathsf{P})_{ij} + \mathrm{Hess}_h(\mathsf{J})_{ij} - (\nabla_i(d\mathsf{J})_j + \nabla_j(d\mathsf{J})_i)\}$$

$$= \frac{1}{2}\Delta_{L,h}(\mathsf{P})_{ij} - \frac{1}{2}\mathrm{Hess}_h(\mathsf{J})_{ij}$$

using $\delta_h(\mathsf{P}) = d\mathsf{J}$ (Lemma 4.2.7). Now for the contribution of the Lichnerowicz-Laplacian, we find

$$\Delta_{L,h}(\mathsf{P})_{ij} = \Delta(\mathsf{P})_{ij} - \{\mathrm{Ric}_{ik}\,\mathsf{P}^k_j + \mathrm{Ric}_{jk}\,\mathsf{P}^k_i\} + 2R_{rij}{}^s\mathsf{P}^r{}_s,$$

where all curvature tensors are with respect to h. Hence we obtain the formula

$$h_{(4)ij} = \frac{1}{n-4}\left[\frac{1}{4}|\mathsf{P}|^2 h_{ij} - \mathsf{P}^k_i\mathsf{P}_{kj} + \frac{1}{4}\mathrm{Hess}_{ij}(\mathsf{J})\right.$$

$$\left. - \frac{1}{4}\Delta(\mathsf{P})_{ij} - \frac{1}{2}R_{rij}{}^s\mathsf{P}^r{}_s + \frac{1}{4}\left(\mathrm{Ric}_{ik}\,\mathsf{P}^k_j + \mathrm{Ric}_{jk}\,\mathsf{P}^k_i\right)\right]. \quad (6.9.15)$$

In the remaining step, we rewrite the Hessian of J using the identity

$$\nabla^k\nabla_j(\mathsf{P})_{ki} = \mathrm{Hess}_{ij}(\mathsf{J}) - \mathcal{K}_{jk}(\mathsf{P})^k_i, \quad (6.9.16)$$

where the curvature contribution $\mathcal{K} \in C^\infty(\wedge^2 T^* M, \mathrm{End}(S^2 T^* M))$ is characterized by

$$(\nabla_X\nabla_Y - \nabla_Y\nabla_X - \nabla_{[X,Y]})(b) = \mathcal{K}(X,Y)(b)$$

for all symmetric bilinear forms b. In order to prove (6.9.16), we use normal coordinates at m and find

$$\sum_k \nabla_k\nabla_j(\mathsf{P})(e_k, e_i) = \sum_k \nabla_j\nabla_k(\mathsf{P})(e_k, e_i) - \mathcal{K}(e_j, e_k)(\mathsf{P})(e_k, e_i)$$

$$= \nabla_j\left(\sum_k \nabla_k(\mathsf{P})(e_k)\right)(e_i) - \mathcal{K}(e_j, e_k)(\mathsf{P})(e_k, e_i)$$

$$= \nabla_j(\delta(\mathsf{P}))(e_i) - \mathcal{K}(e_j, e_k)(\mathsf{P})(e_k, e_i)$$

$$= \nabla_j(d\mathsf{J})(e_i) - \mathcal{K}(e_j, e_k)(\mathsf{P})(e_k, e_i)$$

$$= \mathrm{Hess}(\mathsf{J})(e_i, e_j) - \mathcal{K}(e_j, e_k)(\mathsf{P})(e_k, e_i)$$

using Lemma 4.2.7. Next, we determine an explicit formula for the curvature term. Using the general rule

$$\mathcal{K}(X,Y)(b)(U,V) = -b(R(X,Y)U,V) - b(U,R(X,Y)V), \quad b \in \Gamma(S^2 T^* M),$$

we find

$$\mathcal{K}_{jk}(\mathsf{P})_{ki} = -\mathsf{P}(\sum_l R_{jkkl}e_l, e_i) - \mathsf{P}(e_k, \sum_l R_{jkil}e_l)$$

$$= -\sum_l R_{jkkl}\mathsf{P}_{il} - \sum_l \mathsf{C}_{jkil}\mathsf{P}_{lk} + \sum_l (\mathsf{P} \oslash h)_{jkil}\mathsf{P}_{lk}.$$

Hence

$$\mathrm{Hess}_{ij}(\mathsf{J}) = \nabla^k\nabla_j(\mathsf{P})_{ki} - \mathsf{C}_{jkil}\mathsf{P}^{lk} - \mathrm{Ric}_{jl}\,\mathsf{P}^l_i + (\mathsf{P} \oslash h)_{jkil}\mathsf{P}^{lk}. \qquad (6.9.17)$$

(6.9.15) implies

$$
\begin{aligned}
(n-4)h_{(4)ij} &= \frac{1}{4}|\mathsf{P}|^2 h_{ij} - \mathsf{P}^k_i \mathsf{P}_{kj} \\
&\quad + \frac{1}{4}\left(\nabla^k\nabla_j(\mathsf{P})_{ki} - \nabla^k\nabla_k(\mathsf{P})_{ij}\right) \\
&\quad - \frac{1}{4}\mathsf{C}_{jkil}\mathsf{P}^{kl} - \frac{1}{4}\,\mathrm{Ric}_{jl}\,\mathsf{P}^l_i + \frac{1}{4}(\mathsf{P} \oslash h)_{jkil}\mathsf{P}^{lk} \\
&\quad - \frac{1}{2}\mathsf{C}_{rij}{}^s \mathsf{P}^r_s + \frac{1}{2}(\mathsf{P} \oslash h)_{rijs}\mathsf{P}^{rs} + \frac{1}{4}(\mathrm{Ric}_{ik}\,\mathsf{P}^k_j + \mathrm{Ric}_{jk}\,\mathsf{P}^k_i) \\
&= \frac{1}{4}\left(\nabla^k\nabla_j(\mathsf{P})_{ki} - \nabla^k\nabla_k(\mathsf{P})_{ij}\right) - \frac{1}{4}\mathsf{C}_{kijl}\mathsf{P}^{kl} \\
&\quad + \frac{1}{4}|\mathsf{P}|^2 h_{ij} - \mathsf{P}^k_i \mathsf{P}_{kj} + \frac{1}{4}(\mathsf{P} \oslash h)_{rijs}\mathsf{P}^{rs} + \frac{1}{4}\,\mathrm{Ric}_{ik}\,\mathsf{P}^k_j.
\end{aligned}
$$

Now using the identity

$$(\mathsf{P} \oslash h)_{rijs}\mathsf{P}^{rs} = -\mathsf{P}_{ij}\mathsf{J} + 2\mathsf{P}_{il}\mathsf{P}^l_j - |\mathsf{P}|^2 h_{ij},$$

we find

$$
\begin{aligned}
&\frac{1}{4}|\mathsf{P}|^2 h_{ij} - \mathsf{P}^k_i \mathsf{P}_{kj} + \frac{1}{4}(\mathsf{P} \oslash h)_{rijs}\mathsf{P}^{rs} + \frac{1}{4}\,\mathrm{Ric}_{ik}\,\mathsf{P}^k_j \\
&= \frac{1}{4}|\mathsf{P}|^2 h_{ij} - \mathsf{P}^k_i \mathsf{P}_{kj} - \frac{1}{4}\mathsf{P}_{ij}\mathsf{J} + \frac{1}{2}\mathsf{P}_{il}\mathsf{P}^l_j - \frac{1}{4}|\mathsf{P}|^2 h_{ij} + \frac{1}{4}(n-2)\mathsf{P}_{ik}\mathsf{P}^k_j + \frac{1}{4}\mathsf{J}\mathsf{P}_{ij} \\
&= \frac{1}{4}(n-4)\mathsf{P}^k_i \mathsf{P}_{kj}.
\end{aligned}
$$

Hence

$$(n-4)h_{(4)ij} = \frac{1}{4}\left(\nabla^k\nabla_j(\mathsf{P})_{ki} - \nabla^k\nabla_k(\mathsf{P})_{ij}\right) - \frac{1}{4}\mathsf{C}_{kijl}\mathsf{P}^{kl} + \frac{n-4}{4}\mathsf{P}^k_i \mathsf{P}_{kj}.$$

In order to determine $h_{(2)}$, we have used (6.9.10) at $\rho = 0$. (6.9.11) and (6.9.12) do not impose further conditions on $h_{(2)}$. In fact, (6.9.11) at $\rho = 0$ yields a condition on $h_{(4)}$ and (6.9.12) at $\rho = 0$ states that

$$\nabla^k(\mathsf{P})_{ik} = \nabla_i(\mathsf{J})$$

and thus follows from (4.2.15). In order to determine $h_{(4)}$, we have used the first-order derivative of (6.9.10) at $\rho = 0$ and (6.9.11) at $\rho = 0$. The first-order derivative of (6.9.12) at $\rho = 0$ is satisfied as well. Here we omit the detailed discussion of that point and refer to [122]. This completes the proof of Theorem 6.9.1. □

Theorem 6.9.1 implies that

$$h_{(4)ij} = \frac{1}{4}\mathsf{P}_i^k\mathsf{P}_{kj} \tag{6.9.18}$$

if h is Bach-flat (in dimension $n \geq 6$). By (6.9.4), this is true if h is conformally flat ($\mathsf{C} = 0$).

Now we use Theorem 6.9.1 for the calculation of the coefficient v_6. The main step is the following technical result.

Lemma 6.9.1. *If the matrices A, B and C satisfy the relation*

$$\det(I + tA + t^2B + t^3C + \cdots) = 1 + at + bt^2 + ct^3 + \cdots,$$

then

$$\det\left(I + tA + t^2B + t^3C + \cdots\right)^{\frac{1}{2}}$$
$$= 1 + \frac{a}{2}t + \frac{1}{2}\left(b - \frac{a^2}{4}\right)t^2 + \frac{1}{2}\left[c - \frac{a}{2}\left(b - \frac{a^2}{4}\right)\right]t^3 + \cdots . \tag{6.9.19}$$

and

$$a = \operatorname{tr}(A),$$
$$b = \operatorname{tr}(B) + \frac{1}{2}\left[(\operatorname{tr} A)^2 - \operatorname{tr}(A^2)\right],$$
$$c = \operatorname{tr}(C) + [\operatorname{tr} A \operatorname{tr} B - \operatorname{tr}(AB)] + \frac{1}{6}\operatorname{tr}(A)^3 - \frac{1}{2}\operatorname{tr} A \operatorname{tr}(A^2) + \frac{1}{3}\operatorname{tr}(A^3).$$

Proof. (6.9.19) follows by taking squares. For the proof of the remaining identities we compare the coefficients of the power series of $\log\det = \operatorname{tr}\log$ for the matrix $(I + tA + t^2B + t^3C + \cdots)$ with $\log(1 + at + bt^2 + ct^3 + \cdots)$. □

The relation $\det(I + tA) = \sum_p \operatorname{tr}(\wedge^p A)t^p$ shows that the formulas for the coefficients a, b and c in Lemma 6.9.1 can be written also in the form

$$a = \operatorname{tr}(A),$$
$$b = \operatorname{tr}(B) + \operatorname{tr}(\wedge^2 A),$$
$$c = \operatorname{tr}(C) + [\operatorname{tr} A \operatorname{tr} B - \operatorname{tr}(AB)] + \operatorname{tr}(\wedge^3 A).$$

The relations between both formulas are known as Newton's relations

$$\sigma_1 = s_1,$$
$$2\sigma_2 = s_1^2 - s_2, \tag{6.9.20}$$
$$6\sigma_3 = s_1^3 - 3s_1 s_2 + 2s_3 \tag{6.9.21}$$

between the elementary symmetric functions $\sigma_k = \mathrm{tr}(\wedge^k T)$ and the functions $s_k = \mathrm{tr}(T^k)$ of a linear map T.

Lemma 6.9.1 and the series $h_r = h + r^2 h_{(2)} + r^4 h_{(4)} + \cdots$, i.e.,

$$h_r = 1 - r^2 \mathsf{P} + r^4 \frac{1}{4(n-4)} \left(-\mathcal{B} + (n-4)\mathsf{P}^2 \right) + \cdots$$

imply that the first three coefficients in

$$v(r) = \frac{\mathrm{vol}(h_r)}{\mathrm{vol}(h_0)} = \left(\frac{\det(h_r)}{\det(h_0)} \right)^{\frac{1}{2}} = \det(h_0^{-1} h_r)^{\frac{1}{2}} = 1 + r^2 v_2 + r^4 v_4 + r^6 v_6 + \cdots$$

are given by

$$v_2 = -\frac{1}{2} \mathrm{tr}(\mathsf{P}) = -\frac{1}{2}\mathsf{J},$$

$$v_4 = \frac{1}{2} \left[\frac{1}{4} \mathrm{tr}(\mathsf{P}^2) + \frac{1}{2} \left[\mathrm{tr}(\mathsf{P})^2 - \mathrm{tr}(\mathsf{P}^2) \right] - \frac{1}{4} \mathrm{tr}(\mathsf{P})^2 \right]$$

$$= \frac{1}{2} \left(-\frac{1}{4} \mathrm{tr}(\mathsf{P}^2) + \frac{1}{4} \mathrm{tr}(\mathsf{P})^2 \right)$$

$$= \frac{1}{8} (\mathsf{J}^2 - |\mathsf{P}|^2)$$

$$= \frac{1}{8} (s_1^2(\mathsf{P}) - s_2(\mathsf{P}))$$

$$= \frac{1}{4} \sigma_2(\mathsf{P})$$

$$= \frac{1}{4} \mathrm{tr}(\wedge^2 \mathsf{P})$$

using (6.9.20), and (after some calculation)

$$v_6 = \frac{1}{48} (\mathrm{tr}(h_{(2)}))^3 - \frac{1}{16} \mathrm{tr}(h_{(2)}) \mathrm{tr}(h_{(2)}^2) + \frac{1}{12} \mathrm{tr}(h_{(2)}^3) - \frac{1}{6} \mathrm{tr}(h_{(2)} h_{(4)}) \tag{6.9.22}$$

using

$$\mathrm{tr}(h_{(6)}) = \frac{2}{3} \mathrm{tr}(h_{(2)} h_{(4)}) - \frac{1}{6} \mathrm{tr}(h_{(2)}^3). \tag{6.9.23}$$

The latter formula follows from (6.9.11) by differentiation at $\rho = 0$. In fact, differentiation yields

$$\mathrm{tr}((h^{-1})\dot{}\,\ddot{h}) + \mathrm{tr}(h^{-1}\dddot{h}) = \mathrm{tr}((h^{-1}h)\dot{}\,(h^{-1}h)\dot{}).$$

Hence

$$-2\,\mathrm{tr}\left(h^{-1}h_{(2)}h^{-1}h_{(4)}\right)+6\,\mathrm{tr}\left(h^{-1}h_{(6)}\right)=-\mathrm{tr}\left((h^{-1}h_{(2)})^3\right)+2\,\mathrm{tr}\left(h^{-1}h_{(4)}h^{-1}h_{(2)}\right).$$

This proves the claim. Thus we find

$$v_6 = -\frac{1}{48}\,\mathrm{tr}(\mathsf{P})^3 + \frac{1}{16}\,\mathrm{tr}(\mathsf{P})\,\mathrm{tr}(\mathsf{P}^2) - \frac{1}{12}\,\mathrm{tr}(\mathsf{P}^3) - \frac{1}{24(n-4)}\,\mathrm{tr}(\mathcal{B}\mathsf{P}) + \frac{1}{24}\,\mathrm{tr}(\mathsf{P}^3)$$

i.e.,

$$v_6 = -\frac{1}{24(n-4)}\,\mathrm{tr}(\mathcal{B}\mathsf{P}) + \frac{1}{48}\left(-\mathrm{tr}(\mathsf{P})^3 + 3\,\mathrm{tr}(\mathsf{P})\,\mathrm{tr}(\mathsf{P}^2) - 2\,\mathrm{tr}(\mathsf{P}^3)\right)$$

$$= -\frac{1}{24(n-4)}(\mathcal{B},\mathsf{P}) - \frac{1}{8}\,\mathrm{tr}(\wedge^3\mathsf{P}).$$

Here we have used Newton's relation $6\sigma_3 = s_1^3 - 3s_1 s_2 + 2s_3$. This completes the proof of Theorem 6.9.2. $\qquad\square$

Finally, it is sometimes useful to rewrite the formula for v_6 in Theorem 6.9.2 in terms of Riemannian curvature.

Theorem 6.9.3. *Let $n = 6$. Then*

$$v_6 = \frac{1}{64}\left(\frac{1}{12}\tau|\,\mathrm{Ric}\,|^2 - \frac{1}{100}\tau^3 - \frac{1}{6}R_{ikjl}\,\mathrm{Ric}^{ij}\,\mathrm{Ric}^{kl}\right.$$
$$\left. + \frac{1}{30}(\mathrm{Ric},\mathrm{Hess}(\tau)) - \frac{1}{12}(\Delta\,\mathrm{Ric},\mathrm{Ric}) + \frac{1}{120}\tau\Delta\tau\right). \quad (6.9.24)$$

Since the formula in Theorem 6.9.3 involves the Hessian of τ, the most direct proof of that formula rests on the formula (6.9.14) for the metric $h_{(4)}$ and a calculation as in the proof of Theorem 6.9.2. Here are the details.

Proof. Using (6.9.14) and (6.9.22), we find

$$v_6 = -\frac{1}{48}\,\mathrm{tr}(\mathsf{P})^3 + \frac{1}{16}\,\mathrm{tr}(\mathsf{P})\,\mathrm{tr}(\mathsf{P}^2) - \frac{1}{12}\,\mathrm{tr}(\mathsf{P}^3)$$
$$+ \frac{1}{6(n-4)}\left[\frac{1}{4}|\mathsf{P}|^2\mathsf{J} - \mathrm{tr}(\mathsf{P}^3) + \frac{1}{4}(\mathrm{Hess}(\mathsf{J}),\mathsf{P}) - \frac{1}{4}(\Delta\mathsf{P},\mathsf{P})\right.$$
$$\left. - \frac{1}{2}R^s_{rij}\mathsf{P}^r_s\mathsf{P}^{ij} + \frac{1}{4}\left(\mathrm{Ric}_{ik}\mathsf{P}^k_j\mathsf{P}^{ij} + \mathrm{Ric}_{jk}\mathsf{P}^k_i\mathsf{P}^{ij}\right)\right].$$

Now we write $\frac{1}{n-2}\,\mathrm{Ric} = \mathsf{P} + \frac{1}{n-2}\mathsf{J}h$. Then the contributions of $\mathrm{tr}(\mathsf{P}^3)$ cancel since

$$-\frac{1}{12} + \frac{1}{6(n-4)}\left(-1 + \frac{n-2}{2}\right) = 0.$$

For $n = 6$, we are left with

$$v_6 = -\frac{1}{48}\,\text{tr}(\mathsf{P})^3 + \frac{1}{8}\mathsf{J}|\mathsf{P}|^2 + \frac{1}{48}(\text{Hess}(\mathsf{J}), \mathsf{P}) - \frac{1}{48}(\Delta\mathsf{P}, \mathsf{P})$$
$$-\frac{1}{24}\frac{1}{16}R^s_{rij}\left(\text{Ric}^r_s - \mathsf{J}g^r_s\right)\left(\text{Ric}^{ij} - \mathsf{J}g^{ij}\right)$$
$$= -\frac{1}{48}\,\text{tr}(\mathsf{P})^3 + \frac{1}{8}\mathsf{J}|\mathsf{P}|^2 - \frac{1}{24}\frac{1}{16}\left(R_{rijs}\,\text{Ric}^{rs}\,\text{Ric}^{ij} - 2\mathsf{J}|\text{Ric}|^2 + \mathsf{J}^2\tau\right)$$
$$+\frac{1}{48}(\text{Hess}(\mathsf{J}), \mathsf{P}) - \frac{1}{48}(\Delta\mathsf{P}, \mathsf{P}). \qquad (6.9.25)$$

Now

$$\text{tr}(\mathsf{P}^2) = |\mathsf{P}|^2 = \frac{1}{16}\left(|\text{Ric}|^2 - \frac{7}{50}\tau^2\right),$$
$$(\text{Hess}(\mathsf{J}), \mathsf{P}) = \frac{1}{40}(\text{Hess}(\tau), \text{Ric}) - \frac{1}{400}\tau\Delta\tau$$

and

$$(\Delta\mathsf{P}, \mathsf{P}) = \frac{1}{16}\left((\Delta\,\text{Ric}, \text{Ric}) - \frac{7}{50}\tau\Delta\tau\right).$$

These results suffice to determine the coefficients in (6.9.25). We find

$$
\begin{array}{rcl}
(\Delta\,\text{Ric}, \text{Ric}) &:& -\dfrac{1}{48}\dfrac{1}{16} = -\dfrac{1}{64}\dfrac{1}{12}, \\[2mm]
(\text{Hess}(\tau), \text{Ric}) &:& \dfrac{1}{48}\dfrac{1}{40} = \dfrac{1}{64}\dfrac{1}{30}, \\[2mm]
\tau\Delta\tau &:& -\dfrac{1}{48}\dfrac{1}{400} + \dfrac{1}{48}\dfrac{7}{800} = \dfrac{1}{64}\dfrac{1}{120}, \\[2mm]
\tau|\text{Ric}|^2 &:& \dfrac{1}{8}\dfrac{1}{10}\dfrac{1}{16} + \dfrac{1}{24}\dfrac{1}{16}\dfrac{2}{10} = \dfrac{1}{64}\dfrac{1}{12}, \\[2mm]
\tau^3 &:& -\dfrac{1}{48}\dfrac{1}{1000} - \dfrac{1}{8}\dfrac{1}{10}\dfrac{1}{16}\dfrac{7}{50} - \dfrac{1}{24}\dfrac{1}{16}\dfrac{1}{100} = -\dfrac{1}{64}\dfrac{1}{100}, \\[2mm]
R_{rijs}\,\text{Ric}^{rs}\,\text{Ric}^{ij} &:& -\dfrac{1}{64}\dfrac{1}{6}.
\end{array}
$$

The proof is complete. $\square$

Example 6.9.1. *In order to check signs, we test Theorem 6.9.3 for S^6 with the canonical metric g_c. Then* $\text{Ric} = 5g_c$, $\mathsf{P} = \frac{1}{2}g_c$, $\mathsf{P}^\sharp = \frac{1}{2}\text{id}$, $\mathsf{J} = 3$ *and* $R = -\mathsf{P} \oslash g_c = -\frac{1}{2}(g_c \oslash g_c)$. *Hence the formula in Theorem 6.9.3 gives*

$$v_6 = \frac{1}{64}\left(\frac{1}{12}\cdot 30 \cdot 25 \cdot 6 - \frac{30^3}{100} - \frac{1}{6}\cdot 25 \cdot 30\right) = -\frac{20}{64}$$

using

$$R_{kijl}\,\text{Ric}^{ij}\,\text{Ric}^{kl} = -\frac{1}{2}5^2(g_c \oslash g_c)_{kijl}g_c^{ij}g_c^{kl} = \frac{1}{2}25(6^2 - 6).$$

On the other hand, we have

$$v_6 = -\frac{1}{8} \operatorname{tr} \wedge^3(\mathsf{P}^\sharp) = -\frac{1}{8}\frac{1}{8}\binom{6}{3} = -\frac{20}{64}$$

by Theorem 6.16.1.

It also will be useful to collect the information on traces of $h_{(2)}$, $h_{(4)}$ and $h_{(6)}$.

Lemma 6.9.2.

$$\operatorname{tr}(h_{(2)}) = -\operatorname{tr}(\mathsf{P}),$$

$$\operatorname{tr}(h_{(4)}) = \frac{1}{4}\operatorname{tr}(\mathsf{P}^2) = \frac{1}{4}|\mathsf{P}|^2,$$

$$\operatorname{tr}(h_{(6)}) = \frac{1}{3!}\frac{1}{n-4}(\mathcal{B},\mathsf{P}).$$

Proof. Theorem 6.9.1 directly implies the first and the second equation. Moreover, combined with (6.9.23) it also yields the third equation. $\square$

The following comments should facilitate a comparison with related results in the literature. We do not intend to be exhaustive in any sense and we do not discuss priorities.

The full details of the Graham-Fefferman construction in [99] became available through [96]. Explicit formulas for $h_{(2)}$, $h_{(4)}$ and $h_{(6)}$ are displayed in [96], Chapter 3. In addition, the work [117] describes the full expansion in terms of (extended) obstruction tensors, and derives formulas for the holographic coefficients v_{2j} in these terms. Explicit formulas are given up to v_8.

We continue with comments on some earlier works. The original work [141] states formulas for v_4 (for $n = 4$) as given in Theorem 6.9.2 and v_6 (for $n = 6$) as given in Theorem 6.9.3 (up to normalizing coefficients). Note that in [141], the sign conventions for curvature tensors differ from ours: their curvature R coincides with ours, but Ric and τ have opposite signs. [8] gives formulas for $h_{(2)}$ and $h_{(4)}$ for general dimension n. These results coincide with (6.9.1) and (6.9.2). [8] rewrites the Henningson-Skenderis formula for v_6 ($n = 6$) in terms of the Bach tensor. The result corresponds to Theorem 6.9.2. Again, the sign conventions differ: R is opposite to ours, but Ric and τ coincide. In [8], the authors observe that the *metric* variation of the total holographic anomaly

$$\int_{M^n} v_n \operatorname{vol},$$

or equivalently, the metric variation of the total critical Q-curvature, yields the Fefferman-Graham obstruction tensor. That result is proved in full generality in [122]. [8] also contains an explicit, but structurally unclear formula, for v_8. For a review see also [9]. The seminal work [119] gives formulas for v_4 (for $n = 4$) and v_6 (for $n = 6$). These results correspond to the results in Theorem 6.9.2. Here the sign conventions on R, Ric and τ are as ours, but C has opposite sign. [80] gives a formula for $h_{(4)}$ in terms of R, Ric and τ for general n (see (A.1) in [80]). That formula is equivalent to the corresponding formula in Theorem 6.9.1. The sign conventions in [80] are the same as in [141]. The same formula for $h_{(4)}$ is also presented in [147]. Lemma 6.9.2 is also in [96].

In [175], the authors use the ambient metric to derive formulas for the GJMS-operators and conjecture the existence of a relation between the constant term of P_{2N} and

$$\left(\frac{n}{2}-N\right)v_{2N}$$

in any dimension $n \geq 2N$, up to terms which are $O((\frac{n}{2}-N)^2)$. The following comments should shed light on the situation.

For $N=2$ and $n \geq 4$, we have

$$Q_{4,n} = \frac{n}{2}\mathsf{J}^2 - 2|\mathsf{P}|^2 - \Delta\mathsf{J}$$
$$= 16v_4 + 2\Delta v_2 + \left(\frac{n}{2}-2\right)\mathsf{J}^2.$$

Here the term Δv_2 is an additional divergence term. The appropriate way to write $Q_{4,n}$ seems to be the holographic formula

$$Q_{4,n} = 16v_4 - 2\left(\Delta - \left(\frac{n}{2}-2\right)\mathsf{J}\right)v_2,$$

where the operator $\Delta - (\frac{n}{2}-2)\mathsf{J}$ is interpreted as a multiple of $T_2^*(\frac{n}{2}-2)$.

Similarly, for $N=3$ and $n \geq 6$, (6.10.13) shows that

$$Q_{6,n} = -8 \cdot 48v_6 + \left[(\Delta^2\mathsf{J} - 8\delta(\mathsf{P}\#d\mathsf{J})) + 4\Delta(|\mathsf{P}|^2 - \mathsf{J}^2)\right]$$

up to terms which are $O(\frac{n}{2}-3)$. Theorem 6.10.4 seems to be the appropriate way to express $Q_{6,n}$ in terms of holographic data.

The following formula for the subcritical $Q_{2N,n}$ generalizes these results and the holographic formula (Theorem 6.6.6) for the critical Q-curvature.

Conjecture 6.9.1. *For even n and $2N \leq n$,*

$$4Nc_NQ_{2N,n}(h) = \sum_{j=0}^{N-1}(2N-2j)T_{2j}^*\left(h;\frac{n}{2}-N\right)(v_{2N-2j}), \qquad (6.9.26)$$

where $c_N = (-1)^N(2^{2N}N!(N-1)!)^{-1}$.

It is convenient to summarize the holographic formulas in the matrix form

$$2 \begin{pmatrix} nc_{\frac{n}{2}} Q_n \\ (n-2)c_{\frac{n}{2}-1} Q_{n-2} \\ \vdots \\ 4c_2 Q_4 \\ 2c_1 Q_2 \end{pmatrix}$$

$$= \begin{pmatrix} 1 & T_2^*(0) & T_4^*(0) & \cdots & T_{n-4}^*(0) & T_{n-2}^*(0) \\ & 1 & T_2^*(1) & \cdots & T_{n-6}^*(1) & T_{n-4}^*(1) \\ & & & & \vdots & \vdots \\ & 0 & & & 1 & T_2^*(\frac{n}{2}-2) \\ & & & & & 1 \end{pmatrix} \begin{pmatrix} nv_n \\ (n-2)v_{n-2} \\ \vdots \\ 4v_4 \\ 2v_2 \end{pmatrix}. \qquad (6.9.27)$$

(6.9.27) describes how on any Riemannian manifold (M, h) of even dimension n, the two systems of curvature quantities

$$\{Q_2, Q_4, \ldots, Q_n\} \quad \text{and} \quad \{v_2, v_4, \ldots, v_n\}$$

are related by an invertible upper triangle matrix. Its entries are relatives of the operators $T_{2N}(\frac{n}{2} - N)$ which are constant multiples of the GJMS-operators P_{2N}. For $n = 6$, (6.9.27) will be proved in Section 6.10.

Conjecture 6.9.1 implies that Q_{2N} differs from a multiple of v_{2N} by the sum of the divergence term

$$\sum_{j=1}^{N-1} (2N - 2j) T_{2j}^*(0) (v_{2N-2j})$$

and a multiple of $(\frac{n}{2} - N)$. This refines the idea of [175].

Finally, we prove the following result which has been used in the previous section. Similar results are stated in [80] (Appendix D, without proofs).

Theorem 6.9.4. *Let* $u \sim \sum_{j \geq 0} r^{\lambda+2j} a_{2j}(\lambda)$ *be a formal solution of the eigenequation* $-\Delta_g u = \lambda(n - \lambda)u$ *for the Laplacian of the Poincaré-Einstein metric* $g = r^{-2}(dr^2 + h_r)$. *Then*

$$a_2(\lambda) = \frac{1}{2(n-2-2\lambda)}(\Delta - \lambda J)a_0$$

and

$$a_4(\lambda) = \frac{1}{4(n-4-2\lambda)} \left((\Delta - (\lambda+2)J) a_2(\lambda) - \frac{1}{2}\lambda|P|^2 a_0 - \delta(P\#da_0) - \frac{1}{2}(dJ, da_0) \right).$$

Here P, J, Δ *and the scalar product are to be understood with respect to* $h_0 = h$.

Proof. In terms of the coordinates $r^2 = \rho$, the metric is given by (6.9.9), and the Laplacian reads

$$\Delta_g = \rho \Delta_{h_\rho} + 4\rho^2 \frac{\partial^2}{\partial \rho^2} - 2(n-2)\rho \frac{\partial}{\partial \rho} + 2\rho^2 (\log \det h_\rho)' \frac{\partial}{\partial \rho}.$$

Now for $u(\rho) = \rho^{\frac{\lambda}{2}} v(\rho)$ with $v(\rho) = v_0 + \rho v_1 + \rho^2 v_2 + \cdots$, the equation $-\Delta_g u = \lambda(n-\lambda)u$ is equivalent to

$$4\left(\frac{\lambda}{2}\left(\frac{\lambda}{2}-1\right)\rho^{\frac{\lambda}{2}}v + \lambda\rho^{\frac{\lambda}{2}+1}v' + \rho^{\frac{\lambda}{2}+2}v''\right) - 2(n-2)\left(\frac{\lambda}{2}\rho^{\frac{\lambda}{2}}v + \rho^{\frac{\lambda}{2}+1}v'\right)$$
$$+ 2(\log \det h_\rho)'\left(\frac{\lambda}{2}\rho^{\frac{\lambda}{2}+1}v + \rho^{\frac{\lambda}{2}+2}v'\right) + \rho^{\frac{\lambda}{2}+1}\Delta_{h_\rho}(v) + \lambda(n-\lambda)\rho^{\frac{\lambda}{2}}v = 0,$$

i.e.,

$$4\rho v'' - 2(n-2-2\lambda)v' + \Delta_{h_\rho}(v) + (\log \det h_\rho)'(\lambda v + 2\rho v') = 0. \tag{6.9.28}$$

We compare the coefficients of ρ^0 and ρ^1 and find

$$-2(n-2-2\lambda)v_1 + \Delta_h(v_0) + \lambda(\log \det h_\rho)'\big|_{\rho=0}v_0 = 0 \tag{6.9.29}$$

and

$$-4(n-4-2\lambda)v_2 + [\Delta_h(v_1) + (d/dt)|_0(\Delta_{h-t\mathrm{P}})(v_0)]$$
$$+ \lambda[(\log \det h_\rho)'|_0 v_1 + (\log \det h_\rho)''|_0 v_0] + 2(\log \det h_\rho)'|_0 v_1 = 0. \tag{6.9.30}$$

But using $(\log \det h_\rho)'|_{\rho=0} = (\mathrm{tr} \log h_\rho)'|_{\rho=0} = -\mathrm{tr}(\mathrm{P}) = -\mathrm{J}$, (6.9.29) reads

$$-2(n-2-2\lambda)v_1 + \Delta_h(v_0) - \lambda \mathrm{J} v_0 = 0.$$

This proves the first assertion. In order to prove the second assertion, we apply the identity

$$(\log \det h_\rho)''|_{\rho=0} = -\frac{1}{2}|\mathrm{P}|^2.$$

In fact,

$$(\log \det h_\rho)'' = \frac{(\det h_\rho)''}{\det h_\rho} - \left(\frac{(\det h_\rho)'}{\det h_\rho}\right)^2$$

and

$$\frac{\det h_\rho}{\det h_0} = 1 - \rho \, \mathrm{tr}(\mathrm{P}) + \rho^2\left[\mathrm{tr}(h_{(4)}) + \frac{1}{2}\left(\mathrm{tr}(\mathrm{P})^2 - \mathrm{tr}(\mathrm{P}^2)\right)\right] + \cdots$$
$$= 1 - \rho \mathrm{J} + \rho^2\left[-\frac{1}{4}|\mathrm{P}|^2 + \frac{1}{2}\mathrm{J}^2\right] + \cdots$$

(by Lemma 6.9.1 and (6.9.13)) gives

$$(\log \det h_\rho)''|_{\rho=0} = 2\left(-\frac{1}{4}|\mathsf{P}|^2 + \frac{1}{2}\mathsf{J}^2\right) - \mathsf{J}^2 = -\frac{1}{2}|\mathsf{P}|^2.$$

Then (6.9.30) reads

$$-4(n-4-2\lambda)v_2 + [\Delta_h(v_1) + (d/dt)|_0(\Delta_{h-t\mathsf{P}})(v_0)] - \lambda\mathsf{J}v_1 - \frac{1}{2}\lambda|\mathsf{P}|^2 v_0 - 2\mathsf{J}v_1 = 0.$$

In order to evaluate the variation of the Laplacian, we apply the general formula

$$(d/dt)|_0(\Delta_{g+th})(u) = -(\operatorname{Hess}_g(u), h)_g + (du, \delta_g(h))_g + \frac{1}{2}(du, d(\operatorname{tr}_g(h)))_g \quad (6.9.31)$$

([24], Proposition 1.184). For the convenience of the reader, we sketch a proof of (6.9.31). For the family $G(t) \overset{\text{def}}{=} g + th$, we find

$$d/dt|_{t=0}(G(\nabla^G_X Y, Z)) = h(\nabla^g_X Y, Z) + g(\dot{\nabla}^g_X Y, Z),$$

where $\dot{\nabla}^g_X Y = d/dt|_{t=0}(\nabla^G_X Y)$. Hence

$$
\begin{aligned}
g(\dot{\nabla}^g_X Y, Z) &= \frac{1}{2}\{X(h(Y,Z)) + Y(h(Z,X)) - Z(h(X,Y)) \\
&\quad - h(X,[Y,Z]) + h(Y,[Z,X]) + h(Z,[X,Y])\} - h(\nabla^g_X Y, Z) \\
&= \frac{1}{2}\{\nabla^g_X(h)(Y,Z) + \nabla^g_Y(h)(X,Z) - \nabla^g_Z(h)(X,Y)\}. \quad (6.9.32)
\end{aligned}
$$

Now write

$$\Delta_G = -\delta_G d = \operatorname{tr}_G(\nabla^G d).$$

Then

$$d/dt|_{t=0}(\Delta_G u) = \dot{\operatorname{tr}}_g(\nabla^g du) + \operatorname{tr}_g(\dot{\nabla}^g du).$$

But $\dot{\operatorname{tr}}_g(\omega) = -g(h, \omega)$ for $\omega \in \Omega^1 \otimes \Omega^1$ and (6.9.32) implies, for $\alpha \in \Omega^1$,

$$
\begin{aligned}
\operatorname{tr}_g(\dot{\nabla}^g \alpha) &= -\sum_i \nabla^g_{e_i}(h)(e_i, \alpha^\sharp) + \frac{1}{2}\operatorname{tr}_g(\nabla^g_{\alpha^\sharp}(h)) \\
&= (\delta_g(h), \alpha)_g + \frac{1}{2}\nabla^g_{\alpha^\sharp}\operatorname{tr}_g(h) = (\delta_g(h), \alpha)_g + \frac{1}{2}(\alpha, d(\operatorname{tr}_g(h)))_g.
\end{aligned}
$$

Hence

$$d/dt|_{t=0}(\Delta_G u) = -(h, \nabla^g du)_g + (\delta_g(h), du)_g + \frac{1}{2}(du, d(\operatorname{tr}_g(h)))_g.$$

This proves the variational formula (6.9.31).

In our case, (6.9.31) yields

$$(d/dt)|_0(\Delta_{h-t\mathsf{P}})(u) = (\mathrm{Hess}_h(u),\mathsf{P})_h + (du, d\mathsf{J})_h - \frac{1}{2}(du, d\mathsf{J})_h$$

using $-\delta\mathsf{P} = d\mathsf{J}$ (Lemma 4.2.7). Now Lemma 4.2.6 implies

$$(d/dt)|_0(\Delta_{h-t\mathsf{P}})(u) = -\delta(\mathsf{P}\#du) - \frac{1}{2}(du, d\mathsf{J}). \qquad (6.9.33)$$

The proof is complete. $\square$

6.10   The holographic formula for Q_6

In the present section, we derive explicit formulas for Q_6 in dimension $n \geq 6$ from the asymptotics of eigenfunctions. We use the result to confirm the holographic formula (6.6.24) for the critical Q_6. This also completes the discussion in Example 6.6.5. In particular, we evaluate here the term $\delta_6(1)$ in the formula (6.6.43). The latter result illustrates Theorem 6.6.4. Next, we prove an extension of the holographic formula to *all* subcritical Q-curvatures $Q_{6,n}$. Finally, we relate the explicit formulas for all Q-curvatures Q_6 to those given in [116].

The following calculations rest on the family $P_6(\lambda) : C^\infty(M^n) \to C^\infty(M^n)$ which is defined by the asymptotics

$$u \sim \sum_{j\geq 0} r^{\lambda+2j} a_{2j}(\lambda), \quad -\Delta_g u = \lambda(n-\lambda)u$$

and

$$a_6(\lambda) = \frac{1}{2^3 3!(n-2\lambda-2)(n-2\lambda-4)(n-2\lambda-6)}P_6(\lambda)$$

(see (6.6.3)). In terms of the notation used in the proof of Theorem 6.9.4, we differentiate (6.9.28) twice at $\rho = 0$ and obtain the condition

$$-6(n-6-2\lambda)v_3 + \left[\frac{1}{2}\Delta''(v_0) + \Delta'(v_1) + \Delta(v_2)\right]$$

$$+ \frac{1}{2}\lambda(\log\det h)'''v_0 + (\lambda+2)(\log\det h)''v_1 + (\lambda+4)(\log\det h)'v_2 = 0,$$

where we use the abbreviations

$$\Delta = \Delta_h, \quad \Delta' = (d/d\rho)|_{\rho=0}(\Delta_{h_\rho}), \quad \Delta'' = (d^2/d\rho^2)|_{\rho=0}(\Delta_{h_\rho})$$

and

$$(\log\det h)^{(k)} = (d^k/d\rho^k)|_{\rho=0}(\log\det h_\rho), \ k \geq 1.$$

Using

$$v_1 = \frac{1}{2(n-2-2\lambda)} P_2(\lambda)v_0 \quad \text{and} \quad v_2 = \frac{1}{8(n-4-2\lambda)(n-2-2\lambda)} P_4(\lambda)v_0, \quad (6.10.1)$$

we find

$$\begin{aligned}
v_3 = \; & \frac{1}{48(n-6-2\lambda)(n-4-2\lambda)(n-2-2\lambda)} \\
& \times \Big\{ \left[4(n-4-2\lambda)(n-2-2\lambda)\Delta'' + 4(n-4-2\lambda)\Delta'P_2(\lambda) + \Delta P_4(\lambda)\right](v_0) \\
& + 4\lambda(n-4-2\lambda)(n-2-2\lambda)(\log\det h)'''v_0 \\
& + 4(\lambda+2)(n-4-2\lambda)(\log\det h)''P_2(\lambda)v_0 + (\lambda+4)(\log\det h)'P_4(\lambda)v_0 \Big\},
\end{aligned}$$

i.e.,

$$\begin{aligned}
P_6(\lambda)u = \; & 4(n-4-2\lambda)(n-2-2\lambda)\left[\Delta'' + \lambda(\log\det h)'''\right](u) \\
& + 4(n-4-2\lambda)\left[(\lambda+2)(\log\det h)''P_2(\lambda) + \Delta'P_2(\lambda)\right](u) \\
& + (\Delta - (\lambda+4)J)P_4(\lambda)u. \quad (6.10.2)
\end{aligned}$$

This formula will be the main tool in the proof of the following result.

Theorem 6.10.1 (The explicit formula for Q_6). *On Riemannian manifolds of dimension $n = 6$,*

$$Q_6 = -8 \cdot 48v_6 + \left(\Delta^2 J - 8\delta(P\#dJ)\right) + 4\Delta\left(|P|^2 - J^2\right).$$

Proof. We apply the identity $Q_6 = -\dot{P}_6(0)(1)$ (see (3.2.20)) using (6.10.2) for $n = 6$. First of all, we observe that $\Delta''(1) = 0$ and $P_2(0)(1) = P_4(0)(1) = 0$ (if $n = 6$). Hence

$$\begin{aligned}
-Q_6 &= 32(\log\det h)''' + 16(\log\det h)''\dot{P}_2(0)1 + 8\Delta'\dot{P}_2(0)1 + (\Delta - 4J)\dot{P}_4(0)1 \\
&= 32(\log\det h)''' - 16(\log\det h)''J - 8\Delta'J + (\Delta - 4J)\dot{P}_4(0)1. \quad (6.10.3)
\end{aligned}$$

Now we recall that, by Lemma 6.9.1 and the discussion following it,

$$\begin{aligned}
\frac{\det(h_\rho)}{\det(h_0)} = \; & 1 + \operatorname{tr}(A)\rho + \left[\operatorname{tr}(B) + \operatorname{tr}(\wedge^2 A)\right]\rho^2 \\
& + \left[\operatorname{tr}(C) + \left[\operatorname{tr}(A)\operatorname{tr}(B) - \operatorname{tr}(AB)\right] + \operatorname{tr}(\wedge^3 A)\right]\rho^3 + \cdots
\end{aligned}$$

with

$$A = -\mathsf{P}, \quad B = -\frac{1}{8}\mathcal{B} + \frac{1}{4}\mathsf{P}^2$$

(Theorem 6.9.1) and

$$\operatorname{tr}(C) = \frac{2}{3}\operatorname{tr}(AB) - \frac{1}{6}\operatorname{tr}(A^3). \quad ((6.9.23))$$

Hence for

$$(\log \det h_\rho)'' = \frac{(\det h_\rho)''}{\det h_\rho} - \left(\frac{(\det h_\rho)'}{\det h_\rho}\right)^2$$

and

$$(\log \det h_\rho)''' = \frac{(\det h_\rho)'''}{\det h_\rho} - 3\frac{(\det h_\rho)''(\det h_\rho)'}{(\det h_\rho)^2} + 2\left(\frac{(\det h_\rho)'}{\det h_\rho}\right)^3,$$

we find

$$(\log \det h_\rho)''|_{\rho=0} = 2\left[\mathrm{tr}(B) + \mathrm{tr}(\wedge^2 A)\right] - (\mathrm{tr}\,A)^2 = -\frac{1}{2}|\mathsf{P}|^2$$

(as in the proof of Theorem 6.9.4) and

$$(\log \det h)''' = 6\left[\mathrm{tr}(C) + [\mathrm{tr}(A)\,\mathrm{tr}(B) - \mathrm{tr}(AB)] + \mathrm{tr}(\wedge^3 A)\right]$$
$$- 3\cdot 2\left[\mathrm{tr}(B) + \mathrm{tr}(\wedge^2 A)\right]\mathrm{tr}(A) + 2(\mathrm{tr}\,A)^3.$$

Therefore,

$$-Q_6 = 32\left[6\,\mathrm{tr}(C) - 6\,\mathrm{tr}(AB) + 6\,\mathrm{tr}(\wedge^3 A) - 6\,\mathrm{tr}(\wedge^2 A)\,\mathrm{tr}(A) + 2(\mathrm{tr}\,A)^3\right]$$
$$+ 8|\mathsf{P}|^2\mathsf{J} + 8\left(\delta(\mathsf{P}\#d\mathsf{J}) + \frac{1}{2}(d\mathsf{J}, d\mathsf{J})\right)$$
$$+ (\Delta - 4\mathsf{J})(-\Delta\mathsf{J} + 2\mathsf{J}^2 - 4|\mathsf{P}|^2)$$

using (6.9.33) and

$$\dot{P}_4(0)1 = -\Delta\mathsf{J} + 2\mathsf{J}^2 - 4|\mathsf{P}|^2$$

(see (6.6.41) for $n = 6$). Separating terms without derivatives, we find that $-Q_6$ equals

$$32\left[4\,\mathrm{tr}(AB) - \mathrm{tr}(A^3) - 6\,\mathrm{tr}(AB) + 6\,\mathrm{tr}(\wedge^3 A) - 6\,\mathrm{tr}(\wedge^2 A)\,\mathrm{tr}(A) + 2(\mathrm{tr}\,A)^3\right]$$
$$- 24\,\mathrm{tr}(A)\,\mathrm{tr}(A^2) + 8(\mathrm{tr}\,A)^3$$
$$+ 8\delta(\mathsf{P}\#d\mathsf{J}) + 4(d\mathsf{J}, d\mathsf{J}) - \Delta^2\mathsf{J} + 2\Delta\mathsf{J}^2 - 4\Delta|\mathsf{P}|^2 + 4\mathsf{J}\Delta\mathsf{J}.$$

Now Newton's identities

$$\mathrm{tr}(\wedge^2 A) = \frac{1}{2}\left((\mathrm{tr}\,A)^2 - \mathrm{tr}(A^2)\right),$$
$$\mathrm{tr}(\wedge^3 A) = \frac{1}{6}\left((\mathrm{tr}\,A)^3 - 3\,\mathrm{tr}(A)\,\mathrm{tr}(A^2) + 2\,\mathrm{tr}(A^3)\right)$$

(see (6.9.20), (6.9.21)) show that

$$6\,\mathrm{tr}(\wedge^3 A) - 6\,\mathrm{tr}(\wedge^2 A)\,\mathrm{tr}(A) + 2(\mathrm{tr}\,A)^3 = 2\,\mathrm{tr}(A^3).$$

Hence

$$-Q_6 = 32\left[-2\operatorname{tr}(AB) + \operatorname{tr}(A^3)\right] - 24\operatorname{tr}(A)\operatorname{tr}(A^2) + 8(\operatorname{tr} A)^3 \\ - (\Delta^2 J - 8\delta(P\#dJ)) - 4\Delta(|P|^2 - J^2).$$

Using the explicit formulas for the coefficients A and B, it follows that the latter formula is equivalent to

$$-Q_6 = -6 \cdot 8\operatorname{tr}(\wedge^3 P) - 8(P, B) - (\Delta^2 J - 8\delta(P\#dJ)) - 4\Delta(|P|^2 - J^2).$$

Now

$$v_6 = -\frac{1}{8}\operatorname{tr}(\wedge^3 P) - \frac{1}{48}(B, P)$$

(see Theorem 6.9.2) completes the proof. □

Corollary 6.10.1. Q_6 *can be written in the form*

$$Q_6 = -8 \cdot 48 v_6 - 32 P_2^*(0)(v_4) - 2 P_4^*(0)(v_2) \tag{6.10.4}$$

since $P_2^*(0) = \Delta$ *and* $P_4^*(0)J = \Delta^2 J - 8\delta(P\#dJ)$ *(see (6.6.42)). This shows that Theorem 6.10.1 is a special case of Theorem 6.6.6 stating that*

$$Q_6 = -2^6\left(6v_6 + \frac{1}{2}P_2^*(0)(v_4) + \frac{1}{32}P_4^*(0)(v_2)\right).$$

In order to relate Theorem 6.10.1 and Corollary 6.10.1 to (6.6.43), it remains to determine the quantity $\delta_6(1)$.

Theorem 6.10.2.

$$\delta_6(1) = \left(\dot{P}_6(0) - \dot{P}_6^*(0)\right)(1) = 4\left(\Delta(|P|^2 - J^2) + (\Delta^2 J - 8\delta(P\#dJ))\right) \\ = -32 P_2^*(0)(v_4) - 8 P_4^*(0)(v_2).$$

Theorem 6.10.2 is a special case of Theorem 6.6.4. In fact, Theorem 6.6.4 states that

$$6\delta_6(1) = -2^6 3!2! \left(2T_2^*(0)(v_4) + 4T_4^*(0)(v_2)\right).$$

In view of

$$T_2(0) = \frac{1}{8}P_2(0) \quad \text{and} \quad T_4(0) = \frac{1}{64}P_4(0)$$

(see (6.10.1)), this formula is equivalent to

$$\delta_6(1) = -2^7\left(\frac{1}{4}P_2^*(0)(v_4) + \frac{1}{16}P_4^*(0)(v_2)\right) = -32 P_2^*(0)(v_4) - 8 P_4^*(0)(v_2).$$

Proof. (6.10.2) yields (for $n = 6$)

$$\dot{P}_6(0)u = -48\Delta''u + 32(\log\det h)'''u$$
$$- 8(\log\det h)''P_2(0)u + 16(\log\det h)''\dot{P}_2(0)u$$
$$- 8\Delta'P_2(0)u + 8\Delta'\dot{P}_2(0)u - JP_4(0)u + (\Delta - 4J)\dot{P}_4(0)u. \quad (6.10.5)$$

The last two terms in (6.10.5) contribute to $\delta_6(u)$ by

$$-JP_4(0)u + P_4^*(0)(Ju) + (\Delta - 4J)\dot{P}_4(0)u - \dot{P}_4^*(0)(\Delta - 4J)u,$$

i.e., for $u = 1$ by

$$P_4^*(0)J + (\Delta - 4J)(\dot{P}_4(0)(1)) - \dot{P}_4^*(0)(\Delta - 4J)(1). \quad (6.10.6)$$

In order to find explicit formulas for these terms, we use the formula in Remark 6.6.5. We find

$$P_4^*(0)J = \Delta(\Delta - 2J)J - 8\delta(P\#dJ) - 4\delta(JdJ)$$

and

$$\dot{P}_4(0)u = -\Delta(Ju) - J\Delta u + 2J^2u - 4|P|^2u + 4\delta(P\#du) + 2(dJ, du),$$

i.e.,

$$\dot{P}_4^*(0)u = -J\Delta u - \Delta(Ju) + 2J^2u - 4|P|^2u + 4\delta(P\#du) + 2\delta(udJ).$$

Therefore,

$$(\Delta - 4J)(\dot{P}_4(0)(1)) = (\Delta - 4J)(-\Delta J + 2J^2 - 4|P|^2)$$
$$-\dot{P}_4^*(0)(\Delta - 4J)(1) = 4\dot{P}_4^*(0)J$$
$$= -4J\Delta J - 4\Delta(J^2) + 8J^3 - 16|P|^2J + 16\delta(P\#dJ) + 8\delta(JdJ).$$

Hence (6.10.6) yields

$$\left[\Delta^2(J) - 2\Delta(J^2) - 8\delta(P\#dJ) - 4\delta(JdJ)\right]$$
$$+ \left[-\Delta^2(J) + 2\Delta(J^2) - 4\Delta(|P|^2) + 4J\Delta J - 8J^3 + 16J|P|^2\right]$$
$$- 4J\Delta J - 4\Delta(J^2) + 8J^3 - 16|P|^2J + 16\delta(P\#dJ) + 8\delta(JdJ)$$
$$= -6\Delta J^2 - 4\Delta|P|^2 + 8\delta(P\#dJ).$$

Next, the first two terms in the last line of (6.10.5) contribute to $\delta_6(u)$ by

$$- 8\Delta'P_2(0)u + 8P_2^*(0)\Delta'^*u + 8\Delta'\dot{P}_2(0)u - 8\dot{P}_2^*(0)\Delta'^*u$$
$$= -8\Delta'\Delta u + 8\Delta\Delta'^*u - 8\Delta'Ju + 8J\Delta'^*u.$$

Now by (6.9.33),

$$\Delta' u = -\delta(\mathsf{P}\#du) - \frac{1}{2}(du, d\mathsf{J}) \quad \text{and} \quad \Delta'^* u = -\delta(\mathsf{P}\#du) - \frac{1}{2}\delta(ud\mathsf{J}).$$

It follows that for $u = 1$, the sum $-\Delta'\Delta u + \Delta\Delta'^* u - \Delta'\mathsf{J}u + \mathsf{J}\Delta'^* u$ yields

$$-\frac{1}{2}\Delta(\delta d\mathsf{J}) + \delta(\mathsf{P}\#d\mathsf{J}) + \frac{1}{2}(d\mathsf{J}, d\mathsf{J}) - \frac{1}{2}\mathsf{J}\delta d\mathsf{J} = \frac{1}{2}\Delta^2\mathsf{J} + \delta(\mathsf{P}\#d\mathsf{J}) + \frac{1}{4}\Delta(\mathsf{J}^2),$$

i.e., we find the contribution

$$4\Delta^2\mathsf{J} + 8\delta(\mathsf{P}\#d\mathsf{J}) + 2\Delta\mathsf{J}^2. \tag{6.10.7}$$

The second term in the second line and the second term in the first line of (6.10.5) trivially contribute to $\delta_6(u)$.

The first term in the second line of (6.10.5) contributes to $\delta_6(u)$ by

$$4|\mathsf{P}|^2\Delta u - 4\Delta|\mathsf{P}|^2 u,$$

i.e., for $u = 1$, by

$$-4\Delta|\mathsf{P}|^2. \tag{6.10.8}$$

It remains to determine the contribution of the second variation Δ'', i.e., $\Delta''(1) - \Delta''^*(1)$. However, we avoid determining the second variation Δ'' by using the following argument. The self-adjointness of $P_6(0)$ (known by the general theory) implies an explicit formula for the difference $\Delta''(u) - \Delta''^*(u)$ using (6.10.2). Here are the details.

We have (for $n = 6$)

$$P_6(0)u = 32\Delta'' u - 8|\mathsf{P}|^2\Delta u - 8\delta(\mathsf{P}\#d\Delta u) - 4(d\Delta u, d\mathsf{J}) + (\Delta - 4\mathsf{J})P_4(0)u.$$

Hence

$$P_6^*(0)u = 32\Delta''^* u - 8\Delta(|\mathsf{P}|^2 u) - 8\Delta\delta(\mathsf{P}\#du) - 4\Delta\delta(ud\mathsf{J}) + P_4^*(0)(\Delta - 4\mathsf{J})u.$$

Thus $P_6(0) = P_6^*(0)$ is equivalent to

$$32(\Delta'' - \Delta''^*)u = 8\left(|\mathsf{P}|^2\Delta u - \Delta(|\mathsf{P}|^2 u)\right) + 8\left(\delta(\mathsf{P}\#d\Delta u) - \Delta\delta(\mathsf{P}\#du)\right)$$
$$+ 4\left((d\Delta u, d\mathsf{J}) - \Delta\delta(ud\mathsf{J})\right) - (\Delta - 4\mathsf{J})P_4(0)u + P_4^*(0)(\Delta - 4\mathsf{J})u.$$

In particular, for $u = 1$, we find

$$32(\Delta'' - \Delta''^*)(1) = -8\Delta(|\mathsf{P}|^2) + 4\Delta^2\mathsf{J} - 4P_4^*(0)\mathsf{J}$$
$$= -8\Delta(|\mathsf{P}|^2) + 32\delta(\mathsf{P}\#d\mathsf{J}),$$

i.e.,

$$4(\Delta'' - \Delta''^*)(1) = -\Delta(|\mathsf{P}|^2) + 4\delta(\mathsf{P}\#d\mathsf{J}). \tag{6.10.9}$$

Summarizing the results, we obtain

$$\delta_6(1) = \dot{P}_6(0)1 - \dot{P}_6^*(0)1$$
$$= 12\Delta(|\mathsf{P}|^2) - 48\delta(\mathsf{P}\#d\mathsf{J}) \qquad\qquad\qquad\text{(by (6.10.9))}$$
$$- 4\Delta(|\mathsf{P}|^2) \qquad\qquad\qquad\qquad\qquad\text{(by (6.10.8))}$$
$$+ 4\Delta^2\mathsf{J} + 8\delta(\mathsf{P}\#d\mathsf{J}) + 2\Delta\mathsf{J}^2 \qquad\qquad\text{(by (6.10.7))}$$
$$- 6\Delta\mathsf{J}^2 - 4\Delta|\mathsf{P}|^2 + 8\delta(\mathsf{P}\#d\mathsf{J})$$
$$= 4\Delta^2\mathsf{J} - 4\Delta\mathsf{J}^2 - 32\delta(\mathsf{P}\#d\mathsf{J}) + 4\Delta(|\mathsf{P}|^2)$$
$$= 4\left[\Delta((|\mathsf{P}|^2 - \mathsf{J}^2) + (\Delta^2\mathsf{J} - 8\delta(\mathsf{P}\#d\mathsf{J})\right].$$

This completes the proof.   □

Now using (6.10.4) and Theorem 6.10.2, we find

$$(2Q_6 + 16 \cdot 48v_6) + \delta_6(1)$$
$$= [-64P_2^*(0)(v_4) - 4P_4^*(0)(v_2)] + [-32P_2^*(0)(v_4) - 8P_4^*(0)(v_2)]$$
$$= -96P_2^*(0)(v_4) - 12P_4^*(0)(v_2).$$

The latter formula is equivalent to (6.6.43). This completes the discussion in Example 6.6.5.

These results for Q_6 (for $n = 6$) are completely analogous to results for Q_4 (for $n = 4$). In fact, the relation

$$\delta_4(1) = 2\Delta\mathsf{J} = -4P_2^*(0)(v_2)$$

(see (6.6.40)) implies the equivalence of both formulas

$$2Q_4 = \delta_4(1) + 32v_4 + 8P_2^*(0)(v_2)$$

(see Example 6.6.4) and

$$Q_4 = 16v_4 + 2P_2^*(0)(v_2) \qquad\qquad\qquad\qquad (6.10.10)$$

for the critical Q_4. The latter formula is equivalent to the definition of Q_4 as $2(\mathsf{J}^2 - |\mathsf{P}|^2) - \Delta\mathsf{J}$.

We continue with a discussion of the *subcritical* Q_6. The starting point is the formula

$$P_6\left(\frac{n}{2} - 3\right)1 = -\frac{n-6}{2}Q_{6,n}$$

(see (3.1.3)). For $n \geq 6$, (6.10.2) yields

$$P_6(\lambda)1 = 4(n-4-2\lambda)(n-2-2\lambda)\lambda(\log\det h)'''$$
$$+ 4(n-4-2\lambda)\left[-\lambda(\lambda+2)(\log\det h)''\mathsf{J} - \lambda\Delta'\mathsf{J}\right]$$
$$+ (\Delta - (\lambda+4)\mathsf{J})\left[-\lambda(\Delta - (\lambda+2)\mathsf{J})\mathsf{J} - \lambda|\mathsf{P}|^2(n-2-2\lambda)\right],$$

i.e.,

$$P_6\left(\frac{n}{2}-3\right)1 = \frac{n-6}{2}\Big\{32(\log\det h)''' - 4(n-2)(\log\det h)''\mathsf{J} - 8\Delta'\mathsf{J}$$
$$- \left(\Delta-\left(\frac{n}{2}+1\right)\mathsf{J}\right)\left(\Delta-\left(\frac{n}{2}-1\right)\mathsf{J}\right)\mathsf{J} - 4\left(\Delta-(\tfrac{n}{2}+1)\mathsf{J}\right)|\mathsf{P}|^2\Big\}.$$

Hence

$$- Q_{6,n} = 32(\log\det h)''' - 4(n-2)(\log\det h)''\mathsf{J} - 8\Delta'\mathsf{J}$$
$$- \left(\Delta-\left(\frac{n}{2}+1\right)\mathsf{J}\right)\left(\Delta-\left(\frac{n}{2}-1\right)\mathsf{J}\right)\mathsf{J} - 4\left(\Delta-\left(\frac{n}{2}+1\right)\mathsf{J}\right)|\mathsf{P}|^2. \quad (6.10.11)$$

Note that for $n = 6$, we recover

$$-Q_{6,6} = 32(\log\det h)''' - 16(\log\det h)''\mathsf{J} - 8\Delta'\mathsf{J} - (\Delta-4\mathsf{J})\left(\Delta-2\mathsf{J}^2 + 4|\mathsf{P}|^2\right)$$

which by

$$\dot{P}_4(0)1 = -\Delta\mathsf{J} + 2\mathsf{J}^2 - 4|\mathsf{P}|^2$$

coincides with (6.10.3).

Now let

$$A = -\mathsf{P} \quad \text{and} \quad B = -\frac{1}{4(n-4)}\mathcal{B} + \frac{1}{4}\mathsf{P}^2.$$

Similarly as in the proof of Theorem 6.10.1, (6.10.11) implies that $-Q_{6,n}$ coincides with

$$32\left[6\operatorname{tr}(C) - 6\operatorname{tr}(AB) + 6\operatorname{tr}(\wedge^3 A) - 6\operatorname{tr}(\wedge^2 A)\operatorname{tr}(A) + 2(\operatorname{tr} A)^3\right]$$
$$+ 2(n-2)|\mathsf{P}|^2\mathsf{J} + 8\delta(\mathsf{P}\#d\mathsf{J}) + 4(d\mathsf{J}, d\mathsf{J})$$
$$-\Delta^2\mathsf{J} + \left(\frac{n}{2}-1\right)\Delta\mathsf{J}^2 + \left(\frac{n}{2}+1\right)\mathsf{J}\Delta\mathsf{J} - \left(\frac{n}{2}+1\right)\left(\frac{n}{2}-1\right)\mathsf{J}^3 - 4\Delta|\mathsf{P}|^2 + 2(n+2)\mathsf{J}|\mathsf{P}|^2,$$

i.e., with

$$32\left[4\operatorname{tr}(AB) - \operatorname{tr}(A^3) - 6\operatorname{tr}(AB) + 6\operatorname{tr}(\wedge^3 A) - 6\operatorname{tr}(\wedge^2 A)\operatorname{tr}(A) + 2(\operatorname{tr} A)^3\right]$$
$$- 4n\operatorname{tr}(A)\operatorname{tr}(A^2) + \left(\frac{n}{2}+1\right)\left(\frac{n}{2}-1\right)(\operatorname{tr} A)^3$$
$$+ 8\delta(\mathsf{P}\#d\mathsf{J}) + 4(d\mathsf{J}, d\mathsf{J}) - \Delta^2\mathsf{J} + \left(\frac{n}{2}-1\right)\Delta\mathsf{J}^2 - 4\Delta|\mathsf{P}|^2 + \left(\frac{n}{2}+1\right)\mathsf{J}\Delta\mathsf{J}.$$

The latter result implies

Theorem 6.10.3. *For $n \geq 6$,*

$$Q_{6,n} = \Big\{16\operatorname{tr}(\mathsf{P}^3) - 4n\mathsf{J}|\mathsf{P}|^2 + \frac{n^2-4}{4}\mathsf{J}^3 + \frac{n-6}{2}(d\mathsf{J}, d\mathsf{J}) + \frac{16}{n-4}(\mathcal{B}, \mathsf{P})\Big\}$$
$$+ (\Delta^2\mathsf{J} - 8\delta(\mathsf{P}\#d\mathsf{J})) + \frac{1}{4}\Delta(16|\mathsf{P}|^2 - (3n-2)\mathsf{J}^2).$$

Theorem 6.10.3, obviously, generalizes Theorem 6.10.1 (using (6.9.21)). For closed M^n, Theorem 6.10.3 implies

$$\int_{M^n} Q_{6,n}\,\mathrm{vol}$$
$$= \int_{M^n} \left(16\,\mathrm{tr}(\mathsf{P}^3) - 4n\mathsf{J}|\mathsf{P}|^2 + \frac{n^2-4}{4}\mathsf{J}^3 + \frac{n-6}{2}(d\mathsf{J},d\mathsf{J}) + \frac{16}{n-4}(\mathcal{B},\mathsf{P}) \right)\mathrm{vol}\,.$$
$$(6.10.12)$$

If $\mathsf{C} = 0$ (and hence $\mathcal{B} = 0$), (6.10.12) coincides with Theorem 6.4 in [31] or [32]. Branson's proof rests on invariance arguments and explicit calculations in test cases. In Section 6.11, we will give an alternative proof of the formula for Q_6 for $n = 6$.

The following result extends the holographic formula (6.10.4) to all subcritical $Q_{6,n}$, i.e., confirms Conjecture 6.9.1 for $N = 3$.

Theorem 6.10.4. *For $n \geq 6$,*

$$Q_{6,n} = -8 \cdot 48 v_6 - 32 P_2^* \left(\frac{n}{2} - 3 \right)(v_4) - 2 P_4^* \left(\frac{n}{2} - 3 \right)(v_2).$$

Proof. First of all, a calculation using Theorem 6.9.2 shows that Theorem 6.10.3 can be written in the form

$$Q_{6,n} = \left\{ -8 \cdot 48 v_6 + (\Delta^2 \mathsf{J} - 8\delta(\mathsf{P}\#d\mathsf{J})) + 4\Delta(|\mathsf{P}|^2 - \mathsf{J}^2) \right\}$$
$$+ \frac{n-6}{4} \left(-3\Delta \mathsf{J}^2 + 2(d\mathsf{J},d\mathsf{J}) - 16\mathsf{J}|\mathsf{P}|^2 + (n+6)\mathsf{J}^3 \right). \quad (6.10.13)$$

Thus the assertion is equivalent to the relation

$$\Delta^2 \mathsf{J} - 8\delta(\mathsf{P}\#d\mathsf{J})) + 4\Delta(|\mathsf{P}|^2 - \mathsf{J}^2)$$
$$+ \frac{n-6}{4} \left(-6\mathsf{J}\Delta\mathsf{J} - 4(d\mathsf{J},d\mathsf{J}) - 16\mathsf{J}|\mathsf{P}|^2 + (n+6)\mathsf{J}^3 \right)$$
$$= -4 P_2^* \left(\frac{n}{2} - 3 \right)(\mathsf{J}^2 - |\mathsf{P}|^2) + P_4^* \left(\frac{n}{2} - 3 \right)(\mathsf{J}). \quad (6.10.14)$$

But using $P_2^*(\lambda) = \Delta - \lambda\mathsf{J}$ and

$$P_4^*(\lambda) = (\Delta - \lambda\mathsf{J})(\Delta - (\lambda+2)\mathsf{J}) - \lambda(n-2-2\lambda)|\mathsf{P}|^2$$
$$- 2(n-2-2\lambda)\delta(\mathsf{P}\#d) - (n-2-2\lambda)\delta(\mathsf{J}d)$$

(see Theorem 6.9.4), we find $P_2^*(\frac{n}{2} - 3) = \Delta - (\frac{n}{2} - 3)\mathsf{J}$ and

$$P_4^* \left(\frac{n}{2} - 3 \right)(\mathsf{J})$$
$$= \left(\Delta - \left(\frac{n}{2} - 3 \right)\mathsf{J} \right) \left(\Delta - \left(\frac{n}{2} - 1 \right)\mathsf{J} \right) - 2(n-6)|\mathsf{P}|^2 - 8\delta(\mathsf{P}\#d) - 4\delta(\mathsf{J}d\mathsf{J}).$$

Hence the right-hand side of (6.10.14) equals

$$\left[\Delta^2 J - \left(\frac{n}{2}-1\right)\Delta J^2 - \left(\frac{n}{2}-3\right)J\Delta J + \left(\frac{n}{2}-3\right)\left(\frac{n}{2}-1\right)J^3 - 2(n-6)J|P|^2\right.$$
$$\left. - 8\delta(P\#dJ) + 4J\Delta J + 4(dJ, dJ)\right] - 4\left(\Delta - \left(\frac{n}{2}-3\right)J\right)(J^2 - |P|^2).$$

From here the assertion follows by a direct calculation. $\square$

Remark 6.10.1. *In Theorem 6.11.7 we extend the relation $\dot{D}_6^{\text{res}}(0)(1) = Q_6$ for the critical Q_6 to a relation between $Q_{6,n}$ and the quantity $\dot{D}_6^{\text{res}}(-\frac{n}{2}+3)(1)$.*

Remark 6.10.2. *Theorem 6.10.4 is an analog of the extension*

$$Q_{4,n} = 16v_4 + 2P_2^*\left(\frac{n}{2}-2\right)(v_2) \tag{6.10.15}$$

of the holographic formula for the critical Q_4 to $n \geq 4$. For the proof we observe that the right-hand side of (6.10.15) equals

$$2(J^2 - |P|^2) - \left(\Delta - \left(\frac{n}{2}-2\right)J\right)J = \frac{n}{2}J^2 - 2|P|^2.$$

These results imply that for $n = 6$ the curvature quantities $\{Q_2, Q_4, Q_6\}$ and $\{v_2, v_4, v_6\}$ are related by

$$\begin{pmatrix} -\frac{1}{64}Q_6 \\ \frac{1}{4}Q_4 \\ -Q_2 \end{pmatrix} = \begin{pmatrix} 1 & T_2^*(0) & T_4^*(0) \\ 0 & 1 & T_2^*(1) \\ 0 & 0 & 1 \end{pmatrix} \begin{pmatrix} 6v_6 \\ 4v_4 \\ 2v_2 \end{pmatrix}. \tag{6.10.16}$$

This is a special case of (6.9.27).

We close the present section with a discussion of the relation to a formula of Gover and Peterson ([116]). Using tractor calculus (and computer aided calculations), Gover and Peterson derived formulas for the Q-curvature $Q_{6,n}$ and $Q_{8,n}$, together with formulas for the corresponding GJMS-operators P_6 and P_8, in terms of Riemannian invariants. Their formula for $Q_6 = Q_{6,6}$ is

$$16(P, \Delta P) + 8(\nabla P, \nabla P) - 8J\Delta J + \Delta^2 J$$
$$- 32\operatorname{tr}(P^3) - 16|P|^2 J + 8J^3 + 16P_{ij}P_{kl}C^{kijl}. \tag{6.10.17}$$

Note that in [116] the sign convention for C is opposite to ours. In Table 6.1 we compare the formula for $Q_{6,n}$ (equation (23) in [116]) with Theorem 6.10.3. Both results differ by the sum

$$\frac{16}{n-4}\left(|P|^2 J - n\operatorname{tr}(P^3) - (P, \operatorname{Hess}(J)) + P_{ij}P_{kl}C^{kijl} + \nabla^k\nabla_j(P)_{ik}P^{ij}\right). \tag{6.10.18}$$

The following result shows that both formulas actually coincide.

term	[116]	Theorem 6.10.3
$(P, \Delta P)$	$8 + \frac{16}{n-4}$	$8 + \frac{16}{n-4}$
$\Delta^2 J$	1	1
$(\nabla P, \nabla P)$	8	8
$J \Delta J$	$\frac{2-3n}{2}$	$\frac{2-3n}{2}$
(dJ, dJ)	$6 - n$	$6 - n$
$(P, \mathrm{Hess}(J))$	$8 - \frac{16}{n-4}$	8
J^3	$\frac{n^2-4}{4}$	$\frac{n^2-4}{4}$
$\mathrm{tr}(P^3)$	$-\frac{64}{n-4}$	16
$|P|^2 J$	$\frac{16}{n-4} - 4n$	$-4n$
$P_{ij} P_{kl} C^{kijl}$	$\frac{32}{n-4}$	$\frac{16}{n-4}$
$\nabla^k \nabla_j (P)_{ik} P^{ij}$	0	$-\frac{16}{n-4}$

Table 6.1: $Q_{6,n}$

Lemma 6.10.1. *The sum in* (6.10.18) *vanishes.*

Proof. By (6.9.17), we have

$$\nabla^k \nabla_j (P)_{ki} = \mathrm{Hess}_{ij}(J) + C_{jkil} P^{lk} + \mathrm{Ric}_{jl} P^l_i - (P \oslash h)_{jkil} P^{lk}.$$

Hence

$$\nabla^k \nabla_j (P)_{ki} P^{ij} = (P, \mathrm{Hess}(J)) - C_{kijl} P^{lk} P^{ij}$$
$$+ ((n{-}2)P + Jh)_{jl} P^l_i P^{ij} - (P \oslash h)_{jkil} P^{lk} P^{ij}.$$

But

$$((n{-}2)P + Jh)_{jl} P^l_i P^{ij} = (n-2) \mathrm{tr}(P^3) + J|P|^2$$

and

$$(P \oslash h)_{jkil} P^{lk} P^{ij} = 2J|P|^2 - 2\mathrm{tr}(P^3).$$

Thus

$$\nabla^k \nabla_j (P)_{ki} P^{ij} = (P, \mathrm{Hess}(J)) - P_{ij} P_{kl} C^{kijl} + n\,\mathrm{tr}(P^3) - J|P|^2.$$

The proof is complete.   □

6.11   Factorization identities for residue families. Recursive relations

For $M^n = \mathbb{R}^n$ with the Euclidean metric h_c, the residue family

$$D^{\mathrm{res}}_{2N}(h_c; \lambda) : C^\infty([0, \varepsilon) \times M^n) \to C^\infty(M^n)$$

(Definition 6.6.2) coincides with the differential intertwining operator $D_{2N}^{nc}(\lambda)$ for (the non-compact models of) spherical principal series. It is induced by a homomorphism of Verma modules which is given by (a normalization of) the element $\mathcal{D}_{2N}(\lambda)$. According to Theorem 5.2.2, the family $\mathcal{D}_{2N}(\lambda)$ of order $2N$ satisfies a system of $2N$ factorization relations which induce corresponding factorizations for the differential operator families.

In the present section, we prove that for a *conformally flat* metric h the residue families $D_{2N}^{\mathrm{res}}(h;\lambda)$ satisfy analogous systems of factorization identities. These identities give rise to recursive relations for

$$Q\text{-curvatures}, \quad Q\text{-polynomials} \quad \text{and} \quad \text{GJMS-operators}.$$

The methods are the following. The critical family $D_n^{\mathrm{res}}(h;\lambda)$ satisfies a system of n factorization identities. One half of the system, together with the formula

$$\dot{D}_n^{\mathrm{res}}(h;0)(1) = -(-1)^{\frac{n}{2}} Q_n(h),$$

allows us to determine the critical Q-polynomial $Q_n^{\mathrm{res}}(h;\lambda)$ in terms of $Q_n(h)$, lower order Q-polynomials and lower order GJMS-operators. Applying analogous arguments to the lower order Q-polynomials, yields a formula for $Q_n^{\mathrm{res}}(\lambda)$ in terms of Q_n, lower order Q-curvatures and lower order GJMS-operators. That formula for $Q_n^{\mathrm{res}}(\lambda)$ is equivalent to a set of $\frac{n}{2} - 1$ relatives of the holographic formula for Q_n which relate Q_n to v_n.

Similarly, one half of the system, together with the factorization identity

$$D_n^{\mathrm{res}}\left(h; -\frac{n-1}{2}\right) = D_{n-2}^{\mathrm{res}}\left(h; -\frac{n+3}{2}\right) \circ P_2(dr^2 + h_r), \tag{6.11.1}$$

yields a formula for $\dot{D}_n^{\mathrm{res}}(h;0)(1)$, i.e., $Q_n(h)$, in terms of lower order residue families and lower order GJMS-operators. This formula involves the Yamabe operator $P_2(dr^2 + h_r)$ of the extension $dr^2 + h_r$ of h off M. A repeated application of factorization identities for lower order residue families yields a formula for $Q_n(h)$ in terms of lower order GJMS-operators (acting on 1) and powers of $P_2(dr^2 + h_r)$ (acting on 1). If we express all these powers, except the highest one, in terms of lower order Q-curvatures, we arrive at the recursive formula (6.11.30) for Q_n.

Another source of relations is the fact that the number of identities for $D_{2N}^{\mathrm{res}}(h;\lambda)$ exceeds the number of coefficients of the polynomial (if $N \geq 2$). It yields recursive relations which allow us to determine the critical GJMS-operator $P_n(h)$ using lower order GJMS-operators.

We briefly review the main results. The factorization identities are stated in Theorem 6.11.1. Here we assume that h is conformally flat, but the result is expected to extend to general metrics. For those factorization identities which contain a GJMS-operator on the target manifold M as a factor, this follows from Theorem 6.11.18. However, the remaining identities have not yet been established in full generality. For residue families $D_{2N}^{\mathrm{res}}(h;\lambda)$ of order $2N \leq 6$, Theorem 6.11.2

and Theorem 6.11.4 establish some of these identities. We use these results to
derive a recursive formula for the critical Q_6 (Theorem 6.11.5). Theorem 6.11.6
states that the same formula yields Q_6 in all dimensions $n \geq 6$ (universality).
An extension of the method leads to Conjecture 6.11.1 (for conformally flat h).
Its extension to general metrics would require, in particular, to prove (6.11.1)
in full generality. Theorem 6.11.10 states a recursive formula for the critical Q-
polynomial $Q_n^{\mathrm{res}}(\lambda)$. Its formulation motivates a close consideration of subcritical
Q-polynomials. Such formulas are discussed in Theorem 6.11.11. They generalize
Theorem 6.11.7. The critical Q-polynomial $Q_n(\lambda)$ contains $\frac{n}{2}$ relations between Q_n
and v_n. Among these relations, Corollary 6.11.5 emphasizes the one which arises
by consideration of the leading power of λ. Explicit versions for Q_4 and Q_6 are
given in (6.11.44) and (6.11.48), respectively. The corresponding recursive formula
for Q_8 is discussed in Section 6.13. Finally, Theorem 6.11.15 relates the quadratic
coefficient of the critical Q-polynomial to the renormalized volume, and Theorem
6.11.17 illustrates for P_6 how to derive recursive formulas for GJMS-operators.
Theorem 6.11.17 should be compared with the alternative recursive formula in
Theorem 6.12.1 which rests on (6.11.48).

We start with the discussion of the factorization identities. First of all, we
notice that for $N = 1$, the only factorization identities are formulated in (6.7.2),
i.e., these are

$$D_2^{\mathrm{res}}\left(h; -\frac{n-2}{2}\right) = P_2(h)i^* \quad \text{and} \quad D_2^{\mathrm{res}}\left(h; -\frac{n-1}{2}\right) = i^* P_2(dr^2 + h_r).$$

Both formulas are valid for all metrics h. For the convenience of the reader, we
repeat the argument. Definition 6.6.2 implies

$$D_2^{\mathrm{res}}(\lambda) = -(n-2+2\lambda)i^* \frac{\partial^2}{\partial r^2} + (\Delta + \lambda J)i^*.$$

The first identity is obvious. The second identity asserts that

$$i^* \frac{\partial^2}{\partial r^2} + \Delta_h i^* - \frac{n-1}{2} J(h)i^* = i^* P_2(dr^2 + h_r).$$

But this follows from $H = 0$ and $i^* J(dr^2 + h_r) = J(h)$ (Lemma 6.11.1).

Now we prove the factorization identities for residue families of arbitrary
even order for conformally flat metrics.

Theorem 6.11.1 (Factorization). *Let h be conformally flat. Then the residue family*
$D_{2N}^{\mathrm{res}}(h; \lambda)$ *factorizes for*

$$\lambda \in \left\{ -\frac{n+1}{2}, \dots, -\frac{n+1}{2} + N \right\} \cup \left\{ -\frac{n}{2} + N, \dots, -\frac{n}{2} + 2N \right\}$$

into products of (lower order) residue families and GJMS-operators. More pre-
cisely, for $j = 1, \dots, N$ the identities

$$D_{2N}^{\mathrm{res}}\left(h; -\frac{n}{2} + 2N - j\right) = P_{2j}(h) \circ D_{2N-2j}^{\mathrm{res}}\left(h; -\frac{n}{2} + 2N - j\right) \qquad (6.11.2)$$

and

$$D_{2N}^{\text{res}}\left(h; -\frac{n+1}{2}+j\right) = D_{2N-2j}^{\text{res}}\left(h; -\frac{n+1}{2}-j\right) \circ P_{2j}(dr^2+h_r) \tag{6.11.3}$$

hold true.

Proof. The proof rests on the transformation formula (Theorem 6.6.3)

$$e^{-(\lambda-2L)\varphi} \circ D_{2L}^{\text{res}}(\hat{h}; \lambda) = D_{2L}^{\text{res}}(h; \lambda) \circ \left(\frac{\kappa^*(r)}{r}\right)^{-\lambda} \circ \kappa^*. \tag{6.11.4}$$

Here κ is a diffeomorphism which restricts to the identity on M and satisfies

$$\kappa^*\left(r^{-2}(dr^2+\hat{h}_r)\right) = r^{-2}(dr^2+h_r)$$

and

$$i^*\left(\frac{\kappa^*(r)}{r}\right) = e^{\varphi}, \quad \hat{h} = e^{2\varphi}h.$$

We use local coordinates on M^n so that $h = e^{2\varphi}h_c$ with $h_c = \sum_i dx_i^2$. In order to prove the first set of identities, we apply the transformation formula (6.11.4) and the factorization identities (6.11.2) for $h = h_c$. These hold true in view of $D_{2N}^{\text{res}}(h_c; \lambda) = D_{2N}^{nc}(\lambda)$. We find

$$D_{2N}^{\text{res}}\left(h_c; -\frac{n}{2}+2N-j\right) \circ \left(\frac{\kappa^*(r)}{r}\right)^{-(-\frac{n}{2}+2N-j)} \circ \kappa^*$$

$$= P_{2j}(h_c) \circ D_{2N-2j}^{\text{res}}\left(h_c; -\frac{n}{2}+2N-j\right) \circ \left(\frac{\kappa^*(r)}{r}\right)^{-(-\frac{n}{2}+2N-j)} \circ \kappa^*,$$

i.e.,

$$e^{(\frac{n}{2}+j)\varphi} \circ D_{2N}^{\text{res}}\left(h; -\frac{n}{2}+2N-j\right)$$

$$= P_{2j}(h_c) \circ e^{(\frac{n}{2}-j)\varphi} \circ D_{2N-2j}^{\text{res}}\left(h; -\frac{n}{2}+2N-j\right).$$

Now, by the covariance of P_{2j}, we get

$$D_{2N}^{\text{res}}\left(h; -\frac{n}{2}+2N-j\right) = P_{2j}(h) \circ D_{2N-2j}^{\text{res}}\left(h; -\frac{n}{2}+2N-j\right).$$

This proves (6.11.2) for $h = e^{2\varphi}h_c$. Similarly, we apply (6.11.4) and (6.11.3) for $h = h_c$, and find

$$D_{2N}^{\text{res}}\left(h_c; -\frac{n+1}{2}+j\right) \circ \left(\frac{\kappa^*(r)}{r}\right)^{\frac{n+1}{2}-j} \circ \kappa^*$$

$$= D_{2N-2j}^{\text{res}}\left(h_c; -\frac{n+1}{2}-j\right) \circ P_{2j}(dr^2+h_c) \circ \left(\frac{\kappa^*(r)}{r}\right)^{\frac{n+1}{2}-j} \circ \kappa^*,$$

i.e.,

$$e^{(\frac{n+1}{2}+2N-j)\varphi} \circ D_{2N}^{\mathrm{res}}\left(h; -\frac{n+1}{2}+j\right)$$

$$= D_{2N-2j}^{\mathrm{res}}\left(h_c; -\frac{n+1}{2}-j\right) \circ \left(\frac{\kappa^*(r)}{r}\right)^{\frac{n+1}{2}+j} \circ P_{2j}\left(\left(\frac{\kappa^*(r)}{r}\right)^2(dr^2+h_c)\right) \circ \kappa^*$$

$$= D_{2N-2j}^{\mathrm{res}}\left(h_c; -\frac{n+1}{2}-j\right) \circ \left(\frac{\kappa^*(r)}{r}\right)^{\frac{n+1}{2}+j} \circ \kappa^* \circ \kappa_* \circ P_{2j}(\kappa^*(dr^2+h_r)) \circ \kappa^*$$

$$= e^{(\frac{n+1}{2}+2N-j)\varphi} \circ D_{2N-2j}^{\mathrm{res}}\left(h; -\frac{n+1}{2}-j\right) \circ P_{2j}(dr^2+h_r).$$

This proves (6.11.3) for $h = e^{2\varphi}h_c$. The proof is complete.   □

Remark 6.11.1. *An analogous proof shows factorization identities for odd order families $D_{2N+1}^{\mathrm{res}}(h; \lambda)$ for conformally flat metrics. Here we apply Theorem 5.2.4. The factorizations*

$$D_{2N}^{\mathrm{res}}\left(h; -\frac{n}{2}+N\right) = P_{2N}(h)i^*,$$

$$D_{2N+1}^{\mathrm{res}}\left(h; -\frac{n}{2}+N+1\right) = P_{2N}(h)D_1^{\mathrm{res}}\left(h; -\frac{n}{2}+N+1\right) = P_{2N}(h)i^*\frac{\partial}{\partial r}$$

(*see* (6.6.12)) *directly follow from the definitions and are valid for all metrics h.*

In particular, for a conformally flat metric h, we have the n factorizations

$$D_n^{\mathrm{res}}\left(h; \frac{n}{2}-j\right) = P_{2j}(h) \circ D_{n-2j}^{\mathrm{res}}\left(h; \frac{n}{2}-j\right), \; j = 1,\ldots,\frac{n}{2}$$

and

$$D_n^{\mathrm{res}}\left(h; -\frac{n+1}{2}+j\right) = D_{n-2j}^{\mathrm{res}}\left(h; -\frac{n+1}{2}-j\right) \circ P_{2j}(dr^2+h_r), \; j = 1,\ldots,\frac{n}{2}$$

of the critical residue family.

Since $D_{2N}^{\mathrm{res}}(h; \lambda)$ is a polynomial of degree N, the residue family is completely determined by the $N+1$ operators

$$P_{2j}(h) \circ D_{2N-2j}^{\mathrm{res}}\left(h; -\frac{n}{2}+2N-j\right), \; j = 1,\ldots,N$$

and

$$D_{2N-2}^{\mathrm{res}}\left(h; -\frac{n+3}{2}\right) \circ P_2(dr^2+h_r). \tag{6.11.5}$$

The lower order families $D_{\leq 2N-2}^{\mathrm{res}}(h; \lambda)$ in these compositions, in turn, can be written as linear combinations of compositions of $P_{\leq 2N-2}(h)$ and residue families of order $\leq 2N-4$. Thus $D_{2N}^{\text{res}}(\lambda)$ recursively can be written as a linear combination of compositions of GJMS-operators $P_2(h),\dots,P_{2N}(h)$ and $P_2(dr^2+h_r)$.

In particular, this has the consequence that the critical Q-curvature

$$Q_n(h) = -(-1)^{\frac{n}{2}} \dot{D}_n^{\text{res}}(h;0)(1)$$

(see (6.6.2)) can be written as a linear combination of composition of the GJMS-operators $P_2(h),\dots,P_{n-2}(h)$ and the Yamabe operator $P_2(dr^2+h_r)$ (acting on $u=1$).

Remark 6.11.2. *The only factorization identity in the system (6.11.3) which will be used in connection with the study of the recursive structure of Q-curvature has the right-hand side (6.11.5). This is a curved analog of the identity*

$$\mathcal{D}_{2N}^0\left(-\frac{n-1}{2}\right) = \mathcal{D}_{2N-2}^0\left(-\frac{n+3}{2}\right)\Delta_{n+1}^-,$$

i.e.,

$$\sum_{j=0}^{N} a_j^{(N)}\left(-\frac{n-1}{2}\right)(Y_{n+1}^-)^{2N-2j}(\Delta_n^-)^j$$

$$= \left(\sum_{j=0}^{N-1} a_j^{(N-1)}\left(-\frac{n+3}{2}\right)(Y_{n+1}^-)^{2N-2j}(\Delta_n^-)^j\right)((Y_{n+1}^-)^2 + \Delta_n^-) \quad (6.11.6)$$

in $\mathcal{U}(\mathfrak{n}_{n+2}^-)$, where the coefficients are given by

$$a_j^{(M)}(\lambda) = \frac{M!}{j!(2M-2j)!}(-2)^{M-j}\prod_{k=j}^{M-1}(2\lambda-4M+2k+n+2), \quad j=0,\dots,M-1$$

and $a_0^{(M)}(\lambda) = 1$ (see (5.1.4)). For the convenience of the reader we add a direct proof. It differs from the general (but complicated) argument in the proof of Theorem 5.2.2. (6.11.6) is equivalent to the relations

$$a_j^{(N)}\left(-\frac{n-1}{2}\right) = a_{j-1}^{(N-1)}\left(-\frac{n+3}{2}\right) + a_j^{(N-1)}\left(-\frac{n+3}{2}\right), \quad j=1,\dots,N-1$$

and a corresponding relation for $j=0$ with the undefined term a_{-1} removed. But these identities are equivalent to the obvious relations

$$N(2N-1) = j(4N-2j-1) + \frac{1}{2}(2N-2j-1)(2N-2j).$$

Since $D_{2N}^{\text{res}}(h;\lambda)$ has polynomial degree N, Theorem 6.11.1 implies that there are $2N-(N+1) = N-1$ linear relations among the right-hand sides.

In particular, for $N = 2$ (and $n = 4$) we find a linear relation among the four operators

$$P_4(h)i^*, \quad P_2(h)D_2^{\text{res}}(h;1), \quad D_2^{\text{res}}(h;-7/2)P_2(dr^2+h_r), \quad i^*P_4(dr^2+h_r).$$

Its content will be discussed in Lemma 6.11.6.

Moreover, for even $n \geq 6$, we find a presentation

$$P_n(h)i^* = \Big\{ a_2 P_{n-2}(h)D_2^{\text{res}}(h;1)$$

$$+ a_4 P_{n-4}(h)D_4^{\text{res}}(h;2) + \cdots + a_{n-2}P_2(h)D_{n-2}^{\text{res}}\left(h;\frac{n}{2}-1\right) \Big\}$$

$$+ b_2 D_{n-2}^{\text{res}}\left(h;-\frac{n+3}{2}\right) P_2(dr^2+h_r) + b_4 D_{n-4}^{\text{res}}\left(h;-\frac{n+5}{2}\right) P_4(dr^2+h_r)$$

of the critical GJMS-operator $P_n(h)$ in terms of lower order residue families $D_{\leq n-2}^{\text{res}}(h;\lambda)$, lower order GJMS-operators $P_{n-2}(h), \ldots, P_2(h)$ and $P_4(dr^2+h_r)$, $P_2(dr^2+h_r)$. In turn, all residue families which appear here as factors can be written in terms of $P_{n-2}(h), \ldots, P_2(h)$ and $P_2(dr^2+h_r)$.

Thus we find a formula for $P_n(h)i^*$ ($n \geq 6$) in terms of *lower order* GJMS-operators $P_{n-2}(h), \ldots, P_2(h)$ and $P_2(dr^2+h_r)$, $P_4(dr^2+h_r)$.

It would be interesting to work out the combinatorics of the recursion. The case $n = 6$ will be discussed below.

Next, we consider residue families of order 4 for *arbitrary* metrics h. The following result deals with the critical case.

Theorem 6.11.2. *For any Riemannian four-manifold* (M, h),

$$D_4^{\text{res}}(h;0) = P_4(h)i^*,$$

$$D_4^{\text{res}}(h;1) = P_2(h) \circ D_2^{\text{res}}(h;1),$$

$$D_4^{\text{res}}\left(h;-\frac{3}{2}\right) = D_2^{\text{res}}\left(h;-\frac{7}{2}\right) \circ P_2(dr^2+h_r),$$

$$D_4^{\text{res}}\left(h;-\frac{1}{2}\right) = i^*P_4(dr^2+h_r).$$

In Theorem 6.11.2, the complexity of the identities increases starting with the first identity. The four values $\lambda \in \{-3/2, -1/2, 0, 1\}$ are characterized by the property that the conformal covariance of the compositions on the right-hand sides are consequences of the conformal covariance of the factors.

Theorem 6.11.2 shows that the residue family $D_4^{\text{res}}(h;\lambda)$ interpolate the Paneitz operators $P_4(h)i^*$ and $i^*P_4(dr^2+h_r)$. In that respect, it resembles the family $D_2^{\text{res}}(h;\lambda)$ which interpolates the Yamabe operators $P_2(h)i^*$ and $i^*P_2(dr^2+h_r)$ (see the formulas at the end of Section 6.7).

There is an analogous result for the tractor family $D_4^T(dr^2+h_r;\lambda)$ (Theorem 6.22.4).

For the proof of Theorem 6.11.2 we need the following technical result.

Lemma 6.11.1. *For (M^n, h) and the associated Poincaré-Einstein metric*

$$r^{-2}(dr^2 + h_r),$$

the function $r \mapsto J(dr^2 + h_r)$ satisfies

$$i^* J(dr^2 + h_r) = J(h), \quad \dot{J} = 0 \quad and \quad \ddot{J} = |P|^2.$$

Here the dots denote the derivatives with respect to r at $r = 0$.

Proof. The formula

$$\tau(\hat{g}) = e^{-2\varphi} \left(\tau(g) - 2n\Delta_g(\varphi) - n(n-1)|d\varphi|_g^2 \right)$$

implies

$$\tau(g) = e^{2\varphi} \left(\tau(\hat{g}) + 2n\Delta_{\hat{g}}(\varphi) - n(n-1)|d\varphi|_{\hat{g}}^2 \right),$$

i.e.,

$$\tau(\hat{g}) = e^{-2\varphi} \tau(g) - 2n\Delta_{\hat{g}}(\varphi) + n(n-1)|d\varphi|_{\hat{g}}^2. \tag{6.11.7}$$

We apply (6.11.7) to $\hat{g} = dr^2 + h_r$, $g = r^{-2}(dr^2 + h_r)$ and $\varphi = \log r$. Since $\tau(g) = -n(n+1)$, we find for small $r > 0$

$$\tau(dr^2 + h_r) = -\frac{n(n+1)}{r^2} - 2n\Delta_{dr^2 + h_r}(\log r) + n(n-1)|d\log r|_{dr^2 + h_r}^2$$

$$= -\frac{n(n+1)}{r^2} - 2n\left(\frac{\partial^2}{\partial r^2}(\log r) + \frac{1}{2}\operatorname{tr}(h_r^{-1}\dot{h}_r)\frac{\partial}{\partial r}(\log r) \right) + \frac{n(n-1)}{r^2}$$

$$= -n\operatorname{tr}(h_r^{-1}\dot{h}_r)\frac{1}{r}.$$

Hence

$$J(dr^2 + h_r) = -\frac{1}{2r}\operatorname{tr}(h_r^{-1}\dot{h}_r). \tag{6.11.8}$$

Now $h_r = h_0 + r^2 h_{(2)} + r^4 h_{(4)} + \cdots$ yields

$$h_r^{-1} = h_0^{-1} - r^2 h_0^{-1} h_{(2)} h_0^{-1} + r^4 \left((h_0^{-1} h_{(2)})^2 - h_0^{-1} h_{(4)} \right) h_0^{-1} + \cdots$$

and

$$\operatorname{tr}(h_r^{-1}\dot{h}_r) = \operatorname{tr}\left(2r h_0^{-1} h_{(2)} - 2r^3 (h_0^{-1} h_{(2)})^2 + 4r^3 h_0^{-1} h_{(4)} + \cdots \right)$$

i.e.,

$$-\frac{1}{2}\operatorname{tr}(h_r^{-1}\dot{h}_r)\frac{1}{r} = -\operatorname{tr}(h_0^{-1} h_{(2)}) + r^2 \operatorname{tr}\left((h_0^{-1} h_{(2)})^2 \right) - 2r^2 \operatorname{tr}(h_0^{-1} h_{(4)}) + \cdots.$$

Now (6.11.8) and $h_{(2)} = -P$ imply $i^* J(dr^2 + h_r) = J(h)$ and $\dot{J} = 0$. Moreover, using $\operatorname{tr}(h_{(4)}) = \frac{1}{4}|P|^2$, it follows that

$$i^* \frac{\partial^2}{\partial r^2} \left(J(dr^2 + h_r) \right) = -\frac{1}{2} i^* \frac{\partial^2}{\partial r^2} \left(\operatorname{tr}(h_r^{-1}\dot{h}_r)\frac{1}{r} \right) = |P|^2.$$

This proves $\ddot{J} = |P|^2$. $\qquad\square$

The results of Lemma 6.11.1 can be seen also as consequences of the following useful formula for $P(dr^2 + h_r)$.

Lemma 6.11.2. *In the situation of Lemma 6.11.1,*

$$P(dr^2 + h_r) = -\frac{1}{2r}\dot{h}_r. \tag{6.11.9}$$

In particular,

$$i^* P(dr^2 + h_r) = P(h) \tag{6.11.10}$$

and

$$P_{NN}(dr^2 + h_r) = 0, \quad |P(dr^2 + h_r)|^2 = |P(h)|^2$$

on $r = 0$.

Proof. Applying the conformal transformation law (2.5.9) for the metrics $\hat{g} = dr^2 + h_r$ and $g = r^{-2}(dr^2 + h_r)$ yields

$$P(dr^2 + h_r) = P(g) - \frac{1}{2}|d \log r|_g^2 g - \text{Hess}^g (\log r) + \frac{1}{r^2}dr^2$$

$$= -\frac{1}{2}g - \frac{1}{2}g - \text{Hess}^g (\log r) + \frac{1}{r^2}dr^2$$

$$= -\frac{1}{r^2}h_r - \text{Hess}^g (\log r).$$

In the second line we used that the Schouten tensor of the Einstein metric g is $-\frac{1}{2}g$. Now the tangential components of the Hessian are given by

$$\text{Hess}^g (\log r)_{ij} = \left\langle \nabla_i^g \left(\frac{dr}{r}\right), \frac{\partial}{\partial x_j} \right\rangle = -\frac{1}{r}\Gamma_{ij}^N.$$

The usual formulas for the Christoffel symbols (see (6.8.3)) imply

$$\Gamma_{ij}^N = -\frac{1}{2}(\partial/\partial r)(g_{ij})g^{NN} = -\frac{1}{2}\dot{h}_r + \frac{1}{r}h_r.$$

Hence

$$\text{Hess}^g (\log r)_{ij} = \frac{1}{2r}(\dot{h}_r)_{ij} - \frac{1}{r^2}(h_r)_{ij}.$$

The remaining components of the Hessian vanish:

$$\text{Hess}^g (\log r)_{iN} = -\frac{1}{r}\Gamma_{iN}^N = 0 \quad \text{and} \quad \text{Hess}^g (\log r)_{NN} = -\frac{1}{r^2} - \frac{1}{r}\Gamma_{NN}^N = 0.$$

These results prove (6.11.9). The remaining assertions are direct consequences.
$\square$

We continue with the proof of Theorem 6.11.2. Definition 6.6.1 yields the formula

$$D_4^{\mathrm{res}}(\lambda) = 32(\lambda-1)\lambda$$
$$\times \left\{ \frac{1}{24}i^* \frac{\partial^4}{\partial r^4} + \frac{1}{2}(T_2^*(\lambda) + v_2)i^* \frac{\partial^2}{\partial r^2} + (T_4^*(\lambda) + T_2^*(\lambda)v_2 + v_4)i^* \right\},$$

i.e.,

$$D_4^{\mathrm{res}}(\lambda) = \frac{4}{3}(\lambda-1)\lambda i^* \frac{\partial^4}{\partial r^4} - 4\lambda(\Delta + (\lambda-2)\mathsf{J})i^* \frac{\partial^2}{\partial r^2}$$
$$+ \left[P_4^*(\lambda) + 4\lambda(\Delta - \lambda\mathsf{J})\mathsf{J} + 4\lambda(\lambda-1)(\mathsf{J}^2 - |\mathsf{P}|^2) \right] i^*, \quad (6.11.11)$$

where

$$P_4(\lambda) = (\Delta - (\lambda+2)\mathsf{J})(\Delta - \lambda\mathsf{J}) + 2\lambda(\lambda-1)|\mathsf{P}|^2 + 4(\lambda-1)\delta(\mathsf{P}\#d) + 2(\lambda-1)(d\mathsf{J}, d)$$

(see also Lemma 6.11.8). Here we have used Lemma 6.6.1 and Theorem 6.9.2. In particular, we find $D_4^{\mathrm{res}}(0) = P_4^*(0)i^* = P_4(0)i^*$. This proves the first assertion. It is a special case of (6.6.8). Next, we find

$$D_4^{\mathrm{res}}(h; 1) = -4(\Delta - \mathsf{J})i^* \frac{\partial^2}{\partial r^2} + [(\Delta - \mathsf{J})(\Delta - 3\mathsf{J}) + 4(\Delta - \mathsf{J})\mathsf{J}] i^*$$
$$= (\Delta - \mathsf{J}) \left(-4i^* \frac{\partial^2}{\partial r^2} + (\Delta + \mathsf{J})i^* \right) = P_2(h) \circ D_2^{\mathrm{res}}(h; 1).$$

This proves the second assertion of Theorem 6.11.2. We split the proof of the remaining two identities.

Lemma 6.11.3.

$$D_4^{\mathrm{res}}(h; -3/2) = D_2^{\mathrm{res}}(h; -7/2) \circ P_2(dr^2 + h_r).$$

Proof. By (6.11.11), we find

$$D_4^{\mathrm{res}}(h; -3/2) = 5i^* \frac{\partial^4}{\partial r^4} + 6(\Delta - 7/2\mathsf{J})i^* \frac{\partial^2}{\partial r^2}$$
$$+ \left[P_4^*(-3/2) - 6(\Delta + 3/2\mathsf{J})\mathsf{J} + 15(\mathsf{J}^2 - |\mathsf{P}|^2) \right] i^* \quad (6.11.12)$$

with

$$P_4^*(-3/2)u = (\Delta + 3/2\mathsf{J})(\Delta - 1/2\mathsf{J})u + 15/2|\mathsf{P}|^2u - 10\delta(\mathsf{P}\#du) - 5\delta(ud\mathsf{J});$$

the convention here is $\Delta = -\delta d$. We have to compare this operator with the composition

$$D_2(dr^2 + h_r; -7/2)P_2(dr^2 + h_r)$$
$$= \left(5i^* \frac{\partial^2}{\partial r^2} + \Delta i^* - \frac{7}{2}\mathsf{J}i^* \right) \left(\Delta_{dr^2+h_r} - \frac{3}{2}\mathsf{J}(dr^2 + h_r) \right). \quad (6.11.13)$$

We apply the formula

$$\Delta_{dr^2+h_r} = \frac{\partial^2}{\partial r^2} + \frac{1}{2}\,\mathrm{tr}(h_r^{-1}\dot{h}_r)\frac{\partial}{\partial r} + \Delta_{h_r}.$$

The coefficients of $\partial^4/\partial r^4$ in (6.11.12) and (6.11.13) both are equal to 5. In view of $i^*J(dr^2 + h_r) = J(h)$ (Lemma 6.11.1) and $\ddot{h}_0 = -2\mathsf{P}$ (Theorem 6.9.1), the coefficient of $\partial^2/\partial r^2$ in (6.11.13) is given by

$$-\frac{15}{2}\mathsf{J} + 5\frac{\partial}{\partial r}\Big|_{r=0}\,\mathrm{tr}(h_r^{-1}\dot{h}_r) - \frac{7}{2}\mathsf{J} + 6\Delta = -21\mathsf{J} + 6\Delta.$$

This result coincides with the coefficient of $\partial^2/\partial r^2$ in (6.11.12). It remains to compare the tangential components of both operators. For $u \in C^\infty(M^4)$, (6.11.12) yields the formula

$$\Delta^2 u - \frac{1}{2}\Delta(\mathsf{J}u) + \frac{3}{2}\mathsf{J}\Delta u - \frac{3}{4}\mathsf{J}^2 u + \frac{15}{2}|\mathsf{P}|^2 u - 10\delta(\mathsf{P}\#du) + 5(du, d\mathsf{J}) + 5(\Delta\mathsf{J})u$$
$$- 6\Delta(\mathsf{J}u) - 9\mathsf{J}^2 u + 15\mathsf{J}^2 u - 15|\mathsf{P}|^2 u$$
$$= \Delta^2 u - 5\mathsf{J}\Delta u - 8(du, d\mathsf{J}) - 10\delta(\mathsf{P}\#du) - \frac{3}{2}(\Delta\mathsf{J})u - \frac{15}{2}|\mathsf{P}|^2 u + \frac{21}{4}\mathsf{J}^2 u.$$

Now recall that $\delta(\mathsf{P}\#du) = -(d\mathsf{J}, du) - (\mathsf{P}, \mathrm{Hess}(u))$ (Lemma 4.2.5). Thus we obtain

$$\Delta^2 u - 5\mathsf{J}\Delta u + 10(\mathsf{P}, \mathrm{Hess}(u)) + 2(du, d\mathsf{J}) - \frac{3}{2}(\Delta\mathsf{J})u - \frac{15}{2}|\mathsf{P}|^2 u + \frac{21}{4}\mathsf{J}^2 u.$$

Next, we determine the tangential part of (6.11.13). We find

$$-\frac{15}{2}\ddot{\mathsf{J}}u + \Delta^2 u - \frac{3}{2}\Delta(\mathsf{J})u - \frac{7}{2}\mathsf{J}\Delta u + \frac{21}{4}\mathsf{J}^2 u + 5\frac{\partial^2}{\partial r^2}\Big|_{r=0}(\Delta_{h_r}u),$$

where $\ddot{\mathsf{J}} = (\partial^2/\partial r^2)|_{r=0}(J(dr^2+h_r))$. But using $\ddot{h} = -2\mathsf{P}$, a calculation shows

$$\ddot\Delta u = \frac{\partial^2}{\partial r^2}\Big|_{r=0}(\Delta_{h_r}u)$$
$$= -\mathsf{J}\Delta u - \delta(\mathsf{J}du) + 2(\mathsf{P}, \mathrm{Hess}(u)) = (d\mathsf{J}, du) + 2(\mathsf{P}, \mathrm{Hess}(u)).$$

Moreover, using $\ddot{\mathsf{J}} = |\mathsf{P}|^2$ (Lemma 6.11.1), it follows that the tangential part coincides with

$$\Delta^2 u - 5\mathsf{J}\Delta u + 10(\mathsf{P}, \mathrm{Hess}(u)) + 2(du, d\mathsf{J}) - \frac{3}{2}(\Delta\mathsf{J})u + \frac{21}{4}\mathsf{J}^2 u - \frac{15}{2}|\mathsf{P}|^2 u.$$

The proof is complete.   □

Lemma 6.11.4.
$$D_4^{\mathrm{res}}(h; -1/2) = i^*P_4(dr^2+h_r).$$

Proof. (6.11.11) gives

$$
D_4^{\mathrm{res}}(-1/2)u = i^* \frac{\partial^4 u}{\partial r^4} + (2\Delta - 5\mathsf{J})\, i^* \frac{\partial^2 u}{\partial r^2}
$$
$$
+ \left[P_4^*(-1/2) - 2(\Delta + 1/2\mathsf{J})\mathsf{J} + 3(\mathsf{J}^2 - |\mathsf{P}|^2) \right] i^* u, \quad (6.11.14)
$$

where

$$
P_4^*(-1/2)u = (\Delta + 1/2\mathsf{J})(\Delta - 3/2\mathsf{J})u + 3/2|\mathsf{P}|^2 u - 6\delta(\mathsf{P}\#du) - 3\delta(ud\mathsf{J}).
$$

On the other hand, the Paneitz operator for $dr^2 + h_r$ is given by

$$
P_{4,5} = \Delta^2 + 3\delta(\mathsf{J}d) - 4\mathsf{P}\#d + \frac{1}{2}Q_{4,5}
$$

with

$$
Q_{4,5} = \frac{5}{2}\mathsf{J}^2 - 2|\mathsf{P}|^2 - \Delta\mathsf{J};
$$

here all constructions refer to the metric $dr^2 + h_r$. In order to find a formula for $i^* Q_{4,5}$, we use $i^*\mathsf{J} = \mathsf{J}$, $\dot{\mathsf{J}} = 0$, $\ddot{\mathsf{J}} = |\mathsf{P}|^2$ and $i^*\mathsf{P} = \mathsf{P}$ (Lemma 6.11.2 and Lemma 6.11.1). Hence

$$
i^* Q_{4,5} = \frac{5}{2}\mathsf{J}^2 - 2|\mathsf{P}|^2 - (|\mathsf{P}|^2 + \Delta\mathsf{J}) = \frac{5}{2}\mathsf{J}^2 - 3|\mathsf{P}|^2 - \Delta\mathsf{J}.
$$

We compare the coefficients of $\partial^2/\partial r^2$. On the left-hand side we have $2\Delta - 5\mathsf{J}$ (by (6.11.14)). On the other hand, $i^* P_4(dr^2 + h_r)u$ is given by the restriction of

$$
\left(\frac{\partial^2}{\partial r^2} + \Delta_h \right)\left(\frac{\partial^2}{\partial r^2} + \frac{1}{2}\mathrm{tr}(h_r^{-1}\dot{h}_r)\frac{\partial}{\partial r} + \Delta_{h_r} \right) u
$$
$$
- 3\mathsf{J}\left(\frac{\partial^2 u}{\partial r^2} + \Delta u \right) - 3(d\mathsf{J}, du) - 4\mathsf{P}\#du + \frac{1}{2}Q_{4,5}u \quad (6.11.15)
$$

to $r = 0$; here we used the notation $\Delta = \Delta_h$, $\mathsf{P} = \mathsf{P}(h)$ etc. and we applied the fact that $\mathsf{P}_{NN} = 0$ (by Lemma 6.11.2). Now the latter formula contributes

$$
\left(2\Delta + \frac{\partial}{\partial r}\Big|_{r=0} \mathrm{tr}(h_r^{-1}\dot{h}_r) - 3\mathsf{J} \right) i^* \frac{\partial^2 u}{\partial r^2} = (2\Delta - 5\mathsf{J})i^* \frac{\partial^2 u}{\partial r^2}.
$$

This proves that the coefficients of the normal derivatives on both sides of the assertion coincide. Finally, we verify the coincidence of the tangential parts. The residue family contributes

$$
\left(\Delta + \frac{1}{2}\mathsf{J} \right)\left(\Delta - \frac{3}{2}\mathsf{J} \right) u - 2\left(\Delta + \frac{1}{2}\mathsf{J} \right)\mathsf{J}u - 6\delta(\mathsf{P}\#du) - 3\delta(ud\mathsf{J})
$$
$$
+ \frac{3}{2}|\mathsf{P}|^2 u + 3\left(\mathsf{J}^2 - |\mathsf{P}|^2 \right) u
$$
$$
= \Delta^2 u - 3\mathsf{J}\Delta u + 2(du, d\mathsf{J}) + 6(\mathsf{P}, \mathrm{Hess}(u)) - \frac{1}{2}(\Delta\mathsf{J})u + \left(\frac{5}{4}\mathsf{J}^2 - \frac{3}{2}|\mathsf{P}|^2 \right) u
$$

using $\delta \mathsf{P} = -d\mathsf{J}$. It remains to analyze the tangential part of (6.11.15). Using the formula for $\ddot{\Delta}$ from the proof of Lemma 6.11.3, we find

$$\Delta^2 u + 2(\mathsf{P}, \mathrm{Hess}(u)) + (d\mathsf{J}, du)$$

$$- 3\mathsf{J}\Delta u - 3(d\mathsf{J}, du) + 4(\mathsf{P}, \mathrm{Hess}(u)) + 4(d\mathsf{J}, du) + \frac{1}{2}\left(\frac{5}{2}\mathsf{J}^2 - 3|\mathsf{P}|^2 - \Delta \mathsf{J}\right)u.$$

The proof is complete. $\square$

We emphasize that the proofs of Lemma 6.11.3 and Lemma 6.11.4 rest on the relations

$$i^*\mathsf{P}(dr^2 + h_r) = \begin{pmatrix} 0 & 0 \\ 0 & \mathsf{P}(h) \end{pmatrix}, \quad \mathsf{J} = 0 \quad \text{and} \quad \ddot{\mathsf{J}} = |\mathsf{P}|^2.$$

The family $D_4^{\mathrm{res}}(\lambda)$ is polynomial of degree 2 in λ, i.e., can be written in the form $A\lambda^2 + B\lambda + C$ with three operator-coefficients A, B, C. On the other hand, Theorem 6.11.2 yields four relations, i.e., the right-hand sides of the four relations satisfy one linear relation

$$i^*P_4(dr^2 + h_r) = \alpha P_4(h)i^* + \beta P_2(h)D_2^{\mathrm{res}}(h; 1) + \gamma D_2^{\mathrm{res}}(h; -7/2)P_2(dr^2 + h_r)$$

with rational coefficients. In order to determine (α, β, γ), it suffices to consider the flat case. Then

$$(\alpha, \beta, \gamma) = \frac{1}{5}(5, -1, 1).$$

Thus $D_4^{\mathrm{res}}(h; \lambda)$ can be written as a linear combination of compositions of $P_4(h)$, $P_2(h)$ and $P_2(dr^2 + h_r)$. Moreover, the factorizations

$$D_4^{\mathrm{res}}(h; 0) = P_4(h)i^*,$$
$$D_4^{\mathrm{res}}(h; 1) = P_2(h)D_2^{\mathrm{res}}(h; 1),$$
$$D_4^{\mathrm{res}}\left(h; -\frac{3}{2}\right) = D_2^{\mathrm{res}}\left(h; -\frac{7}{2}\right)P_2(dr^2 + h_r)$$

imply $C(h) = P_4(h)i^*$ and

$$\begin{pmatrix} (-\frac{3}{2})^2 & -\frac{3}{2} \\ 1 & 1 \end{pmatrix} \begin{pmatrix} A(h)(1) \\ B(h)(1) \end{pmatrix} = \begin{pmatrix} D_2^{\mathrm{res}}(h; -\frac{7}{2})P_2(dr^2 + h_r)(1) \\ P_2(h)D_2^{\mathrm{res}}(h; 1)(1) \end{pmatrix},$$

i.e.,

$$\begin{pmatrix} A(h)(1) \\ B(h)(1) \end{pmatrix} = \begin{pmatrix} \frac{4}{15} & \frac{2}{5} \\ -\frac{4}{15} & \frac{3}{5} \end{pmatrix} \begin{pmatrix} D_2^{\mathrm{res}}(h; -\frac{7}{2})P_2(dr^2 + h_r)(1) \\ P_2(h)D_2^{\mathrm{res}}(h; 1)(1) \end{pmatrix}.$$

In particular, we can write $Q_4(h) = -B(h)(1)$ as the linear combination

$$Q_4(h) = -\frac{4}{15}D_2^{\mathrm{res}}\left(h; -\frac{7}{2}\right)P_2(dr^2 + h_r)(1) + \frac{3}{5}P_2(h)D_2^{\mathrm{res}}(h; 1)(1). \quad (6.11.16)$$

But using

$$D_2^{\mathrm{res}}\left(h; -\frac{7}{2}\right) = 5i^* P_2(dr^2 + h_r) - 4P_2(h)i^*,$$

$$D_2^{\mathrm{res}}(h; 1) = -4i^* P_2(dr^2 + h_r) + 5P_2(h)i^*,$$

we find the following formula for Q_4 in terms of Yamabe operators P_2 and Q-curvatures $Q_2 = \mathsf{J}$.

Corollary 6.11.1. *For any Riemannian four-manifold* (M, h),

$$Q_4(h) = P_2(h)Q_2(h) + \frac{4}{3}i^* P_2^2(dr^2 + h_r)(1)$$

$$= P_2(h)Q_2(h) - 2i^* P_2(dr^2 + h_r)Q_2(dr^2 + h_r).$$

Proof. (6.11.16) yields

$$Q_4(h) = -3P_2^2(h)(1) + \frac{4}{3}\left(P_2(h)i^* + i^* P_2(dr^2 + h_r)\right) P_2(dr^2 + h_r)(1).$$

Now using $P_2(h)(1) = -\mathsf{J}(h)$, $P_2(dr^2 + h_r)(1) = -\frac{3}{2}\mathsf{J}(dr^2 + h_r)$ and $i^* \mathsf{J}(dr^2 + h_r) = \mathsf{J}(h)$, we find

$$Q_4(h) = 3P_2(h)\mathsf{J}(h) - 2\left(P_2(h)i^* + i^* P_2(dr^2 + h_r)\right) \mathsf{J}(dr^2 + h_r)$$

$$= P_2(h)Q_2(h) - 2i^* P_2(dr^2 + h_r)\mathsf{J}(dr^2 + h_r).$$

The proof is complete. □

It is instructive to observe how the various terms in Corollary 6.11.1 are organized. The proof of the following somewhat more general result rests on such explicit formulas.

Lemma 6.11.5. *The second identity in Corollary* 6.11.1, *i.e.,*

$$Q_4(h) = P_2(h)Q_2(h) - 2i^* P_2(dr^2 + h_r)Q_2(dr^2 + h_r)$$

holds true in all dimensions $n \geq 3$.

Proof. We recall that $Q_{4,n} = \frac{n}{2}\mathsf{J}^2 - 2|\mathsf{P}|^2 - \Delta\mathsf{J}$ (see (4.1.7)). On the other hand, the right-hand side of the assertion is given by

$$\left(\Delta - (\frac{n}{2} - 1)\mathsf{J}\right)\mathsf{J} - 2i^*\left(\Delta_{dr^2 + h_r} - \frac{n-1}{2}\mathsf{J}(dr^2 + h_r)\right)\mathsf{J}(dr^2 + h_r)$$

$$= \Delta\mathsf{J} - (\frac{n}{2} - 1)\mathsf{J}^2 - 2\left(\ddot{\mathsf{J}} + \Delta\mathsf{J} - \frac{n-1}{2}\mathsf{J}^2\right) = -\Delta\mathsf{J} - 2|\mathsf{P}|^2 + \frac{n}{2}\mathsf{J}^2.$$

The proof is complete. □

The following result extends Theorem 6.11.2.

Theorem 6.11.3. *For (M^n, h) $(n \geq 3)$,*

$$D_4^{\text{res}}\left(h; -\frac{n}{2}+2\right) = P_4(h)i^*,$$

$$D_4^{\text{res}}\left(h; -\frac{n}{2}+3\right) = P_2(h) \circ D_2^{\text{res}}\left(h; -\frac{n}{2}+3\right),$$

$$D_4^{\text{res}}\left(h; -\frac{n-1}{2}\right) = D_2^{\text{res}}\left(h; -\frac{n+3}{2}\right) \circ P_2(dr^2 + h_r),$$

$$D_4^{\text{res}}\left(h; -\frac{n-3}{2}\right) = i^* P_4(dr^2 + h_r).$$

We omit the analogous proof.

For the remainder of the present section it will be convenient to introduce the abbreviations

$$\bar{P}_{2j}(h) = P_{2j}(dr^2 + h_r) \quad \text{and} \quad \bar{Q}_{2j}(h) = Q_{2j}(dr^2 + h_r). \tag{6.11.17}$$

The four right-hand sides in Theorem 6.11.3 satisfy *one* linear relation. The following result is equivalent to that relation.

Lemma 6.11.6. *For any Riemannian manifold (M^n, h) $(n \geq 3)$,*

$$i^*\left(\bar{P}_4(h) - \bar{P}_2^2(h)\right) = \left(P_4(h) - P_2^2(h)\right) i^*. \tag{6.11.18}$$

Proof. The formula

$$D_2^{\text{res}}(\lambda) = -(n-2+2\lambda)i^*\bar{P}_2 + (n-1+2\lambda)P_2i^*$$

yields

$$D_2^{\text{res}}\left(-\frac{n}{2}+3\right) = -4i^*\bar{P}_2 + 5P_2i^* \quad \text{and} \quad D_2^{\text{res}}\left(-\frac{n+3}{2}\right) = 5i^*\bar{P}_2 - 4P_2i^*.$$

Now the right-hand sides of the identities in Theorem 6.11.3 satisfy the linear relation

$$P_4i^* = \frac{1}{5}P_2D_2^{\text{res}}\left(-\frac{n}{2}+3\right) - \frac{1}{5}D_2^{\text{res}}\left(-\frac{n+3}{2}\right)\bar{P}_2 + i^*\bar{P}_4.$$

In fact, for the flat metric $h = h_c$ on $\mathbb{R}^n$ it specializes to the obvious relation

$$\Delta_n^2 i^* = \frac{1}{5}\Delta_n(-4i^*\Delta_{n+1} + 5\Delta_n i^*) - \frac{1}{5}(5i^*\Delta_{n+1} - 4\Delta_n i^*)\Delta_{n+1} + i^*\Delta_{n+1}^2.$$

But for a general metric it implies

$$P_4i^* = \frac{1}{5}P_2\left(-4i^*\bar{P}_2 + 5P_2i^*\right) - \frac{1}{5}\left(5i^*\bar{P}_2 - 4P_2i^*\right)\bar{P}_2 + i^*\bar{P}_4$$

$$= P_2^2 i^* - i^*\bar{P}_2^2 + i^*\bar{P}_4.$$

The proof is complete. $\square$

A direct calculation shows that for all n,

$$P_4 - P_2^2 = -4\delta(\mathsf{P}\#)d - \left((n-4)|\mathsf{P}|^2 + \mathsf{J}^2 - \Delta\mathsf{J}\right).$$

The latter identity can be used for a more direct proof of Lemma 6.11.6.

We continue with the discussion of the critical case of order 6 residue families, i.e., of $D_6^{\mathrm{res}}(h;\lambda)$ for (M^6, h).

Theorem 6.11.4. *For any Riemannian 6-manifold (M, h),*

$$D_6^{\mathrm{res}}(h;0) = P_6(h)i^*,$$
$$D_6^{\mathrm{res}}(h;1) = P_4(h) \circ D_2^{\mathrm{res}}(h;1),$$
$$D_6^{\mathrm{res}}(h;2) = P_2(h) \circ D_4^{\mathrm{res}}(h;2),$$
$$D_6^{\mathrm{res}}\left(h; -\frac{5}{2}\right) = D_4^{\mathrm{res}}\left(h; -\frac{9}{2}\right) \circ P_2(dr^2 + h_r).$$

Proof. The complexity of the identities increases starting with the first one. The first assertion directly follows from the definition of $D_6^{\mathrm{res}}(h;\lambda)$ using the self-adjointness of $P_6 = P_6(0)$. Definition 6.6.1 yields the formula

$$D_6^{\mathrm{res}}(\lambda) = 2^6 3! (2-\lambda)(1-\lambda)(-\lambda)$$

$$\times \left\{ \frac{1}{6!} i^* \left(\frac{\partial}{\partial r}\right)^6 + \frac{1}{4!} \left[T_2^*(\lambda) + v_2\right] i^* \left(\frac{\partial}{\partial r}\right)^4 \right.$$

$$+ \frac{1}{2!} \left[T_4^*(\lambda) + T_2^*(\lambda)v_2 + v_4\right] i^* \left(\frac{\partial}{\partial r}\right)^2 + \left. \left[T_6^*(\lambda) + T_4^*(\lambda)v_2 + T_2^*(\lambda)v_4 + v_6\right] i^*\right\}$$

with

$$T_2(\lambda) = \frac{1}{4(2-\lambda)}(\Delta - \lambda\mathsf{J})$$

and

$$T_4(\lambda) = \frac{1}{32(2-\lambda)(1-\lambda)}$$
$$\times \left[(\Delta - (\lambda+2)\mathsf{J})(\Delta - \lambda\mathsf{J}) + 4(2-\lambda)\left(-\lambda/2|\mathsf{P}|^2 - \delta(\mathsf{P}\#d) - 1/2(d\mathsf{J}, d)\right)\right].$$

Now (6.10.2) implies

$$D_6^{\mathrm{res}}(1) = -6P_4^* \circ i^* \frac{\partial^2}{\partial r^2} + \left(P_4^* \circ P_2^*(5) + 6P_4^* \circ \mathsf{J}\right) i^*$$

$$= P_4 \circ \left(-6i^* \frac{\partial^2}{\partial r^2} + (\Delta + \mathsf{J})i^*\right)$$

$$= P_4 \circ D_2^{\mathrm{res}}(1).$$

We omit the details of the proof of the third identity. It rests on the following result (for $n = 6$).

Lemma 6.11.7. *For (M^n, h) we have the factorizations*

$$P_4\left(h; \frac{n}{2}-1\right) = P_2\left(h; \frac{n}{2}+1\right) \circ P_2(h)$$

(if $n \geq 4$) and

$$P_6\left(h; \frac{n}{2}-2\right) = P_2\left(h; \frac{n}{2}+2\right) \circ P_4(h),$$
$$P_6\left(h; \frac{n}{2}-1\right) = P_4\left(h; \frac{n}{2}+1\right) \circ P_2(h)$$

(if $n \geq 6$).

Proof. The factorization of $P_4(\lambda)$ is a direct consequence of Theorem 6.9.4. For $\lambda = \frac{n}{2}-2$, (6.10.2) yields

$$P_6\left(\frac{n}{2}-2\right) = \left(\Delta - \left(\frac{n}{2}+2\right)\mathsf{J}\right) P_4\left(\frac{n}{2}-2\right) = P_2\left(\frac{n}{2}+2\right) P_4.$$

This proves the first factorization for $P_6(\lambda)$. For $\lambda = \frac{n}{2}-1$, (6.10.2) yields

$$P_6\left(\frac{n}{2}-1\right)$$
$$= -8\left[\left(\frac{n}{2}+1\right)(\log \det h)'' + \Delta'\right] P_2\left(\frac{n}{2}-1\right) + \left(\Delta - (\frac{n}{2}+3)\mathsf{J}\right) P_4\left(\frac{n}{2}-1\right).$$

But using $(\log \det h)'' = -\frac{1}{2}|\mathsf{P}|^2$ and

$$\Delta' = -\delta(\mathsf{P}\#d\cdot) - \frac{1}{2}(d\cdot, d\mathsf{J})$$

(these formulas extend the corresponding results in Section 6.10 to general dimension), the formula simplifies to

$$\left[4\left(\frac{n}{2}+1\right)|\mathsf{P}|^2 + 8\delta(\mathsf{P}\#d\cdot) + 4(d\cdot, d\mathsf{J})\right] P_2 + \left(\Delta - (\frac{n}{2}+3)\mathsf{J}\right)\left(\Delta - (\frac{n}{2}+1)\mathsf{J}\right) P_2.$$

Here we have used the factorization of $P_4(\lambda)$. Now it suffices to apply Theorem 6.9.4. $\square$

Lemma 6.11.7 is a special case of Theorem 6.11.18.

The remaining part of the proof of Theorem 6.11.4 concerns the proof of the last identity. We verify that the normal components of both sides coincide. A calculation shows that for $\lambda = -5/2$, the respective coefficients of $i^*\partial^6/\partial r^6$, $i^*\partial^4/\partial r^4$ and $i^*\partial^2/\partial r^2$ are given by the operators 21,

$$18 \cdot 35\left[1/18(\Delta + 5/2\mathsf{J}) - 1/2\mathsf{J}\right] = 35(\Delta - 13/2\mathsf{J}) \tag{6.11.19}$$

and

$$15 \left[(\Delta + 5/2J)(\Delta + 1/2J) + 18(5/4|P|^2 - \delta(P\#d) - 1/2\delta(\cdot dJ)) \right]$$
$$- 14 \cdot 15(\Delta + 5/2J)J + 63 \cdot 15(J^2 - |P|^2). \quad (6.11.20)$$

These operators are to be compared with the respective coefficients of $i^*\partial^6/\partial r^6$, $i^*\partial^4/\partial r^4$ and $i^*\partial^2/\partial r^2$ in the composition $D_4^{res}(h; -9/2) \circ P_2(dr^2 + h_r)$. For that purpose, we apply the following formula for $D_4^{res}(h; \lambda)$.

Lemma 6.11.8. *For* (M^n, h) $(n \geq 3)$,

$$D_4^{res}(\lambda) = \frac{1}{3}(n+2\lambda-4)(n+2\lambda-6)i^*\frac{\partial^4}{\partial r^4} - 2(n+2\lambda-4)(\Delta+(\lambda-2)J)i^*\frac{\partial^2}{\partial r^2}$$
$$+ \left[P_4^*(n+\lambda-4) + 2(n+2\lambda-4)(\Delta-(n+\lambda-4)J)J \right.$$
$$\left. + (n+2\lambda-4)(n+2\lambda-6)(J^2-|P|^2) \right] i^*.$$

We omit the proof. Now Remark 6.6.5 and Lemma 6.11.8 yield

$$D_4^{res}(-9/2) = 21i^*\frac{\partial^4}{\partial r^4} + 14(\Delta - 13/2J)i^*\frac{\partial^2}{\partial r^2}$$
$$+ \left[(\Delta + 5/2J)(\Delta + 1/2J) + 18 \left(5/4|P|^2 - \delta(P\#d) - 1/2\delta(\cdot dJ) \right) \right.$$
$$\left. - 14(\Delta + 5/2J)J + 63(J^2 - |P|^2) \right] i^*.$$

Using

$$P_2(dr^2 + h_r) = \frac{\partial^2}{\partial r^2} + \frac{1}{2} \text{tr}(h_r^{-1}\dot{h}_r)\frac{\partial}{\partial r} + \Delta_{h_r} - \frac{5}{2}J(dr^2 + h_r),$$

we find that the respective coefficients of $i^*(\partial/\partial r)^6$ and $i^*(\partial/\partial r)^4$ in the composition are 21 and

$$42(\partial/\partial r)\big|_{r=0} \text{tr}(h_r^{-1}\dot{h}_r) + 21(\Delta - 5/2J) + 14(\Delta - 13/2J) = 35(\Delta - 13/2J).$$

The latter result coincides with (6.11.19). Finally, it remains to compare (6.11.20) with the coefficient of $i^*(\partial/\partial r)^2$ in the composition $D_4^{res}(-9/2)P_2$, i.e., with

$$21 \left(2(\partial^3/\partial r)^3\big|_{r=0} \text{tr}(h_r^{-1}\dot{h}_r) + 6\ddot{\Delta} - 15\dddot{J} \right) + 14(\Delta - 13/2J)(\Delta - 9/2J)$$
$$+ \left[(\Delta + 5/2J)(\Delta + 1/2J) + 18 \left(5/4|P|^2 - \delta(P\#d) - 1/2\delta(\cdot dJ) \right) \right.$$
$$\left. - 14(\Delta + 5/2J)J + 63(J^2 - |P|^2) \right] i^*. \quad (6.11.21)$$

Now (6.11.20) coincides with (6.11.21) iff

$$(\Delta + 5/2\mathsf{J})(\Delta + 1/2\mathsf{J}) + 18\left(5/4|\mathsf{P}|^2 - \delta(\mathsf{P}\#d) - 1/2\delta(\cdot d\mathsf{J})\right)$$
$$- 14(\Delta + 5/2\mathsf{J})\mathsf{J} + 63(\mathsf{J}^2 - |\mathsf{P}|^2) = (\Delta - 13/2\mathsf{J})(\Delta - 9/2\mathsf{J})$$
$$+ 3\left((\partial^3/\partial r)^3\big|_{r=0}\,\mathrm{tr}(h_r^{-1}\dot{h}_r) + 3\ddot{\Delta} - 15/2\ddot{\mathsf{J}}\right). \quad (6.11.22)$$

Using the formulas

$$(\partial^3/\partial r)^3\big|_{r=0}\,\mathrm{tr}(h_r^{-1}\dot{h}_r) = -6|\mathsf{P}|^2, \quad \ddot{\mathsf{J}} = |\mathsf{P}|^2$$

(by Lemma 6.11.1 and its proof) and

$$\ddot{\Delta} = 2(\mathsf{P}, \mathrm{Hess}) + (d\mathsf{J}, d)$$

(see the proof of Lemma 6.11.3), we determine the coefficients of $\Delta^2 u$, $\mathsf{J}(\Delta u)$, $u(\Delta \mathsf{J})$, $(\mathsf{P}, \mathrm{Hess}(\mathsf{J}))$, $\mathsf{J}^2 u$, $|\mathsf{P}|^2 u$ and $(d\mathsf{J}, du)$ on both sides of (6.11.22). The results are listed in Table 6.2. It shows that all coefficients coincide.

term	left-hand side	right-hand side
$\Delta^2 u$	1	1
$u(\Delta \mathsf{J})$	$\frac{1}{2} + 9 - 14$	$-\frac{9}{2}$
$\mathsf{J}\Delta u$	$3 - 14$	-11
$(\mathsf{P}, \mathrm{Hess}(u))$	18	18
J^2	$\frac{5}{4} - 35 + 63$	$\frac{117}{4}$
$|\mathsf{P}|^2$	$\frac{45}{2} - 63$	$3(-6 - \frac{15}{2})$
$(d\mathsf{J}, du)$	$1 + 18 + 9 - 28$	$-9 + 9$

Table 6.2: Evaluation of (6.11.22)

It remains to prove the coincidence of the tangential components of both sides of the last factorization identity. That proof, in particular, involves the operator $\mathcal{T}_6(-\frac{5}{2})$. We omit the details of the direct calculation. An alternative proof of the fact that the constant terms of both operators coincide can be derived from the observation below that the factorization identities imply the correct formula for Q_6.   □

We do not discuss here the remaining two factorization identities

$$D_6^{\mathrm{res}}(h; -1/2) = i^* P_6(dr^2 + h_r),$$
$$D_6^{\mathrm{res}}(h; -3/2) = D_2^{\mathrm{res}}(h; -11/2) \circ P_4(dr^2 + h_r) \quad (6.11.23)$$

for metrics which are *not* conformally flat. These hold true for conformally flat metrics by Theorem 6.11.1.

(6.11.23) has the consequence that the five operators $P_6(h)i^*$ and

$$P_4(h) \circ D_2^{\mathrm{res}}(h;1), \qquad\qquad P_2(h) \circ D_4^{\mathrm{res}}(h;2),$$
$$D_4^{\mathrm{res}}(h;-9/2) \circ P_2(dr^2+h_r), \qquad D_2^{\mathrm{res}}(h;-11/2) \circ P_4(dr^2+h_r)$$

are linearly dependent. More precisely, $P_6(h)i^*$ can be presented as a linear combination of the last four, i.e.,

$$P_6(h)i^* = \alpha P_4(h)D_2^{\mathrm{res}}(h;1) + \beta P_2(h)D_4^{\mathrm{res}}(h;2)$$
$$+ \gamma D_4^{\mathrm{res}}(h;-9/2)P_2(dr^2+h_r) + \delta D_2^{\mathrm{res}}(h;-11/2)P_4(dr^2+h_r) \quad (6.11.24)$$

for certain rational coefficients $\alpha, \beta, \gamma, \delta$. In order to determine the numerical coefficients, it suffices to choose them in the flat case. The result is

$$(\alpha, \beta, \gamma, \delta) = \frac{1}{21}(18, -5, -4, 12).$$

In turn, the families $D_4^{\mathrm{res}}(h; \lambda)$ and $D_2^{\mathrm{res}}(h; \lambda)$ can be written as linear combinations of compositions of the Yamabe operators $P_2(h)$, $P_2(dr^2+h_r)$ and the Paneitz operator $P_4(h)$. Hence $P_6(h)i^*$ is a linear combination of compositions of the *lower order* GJMS-operators

$$P_2(h), \ P_4(h), \ P_2(dr^2+h_r) \text{ and } P_4(dr^2+h_r).$$

The explicit result is given in Theorem 6.11.17 and Corollary 6.11.6.

On the other hand, the four identities in Theorem 6.11.4 imply that the family $D_6^{\mathrm{res}}(h; \lambda)$ can be written as a linear combination of the right-hand sides, i.e., as a linear combination of compositions involving $P_2(dr^2+h_r)$, $P_2(h)$, $P_4(h)$ and $P_6(h)$.

Finally, we show how Theorem 6.11.4 can be applied to determine Q_6. The family $D_6^{\mathrm{res}}(h; \lambda)$ is a polynomial of degree 3 in λ, i.e., can be written in the form $A\lambda^3 + B\lambda^2 + C\lambda + D$ with four operator-coefficients A, B, C, D. Now, by Theorem 6.11.4, we have $D = P_6(h)i^*$. We determine Q_6 by combining the formula

$$Q_6(h) = \dot{D}_6^{\mathrm{res}}(h;0)(1) = C(h)(1)$$

with the formula for $C(1)$ which follows from the last three factorization identities. In other words, we solve the equation

$$\begin{pmatrix} (-\frac{5}{2})^3 & (\frac{5}{2})^2 & -\frac{5}{2} \\ 2^3 & 2^2 & 2 \\ 1 & 1 & 1 \end{pmatrix} \begin{pmatrix} A(1) \\ B(1) \\ C(1) \end{pmatrix} = \begin{pmatrix} D_4^{\mathrm{res}}(h;-\frac{9}{2})P_2(dr^2+h_r)(1) \\ P_2(h)D_4^{\mathrm{res}}(h;2)(1) \\ P_4(h)D_2^{\mathrm{res}}(h;1)(1) \end{pmatrix}$$

for $C(1)$. We find

$$\begin{pmatrix} A(1) \\ B(1) \\ C(1) \end{pmatrix} = -\frac{4}{315} \begin{pmatrix} * & * & * \\ * & * & * \\ 4 & \frac{175}{8} & -\frac{225}{2} \end{pmatrix} \begin{pmatrix} D_4^{\mathrm{res}}(h;-\frac{9}{2})P_2(dr^2+h_r)(1) \\ P_2(h)D_4^{\mathrm{res}}(h;2)(1) \\ P_4(h)D_2^{\mathrm{res}}(h;1)(1) \end{pmatrix}.$$

In particular, $Q_6(h)$ can be written as the linear combination

$$-\frac{16}{9 \cdot 7 \cdot 5} D_4^{\text{res}}\left(h; -\frac{9}{2}\right) P_2(dr^2 + h_r)(1)$$

$$-\frac{5}{9 \cdot 2} P_2(h) D_4^{\text{res}}(h; 2)(1) + \frac{10}{7} P_4(h) D_2^{\text{res}}(h; 1)(1). \quad (6.11.25)$$

The analogous formula for $Q_4(h)$ is (6.11.16). But the families $D_4^{\text{res}}(h; \lambda)$ and $D_2^{\text{res}}(h; \lambda)$, in turn, can be written as compositions of

$$P_4(h), \quad P_2(h) \quad \text{and} \quad P_2(dr^2 + h_r).$$

This shows that $Q_6(h)$ can be written as a composition of $P_4(h)$, $P_2(h)$ and $P_2(dr^2 + h_r)$ (acting on $u = 1$). This leads to a recursive formula for Q_6 in terms of Q_4, Q_2 and the GJMS-operators P_4, P_2 (Theorem 6.11.5).

The results of the evaluation of the terms in (6.11.25) and the resulting formula for $C(h)(1)$ are listed in Table 6.3.

term	$P_4 D_2^{\text{res}}(1)(1)$	$P_2 D_4^{\text{res}}(2)(1)$	$D_4^{\text{res}}(-\frac{9}{2})P_2(1)$	$C(1)$
$\Delta^2 \mathsf{J}$	1	2	$-\frac{5}{2}$	1
$\Delta(|\mathsf{P}|^2)$	0	-8	$(-\frac{5}{2})14$	4
$\mathsf{J}\Delta\mathsf{J}$	-5	-4	$(-\frac{5}{2})(-\frac{31}{2})$	-8
$(d\mathsf{J}, d\mathsf{J})$	0	0	0	0
$(\mathsf{P}, \text{Hess}(\mathsf{J}))$	4	0	$(-\frac{5}{2})18$	8
J^3	3	0	$(-\frac{5}{2})\frac{117}{4}$	8
$|\mathsf{P}|^2\mathsf{J}$	-2	16	$(-\frac{5}{2})(-\frac{263}{2})$	-24
$\text{tr}(\mathsf{P}^3)$	0	0	$21(-\frac{5}{2})6$	16
$(\mathcal{B}, \mathsf{P})$	0	0	$21(-\frac{5}{2})3$	8

Table 6.3: Evaluation of right-hand sides of factorization identities and $C(1)$

The result for Q_6 fits with Table 6.1 for $n = 6$.

Among the calculations that lead to Table 6.3, only those which yield the last two contributions are interesting enough to be given explicitly. These two terms arise from the contribution

$$21\left(-\frac{5}{2}\right)(\partial/\partial r)^4\big|_{r=0}\mathsf{J}(dr^2 + h_r)$$

in the composition $D_4^{\text{res}}(h; -\frac{9}{2})P_2(dr^2 + h_r)(1)$.

Lemma 6.11.9. *For* (M^6, h),

$$(\partial/\partial r)^4\big|_{r=0}\mathsf{J}(dr^2 + h_r) = 3(\mathcal{B}, \mathsf{P}) + 6\,\text{tr}(\mathsf{P}^3).$$

Proof. From the proof of Lemma 6.11.1 we recall that

$$\mathsf{J}(dr^2 + h_r) = -\frac{1}{2r}\operatorname{tr}(h_r^{-1}\dot{h}_r)$$

(see (6.11.8)). But the formula for h_r^{-1} in the proof of Lemma 6.11.1 and

$$\dot{h}_r = 2rh_{(2)} + 4r^3 h_{(4)} + 6r^5 h_{(6)} + \cdots$$

imply

$$\frac{1}{r}\operatorname{tr}(h_r^{-1}\dot{h}_r) = \cdots + r^4\Big[6\operatorname{tr}(h_0^{-1}h_{(6)}) - 4\operatorname{tr}(h_0^{-1}h_{(2)}h_0^{-1}h_{(4)})$$
$$+ 2\operatorname{tr}((h_0^{-1}h_{(2)})^3) - 2\operatorname{tr}(h_0^{-1}h_{(4)}h_0^{-1}h_{(2)})\Big] + \cdots .$$

Hence for $(\partial/\partial r)^4|_{r=0}(\mathsf{J}(dr^2 + h_r))$ we find the formula

$$4!\left(-3\operatorname{tr}(h_0^{-1}h_{(6)}) + 3\operatorname{tr}(h_0^{-1}h_{(2)}h_0^{-1}h_{(4)}) - \operatorname{tr}((h_0^{-1}h_{(2)})^3)\right)$$
$$= 4!\left(-3\operatorname{tr}(h_{(6)}^{\sharp}) + 3\operatorname{tr}(h_{(2)}^{\sharp}h_{(4)}^{\sharp}) - \operatorname{tr}((h_{(2)}^{\sharp})^3)\right).$$

Using the relation (6.9.23) and the explicit formulas for $h_{(2)}$, $h_{(4)}$ (Theorem 6.9.1), the latter sum simplifies to

$$4!\left(\operatorname{tr}(h_{(2)}^{\sharp}h_{(4)}^{\sharp}) - \frac{1}{2}\operatorname{tr}((h_{(2)}^{\sharp})^3)\right) = 4!\left(\frac{1}{8}(\mathcal{B},\mathsf{P}) + \frac{1}{4}\operatorname{tr}((\mathsf{P}^{\sharp})^3)\right).$$

The proof is complete. $\qquad\square$

Thus in $C(h)(1)$ we find the contribution

$$-\frac{4}{315}4\left(-\frac{5}{2}\right)21\left(3(\mathcal{B},\mathsf{P}) + 6\operatorname{tr}(\mathsf{P}^3)\right) = 8(\mathcal{B},\mathsf{P}) + 16\operatorname{tr}(\mathsf{P}^3).$$

These terms yield the last two terms in Table 6.3.

Next, we use (6.11.25) to derive a recursive formula for Q_6 which generalizes Corollary 6.11.1 for Q_4.

Theorem 6.11.5 (Recursive formula for Q_6). *For any Riemannian 6-manifold* (M, h),

$$Q_6 = \frac{2}{3}(P_2 Q_4 + P_4 Q_2) - \frac{5}{3}P_2^2 Q_2 + \frac{8}{3}i^*\bar{P}_2^2\bar{Q}_2,$$

where

$$\bar{P}_2(h) = P_2(dr^2 + h_r), \quad \bar{Q}_2(h) = Q_2(dr^2 + h_r),$$

and all other quantities are understood with respect to h.

Proof. First, we determine a formula for $D_4^{\mathrm{res}}(\lambda)$ in terms of Yamabe and Paneitz operators. In order to find such a formula, we use the factorization identities

$$D_4^{\mathrm{res}}(-1) = P_4 i^*, \quad D_4^{\mathrm{res}}(0) = P_2 D_2^{\mathrm{res}}(0), \quad D_4^{\mathrm{res}}(-5/2) = D_2^{\mathrm{res}}(-9/2)\bar{P}_2.$$

These are special cases of Theorem 6.11.3 for $n = 6$. We find

$$D_4^{\mathrm{res}}(\lambda) = \left[-\frac{2}{3}P_4 + 2P_2^2 - \frac{8}{3}P_2\bar{P}_2 + \frac{4}{3}\bar{P}_2^2 \right] \lambda^2$$
$$+ \left[-\frac{5}{3}P_4 + 7P_2^2 - \frac{20}{3}P_2\bar{P}_2 + \frac{4}{3}\bar{P}_2^2 \right] \lambda + \left(5P_2^2 - 4P_2\bar{P}_2 \right);$$

here we omit the restrictions i^* in order to simplify the formulas. It follows that

$$D_4^{\mathrm{res}}(2) = -6P_4 + 27P_2^2 - 28P_2\bar{P}_2 + 8\bar{P}_2^2, \tag{6.11.26}$$
$$D_4^{\mathrm{res}}(-9/2) = -6P_4 + 14P_2^2 - 28P_2\bar{P}_2 + 21\bar{P}_2^2. \tag{6.11.27}$$

Now

$$D_2^{\mathrm{res}}(\lambda) = -(2\lambda+4)i^*\bar{P}_2 + (2\lambda+5)P_2 i^*$$

and (6.11.25) imply

$$Q_6 = \left(-\frac{15}{2}P_2^3 + \frac{5}{3}P_2 P_4 + 10P_4 P_2 \right)(1)$$
$$+ \left(-\frac{124}{15}P_4 + \frac{106}{15}P_2^2 \right) i^*\bar{P}_2(1) - \frac{4}{5}P_2 i^*\bar{P}_2^2(1) - \frac{16}{15}i^*\bar{P}_2^3(1).$$

But using

$$P_2(1) = -2\mathsf{J}, \quad i^*\bar{P}_2(1) = -\frac{5}{2}\mathsf{J}, \quad P_4(1) = Q_4,$$

the latter formula yields

$$Q_6 = \frac{5}{3}P_2 Q_4 + \frac{2}{3}P_4 Q_2 - \frac{8}{3}P_2^2 Q_2 - \frac{4}{5}P_2 i^*\bar{P}_2^2(1) - \frac{16}{15}i^*\bar{P}_2^3(1).$$

Now combining this identity with the relation

$$Q_4 = P_2 Q_2 - 2i^*\bar{P}_2\bar{Q}_2 = P_2 Q_2 + \frac{4}{5}i^*\bar{P}_2^2(1)$$

(Lemma 6.11.5 for $n = 6$), we find

$$Q_6 = \frac{5}{3}P_2 Q_4 + \frac{2}{3}P_4 Q_2 - \frac{8}{3}P_2^2 Q_2 + \left[P_2^2 Q_2 - P_2 Q_4 \right] + \frac{8}{3}i^*\bar{P}_2^2\bar{Q}_2.$$

This yields the assertion. $\square$

In Theorem 6.11.5, the term $i^* \bar{P}_2 \bar{Q}_2$ is responsible for the contribution $(\mathcal{B}, \mathsf{P})$ to Q_6.

Example 6.11.1. *Let $n = 6$ and h be Einstein. Then*

$$h_r = (1 - cr^2)^2 h, \quad c = \mathsf{J}(h)/2$$

(*Theorem 6.16.1*) *and*

$$\mathsf{J}(dr^2 + h_r) = -\frac{1}{2r} \operatorname{tr}(h_r^{-1} \dot{h}_r) = \mathsf{J}(1 - cr^2)^{-1}$$

(*by (6.11.8)*). *Moreover, for the Yamabe operator $\bar{P}_2$ we find*

$$\bar{P}_2(dr^2 + h_r) = \frac{\partial^2}{\partial r^2} - \mathsf{J}r(1 - cr^2)^{-1} \frac{\partial}{\partial r} + (1 - cr^2)^{-2} \Delta - \frac{5}{2} \mathsf{J}(1 - cr^2)^{-1}.$$

Now on (M^6, h), GJMS-operators of order up to 6 are given by

$$P_2 = \Delta - 2\mathsf{J}, \quad P_4 = (\Delta - 2\mathsf{J})(\Delta - 4/3\mathsf{J}), \quad P_6 = (\Delta - 2\mathsf{J})(\Delta - 4/3\mathsf{J})\Delta.$$

Hence

$$Q_2 = \mathsf{J}, \quad Q_4 = \frac{8}{3} \mathsf{J}^2 \quad and \quad Q_6 = 5! \left(\frac{\mathsf{J}}{3}\right)^3 = \frac{40}{9} \mathsf{J}^3$$

(*see also (6.16.1)*). *Some calculations show that*

$$\bar{P}_2^2(1) = \frac{5}{2}(28c - 12c^2 r^2)(1 - cr^2)^{-3}$$

and hence

$$i^* \bar{P}_2^2 \bar{Q}_2 = \frac{29}{6} \mathsf{J}^3.$$

Thus the recursive formula for Q_6 reads

$$Q_6 = \frac{2}{3}\left(-\frac{16}{3} + \frac{8}{3}\right) \mathsf{J}^3 - \frac{5}{3} 4\mathsf{J}^3 + \frac{8}{3} \frac{29}{6} \mathsf{J}^3 = -\frac{16}{9} \mathsf{J}^3 - \frac{60}{9} \mathsf{J}^3 + \frac{116}{9} \mathsf{J}^3 = \frac{40}{9} \mathsf{J}^3.$$

Next, recall that by Lemma 6.11.5 the formula

$$Q_4 = P_2 Q_2 - 2i^* \bar{P}_2 \bar{Q}_2$$

(Corollary 6.11.1) for the critical Q-curvature of a Riemannian four-manifold (M, h) continues to hold true in all dimensions $n \geq 4$. The following result claims an analogous effect for the recursive formula for Q_6.

Theorem 6.11.6. *For a Riemannian manifold (M, h) of dimension $n \geq 6$,*

$$Q_6 = \frac{2}{3}(P_2 Q_4 + P_4 Q_2) - \frac{5}{3} P_2^2 Q_2 + \frac{8}{3} i^* \bar{P}_2^2 \bar{Q}_2. \tag{6.11.28}$$

term	contribution in $Q_{6,n}$
$\Delta(|P|^2)$	4
$\Delta^2 J$	1
$J\Delta J$	$\frac{2-3n}{2}$
(dJ, dJ)	$6 - n$
$(P, \text{Hess}(J))$	8
J^3	$\frac{n^2-4}{4}$
$\text{tr}(P^3)$	16
$|P|^2 J$	$-4n$
$(\mathcal{B}, P)$	$\frac{16}{n-4}$

Table 6.4: The structure of $Q_{6,n}$

We emphasize that the numerical coefficients in (6.11.28) do *not* depend on the dimension of M. We rephrase this observation by saying that (6.11.28) is *universal*.

Proof. We prove (6.11.28) by direct calculation of the right-hand side and comparison with Table 6.1, i.e., with the equivalent Table 6.4.

The right-hand side of (6.11.28) is the sum of

$$\frac{2}{3}\left(\Delta - \frac{n-2}{2}J\right)\left(\frac{n}{2}J^2 - 2|P|^2 - \Delta J\right)$$
$$+ \frac{2}{3}\left(\Delta^2 + \delta((n-2)Jh - 4P)\#dJ + \frac{n-4}{2}\left(\frac{n}{2}J^2 - 2|P|^2 - \Delta J\right)\right)J$$
$$- \frac{5}{3}\left(\Delta - \frac{n-2}{2}J\right)^2 J$$

and

$$\frac{8}{3}\Big(J^{(4)} - 2J\ddot{J} + \Delta\ddot{J} + \ddot{\Delta}J - (n-1)J\ddot{J} + \Delta\ddot{J} + \Delta^2 J - \frac{n-1}{2}\Delta(J^2)$$
$$- \frac{n-1}{2}J\ddot{J} - \frac{n-1}{2}J\Delta J + \left(\frac{n-1}{2}\right)^2 J^3\Big).$$

We evaluate these terms by using

$$J^{(4)} = 4!\left(\text{tr}(h_{(2)}h_{(4)}) - \frac{1}{2}\text{tr}(h_{(2)}^3)\right) = 3!\left(\frac{1}{n-4}(\mathcal{B}, P) + \text{tr}(P^3)\right) \qquad (6.11.29)$$

(extending Lemma 6.11.9), $\ddot{J} = |P|^2$ and $\ddot{A}u = (dJ, du) + 2(P, \text{Hess}(u))$. We find that the respective coefficients of $J\Delta J$ and (dJ, dJ) are given by

$$\frac{2}{3}\frac{3n-2}{2} - \frac{2}{3}\frac{3n-8}{2} + \frac{5}{3}\frac{3}{2}(n-2) - \frac{8}{3}\frac{3}{2}(n-1) = 1 - \frac{3n}{2}$$

and

$$\frac{2}{3}n - \frac{2}{3}(n-6) + \frac{5}{3}(n-2) - \frac{8}{3}(n-2) = 6 - n.$$

Next, the respective coefficients of J^3 and $|P|^2 J$ are

$$-\frac{2}{3}\frac{n(n-2)}{4} + \frac{2}{3}\frac{n(n-4)}{4} - \frac{5}{3}\left(\frac{n-2}{2}\right)^2 + \frac{8}{3}\left(\frac{n-1}{2}\right)^2 = \frac{n^2-4}{4}$$

and

$$\frac{2}{3}(n-2) - \frac{2}{3}(n-4) - \frac{8}{3}\frac{3n+1}{2} = -4n.$$

Finally, (6.11.29) yields the correct contributions of $\text{tr}(P^3)$ and $(\mathcal{B}, P)$. We omit the details concerning the coefficients of $\Delta(|P|^2)$, $\Delta^2 J$ and $(P, \text{Hess}(J))$. $\square$

Combining Theorem 6.11.6 with (6.11.18), yields a formula for $Q_{6,n}$ which contains the Paneitz operator $\bar{P}_4$ instead of $\bar{P}_2^2$.

Corollary 6.11.2. $Q_6 = \frac{2}{3}P_2 Q_4 + (P_2^2 - 2P_4)Q_2 + \frac{8}{3}i^* \bar{P}_4 \bar{Q}_2$.

Now we generalize. The method in the proof of Theorem 6.11.5 is expected to generate the following recursive formula for Q_n in terms of lower order Q-curvatures, lower order GJMS-operators *and* powers of the Yamabe operator $\bar{P}_2$.

Conjecture 6.11.1 (Recursive formula for Q_n). *For a conformally flat metric, the critical Q-curvature Q_n admits a representation*

$$Q_n = \sum_I a_I^{(n)} P_{2I}(Q_{n-2|I|}) + (-1)^{\frac{n}{2}-1} \frac{(n-2)(n-4)\cdots 2}{(n-3)(n-5)\cdots 1} i^* \bar{P}_2^{\frac{n}{2}-1}(\bar{Q}_2) \quad (6.11.30)$$

with rational coefficients $a_I^{(n)}$. In (6.11.30) the sum runs over all ordered partitions I of integers in the interval $[1, \frac{n}{2} - 1]$ as sums of natural numbers. For $I = (I_1, \ldots, I_m)$, the operator P_{2I} is defined as the composition $P_{2I_1} \circ \cdots \circ P_{2I_m}$ of GJMS-operators and $|I| = \sum_i I_i$.

A number of comments are in order.

An additional important feature of the representation (6.11.30) is its conjectural *universality*. It states that in the critical case, i.e., for the critical Q-curvature Q_n in dimension n, the formula is also valid in all higher even dimensions, i.e., represents the subcritical Q-curvature Q_n in even dimensions $\geq n + 2$. Here the point is that the coefficients a_I depend only on the order of the curvature invariant but not on the dimension of the underlying space. This extends the observed universality in Lemma 6.11.5 and Theorem 6.11.6.

The proof of Theorem 6.11.5 uses the universality of the recursive formula $Q_4 = P_2 Q_2 - 2i^* \bar{P}_2 \bar{Q}_2$ (in dimension 6) in order to replace the term $\bar{P}_2^2(1)$. In a similar way, universality plays a central role in the derivation of the representations (6.11.30) for all n.

Theorem 6.11.6 supports the expectation that finally the assumption of conformal flatness can be removed from the scene. It appears here since the factorization identities which involve the factor $\bar{P}_2$ are (presently) not known for general metrics.

We count the number of terms in (6.11.30). The number of solutions of the equation

$$I_1 + \cdots + I_r = m, \ m \in \mathbb{N}_0 \tag{6.11.31}$$

with $I_i \in \mathbb{N}_0$ is

$$\binom{m+r-1}{r-1}.$$

Hence the number of solutions of (6.11.31) with $m \geq r$ and $I_i \in \mathbb{N}$ is

$$\binom{m-r+r-1}{r-1} = \binom{m-1}{r-1}.$$

In particular, $m \in [1, \frac{n}{2}-1]$ can be written in

$$\binom{m-1}{0} + \binom{m-1}{1} + \cdots + \binom{m-1}{m-1} = 2^{m-1}$$

ways as the sum of natural numbers (here we take the order into account). Thus there are

$$\left(2^0 + 2^1 + \cdots + 2^{\frac{n}{2}-2} \right) + 1 = 2^{\frac{n}{2}-1}$$

terms on the right-hand side of (6.11.30).

The coefficient of the term $i^* \bar{P}_2^{\frac{n}{2}-1}(\bar{Q}_2)$ can be determined as follows by tracing the contribution of $\mathrm{tr}(\mathsf{P}^{\frac{n}{2}})$. On the one hand, using Lemma 6.14.1, we determine the coefficient of $\mathrm{tr}(\mathsf{P}^{\frac{n}{2}})$ in the holographic formula for Q_n. By Newton's formula,

$$\mathrm{tr}(\wedge^m T) = \frac{1}{m!} \left(\mathrm{tr}(T)^m + \cdots + (m-1)!(-1)^{m-1} \mathrm{tr}(T^m) \right).$$

Thus we find

$$\left\{ 2^{\frac{n}{2}-1} \left(\frac{n}{2} \right)! \left(\frac{n}{2}-1 \right)! \right\} \frac{1}{\left(\frac{n}{2} \right)!} (-1)^{\frac{n}{2}-1} \left(\frac{n}{2}-1 \right)! = (-2)^{\frac{n}{2}-1} \left(\frac{n}{2}-1 \right)!^2 \tag{6.11.32}$$

for this coefficient. The latter result is confirmed for $n \leq 6$ by the explicit formulas for Q_n. On the other hand, $\bar{P}_2^{\frac{n}{2}-1}(\bar{Q}_2)$ contains a multiple of $\mathrm{tr}(\mathsf{P}^{\frac{n}{2}})$ in

$$i^* (\partial/\partial r)^{n-2} \mathsf{J}(dr^2 + h_r).$$

Similarly as in the proof of Lemma 6.11.9, we find that this coefficient is

$$\frac{(n-2)!}{2^{\frac{n}{2}-1}}.$$

Now

$$(-1)^{\frac{n}{2}-1}\frac{(n-2)(n-4)\cdots 2\,(n-2)!}{(n-3)(n-5)\cdots 1\;2^{\frac{n}{2}-1}} = (-2)^{\frac{n}{2}-1}\left(\frac{n}{2}-1\right)!^2$$

confirms the coefficient of $\bar{P}_2^{\frac{n}{2}-1}(\bar{Q}_2)$ in (6.11.30).

It is clearly of interest to make the rational coefficients a_I more explicit. This problem is discussed in [95].

The fact that the $2N$ right-hand sides of the system of factorization identities of $D_{2N}^{\mathrm{res}}(\lambda)$ satisfy $N-1$ linear relations can be used to replace $\bar{P}_2^{\frac{n}{2}-1}(\bar{Q}_2)$ in (6.11.30) by similar compositions which involve higher order GJMS-operators of dr^2+h_r. For Q_6, such a result is given in Corollary 6.11.2.

As explained above, the representation (6.11.30) of the critical Q-curvature follows by combining the relation

$$\dot{D}_n^{\mathrm{res}}(0)(1) = -(-1)^{\frac{n}{2}}Q_n$$

with recursive formulas for subcritical Q-curvatures. It is natural to ask for a similar device to derive recursive formulas in the subcritical case. Now the holographic formula in Conjecture 6.9.1 actually may be regarded as pointing to the existence of a relation between the quantities

$$Q_{2N} \quad \text{and} \quad \dot{D}_{2N}^{\mathrm{res}}\left(-\frac{n}{2}+N\right)(1).$$

Although for $2N < n$, the value $\dot{D}_{2N}^{\mathrm{res}}(-\frac{n}{2}+N)(1)$ does *not* coincide with Q_{2N}, Theorem 6.11.7 establishes such relations for $N = 2$ and $N = 3$. The proof shows that the relations are *equivalent* to the respective recursive formulas (see Remark 6.11.3). On the other hand, the relations in Theorem 6.11.7 can be shown to be equivalent to the corresponding holographic formula in Conjecture 6.9.1.

Theorem 6.11.7 will be used to derive explicit formulas for the Q-polynomials $Q_4^{\mathrm{res}}(\lambda)$ and $Q_6^{\mathrm{res}}(\lambda)$ which generalize the holographic formulas for the respective Q-curvatures Q_4 and Q_6.

Theorem 6.11.7. *For any Riemannian n-manifold (M, h),*

$$\dot{D}_4^{\mathrm{res}}\left(h; -\frac{n}{2}+2\right)(1) = -Q_{4,n} - \frac{n-4}{2}(Q_{4,n}+P_2Q_{2,n})$$

for $n \geq 4$, and

$$\dot{D}_6^{\mathrm{res}}\left(h; -\frac{n}{2}+3\right)(1) = Q_{6,n} + \frac{n-6}{2}\left(\frac{3}{2}Q_{6,n}+P_2Q_{4,n}-2P_4Q_{2,n}+\frac{3}{2}P_2^2Q_{2,n}\right)$$

for $n \geq 6$.

Theorem 6.11.7 shows that for $N \leq 3$ the linear term of the polynomial $D_{2N}^{\mathrm{res}}(h; \lambda)(1)$ is a *combination* of $Q_{2N,n}$ and contributions in terms of lower order Q-curvatures and GJMS-operators. In particular, it confirms that in the respective critical cases

$$\dot{D}_4^{\mathrm{res}}(0)(1) = -Q_4 \quad \text{and} \quad \dot{D}_6^{\mathrm{res}}(0)(1) = Q_6.$$

The general picture will be discussed below.

Proof. The factorization identities

$$D_4^{\mathrm{res}}\left(h; -\frac{n}{2}+2\right) = P_4(h)i^*,$$

$$D_4^{\mathrm{res}}\left(h; -\frac{n}{2}+3\right) = P_2(h) \circ D_2^{\mathrm{res}}\left(h; -\frac{n}{2}+3\right),$$

$$D_4^{\mathrm{res}}\left(h; -\frac{n-1}{2}\right) = D_2^{\mathrm{res}}\left(h; -\frac{n+3}{2}\right) \circ P_2(dr^2+h_r)$$

(Theorem 6.11.3) imply that the coefficients of

$$D_4^{\mathrm{res}}(h; \lambda) = A\left(\lambda+\frac{n}{2}-2\right)^2 + B\left(\lambda+\frac{n}{2}-2\right) + C$$

are determined by

$$\begin{pmatrix} A \\ B \\ C \end{pmatrix} = \begin{pmatrix} \frac{4}{15} & \frac{2}{5} & -\frac{2}{3} \\ -\frac{4}{15} & \frac{3}{5} & -\frac{1}{3} \\ 0 & 0 & 1 \end{pmatrix} \begin{pmatrix} D_2^{\mathrm{res}}\left(-\frac{n+3}{2}\right)\bar{P}_2 \\ P_2 D_2^{\mathrm{res}}\left(-\frac{n}{2}+3\right) \\ P_4 i^* \end{pmatrix}.$$

Here the arguments are analogous to those which yield (6.11.16). Hence

$$\dot{D}_4^{\mathrm{res}}\left(-\frac{n}{2}+2\right)(1) = B(1)$$

$$= -\frac{1}{3}P_4(1) + \frac{3}{5}P_2 D_2^{\mathrm{res}}\left(-\frac{n}{2}+3\right)(1) - \frac{4}{15}D_2^{\mathrm{res}}\left(-\frac{n+3}{2}\right)\bar{P}_2(1).$$

But using

$$D_2^{\mathrm{res}}\left(-\frac{n}{2}+3\right) = -4i^*\bar{P}_2 + 5P_2 i^* \quad \text{and} \quad D_2^{\mathrm{res}}\left(-\frac{n+3}{2}\right) = 5i^*\bar{P}_2 - 4P_2 i^*,$$

we find

$$\dot{D}_4^{\mathrm{res}}\left(-\frac{n}{2}+2\right)(1) = -\frac{1}{3}P_4(1) + 3P_2^2(1) - \frac{4}{3}\bar{P}_2^2(1) - \frac{4}{3}P_2 i^*\bar{P}_2(1)$$

$$= -\frac{1}{3}\frac{n-4}{2}Q_4 - 3\frac{n-2}{2}P_2 Q_2 + \frac{4}{3}\frac{n-1}{2}i^*\bar{P}_2\bar{Q}_2 + \frac{4}{3}\frac{n-1}{2}P_2 i^*\bar{Q}_2.$$

Thus, in view of $i^*\bar{Q}_2 = Q_2$ (Lemma 6.11.1), we obtain

$$\dot{D}_4^{\text{res}}\left(-\frac{n}{2}+2\right)(1) = -\frac{1}{3}\frac{n-4}{2}Q_4 - \frac{5n-14}{6}P_2Q_2 + \frac{2n-2}{3}i^*\bar{P}_2\bar{Q}_2.$$

Next, we apply Lemma 6.11.5, i.e.,

$$Q_4 = P_2Q_2 - 2i^*\bar{P}_2\bar{Q}_2,$$

and find

$$\dot{D}_4^{\text{res}}\left(-\frac{n}{2}+2\right)(1) = -\frac{n-4}{6}Q_4 - \frac{5n-14}{6}P_2Q_2 + \frac{n-1}{3}(P_2Q_2 - Q_4)$$
$$= -\frac{n-2}{2}Q_4 - \frac{n-4}{2}P_2Q_2.$$

This proves the first assertion.

In order to prove the second identity, we first observe that

$$D_6^{\text{res}}(\lambda) = A\left(\lambda+\frac{n}{2}-3\right)^3 + B\left(\lambda+\frac{n}{2}-3\right)^2 + C\left(\lambda+\frac{n}{2}-3\right) + D$$

with

$$\begin{pmatrix} A \\ B \\ C \\ D \end{pmatrix} = -\frac{4}{315}\begin{pmatrix} 2 & -\frac{35}{4} & \frac{45}{2} & -\frac{63}{4} \\ -6 & -\frac{105}{8} & \frac{45}{4} & \frac{63}{8} \\ 4 & \frac{175}{8} & -\frac{225}{2} & \frac{693}{8} \\ 0 & 0 & 0 & -\frac{315}{4} \end{pmatrix}\begin{pmatrix} D_4^{\text{res}}(-\frac{n+3}{2})\bar{P}_2 \\ P_2 D_4^{\text{res}}\left(-\frac{n}{2}+5\right) \\ P_4 D_2^{\text{res}}\left(-\frac{n}{2}+4\right) \\ P_6 i^* \end{pmatrix}.$$

The latter formula follows from an extension to general dimension of the four factorization identities in Theorem 6.11.4. Hence for

$$\dot{D}_6^{\text{res}}\left(-\frac{n}{2}+3\right)(1) = C(1),$$

we find

$$-\frac{16}{9\cdot7\cdot5}D_4^{\text{res}}\left(-\frac{n+3}{2}\right)\bar{P}_2(1)$$
$$-\frac{5}{9\cdot2}P_2 D_4^{\text{res}}\left(-\frac{n}{2}+5\right)(1) + \frac{10}{7}P_4 D_2^{\text{res}}\left(-\frac{n}{2}+4\right)(1) - \frac{11}{10}P_6(1).$$

The same arguments as in the proof of Theorem 6.11.5 simplify the sum to

$$\left(-\frac{15}{2}P_2^3 + \frac{5}{3}P_2P_4 + 10P_4P_2\right)(1)$$
$$+\left(-\frac{124}{15}P_4 + \frac{106}{15}P_2^2\right)i^*\bar{P}_2(1) - \frac{4}{5}P_2i^*\bar{P}_2^2(1) - \frac{16}{15}i^*\bar{P}_2^3(1) - \frac{11}{10}P_6(1).$$

Now using

$$P_2(1) = -\frac{n-2}{2}Q_2, \quad i^*\bar{P}_2(1) = -\frac{n-1}{2}i^*\bar{Q}_2 = -\frac{n-1}{2}Q_2, \quad P_4(1) = \frac{n-4}{2}Q_4,$$

the latter sum equals

$$\frac{5(n-4)}{6}P_2Q_4 - \frac{13n-88}{15}P_4Q_2 - \left(\frac{106(n-1)}{30} - \frac{15(n-2)}{4}\right)P_2^2Q_2$$
$$- \frac{4}{5}P_2i^*\bar{P}_2^2(1) + \frac{16(n-1)}{30}i^*\bar{P}_2^2\bar{Q}_2 + \frac{11}{10}\frac{n-6}{2}Q_6.$$

But the relation

$$i^*\bar{P}_2^2(1) = \frac{n-1}{2}i^*\bar{P}_2\bar{Q}_2 = -\frac{n-1}{4}(-P_2Q_2 + Q_4)$$

(Lemma 6.11.5) shows that the sum can be written in the form

$$\frac{11}{10}\frac{n-6}{2}Q_6 + \frac{n-1}{15}\left(2P_2Q_4 + 2P_4Q_2 - 5P_2^2Q_2 + 8i^*\bar{P}_2^2\bar{Q}_2\right)$$
$$+ \frac{n-6}{2}\left(P_2Q_4 - 2P_4Q_2 + \frac{3}{2}P_2^2Q_2\right).$$

Now Theorem 6.11.6 completes the proof.   □

Remark 6.11.3. *The proof of Theorem 6.11.7 also shows that the recursive formula*

$$Q_4 = P_2Q_2 - 2i^*\bar{P}_2\bar{Q}_2$$

(in any dimension $n \geq 3$) follows from the first assertion of Theorem 6.11.7 and the factorization identities (used to evaluate $\dot{D}_4^{\mathrm{res}}(-\frac{n}{2}+2)(1)$). A similar comment concerns the recursive formula (6.11.28) for Q_6.

Now we apply Theorem 6.11.7 in order to derive explicit formulas for the Q-polynomials

$$Q_{2N}^{\mathrm{res}}(h;\lambda) = -(-1)^N D_{2N}^{\mathrm{res}}(h;\lambda)(1)$$

for $N = 2$ and $N = 3$ in terms of Q-curvatures and GJMS-operators. We recall that

$$Q_{2N}^{\mathrm{res}}(\lambda) = -(-1)^N 2^{2N} N! \left[\left(-\frac{n}{2}-\lambda+2N-1\right)\cdots\left(-\frac{n}{2}-\lambda+N\right)\right]$$
$$\times [T_{2N}^*(\lambda+n-2N)(v_0) + \cdots + T_0^*(\lambda+n-2N)(v_{2N})]$$

(Definition 6.6.3). In particular, the critical Q-polynomial is given by

$$Q_n^{\mathrm{res}}(\lambda) = -2^n\left(\frac{n}{2}\right)!\left[\left(\lambda-\frac{n}{2}+1\right)\cdots\lambda\right][T_n^*(\lambda)(v_0) + \cdots + T_0^*(\lambda)(v_n)].$$

Theorem 6.11.8. *For all metrics,*

$$Q_4^{\mathrm{res}}(\lambda) = -\lambda\left(\lambda+\frac{n}{2}-3\right)Q_4 - \lambda\left(\lambda+\frac{n}{2}-2\right)P_2Q_2 \qquad (6.11.33)$$

for even $n \geq 4$, and

$$Q_6^{\mathrm{res}}(\lambda) = \frac{1}{2}\lambda\left(\lambda+\frac{n}{2}-4\right)\left(\lambda+\frac{n}{2}-5\right)Q_6$$

$$+ \lambda\left(\lambda+\frac{n}{2}-3\right)\left(\lambda+\frac{n}{2}-4\right)P_2\left(Q_4 + \frac{3}{2}P_2Q_2\right)$$

$$- \lambda\left(\lambda+\frac{n}{2}-3\right)\left(\lambda+\frac{n}{2}-5\right)P_4Q_2 \qquad (6.11.34)$$

for even $n \geq 6$.

Proof. The proof of the first assertion rests on the characterizing properties

$$Q_4^{\mathrm{res}}\left(-\frac{n}{2}+2\right) = -P_4(1) = -\frac{n-4}{2}Q_4,$$

$$Q_4^{\mathrm{res}}\left(-\frac{n}{2}+3\right) = -P_2 D_2^{\mathrm{res}}\left(-\frac{n}{2}+3\right)(1),$$

$$\dot{Q}_4^{\mathrm{res}}\left(-\frac{n}{2}+2\right) = -\dot{D}_4^{\mathrm{res}}\left(-\frac{n}{2}+2\right)(1)$$

of the Q-polynomial. We verify that the polynomial on the right-hand side of (6.11.33), in fact, has these three properties. The first identity is obvious. The value of that polynomial at $\lambda = -\frac{n}{2}+3$ is

$$\left(\frac{n}{2}-3\right)P_2Q_2.$$

But

$$-D_2^{\mathrm{res}}\left(-\frac{n}{2}+3\right)(1) = 4i^*\bar{P}_2(1) - 5P_2(1) = -4\frac{n-1}{2}Q_2 + 5\frac{n-2}{2}Q_2 = \left(\frac{n}{2}-3\right)Q_2.$$

This proves the second property. The third property is a consequence of Theorem 6.11.7.

Similarly, the proof of the formula for $Q_6^{\mathrm{res}}(\lambda)$ rests on the characterizing properties

$$Q_6^{\mathrm{res}}\left(-\frac{n}{2}+3\right) = -\frac{n-6}{2}Q_6,$$

$$Q_6^{\mathrm{res}}\left(-\frac{n}{2}+4\right) = P_4 D_2^{\mathrm{res}}\left(-\frac{n}{2}+4\right)(1),$$

$$Q_6^{\mathrm{res}}\left(-\frac{n}{2}+5\right) = P_2 D_4^{\mathrm{res}}\left(-\frac{n}{2}+5\right)(1),$$

$$\dot{Q}_6^{\mathrm{res}}\left(-\frac{n}{2}+3\right) = \dot{D}_6^{\mathrm{res}}\left(-\frac{n}{2}+3\right)(1)$$

of the Q-polynomial. We verify that the polynomial on the right-hand side of (6.11.34), in fact, has these four properties. The first property is obvious. The last property follows from Theorem 6.11.7. The remaining two identities assert that

$$-\left(\frac{n}{2}-4\right)Q_2 = D_2^{\mathrm{res}}\left(-\frac{n}{2}+4\right)(1),$$

$$-\left(\frac{n}{2}-5\right)2\left(Q_4+\frac{3}{2}Q_2\right) = D_4^{\mathrm{res}}\left(-\frac{n}{2}+5\right)(1),$$

respectively. But

$$D_2^{\mathrm{res}}\left(-\frac{n}{2}+4\right)(1) = -6i^*\bar{P}_2(1)+7P_2(1) = 6\frac{n-1}{2}i^*\bar{Q}_2 - 7\frac{n-2}{2}Q_2 = -\left(\frac{n}{2}-4\right)Q_2$$

and

$$D_4^{\mathrm{res}}\left(-\frac{n}{2}+5\right)(1) = -Q_4^{\mathrm{res}}\left(-\frac{n}{2}+5\right)(1) = -\left(\frac{n}{2}-5\right)(2Q_4+3P_2Q_2)$$

by (6.11.33). The proof is complete.   □

Note that Example 6.6.7 is the special case $n = 4$ of (6.11.33).

The formula for the polynomial $Q_4^{\mathrm{res}}(\lambda)$ has an interesting consequence.

Corollary 6.11.3. *In all dimensions,*

$$i^*\bar{Q}_4 = \frac{3}{2}Q_4 + \frac{1}{2}P_2Q_2 \quad and \quad i^*\bar{P}_2\bar{Q}_2 = \frac{1}{2}P_2Q_2 - \frac{1}{2}Q_4.$$

Proof. The identity

$$D_4^{\mathrm{res}}\left(-\frac{n-3}{2}\right) = i^*\bar{P}_4$$

(Theorem 6.11.3) implies

$$-Q_4^{\mathrm{res}}\left(-\frac{n-3}{2}\right) = \frac{n-3}{2}i^*\bar{Q}_4.$$

But (6.11.33) yields

$$-Q_4^{\mathrm{res}}\left(-\frac{n-3}{2}\right) = \frac{n-3}{2}\frac{3}{2}Q_4 + \frac{n-3}{2}\frac{1}{2}P_2Q_2.$$

Hence

$$i^*\bar{Q}_4 = \frac{3}{2}Q_4 + \frac{1}{2}P_2Q_2.$$

Similarly, the second assertion is a consequence of the factorization identity

$$D_4^{\mathrm{res}}\left(-\frac{n-1}{2}\right) = D_2^{\mathrm{res}}\left(-\frac{n-3}{2}\right)\bar{P}_2$$

(Theorem 6.11.3). In fact, we get

$$Q_4^{\mathrm{res}}\left(-\frac{n-1}{2}\right) = D_2^{\mathrm{res}}\left(-\frac{n-3}{2}\right)\frac{n-1}{2}\bar{Q}_2$$

and (6.11.33) implies the assertion. $\qquad$ □

Note that the first identity in Corollary 6.11.3 already appeared in the proof of Lemma 6.11.4 in the case $n = 4$. The second identity coincides with Lemma 6.11.5.

For the convenience of later reference, we summarize the critical cases of Theorem 6.11.8.

Theorem 6.11.9. *For (M^4, h), the critical Q-polynomial $Q_4^{\mathrm{res}}(h; \lambda)$ is given by*

$$Q_4^{\mathrm{res}}(\lambda) = -\lambda(\lambda-1)Q_4 - \lambda^2 P_2 Q_2.$$

Similarly, for (M^6, h) the critical Q-polynomial $Q_6^{\mathrm{res}}(h; \lambda)$ is given by

$$Q_6^{\mathrm{res}}(\lambda) = \frac{1}{2}\lambda(\lambda-1)(\lambda-2)Q_6 + \lambda^2(\lambda-1)P_2\left(Q_4 + \frac{3}{2}P_2 Q_2\right) - \lambda^2(\lambda-2)P_4 Q_2.$$

If h is Einstein, the formulas for the critical Q-polynomials $Q_n^{\mathrm{res}}(h; \lambda)$ $(n \leq 6)$ further simplify.

Corollary 6.11.4. *Let h be Einstein. Then the respective critical Q-polynomials factorize as*

$$Q_2^{\mathrm{res}}(h; \lambda) = \lambda Q_2(h),$$

$$Q_4^{\mathrm{res}}(h; \lambda) = -\frac{\lambda(\lambda-3)}{3}Q_4(h),$$

$$Q_6^{\mathrm{res}}(h; \lambda) = \frac{\lambda(\lambda-4)(\lambda-5)}{4\cdot 5}Q_6(h).$$

Proof. The first assertion is trivial by Example 6.6.6. For the proof of the second assertion we use

$$P_2 = \Delta - \mathsf{J}, \quad Q_4 = 3!\left(\frac{\mathsf{J}}{2}\right)^2 = \frac{3}{2}\mathsf{J}^2$$

(Theorem 3.2.3). Hence

$$Q_4^{\mathrm{res}}(\lambda) = -\frac{3}{2}\lambda(\lambda-1)\mathsf{J}^2 + \lambda^2\mathsf{J}^2 = -\frac{1}{2}\lambda(\lambda-3)\mathsf{J}^2.$$

Similarly, for the proof of the last assertion we apply

$$P_2 = \Delta - 2\mathsf{J}, \quad P_4 = (\Delta - 2\mathsf{J})(\Delta - 4/3\mathsf{J})$$

and

$$Q_4 = \frac{8}{3}\mathsf{J}^2, \quad Q_6 = 5!\left(\frac{\mathsf{J}}{3}\right)^3 = \frac{40}{9}\mathsf{J}^3.$$

Hence

$$Q_6^{\text{res}}(\lambda) = \frac{20}{9}\lambda(\lambda-1)(\lambda-2)\mathsf{J}^3 + \frac{2}{3}\lambda^2(\lambda-1)\mathsf{J}^3 - \frac{8}{3}\lambda^2(\lambda-2)\mathsf{J}^3$$
$$= \frac{2}{9}\lambda(\lambda-4)(\lambda-5)\mathsf{J}^3.$$

The proof is complete.   □

It is open whether Corollary 6.11.4 extends to general even dimensions.

Moreover, the above techniques yield the following generalization of Theorem 6.11.9 for conformally flat metrics.

Theorem 6.11.10. *Let (M, h) be conformally flat of even dimension n. Assume that*

$$Q_{2j}^{\text{res}}(0) = 0 \quad for \quad j = 1, \dots, \frac{n}{2}-1, \tag{6.11.35}$$

and define the polynomials $\mathcal{Q}_{2j}^{\text{res}}(\lambda)$ by

$$Q_{2j}^{\text{res}}(\lambda) = \lambda \mathcal{Q}_{2j}^{\text{res}}(\lambda).$$

Then

$$Q_n^{\text{res}}(\lambda) = (-1)^{\frac{n}{2}-1}\lambda \prod_{k=1}^{\frac{n}{2}-1}\left(\frac{\lambda - \frac{n}{2} + k}{k}\right)Q_n$$
$$+ \lambda \sum_{j=1}^{\frac{n}{2}-1}(-1)^j \prod_{\substack{k=1 \\ k\neq j}}^{\frac{n}{2}}\left(\frac{\lambda - \frac{n}{2} + k}{k - j}\right)P_{2j}\mathcal{Q}_{n-2j}^{\text{res}}\left(\frac{n}{2}-j\right). \tag{6.11.36}$$

Proof. The $\frac{n}{2}+1$ relations

$$Q_n^{\text{res}}\left(\frac{n}{2}-j\right) = (-1)^j P_{2j}Q_{n-2j}^{\text{res}}\left(\frac{n}{2}-j\right), \quad j = 1, \dots, \frac{n}{2}$$

and

$$\dot{Q}_n^{\text{res}}(0) = -(-1)^{\frac{n}{2}}\dot{D}_n^{\text{res}}(0)(1)$$

characterize the Q-polynomial $Q_n^{\text{res}}(\lambda)$. We verify that the given polynomial of degree $\frac{n}{2}$ has these properties. The definition immediately implies that for $1 \le N \le \frac{n}{2}$ its value at $\lambda = \frac{n}{2} - N$ is

$$\left(\frac{n}{2}-N\right)(-1)^N P_{2N}\mathcal{Q}_{n-2N}^{\text{res}}\left(\frac{n}{2}-N\right) = (-1)^N P_{2N}Q_{n-2N}^{\text{res}}\left(\frac{n}{2}-N\right).$$

Next, the value at $\lambda = 0$ of its derivative is Q_n. Thus it suffices to apply the holographic formula $Q_n = -(-1)^{\frac{n}{2}}\dot{D}_n^{\text{res}}(0)(1)$.   □

Remark 6.11.4. *Theorem 6.11.10 can be regarded as a generalization of the holographic formula* $Q_n = -(-1)^{\frac{n}{2}} \dot{D}_n^{res}(0)(1)$. *The latter identity is valid for all metrics. Since the proof of Theorem 6.11.10 rests only on that system of factorization identities which extends to general metrics (we refer to the comments at the end of the present section), (6.11.36) also continues to hold true for all metrics (under the assumption (6.11.35)).*

Remark 6.11.5. *The condition* $Q_{2N}^{res}(0) = 0$ *means that the polynomial*

$$\left[\left(-\frac{n}{2}-\lambda+2N-1\right)\cdots\left(-\frac{n}{2}-\lambda+N\right)\right]$$
$$\times \left[\mathcal{T}_{2N}^*(\lambda+n-2N)(v_0) + \cdots + \mathcal{T}_0^*(\lambda+n-2N)(v_{2N})\right] \quad (6.11.37)$$

vanishes at $\lambda = 0$. *For* $N = \frac{n}{2}$, *this is equivalent to the condition* $P_n^*(0)(1) = 0$ *which is satisfied since* P_n *is self-adjoint and annihilates constants. Moreover, by Theorem 6.11.8, the vanishing holds true for* $N = 1, 2, 3$. *We expect that the vanishing holds true in general. In fact, this would be a consequence of the general formula (6.11.38).*

We work out the meaning of the condition for $N = 1$ *and* $N = 2$. *For* $N = 1$, *it is equivalent to the fact that*

$$-2(n-2+2\lambda)\left(\mathcal{T}_2^*(\lambda+n-2)(1) + v_2\right) = (\Delta - (\lambda+n-2)J)(1) - 2(n-2+2\lambda)v_2$$

vanishes at $\lambda = 0$. *For* $N = 2$, *the condition states that*

$$(2\lambda+n-6)(2\lambda+n-4)\left[\mathcal{T}_4^*(\lambda+n-4)(v_0) + \mathcal{T}_2^*(\lambda+n-4)(v_2) + v_4\right]$$

vanishes at $\lambda = 0$. *Using Remark 6.6.5, the latter assertion is equivalent to the (obvious) relation*

$$-(\Delta - (n-4)J)(n-2)J + (n-4)(n-6)|P|^2 - (n-6)\Delta J$$
$$- 4(n-4)(\Delta - (n-4)J)(v_2) + 8(n-6)(n-4)v_4 = 0.$$

In the recursive formula (6.11.36), the lower order Q-polynomials play an important role. For these polynomials, the following analogous formula would generalize (6.11.36):

$$Q_{2N}^{res}(\lambda) = (-1)^{N-1}\lambda \prod_{k=1}^{N-1}\left(\frac{\lambda+\frac{n}{2}-2N+k}{k}\right)Q_{2N}$$

$$+ \lambda \sum_{j=1}^{N-1}(-1)^j \prod_{\substack{k=1\\k\neq j}}^{N}\left(\frac{\lambda+\frac{n}{2}-2N+k}{k-j}\right)P_{2j}Q_{2N-2j}^{res}\left(-\frac{n}{2}+2N-j\right). \quad (6.11.38)$$

Note that (6.11.38) implies $Q_{2N}^{res}(0) = 0$, i.e., shows that the assumption (6.11.35) in Theorem 6.11.10 is always satisfied.

Now Theorem 6.11.10, together with (6.11.38), yields a formula of the form

$$Q_n^{\mathrm{res}}(\lambda) = \sum_I c_I(\lambda) P_{2I} Q_{n-2|I|}, \qquad (6.11.39)$$

where the sum runs over all partitions I of integers in $[0, \frac{n}{2}-1]$ and the coefficients $c_I(\lambda)$ are polynomials of degree $\frac{n}{2}$. Theorem 6.11.8 covers the special cases $n=4$ and $n=6$.

The value of the polynomial on the right-hand side of (6.11.38) at $\lambda = -\frac{n}{2}+2N-L$ $(1 \le L \le N-1)$ is given by

$$\left(-\frac{n}{2}+2N-L\right)(-1)^L P_{2L} \mathcal{Q}_{2N-2L}^{\mathrm{res}}\left(-\frac{n}{2}+2N-L\right)$$
$$= (-1)^L P_{2L} Q_{2N-2L}^{\mathrm{res}}\left(-\frac{n}{2}+2N-L\right).$$

Similarly, for $\lambda = -\frac{n}{2}+N$, we find the value

$$\left(-\frac{n}{2}+N\right)(-1)^{N-1}\prod_{k=1}^{N-1}\left(\frac{-N+k}{k}\right)Q_{2N} = -\left(\frac{n}{2}-N\right)Q_{2N}.$$

These observations show that the polynomial on the right hand side of (6.11.38) satisfies the same N factorization identities as those which follow for $Q_{2N}^{\mathrm{res}}(\lambda)$ from the first N factorization identities in Theorem 6.11.1. Moreover, the validity of implies a formula for

$$\dot{Q}_{2N}^{\mathrm{res}}\left(-\frac{n}{2}+N\right) = -(-1)^N \dot{D}_{2N}^{\mathrm{res}}\left(-\frac{n}{2}+N\right)(1),$$

and, by the above arguments, an independent proof of that formula would be enough to prove (6.11.38). The following result makes the missing identity explicit.

Theorem 6.11.11 (Holographic formula for Q_{2N}). *The identity* (6.11.38) *implies*

$$\left[1+\left(\frac{n}{2}-N\right)\sum_{k=1}^{N-1}\frac{1}{k}\right]Q_{2N} = \dot{Q}_{2N}^{\mathrm{res}}\left(-\frac{n}{2}+N\right)$$
$$-(-1)^N\left(\frac{n}{2}-N\right)\sum_{j=1}^{N-1}\frac{1}{N-j}\binom{N-1}{j-1}P_{2j}\mathcal{Q}_{2N-2j}^{\mathrm{res}}\left(-\frac{n}{2}+2N-j\right). \quad (6.11.40)$$

For $2N = n$, (6.11.40) specializes to the holographic formula

$$Q_n = -(-1)^{\frac{n}{2}}\dot{D}_n^{\mathrm{res}}(0)(1) = \dot{Q}_n^{\mathrm{res}}(0).$$

All terms on the right-hand side of (6.11.40) are given by holographic data, i.e., holographic coefficients and the asymptotics of eigenfunctions of the Laplacian of the Poincaré-Einstein metric. Theorem 6.11.11 emphasizes the role of the holographic formula (6.11.40) for the subcritical Q-curvatures in the proof of the formula (6.11.38) for the subcritical Q-polynomials. It remains to find an independent proof of (6.11.40).

Example 6.11.2. *The cases $N = 2$ and $N = 3$ of (6.11.40) are covered by Theorem 6.11.7. In fact, for $N = 2$, (6.11.40) reads*

$$\dot{Q}_4^{\mathrm{res}}\left(-\frac{n}{2}+2\right) = \left[1+\left(\frac{n}{2}-2\right)\right]Q_4 + \left(\frac{n}{2}-2\right)P_2\mathcal{Q}_2^{\mathrm{res}}\left(-\frac{n}{2}+3\right).$$

In view of $\mathcal{Q}_2^{\mathrm{res}}(\lambda) = Q_2$, this formula coincides with the first identity of Theorem 6.11.7. Similarly, for $N = 3$, (6.11.40) states

$$\dot{Q}_6^{\mathrm{res}}\left(-\frac{n}{2}+3\right) = \left[1+\frac{n-6}{2}\cdot\frac{3}{2}\right]Q_6$$

$$-\left(\frac{n}{2}-3\right)\left(\frac{1}{2}P_2\mathcal{Q}_4^{\mathrm{res}}\left(-\frac{n}{2}+5\right) + 2P_4\mathcal{Q}_2^{\mathrm{res}}\left(-\frac{n}{2}+4\right)\right).$$

Now using $\mathcal{Q}_2^{\mathrm{res}}(\lambda) = Q_2$ and

$$\mathcal{Q}_4^{\mathrm{res}}\left(-\frac{n}{2}+5\right) = -2Q_4 - 3P_2Q_2,$$

the latter formula reads

$$\dot{Q}_6^{\mathrm{res}}\left(-\frac{n}{2}+3\right) = \left[1+\frac{n-6}{2}\cdot\frac{3}{2}\right]Q_6 + \left(\frac{n}{2}-3\right)\left(P_2Q_4 + \frac{3}{2}P_2^2Q_2 - 2P_4Q_2\right).$$

This formula coincides with the second identity of Theorem 6.11.7.

Proof of Theorem 6.11.11. Differentiating the first term in (6.11.38) at $\lambda = -\frac{n}{2} + N$, yields

$$(-1)^{N-1}\prod_{k=1}^{N-1}\left(\frac{-N+k}{k}\right)Q_{2N} + \left(-\frac{n}{2}+N\right)(-1)^{N-1}\sum_{r=1}^{N-1}\frac{1}{r}\prod_{\substack{k=1\\k\neq r}}^{N-1}\left(\frac{-N+k}{k}\right)Q_{2N}$$

$$= Q_{2N} - \left(\frac{n}{2}-N\right)\sum_{r=1}^{N-1}\frac{1}{-N+r}Q_{2N}.$$

Next, by differentiation of the second term, we obtain

$$\left(-\frac{n}{2}+N\right)\sum_{j=1}^{N-1}(-1)^j\frac{1}{N-j}\prod_{\substack{k=1\\k\neq j}}^{N-1}\left(\frac{-N+k}{k-j}\right)P_{2j}\mathcal{Q}_{2N-2j}^{\mathrm{res}}(\cdots).$$

But since

$$\prod_{\substack{k=1\\k\neq j}}^{N-1}\left(\frac{-N+k}{k-j}\right) = -(-1)^{N-j}\binom{N-1}{j-1},$$

the sum equals

$$\left(\frac{n}{2}-N\right)(-1)^N \sum_{j=1}^{N-1} \frac{1}{N-j}\binom{N-1}{j-1} P_{2j}\mathcal{Q}^{\mathrm{res}}_{2N-2j}(\cdots).$$

The proof is complete.   □

The polynomial identity (6.11.38) is equivalent to a *set* of N identities for Q_{2N} which arise by comparing coefficients of powers of λ on both sides. Theorem 6.11.11 deals with the coefficient of λ. In the critical case $2N = n$, it leads to the holographic formula for Q_n. A second case of substantial interest concerns the coefficients of the leading power. In fact, comparing coefficients of λ^N in (6.11.38) implies

$$(N-1)!(Q^{\mathrm{res}}_{2N})^{\mathrm{top}} = (-1)^{N-1}Q_{2N}$$
$$- \sum_{j=1}^{N-1}\binom{N-1}{j-1} P_{2j}\mathcal{Q}^{\mathrm{res}}_{2N-2j}\left(-\frac{n}{2}+2N-j\right). \quad (6.11.41)$$

Here $(Q^{\mathrm{res}}_{2N})^{\mathrm{top}}$ denotes the coefficient of λ^N in the Q-polynomial $Q^{\mathrm{res}}_{2N}(\lambda)$. In the critical case $2N = n$, we find the following result.

Corollary 6.11.5. *Under the assumption* (6.11.35),

$$\left(\frac{n}{2}-1\right)!(Q^{\mathrm{res}}_n)^{\mathrm{top}} = (-1)^{\frac{n}{2}-1}Q_n - \sum_{j=1}^{\frac{n}{2}-1}\binom{\frac{n}{2}-1}{j-1} P_{2j}\mathcal{Q}^{\mathrm{res}}_{n-2j}\left(\frac{n}{2}-j\right). \quad (6.11.42)$$

As described above, the terms which involve the values of the polynomials $\mathcal{Q}$ again can be written in terms of GJMS-operators and Q-curvatures. Since the left-hand side is defined in terms of holographic data, (6.11.42) is a second holographic formula for Q_n (see also Remark 6.11.6). The main difference between that formula and the holographic formula of Section 6.6 is that the structure of $(Q^{\mathrm{res}}_n)^{\mathrm{top}}$ is much simpler than that of $\dot{Q}^{\mathrm{res}}_n(0)$. We illustrate the relations between both holographic formulas by two examples.

Example 6.11.3. *Let* $n = 4$. *The following discussion is valid for all metrics.* (6.11.42) *states that*

$$-(Q_4 + P_2\mathcal{Q}_2(1)) = -(Q_4 + P_2Q_2)$$

coincides with the coefficient of λ^2 *in* $Q^{\mathrm{res}}_4(\lambda) = -D^{\mathrm{res}}_4(\lambda)(1)$, *i.e., in*

$$-2^4 2!(\lambda-1)\lambda\left[T^*_4(\lambda)(v_0) + T^*_2(\lambda)(v_2) + v_4\right]$$
$$= -P^*_4(\lambda)(v_0) + 8\lambda P^*_2(\lambda)(v_2) - 32\lambda(\lambda-1)v_4.$$

Now we observe

Lemma 6.11.10.
$$P_4^*(\lambda)(1) = 6\lambda P_2^*(\lambda)(v_2) - 16\lambda(\lambda-1)v_4.$$

Proof. We recall that

$$P_4^*(\lambda)(1) = \lambda^2(J^2 + 2|\mathsf{P}|^2) + \lambda(-3\Delta J + 2J^2 - 2|\mathsf{P}|^2)$$

(for the discussion of the linear term see Example 6.6.4). For the right-hand side we find the formula

$$6\lambda(\Delta - \lambda J)\left(-\frac{1}{2}J\right) - 2\lambda(\lambda-1)(J^2 - |\mathsf{P}|^2)$$

and the assertion follows by an easy calculation. $\qquad\square$

Hence (6.11.42) *reads*

$$
\begin{aligned}
-(Q_4 + P_2Q_2) &= \left(-P_4^*(\lambda)(1) + 8\lambda P_2^*(\lambda)(v_2) - 32\lambda(\lambda-1)v_4\right)^{\text{top}} \\
&= \left(2\lambda P_2^*(\lambda)(v_2) - 16\lambda(\lambda-1)v_4\right)^{\text{top}} \\
&= -2Jv_2 - 16v_4 \qquad\qquad\qquad\qquad (6.11.43)
\end{aligned}
$$

or, equivalently,

$$Q_4 = -P_2(Q_2) - Q_2^2 + 16v_4. \qquad\qquad (6.11.44)$$

The latter formula will have an analog for Q_6 (see (6.11.48)) and Q_8 (see Theorem 6.13.1). (6.11.44) should be compared with the holographic formula

$$Q_4 = 16v_4 + 2\Delta v_2$$

(Theorem 6.6.6, see also Example 6.6.4). Both formulas relate Q_4 to $16v_4$ and are easily seen to be equivalent.

Note that Lemma 6.11.10 shows

$$
\begin{aligned}
Q_4^{\text{res}}(\lambda) &= -P_4^*(\lambda)(v_0) + 8\lambda P_2^*(\lambda)(v_2) - 32\lambda(\lambda-1)v_4 \\
&= 2\lambda P_2^*(\lambda)(v_2) - 16\lambda(\lambda-1)v_4.
\end{aligned}
$$

We continue with a detailed discussion of the case $n = 6$.

Example 6.11.4. Let $n = 6$. The following discussion is valid for all metrics. In view of Theorem 6.11.8, the relation (6.11.42) states that

$$
Q_6 - P_2\mathcal{Q}_4^{\text{res}}(2) - 2P_4\mathcal{Q}_2^{\text{res}}(1)
$$
$$
= Q_6 - P_2(-2Q_4 - 3P_2Q_2) - 2P_4Q_2 = Q_6 + 2P_2Q_4 + 3P_2^2Q_2 - 2P_4Q_2
$$

coincides with the coefficient of λ^3 in the polynomial

$$2Q_6^{\text{res}}(\lambda) = 2D_6^{\text{res}}(\lambda)(1).$$

Here $D_6^{\mathrm{res}}(\lambda)(1)$ is given by

$$2^6 3!(-\lambda+2)(-\lambda+1)(-\lambda)\left[\mathcal{T}_6^*(\lambda)(v_0) + \mathcal{T}_4^*(\lambda)(v_2) + \mathcal{T}_2^*(\lambda)(v_4) + v_6\right],$$

i.e., equals

$$P_6^*(\lambda)(1) - 2^1 3!\lambda P_4^*(\lambda)(v_2) + 2^4 3!\lambda(\lambda-1)P_2^*(\lambda)(v_4) - 2^6 3!\lambda(\lambda-1)(\lambda-2)v_6.$$

Similarly as in Example 6.11.3, we observe that in the latter sum the first term can be written as a linear combination of the other terms.

Lemma 6.11.11.

$$P_6^*(\lambda)(1) = 10\lambda P_4^*(\lambda)(v_2) - 2^6\lambda(\lambda-1)P_2^*(\lambda)(v_4) + 2^6 3\lambda(\lambda-1)(\lambda-2)v_6.$$

Proof. We compare coefficients of powers of λ. By the self-adjointness of $P_6 = P_6(0)$, we have $P_6^*(0)(1) = P_6(0)(1) = 0$. This implies that the constant terms of the polynomials on both sides vanish. For the coefficients of λ the claim is that

$$\dot{P}_6^*(0)(1) = 10\dot{P}_4^*(0)(v_2) + 2^6 \dot{P}_2^*(0)(v_4) + 2^6 3!v_6. \qquad (6.11.45)$$

But using $\dot{P}_6(0)(1) = -Q_6$, Theorem 6.10.2 yields

$$\dot{P}_6^*(0)(1) = -Q_6 + 32\dot{P}_2^*(0)(v_4) + 8\dot{P}_2^*(0)(v_2).$$

Now substituting

$$Q_6 = -8 \cdot 48v_6 - 32\dot{P}_2^*(0)(v_4) - 2\dot{P}_4^*(0)(v_2)$$

(Corollary 6.10.1), completes the proof of (6.11.45). In order to compare the coefficients of λ^3, we apply (6.6.41) and (6.10.2). We find that on the left-hand side the coefficient is given by

$$2^4(\log\det h)''' + 2^3(\log\det h)''\mathsf{J} - \mathsf{J}(\mathsf{J}^2 + 2|\mathsf{P}|^2).$$

In order to simplify the sum, we observe that

$$(\log\det h)''' = -\frac{1}{4}(\mathcal{B},\mathsf{P}) - \frac{1}{2}\operatorname{tr}(\mathsf{P}^3) \quad\text{and}\quad (\log\det h)'' = -\frac{1}{2}|\mathsf{P}|^2.$$

The latter formula has been derived in the proof of Theorem 6.10.1. Similarly, the first formula can be proved on the basis of the formulas established in the proof of Theorem 6.10.1. We omit the details. Hence

$$2^4(\log\det h)''' + 2^3(\log\det h)''\mathsf{J} - \mathsf{J}(\mathsf{J}^2 + 2|\mathsf{P}|^2) = -4(\mathcal{B},\mathsf{P}) - 8\operatorname{tr}(\mathsf{P}^3) - 6\mathsf{J}|\mathsf{P}|^2 - \mathsf{J}^3.$$

But using

$$v_6 = -\frac{1}{8}\operatorname{tr}(\wedge^3\mathsf{P}) - \frac{1}{48}(\mathcal{B},\mathsf{P}) = -\frac{1}{48}\left(\mathsf{J}^3 - 3\mathsf{J}|\mathsf{P}|^2 + 2\operatorname{tr}(\mathsf{P}^3)\right) - \frac{1}{48}(\mathcal{B},\mathsf{P})$$

(Theorem 6.9.2), it follows that the sum can be written in the form

$$2^6 3v_6 + 3J^3 - 18|P|^2 J.$$

Now it only remains to verify that

$$(10\lambda P_4^*(\lambda)(v_2) - 2^6\lambda(\lambda-1)P_2^*(\lambda)(v_4))^{\text{top}} = 3J^3 - 18J|P|^2.$$

But it is easy to see that the left-hand side equals

$$-5(J^2 + 2|P|^2)J + 8J(J^2 - |P|^2).$$

We omit the detailed discussion of the quadratic term and indicate only, where the contribution $12((\mathcal{B}, P) + 2\operatorname{tr}(P^3))$ appears on both sides. By (6.10.2), the coefficient of λ^2 on the left-hand side contains the term $-2^4 3(\log\det h)'''$, i.e., $12((\mathcal{B}, P) + 2\operatorname{tr}(P^3))$. On the other hand, the coefficient of λ^2 on the right-hand side contains the term $-2^6 9v_6$, i.e., in particular, the term $12((\mathcal{B}, P) + 2\operatorname{tr}(P^3))$. The discussion of the remaining terms can be done on the basis of the results of Section 6.10. The proof is complete. □

Lemma 6.11.11 shows that the Q-polynomial is given by the formula

$$Q_6^{\text{res}}(\lambda) = -2\lambda P_4^*(\lambda)(v_2) + 2^5\lambda(\lambda-1)P_2^*(\lambda)(v_4) - 2^6 3\lambda(\lambda-1)(\lambda-2)v_6,$$

and that (6.11.42) is equivalent to

$$Q_6 + \left[2P_2Q_4 + 3P_2^2Q_2 - 2P_4Q_2\right]$$
$$= \left(-4\lambda P_4^*(\lambda)(v_2) + 2^6\lambda(\lambda-1)P_2^*(\lambda)(v_4)\right)^{\text{top}} - 2^6 3!v_6. \quad (6.11.46)$$

Thus we obtain the formula

$$Q_6 + \left[2P_2Q_4 + 3P_2^2Q_2 - 2P_4Q_2\right] = -4(J^2 + 2|P|^2)(v_2) - 64Jv_4 - 2^6 3!v_6$$
$$= -6J^3 + 12|P|^2J - 2^6 3!v_6. \quad (6.11.47)$$

Now using

$$Q_4 + P_2Q_2 = J^2 - 2|P|^2,$$

we can write (6.11.47) in the form

$$Q_6 = [P_2(Q_4^{\text{res}}(2)) + 2P_4(Q_2^{\text{res}}(1))] - 6[Q_4 + P_2(Q_2)]Q_2 - 2^6 3!v_6. \quad (6.11.48)$$

This is an analog of (6.11.44). The structure of (6.11.48) substantially differs from that of the holographic formula

$$Q_6 = -2^6 3!v_6 - 32P_2^*(0)(v_4) - 2P_4^*(0)(v_2) \quad (6.11.49)$$

(see (6.10.4)) although both formulas express a relation between the quantities Q_6 and $-2^6 6 v_6$. The main difference concerns the way in which both formulas explain the divergence terms. Whereas in (6.11.49) these contributions are formulated in terms of the operators $P_{2j}^*(0)$ (which describe the asymptotics of harmonic functions of the Poincaré-Einstein metric) acting on holographic coefficients, in (6.11.48) these are covered by lower order GJMS-operators acting on lower order Q-curvature. In particular, (6.11.48) is recursive.

Remark 6.11.6. *Comparing coefficients of powers of λ in (6.11.36) actually implies a set of $\frac{n}{2}$ identities. The respective coefficients of v_n (on the left-hand side) and Q_n (on the right-hand side) are*

$$-2^n \left(\frac{n}{2}\right)! \prod_{k=1}^{\frac{n}{2}} \left(\lambda - \frac{n}{2} + k\right) \tag{6.11.50}$$

and

$$(-1)^{\frac{n}{2}-1} \frac{1}{\left(\frac{n}{2}-1\right)!} \prod_{k=1}^{\frac{n}{2}} \left(\lambda - \frac{n}{2} + k\right). \tag{6.11.51}$$

The polynomial $P_n^*(\lambda)(1)$ on the left-hand side plays a special role. It is conjectured to be a linear combination of the other terms so that v_n contributes again with the coefficient

$$(-1)^{\frac{n}{2}-1} 2^{n-1} \left(\frac{n}{2}\right)! \prod_{k=1}^{\frac{n}{2}} \left(\lambda - \frac{n}{2} + k\right)$$

(see (6.11.56)); the cases $n = 4$ and $n = 6$ are covered by Lemma 6.11.10 and Lemma 6.11.11. On the left-hand side of (6.11.36) this yields the additional contribution

$$2^{n-1} \left(\frac{n}{2}\right)! \prod_{k=1}^{\frac{n}{2}} \left(\lambda - \frac{n}{2} + k\right) v_n.$$

Together with (6.11.50) we find

$$-2^{n-1} \left(\frac{n}{2}\right)! \prod_{k=1}^{\frac{n}{2}} \left(\lambda - \frac{n}{2} + k\right) v_n$$

on the left-hand side of (6.11.36). Using (6.11.51), it follows that all identities relate

$$(-1)^{\frac{n}{2}} 2^{n-1} \left(\frac{n}{2}\right)! \left(\frac{n}{2}-1\right)! v_n \quad \text{to} \quad Q_n.$$

These identities contain additional holographic terms and lower order GJMS-operators and Q-curvatures.

Lemma 6.11.10 and Lemma 6.11.11 state that for $n = 4$ and $n = 6$ the respective leading term $P_n^*(\lambda)(1)$ in $Q_n^{res}(\lambda)$ can be written as a linear combination of the other terms which enter into $Q_n^{res}(\lambda)$. The following result and the accompanying conjecture provide a uniform formulation of these relations for general n.

Theorem 6.11.12. *For a Riemannian metric h on a manifold of even dimension n, the function*

$$\mathcal{V}_n(h; \lambda) \stackrel{\text{def}}{=} \lambda(\lambda - 1) \cdots \left(\lambda - \frac{n}{2} + 1\right) \sum_{j=0}^{\frac{n}{2}} (n+2j) T_{2j}^*(h; \lambda)(v_{n-2j}) \qquad (6.11.52)$$

is a polynomial with vanishing constant and linear terms.

Proof. The families $T_{2j}(\lambda)$, $j = 1, \ldots, \frac{n}{2} - 1$ are regular at $\lambda = 0$. Hence the relation

$$P_n(\lambda) = 2^n \left(\frac{n}{2}\right)!(-1)^{\frac{n}{2}} \left[\lambda(\lambda - 1) \cdots \left(\lambda - \frac{n}{2} + 1\right)\right] T_n(\lambda) \qquad (6.11.53)$$

(see (6.6.3)) implies that the constant term of $\mathcal{V}_n(\lambda)$ is a multiple of $P_n^*(0)(1) = P_n^*(1) = P_n(1) = 0$. In order to determine the coefficient of λ, we apply (6.11.53). We find

$$(-1)^{\frac{n}{2}-1}\left(\frac{n}{2}-1\right)! \sum_{j=0}^{\frac{n}{2}-1} (n+2j) T_{2j}^*(0)(v_{n-2j}) + 2n \left(2^n \left(\frac{n}{2}\right)!\right)^{-1} (-1)^{\frac{n}{2}} \dot{P}_n^*(0)(1),$$

i.e., the product of $(-1)^{\frac{n}{2}} 2^{-(n-1)} \left(\frac{n}{2}\right)^{-1}$ and

$$n\dot{P}_n^*(0)(1) - \left(\frac{n}{2}-1\right)! \left(\frac{n}{2}\right) 2^{n-1} \sum_{j=0}^{\frac{n}{2}-1} (n+2j) T_{2j}^*(0)(v_{n-2j}). \qquad (6.11.54)$$

In order to prove that (6.11.54) vanishes, we recall that by Theorem 6.6.4,

$$n\dot{P}_n^*(0)(1) = n\dot{P}_n(0)(1) + 2^n \left(\frac{n}{2}\right)! \left(\frac{n}{2}-1\right)! \sum_{j=0}^{\frac{n}{2}-1} 2j T_{2j}^*(0)(v_{n-2j}).$$

But $\dot{P}_n(0)(1) = (-1)^{\frac{n}{2}} Q_n$ and Theorem 6.6.6 imply

$$n\dot{P}_n^*(0)(1) = 2^{n-1} \left(\frac{n}{2}\right)! \left(\frac{n}{2}-1\right)! \sum_{j=0}^{\frac{n}{2}-1} (n-2j) T_{2j}^*(0)(v_{n-2j})$$

$$+ 2^n \left(\frac{n}{2}\right)! \left(\frac{n}{2}-1\right)! \sum_{j=0}^{\frac{n}{2}-1} 2j T_{2j}^*(0)(v_{n-2j})$$

$$= 2^{n-1} \left(\frac{n}{2}\right)! \left(\frac{n}{2}-1\right)! \sum_{j=0}^{\frac{n}{2}-1} (n+2j) T_{2j}^*(0)(v_{n-2j}).$$

The latter identity means that (6.11.54) vanishes. $\qquad \square$

Conjecture 6.11.2. *For any Riemannian manifold* (M, h) *of dimension* n, *the polynomial* $\mathcal{V}_n(h; \lambda)$ *vanishes identically.*

Of course, Conjecture 6.11.2 is equivalent to the vanishing of the rational function

$$\sum_{j=0}^{\frac{n}{2}} (n+2j) \mathcal{T}_{2j}^*(h; \lambda)(v_{n-2j}). \tag{6.11.55}$$

Lemma 6.11.10 and Lemma 6.11.11 are the special cases $n = 4$ and $n = 6$ of Conjecture 6.11.2. In fact, $\mathcal{V}_n(\lambda) \equiv 0$ is equivalent to

$$P_n^*(\lambda)(1) = (-1)^{\frac{n}{2}-1} \left(\frac{n}{2}-1\right)! 2^{n-2}$$

$$\times \sum_{j=0}^{\frac{n}{2}-1} (-1)^j (n+2j) \frac{\lambda(\lambda-1)\dots(\lambda-\frac{n}{2}+j+1)}{2^{2j} j!} P_{2j}^*(\lambda)(v_{n-2j}). \tag{6.11.56}$$

It suffices to specialize that formula to $n = 4$ and $n = 6$.

Conjecture 6.11.2 yields an alternative formula for the Q-polynomial.

Theorem 6.11.13. $\mathcal{V}_n \equiv 0$ *is equivalent to*

$$n Q_n^{\text{res}}(\lambda) = 2^n \left(\frac{n}{2}\right)! \left[\lambda(\lambda-1)\dots\left(\lambda-\frac{n}{2}+1\right)\right] \sum_{j=0}^{\frac{n}{2}} 2j \mathcal{T}_{2j}^*(\lambda)(v_{n-2j}). \tag{6.11.57}$$

Proof. It suffices to notice that

$$2^n \left(\frac{n}{2}\right)! \left[\lambda(\lambda-1)\dots\left(\lambda-\frac{n}{2}+1\right)\right] \sum_{j=0}^{\frac{n}{2}} \mathcal{T}_{2j}^*(\lambda)(v_{n-2j})$$

$$= (-1)^{\frac{n}{2}} D_n^{\text{res}}(\lambda)(1) = -Q_n^{\text{res}}(\lambda).$$

$\square$

Example 6.11.5. *For* $n = 4$, (6.11.57) *states that*

$$4 Q_4^{\text{res}}(\lambda) = 2^5 \lambda(\lambda-1) \left[2\mathcal{T}_2^*(\lambda)(v_2) + 4\mathcal{T}_4^*(\lambda)(v_0)\right].$$

Using

$$\mathcal{T}_2(\lambda) = -\frac{1}{4(\lambda-1)} P_2(\lambda) \quad \text{and} \quad \mathcal{T}_4(\lambda) = \frac{1}{32\lambda(\lambda-1)} P_4(\lambda),$$

this formula is equivalent to

$$Q_4^{\text{res}}(\lambda) = -4\lambda P_2^*(\lambda)(v_2) + P_4^*(\lambda)(v_0).$$

The latter identity can be checked directly using the explicit formulas for the terms involved. For the right-hand side we find

$$2\lambda(\Delta - \lambda J)J + (\Delta - \lambda J)(\Delta - (\lambda+2)J)(1) + 2\lambda(\lambda-1)|P|^2 - 2(\lambda-1)\Delta J$$
$$= \lambda^2(2|P|^2 - J^2) + \lambda(-\Delta J - 2|P|^2 + 2J^2)$$
$$= -\lambda^2(Q_4 + P_2 Q_2) + \lambda Q_4.$$

By Theorem 6.11.9, this is $Q_4^{\mathrm{res}}(\lambda)$.

Next, we discuss a consequence of (6.11.57) for *global* conformally compact Einstein metrics. The setting is the following. Let X^{n+1} be a compact Riemannian manifold with boundary $\partial X = M$. Assume that g_E is a global conformally compact Einstein metric on $X \setminus \partial X$ with conformal infinity $[h]$. A choice of a representing metric h on M gives rise to a defining function r such that $|dr|^2_{r^2 g_E} = 1$ near ∂X. Near the boundary g_E can be written in the normal form $r^{-2}(dr^2 + h_r)$, and for even n we have the asymptotics

$$\int_{r \geq \varepsilon} \mathrm{vol}(g_E) = c_0 \varepsilon^{-n} + c_2 \varepsilon^{-(n-2)} + \cdots + c_{n-2}\varepsilon^{-2} - L \log \varepsilon + V + o(1) \quad (6.11.58)$$

for $\varepsilon \to 0$. Here $V = V(g_E, h)$ is the asymptotic or renormalized volume of g_E with respect to h.

Lemma 6.11.12. $\ddot{V}_n(0) = 0$ *(i.e., Conjecture 6.11.2 for the quadratic coefficient) implies*

$$(-1)^{\frac{n}{2}} c_{\frac{n}{2}}^{-1} \frac{1}{n} \int_M \sum_{j=1}^{\frac{n}{2}} 2j \dot{T}_{2j}^*(0)(v_{n-2j}) \,\mathrm{vol} = -\frac{1}{2} \int_M \ddot{Q}_n^{\mathrm{res}}(0) \,\mathrm{vol} - \alpha \int_M Q_n \,\mathrm{vol},$$

where $c_{\frac{n}{2}} = \left(2^n \left(\frac{n}{2}\right)! \left(\frac{n}{2}-1\right)!\right)^{-1}$ *and* $\alpha = \sum_{k=1}^{n/2-1} \frac{1}{k}$.

Proof. By Theorem 6.11.13, the condition $\ddot{V}_n(0) = 0$ is equivalent to

$$\frac{1}{2} n \ddot{Q}_n^{\mathrm{res}}(0) = 2^n \left(\frac{n}{2}\right)! \left(\frac{n}{2}-1\right)!(-1)^{\frac{n}{2}-1} \sum_{j=0}^{\frac{n}{2}} 2j \left(-\alpha T_{2j}^*(0) + \dot{T}_{2j}^*(0)\right)(v_{n-2j});$$

recall that $T_n(\lambda)(1)$ is holomorphic at $\lambda = 0$. But

$$\int_M T_{2j}^*(0)(v_{n-2j}) \,\mathrm{vol} = \int_M v_{n-2j} T_{2j}(0)(1) \,\mathrm{vol} = 0$$

for $j = 1, \ldots, \frac{n}{2} - 1$. Moreover, (6.11.53) implies

$$-2^n \left(\frac{n}{2}\right)! \left(\frac{n}{2}-1\right)! T_n^*(0)(1) = \dot{P}_n^*(0)(1).$$

Hence

$$2^n \left(\frac{n}{2}\right)! \left(\frac{n}{2}-1\right)!(-1)^{\frac{n}{2}-1} \int_M \mathcal{T}_n^*(0)(1) \, \mathrm{vol}$$

$$= (-1)^{\frac{n}{2}} \int_M \dot{P}_n^*(0)(1) \, \mathrm{vol} = (-1)^{\frac{n}{2}} \int_M \dot{P}_n(0)(1) \, \mathrm{vol} = \int_M Q_n \, \mathrm{vol}$$

and thus we obtain

$$\frac{1}{2} n \int_M \ddot{Q}_n^{\mathrm{res}}(0) \, \mathrm{vol}$$

$$= -\alpha n \int_M Q_n \, \mathrm{vol} - 2^n \left(\frac{n}{2}\right)! \left(\frac{n}{2}-1\right)!(-1)^{\frac{n}{2}} \int_M \sum_{j=0}^{\frac{n}{2}} 2j \dot{\mathcal{T}}_{2j}^*(0)(v_{n-2j}) \, \mathrm{vol},$$

i.e.,

$$(-1)^{\frac{n}{2}} c_{\frac{n}{2}}^{-1} \frac{1}{n} \int_M \sum_{j=0}^{\frac{n}{2}} 2j \dot{\mathcal{T}}_{2j}^*(0)(v_{n-2j}) \, \mathrm{vol} = -\alpha \int_M Q_n \, \mathrm{vol} - \frac{1}{2} \int_M \ddot{Q}_n^{\mathrm{res}}(\lambda) \, \mathrm{vol}.$$

The proof is complete.   □

Combined with $Q_n^{\mathrm{res}}(0) = 0$ and $\dot{Q}_n^{\mathrm{res}}(0) = Q_n$, we get

$$\int_M Q_n^{\mathrm{res}}(\lambda) \, \mathrm{vol} = \lambda \int_M Q_n \, \mathrm{vol}$$

$$- \lambda^2 \left(\alpha \int_M Q_n \, \mathrm{vol} + (-1)^{\frac{n}{2}} c_{\frac{n}{2}}^{-1} \frac{1}{n} \sum_{j=0}^{\frac{n}{2}} 2j \dot{\mathcal{T}}_{2j}^*(0)(v_{n-2j}) \, \mathrm{vol} \right) + \cdots . \quad (6.11.59)$$

Now we need

Theorem 6.11.14 ([70]). *Let $n \geq 2$ be even. For a global conformally compact Einstein metric g_E and a representing metric h in its conformal infinity,*

$$V = -\int_M \dot{\mathcal{S}}(n)(1) \, \mathrm{vol} + \frac{1}{n} \int_M \sum_{j=1}^{\frac{n}{2}} 2j \dot{\mathcal{T}}_{2j}^*(0)(v_{n-2j}) \, \mathrm{vol}.$$

Here V denotes the renormalized volume as in (6.11.58), $\mathcal{S}$ is the scattering operator and the local terms on the right-hand side are to be understood with respect to the metric h.

The quantity $\dot{\mathcal{S}}(n)(1)$ is well defined since $\mathcal{S}(\lambda)(1)$ is holomorphic near $\lambda = n$. V and $\dot{\mathcal{S}}(n)(1)$ both depend on the global metric g_E. For odd n, Fefferman and Graham ([97]) proved that

$$V = \int_M \dot{\mathcal{S}}(n)(1) \, \mathrm{vol}.$$

Theorem 6.11.14 is an analog of that relation for even n.

Example 6.11.6. *For $n = 2$, we have $\mathcal{T}_2(\lambda)(1) = \frac{1}{4}\mathsf{J}$ and Theorem 6.11.14 states that*

$$V + \int_M \dot{\mathcal{S}}(2)(1)\,\mathrm{vol} = 0.$$

In dimension $n = 4$, Theorem 6.11.14 states that

$$V + \int_M \dot{\mathcal{S}}(4)(1)\,\mathrm{vol} = \frac{1}{4}\int_M \sum_{j=1}^2 2j T_{2j}^*(0)(v_{4-2j})\,\mathrm{vol}.$$

By Example 6.11.5, the right-hand side coincides with

$$\frac{1}{32}\int_M \left(\frac{Q_4^{\mathrm{res}}(\lambda)}{\lambda(\lambda-1)}\right)\Big|_{\lambda=0}\,\mathrm{vol}.$$

But Theorem 6.11.9 implies that the latter integral equals

$$\frac{1}{32}\int_M P_2 Q_2\,\mathrm{vol} = -\frac{1}{32}\int_M \mathsf{J}^2\,\mathrm{vol}.$$

Hence

$$V + \int_M \dot{\mathcal{S}}(4)(1)\,\mathrm{vol} = -\frac{1}{32}\int_M \mathsf{J}^2\,\mathrm{vol}.$$

We use Theorem 6.11.14 to rewrite the expansion (6.11.59) in the form

$$\lambda \int_M Q_n\,\mathrm{vol} - \lambda^2\alpha \int_M Q_n\,\mathrm{vol} - \lambda^2(-1)^{\frac{n}{2}}c_{\frac{n}{2}}^{-1}\left(\int_M \dot{\mathcal{S}}(n)(1)\,\mathrm{vol} + V\right) + \cdots.$$

But

$$(-1)^{\frac{n}{2}}c_{\frac{n}{2}}\int_M Q_n\,\mathrm{vol} = \int_M \mathcal{S}(n)(1)\,\mathrm{vol}$$

(Theorem 3.2.2) yields the equivalent formula

$$\lambda(-1)^{\frac{n}{2}}c_{\frac{n}{2}}^{-1}\int_M \mathcal{S}(n)(1)\,\mathrm{vol}$$

$$- \lambda^2(-1)^{\frac{n}{2}}c_{\frac{n}{2}}^{-1}\left[\alpha\int_M \mathcal{S}(n)(1)\,\mathrm{vol} + \int_M \dot{\mathcal{S}}(n)(1)\,\mathrm{vol}\right] - \lambda^2(-1)^{\frac{n}{2}}c_{\frac{n}{2}}^{-1}V + \cdots.$$

Since for the polynomial

$$\alpha(\lambda) = (-1)^{\frac{n}{2}}\frac{(\lambda-1)\dots(\lambda-\frac{n}{2}+1)}{(\frac{n}{2}-1)!}$$

we have $\alpha(0) = -1$ and $\dot{\alpha}(0) = \alpha$, we can write the expansion in the more compressed form

$$-\lambda\alpha(\lambda)(-1)^{\frac{n}{2}}c_{\frac{n}{2}}^{-1}\int_M \mathcal{S}(n-\lambda)(1)\,\mathrm{vol} - \lambda^2(-1)^{\frac{n}{2}}c_{\frac{n}{2}}^{-1}V + \cdots.$$

Thus we have proved

Theorem 6.11.15. *In the situation of Theorem 6.11.14, the relation* $\ddot{\mathcal{V}}_n(0) = 0$ *implies*

$$c_{\frac{n}{2}}(-1)^{\frac{n}{2}} \int_M Q_n^{\mathrm{res}}(\lambda) \, \mathrm{vol} + (-1)^{\frac{n}{2}} \frac{\lambda(\lambda-1)\dots(\lambda-\frac{n}{2}+1)}{(\frac{n}{2}-1)!} \int_M \mathcal{S}(n-\lambda)(1) \, \mathrm{vol}$$
$$= -\lambda^2 V + O(\lambda^3).$$

Theorem 6.11.15 relates the quadratic coefficient in the total Q-polynomial $\int_M Q_n^{\mathrm{res}}(\lambda) \, \mathrm{vol}$ to the scattering operator and the asymptotic volume. That relation, in particular, allows us to determine its infinitesimal conformal variation.

Theorem 6.11.16. *In the situation of Theorem 6.11.15,*

$$\frac{1}{2} \left(\int_M \ddot{Q}_n^{\mathrm{res}}(0) \, \mathrm{vol} \right)^\bullet [\varphi] = \int_M \varphi \left[2Q_n - (-1)^{\frac{n}{2}} c_{\frac{n}{2}}^{-1} v_n \right] \mathrm{vol}.$$

Here the bullet denotes the infinitesimal conformal variation

$$\mathcal{F}^\bullet(h)[\varphi] = d/dt|_{t=0}(\mathcal{F}(e^{2t\varphi}h))$$

of the functional $\mathcal{F}$.

Proof. Lemma 6.11.12 and Theorem 6.11.14 imply

$$V + \int_M \dot{S}(n)(1) \, \mathrm{vol} = -\frac{1}{2}(-1)^{\frac{n}{2}} c_{\frac{n}{2}} \int_M \ddot{Q}_n^{\mathrm{res}}(0) \, \mathrm{vol} - \alpha(-1)^{\frac{n}{2}} c_{\frac{n}{2}} \int_M Q_n \, \mathrm{vol}.$$

Now we have

$$\left(\int_M Q_n \, \mathrm{vol} \right)^\bullet [\varphi] = 0, \qquad V^\bullet[\varphi] = \int_M \varphi v_n \, \mathrm{vol}$$

and

$$\left(\int_M \dot{S}(n)(1) \, \mathrm{vol} \right)^\bullet [\varphi] = -2(-1)^{\frac{n}{2}} c_{\frac{n}{2}} \int_M \varphi Q_n \, \mathrm{vol}$$

(by Theorem 4.3 in [70]). The assertion follows by combining these results. $\qquad\square$

Note that, by Theorem 6.6.6, the conformal anomaly on the right-hand side is the divergence of a natural one-form. It follows that

$$d/dt|_{t=0} \int_M \ddot{Q}_n^{\mathrm{res}}(th; 0) \, \mathrm{vol}(th) = 0.$$

The latter observation actually is a consequence of the *scaling property*

$$Q_n^{\mathrm{res}}(c^2 h; \lambda) = c^{-n} Q_n^{\mathrm{res}}(h; \lambda), \ c \in \mathbb{R}^+ \tag{6.11.60}$$

of the Q-polynomial. In order to prove (6.11.60), it suffices to observe that

$$T_{2j}(c^2 h; \lambda) = c^{-2j} T_{2j}(h; \lambda) \quad \text{and} \quad v_{2j}(c^2 h) = c^{-2j} v_{2j}(h).$$

The proof of Theorem 6.11.16 uses global arguments (involving V and S) to prove a local result. Therefore, it is natural to ask for a local proof. In the following local proof, we do *not* use the relation $\ddot{V}_n(0) = 0$.

Local proof of Theorem 6.11.16. The transformation formula

$$D_n^{\mathrm{res}}(\hat{h}; \lambda) = e^{(\lambda-n)\varphi} \circ D_n^{\mathrm{res}}(h; \lambda) \circ \kappa_* \circ (\kappa^*(r)/r)^\lambda, \quad \hat{h} = e^{2\varphi} h$$

(see Theorem 6.6.3) implies

$$D_n^{\mathrm{res}}(\hat{h}; \lambda)(1) = e^{(\lambda-n)\varphi} D_n^{\mathrm{res}}(h; \lambda) \left(\kappa_*(\kappa^*(r)/r)^\lambda \right).$$

In order to determine the quantity

$$-(-1)^{\frac{n}{2}} \frac{1}{2} \ddot{Q}_n^{\mathrm{res}}(h; 0),$$

we differentiate this identity twice at $\lambda = 0$ and find the formula

$$\varphi^2 e^{-n\varphi} D_n^{\mathrm{res}}(h; 0)(1) + e^{-n\varphi} \ddot{D}_n^{\mathrm{res}}(h; 0)(1) + e^{-n\varphi} D_n^{\mathrm{res}}(h; 0) \left(\kappa_* \log^2(\kappa^*(r)/r) \right)$$

$$+ \varphi e^{-n\varphi} \dot{D}_n^{\mathrm{res}}(h; 0)(1) + e^{-n\varphi} \dot{D}_n^{\mathrm{res}}(\kappa_* \log(\kappa^*(r)/r)).$$

Now using $D_n^{\mathrm{res}}(h; 0) = P_n(h) i^*$ and $i^* \log(\kappa^*(r)/r) = -\varphi$, the first three items simplify to

$$\varphi^2 e^{-n\varphi} P_n(h) + e^{-n\varphi} \ddot{D}_n^{\mathrm{res}}(h; 0)(1) + e^{-n\varphi} P_n(h)(\varphi^2).$$

The remaining two items give

$$-(-1)^{\frac{n}{2}} \varphi e^{-n\varphi} \dot{Q}_n^{\mathrm{res}}(h; 0) + e^{-n\varphi} \dot{D}_n^{\mathrm{res}}(h; 0)(\kappa_* \log(\kappa^*(r)/r)).$$

It follows that

$$\frac{1}{2} \left(\int_M \ddot{Q}_n^{\mathrm{res}}(h; 0) \, \mathrm{vol}(h) \right)^{\bullet} [\varphi]$$

equals the sum of

$$\int_M \varphi Q_n(h) \, \mathrm{vol}(h)$$

(Theorem 6.6.1) and

$$-(-1)^{\frac{n}{2}} \int_M \dot{D}_n^{\mathrm{res}}(h; 0) \left((\kappa_* \log(\kappa^*(r)/r))^{\bullet} [\varphi] \right) \mathrm{vol}(h). \tag{6.11.61}$$

Now we observe that in the latter contribution the conformal variations of all terms which involve normal derivatives vanish. Thus using $i^* \log(\kappa^*(r)/r) = -\varphi$, (6.11.61) equals

$$(-1)^{\frac{n}{2}} \int_M \dot{P}_n^*(h; 0)(v_0 \varphi) \, \mathrm{vol}(h)$$

$$- (-1)^{\frac{n}{2}} 2^n \left(\frac{n}{2} \right)! \left(\frac{n}{2} - 1 \right)! \int_M \sum_{j=0}^{\frac{n}{2}-1} T_{2j}^*(h; 0)(v_{n-2j} \varphi) \, \mathrm{vol}(h),$$

i.e.,

$$(-1)^{\frac{n}{2}} \int_M \varphi \dot{P}_n(h;0)(1) \, \mathrm{vol}(h)$$

$$- (-1)^{\frac{n}{2}} 2^n \left(\frac{n}{2}\right)! \left(\frac{n}{2}-1\right)! \int_M \varphi \sum_{j=0}^{\frac{n}{2}-1} \mathcal{T}_{2j}(h;0)(1) v_{n-2j} \, \mathrm{vol}(h)$$

$$= (-1)^{\frac{n}{2}} \int_M \varphi \dot{P}_n(h;0)(1) \, \mathrm{vol}(h) - (-1)^{\frac{n}{2}} 2^n \left(\frac{n}{2}\right)! \left(\frac{n}{2}-1\right)! \int_M \varphi v_n \, \mathrm{vol}(h).$$

Here we have used that $\mathcal{T}_{2j}(0)(1) = 0$ for $j = 1, \ldots, \frac{n}{2} - 1$. Now $\dot{P}_n(0)(1) = (-1)^{\frac{n}{2}} Q_n$ implies that (6.11.61) is given by

$$\int_M \varphi \left[Q_n - (-1)^{\frac{n}{2}} c_{\frac{n}{2}}^{-1} v_n \right] \mathrm{vol}.$$

This completes the proof.   □

Example 6.11.7. *For $n = 4$, we have*

$$\frac{1}{2} \ddot{Q}_4^{\mathrm{res}}(0) = -2(Q_4 + P_2 Q_2) = -16 v_4 - 2 \mathsf{J} v_2$$

(see Example 6.11.3) and $Q_4 - 16 v_4 = -\Delta \mathsf{J}$. Thus Theorem 6.11.16 states that

$$- \left(\int_M (16 v_4 + 2 \mathsf{J} v_2) \, \mathrm{vol} \right)^{\bullet} [\varphi] = -2 \int_M \varphi \Delta \mathsf{J} \, \mathrm{vol}.$$

By the conformal invariance of $\int_M v_4 \, \mathrm{vol}$, this is equivalent to

$$\left(\int_M \mathsf{J}^2 \, \mathrm{vol} \right)^{\bullet} [\varphi] = -2 \int_M \varphi \Delta \mathsf{J} \, \mathrm{vol}.$$

The latter identity is a consequence of the transformation formula (2.5.7) for J.

We continue with a derivation of a *recursive* formula for the GJMS-operator P_6 for conformally flat metrics. An alternative formula will be derived in Section 6.12.

Theorem 6.11.17. *On conformally flat manifolds of dimension 6, the critical GJMS-operator P_6 can be written in the form*

$$P_6 i^* = \left(6 P_4 P_2 - 2 P_2 P_4 - 3 P_2^3\right) i^* + 4 i^* \left[\bar{P}_2, \bar{P}_4\right]$$
$$= 2 \left(P_4 P_2 + P_2 P_4\right) i^* - 3 P_2^3 i^* + 4 \left(i^* \left[\bar{P}_2, \bar{P}_4\right] - [P_2, P_4] i^*\right). \qquad (6.11.62)$$

Proof. We combine the presentation (6.11.24), i.e.,

$$P_6 i^* = \alpha P_4 D_2^{\text{res}}(1) + \beta P_2 D_4^{\text{res}}(2) + \gamma D_4^{\text{res}}(-9/2)\bar{P}_2 + \delta D_2^{\text{res}}(-11/2)\bar{P}_4,$$

where

$$(\alpha, \beta, \gamma, \delta) = \frac{1}{21}(18, -5, -4, 12),$$

with

$$D_2^{\text{res}}(1) = -6i^*\bar{P}_2 + 7P_2 i^*, \quad D_2^{\text{res}}(-11/2) = 7i^*\bar{P}_2 - 6P_2 i^*$$

and (6.11.26), (6.11.27). We find (omitting i^*)

$$\begin{aligned}
P_6 = {}& \alpha P_4 \left(-6\bar{P}_2 + 7P_2\right) \\
& + \beta P_2 \left(-6P_4 + 27P_2^2 - 28P_2\bar{P}_2 + 8\bar{P}_2^2\right) \\
& + \gamma \left(-6P_4 + 14P_2^2 - 28P_2\bar{P}_2 + 21\bar{P}_2^2\right)\bar{P}_2 \\
& + \delta \left(7\bar{P}_2 - 6P_2\right)\bar{P}_4.
\end{aligned}$$

Hence

$$\begin{aligned}
21P_6 i^* = {}& \left(126P_4 P_2 + 30P_2 P_4 - 135P_2^3\right)i^* \\
& + 84\left(P_2^2 - P_4\right)i^*\bar{P}_2 \\
& + 72P_2 i^* \left(\bar{P}_2^2 - \bar{P}_4\right) \\
& + 84i^* \left(\bar{P}_2\bar{P}_4 - \bar{P}_2^3\right).
\end{aligned}$$

Now we use the identity $\left(P_2^2 - P_4\right)i^* = i^*\left(\bar{P}_2^2 - \bar{P}_4\right)$ (Lemma 6.11.6 for $n = 6$) twice and find

$$\begin{aligned}
21P_6 i^* = {}& \left(126P_4 P_2 + 30P_2 P_4 - 135P_2^3\right)i^* \\
& + 84i^* \left(\bar{P}_2^2 - \bar{P}_4\right)\bar{P}_2 \\
& + 72P_2 \left(P_2^2 - P_4\right)i^* \\
& + 84i^* \left(\bar{P}_2\bar{P}_4 - \bar{P}_2^3\right) \\
= {}& \left(126P_4 P_2 - 42P_2 P_4 - 63P_2^3\right)i^* + 84i^* \left[\bar{P}_2, \bar{P}_4\right].
\end{aligned}$$

This implies the first assertion. The second assertion is a direct consequence. $\quad\square$

Remark 6.11.7. *Analogous arguments show that Theorem 6.11.17 holds true for conformally flat metrics in any dimension $n \geq 3$, i.e., (6.11.62) is universal.*

(6.11.62) implies that the operator $i^*\left[\bar{P}_2, \bar{P}_4\right]$ is tangential, i.e., can be written in the form

$$i^*\left[\bar{P}_2, \bar{P}_4\right] = \mathcal{L}i^*$$

with a natural operator $\mathcal{L}$. In these terms, we find

Corollary 6.11.6. *On conformally flat manifolds of dimension $n = 6$,*

$$P_6 = 2(P_4 P_2 + P_2 P_4) - 3P_2^3 + 2(\mathcal{L} + \mathcal{L}^*). \qquad (6.11.63)$$

Proof. We apply the fact that the operators P_6, P_4 and P_2 are selfadjoint. Hence (6.11.62) implies

$$\mathcal{L}^* - [P_2, P_4]^* = \mathcal{L} - [P_2, P_4].$$

In particular, $\mathcal{L} - \mathcal{L}^* = 2[P_2, P_4]$ and

$$\mathcal{L} - [P_2, P_4] = \frac{1}{2} \left((\mathcal{L} - [P_2, P_4]) + (\mathcal{L} - [P_2, P_4])^* \right) = \frac{1}{2}(\mathcal{L} + \mathcal{L}^*).$$

This implies the assertion. $\square$

Remark 6.11.8. *On the round sphere S^n, the operator $\mathcal{L}$ is given by $\mathcal{L} = 3P_2$. Similarly, for Einstein metrics, $\mathcal{L} = 3(4c)^2 P_2$, where $4c = \tau/n(n-1)$ (see (6.16.13)). In these cases, Theorem 6.11.17 can be verified directly using (6.16.13). For general metrics, P_6 differs from $2(P_4 P_2 + P_2 P_4) - 3P_2^3$ by a second-order operator (Corollary 6.12.1).*

In the following result, the terms with at least three derivatives in Corollary 6.11.6 (and its generalization to general dimensions) are made explicit.

Corollary 6.11.7. *On conformally flat manifolds of dimension $n \geq 3$,*

$$P_6 = \Delta^3 - \frac{3}{2}(n-2)\mathsf{J}\Delta^2 + 16(\mathsf{P}, \operatorname{Hess}\Delta) + 16(\nabla\mathsf{P}, \nabla\operatorname{Hess}) + (22-3n)(d\mathsf{J}, d\Delta),$$

up to terms of order ≤ 2.

Alternatively, Corollary 6.11.7 follows from (6.10.2). We omit the details.

Corollary 6.11.7 fits with Lemma 2.2 in [118] describing the structure of any conformally covariant cube of the Laplacian. The structure of the suppressed terms is more subtle and will be discussed in Section 6.12.

For the convenience of the reader, we finally summarize the main formulas for Q_6.

- **Holographic formula.** For $n \geq 6$,

$$Q_6 = -2^6 3! v_6 - 32 P_2^* \left(\frac{n}{2} - 3 \right)(v_4) - 2P_4^* \left(\frac{n}{2} - 3 \right)(v_2)$$

  (Corollary 6.10.1 for $n = 6$, Theorem 6.10.4 in general).

- **Universal recursive formula.** For $n \geq 6$,

$$Q_6 = \frac{2}{3} P_2(Q_4) + \frac{2}{3} P_4(Q_2) - \frac{5}{3} P_2^2(Q_2) + \frac{8}{3} i^* \bar{P}_2^2(\bar{Q}_2)$$

  (Theorem 6.11.6).

- For $n \geq 6$,

$$Q_6 = - \left[3P_2^2(Q_2) + 2P_2(Q_4) - 2P_4(Q_2) \right] - 6[Q_4 + P_2(Q_2)]Q_2 - 2^6 3! v_6$$

((6.11.48) for $n = 6$).

These formulas are the analogs of the following results for Q_4.

- **Holographic formula.** For $n \geq 4$,

$$Q_4 = 16v_4 + 2P_2^* \left(\frac{n}{2} - 2 \right) (v_2)$$

(Example 6.6.4 for $n = 4$, (6.10.15) in general).

- **Universal recursive formula.** For $n \geq 4$,

$$Q_4 = P_2(Q_2) - 2i^* \bar{P}_2(\bar{Q}_2)$$

(Lemma 6.11.1).

- For $n \geq 4$,

$$Q_4 = -P_2(Q_2) - Q_2^2 + 16v_4$$

((6.11.44) for $n = 4$).

The respective last formulas for Q_4 and Q_6 were derived above in the respective critical dimensions $n = 4$ and $n = 6$. For more details concerning the extension of this formula for Q_6 to general dimensions we refer to Section 6.12, where it is used to derive a universal recursive formula for the GJMS-operator P_6.

We finish the present section with some comments concerning the extension of the methods beyond the conformally flat case. In that connection, it is crucial to establish the factorization identities for residue families in full generality.

For a conformally flat metric h, the set of $2N$ factorization identities for $D_{2N}^{\mathrm{res}}(h; \lambda)$ consists of *two* subsets of N identities of different nature. The N identities in the first set (see (6.11.2)) factorize $D_{2N}^{\mathrm{res}}(h; \lambda)$, for certain values of λ, into products of lower order residue families and GJMS-operators on M. In the proofs of Theorem 6.11.2 and Theorem 6.11.4, we have seen that the identities in this set follow from factorizations of the families $P_{2N}(\lambda)$ (Lemma 6.11.7). In fact, this is a general phenomenon. The following result actually suffices to prove the first system of factorization identities for general metrics.

Theorem 6.11.18. *For* (M^n, h), $k = 0, \ldots, N$ *and* $2N \leq n$,

$$P_{2N} \left(h; \frac{n}{2} - k \right) = P_{2N-2k} \left(h; \frac{n}{2} + k \right) \circ P_{2k}(h). \tag{6.11.64}$$

For the proof of Theorem 6.11.18, we recall the identification of GJMS-operators as residues of the scattering operator (Chapter 3). Now the assertion follows from the fact that the scattering operator relates the leading terms of both

ladders in the asymptotics of an eigenfunction for the Laplacian of the corresponding Poincaré-Einstein metric. In the flat case, this argument was used in the proof
of Theorem 5.2.7.

The remaining N factorization identities for $D_{2N}^{\mathrm{res}}(h; \lambda)$ (see (6.11.3)) are of
different nature. For $N = 2$ and $N = 3$, the difference is clearly indicated by the
corresponding proofs of Theorem 6.11.2 and Theorem 6.11.4.

For the proof of Conjecture 6.11.1, it would suffice to establish the factorization identities with the factor $\bar{P}_2$. But even for conformally flat metrics, these
identities only follow by using the full power of the arguments of Theorem 6.11.1.
The remaining identities in the second set are even more mysterious.

However, the situation is different in connection with the discussion of the
structure of the critical Q-polynomial. The proof of Theorem 6.11.10 uses only the
first set of the factorization identities. Therefore, by the above remarks, it extends
to the general case (see Remark 6.11.4). In particular, Corollary 6.11.5 holds true
in the general case. This will be used in Section 6.12 and Section 6.13.

6.12   A recursive formula for P_6. Universality

In Section 6.11, we have seen that the known structure of the Q-polynomial $Q_6^{\mathrm{res}}(\lambda)$
(see Theorem 6.11.9) implies the formula

$$Q_6 = - \left[2P_2(Q_4) - 2P_4(Q_2) + 3P_2^2(Q_2)\right] - 6[Q_4 + P_2(Q_2)]Q_2 - 2^6 3! v_6 \quad (6.12.1)$$

(see (6.11.48)) for the critical Q_6. Now infinitesimal conformal variation of (6.12.1)
yields the following recursive formula for the critical GJMS-operator P_6.

Theorem 6.12.1. *The critical GJMS-operator P_6 can be written in the form*

$$\begin{aligned}
P_6 u = {}& [2(P_2P_4 + P_4P_2) - 3P_2^3]^0 u \\
& - 24(\mathsf{J}^2 - |\mathsf{P}|^2)\Delta u + 48\mathsf{J}(\mathsf{P}, \mathrm{Hess}(u)) + 24(d|\mathsf{P}|^2, du) + 48(\mathsf{P}\#d\mathsf{J}, du) \\
& + 2^6 3! [v_6^\bullet[u] + 6uv_6]. \quad (6.12.2)
\end{aligned}$$

Here $[\cdot]^0$ denotes the non-constant part of the respective operator.

Some comments are in order. By the discussion in Section 6.11, the relation
(6.12.1) holds true for all metrics. The same is true for Theorem 6.12.1. The terms
in the second and the third line of (6.12.2) define a *second*-order linear differential
operator with vanishing constant term. This shows that the operator $\mathcal{L} + \mathcal{L}^*$ in
(6.11.63) is of second order (see (6.12.5)).

Proof. We combine the identity $-P_6(h)(\varphi) = 6Q_6(h)\varphi + Q_6^\bullet(h)[\varphi]$ (see (3.1.7))
with (6.12.1). The calculation rests on the infinitesimal conformal transformation

laws

$$P_4^0(h)(\varphi) - 4Q_4(h)\varphi = (d/dt)|_0 \left(Q_4(e^{2t\varphi}h)\right) = Q_4^\bullet(h)[\varphi],$$
$$-P_2^0(h)(\varphi) - 2Q_2(h)\varphi = (d/dt)|_0 \left(Q_2(e^{2t\varphi}h)\right) = Q_2^\bullet(h)[\varphi]$$

(see (3.1.7)), and the conformal covariance of P_4 and P_2. Since

$$e^{6t\varphi}P_2(e^{2t\varphi}h)(Q_4(e^{2t\varphi}h)) = e^{2t\varphi}P_2(h)(e^{2t\varphi}Q_4(e^{2t\varphi}h)),$$

we find

$$P_2(Q_4)^\bullet[\varphi] + 6\varphi P_2(Q_4) = -2[P_2^0, \varphi](Q_4) + P_2 P_4^0(\varphi).$$

Similarly,

$$P_4(Q_2)^\bullet[\varphi] + 6\varphi P_4(Q_2) = -[P_4^0, \varphi](Q_2) - P_4 P_2^0(\varphi),$$
$$P_2^2(Q_2)^\bullet[\varphi] + 6\varphi P_2^2(Q_2) = -2[P_2^0, \varphi](P_2(Q_2)) - P_2^2 P_2^0(\varphi)$$

and

$$(Q_4 Q_2)^\bullet[\varphi] + 6\varphi Q_4 Q_2 = P_4^0(\varphi)Q_2 - P_2^0(\varphi)Q_4,$$
$$(P_2(Q_2)Q_2)^\bullet[\varphi] + 6\varphi P_2(Q_2)Q_2 = -P_2(P_2^0(\varphi))Q_2 - P_2(Q_2)P_2^0(\varphi).$$

These results yield

$$P_6(\varphi) = \left(-4[P_2^0, \varphi](Q_4) + 2P_2 P_4^0(\varphi)\right) + \left(2[P_4^0, \varphi](Q_2) + 2P_4 P_2^0(\varphi)\right)$$
$$- \left(6[P_2^0, \varphi](P_2(Q_2)) + 3P_2^2 P_2^0(\varphi)\right) + \left(6P_4^0(\varphi)Q_2 - 6P_2^0(\varphi)Q_4\right)$$
$$- \left(6P_2(P_2^0(\varphi))Q_2 + 6P_2(Q_2)P_2^0(\varphi)\right) + 2^6 3! \left(v_6^\bullet[\varphi] + 6\varphi v_6\right).$$

Now reordering gives

$$P_6(u) = \left[2P_2 P_4 + 2P_4 P_2 - 3P_2^3\right](u) - 2P_2(Q_4 u) + 4P_4(Q_2 u) - 6P_2^2(Q_2 u)$$
$$- [P_2^0, u] (4Q_4 + 6P_2(Q_2)) + 2[P_4^0, u](Q_2)$$
$$- 6(P_2 P_2^0 - P_4^0)(u)Q_2 - 6P_2^0(u)(Q_4 + P_2(Q_2)) + 2^6 3! \left(v_6^\bullet[u] + 6u v_6\right)$$

by using $P_2 = P_2^0 - 2Q_2$ and $P_4 = P_4^0 + Q_4$. In order to evaluate the last sum, we apply the explicit formulas

$$P_2 = \Delta - 2\mathsf{J} \quad \text{and} \quad P_4 = \Delta^2 + \delta(4\mathsf{J} - 4\mathsf{P}\#)d + (3\mathsf{J}^2 - 2|\mathsf{P}|^2 - \Delta\mathsf{J}).$$

Since the constant term of P_6 vanishes, it suffices to determine the constant term of the sum. A calculation yields the result

$$P_6(u) = [2(P_2 P_4 + P_4 P_2) - 3P_2^3]^0(u)$$
$$- 24(d(\mathsf{J}^2 - |\mathsf{P}|^2), du) - 24(\mathsf{J}^2 - |\mathsf{P}|^2)\Delta(u)$$
$$- 24\delta(\mathsf{P}\#d)(u\mathsf{J}) + 24\delta(\mathsf{P}\#d)(\mathsf{J})u - 24\delta(\mathsf{P}\#d)(u)\mathsf{J}$$
$$+ 2^6 3! \left(v_6^\bullet[u] + 6u v_6\right).$$

The assertion follows from here using $\delta(\mathsf{P}) = -d\mathsf{J}$. $\qquad \square$

Using Theorem 6.9.2 and the infinitesimal conformal transformation laws of P and $\mathcal{B}$, the formula (6.12.2) can be simplified.

Corollary 6.12.1. *The critical P_6 can be written in the form*

$$P_6 u = [2(P_2 P_4 + P_4 P_2) - 3P_2^3]^0 u - 48\delta(\mathsf{P}^2 \# du) - 8\delta(\mathcal{B} \# du). \qquad (6.12.3)$$

Proof. By Theorem 6.9.2 and Newton's identity $\operatorname{tr}(\wedge^3 \mathsf{P}) = \frac{1}{6}\mathsf{J}^3 - \frac{1}{2}\mathsf{J}|\mathsf{P}|^2 + \frac{1}{3}\operatorname{tr}(\mathsf{P}^3)$,

$$v_6 = -\frac{1}{48}\mathsf{J}^3 + \frac{1}{16}\mathsf{J}|\mathsf{P}|^2 - \frac{1}{24}\operatorname{tr}(\mathsf{P}^3) - \frac{1}{48}(\mathcal{B}, \mathsf{P}).$$

Hence by the conformal transformation law of P, the term $v_6^\bullet[u] + 6uv_6$ equals

$$\frac{1}{16}\mathsf{J}^2 \Delta u - \frac{1}{16}|\mathsf{P}|^2 \Delta u - \frac{1}{8}\mathsf{J}(\mathsf{P}, \operatorname{Hess}(u)) + \frac{1}{8}\operatorname{tr}(\mathsf{P}^2 \operatorname{Hess}(u)) - \frac{1}{48}\left[\mathfrak{b}^\bullet[u] + 6u\mathfrak{b}\right],$$

where $\mathfrak{b} = (\mathcal{B}, \mathsf{P})$. It follows that (6.12.2) simplifies to

$$\begin{aligned}
P_6 u = {}&[2(P_2 P_4 + P_4 P_2) - 3P_2^3]^0 u \\
&+ 24(d|\mathsf{P}|^2, du) + 48(\mathsf{P} \# d\mathsf{J}, du) + 48(\mathsf{P}^2, \operatorname{Hess}(u)) \\
&\qquad\qquad\qquad\qquad - 8\left[\mathfrak{b}^\bullet[u] + 6u\mathfrak{b}\right]. \quad (6.12.4)
\end{aligned}$$

Now

$$\begin{aligned}
\delta(\mathsf{P}^2 \# du) &= (\delta(\mathsf{P}^2), du) - (\mathsf{P}^2, \operatorname{Hess}(u)) \\
&= -(\mathsf{P} \# d\mathsf{J}, du) - \mathcal{C}_{jik}\mathsf{P}^{jk} u^i - \frac{1}{2}(d|\mathsf{P}|^2, du) - (\mathsf{P}^2, \operatorname{Hess}(u))
\end{aligned}$$

and

$$[\mathfrak{b}^\bullet[u] + 6u\mathfrak{b}] = -2(n-4)\mathcal{C}_{ikj}\mathsf{P}^{ij} u^k - (\mathcal{B}, \operatorname{Hess}(u))$$

(see (2.5.9), (6.9.6) and (6.9.5)). Hence the last two lines of (6.12.4) yield

$$\begin{aligned}
&24(d|\mathsf{P}|^2, du) + 48(\mathsf{P} \# d\mathsf{J}, du) + 48(\mathsf{P}^2, \operatorname{Hess}(u)) + 32\mathcal{C}_{ikj}\mathsf{P}^{ij} u^k + 8(\mathcal{B}, \operatorname{Hess}(u)) \\
&= -48\delta(\mathsf{P}^2 \# du) - 16\mathcal{C}_{ikj}\mathsf{P}^{ij} u^k + 8(\mathcal{B}, \operatorname{Hess}(u)) \\
&= -48\delta(\mathsf{P}^2 \# du) - 8\delta(\mathcal{B} \# du) - 16\mathcal{C}_{ikj}\mathsf{P}^{ij} u^k + 8(\delta(\mathcal{B}), du).
\end{aligned}$$

Now it suffices to apply the identity $\delta(\mathcal{B})_k = (n-4)\mathcal{C}_{ikj}\mathsf{P}^{ij}$. For $n = 6$, the latter relation is a consequence of the self-adjointness of the GJMS-operators. In fact, combining their self-adjointness with that of the second-order operators $\delta(\mathsf{P}^2 \# du)$, $\delta(\mathcal{B} \# du)$, the above calculation shows that the first-order differential operator $2\mathcal{C}_{ikj}\mathsf{P}^{ij} u^k - 8(\delta(\mathcal{B}), du)$ is self-adjoint, i.e., vanishes. □

Note that (6.12.3) makes the self-adjointness of P_6 obvious.

The Bach tensor $\mathcal{B}$ vanishes on the conformal class of flat metrics (although for $n \neq 4$ it is not conformally invariant). Hence for conformally flat metrics, the contribution $\delta(\mathcal{B}\#du)$ in (6.12.3) vanishes, and by comparing Corollary 6.12.1 with Corollary 6.11.6, we find the following explicit formula for the self-adjoint part of $\mathcal{L}$:

$$\frac{1}{2}(\mathcal{L}+\mathcal{L}^*)^0 u = -12\delta(\mathsf{P}^2\#du). \tag{6.12.5}$$

A closely related formula for a conformally covariant operator of the form $\Delta^3 + LOT$ can be found in [30] (Theorem 2.8).

Example 6.12.1. *For the round sphere S^6, we have* $\mathsf{P} = \frac{1}{2}g_c$ *and* $\mathsf{C} = 0$. *Therefore,* (6.12.3) *reads*

$$P_6 u = [2(P_2 P_4 + P_4 P_2) - 3P_2^3]^0 u + 12\Delta u. \tag{6.12.6}$$

This formula also follows from the product formulas for the GJMS-operators on the sphere. In fact, by (1.4.1), *the right-hand side of* (6.12.6) *equals*

$$\begin{aligned}[4(\Delta-6)(\Delta-6)(\Delta-4) - 3(\Delta-6)^3]^0 + 12\Delta &= (\Delta^3 - 10\Delta^2 + 12\Delta) + 12\Delta \\ &= (\Delta-6)(\Delta-4)\Delta \\ &= P_6.\end{aligned}$$

Next we discuss *universality*. The presentation (6.12.1) has the distinguished property of being universal. More precisely,

Theorem 6.12.2. (6.12.1) *holds true for all $n \geq 6$.*

Proof. We recall the formula

$$\begin{aligned}Q_6 = \Delta^2 \mathsf{J} &- 8\delta(\mathsf{P}\#d\mathsf{J}) - 4\Delta(\mathsf{J}^2 - |\mathsf{P}|^2) \\ &+ \frac{n-6}{4}\left(-6\mathsf{J}\Delta\mathsf{J} - 4(d\mathsf{J}, d\mathsf{J}) - 16\mathsf{J}|\mathsf{P}|^2 + (n+6)\mathsf{J}^3\right) - 2^6 3! v_6 \tag{6.12.7}\end{aligned}$$

(see (6.10.13)). Now

$$-2P_2(Q_4) = -2\left(\Delta - \left(\frac{n}{2} - 1\right)\mathsf{J}\right)\left(\frac{n}{2}\mathsf{J}^2 - 2|\mathsf{P}|^2 - \Delta\mathsf{J}\right),$$
$$-3P_2^2(Q_2) = -3\left(\Delta - \left(\frac{n}{2} - 1\right)\mathsf{J}\right)\left(\Delta - \left(\frac{n}{2} - 1\right)\mathsf{J}\right)\mathsf{J},$$
$$2P_4(Q_2) = 2\left(\Delta^2 + \delta((n-2)\mathsf{J}^2 - 4\mathsf{P}\#)d + \frac{n-4}{2}\left(\frac{n}{2}\mathsf{J}^2 - 2|\mathsf{P}|^2 - \Delta\mathsf{J}\right)\right)\mathsf{J}$$

and

$$\begin{aligned}-6(Q_4 + P_2(Q_2))Q_2 &= -6\left(\frac{n}{2}\mathsf{J}^2 - 2|\mathsf{P}|^2 - \Delta\mathsf{J} + \left(\Delta - \left(\frac{n}{2} - 1\right)\mathsf{J}\right)\mathsf{J}\right)\mathsf{J} \\ &= -6\mathsf{J}^3 - 12|\mathsf{P}|^2\mathsf{J}.\end{aligned}$$

A direct calculation shows that the sum of these four terms coincides with the right-hand side of (6.12.7), up to the v_6-term. $\hspace{2cm}\square$

In all cases, infinitesimal conformal variation of Q_6 yields P_6^0 (see (3.1.7)). Using (6.12.1) in the subcritical cases, we find the following result. It states that the formula for P_6 in Theorem 6.12.1 is universal, too.

Theorem 6.12.3 (Universal recursive formula for P_6). *On manifolds of dimension $n \geq 6$, the GJMS-operator P_6 satisfies*

$$P_6^0 u = [2(P_2 P_4 + P_4 P_2) - 3P_2^3]^0 u$$
$$- 24(\mathsf{J}^2 - |\mathsf{P}|^2)\Delta u + 48\mathsf{J}(\mathsf{P}, \mathrm{Hess}(u)) + 24(d|\mathsf{P}|^2, du) + 48(\mathsf{P}\#d\mathsf{J}, du)$$
$$+ 2^6 3![v_6^\bullet[u] + 6uv_6]. \quad (6.12.8)$$

The central point is that the coefficients in (6.12.8) do *not* depend on the dimension.

Proof. The calculation rests on the infinitesimal conformal transformation laws

$$P_2(Q_4)^\bullet[\varphi] + 6\varphi P_2(Q_4) = \left(\frac{n}{2} - 5\right)[P_2^0, \varphi](Q_4) + P_2 P_4^0(\varphi),$$
$$P_4(Q_2)^\bullet[\varphi] + 6\varphi P_4(Q_2) = \left(\frac{n}{2} - 4\right)[P_4^0, \varphi](Q_2) - P_4 P_2^0(\varphi),$$

$$P_2^2(Q_2)^\bullet[\varphi] + 6\varphi P_2^2(Q_2)$$
$$= \left(\frac{n}{2} - 5\right)[P_2^0, \varphi](P_2(Q_2)) + \left(\frac{n}{2} - 3\right)P_2[P_2^0, \varphi](Q_2) - P_2^2 P_2^0(\varphi)$$

and

$$(P_2(Q_2)Q_2)^\bullet[\varphi] + 6\varphi P_2(Q_2)Q_2$$
$$= \left(\frac{n}{2} - 3\right)[P_2^0, \varphi](Q_2)Q_2 - P_2(P_2^0(\varphi))Q_2 - P_2(Q_2)P_2^0(\varphi).$$

The calculations are analogous to those in the proof of Theorem 6.12.1. We omit the details. □

As above, we simplify (6.12.8) by using the known structure of v_6. The arguments in the proof of Corollary 6.12.1 yield the following result.

Corollary 6.12.2 (Explicit recursive formula for P_6). *On manifolds of dimension $n \geq 6$, the GJMS-operator P_6 is given by*

$$P_6^0 u = [2(P_2 P_4 + P_4 P_2) - 3P_2^3]^0 u - 48\delta(\mathsf{P}^2 \# du) - \frac{16}{n-4}\delta(\mathcal{B}\#du) \quad (6.12.9)$$

and

$$P_6 1 = -\left(\frac{n}{2} - 3\right)Q_6$$

with Q_6 as in (6.12.1).

In dimension $n = 4$, the right-hand side of (6.12.9) is well defined only for Bach-flat metrics. This reflects the non-existence ([118]) of a conformally covariant operator of the form $\Delta^3 + LOT$ in dimension 4 and, more precisely, shows the role of the Bach tensor as an obstruction. In [118], the non-existence is derived from the fact that the weight of the second-order operator $(\Delta P, \text{Hess})$ in *any* conformally covariant cube of the Laplacian is given by $\frac{8(n-2)}{n-4}$ (see the calculations on p. 574). It is easy to see that the weight of the corresponding contribution in (6.12.9) is given by the sum

$$8 + \frac{16}{n-4} = \frac{8(n-2)}{n-4}.$$

The method of curved Casimir operators ([56]) yields an alternative construction of a conformally covariant cube of the Laplacian in all dimensions $n \geq 5$ ($n \neq 6$). In dimension $n = 4$ it works only for conformally flat metrics. However, no explicit formulas are given in [56].

Using Theorem 6.10.3 and analytic continuation in dimension, (6.12.9) suggests the following result.

Lemma 6.12.1. *On four-manifolds, the second-order operator*

$$\mathcal{R} = -\delta(\mathcal{B}\#d) + (\mathcal{B}, \mathsf{P})$$

is conformally covariant:

$$e^{5\varphi} \circ \hat{\mathcal{R}} = \mathcal{R} \circ e^{-\varphi}.$$

Proof. In dimension 4, the Bach tensor is divergence free. Hence

$$\mathcal{R}(u) = (\mathcal{B}, \text{Hess}(u)) + (\mathcal{B}, \mathsf{P})u.$$

The assertion follows by direct calculation using the conformal invariance of $\mathcal{B}$ (see (6.9.7)) and the conformal transformation laws for Hess and P. We omit the details. $\qquad\square$

Note that even if the obstruction operator $\mathcal{R}$ vanishes in dimension $n = 4$, i.e., iff the Bach tensor vanishes, deriving a conformally covariant cube of the Laplacian in $n = 4$ from (6.12.9) is more subtle. In fact, this requires that

$$\lim_{n \to 4} \frac{e^{2\varphi}\hat{\mathcal{B}} - \mathcal{B}}{n - 4} = 0.$$

By the conformal transformation law (6.9.5) of $\mathcal{B}$, this condition is equivalent to $\mathsf{C} = 0$. This observation is parallel to the observation that the curved Casimir construction of [56] works in $n = 4$ only if $\mathsf{C} = 0$.

Corollary 6.12.2 is an analog of the universal formula

$$
\begin{aligned}
P_4^0 u &= (P_2^2)^0 u + 4\mathsf{J}\Delta u + 4(d\mathsf{J}, du) + 16[v_4^\bullet[u] + 4uv_4] \\
&= (P_2^2)^0 u + 4(d\mathsf{J}, du) + 4(\mathsf{P}, \text{Hess}(u)) \\
&= (P_2^2)^0 u - 4\delta(\mathsf{P}\#du)
\end{aligned}
$$

for P_4, which follows from $Q_4 = -P_2(Q_2) - Q_2^2 + 16v_4$ by infinitesimal conformal variation (see also (4.3.2)).

Example 6.12.2. *On the round sphere S^n, we have* $\mathsf{P} = \frac{1}{2}g_c$ *and* $\mathcal{B} = 0$. *Thus* (6.12.9) *states that*

$$P_6^0 = [2(P_2P_4 + P_4P_2) - 3P_2^3]^0 + 12\Delta.$$

By the product formula (1.4.1), *the latter identity follows from the polynomial identity*

$$[(x-a(a-1))(x-(a+1)(a-2))(x-(a+2)(a+3))]$$
$$= [4(x-a(a-1))(x-a(a-1))(x-(a+1)(a-2)) - 3(x-a(a-1))^3]$$
$$+ 12(x-a(a-1)).$$

In fact, this argument proves the stronger relation

$$P_6 = 2(P_2P_4 + P_4P_2) - 3P_2^3 + 12P_2 \tag{6.12.10}$$

which generalizes

$$P_4 = P_2^2 + 2P_2.$$

We recall that the recursive formula

$$P_6 = 2(P_2P_4 + P_4P_2) - 3P_2^3 + 2(\mathcal{L} + \mathcal{L}^*) \tag{6.12.11}$$

(see Corollary 6.11.6) is universal, too. This formula was derived in Section 6.11 under the assumption that h is conformally flat. The conformal flatness of the metric is needed here in order to apply the factorization identity of $D_6^{\mathrm{res}}(\lambda)$ which contains the factor $\bar{P}_4$. We finish the present section with an observation which supports the conjecture that (6.12.11) holds true *without* additional assumptions on the metric. In turn, this supports the conjecture that the mentioned factorization identity is valid for general metrics.

In the following we confirm the coincidence of the contributions of the Bach tensor $\mathcal{B}$ in (6.12.9) and (6.12.11). In order to determine the contribution of $\mathcal{B}$ to the commutator $i^*[\bar{P}_2, \bar{P}_4]$, we use the explicit formulas

$$\bar{P}_2 = \bar{\Delta} - \frac{n-1}{2}\bar{\mathsf{J}}$$

and

$$\bar{P}_4 = \bar{\Delta}^2 + \bar{\delta}((n-1)\bar{\mathsf{J}} - 4\bar{\mathsf{P}}\#)d + \frac{n-3}{2}\left(\frac{n+1}{2}\bar{\mathsf{J}}^2 - 2|\bar{\mathsf{P}}|^2 - \bar{\Delta}(\bar{\mathsf{J}})\right)$$

(with obvious notations). It follows that all contributions of $\mathcal{B}$ are contained in

$$-4\delta(\ddot{\mathsf{P}}\#d) - \frac{n-3}{2}\mathsf{J}^{(4)} + \frac{n-1}{2}\mathsf{J}^{(4)} - (n-3)|\ddot{\mathsf{P}}|^2, \tag{6.12.12}$$

where $\ddot{\mathsf{P}} = i^*(\partial/\partial r)^2|_0(\bar{\mathsf{P}})$ and $\mathsf{J}^{(k)} = i^*(\partial/\partial r)^k(\bar{\mathsf{J}})$. Now (6.11.9) implies

$$\ddot{\mathsf{P}} = -\frac{\partial^2}{\partial r^2}\Big|_0\left(\frac{1}{2r}h_r\right) = -4h_{(4)} = -\mathsf{P}^2 + \frac{\mathcal{B}}{n-4}$$

by Theorem 6.9.1. Hence the first term in (6.12.12) contributes

$$-\frac{4}{n-4}\delta(\mathcal{B}\#d).$$

Next, the generalization

$$\mathsf{J}^{(4)} = \frac{6}{n-4}(\mathcal{B},\mathsf{P}) + \cdots$$

of (6.11.9) implies that the remaining terms in (6.12.12) are given by

$$\frac{6}{n-4}(\mathcal{B},\mathsf{P}) - 2(n-3)(\mathsf{P},\ddot{\mathsf{P}}).$$

This sum contributes

$$\left(\frac{6}{n-4} - \frac{2(n-3)}{n-4}\right)(\mathcal{B},\mathsf{P}) = -2\frac{n-6}{n-4}(\mathcal{B},\mathsf{P}).$$

These results show that $\mathcal{B}$ contributes to $2(\mathcal{L}+\mathcal{L}^*)$ by

$$-\frac{16}{n-4}\delta(\mathcal{B}\#d) - \frac{n-6}{2}\frac{16}{n-4}(\mathcal{B},\mathsf{P}). \tag{6.12.13}$$

The latter sum coincides with the contributions of $\mathcal{B}$ in (6.12.9) (see Theorem 6.10.3).

Finally, we note that an explicit, but less structured, formula for a conformally covariant cube of the Laplacian can be extracted from the work [176] concerning the construction of actions of the form

$$S[g,u] = \int_M u(\Delta_g^k u + \cdots)\operatorname{vol}(g)$$

with the conformal invariance

$$S[e^{2\varphi}g, e^{(k-\frac{n}{2})\varphi}u] = S[g,u].$$

6.13 Recursive formulas for Q_8 and P_8

We consider a manifold of dimension $n = 8$. The following identity is the explicit version of the recursive formula (6.11.1) for the critical Q-curvature Q_8.

$$Q_8 = \frac{3}{5}P_2(Q_6) + \left[-4P_2^2 + \frac{17}{5}P_4\right](Q_4)$$

$$+ \left[-\frac{22}{5}P_2^3 + \frac{8}{5}P_2P_4 + \frac{28}{5}P_4P_2 - \frac{9}{5}P_6\right](Q_2) - \frac{16}{5}i^*\bar{P}_2^3(\bar{Q}_2). \tag{6.13.1}$$

As described in Section 6.11, its proof rests on the identity $-Q_8 = \dot{D}_8^{\text{res}}(0)(1)$, certain factorization identities for the families $D_{2N}^{\text{res}}(\lambda)$ for $N \leq 4$, and the universality of the recursive formulas for Q_6 and Q_4. Since the factorization identity with the second factor $\bar{P}_2$ is only known for conformally flat metrics, (6.13.1) is also only known for such metrics (see the comments after Conjecture 6.11.1). For more details we refer to [95].

An unfortunate aspect of (6.13.1) is that the contribution $i^* \bar{P}_2^3(\bar{Q}_2)$ is less understood. The following alternative recursive formula for Q_8 generalizes (6.11.44) and (6.11.48), and holds in full generality. The result stresses the significance of the Q-polynomials.

Theorem 6.13.1. *Assume that $\mathcal{V}_8(\lambda) \equiv 0$ (Conjecture 6.11.2). Then the critical Q_8 admits the presentation*

$$Q_8 = -\big[P_2(\mathcal{Q}_6^{\text{res}}(3)) + 3P_4(\mathcal{Q}_4^{\text{res}}(2)) + 3P_6(Q_2)\big]$$
$$- 12\big[Q_6 - P_2(\mathcal{Q}_4^{\text{res}}(1)) - 2P_4(Q_2)\big]Q_2 - 18\big[Q_4 + P_2(Q_2)\big]^2$$
$$+ 4!3!2^7 v_8. \quad (6.13.2)$$

Formula (6.13.2) yields an explicit presentation of Q_8 when combined with the following results for the individual terms. By Theorem 6.11.8, the coefficients of the Q-polynomials $\mathcal{Q}_4^{\text{res}}(\lambda)$ and $\mathcal{Q}_6^{\text{res}}(\lambda)$ can be expressed in terms of GJMS-operators and Q-curvatures (see (6.13.7)). Corollary 6.12.2 gives a recursive formula for P_6 in dimension $n = 8$. Lemma 6.14.1 provides a formula for v_8 for (locally) conformally flat metrics. For general metrics, the holographic coefficient v_8 is more complicated. By [117], it can be written in terms of P and the extended obstruction tensors $\Omega^{(1)}$ and $\Omega^{(2)}$.

Proof. We start from the identity

$$3!(Q_8^{\text{res}})^{\text{top}} = -\big[Q_8 + P_2\mathcal{Q}_6^{\text{res}}(3) + 3P_4\mathcal{Q}_4^{\text{res}}(2) + 3P_6\mathcal{Q}_2^{\text{res}}(1)\big] \quad (6.13.3)$$

(see Corollary 6.11.5 and the comments at the end of Section 6.11); the vanishing of $Q_6^{\text{res}}(0)$, $Q_4^{\text{res}}(0)$ and $Q_2^{\text{res}}(0)$ follows from Theorem 6.11.8. The left-hand side of (6.13.3) is given by the coefficient of λ^4 in

$$D_8^{\text{res}}(\lambda)(1) = P_8^*(\lambda)(1) - 16\lambda P_6^*(\lambda)(v_2) + 2^6 3\lambda(\lambda-1)P_4^*(\lambda)(v_4)$$
$$- 2^9 3\lambda(\lambda-1)(\lambda-2)P_2^*(\lambda)(v_6) + 2^{10} 3\lambda(\lambda-1)(\lambda-2)(\lambda-3)v_8.$$

Now $\mathcal{V}_8(\lambda) \equiv 0$ is equivalent to

$$P_8^*(\lambda)(1) = 2^{10} 3\lambda(\lambda-1)(\lambda-2)(\lambda-3)v_8 + 2^6 15\lambda(\lambda-1)(\lambda-2)P_2^*(\lambda)(v_6)$$
$$- 2^4 9\lambda(\lambda-1)P_4^*(\lambda)(v_4) + 14\lambda P_6^*(\lambda)(v_2)$$

(see (6.11.56)). Hence

$$D_8^{\text{res}}(\lambda)(1) = -2\lambda P_6^*(\lambda)(v_2) + 48\lambda(\lambda-1)P_4^*(\lambda)(v_4)$$
$$- 2^6 9\lambda(\lambda-1)(\lambda-2)P_2^*(\lambda)(v_6) + 2^{10}3\lambda(\lambda-1)(\lambda-2)(\lambda-3)v_8.$$

It follows that the left-hand side of (6.13.3) is given by

$$6\left[-2P_6^*(\lambda)^{[3]}(v_2) + 48P_4^*(\lambda)^{[2]}(v_4) - 2^6 9P_2^*(\lambda)^{[1]}(v_6) + 2^{10}3v_8\right], \qquad (6.13.4)$$

where the superscripts indicate the coefficients of the respective powers of λ. Now we have

$$P_2^*(\lambda)^{[2]} = -\mathsf{J}, \quad P_4^*(\lambda)^{[2]} = \mathsf{J}^2 + 2|\mathsf{P}|^2$$

and

$$P_6^*(\lambda)^{[3]} = 16(\log \det h)''' + 8(\log \det h)'' - \mathsf{J}(\mathsf{J}^2 + 2|\mathsf{P}|^2)$$

by (6.10.2). A calculation shows that

$$\mathsf{J}^2 + 2|\mathsf{P}|^2 = 12v_6 - 12v_2v_4 + 4v_2^3 \quad \text{and} \quad (\log \det h)'' = -\frac{1}{2}|\mathsf{P}|^2.$$

Hence

$$-2P_6^*(\lambda)^{[3]}(v_2) = 48(-8v_2v_6 + 12v_2^2v_4 - 5v_2^4),$$

and it follows that (6.13.4) is given by the sum of

$$6 \cdot 48\left(-32v_2v_6 + 24v_2^2v_4 - 5v_2^4 - 16v_4^2\right) \qquad (6.13.5)$$

and $4!3!2^7v_8$. Now we write (6.13.5) in terms of GJMS-operators and Q-curvatures. For that purpose, we use the relations

$$Q_2 = -2v_2, \quad Q_4 = -P_2Q_2 - Q_2^2 + 16v_4 \qquad (6.13.6)$$

and

$$Q_6 = -\left[3P_2^2(Q_2) - 2P_4(Q_2) + 2P_2(Q_4)\right] - 6\left[Q_4 + P_2(Q_2)\right]Q_2 - 16 \cdot 24v_6$$
$$= \left[P_2(Q_4^{\text{res}}(1) + 2P_4(Q_2)\right] - 6\left[Q_4 + P_2(Q_2)\right]Q_2 - 16 \cdot 24v_6.$$

The version of the second identity in (6.13.6) in dimension $n = 4$ was proved in (6.11.44). A direct calculation shows that it is universal. Similarly, the version of the first identity for Q_6 in $n = 6$ was proved in (6.11.48). A direct calculation shows that it is universal. Thus (6.13.5) equals

$$6 \cdot 48\left(16v_6Q_2 + \frac{6}{16}(Q_4 + P_2Q_2 + Q_2^2)Q_2^2 - \frac{1}{16}(Q_4 + P_2Q_2 + Q_2^2)^2 - \frac{5}{16}Q_2^4\right)$$

$$= 12 \cdot 24 \cdot 16v_6Q_2$$
$$+ 18\left[-(Q_4 + P_2Q_2 + Q_2^2)^2 + 6(Q_4 + P_2Q_2 + Q_2^2)Q_2^2 - 5Q_2^4\right]$$
$$= 12\left[-Q_6 + (P_2Q_4^{\text{res}}(1) + 2P_4Q_2) - 6(Q_4 + P_2Q_2)Q_2\right]Q_2$$
$$+ 18\left[-(Q_4 + P_2Q_2 + Q_2^2)(Q_4 + P_2Q_2 - 5Q_2^2) - 5Q_2^4\right]$$
$$= 12\left[-Q_6 + (P_2Q_4^{\text{res}}(1) + 2P_4Q_2)\right] - 18(Q_4 + P_2Q_2)^2.$$

The proof is complete. $\qquad \square$

Combining the formula in Theorem 6.13.1 with the holographic formula for Q_8 yields a formula for the divergence term

$$2\mathcal{T}_6^*(0)(v_2) + 4\mathcal{T}_4^*(0)(v_4) + 6\mathcal{T}_2^*(0)(v_6)$$

in terms of GJMS-operators and Q-curvatures. It would be of interest to establish such a result in full generality.

Infinitesimal conformal variation of (6.13.2) yields a recursive formula for the critical GJMS-operator P_8. We finish with some comments on that formula. In more explicit terms, the first bracket in (6.13.2) equals

$$P_2\left(3Q_6 + 12P_2\left(Q_4 + \frac{3}{2}P_2Q_2\right) - 8P_4Q_2\right)$$

$$+ 3P_4(-3Q_4 - 4P_2Q_2) + 3P_6Q_2. \quad (6.13.7)$$

The remaining terms are *non-linear* in lower order Q-curvatures. It follows that the resulting recursive formula for the critical P_8 expresses the operator as the sum of the *main part*

$$[3(P_2P_6 + P_6P_2) - 12(P_2^2P_4 + P_4P_2^2) - 8P_2P_4P_2 + 9P_4^2 + 18P_2^4] \quad (6.13.8)$$

and some low order operators.

The following result shows that, on the round sphere S^8, the low order correction terms in that formula are given by a multiple of the second-order operator P_2. In that case, the GJMS-operators commute, and the main part (6.13.8) coincides with the main part in

Lemma 6.13.1. *On S^8, we have*

$$P_8 = \left[6P_2P_6 - 32P_2^2P_4 + 9P_4^2 + 18P_2^4\right] + 3!4!P_2. \quad (6.13.9)$$

Lemma 6.13.1 follows from the product formulas for GJMS-operators on spheres. It is the analog of (6.12.6). Moreover, (6.13.9) is *universal*, i.e., holds true on all spheres S^n, $n \geq 8$. This generalizes (6.12.10).

In view of the latter observation, it is natural to ask whether in the curved case the low order correction terms are only of second order, too. In fact, this is what happens for P_6: formula (6.12.9) is the curved analog of (6.12.10).

We recall that Lemma 6.12.1 relates the non-existence of P_6 in dimension 4 to the conformally covariant second-order operator $\mathcal{R}$ which is defined by the Bach tensor. In a similar way, it would be natural to relate the non-existence of P_8 in dimension 4 and 6 ([112]) explicitly to the first two (extended) Fefferman-Graham obstruction tensors (in the sense of [117]). In particular, we expect that the existence of P_8 in dimension 6 is obstructed by the conformally covariant second-order operator $-\delta(\mathcal{O}\#d) + (\mathcal{O}, \mathsf{P})$ defined by the Fefferman-Graham obstruction tensor $\mathcal{O}$ in dimension 6. Similarly, the existence of P_8 in dimension 4 should be obstructed by a fourth-order conformally covariant operator.

Finally, it would be of interest to derive a presentation of P_8 with the main part (6.13.8) from the factorization identities (generalizing Theorem 6.11.17).

6.14 The holographic formula for conformally flat metrics and tube formulas

The theory of the residue families $D_{2N}^{\mathrm{res}}(\lambda)$ emphasizes the significance of *all* holographic coefficients v_{2j}, $2j \leq n$. In the present section, we determine the holographic coefficients v_{2j} for conformally flat metrics. The result suggests that we compare the situation with Weyl's tube formula ([130]). Moreover, we discuss the consequences for the holographic formulas for Q-curvatures.

Lemma 6.14.1. *The holographic coefficients of a conformally flat metric are given by*

$$v_{2j} = (-2)^{-j} \operatorname{tr}\left(\wedge^j(\mathsf{P})\right), \; j = 1, \ldots, \frac{n}{2}. \tag{6.14.1}$$

Proof. The result follows from the formula (see (6.9.8))

$$g = r^{-2}(dr^2 + h_r) = \frac{1}{4\rho^2} d\rho^2 + \frac{1}{\rho} h_\rho$$

with

$$h_\rho = \left(1 - \frac{\rho}{2}\mathsf{P}(h)\right)^2, \quad h_r = \left(1 - \frac{r^2}{2}\mathsf{P}(h)\right)^2 \tag{6.14.2}$$

for the Poincaré metric of a conformally flat metric h. For the flat metric $h_0 = \sum_1^n dx_i^2$, the metric $g_0 = r^{-2}(dr^2 + h_0)$ satisfies $\operatorname{Ric}(g_0) + ng_0 = 0$ and $i^*(r^2 g_0) = h_0$. g_0 is conformally flat. For $h = e^{2\varphi} h_0 \in [h_0]$, the solution $g = r^{-2}(dr^2 + h_r)$ of $\operatorname{Ric}(g) + ng = 0$ with $h(0) = h$ is the pull-back $\kappa^*(g_0)$ under a (local) diffeomorphism κ (fixing the boundary $r = 0$). In particular, g is conformally flat. It follows that

$$R(g) = -\mathsf{P}(g) \oslash g = \frac{1}{2}(g \oslash g).$$

By Lemma 6.14.2, the latter relation has the consequence

$$\ddot{h} = \frac{1}{2}\dot{h}h^{-1}\dot{h} \tag{6.14.3}$$

for small ρ; here derivatives are taken with respect to ρ (compare also with (6.9.11)). We claim that (6.14.3), in turn, yields

$$\dddot{h} = 0 \tag{6.14.4}$$

for small ρ. Since $h_{(2)}$ and $h_{(4)}$ are known by Theorem 6.9.1, formula (6.14.2) follows. It remains to verify (6.14.4). We differentiate (6.14.3) and find

$$\dddot{h} = \frac{1}{2}\dddot{h}h^{-1}\dot{h} + \frac{1}{2}\dot{h}(h^{-1})\dot{}\dot{h} + \frac{1}{2}\dot{h}h^{-1}\ddot{h}.$$

Using $(\dot{h^{-1}}) = -h^{-1}\dot{h}h^{-1}$ in the latter formula, gives

$$\dddot{h} = \frac{1}{2}\ddot{h}h^{-1}\dot{h} - \frac{1}{2}\dot{h}h^{-1}\dot{h}h^{-1}\dot{h} + \frac{1}{2}\dot{h}h^{-1}\ddot{h}.$$

But using (6.14.3) again, yields

$$\dddot{h} = \frac{1}{4}\dot{h}h^{-1}\dot{h}h^{-1}\dot{h} - \frac{1}{2}\dot{h}h^{-1}\dot{h}h^{-1}\dot{h} + \frac{1}{4}\dot{h}h^{-1}\dot{h}h^{-1}\dot{h} = 0.$$

This proves (6.14.4). $\square$

Lemma 6.14.2. *If the metric* $g = \frac{1}{4\rho^2}d\rho^2 + \frac{1}{\rho}h_\rho$ *satisfies* $R(g) = \frac{1}{2}(g \oslash g)$, *then*

$$\ddot{h} = \frac{1}{2}\dot{h}h^{-1}\dot{h}.$$

Proof. The assumption implies

$$R_{iNNj} = R(e_i, N, N, e_j) = -g_{ij}g_{NN} = -\frac{1}{4\rho^3}h_{ij}, \; N = \partial/\partial\rho.$$

On the other hand, the curvature is given by

$$R_{ijkl} = \frac{1}{2}\left((g_{ik})_{jl} + (g_{jl})_{ik} - (g_{jk})_{il} - (g_{il})_{jk}\right) + \sum_{p,q}g_{pq}\left(\Gamma^p_{ik}\Gamma^q_{jl} - \Gamma^p_{il}\Gamma^q_{jk}\right)$$

with

$$\Gamma^k_{ij} = \frac{1}{2}\sum_r\left((g_{jr})_i + (g_{ri})_j - (g_{ij})_r\right)g^{rk}.$$

For the present metric g, we find

$$R_{iNNj} = -\frac{1}{2\rho}\ddot{h}_{ij} + \sum_{p,q}g_{pq}\left(\Gamma^p_{iN}\Gamma^q_{Nj} - \Gamma^p_{ij}\Gamma^q_{NN}\right).$$

But

$$\Gamma^k_{NN} = \begin{cases} \frac{1}{2}(g_{NN})_N g^{NN} = -\frac{1}{\rho} & \text{for } k = N \\ 0 & \text{for } k \neq N \end{cases},$$

$$\Gamma^N_{ij} = -\frac{1}{2}(g_{ij})_N g^{NN} = -2\rho\dot{h} + 2h$$

and

$$\Gamma^k_{iN} = \begin{cases} 0 & \text{for } k = N \\ \frac{1}{2}\sum_r(g_{ri})_N g^{rk} = \frac{1}{2}(-\frac{1}{\rho}\delta_{ik} + (\dot{h}h^{-1})_{ik}) & \text{for } k \neq N \end{cases}.$$

These results (and some calculation) yield

$$R_{iNNj} = -\frac{1}{2\rho}\ddot{h}_{ij} + \frac{1}{4\rho}\sum_{p,q} h_{pq}\left(-\frac{1}{\rho}\delta_{ip} + (\dot{h}h^{-1})_{ip}\right)\left(-\frac{1}{\rho}\delta_{jq} + (\dot{h}h^{-1})_{jq}\right)$$

$$+ \frac{1}{2\rho^2}\dot{h}_{ij} - \frac{1}{2\rho^3}h_{ij} = -\frac{1}{2\rho}\ddot{h}_{ij} + \frac{1}{4\rho}(\dot{h}h^{-1}\dot{h})_{ij} - \frac{1}{4\rho^3}h_{ij}.$$

The combination of both formulas for R_{iNNj} implies the assertion. $\qquad\square$

The observation that for conformally flat h, (6.14.2) yields the *exact* Poincaré-Einstein metric, i.e., that the Fefferman-Graham series terminates at r^4, goes back to [220].

Remark 6.14.1. *A version of the formula*

$$g = r^{-2}(dr^2 + h_r), \quad h_r = \left(1 - \frac{r^2}{2}\mathsf{P}(h)\right)^2 = 1 - r^2\mathsf{P}(h) + \frac{r^4}{4}\mathsf{P}(h)^2$$

for a conformally flat metric h appears also in [195]. Let Γ be a convex-cocompact Kleinian group. It gives rise to the Kleinian manifold $X = \Gamma\backslash\mathbb{H}^{n+1}$ with boundary $\Gamma\backslash\Omega(\Gamma)$. In Epstein's appendix of [195] it is shown that the hyperbolic metric on an end $(0,\infty) \times \Sigma^n$ of X (Σ is a horospherically convex smooth hypersurface of X) is of the form

$$dt^2 + e^{-2t}h_- + h_0 + e^{2t}h_+,$$

where h_0, $h_\pm$ are symmetric bilinear forms on Σ. Moreover, the pull-back g_t of the hyperbolic metric to the parallel hypersurface Σ_t of (hyperbolic) distance t satisfies

$$g_t(X,Y) = g_0\left((\cosh tI + \sinh tS)(X), (\cosh tI + \sinh tS)(Y)\right),$$

when viewed as a metric on Σ. Here S is the shape operator on $T(\Sigma)$. Hence we find the relations

$$4h_\pm(X,Y) = g_0((I \pm S)(X), (I \pm S)(Y)),$$
$$2h_0(X,Y) = g_0(X,Y) - g_0(SX, SY).$$

It follows that the linear operator $T : T\Sigma \to T\Sigma$ which is defined by

$$h_0(X,Y) = h_+(TX,Y) \qquad\qquad (6.14.5)$$

is given by

$$T = 2(I - S)(I + S)^{-1}.$$

In fact, (6.14.5) is equivalent to

$$2\left[g_0(X,Y) - g_0(X, S^2(Y))\right] = g_0\left(TX, (I + S)^2(Y)\right),$$

i.e.,

$$g_0(TX, Y) = 2\left[g_0(X, (I + S)^{-2}(Y)) - g_0(X, S^2(I + S)^{-2}(Y))\right]$$
$$= 2g_0(X, (I - S)(I + S)^{-1}(Y)).$$

Similarly, for the linear map $R : T\Sigma \to T\Sigma$ so that

$$h_-(X, Y) = h_+(RX, Y), \tag{6.14.6}$$

we find

$$g_0((I - S)(X), (I - S)(Y)) = g_0((I + S)RX, (I + S)(Y)),$$

i.e.,

$$g_0(RX, Y) = g_0((I - S)(X), (I - S)(I + S)^{-2}(Y)),$$

i.e.,

$$R = (I + S)^{-2}(I - S)^2 = \frac{1}{4}T^2.$$

In other words (using $r = e^{-t}$), we can write the metric in the form

$$r^{-2}\left(dr^2 + h_+ + r^2 h_0 + r^4 h_-\right)$$

with

$$h_0^\sharp = 2(I - S)(I + S)^{-1} \quad and \quad h_-^\sharp = \frac{1}{4}(h_0^\sharp)^2,$$

where $\sharp$ is understood with respect to the metric h_+. Finally, note that h_+ can be viewed as a metric on the boundary at infinity. Similarly, h_- and h_0 can be viewed as symmetric bilinear forms on the boundary at infinity. For a related discussion we refer to [161].

Now combining Lemma 6.14.1 with the holographic formula (6.6.24) yields the following result.

Theorem 6.14.1 (Q-curvature for conformally flat metrics). *For conformally flat metrics on M^n ($n \geq 4$ even),*

$$Q_n = 2^{\frac{n}{2}-1}\left(\frac{n}{2}-1\right)!\sum_{j=0}^{\frac{n}{2}-1}(-1)^j 2^{-j}\frac{(\frac{n}{2}-j)!}{j!}P_{2j}^*(0)\,\mathrm{tr}\left(\wedge^{\frac{n}{2}-j}(\mathsf{P})\right).$$

In particular, we can write Q_n in the form

$$Q_n = 2^{\frac{n}{2}-1}\left(\frac{n}{2}\right)!\left(\frac{n}{2}-1\right)!\,\mathrm{tr}(\wedge^{\frac{n}{2}}\mathsf{P}) + \delta\omega \tag{6.14.7}$$

with a natural 1-form ω. The first term on the right-hand side of (6.14.7) is related to the integrand in the Chern-Gauß-Bonnet formula (Euler form).

We briefly recall the formulation of that formula. It will be convenient to denote the even dimension on M by $2n$. Let

$$E_{2n} = \frac{1}{(2\pi)^n} (-1)^n \operatorname{Pf}_{2n} \in \Omega^{2n}(M),$$

where the Pfaffian form is defined by

$$\operatorname{Pf}_{2n} = \frac{1}{2^n n!} \sum_{\sigma \in S_{2n}} \epsilon(\sigma) \Omega_{\sigma_1 \sigma_2} \wedge \cdots \wedge \Omega_{\sigma_{2n-1} \sigma_{2n}}$$

in terms of the curvature forms $\Omega_{ij} = g(R(\cdot, \cdot)e_i, e_j)$ with respect to a local orthonormal basis $\{e_i\}$. Then the Euler form E_{2n} has the density

$$\frac{1}{(2\pi)^n} (-1)^n \frac{1}{2^{2n} n!} \sum_{\sigma, \rho \in S_{2n}} \epsilon(\sigma) \epsilon(\rho) R_{\sigma_1 \sigma_2 \rho_1 \rho_2} \cdots R_{\sigma_{2n-1} \sigma_{2n} \rho_{2n-1} \rho_{2n}} \qquad (6.14.8)$$

with respect to the corresponding Riemannian volume. In the following, we shall often identify E_{2n} and Pf_{2n} with its respective density with respect to the Riemannian volume form. In particular, we have

$$-2\pi E_2 = R_{1212}$$

and

$$8(2\pi)^2 E_4 = \sum_{i,j,k,l} (R_{ijkl} R_{ijkl} - 4 R_{ijik} R_{ljlk} + R_{ijij} R_{klkl}) = |R|^2 - 4|\operatorname{Ric}|^2 + \tau^2.$$

The Chern-Gauß-Bonnet theorem ([130], [107]) states that

$$\int_{M^{2n}} E_{2n} = \chi(M^{2n}). \qquad (6.14.9)$$

Now for a conformally flat metric g,

$$\operatorname{Pf}_n(R) = \operatorname{Pf}_n(-\mathsf{P} \oslash g) = (-1)^{\frac{n}{2}} \left(\frac{n}{2}\right)! \operatorname{tr}(\wedge^{\frac{n}{2}} \mathsf{P}) \qquad (6.14.10)$$

using $C(g) = 0$ ([235], Proposition 8; the present formula differs from the one in [235] due to a missing coefficient $\frac{1}{2}$ in the definition of the curvature form). (6.14.7) and (6.14.10) imply the relation

$$Q_n = 2^{\frac{n}{2}-1} (-1)^{\frac{n}{2}} \left(\frac{n}{2} - 1\right)! \operatorname{Pf}_n + \delta \omega. \qquad (6.14.11)$$

The latter formula directly links Q_n to the Pfaffian. It refines the result ([39]) that for a conformally flat metric, the total integral of its Q-curvature is a multiple of the Euler characteristic.

(6.14.11) is of interest also in connection with the following classification result.

Theorem 6.14.2 ([81], [3]). *Let S_n be a scalar Riemannian curvature invariant of weight $-n$ on M^n (n even), i.e., $S_n(t^2 g) = t^{-n} S_n(g)$ for $t > 0$ and all metrics g. Assume that, for closed M, the integrals*

$$\int_M S_n \, \text{vol}$$

are conformally invariant. Then S_n can be decomposed as the sum

$$c_n \, \text{Pf}_n + C + \delta\omega \tag{6.14.12}$$

of a multiple of the Pfaffian, a local conformal invariant C and an exact divergence $\delta\omega$.

The decomposition in Theorem 6.14.2 was proposed by Deser and Schwimmer in [81]. Alexakis' monumental work [3] (combined with preparations in [5] and [6]) contains the first mathematically rigorous proof of this important fact. For the formulation of a number of related conjectures we refer to [41].

In Section 6.15, we shall decompose conformal anomalies of certain functional determinants in the form (6.14.12). These examples illustrate Theorem 6.14.2.

The critical Q-curvature Q_n is a local scalar Riemannian curvature invariant of weight $-n$ with a conformally invariant integral

$$\int_M Q_n \, \text{vol} \, .$$

Thus Theorem 6.14.2 yields a decomposition

$$Q_n = (-1)^{\frac{n}{2}} \left[(n-2)(n-4) \cdots 2 \right] \text{Pf}_n + C + \delta\omega \tag{6.14.13}$$

with a generally unknown local conformal invariant C: $e^{n\varphi} C(e^{2\varphi} g) = C(g)$ and a generally unknown $\omega \in \Omega^1(M)$.

Concerning the nature of the decomposition (6.14.13), we recall from Section 6.6 that the holographic formula for Q_n also shows that the quantity

$$(-1)^{\frac{n}{2}-1} \Delta^{\frac{n}{2}-1}(J)$$

contributes to Q_n. That contribution is the one which contains the *maximal* number $n-2$ of covariant derivatives ∇ acting on the curvature tensor R. On the other hand, Pf_n is a contribution which can be written *without* covariant derivatives acting on the curvature tensor R. In that sense both terms are extreme cases.

Note that the coefficient of Pf_n in (6.14.11) and (6.14.13) can be verified by the following argument. For $M = S^n$ (n even) with the round metric g_c, we have $Q_n = (n-1)!$ (see Theorem 3.2.3). Hence

$$\int_{S^n} Q_n \, \text{vol} = \text{vol}(S^n)(n-1)! = (2\pi)^{\frac{n}{2}} \frac{2}{(n-1)(n-3)\cdots 1}(n-1)!$$

$$= (2\pi)^{\frac{n}{2}} 2 \left[(n-2)(n-4) \cdots 2 \right].$$

On the other hand, by Chern-Gauß-Bonnet,

$$\int_{S^n} \mathrm{Pf}_n = (-2\pi)^{\frac{n}{2}} \chi(S^n) = 2(-2\pi)^{\frac{n}{2}}.$$

Thus

$$Q_n(S^n, g_c) = (-1)^{\frac{n}{2}} \left[(n-2)(n-4) \cdots 2 \right] \mathrm{Pf}(S^n, g_c).$$

There is a corresponding decomposition

$$v_n = \frac{1}{(n(n-2)\cdots 2)} \mathrm{Pf}_n + C + \delta\omega \qquad (6.14.14)$$

of the curvature invariant v_n. In the decompositions of $2Q_n$ and v_n, the respective coefficients of Pf_n differ by the factor $(-1)^{\frac{n}{2}} 2^n (\frac{n}{2})! (\frac{n}{2} - 1)!$. This is the coefficient of v_n in (6.6.24).

In dimension $n = 4$, the quantities Q_4 and v_4 are given by explicit formulas and the decompositions (6.14.13) and (6.14.14) can be made explicit. First of all, we have

$$Q_4 = 2J^2 - 2|P|^2 - \Delta J, \quad -\Delta = \delta d$$

(see (4.1.1)) and

$$v_4 = \frac{1}{8}(J^2 - |P|^2)$$

(Theorem 6.9.2), i.e.,

$$2Q_4 = 32v_4 - 2\Delta J.$$

Now the quantity $J^2 - |P|^2$ satisfies the identity

$$(J^2 - |P|^2) + \frac{1}{8}|C|^2 = \mathrm{Pf}_4. \qquad (6.14.15)$$

Hence

$$Q_4 = 2\,\mathrm{Pf}_4 - \frac{1}{4}|C|^2 - \Delta J, \qquad v_4 = \frac{1}{8}\mathrm{Pf}_4 - \frac{1}{64}|C|^2. \qquad (6.14.16)$$

In order to prove (6.14.15), we rewrite both sides as follows. (6.14.8) implies that

$$\mathrm{Pf}_4 = \frac{1}{8}\left(|R|^2 - 4|\,\mathrm{Ric}\,|^2 + \tau^2 \right).$$

On the other hand, the identities

$$J^2 - |P|^2 = -\frac{1}{4}|\,\mathrm{Ric}\,|^2 + \frac{1}{12}\tau^2 \qquad (6.14.17)$$

and

$$|C|^2 = |R|^2 - 2|\,\mathrm{Ric}\,|^2 + \frac{1}{3}\tau^2 \qquad (6.14.18)$$

(see the proof of Theorem 6.14.3) yield the result $\frac{1}{8}(|R|^2 - 4|\,\mathrm{Ric}\,|^2 + \tau^2)$ for the left-hand side of (6.14.15).

The formula

$$16v_4 = 2\,\mathrm{Pf}_4 - \frac{1}{4}|\mathsf{C}|^2$$

(see (6.14.16)) plays a key role in testing the AdS/CFT correspondence (see Section 6.15).

Conjecture 6.9.1 formulates an analog of the holographic formula (6.6.24) for all subcritical Q-curvatures. In the conformally flat case, combining it with Lemma 6.14.1 gives

$$Q_{2N} = 2^{N-1}N!(N-1)!\,\mathrm{tr}(\wedge^N \mathsf{P}) + \delta\omega, \tag{6.14.19}$$

up to terms of the form $\left(\frac{n}{2} - N\right)(\cdot)$.

Now $\mathsf{C} = 0$ implies that $\mathrm{tr}(\wedge^N \mathsf{P})$ and $\mathrm{tr}(\wedge^N \Omega)$ are proportional. Thus the latter formula relates Q_{2N} to a multiple of $\mathrm{tr}(\wedge^N \Omega)$. In other words, in the conformally flat category, the constant term of P_{2N} naturally splits off the quantity $\mathrm{tr}(\wedge^N \Omega)$. Similar observations have been discussed in the physical literature ([176]).

The quantities $\mathrm{tr}(\wedge^N \Omega)$ also appear in the coefficients in Weyl's formula for the volume of tubes ([130]). We describe that relation. Let $M^{n+1} \subset \mathbb{R}^{n+1}$ (n even) be an open set with compact closure and smooth boundary Σ^n. Let

$$M(\varepsilon) = \left\{x \in \mathbb{R}^{n+1} \mid \mathrm{dist}(x, M) \leq \varepsilon\right\}.$$

Then the volume

$$\mathrm{vol}(M^{n+1}(\varepsilon)) - \mathrm{vol}(M^{n+1})$$

is given by the formula

$$\varepsilon k_0 + \frac{\varepsilon^3}{3}k_2 + \cdots + \frac{\varepsilon^{n+1}}{(n+1)\cdots 3}k_n \tag{6.14.20}$$

(up to a linear combination of the powers $\varepsilon^2, \varepsilon^4, \dots, \varepsilon^n$), where the coefficients k_{2j} are defined by

$$k_{2j} = \frac{(-1)^j}{j!} \int_{\Sigma^n} \mathrm{tr}(\Lambda^j \Omega)\,\mathrm{vol},\ j = 0, \dots, \frac{n}{2}. \tag{6.14.21}$$

Here $\Omega : \Lambda^2(T\Sigma) \to \Lambda^2(T\Sigma)$ denotes the Riemann curvature endomorphism of Σ. In particular, k_{2j} only depends on the inner geometry of Σ^n. The coefficients of the even powers of ε depend on the embedding. More precisely, these coefficients depend linearly on the second fundamental form L; they vanish if L vanishes. For details see [130], Chapter 10.

The special case

$$\mathrm{vol}(M^3(\varepsilon)) = \mathrm{vol}(M^3) + \varepsilon\,\mathrm{area}\,(\Sigma^2) + \varepsilon^2 \int_{\Sigma^2} H\,\mathrm{vol} + \frac{\varepsilon^3}{3}4\pi$$

for a convex $M^3 \subset \mathbb{R}^3$ with boundary Σ^2 goes back to Steiner (see [130] for more details).

The numbers $k_0, k_2, \ldots, k_n$ also appear as coefficients in the formula for the volume of the ε-tube

$$T(\Sigma, \varepsilon) = \{x \in \mathbb{R}^{n+1} \mid \text{dist}(x, \Sigma) \le \varepsilon\}$$

of the closed hypersurface $\Sigma^n \subset \mathbb{R}^{n+1}$:

$$\text{vol}(T(\Sigma, \varepsilon)) = 2 \left(\varepsilon k_0 + \frac{\varepsilon^3}{3} k_2 + \cdots + \frac{\varepsilon^{n+1}}{(n+1) \cdots 3} k_n \right). \tag{6.14.22}$$

For a closed surface Σ^2 in $\mathbb{R}^3$, that formula reads

$$\text{vol}(T(\Sigma^2, \varepsilon)) = 2\varepsilon \, \text{area} \, (\Sigma^2) + \frac{4\pi}{3} \varepsilon^3 \chi(\Sigma^2)$$

by Gauß-Bonnet. For details see [130], Chapter 4.

Note also that the coefficients $k_{2j}, j = 1, \ldots, \frac{n}{2}$ are naturally related to heat-coefficients of Hodge-Laplacians ([86]). The relation between v_4 in dimension $n = 4$ to heat-coefficients will be the subject of Section 6.15.

There are analogous tube formulas for higher codimension submanifolds of the flat space $\mathbb{R}^n$, but no general results for submanifolds in a curved background. However, (6.14.20) should be regarded as a relative of the following volume formula for the half-tube $[0, \varepsilon] \times M^n$ with respect to the Riemannian volume of the conformal compactification $dr^2 + h_r$ of the Einstein metric $r^{-2}(dr^2 + h_r)$. Let $h = h_0$. Then

$$\text{vol}(\{r < \varepsilon\}; dr^2 + h_r)$$
$$= \varepsilon \int_M v_0 \, \text{vol}(h) + \frac{\varepsilon^3}{3} \int_M v_2 \, \text{vol}(h) + \cdots + \frac{\varepsilon^{n+1}}{n+1} \int_M v_n \, \text{vol}(h) + \cdots. \tag{6.14.23}$$

The structure of (6.14.23) resembles the structure of the half-tube formula (6.14.20). We refer to Section 6.16 for more details in the case of an Einstein metric h.

In the conformally flat case, the holographic coefficients are given by Lemma 6.14.1, i.e., are proportional to the quantities $\text{tr}(\wedge^j \Omega)$. For general metrics, it seems reasonable to ask for the existence of decompositions of the holographic coefficients v_{2j}, $2j \le n$ which naturally split off a multiple of $\text{tr}(\wedge^j \Omega)$. The case v_0 is trivial. The following result establishes such formulas for v_2 and v_4 (for general n).

Lemma 6.14.3.

$$v_2 = \frac{1}{2(n-1)} \text{tr}(\Omega)$$

for $n \ge 2$ and

$$v_4 = \frac{1}{8} \frac{1}{(n-2)(n-3)} \left(\text{tr}(\wedge^2 \Omega) - \frac{1}{4} |C|^2 \right)$$

for $n \ge 4$.

Proof. The observations

$$v_2 = \frac{1}{2}\operatorname{tr}(h_{(2)}) = -\frac{1}{2}J = -\frac{1}{4(n-1)}\tau = \frac{1}{2(n-1)}\operatorname{tr}(\Omega)$$

prove the first assertion. In order to prove the assertion for v_4, we recall the formula

$$v_4 = \frac{1}{8}\left((\operatorname{tr}(P))^2 - |P|^2\right) \tag{6.14.24}$$

for all $n \geq 4$ (Theorem 6.9.1). The definition (2.5.10) yields

$$|P|^2 = \frac{1}{(n-2)^2}\left(|\operatorname{Ric}|^2 + \frac{4-3n}{4(n-1)^2}\tau^2\right).$$

Hence

$$v_4 = \frac{1}{8(n-2)^2}\left[-|\operatorname{Ric}|^2 + \frac{n}{4(n-1)}\tau^2\right]. \tag{6.14.25}$$

On the other hand, for all $n \geq 4$,

$$\frac{1}{2}\operatorname{tr}(\wedge^2\Omega) = \frac{1}{8}\left(|R|^2 - 4|\operatorname{Ric}|^2 + \tau^2\right)$$

(see [130], Lemma 4.2). For $n = 4$, that expression coincides with Pf_4. Finally, we determine $|C|^2$. Using $C = R + P \oslash g$ (see (2.5.13)) and the identities

$$(R, P \oslash g) = -4(\operatorname{Ric}, P), \quad |P \oslash g|^2 = 4(n-2)|P|^2 + 4J^2,$$

we obtain

$$\begin{aligned}|C|^2 &= |R|^2 + 2(R, P \oslash g) + |P \oslash g|^2 \\ &= |R|^2 - 8(P, \operatorname{Ric}) + 4(n-2)|P|^2 + 4J^2 \\ &= |R|^2 - \frac{4}{n-2}|\operatorname{Ric}|^2 + \frac{2}{(n-2)(n-1)}\tau^2.\end{aligned}$$

Hence

$$\operatorname{tr}(\wedge^2(\Omega)) - \frac{1}{4}|C|^2 = \frac{n-3}{n-2}\left[-|\operatorname{Ric}|^2 + \frac{n}{4(n-1)}\tau^2\right]. \tag{6.14.26}$$

(6.14.25) and (6.14.26) imply the assertion. $\qquad\square$

On the other hand, it also seems to be reasonable to find generalizations of Lemma 6.14.1 in the form of natural decompositions of v_{2j} which split off a multiple of $\operatorname{tr}(\wedge^j P)$. In the special cases $j = 1, 2, 3$, such formulas easily follow from the explicit formulas for v_{2j}.

We close the present section with a comment concerning the relation to Anselmi's work [12]. Anselmi studied the problem of finding scalar Riemannian invariants of the form $\mathrm{Pf}_n + \delta \omega$ (up to local conformal invariants) with the property that their transformation law under conformal changes $e^{2\varphi} h$ is governed by a *linear* differential operator. In dimension $n = 4$, this question led Riegert ([208]) to the discovery of a version of Q-curvature and the associated operator P_4 (see the remarks in the introduction of Chapter 4). Anselmi called such quantities *pondered Euler densities*, and found that, in dimension $n = 6$, the quantity

$$G_6 = 2^6 3! \, \mathrm{Pf}_6 + \delta \omega$$

with

$$\delta \omega = \frac{48}{5} \delta(\mathrm{Ric} \, \# d\tau) + \frac{102}{25} \Delta(\tau^2) - 12\Delta(|\mathrm{Ric}|^2) - \frac{24}{5} \Delta^2 \tau \qquad (6.14.27)$$

is a pondered Euler density for conformally flat metrics. This result, in fact, is a consequence of the following observation.

Lemma 6.14.4. *For conformally flat metrics, $G_6 = -48 Q_6$.*

Proof. Using $\mathsf{P} = \frac{1}{4}(\mathrm{Ric} - \mathsf{J}h)$ and $\mathsf{J} = \frac{\tau}{10}$, we find $|\mathrm{Ric}|^2 = 16|\mathsf{P}|^2 + 14\mathsf{J}^2$. Hence $\frac{1}{48} \times (6.14.27)$ coincides with

$$2\delta(\mathrm{Ric} \, \# d\mathsf{J}) + \frac{17}{2}\Delta(\mathsf{J}^2) - \frac{1}{4}\Delta(|\mathrm{Ric}|^2) - \Delta^2 \mathsf{J}$$
$$= 8\delta(\mathsf{P} \# d\mathsf{J}) - 2\delta(\mathsf{J}d\mathsf{J}) + 5\Delta(\mathsf{J}^2) - 4\Delta(|\mathsf{P}|^2) - \Delta^2 \mathsf{J}$$
$$= 8\delta(\mathsf{P} \# d\mathsf{J}) + 4\Delta(\mathsf{J}^2) - 4\Delta(|\mathsf{P}|^2) - \Delta^2 \mathsf{J}.$$

Therefore,

$$\frac{1}{48}G_6 = 8\,\mathrm{Pf}_6 + 8\delta(\mathsf{P} \# d\mathsf{J}) + 4\Delta(\mathsf{J}^2 - |\mathsf{P}|^2) - \Delta^2 \mathsf{J} = -Q_6$$

by (6.14.11) and Theorem 6.10.1. □

Similarly, in dimension $n = 4$, Anselmi's pondered Euler density G_4 is given by

$$G_4 = 2^4 2! \, \mathrm{Pf}_4 - \frac{8}{3}\Delta\tau = 32\,\mathrm{Pf}_4 - 16\Delta\mathsf{J}.$$

Hence $G_4 = 16 Q_4 + 4|\mathsf{C}|^2$ by (6.14.16).

6.15 v_4 as a conformal index density

In the present section, we describe a relation between the holographic anomaly v_4 and certain conformal anomalies in spectral theory. This relation is considered as one of the earliest successful tests of the AdS/CFT-duality.

The setting is a closed oriented spin manifold (M, h) of dimension 4 with the spinor bundle S. Associated to (M, h) are the holographic coefficients

$$v_0 = 1, \quad v_2 = -\frac{1}{2}\operatorname{tr}(\mathsf{P}) = -\frac{1}{2}\mathsf{J}, \quad v_4 = \frac{1}{4}\operatorname{tr}(\wedge^2\mathsf{P}) = \frac{1}{8}(\mathsf{J}^2 - |\mathsf{P}|^2)$$

(Theorem 6.9.2). We relate the conformal anomaly v_4 to a combination of conformal anomalies of functional determinants of differential operators acting on certain types of fields on M. More precisely, we consider

- The Yamabe operator $-Y = -\Delta + \mathsf{J} = -\Delta + \frac{1}{6}\tau$.
- The square of the Dirac operator $\slashed{\nabla} : \Gamma(S) \to \Gamma(S)$.
- The form Laplacians $-\Delta_p$ on functions and 1-forms.

The Yamabe operator and the Dirac operator are conformally covariant:

$$e^{3\varphi} \circ Y(\hat{h}) \circ e^{-\varphi} = Y(h), \quad e^{\frac{5}{2}\varphi} \circ \slashed{\nabla}(\hat{h}) \circ e^{-\frac{3}{2}\varphi} = \slashed{\nabla}(h).$$

The square $\slashed{\nabla}^2$ and the Laplacians Δ_0 and Δ_1 are not conformally covariant. However, the operator

$$\delta d : \Omega^1(M^4) \to \Omega^1(M^4)$$

is conformally covariant: $e^{2\varphi} \circ \hat{\delta}d = \delta d$ by Lemma 4.2.1. In contrast to Y and $\slashed{\nabla}^2$, the operator δd is not elliptic.

Now if $D = -Y$ is positive, we define its determinant by

$$\det(D) = e^{-\zeta'(0, D)},$$

where

$$\zeta(s, D) = \sum_j \lambda_j^{-s} = \operatorname{tr}(D^{-s}), \quad \Re(s) > 2$$

is the spectral zeta function of D. Then we have the Polyakov-formula

$$-\log \det(-Y(h))^\bullet[\varphi] \overset{\text{def}}{=} -(d/ds)|_0 \log \det(-Y(e^{2s\varphi}h))$$

$$= 2 \int_M \varphi a_4(-Y(h)) \operatorname{vol}(h), \qquad (6.15.1)$$

where $a_4 \in C^\infty(M)$ appears in the coefficient of t^0 in the heat asymptotics

$$\operatorname{tr}(\varphi e^{-tD}) \sim \sum_{j \geq 0} t^{-2+\frac{j}{2}} \int_M \varphi a_j(D) \operatorname{vol}, \quad t \to 0$$

(see Theorem 4.10 in [31]). Moreover, the zeta-value

$$\zeta(0, D) = \int_M a_4(D) \operatorname{vol}$$

is a conformal invariant. $a_4(D)$ is the *conformal index density* of D (see [45] and Theorem 4.8 in [31]).

Now by general results of Gilkey [106] (see also [194]),

$$\mathcal{A}(Y) = 180(4\pi)^2 a_4(-Y) = \Delta\tau + |R|^2 - |\operatorname{Ric}|^2$$
$$= 6\Delta\mathsf{J} + 4(|\mathsf{P}|^2 - \mathsf{J}^2) + |\mathsf{C}|^2 \qquad (6.15.2)$$

for the Yamabe operator. Here we have used the identity

$$|R|^2 - |\operatorname{Ric}|^2 = |\mathsf{C}|^2 + 4(|\mathsf{P}|^2 - \mathsf{J}^2)$$

(see (6.14.17) and (6.14.18)). It follows that the conformal invariance of the conformal index $\int_M a_4$ vol is equivalent to the conformal invariance of the integral $\int_M v_4$ vol.

Next, for the square $\not\!\nabla^2$, we have (by [106], [46], Section 5 and [48])

$$\mathcal{A}(\not\!\nabla^2) = 180(4\pi)^2 a_4(-\not\!\nabla^2) = -6\Delta\tau - \frac{7}{2}|R|^2 - 4|\operatorname{Ric}|^2 + \frac{5}{2}\tau^2$$
$$= -36\Delta\mathsf{J} - 44(|\mathsf{P}|^2 - \mathsf{J}^2) - \frac{7}{2}|\mathsf{C}|^2 \qquad (6.15.3)$$

using (6.14.17) and (6.14.18). The latter formula shows the conformal invariance of the conformal index $\int_M a_4(-\not\!\nabla^2)$ vol.

Finally, for the Laplacians $-\Delta_0$ and $-\Delta_1$, we again use [106] and (6.14.17), (6.14.18) to find that the respective coefficients a_4 are given by

$$\mathcal{A}(\Delta_0) = 180(4\pi)^2 a_4(-\Delta_0) = 6\Delta\tau + |R|^2 - |\operatorname{Ric}|^2 + \frac{5}{2}\tau^2$$
$$= 36\Delta\mathsf{J} + 4(|\mathsf{P}|^2 - \mathsf{J}^2) + |\mathsf{C}|^2 + \frac{5}{2}\tau^2 \qquad (6.15.4)$$

and

$$\mathcal{A}(\Delta_1) = 180(4\pi)^2 a_4(-\Delta_1) = -6\Delta\tau - 11|R|^2 + 86|\operatorname{Ric}|^2 - 20\tau^2$$
$$= -36\Delta\mathsf{J} + 256(|\mathsf{P}|^2 - \mathsf{J}^2) - 11|\mathsf{C}|^2 + 5\tau^2. \qquad (6.15.5)$$

The integrals of $\mathcal{A}(-\Delta_0)$ and $\mathcal{A}(-\Delta_1)$ are not conformally invariant. However, we observe that the difference

$$\mathcal{A}(\Delta_1) - 2\mathcal{A}(\Delta_0) = -108\Delta\mathsf{J} + 248(|\mathsf{P}|^2 - \mathsf{J}^2) - 13|\mathsf{C}|^2 \qquad (6.15.6)$$

integrates to a multiple of the conformal invariant

$$\zeta(0, -\Delta_1) - 2\zeta(0, -\Delta_0).$$

Here we should emphasize that although the operator $(\delta d)_1 : \Omega^1(M) \to \Omega^1(M)$ is conformally covariant, it is not true that the zeta-value

$$\zeta(0, (\delta d)_1) = \zeta(0, -\Delta_1) - \zeta(0, -\Delta_0)$$

is conformally invariant. However, the function

$$\rho(s) \overset{\text{def}}{=} \zeta(s, -\Delta_1) - 2\zeta(s, -\Delta_0)$$

is regular at $s = 0$ and the value is conformally invariant. The quantity

$$\frac{\det(-\Delta_0)^4}{\det(-\Delta_1)^2} \tag{6.15.7}$$

is a special case of the *Cheeger half-torsion* (see [35]).

Now the formulas (6.15.2), (6.15.3), (6.15.4) and (6.15.5) imply the miraculous identity

$$6\mathcal{A}(Y) - 2\mathcal{A}(\overset{2}{\nabla}) + [\mathcal{A}(\Delta_1) - 2\mathcal{A}(\Delta_0)] = 360(|\mathsf{P}|^2 - \mathsf{J}^2), \tag{6.15.8}$$

and a consequence is the Polyakov-formula

$$-\frac{1}{2} \log \left(\frac{\det(-Y)^6 \det(-\Delta_1)}{\det(-\overset{2}{\nabla})^2 \det(-\Delta_0)^2} \right)^{\bullet} = -\frac{1}{\pi^2} \int_M \varphi v_4 \, \text{vol} \tag{6.15.9}$$

(ignoring complications due to non-trivial kernels).

The explicit formulas of the individual spectral anomalies on the left-hand side of (6.15.8) have been known in the physics literature since the middle of the 1970s (see [87] for a historical account). (6.15.2) can be found, for instance, in [136] and [137] (Chapter 15). We refer also to the useful review [234].

(6.15.9) can be interpreted as follows as a confirmation of the AdS/CFT-duality on the level of conformal anomalies. In [241], Witten proposed to formulate the duality as the identity

$$Z_{CFT}[h] = \exp(-S[g]) \tag{6.15.10}$$

of partition functions, where g is a critical point of the Einstein-Hilbert action S with $[h]$ as conformal infinity.

We consider a multiplet of six (bosonic) scalar fields, two (fermionic) spinors and one (bosonic) vector field. This is the field content of the super Yang-Mills Lagrangian on M^4 ([82]). Henningson and Skenderis ([141]) observed that the resulting conformal anomaly, i.e., the linear combination on the left-hand side of (6.15.8), is given by the conformal anomaly v_4 of the volume functional which is defined by holographic renormalization. In fact, the conformal anomaly of the left-hand side of (6.15.10) is given by

$$(d/ds)|_0 \left(D_{HS}^{-1/2}(e^{2s\varphi}h) \right),$$

where

$$D_{HS} = \frac{\det(-Y)^6 \det(-\Delta_1)}{\det(-\overset{2}{\nabla})^2 \det(-\Delta_0)^2}.$$

Note that the Dirac operator contributes to the denominator in view of the fermionic character of spinors. The contribution of $\det(-\Delta_0)$ is usually attributed to ghost fields. On the other hand, holographic renormalization of the right-hand side of (6.15.10) yields $\exp(-2nV(g;h))$ (see 1.6.21) with the conformal anomaly

$$\exp(-2n \int_M \varphi v_n \text{ vol}).$$

It remains to choose appropriate units.

Using the identity

$$8(\mathsf{J}^2 - |\mathsf{P}|^2) = 8\,\mathrm{Pf}_4 - |\mathsf{C}|^2, \quad \mathrm{Pf}_4 = \frac{1}{2}\,\mathrm{tr}(\wedge^2\Omega)$$

(see Theorem 6.14.3), the formulas (6.15.2), (6.15.3) and (6.15.6) can be written in the form

$$\mathcal{A}(Y) = 6\Delta\mathsf{J} + \frac{3}{2}|\mathsf{C}|^2 - 4\,\mathrm{Pf}_4,$$

$$\mathcal{A}(\nabla^2) = -36\Delta\mathsf{J} - 9|\mathsf{C}|^2 + 44\,\mathrm{Pf}_4,$$

$$\mathcal{A}(\Delta_1) - 2\mathcal{A}(\Delta_0) = -108\Delta\mathsf{J} + 18|\mathsf{C}|^2 - 248\,\mathrm{Pf}_4.$$

In these terms, the results appear in [141], up to the contributions of $\Delta\mathsf{J}$. The idea of the latter decompositions is to write the anomalies as sums of a multiple of Pf_4, a local scalar invariant and a divergence term (which is possible by Theorem 6.14.2). It follows that

$$6\mathcal{A}(Y) - 2\mathcal{A}(\nabla^2) + [\mathcal{A}(\Delta_1) - 2\mathcal{A}(\Delta_0)] = 45(|\mathsf{C}|^2 - 8\,\mathrm{Pf}_4).$$

The cancellation of the contributions of $\Delta\mathsf{J}$ in the latter sum was ignored in [141] since from the physical point these terms are considered to be of no importance.

The above calculations involve only *free* fields. A more refined version of the AdS/CFT-duality, however, would involve an interacting $SU(N)$ super Yang-Mills multiplet.

In dimension $n = 6$ the relation between v_6 and the conformal index densities of the free $(2,0)$ tensor multiplet was studied in [18].

6.16 The holographic formula for Einstein metrics

Here we demonstrate that the holographic formula (6.6.20) implies Gover's formula

$$Q_n(h) = (n-1)! \left(\frac{\tau(h)}{n(n-1)}\right)^{\frac{n}{2}} \tag{6.16.1}$$

(Theorem 3.2.3) for the Q-curvature of an Einstein metric h on Σ^n.

The first step is

Theorem 6.16.1. *Let h be an Einstein metric on Σ^n (n even). Then*

$$g = \frac{1}{r^2}\left(dr^2 + (1-cr^2)^2 h\right), \quad c = \frac{\tau(h)}{4n(n-1)} \tag{6.16.2}$$

is the Poincaré-Einstein metric on $(0,\varepsilon) \times \Sigma^n$ ($\varepsilon < 1/\sqrt{c}$ if $c > 0$ and $\varepsilon = \infty$ if $c \leq 0$) with conformal infinity $[h]$, i.e.,

$$\mathrm{Ric}(g) + ng = 0 \tag{6.16.3}$$

and the restriction of $r^2 g$ to $r = 0$ coincides with h. In particular,

$$v_{2j} = (-1)^j 2^{-j} \sigma_j(\mathsf{P}(h)) = (-1)^j 2^{-j} \operatorname{tr}(\wedge^j(\mathsf{P}(h))),$$

i.e.,

$$v(r) = \sum_j r^{2j} v_{2j} = \det\left(1 - \frac{r^2}{2}\mathsf{P}(h)\right).$$

Here we use the same symbol for the linear map $\mathsf{P}_h^\sharp : T(\Sigma) \to T(\Sigma)$ and the symmetric bilinear form $\mathsf{P}(h)$ (Schouten tensor) inducing it.

Note that if $\tau(h) > 0$ the metric in Theorem 6.16.1 exists only for $0 < r < 1/\sqrt{c}$. But the relation $\tau_{\lambda h} = \lambda^{-1}\tau(h)$ implies that the Poincaré-Einstein metric $g_{\lambda h}$ with the property $(r^2 g_{\lambda h})|_{r=0} = \lambda h \in [h]$ is given by

$$\frac{1}{r^2}\left(dr^2 + (1-\frac{c}{\lambda}r^2)^2 \lambda h\right)$$

for $0 < r < \sqrt{\lambda/c}$. In particular, for $\Sigma^n = S^n$ and $\lambda \geq c$ $g_{\lambda h}$ extends to a metric on the ball $\mathbb{B}^{n+1}$.

Proof. The construction of g is a special case of the warped product metric constructions in [24] (Section 9.J). We give an independent proof. h Einstein means

$$\mathrm{Ric}(h) = \frac{\tau(h)}{n}h.$$

In order to calculate $\mathrm{Ric}(g)$, we write g using the coordinates $r = e^s$. Then g is the warped product metric

$$g = ds^2 + \left(e^{-s} - ce^s\right)^2 h = ds^2 + f^2(s)h.$$

Since $g = f^2(f^{-2}ds^2 + h)$, where f only depends on s, it is enough to combine the transformation rule (2.5.4) with the explicit formula

$$\mathrm{Ric}\left(f^{-2}ds^2 + h\right)(X,Y) = \mathrm{Ric}(h)(X,Y)$$

for $X, Y \in \mathcal{X}(\Sigma)$; all other terms vanish. We find

$$\mathrm{Ric}(g)\left(\frac{\partial}{\partial s}, \frac{\partial}{\partial s}\right)$$

$$= -nf^{-2}\Delta_{f^{-2}ds^2}(\log f) - (n-1)f^{-2}|d\log f|^2 + (n-1)\left(\frac{\partial \log f}{\partial s}\right)^2$$

$$= -nf^{-2}(ff'') = -n = -ng\left(\frac{\partial}{\partial s}, \frac{\partial}{\partial s}\right),$$

$$\mathrm{Ric}(g)\left(\frac{\partial}{\partial s}, X\right) = 0, \ X \in \mathcal{X}(\Sigma)$$

and

$$\mathrm{Ric}(g)(X,Y) = \mathrm{Ric}(h)(X,Y) - \Delta_{f^{-2}ds^2}(\log f)h(X,Y) - (n-1)|d\log f|^2 h(X,Y)$$

$$= \frac{\tau(h)}{n}h(X,Y) - ff''h(X,Y) - (n-1)(f')^2 h(X,Y)$$

$$= \left(\frac{\tau(h)}{n} - ff'' - (n-1)(f')^2\right)h(X,Y).$$

For $f(s) = e^{-s} - ce^s$, we find for the coefficient in the latter formula

$$\frac{\tau(h)}{n} - \left(e^{-2s} + c^2 e^{2s} - 2c\right) - (n-1)\left(e^{-2s} + c^2 e^{2s} + 2c\right)$$

$$= -ne^{-2s} - nc^2 e^{2s} + \frac{\tau(h)}{n} - 2c(n-2) = -n\left(e^{-s} - ce^s\right)^2$$

using the definition of c. Hence

$$\mathrm{Ric}(g)(X,Y) = -nf^2 h(X,Y) = -ng(X,Y).$$

This proves (6.16.3). It follows that

$$v(r) = (1 - cr^2)^n = \sum_{j=0}^{n}\binom{n}{j}(-1)^j c^j r^{2j},$$

i.e.,

$$v_{2j} = (-1)^j c^j \binom{n}{j} = (-1)^j 2^{-2j}\binom{n}{j}\left(\frac{\tau(h)}{n(n-1)}\right)^j$$

$$= (-1)^j 2^{-j}\binom{n}{j}\left(\frac{\tau(h)}{2n(n-1)}\right)^j = (-1)^j 2^{-j}\sigma_j(P(h))$$

since

$$P(h) = \frac{\tau(h)}{2n(n-1)}h \qquad (6.16.4)$$

for the Einstein metric h. The proof is complete. $\qquad\square$

We have seen that g can be written as a warped product metric $ds^2 + f^2(s)h$. Hence its Ricci curvature can be determined by applying the following well-known general formula for the Ricci curvature of such a metric $G = g + f^2h$ on $M^m \times N^n$, $f \in C^\infty(M)$, $f > 0$ (see [24], Proposition 9.106).

$$\mathrm{Ric}(G)(X,Y) = \mathrm{Ric}(g)(X,Y) - \frac{n}{f} \, \mathrm{Hess}(f)(X,Y),$$

$$\mathrm{Ric}(G)(X,V) = 0,$$

$$\mathrm{Ric}(G)(V,W) = \mathrm{Ric}(h)(V,W) - f^2 h(V,W)\left[\frac{\Delta_g f}{f} + (n-1)\frac{|\,\mathrm{grad}\, f|^2}{f^2}\right]$$

for $X,Y \in \mathcal{X}(M)$ and $V,W \in \mathcal{X}(N)$. Such an argument is equivalent to the self-contained proof given above.

Note that if, in addition, the boundary metric h is conformally *flat*, then the associated Poincaré-Einstein metric g constructed in Theorem 6.16.1 is conformally *flat*, too. In order to prove the claim, we first note that $C(h) = 0$ implies $R(h) = -\mathsf{P}(h) \oslash h = -2c(h \oslash h)$ using $\mathsf{P}(h) = 2ch$. Now as in the above proof we find

$$
\begin{aligned}
R(g) &= R(ds^2 + f^2 h) \\
&= f^2 \left[R(f^{-2}ds^2 + h) + \Xi \oslash (f^{-2}ds^2 + h) \right] \\
&= f^2 R(h) + \Xi \oslash (ds^2 + f^2 h) \\
&= -2cf^2(h \oslash h) + \Xi \oslash (ds^2 + f^2 h),
\end{aligned}
$$

where $\Xi = \Xi_{(f^{-2}ds^2 + h, \log f)}$, i.e.,

$$
\begin{aligned}
\Xi &= \Delta_{f^{-2}ds^2}(\log f) f^{-2}ds^2 - d\log f \otimes d\log f + \frac{1}{2}f^2|d\log f|^2(f^{-2}ds^2 + h) \\
&= \frac{\ddot{f}}{f}ds^2 - \frac{1}{2}\left(\frac{\dot{f}}{f}\right)^2 ds^2 + \frac{1}{2}\dot{f}^2 h.
\end{aligned}
$$

Hence

$$R(g) = \left(-2cf^2 + \frac{1}{2}\dot{f}^2 f^2\right)(h \oslash h) + \left(\left(\ddot{f}f - \frac{1}{2}\dot{f}^2\right) + \frac{1}{2}\dot{f}^2\right)(ds^2 \oslash h).$$

Now for $f(s) = (e^{-s} - ce^s)$, we find $\ddot{f}f = f^2$ and $\dot{f}^2 - 4c = f^2$, i.e.,

$$R(g) = \frac{1}{2}f^4(h \oslash h) + f^2(ds^2 \oslash h) = \frac{1}{2}(g \oslash g).$$

But $\mathrm{Ric}(g) = -ng$ implies $\mathsf{P}(g) = -\frac{1}{2}g$, i.e., $R(g) = -\mathsf{P}(g) \oslash g$. Hence $C(g) = 0$. This proves the conformal flatness of g.

Now we combine Theorem 6.16.1 with the holographic formula

$$2Q_n(h) = (-1)^{\frac{n}{2}} \left(\dot{P}_n(h;0) - \dot{P}_n^*(h;0) \right) (1)$$

$$+ (-1)^{\frac{n}{2}} 2^n \left(\frac{n}{2} \right)! \sum_{j=0}^{\frac{n}{2}-1} \frac{(\frac{n}{2}-j-1)!}{2^{2j} j!} P_{2j}^*(h;0)(v_{n-2j}).$$

We apply the following two facts.

(i) The operator $\dot{P}_n(h;0) - \dot{P}_n^*(h;0)$ annihilates constants.

(ii) The operators $P_{2j}^*(h;0)$, $j = 1, \dots, \frac{n}{2} - 1$ annihilate constants.

In fact, the operators $P_{2j}(h;\lambda)$ are polynomials in the Laplacian Δ_h with constant coefficients. In particular, these operators are self-adjoint. But $P_{2j}(h;0)$ annihilates constants since the harmonic function $u = 1$ has a trivial asymptotics. Hence (ii) holds true. Similarly, (i) follows from the self-adjointness of $P_n(h;\lambda)$. It follows that the holographic formula simplifies to

$$2Q_n(h) = (-1)^{\frac{n}{2}} 2^n \left(\frac{n}{2} \right)! \left(\frac{n}{2} - 1 \right)! v_n$$

$$= (-1)^{\frac{n}{2}} 2^n \left(\frac{n}{2} \right)! \left(\frac{n}{2} - 1 \right)! \left\{ (-1)^{\frac{n}{2}} 2^{-\frac{n}{2}} \left(\frac{n}{\frac{n}{2}} \right) \left(\frac{\tau(h)}{2n(n-1)} \right)^{\frac{n}{2}} \right\}$$

$$= \frac{n!}{\frac{n}{2}} \left(\frac{\tau(h)}{n(n-1)} \right)^{\frac{n}{2}},$$

i.e.,

$$Q_n(h) = (n-1)! \left(\frac{\tau(h)}{n(n-1)} \right)^{\frac{n}{2}}.$$

This is Gover's formula (6.16.1) (Theorem 3.2.3, [110]).

Note that, alternatively, we can use Theorem 6.6.6 and (ii) to conclude more directly that

$$Q_n(h) = (-1)^{\frac{n}{2}} 2^{n-1} \left(\frac{n}{2} - 1 \right)! \left(\frac{n}{2} \right)! v_n.$$

Remark 6.16.1. *More generally, for $M^m = M^{m_1} \times M^{m_2}$ with the product metric $h = h_1 + h_2$ of two Einstein metrics h_i so that*

$$\frac{\tau_1}{2m_1(m_1-1)} = -\frac{\tau_2}{2m_2(m_2-1)}, \quad \tau_i = \tau(h_i) \qquad (6.16.5)$$

the associated Poincaré-Einstein metric is given by

$$\frac{1}{r^2} \left(dr^2 + (1-cr^2)^2 h_1 + (1+cr^2)^2 h_2 \right), \quad c = \frac{\tau_1}{4m_1(m_1-1)}$$

for sufficiently small r ([114]). In that case, the holographic formula says that

$$2Q_m(h_1+h_2) = (-1)^{\frac{m}{2}} 2^m \left(\frac{m}{2}\right)! \left(\frac{m}{2}-1\right)! v_m,$$

where v_m is the coefficient of r^m in

$$v(r) = (1 - cr^2)^{m_1} (1 + cr^2)^{m_2}.$$

Note that the formula

$$v(r) = \det\left(1 - \frac{r^2}{2}P(h)\right)$$

extends to the product case since

$$P(h_1+h_2) = P(h_1) + P(h_2).$$

In fact, for $P(h_1+h_2)$ we find

$$\frac{1}{m-2}\left(\frac{\tau_1}{m_1}h_1 + \frac{\tau_2}{m_2}h_2 - \frac{\tau_1+\tau_2}{2(m-1)}(h_1+h_2)\right).$$

But the relation (6.16.5) implies

$$\frac{1}{m-2}\left(\frac{\tau_i}{m_i} - \frac{\tau_1+\tau_2}{2(m-1)}\right) = \frac{\tau_i}{2m_i(m_i-1)}.$$

Using

$$P(h_i) = \frac{\tau_i}{2m_i(m_i-1)}h_i,$$

this yields the asserted formula for P.

Note that the formula for the Schouten tensor of the product metric, when combined with formula (6.9.3), shows the vanishing of the Bach tensor of the product metric. Hence Theorem 6.9.1 yields

$$h_{(4)} = \frac{1}{4}P(h_1+h_2)^2 = \frac{1}{4}(P(h_1)^2 + P(h_2)^2)$$

$$= \frac{1}{4}\left(\left(\frac{\tau_1}{2m_1(m_1-1)}\right)^2 h_1 + \left(\frac{\tau_2}{2m_2(m_2-1)}\right)^2 h_2\right) = c^2(h_1+h_2).$$

Thus the above result is compatible with Theorem 6.9.1.

Remark 6.16.2. *In the situation of Remark 6.16.1, with $(M_1, M_2) = (S^k, N^{n-k})$ and the product of the round metric on the sphere and an Einstein metric on N so that $\mathrm{Ric} + (n-k+1) = 0$, the formula for the corresponding Poincaré-Einstein metric was also mentioned in [238]. For $N = \Gamma \backslash \mathbb{H}^{n-k}$ with a cocompact discrete $\Gamma \subset SO(1, n-k)^\circ$, the product $M_1 \times M_2$ is the boundary of the Kleinian manifold*

$$\Gamma \backslash \mathbb{H}^{n+1} \simeq \mathbb{H}^{k+1} \times \Gamma \backslash \mathbb{H}^{n-k}$$

(here Γ is viewed as a subgroup of $SO(1, n+1)^\circ$).

The present context is a good case to illustrate again the relation to Weyl's tube formula. The formula

$$\text{vol}(\{t < r\}; dt^2 + h_t) = \int_0^r \int_\Sigma v(t) \, \text{vol}(\Sigma; h) dt$$

$$= \int_0^r \int_\Sigma \det\left(1 - \frac{t^2}{2} \mathsf{P}(h)\right) \text{vol}(\Sigma; h) dt \qquad (6.16.6)$$

for an Einstein metric h *and* the formula

$$\text{vol}(T(\Sigma, r); g_c) = \sum_\pm \int_0^r \int_\Sigma \det(1 \pm tS) \, \text{vol}(\Sigma; g_c) dt \qquad (6.16.7)$$

for the volume of a tube of radius r around a closed hypersurface Σ of the Euclidean space $(\mathbb{R}^n, g_c)$ both have a common origin. In (6.16.7) the operator $S : T(\Sigma) \to T(\Sigma)$ is the shape operator of Σ, i.e., $g(S(X), Y) = L(X, Y)$. Weyl's volume formula (6.14.22) is a consequence of (6.16.7) ([130]).

First of all, note that both formulas deal with the volumes of (geodesic) tubes. But whereas (6.16.7) deals with a flat background metric, in (6.16.6) the background metric is $dt^2 + h_t$ so that $t^{-2}(dt^2 + h_t)$ is Einstein.

Now the volume $\text{vol}(T(\Sigma, r); g)$ of a general tube

$$T(\Sigma, r) = \{x \in M \mid \text{dist}(x, \Sigma) \le r\}$$

around a closed hypersurface $\Sigma \subset M$ with respect to a background metric g is given by the formula

$$\text{vol}(T(\Sigma, r); g) = \sum_\pm \int_0^r \int_\Sigma v_\pm \, \text{vol}(\Sigma; g) dt,$$

where $v_\pm$ satisfies the differential equation

$$\frac{\dot{v}_\pm(t)}{v_\pm(t)} = \text{tr}(S_\pm(t)). \qquad (6.16.8)$$

Here $S_\pm(t) : T(\Sigma_t) \to T(\Sigma_t)$ are the shape operators of the two components $\Sigma_t^\pm$ of the parallel hypersurface with geodesic distance t from Σ. $S(t)$ itself satisfies the Ricatti differential equation

$$\dot{S} + S^2 + G = 0, \quad G(X) = R(X, N)N, \qquad (6.16.9)$$

where N denotes a geodesic normal field. Gray's studies of tubes ([130]) rest on these two fundamental differential equations.

In Weyl's case $G \equiv 0$ and

$$S_\pm(t) = \pm \frac{S(0)}{\det(1 \pm tS(0))}, \quad v_\pm(t) = \det(1 \pm tS(0))$$

is the solution of (6.16.8), (6.16.9).

In the case (6.16.6), the background curvature, i.e., G, is non-trivial and the equations (6.16.8), (6.16.9) are more complicated. For the warped product metric $dr^2 + \varphi^2(r)h$, we find

$$G = -\left(\left(\frac{\dot{\varphi}}{\varphi}\right)^2 + \left(\frac{\dot{\varphi}}{\varphi}\right)^{\cdot}\right) \mathrm{id}$$

and

$$S(t) = \frac{\dot{\varphi}}{\varphi}(t)\,\mathrm{id}$$

solves (6.16.9). (6.16.8) yields $v = \varphi^n$. For $\varphi(r) = 1 - cr^2$ (as in Theorem 6.16.1) we get

$$G(t) = \frac{2c}{1-ct^2}\,\mathrm{id}, \quad S(t) = -\frac{2ct}{1-ct^2}\,\mathrm{id}, \quad v(t) = (1-ct^2)^n.$$

Note that

$$G(0) = 2c\,\mathrm{id} = \frac{\tau(h)}{2n(n-1)}\,\mathrm{id} = \mathsf{P}(h).$$

The arguments in the proof of Lemma 6.8.1 can be used to prove the latter relation for *all* metrics h. Moreover, for Einstein h, we find the formula

$$G(t) = \mathsf{P}(h)\left(\mathrm{id} - \frac{t^2}{2}\mathsf{P}(h)\right)^{-1}.$$

In general, the equations (6.16.8), (6.16.9), however, do not yield new insight without further information on h_r. In fact, for the background metric $dr^2 + h_r$, we find

$$L(t) = \frac{1}{2}\dot{h}_t. \qquad (6.16.10)$$

In particular, $L(0) = 0$. (6.16.9) can be verified by an elementary calculation. (6.16.8) and (6.16.10) imply

$$\frac{\dot{v}}{v}(t) = \mathrm{tr}(S(t)) = \frac{1}{2}\mathrm{tr}(\dot{h}_t h_t^{-1}).$$

Hence using the initial condition $v(0) = 1$, we obtain

$$v(r) = \exp\int_0^r \frac{\dot{v}}{v}(t)dt = \exp\frac{1}{2}\int_0^r \mathrm{tr}(\dot{h}_t h_t^{-1})dt$$

$$= \exp\frac{1}{2}\int_0^r (d/dt)\,\mathrm{tr}(\log(h_t))dt = \exp\frac{1}{2}\int_0^r (d/dt)\log\det(h_t)dt$$

$$= (\det(h_r)/\det(h_0))^{1/2} = \mathrm{vol}(h_r)/\mathrm{vol}(h_0),$$

i.e., (6.16.8) is just the infinitesimal form of the definition of $v(r)$.

Now we add the Einstein condition

$$\mathrm{Ric}(g) + ng = 0, \quad g = r^{-2}(dr^2 + h_r).$$

Lemma 6.16.1.
$$\operatorname{tr}(G(r)) = -\frac{1}{r}\operatorname{tr}(S(r)).$$

Proof. We use the identity

$$-ng(X,Y) = \operatorname{Ric}(g)(X,Y) = \operatorname{Ric}(\bar{g})(X,Y) + (n-1)\bar{g}(\bar{\nabla}_X \overline{\operatorname{grad}}\varphi, Y)$$
$$+ \bar{g}(X,Y)\Delta_{\bar{g}}(\varphi) - (n-1)|d\varphi|^2_{\bar{g}}\bar{g}(X,Y) + (n-1)X(\varphi)Y(\varphi)$$

for $\varphi = \log r$, $\bar{g} = e^{2\varphi}g = dr^2 + h_r$ and $X = Y = \partial/\partial r$. Then

$$-\frac{n}{r^2} = \operatorname{Ric}(\bar{g})(\partial/\partial s, \partial/\partial s) - \frac{n-1}{r^2} + \Delta_{dr^2+h_r}(\log r) - \frac{n-1}{r^2} + \frac{n-1}{r^2}$$

$$= \operatorname{tr}(G) - \frac{n-1}{r^2} + \frac{1}{\sqrt{\det(h_r)}}\frac{\partial}{\partial r}\left(\sqrt{\det(h_r)}\frac{1}{r}\right),$$

i.e.,

$$\operatorname{tr}(G(r)) = -\frac{1}{2r}\frac{(\det(h_r))^\cdot}{\det(h_r)} = -\frac{1}{r}\operatorname{tr}(S(r)).$$

The proof is complete. □

It is obvious that for *any* metric h, the families

$$S(t) \stackrel{\text{def}}{=} -tP(h)\left(\operatorname{id} - \frac{t^2}{2}P(h)\right)^{-1}, \quad G(t) \stackrel{\text{def}}{=} P(h)\left(\operatorname{id} - \frac{t^2}{2}P(h)\right)^{-1}$$

(for sufficiently small t) provide a solution of the Ricatti equation

$$\dot{S} + S^2 + G = 0 \tag{6.16.11}$$

with the initial condition $S(0) = 0$. In addition, the pair (S, G) satisfies the necessary relation (Lemma 6.16.1)

$$G(t) = -\frac{1}{t}S(t)$$

for $r^{-2}(dr^2 + h_r)$ being Einstein. These data correspond to the volume function $v(r) = \det(\operatorname{id} - \frac{r^2}{2}P)$. For Einstein h, this is the *exact* solution and it is an *open problem* to describe how the *true* volume function deviates from the latter one.

Note that by taking traces (6.16.11) and Lemma 6.16.1 imply

$$\operatorname{tr}(\dot{S}(t)) + |S^2(t)|^2 - \frac{1}{t}\operatorname{tr}(S(t)) = 0.$$

The latter relation, in turn, yields the relations

$$\operatorname{tr}(S_j) = -\frac{1}{j-1}\operatorname{tr}(S_1 S_{j-1} + \cdots + S_{j-1}S_1), \ j \geq 2$$

for the coefficients S_j in the series $S(t) = \sum_{j\geq 0} t^j S_j$, $S_0 = 0$.

Note that the Ricatti equation also plays a central role in the so-called Hamilton-Jacobi method in holographic renormalization ([193], [174]).

We close the present section with an *alternative* more direct proof of (6.16.1). Since

$$Q_n(\Sigma^n; h) = \frac{1}{2^n} \left(\frac{\tau(h)}{n(n-1)} \right)^{\frac{n}{2}} Q_n(\Sigma^n; ch)$$

with c as in Theorem 6.16.1 and $\tau(ch) = c^{-1}\tau(h) = 4n(n-1) = \tau(S^n; \frac{1}{4}g_c)$, it is enough to prove

$$Q_n(\Sigma^n; h) = 2^n (n-1)! \tag{6.16.12}$$

for an Einstein metric h so that $\tau(h) = 4n(n-1)$. In order to prove (6.16.12), we use the identity $Q_n = (-1)^{\frac{n}{2}} \dot{P}_n(0)(1)$ (see (3.2.20)). For that purpose, we consider eigenfunctions of the Laplacian of the Poincaré-Einstein metric

$$g_h = \frac{1}{r^2} \left(dr^2 + (1-r^2)h \right).$$

The latter metric generalizes the Poincaré metric on the ball $\mathbb{B}^{n+1}$,

$$g_E = \frac{1}{r^2} \left(dr^2 + (1-r^2)^2 \frac{1}{4} h_{S^n} \right)$$

(see (6.5.9)). The Laplacian for g_h is readily obtained by replacing in (6.5.10) the Laplacian of $\frac{1}{4}h_{S^n}$ by the Laplacian of h. Moreover, the asymptotics of eigenfunctions for g_h can be obtained directly from the asymptotics of eigenfunctions for g_E just by replacing Laplacians (see Section 6.5 for details on the asymptotics). In particular, it follows that the functions

$$(-1)^{\frac{n}{2}} \dot{P}_n(\Sigma; h; 0)(1) \quad \text{and} \quad (-1)^{\frac{n}{2}} \dot{P}_n(S^n; \frac{1}{4}h_{S^n}; 0)(1)$$

coincide. Since the latter is given by

$$Q_n(S^n; \frac{1}{4}h_{S^n}) = 2^n Q_n(S^n; h_{S^n}) = 2^n (n-1)!,$$

this proves (6.16.12).

Combining the above arguments with (3.2.18), it follows also that the critical GJMS-operator $P_n(\Sigma^n, h)$ for (normalized) Einstein h (i.e., $\tau(h) = 4n(n-1)$), readily can be obtained from $P_n(S^n; \frac{1}{4}h_{S^n})$ just by replacing Laplacians in the product formula (3.2.25) on the standard sphere. We find

$$P_n(\Sigma^n; h) = \prod_{j=\frac{n}{2}}^{n-1} (\Delta_h - 4j(n-1-j)).$$

Hence, for (not normalized) Einstein h, we find

$$P_n(\Sigma^n; h) = P_n(\Sigma^n; c^{-1}(ch)) = c^{\frac{n}{2}} P_n(\Sigma^n; ch)$$

$$= c^{\frac{n}{2}} \prod_{j=\frac{n}{2}}^{n-1} (\Delta_{ch} - 4j(n-1-j)) = \prod_{j=\frac{n}{2}}^{n-1} (\Delta_h - 4j(n-1-j)c) \quad (6.16.13)$$

$$= \prod_{j=\frac{n}{2}}^{n-1} \left(\Delta_h - \frac{j(n-1-j)}{n(n-1)} \tau(h)\right).$$

This proves Gover's product formula (Theorem 3.2.3). The argument extends to the non-critical GJMS-operators.

In certain special cases, the product formula (6.16.13) can be proved by more elementary methods. Graham ([120]) gave such an argument proving (6.16.13) for the round sphere. Here we add a similar argument for the real hyperbolic space. A consequence of that formula naturally leads to the families $D_N^{nc}(\lambda)$ for certain parameters N and λ. The following arguments are taken from [150].

Let $(\mathbb{H}^n, g_c)$ be the upper half-space model of hyperbolic space with the metric

$$g_c = x_n^{-2}(dx_1^2 + \cdots + dx_n^2).$$

Theorem 6.16.2. *Let n be even and $N \geq 1$. Then*

$$P_{2N}((\mathbb{H}^n, g_c)) = x_n^{\frac{n}{2}+N} \Delta_{\mathbb{R}^n}^N x_n^{-\frac{n}{2}+N} = \prod_{j=\frac{n}{2}}^{\frac{n}{2}+N-1} (\Delta_{\mathbb{H}^n} + \kappa_j),$$

where $\kappa_j = j(n-1-j)$.

The first line is a consequence of the conformal covariance of P_{2N}. For the proof of the factorization, we recall that

$$\Delta_{\mathbb{H}^n} = x_n^2 \sum_{i=1}^{n} \frac{\partial^2}{\partial x_i^2} - (n-2)x_n \frac{\partial}{\partial x_n},$$

and apply a partial Fourier transform in the variable x'. It follows that Theorem 6.16.2 is a consequence of

Theorem 6.16.3. *On $\mathbb{R}^+$,*

$$y^{\frac{n}{2}+N} \left(\frac{d^2}{dy^2} + c^2\right)^N y^{-\frac{n}{2}+N} = \prod_{j=\frac{n}{2}}^{\frac{n}{2}+N-1} \left(y^2 \left(\frac{d^2}{dy^2} + c^2\right) - (n-2)y\frac{d}{dy} + \kappa_j\right)$$

for $c \in \mathbb{C}$.

For the proof of Theorem 6.16.3, we introduce the abbreviations

$$L = \frac{d^2}{dy^2} + c^2, \quad T = y^2 L - (n-2)y\frac{d}{dy} + \left(\frac{n}{2}-1\right)\frac{n}{2}. \tag{6.16.14}$$

Lemma 6.16.2.

(i) $L \circ y^m - y^m \circ L = m(m-1)y^{m-2} + 2my^{m-1} \circ \frac{d}{dy}$ for $m \in \mathbb{N}$.

(ii) $L^m \circ y - y \circ L^m = 2mL^{m-1} \circ \frac{d}{dy}$ for $m \in \mathbb{N}$.

Proof. (i) is obvious. We prove (ii) by induction. For $m = 1$, it is a special case of (i). Thus assuming

$$L^m \circ y - y \circ L^m = 2mL^{m-1} \circ \frac{d}{dy}$$

we calculate

$$L^{m+1} \circ y - y \circ L^{m+1} = L \circ (L^m \circ y - y \circ L^m) + L \circ y \circ L^m - y \circ L^{m+1}$$

$$= L \circ \left(2mL^{m-1} \circ \frac{d}{dy}\right) + (L \circ y - y \circ L) \circ L^m$$

$$= 2mL^m \circ \frac{d}{dy} + 2\frac{d}{dy} \circ L^m = 2(m+1)L^m \circ \frac{d}{dy}.$$

The proof is complete. $\qquad\qquad\qquad\qquad\qquad\qquad\qquad\qquad\qquad\qquad\qquad\square$

Lemma 6.16.3. $y \circ L^m \circ y^{m-1} = L^{m-1} \circ y^m \circ L - m(m-1)L^{m-1} \circ y^{m-2}$ for $m \in \mathbb{N}$.

Proof. We calculate

$$y \circ L^m \circ y^{m-1} = y \circ L^{m-1} \circ L \circ y^{m-1}$$

$$= L^{m-1} \circ y \circ L \circ y^{m-1} - 2(m-1)L^{m-1} \circ \frac{d}{dy} \circ y^{m-1} \quad \text{(by Lemma 6.16.2/(ii))}$$

$$= L^{m-1} \circ y \circ \left(y^{m-1} \circ L + (m-1)(m-2)y^{m-3} + 2(m-1)y^{m-2} \circ \frac{d}{dy}\right)$$

$$\quad - 2(m-1)L^{m-1} \circ \frac{d}{dy} \circ y^{m-1} \qquad\qquad\qquad\qquad\qquad \text{(by Lemma 6.16.2/(i))}$$

$$= L^{m-1} \circ y^m \circ L + (m-1)(m-2)L^{m-1} \circ y^{m-2} + 2(m-1)L^{m-1} \circ y^{m-1} \circ \frac{d}{dy}$$

$$\quad - 2(m-1)L^{m-1} \circ \frac{d}{dy} \circ y^{m-1}$$

$$= L^{m-1} \circ y^m \circ L + (m-1)(m-2)L^{m-1} \circ y^{m-2} - 2(m-1)(m-1)L^{m-1} \circ y^{m-2}$$

$$= L^{m-1} \circ y^m \circ L - m(m-1)L^{m-1} \circ y^{m-2}.$$

The proof is complete. $\qquad\qquad\qquad\qquad\qquad\qquad\qquad\qquad\qquad\qquad\qquad\square$

Lemma 6.16.4. $T = y^{\frac{n}{2}+1} \circ L \circ y^{-\frac{n}{2}+1}$.

Proof. Direct calculation.   □

Theorem 6.16.4. *For $N \in \mathbb{N}_0$,*

$$T \circ (T-2) \circ \cdots \circ (T - N(N+1)) = y^{\frac{n}{2}+N+1} \circ L^{N+1} \circ y^{-\frac{n}{2}+N+1}.$$

Proof. We use induction. For $N = 0$, Lemma 6.16.4 proves the assertion. Now

$$\begin{aligned}
&T(T-2)\cdots(T-N(N+1))\\
&= T(T-2)\cdots(T-(N-1)N)(T-N(N+1))\\
&= y^{\frac{n}{2}+N}L^N y^{-\frac{n}{2}+N}(T-N(N+1)) \quad \text{(by assumption)}\\
&= y^{\frac{n}{2}+N}L^N y^{-\frac{n}{2}+N}\left(y^{\frac{n}{2}+1}Ly^{-\frac{n}{2}+1} - N(N+1)\right) \quad \text{(by Lemma 6.16.4)}\\
&= y^{\frac{n}{2}+N}(L^N y^{N+1}L)y^{-\frac{n}{2}+1} - N(N+1)y^{\frac{n}{2}+N}L^N y^{-\frac{n}{2}+N}\\
&= y^{\frac{n}{2}+N+1}L^{N+1}y^{-\frac{n}{2}+N+1} \quad \text{(by Lemma 6.16.3)}.
\end{aligned}$$

The proof is complete.   □

As a corollary of Theorem 6.16.4, we obtain for $N \geq 1$,

$$\begin{aligned}
y^{\frac{n}{2}+N}L^N y^{-\frac{n}{2}+N} &= T(T-2)\cdots(T-(N-1)N)\\
&= \prod_{j=0}^{N-1}\left(y^2 L - (n-2)y\frac{d}{dy} + \kappa_{\frac{n}{2}-1-j}\right) = \prod_{j=\frac{n}{2}}^{\frac{n}{2}+N-1}\left(y^2 L - (n-2)y\frac{d}{dy} + \kappa_j\right),
\end{aligned}$$

i.e., the proof of Theorem 6.16.3 is complete.

Corollary 6.16.1. *Let $u \in C^\infty(\mathbb{H}^n)$ satisfy*

$$-\Delta_{\mathbb{H}^n} u = \mu(n-1-\mu)u.$$

Then

$$x_n^n \Delta_{\mathbb{R}^n}^{\frac{n}{2}} u = (-\mu)\cdots(n-1-\mu)u.$$

Proof. Theorem 5.2.2 (for $N = \frac{n}{2}$) implies

$$x_n^n \Delta_{\bar{E}}^{\frac{n}{2}}\omega = \prod_{j=0}^{\frac{n}{2}-1}\left(-\mu(n-1-\mu)+\kappa_j\right)\omega = \prod_{j=0}^{\frac{n}{2}-1}(\mu-j)(\mu-(n-1-j))\omega = \prod_{j=0}^{n-1}(\mu-j)\omega$$

by using the identity

$$-\mu(n-1-\mu) + j(n-1-j) = (\mu-j)(\mu+j) - (\mu-j)(n-1).$$

Since n is even the proof is complete.   □

Corollary 6.16.2. *Let n be even. Any eigenfunction $u \in C^\infty(\mathbb{R}^n_+)$ such that*

$$-\Delta_{\mathbb{H}^n} u = \lambda(n - \lambda)u$$

with $\lambda \in [0, n-1] \cap \mathbb{Z}$ satisfies

$$\Delta_{\mathbb{R}^n}^{\frac{n}{2}} u = 0.$$

Corollary 6.16.2 has an interesting consequence. A bounded eigenfunction u as in Corollary 6.16.2 can be considered as a measure $M(u)$ on $\mathbb{R}^n$ by extending it by 0 to the lower half plane, i.e., we set

$$\langle M(u), \varphi \rangle = \int_{\mathbb{R}^n_+} u\varphi dx, \quad \varphi \in C_0^\infty(\mathbb{R}^n).$$

Now we form the distribution

$$\Delta_{\mathbb{R}^n}^{\frac{n}{2}}(M(u)) \in C^{-\infty}(\mathbb{R}^n).$$

Corollary 6.16.2 implies that

$$\mathrm{supp}(\Delta_{\mathbb{R}^n}^{\frac{n}{2}}(M(u))) \subset \{x_n = 0\}.$$

A closer analysis shows that for $\lambda = N \in [0, \dots, n-1] \cap \mathbb{Z}$, the latter distribution can be described in terms of the adjoint operators

$$D_N^{nc}(\lambda)^*, \; D_{n-1-N}^{nc}(\lambda)^* : C^{-\infty}(\mathbb{R}^{n-1}) \to C^{-\infty}(\mathbb{R}^n)$$

for appropriate values of λ, acting on the boundary distributions of u.

6.17   Semi-holonomic Verma modules and their role

We recall that the universal enveloping algebra $\mathcal{U}(\mathfrak{g})$ of a Lie algebra $\mathfrak{g}$ is defined as the quotient of the tensor algebra $T(\mathfrak{g})$ by the two-sided ideal which is generated by the elements $X \otimes Y - Y \otimes X - [X, Y]$ for $X, Y \in \mathfrak{g}$. Given a representation (σ, V_σ) of a subgroup P let

$$\mathcal{M}(V_\sigma) \stackrel{\mathrm{def}}{=} (\mathcal{U}(\mathfrak{g}) \otimes V_\sigma) / \langle X \otimes v - 1 \otimes d\sigma(X)v, \; X \in \mathfrak{p} \rangle = \mathcal{U}(\mathfrak{g}) \otimes_{\mathcal{U}(\mathfrak{p})} V_\sigma,$$

where $\langle \cdot \rangle$ denotes the left $\mathcal{U}(\mathfrak{g})$-ideal generated by the elements indicated.

In the following, let $\mathfrak{g}$ be the Lie algebra of $G = SO(1, n)^\circ$, $\mathfrak{p}$ the Lie algebra of a parabolic subgroup $P \subset G$ and (σ, V_σ) an irreducible representation of P. Then the $\mathcal{U}(\mathfrak{g})$-module $\mathcal{M}(V_\sigma)$ is called the (generalized) Verma module induced by (σ, V_σ).

Following [19], we define the algebra $\mathcal{A}(\mathfrak{g})$ as the quotient of the tensor algebra $T(\mathfrak{g})$ by the two-sided ideal which is generated by the elements $X \otimes Y - Y \otimes$

$X - [X, Y]$ for $X \in \mathfrak{p}$ and $Y \in \mathfrak{g}$. In other words, in $\mathcal{A}(\mathfrak{g})$ it is only allowed to commute two elements if at least one is in $\mathfrak{p}$ or, equivalently, in $\mathcal{A}(\mathfrak{g})$ the relations $X \otimes Y - Y \otimes X = 0$ for $X, Y \in \mathfrak{n}^-$ do not hold true.

For a representation (σ, V_σ) of P, we define

$$\mathcal{N}(V_\sigma) \overset{\text{def}}{=} (\mathcal{A}(\mathfrak{g}) \otimes V_\sigma) / \langle X \otimes v - 1 \otimes d\sigma(X)v, \ X \in \mathfrak{p} \rangle,$$

where $\langle \cdot \rangle$ denotes the left $\mathcal{A}(\mathfrak{g})$-ideal generated by the elements indicated. The $\mathcal{A}(\mathfrak{g})$-module $\mathcal{N}(V_\sigma)$ is called the *semi-holonomic* Verma module induced by (σ, V_σ) ([92]).

The canonical projection $\pi : \mathcal{A}(\mathfrak{g}) \to T(\mathfrak{g})$ induces a canonical projection $\pi : \mathcal{N}(V_\sigma) \to \mathcal{M}(V_\sigma)$.

Next, we recall the correspondence between G-equivariant differential operators

$$D : C^\infty(G, E_\sigma)^P \to C^\infty(G, F_\eta)^P$$

and $\mathcal{U}(\mathfrak{g})$-module homomorphisms

$$\mathcal{D} : \mathcal{M}(F_\eta^*) \to \mathcal{M}(E_\sigma^*)$$

of generalized Verma modules. D induces a map

$$\mathcal{D} : \mathcal{U}(\mathfrak{g}) \otimes F_\eta^* \ni V \otimes f^* \mapsto \sum_i VT_i \otimes L_i(f^*) \in \mathcal{U}(\mathfrak{g}) \otimes E_\sigma^*,$$

where $T_i \in \mathcal{U}(\mathfrak{g})$ and the linear maps $L_i \in \text{Hom}(F_\eta^*, E_\sigma^*)$ are characterized by the condition

$$\langle D(u)(e), f^* \rangle = \sum_i \langle T_i(u)(e), L_i(f^*) \rangle \tag{6.17.1}$$

for $u \in C^\infty(G, E_\sigma)^P$. Here $\mathcal{U}(\mathfrak{g})$ acts from the right on $C^\infty(G)$. In order to prove that $\mathcal{D}$ induces a homomorphism of generalized Verma modules, we verify that $\mathcal{D}$ induces a map

$$\langle X \otimes f^* - 1 \otimes d\eta^*(X)f^* \mid X \in \mathfrak{p}, f^* \in F_\eta^* \rangle$$
$$\longrightarrow \langle X \otimes e^* - 1 \otimes d\sigma^*(X)e^* \mid X \in \mathfrak{p}, e^* \in E_\sigma^* \rangle$$

of left $\mathcal{U}(\mathfrak{g})$-ideals. In fact, (6.17.1) implies

$$\mathcal{D}(1 \otimes d\eta^*(X)f^*) = \sum_i (\text{ad}(X)T_i \otimes e_i^* + T_i \otimes d\sigma^*(X)e_i^*), \ X \in \mathfrak{p} \tag{6.17.2}$$

with $e_i^* = L_i(f^*)$. Hence

$$\mathcal{D}(X \otimes f^*) - \mathcal{D}(1 \otimes d\eta^*(X)f^*)$$
$$= \sum_i XT_i \otimes e_i^* - \sum_i (\text{ad}(X)T_i \otimes e_i^* + T_i \otimes d\sigma^*(X)e_i^*)$$
$$= \sum_i T_i (X \otimes e_i^* - 1 \otimes d\sigma^*(X)e_i^*), \ X \in \mathfrak{p}.$$

This proves the claim. It remains to verify the relation (6.17.2). But for $p \in P$, we have

$$\begin{aligned}
\langle D(u)(e), \eta^*(p)f^* \rangle &= \langle \eta(p^{-1})D(u)(e), f^* \rangle \\
&= \langle L_{p^{-1}}(D(u))(e), f^* \rangle \\
&= \langle D(L_{p^{-1}}(u))(e), f^* \rangle \\
&= \sum_i \langle T_i(L_{p^{-1}}(u))(e), e_i^* \rangle \\
&= \langle (\mathrm{Ad}(p)T_i)(u)(e), \sigma^*(p)e_i^* \rangle
\end{aligned}$$

by left G-equivariance of D. Here L_g denotes left translation by g. (6.17.2) follows by differentiation. In the reverse direction, we reconstruct D from $\mathcal{D}$ by

$$g \mapsto \left(F_\eta^* \ni f^* \mapsto \mathcal{D}(1 \otimes f^*)(u)(g) \right) \in F_\eta.$$

Here $T \otimes e^* \in \mathcal{U}(\mathfrak{g}) \otimes E_\sigma^*$ acts on $u \in C^\infty(G, E_\sigma)^P$ by $\langle T(u)(g), e^* \rangle$.

The following special case is the one which is of most significance here. Let $\mathbb{C}(\lambda)$ denote the 1-dimensional P-module with the action $a_t \mapsto a_t^\lambda = e^\lambda t$. Then $\mathbb{C}(\lambda)^* = \mathbb{C}(-\lambda)$. We use the notation

$$\mathcal{M}_\lambda(\mathfrak{g}) = \mathcal{U}(\mathfrak{g}) \otimes_{\mathcal{U}(\mathfrak{p})} \mathbb{C}(\lambda)$$

for the corresponding Verma module. Note that $C^\infty(G, \mathbb{C}(\lambda)^*)^P \simeq \mathrm{Ind}_P^G(\xi_\lambda)$ (Section 5.2).

In order to construct conformally covariant curved analogs of G-equivariant differential operators on sections of homogeneous vector bundles on spheres G/P, Baston ([19]) introduced an induction mechanism which combines Cartan's conformal connection with homomorphism of semi-holonomic Verma modules. Here it is necessary to go to the *semi-holonomic* category since the curvature of Cartan's conformal connection does not vanish, in general. We briefly explain the idea. For a conformal manifold $(M, [g])$, there is a replacement of the principal P-fibre bundle $G \to G/P$ together with the Maurer-Cartan form ω. More precisely, there exists a principal P-fibre bundle $\mathcal{G} \to M$ (P is a parabolic subgroup in $G = SO(1, n+1)$ if M has dimension n) and a one-form $\omega \in \Omega^1(\mathcal{G}, \mathfrak{g})$ so that

(i) $\omega_m : T_m\mathcal{G} \to \mathfrak{g}$ is an isomorphism for all $m \in \mathcal{G}$.

(ii) $\omega^{-1}(X) = \zeta_X$ for $X \in \mathfrak{p}$, where ζ_X denotes the corresponding fundamental vector field of the P-action.

(iii) $p_*(\omega) = \mathrm{Ad}(p)\omega$ for $p \in P$.

By (i), the horizontal vector fields $\omega^{-1}(X) \in \mathcal{X}(\mathcal{G})$, $X \in \mathfrak{g}$ are well defined. For $X \in \mathfrak{p}$, these are the fundamental vector fields of the P-action. For details on Cartan connections and its applications to constructions of invariant differential operators we refer to [57], [58], [59], [22].

Now an $\mathcal{A}(\mathfrak{g})$-module homomorphism $\mathcal{D} : \mathcal{N}(F_\eta^*) \to \mathcal{N}(E_\sigma^*)$ induces a differential operator

$$D : C^\infty(\mathcal{G}, E_\sigma)^P \to C^\infty(\mathcal{G}, F_\eta)^P$$

as follows. For $u \in C^\infty(\mathcal{G}, E_\sigma)^P = \{u \in C^\infty(\mathcal{G}, E_\sigma) \mid u(gp) = \sigma^{-1}(p)u(g), \; p \in P\}$, we define $Du(m)$ by

$$\left(F_\eta^* \ni f^* \mapsto \mathcal{D}(1 \otimes f^*)(u)(g)\right) \in F_\eta.$$

Here an element $T \otimes e^* \in \mathcal{A}(\mathfrak{g}) \otimes E_\sigma^*$ acts on u by $\langle \omega^{-1}(T)u, e^* \rangle$. The fact that $Du \in C^\infty(\mathcal{G}, F_\eta)^P$ is a consequence of (iii). The infinitesimal form of (iii) states that

$$\omega^{-1}([X, Y]) = \left[\omega^{-1}(X), \omega^{-1}(Y)\right]$$

for $X \in \mathfrak{p}, Y \in \mathfrak{g}$. But other commutators are *not* preserved by ω^{-1}. That is the reason to go from $\mathcal{U}(\mathfrak{g})$ to $\mathcal{A}(\mathfrak{g})$.

Now the problem of constructing a conformally covariant curved analog of an invariant differential operator on homogeneous vector bundles on $G/P \simeq S^n$ can be treated in the following three steps.

1. Identify the invariant differential operator with a homomorphism of (generalized) Verma modules.

2. Lift the homomorphism of Verma modules to a homomorphism of semi-holonomic Verma modules.

3. Use the lift to induce a differential operator on associated vector bundles on $\mathcal{G}/P \simeq M$.

An unfortunate aspect of that procedure is that the lifts required by the second step do not always exist. In particular, Eastwood and Slovak ([92]) proved that the Verma module homomorphisms which correspond to the critical GJMS-operators on the spheres (viewed as conformally invariant operators on densities) do *not* lift to the semi-holonomic category. Thus the above method cannot be used to induce the critical GJMS-operators.

Now the fact that the equivariant families $D_N^c(\lambda) : C^\infty(S^n) \to C^\infty(S^{n-1})$ are induced by the families $\mathcal{D}_N(\lambda)$ of homomorphisms of Verma modules (Theorem 5.1.5) suggests that we

1. lift the family $\mathcal{D}_N(\lambda)$ to the semi-holonomic category and
2. use the lifts to induce curved analogs of $D_N^c(\lambda)$.

These problems will be discussed in the following sections. Although we shall be able to construct certain rational lifts of all polynomial families $\mathcal{D}_N(\lambda)$ (Section 6.18), the curved analogs of $D_N^c(\lambda)$ will *not* arise by an induction procedure in terms of conformal connections. Instead, the specific form of the constructed semi-holonomic lifts will allow us to apply tractor calculus for that purpose. The consequences of the non-uniqueness of semi-holonomic lifts remain to be understood.

6.18   Zuckerman translation and $\mathcal{D}_N(\lambda)$

In the present section, we lift the families $\mathcal{D}_N(\lambda)$ to *rational* families of homomorphisms of semi-holonomic Verma modules using Zuckerman translation.

We first recall the construction of GJMS-operators by Zuckerman translation of the Yamabe operator. The idea is as follows. The Yamabe operator $P_2(\cdot)$: $C^\infty(M^n) \to C^\infty(M^n)$ satisfies the relation

$$P_2(e^{2\varphi}g) = e^{-(\frac{n}{2}+1)\varphi} \circ P_2(g) \circ e^{(\frac{n}{2}-1)\varphi}.$$

Using Cartan's conformal connection, it is induced by a homomorphism

$$\mathcal{N}_{-\frac{n}{2}-1}(\mathfrak{g}_{n+1}) \to \mathcal{N}_{-\frac{n}{2}+1}(\mathfrak{g}_{n+1}) \tag{6.18.1}$$

of semi-holonomic Verma modules

$$\mathcal{N}_\lambda(\mathfrak{g}_{n+1}) = \mathcal{A}(\mathfrak{g}_{n+1}) \otimes \mathbb{C}(\lambda)/\langle X \otimes 1 - 1 \otimes d\xi_\lambda(X)1, X \in \mathfrak{p}\rangle,$$

where ξ_λ denotes the irreducible P-representation on $\mathbb{C}(\lambda)$ defined by $\xi_\lambda(ma_t)z = e^{\lambda t}z$. In explicit terms that homomorphism is induced by the map

$$\mathcal{A}(\mathfrak{g}_{n+1}) \otimes \mathbb{C}\left(-\frac{n}{2}-1\right) \ni T \otimes 1 \mapsto T\Delta_n^- \otimes 1 \in \mathcal{A}(\mathfrak{g}_{n+1}) \otimes \mathbb{C}\left(-\frac{n}{2}+1\right).$$

The fact that (6.18.1) is well defined is a consequence of the commutator relations (6.18.3), $[H_0, \Delta_n^-] = -2\Delta_n^-$ and $[M_{ij}, \Delta_n^-] = 0$. The homomorphism (6.18.1) covers the corresponding homomorphism

$$\Delta_n^- : \mathcal{M}_{-\frac{n}{2}-1}(\mathfrak{g}_{n+1}) \to \mathcal{M}_{-\frac{n}{2}+1}(\mathfrak{g}_{n+1})$$

of generalized Verma modules for $\mathfrak{g}_{n+1}$.

Zuckerman translation constructs a new homomorphism using a given one. In the present context, let (π, F) be the fundamental representation of $\mathfrak{g}_{n+1}$ and consider the composition of the homomorphism

$$\Delta_n^- \otimes I : \mathcal{M}_{-\frac{n}{2}-1}(\mathfrak{g}_{n+1}) \otimes F \to \mathcal{M}_{-\frac{n}{2}+1}(\mathfrak{g}_{n+1}) \otimes F$$

with an embedding

$$i : \mathcal{M}_{-\frac{n}{2}-2} \hookrightarrow \mathcal{M}_{-\frac{n}{2}-1} \otimes F$$

and a projection

$$p : \mathcal{M}_{-\frac{n}{2}+1} \otimes F \to \mathcal{M}_{-\frac{n}{2}+2}$$

of modules. Here $X \in \mathfrak{g}$ acts on $\mathcal{M}_\lambda \otimes F$ by

$$T \otimes 1 \otimes v \mapsto XT \otimes 1 \otimes v + T \otimes 1 \otimes \pi(X)v.$$

For (π, F) we use the model given by the defining representation of $SO(1, n+1)$ on $\mathbb{R}^{1,n+1}$. The intertwining map

$$p \circ (\Delta_n^- \otimes I) \circ i : \mathcal{M}_{-\frac{n}{2}-2} \to \mathcal{M}_{-\frac{n}{2}+2}$$

is the Zuckerman translate of Δ_n^-. The procedure can be iterated, and yields non-trivial intertwining maps

$$\mathcal{M}_{-\frac{n}{2}-j}(\mathfrak{g}_{n+1}) \to \mathcal{M}_{-\frac{n}{2}+j}(\mathfrak{g}_{n+1})$$

for $j = 1, \ldots, \frac{n}{2} - 1$. However, it fails to construct a non-trivial intertwining map

$$\mathcal{M}_{-n}(\mathfrak{g}_{n+1}) \to \mathcal{M}_0(\mathfrak{g}_{n+1}).$$

The method can be used in the semi-holonomic category as well since the corresponding embedding i and projection p can be lifted.

For later use, it will be important to have explicit formulas for these maps. We introduce some more notation. We choose a basis $\{v_+, v_i, v_-, i = 1, \ldots, n\}$ of $F \simeq \mathbb{R}^{1,n+1}$ by

$$v_+ = (1, 1, 0, \ldots, 0)^t, \quad v_- = (1, -1, 0, \ldots, 0)^t, \quad v_i = (0, 0, e_i)^t, \ i = 1, \ldots, n.$$

Then

$$\begin{aligned}
Y_i^+ v_- &= 2v_i, & Y_i^+ v_j &= \delta_{ij} v_+, & Y_i^+ v_+ &= 0, \\
Y_i^- v_+ &= 2v_i, & Y_i^- v_j &= \delta_{ij} v_-, & Y_i^- v_- &= 0.
\end{aligned} \tag{6.18.2}$$

Moreover, the decomposition

$$\mathbb{R}^{1,n+1} \simeq \mathbb{R}v_- \oplus \langle v_i \rangle \oplus \mathbb{R}v_+ = V_{-1} \oplus V_0 \oplus V_1$$

is MA-equivariant. Here M acts trivially on $V_{\pm 1}$, by rotation on the n-dimensional space V_0, and A acts by $a_t : v_\pm \mapsto e^{\pm t} v_\pm$, $v_i \mapsto v_i$. Let $J(\lambda)$ be the left ideal generated by the elements $X \otimes 1 - 1 \otimes \xi_\lambda(X)1$, $X \in \mathfrak{p}$.

Lemma 6.18.1. *The homomorphism* $i : \mathcal{M}_{-\frac{n}{2}-2}(\mathfrak{g}) \to \mathcal{M}_{-\frac{n}{2}-1}(\mathfrak{g}) \otimes F$ *of* $\mathcal{U}(\mathfrak{g})$-*modules is induced by the* $\mathcal{U}(\mathfrak{g})$-*module map*

$$\mathcal{U}(\mathfrak{g}) \otimes \mathbb{C}\left(-\frac{n}{2}-2\right) \to \mathcal{U}(\mathfrak{g}) \otimes \mathbb{C}\left(-\frac{n}{2}-1\right) \otimes F$$

which is determined by

$$i : 1 \otimes 1 \mapsto \left(\frac{n}{2}+1\right)(1 \otimes 1 \otimes v_-) + \sum_i Y_i^- \otimes 1 \otimes v_i + \frac{1}{8}\Delta_n^- \otimes 1 \otimes v_+.$$

Similarly, the homomorphism $p : \mathcal{M}_{-\frac{n}{2}+1}(\mathfrak{g}) \otimes F \to \mathcal{M}_{-\frac{n}{2}+2}(\mathfrak{g})$ *of* $\mathcal{U}(\mathfrak{g})$-*modules is induced by the* $\mathcal{U}(\mathfrak{g})$-*module map*

$$p : \mathcal{U}(\mathfrak{g}) \otimes \mathbb{C}\left(-\frac{n}{2}+1\right) \otimes F \to \mathcal{U}(\mathfrak{g}) \otimes \mathbb{C}\left(-\frac{n}{2}+2\right)$$

which is determined by

$$1 \otimes 1 \otimes v_- \mapsto \frac{1}{2}(\Delta_n^- \otimes 1), \quad 1 \otimes 1 \otimes v_i \mapsto Y_i^- \otimes 1, \quad 1 \otimes 1 \otimes v_+ \mapsto (4-n)(1 \otimes 1).$$

Proof. In order to prove the assertion for i, it is enough to verify that

$$i : J\left(-\frac{n}{2}-2\right) \to J\left(-\frac{n}{2}-1\right) \otimes F.$$

But (6.18.2) says that modulo terms in $J(-\frac{n}{2}-1) \otimes F$

$$i(Y_j^+ \otimes 1) = \left(\frac{n}{2}+1\right)\left(Y_j^+ \otimes 1 \otimes v_- + 1 \otimes 1 \otimes Y_j^+ v_-\right)$$
$$+ \sum_i \left(Y_j^+ Y_i^- \otimes 1 \otimes v_i + Y_i^- \otimes 1 \otimes Y_j^+ v_i\right)$$
$$+ \frac{1}{8}\left(Y_j^+ \Delta_n^- \otimes 1 \otimes v_+ + \Delta_n^- \otimes 1 \otimes Y_j^+ v_+\right)$$
$$\equiv \left(\frac{n}{2}+1\right)\left(1 \otimes 1 \otimes 2v_j\right)$$
$$+ 2H_0 \otimes 1 \otimes v_j + 2\sum_{i \neq j} M_{ji} \otimes 1 \otimes v_i + \sum_i Y_i^- \otimes 1 \otimes \delta_{ij}v_+$$
$$+ \frac{1}{8}\left(2(n-2)Y_j^- + 4Y_j^- H_0\right) \otimes 1 \otimes v_+$$
$$\equiv 0.$$

Here we have used the commutator relation

$$\left[Y_j^+, \Delta_n^-\right] = 2(n-2)Y_j^- + 4Y_j^- H_0 + 4\sum_{r \neq j} Y_r^- M_{jr} \tag{6.18.3}$$

which follows from Lemma 5.1.3 for $j = 1$, and by a rotation for all j. Similarly, we calculate

$$i\left(H_0 \otimes 1 + \left(\frac{n}{2}+2\right) \otimes 1\right) = \left(\frac{n}{2}+1\right)\left(H_0 \otimes 1 \otimes v_- - 1 \otimes 1 \otimes v_-\right)$$
$$+ \left(\frac{n}{2}+2\right)\left(\left(\frac{n}{2}+1\right) \otimes 1 \otimes v_-\right) + \sum_i H_0 Y_i^- \otimes 1 \otimes v_i + \left(\frac{n}{2}+2\right) Y_i^- \otimes 1 \otimes v_i$$
$$+ \frac{1}{8}\left(H_0 \Delta_n^- \otimes 1 \otimes v_+ + \Delta_n^- \otimes 1 \otimes v_+ + \left(\frac{n}{2}+2\right) \Delta_n^- \otimes 1 \otimes v_+\right),$$

i.e.,

$$i\left(H_0 \otimes 1 + \left(\frac{n}{2}+2\right) \otimes 1\right) = \left(\frac{n}{2}+1\right)\left(H_0 \otimes 1 \otimes v_- + \left(\frac{n}{2}+1\right) \otimes 1 \otimes v_-\right)$$
$$+ \sum_i Y_i^- H_0 \otimes 1 \otimes v_i + \left(\frac{n}{2}+1\right) Y_i^- \otimes 1 \otimes v_i$$
$$+ \frac{1}{8}\left(\Delta_n^- H_0 \otimes 1 \otimes v_+ + \left(\frac{n}{2}+1\right) \Delta_n^- \otimes 1 \otimes v_+\right) \equiv 0$$

using $[H_0, \Delta_n^-] = -2\Delta_n^-$. Finally, $i(M_{ij} \otimes 1) \equiv 0$ since M acts trivially on $v_\pm$ and Δ_n^- and $\mathfrak{n}^- \simeq V_0$ as a M-module.

We continue with the proof for p. In order to verify that p defines a map $J(-\frac{n}{2}+1) \otimes F \to J(-\frac{n}{2}+2)$, we first calculate using (6.18.3)

$$p(Y_j^+ \otimes 1 \otimes v_-) = p(Y_j^+(1 \otimes 1 \otimes v_-) - 2 \otimes 1 \otimes v_j)$$
$$= \frac{1}{2}Y_j^+\Delta_n^- \otimes 1 - 2Y_j^- \otimes 1 \equiv ((n-2)Y_j^- + 2Y_j^- H_0 - 2Y_j^-) \otimes 1 \equiv 0$$

modulo terms in $J(-\frac{n}{2}+2)$. Similarly, we find

$$p(Y_j^+ \otimes 1 \otimes v_i) = p(Y_j^+(1 \otimes 1 \otimes v_i) - 1 \otimes 1 \otimes \delta_{ij}v_+)$$
$$= Y_j^+Y_i^- \otimes 1 + (n-4)\delta_{ij}(1 \otimes 1) \equiv 2H_0\delta_{ij} \otimes 1 + (n-4)\delta_{ij}(1 \otimes 1) \equiv 0$$

modulo $J(-\frac{n}{2}+2)$ and

$$p(Y_j^+ \otimes 1 \otimes v_+) = p(Y_j^+(1 \otimes 1 \otimes v_+)) = (4-n)Y_j^+ \otimes 1 \equiv 0.$$

The proof is complete. □

The proof of Theorem 6.18.1 shows that the maps i and p also induce homomorphisms of semi-holonomic Verma modules.

Next, we determine the translation $p \circ (\Delta_n^- \otimes I) \circ i$ of Δ_n^-.

Theorem 6.18.1. *The composition* $p \circ (\Delta_n^- \otimes I) \circ i : \mathcal{N}_{-\frac{n}{2}-2}(\mathfrak{g}) \to \mathcal{N}_{-\frac{n}{2}+2}(\mathfrak{g})$ *is given by right multiplication with*

$$\frac{1}{8}\left(4L_{4,n} - n(\Delta_n^-)^2\right),$$

where

$$L_{4,n} = \sum_i Y_i^- \Delta_n^- Y_i^-.$$

Since $\pi(L_{4,n}) = (\Delta_n^-)^2 \in \mathcal{U}(\mathfrak{n}^-)$, the homomorphism $p \circ (\Delta_n^- \otimes I) \circ i : M_{-\frac{n}{2}-2}(\mathfrak{g}) \to M_{-\frac{n}{2}+2}(\mathfrak{g})$ of Verma modules is induced by right multiplication with $\frac{1}{8}(4 - n)(\Delta_n^-)^2$. In particular, for $n = 4$, the latter operator vanishes. Notice, however, that it is still non-trivial in the semi-holonomic category since $L_{4,n} \neq (\Delta_n^-)^2$.

Proof. The proof is a straightforward calculation. We have to calculate the image of

$$\left(\frac{n}{2}+1\right)(\Delta_n^- \otimes 1 \otimes v_-) + \sum_i Y_i^-\Delta_n^- \otimes 1 \otimes v_i + \frac{1}{8}((\Delta_n^-)^2 \otimes 1 \otimes v_+)$$

under p. For that purpose, we use the formulas

$$\sum_i Y_i^-\Delta_n^-(1 \otimes 1 \otimes v_i) = \sum_i Y_i^-\Delta_n^- \otimes 1 \otimes v_i + (n+2)\Delta_n^-(1 \otimes 1 \otimes v_-) \quad (6.18.4)$$

and

$$(\Delta_n^-)^2(1 \otimes 1 \otimes v_+)$$
$$= (\Delta_n^-)^2 \otimes 1 \otimes v_+ + 4\sum_i Y_i^- \Delta_n^- \otimes 1 \otimes v_i + 4\sum_i \Delta_n^- Y_i^-(1 \otimes 1 \otimes v_i)$$
$$= (\Delta_n^-)^2 \otimes 1 \otimes v_+ + 4\sum_i Y_i^- \Delta_n^-(1 \otimes 1 \otimes v_i)$$
$$- 4(n+2)\Delta_n^-(1 \otimes 1 \otimes v_-) + 4\sum_i \Delta_n^- Y_i^-(1 \otimes 1 \otimes v_i) \quad \text{(by (6.18.4))},$$

the proofs of which are left to the reader. Now for the composition we obtain

$$\frac{n+2}{4}((\Delta_n^-)^2 \otimes 1) + \left(L_{4,n} \otimes 1 - \frac{n+2}{2}(\Delta_n^-)^2 \otimes 1\right)$$
$$+ \frac{1}{8}\left((4-n)(\Delta_n^-)^2 - 4L_{4,n} + 2(n+2)(\Delta_n^-)^2 - 4(\Delta_n^-)^2\right) \otimes 1$$
$$= \left(\frac{1}{2}L_{4,n} - \frac{n}{8}(\Delta_n^-)^2\right) \otimes 1.$$

The proof is complete.   □

Lemma 6.18.1 admits the following *family version*.

Lemma 6.18.2. *The family of* $\mathcal{U}(\mathfrak{g})$-*module maps*

$$i(\lambda) : \mathcal{U}(\mathfrak{g}) \otimes \mathbb{C}(\lambda-1) \to \mathcal{U}(\mathfrak{g}) \otimes \mathbb{C}(\lambda) \otimes F$$

which is determined by

$$1 \otimes 1 \mapsto -\lambda(1 \otimes 1 \otimes v_-) + \sum_i Y_i^- \otimes 1 \otimes v_i - \frac{1}{2(n-2+2\lambda)}(\Delta_n^- \otimes 1 \otimes v_+)$$

$$(6.18.5)$$

induces a family of homomorphisms $\mathcal{M}_{\lambda-1}(\mathfrak{g}) \to \mathcal{M}_\lambda(\mathfrak{g}) \otimes F$ *of* $\mathcal{U}(\mathfrak{g})$-*modules. Similarly, the family of* $\mathcal{U}(\mathfrak{g})$-*module maps*

$$p(\lambda) : \mathcal{U}(\mathfrak{g}) \otimes \mathbb{C}(\lambda-1) \otimes F \to \mathcal{U}(\mathfrak{g}) \otimes \mathbb{C}(\lambda)$$

which is determined by

$$1 \otimes 1 \otimes v_- \mapsto \frac{1}{n-2+2\lambda}(\Delta_n^- \otimes 1), \ 1 \otimes 1 \otimes v_i \mapsto Y_i^- \otimes 1, \ 1 \otimes 1 \otimes v_+ \mapsto 2\lambda(1 \otimes 1)$$

induces a family of homomorphisms $\mathcal{M}_{\lambda-1}(\mathfrak{g}) \otimes F \to \mathcal{M}_\lambda(\mathfrak{g})$ *of* $\mathcal{U}(\mathfrak{g})$-*modules.* $i(\lambda)$ *and* $p(\lambda)$ *induce analogous homomorphisms in the semi-holonomic category.*

The proof is analogous to that of Lemma 6.18.1 and we omit the details. We will also need the following generalization of Lemma 6.18.2.

Lemma 6.18.3. *For any $\mathfrak{g}$-module W, the family*

$$i(\lambda) : \mathcal{U}(\mathfrak{g}) \otimes \mathbb{C}(\lambda-1) \otimes W \to \mathcal{U}(\mathfrak{g}) \otimes \mathbb{C}(\lambda) \otimes W \otimes F$$

of $\mathcal{U}(\mathfrak{g})$-module maps defined by

$$1\otimes 1\otimes w \to -\lambda(1\otimes 1\otimes w\otimes v_-)+\sum_i Y_i^-\otimes 1\otimes w\otimes v_i-\frac{1}{2(n-2+2\lambda)}(\Delta_n^-\otimes 1\otimes w\otimes v_+)$$

descends to a family

$$\mathcal{M}_{\lambda-1}(\mathfrak{g}) \otimes W \to \mathcal{M}_\lambda(\mathfrak{g}) \otimes W \otimes F$$

of homomorphisms of $\mathcal{U}(\mathfrak{g})$-modules. Similarly, the family of $\mathcal{U}(\mathfrak{g})$-module maps

$$p(\lambda) : \mathcal{U}(\mathfrak{g}) \otimes \mathbb{C}(\lambda-1) \otimes W \otimes F \to \mathcal{U}(\mathfrak{g}) \otimes \mathbb{C}(\lambda) \otimes W$$

defined by

$$1 \otimes 1 \otimes w \otimes v_- \mapsto \frac{1}{n-2+2\lambda}(\Delta_n^- \otimes 1 \otimes w),$$

$$1 \otimes 1 \otimes w \otimes v_i \mapsto Y_i^- \otimes 1 \otimes w, \quad 1 \otimes 1 \otimes w \otimes v_+ \mapsto 2\lambda(1 \otimes 1 \otimes w)$$

induces a family

$$\mathcal{M}_{\lambda-1}(\mathfrak{g}) \otimes W \otimes F \to \mathcal{M}_\lambda(\mathfrak{g}) \otimes W$$

of homomorphisms of $\mathcal{U}(\mathfrak{g})$-modules. $i(\lambda)$ and $p(\lambda)$ induce analogous homomorphisms in the semi-holonomic category.

Proof. The proof is analogous to that of Lemma 6.18.2. In particular, we have to prove that $i(\lambda)(Y_j^+\otimes 1\otimes w) \in J(\lambda)\otimes W$, where $J(\lambda) = \langle X\otimes 1-1\otimes\xi_\lambda(X)1, X \in \mathfrak{p}\rangle$. Now $Y_j^+ \otimes 1 \otimes w = Y_j^+(1 \otimes 1 \otimes w) - 1 \otimes 1 \otimes Y_j^+w$ implies

$$i(\lambda)(Y_j^+\otimes 1\otimes w)=-\lambda\big[Y_j^+\otimes 1\otimes w\otimes v_-+1\otimes 1\otimes Y_j^+(w\otimes v_-)\big]$$

$$+\sum_i Y_j^+Y_i^-\otimes 1\otimes w\otimes v_i+\sum_i Y_i^-\otimes 1\otimes Y_j^+(w\otimes v_i)$$

$$-\frac{1}{2(n-2+2\lambda)}\big(Y_j^+\Delta_n^-\otimes 1\otimes w\otimes v_++\Delta_n^-\otimes 1\otimes Y_j^+(w\otimes v_+)\big)$$

$$+\lambda(1\otimes 1\otimes Y_j^+(w)\otimes v_-)-\sum_i Y_j^-\otimes 1\otimes Y_j^+(w)\otimes v_i$$

$$+\frac{1}{2(n-2+2\lambda)}(\Delta_n^-\otimes 1\otimes Y_j^+(w)\otimes v_+)$$

$$= -\lambda \left[Y_j^+ \otimes 1 \otimes w \otimes v_- + 1 \otimes 1 \otimes w \otimes Y_j^+(v_-) \right]$$
$$+ \sum_i Y_j^+ Y_i^- \otimes 1 \otimes w \otimes v_i + \sum_i Y_i^- \otimes 1 \otimes w \otimes Y_j^+(v_i)$$
$$- \frac{1}{2(n-2+2\lambda)} \left(Y_j^+ \Delta_n^- \otimes 1 \otimes w \otimes v_+ + \Delta_n^- \otimes 1 \otimes w \otimes Y_j^+(v_+) \right).$$

Modulo terms in $(J(\lambda) \otimes W) \otimes F$, the latter sum equals

$$- \lambda (1 \otimes 1 \otimes w \otimes Y_j^+(v_-)) + 2\lambda (1 \otimes 1 \otimes w \otimes v_j) + Y_j^- \otimes 1 \otimes w \otimes v_+$$
$$- \frac{1}{2(n-2+2\lambda)} \left[Y_j^+, \Delta_n^- \right] \otimes 1 \otimes w \otimes v_+$$
$$\equiv Y_j^- \otimes 1 \otimes w \otimes v_+ - \frac{1}{2(n-2+2\lambda)} \left(2(n-2)Y_j^- + 4\lambda Y_j^- \right) \otimes 1 \otimes w \otimes v_+$$
$$\equiv 0$$

using (6.18.3). The remaining parts of the proof are left to the reader. □

Lemma 6.18.3 allows us to relate $i(\lambda)$ and $p(\lambda)$ as follows.

Corollary 6.18.1. $p(\lambda) = \mathrm{tr}_F \circ i(\lambda)$, where $i(\lambda)$ is viewed as a homomorphism

$$\mathcal{U}(\mathfrak{g}) \otimes \mathbb{C}(\lambda-1) \otimes F \to \mathcal{U}(\mathfrak{g}) \otimes \mathbb{C}(\lambda) \otimes F \otimes F,$$

and $\mathrm{tr}_F(v \otimes w) = (v, w)$ is the contraction defined by the scalar product on $F \simeq \mathbb{R}^{1,n+1}$.

Proof. It suffices to notice that for $v \in F$,

$$\mathrm{tr}_F\, i(\lambda)(1 \otimes 1 \otimes v) = -\lambda \otimes (v, v_-) + \sum_i Y_i^- \otimes (v, v_i) - \frac{1}{2(n-2+2\lambda)} \Delta_n^- \otimes (v, v_+).$$

Thus using $(v_\pm, v_\pm) = 0$, $(v_\pm, v_\mp) = -2$, $(v_\pm, v_i) = 0$ and $(v_i, v_j) = \delta_{ij}$, we get the assertion. □

Corollary 6.18.1 shows that the mapping properties of $p(\lambda)$ are consequences of those of $i(\lambda)$. In fact, $\mathrm{tr}_F \circ i(\lambda)$ defines a $\mathfrak{g}$-module map $\mathcal{U}(\mathfrak{g}) \otimes \mathbb{C}(\lambda - 1) \otimes F \to \mathcal{U}(\mathfrak{g}) \otimes \mathbb{C}(\lambda)$ since $(Xv, w) + (v, Xw) = 0$ for $X \in \mathfrak{g}$ and $v, w \in F$. Moreover, the composition descends to a homomorphism $\mathcal{M}_{\lambda-1}(\mathfrak{g}) \otimes F \to \mathcal{M}_\lambda(\mathfrak{g})$.

Note that $i(\lambda)$ and $p(\lambda)$ are *rational* in λ. A calculation shows that the composition

$$\mathcal{M}_{\lambda-2}(\mathfrak{g}_{n+1}) \xrightarrow{i(\lambda-1)} \mathcal{M}_{\lambda-1}(\mathfrak{g}_{n+1}) \otimes F \xrightarrow{p(\lambda)} \mathcal{M}_\lambda(\mathfrak{g}_{n+1})$$

vanishes.

Definition 6.18.1. *The formulas*

$$\Theta_{2N}(\lambda) = p(\lambda) \circ (\Theta_{2N-2}(\lambda-1) \otimes I_n) \circ i(\lambda-(2N-1)), \quad \Theta_0(\lambda) = \mathcal{D}_0(\lambda) = 1,$$

and

$$\Theta_{2N+1}(\lambda) = p(\lambda) \circ (\Theta_{2N-1}(\lambda-1) \otimes I_n) \circ i(\lambda-2N), \quad \Theta_1(\lambda) = \mathcal{D}_1(\lambda)$$

define rational families

$$\Theta_N(\lambda) : \mathcal{M}_{\lambda-N}(\mathfrak{g}_n) \to \mathcal{M}_\lambda(\mathfrak{g}_{n+1})$$

of homomorphisms of Verma modules and rational families

$$\tilde{\Theta}_N(\lambda) : \mathcal{N}_{\lambda-N}(\mathfrak{g}_n) \to \mathcal{N}_\lambda(\mathfrak{g}_{n+1})$$

of homomorphisms of semi-holonomic Verma modules. $\tilde{\Theta}_N(\lambda)$ lifts $\Theta_N(\lambda)$.

Here we use the G^n-equivariant map $I_n : F_n \to F_{n+1}$ which corresponds to the embedding $SO(1,n) \hookrightarrow SO(1, n+1)$, i.e., $SO(1,n)$ is the subgroup which preserves the subspace $\mathbb{R}^{1,n} \hookrightarrow \mathbb{R}^{1,n+1}$ defined by $x_{n+1} = 0$. Then v_-, v_i ($i = 1, \ldots, n-1$) and $v_+ \in F_n$ will be viewed as elements of F_{n+1}.

The definition of $\Theta_N(\lambda)$ is correct since $\mathcal{D}_0$ and $\mathcal{D}_1(\lambda) = Y_n^-$ induce homomorphisms of semi-holonomic Verma modules. In fact, it suffices to note that $[Y_i^+, Y_n^-] = 2M_{in}$ for $i = 1, \ldots, n-1$.

The poles of the families $\Theta_N(\lambda)$ are caused only by the poles of the coefficients of $i(\lambda)$ and $p(\lambda)$. An obvious question is to relate $\Theta_N(\lambda)$ and $\mathcal{D}_N(\lambda)$. The answer is given by Theorem 6.18.7 and Theorem 6.18.8.

In addition, it is instructive and useful to find explicit formulas for low order translates. The first non-trivial translate is $\Theta_2(\lambda)$. Its discussion will be followed by establishing formulas for $\Theta_3(\lambda)$ and $\Theta_4(\lambda)$.

Theorem 6.18.2. *The homomorphism $\Theta_2(\lambda) : \mathcal{N}_{\lambda-2}(\mathfrak{g}_n) \to \mathcal{N}_\lambda(\mathfrak{g}_{n+1})$ is induced by right multiplication with*

$$\frac{(n-3+\lambda)}{(n-5+2\lambda)(n-2+2\lambda)} \mathcal{D}_2(\lambda), \tag{6.18.6}$$

where

$$\mathcal{D}_2(\lambda) = -(n-3+2\lambda)\Delta_n^- + (n-2+2\lambda)\Delta_{n-1}^-.$$

Proof. We recall that $\Theta_2(\lambda) = p(\lambda) \circ (1 \otimes I_n) \otimes i(\lambda-1)$. Thus we have to calculate the image of $i(\lambda-1)(1 \otimes 1) \in \mathcal{A}(\mathfrak{g}_n) \otimes F_{n+1}$, i.e., of

$$-(\lambda-1) \otimes 1 \otimes v_- + \sum_{i=1}^{n-1} Y_i^- \otimes 1 \otimes v_i - \frac{1}{2(n-5+2\lambda)}(\Delta_{n-1}^- \otimes 1 \otimes v_+)$$

under $p(\lambda)$. For that purpose, we use that $\sum_1^{n-1} Y_i^- \otimes v_i = \sum_1^{n-1} Y_i^-(1 \otimes v_i) - (n-1)(1 \otimes v_-)$ and

$$\Delta_{n-1}^- \otimes v_+ = \Delta_{n-1}^-(1 \otimes v_+) - 4\sum_1^{n-1} Y_i^-(1 \otimes v_i) + 2(n-1)(1 \otimes v_-).$$

We apply $p(\lambda)$ and obtain (for $\beta^{-1} = n - 2 + 2\lambda$)

$$-(\lambda-1)\beta\Delta_n^- \otimes 1 + \Delta_{n-1}^- \otimes 1 - (n-1)\beta\Delta_n^- \otimes 1$$
$$-\frac{1}{2(n-5+2\lambda)}\left[2\lambda\Delta_{n-1}^- - 4\Delta_{n-1}^- + 2(n-1)\beta\Delta_n^-\right] \otimes 1$$
$$= -\frac{1}{(n-2+2\lambda)(n-5+2\lambda)}\left[(n-2+\lambda)(n-5+2\lambda) + (n-1)\right]\Delta_n^- \otimes 1$$
$$+ \frac{n-3+\lambda}{n-5+2\lambda}\Delta_{n-1}^- \otimes 1$$
$$= \frac{(n-3+\lambda)}{(n-2+2\lambda)(n-5+2\lambda)}\left[-(n-3+2\lambda)\Delta_n^- + (n-2+2\lambda)\Delta_{n-1}^-\right].$$

The proof is complete.   □

We continue with a discussion of $\Theta_3(\lambda)$.

Theorem 6.18.3. *The homomorphism* $\Theta_3(\lambda) : \mathcal{N}_{\lambda-3}(\mathfrak{g}_n) \to \mathcal{N}_\lambda(\mathfrak{g}_{n+1})$ *is induced by right multiplication with*

$$\frac{3}{(n-7+2\lambda)(n-2+2\lambda)}(n-4+\lambda)\mathcal{D}_3^T(\lambda), \tag{6.18.7}$$

where

$$\mathcal{D}_3^T(\lambda) = -\frac{1}{3}(n-5+2\lambda)\left((Y_n^-)^3 + Y_n^-\Delta_{n-1}^-\right)$$
$$+ \frac{1}{3}\frac{(n-2+2\lambda)}{(n-4+\lambda)}\left((n-5+2\lambda)\sum_i Y_i^- Y_n^- Y_i^- - (\lambda-1)\Delta_{n-1}^- Y_n^-\right). \tag{6.18.8}$$

Note that $\pi\mathcal{D}_3^T(\lambda) = -\frac{1}{3}(n-5+2\lambda)(Y_n^-)^3 + Y_n^-\Delta_{n-1}^- = \mathcal{D}_3(\lambda)$.

In other words, $\mathcal{D}_3^T(\lambda)$ is a lift of $\mathcal{D}_3(\lambda)$ which defines a homomorphism of semi-holonomic Verma modules. It arises by translation of $\Theta_1(\lambda) = \mathcal{D}_1(\lambda) = Y_n^+$. Note that the factorization identities

$$\mathcal{D}_3\left(-\frac{n}{2}+1\right) = Y_n^-\Delta_n^- \quad \text{and} \quad \mathcal{D}_3\left(-\frac{n-5}{2}\right) = \Delta_{n-1}^- Y_n^-$$

continue to hold true for the lift $\mathcal{D}_3^T(\lambda)$.

Although $\mathcal{D}_3(\lambda)$ is polynomial, its lift has a simple pole at $\lambda = 4 - n$ with residue

$$\mathrm{Res}_{4-n}(\mathcal{D}_3^T(\lambda)) = \frac{1}{3}(n-3)(n-6)\sum_{1}^{n-1} Y_i^-\left(Y_n^- Y_i^- - Y_i^- Y_n^-\right) \in \ker\pi. \qquad (6.18.9)$$

In particular, the residue defines a homomorphism $\mathcal{N}_{1-n}(\mathfrak{g}_n) \to \mathcal{N}_{4-n}(\mathfrak{g}_{n+1})$. It is of interest to determine the conformally covariant differential operator $C^\infty(M^n) \to C^\infty(\Sigma^{n-1})$ which is induced by that homomorphism.

The critical case $n = 4$, $\lambda = 0$ is of particular interest. We see that, for $n = 4$, the family $\mathcal{D}_3^T(\lambda)$ has a simple pole at $\lambda = 0$ with residue

$$-\frac{2}{3}\sum_{1}^{3} Y_i^-\left(Y_4^- Y_i^- - Y_i^- Y_4^-\right) \in \ker\pi$$

and constant term

$$CT_0(\mathcal{D}_3^T) = \frac{1}{3}\left((Y_4^-)^3 + Y_4^- \Delta_3^- + 2\sum_{1}^{3} Y_i^- Y_4^- Y_i^-\right)$$

so that $\pi CT_0(\mathcal{D}_3^T) = \frac{1}{3}(Y_4^-)^3 + Y_4^- \Delta_3^- = \mathcal{D}_3(0)$.

Note that the coefficients $(n - 7 + 2\lambda)^{-1}$ and $(n - 2 + 2\lambda)^{-1}$ are caused by the poles of i and p. On the other hand, the pole of $\mathcal{D}_3^T(\lambda)$ at $\lambda = 4 - n$ is due to the *overall* coefficient $n - 4 + \lambda$ in (6.18.7); $\Theta_3(\lambda)$ is regular at $\lambda = 4 - n$. More precisely,

$$\Theta_3(4 - n) = \frac{n-3}{n-1}\sum_{1}^{n-1} Y_i^-\left(Y_n^- Y_i^- - Y_i^- Y_n^-\right) \in \ker\pi.$$

Proof. We have to calculate the image of

$$-(\lambda-2)(Y_n^- \otimes 1 \otimes v_-) + \sum_{1}^{n-1} Y_i^- Y_n^- \otimes 1 \otimes v_i - \frac{1}{2(n-7+2\lambda)}(\Delta_{n-1}^- Y_n^- \otimes 1 \otimes v_+)$$

$$(6.18.10)$$

under $p(\lambda)$. For that purpose, we use the formulas

$$\sum_{1}^{n-1} Y_i^- Y_n^-(1 \otimes 1 \otimes v_i) = \sum_{1}^{n-1} Y_i^- Y_n^- \otimes 1 \otimes v_i + (n-1)Y_n^-(1 \otimes 1 \otimes v_-)$$

and

$$\Delta_{n-1}^- Y_n^-(1 \otimes 1 \otimes v_+) = \Delta_{n-1}^- Y_n^- \otimes 1 \otimes v_+$$

$$+ 2\Delta_{n-1}^-(1 \otimes 1 \otimes v_n) - 2(n-1)Y_n^-(1 \otimes 1 \otimes v_-) + 4\sum_{1}^{n-1} Y_i^- Y_n^-(1 \otimes 1 \otimes v_i)$$

the proofs of which are left to the reader. We apply $p(\lambda)$ to (6.18.10) and obtain

$$- \left[(\lambda-2)+(n-1)\right]\beta Y_n^- \Delta_n^- \otimes 1 + \left(1 + \frac{2}{n-7+2\lambda}\right) \sum_i Y_i^- Y_n^- Y_i^- \otimes 1$$

$$- \frac{1}{2(n-7+2\lambda)} \left(2\lambda\Delta_{n-1}^- Y_n^- - 2\Delta_{n-1}^- Y_n^- + 2(n-1)\beta Y_n^- \Delta_n^-\right) \otimes 1,$$

where $\beta^{-1} = (n-2+2\lambda)$. A direct calculation yields the assertion. $\qquad\square$

It is interesting to notice that $\mathcal{D}_3^T(\lambda)$ is an element in a *one-parameter family* of lifts of $\mathcal{D}_3(\lambda)$.

Theorem 6.18.4. *Let*

$$\bar{\mathcal{D}}_3(\lambda) = -\frac{1}{3}(2\lambda+n-5)(Y_n^-)^3 + \left(a\Delta_{n-1}^- Y_n^- + b\sum_1^{n-1} Y_j^- Y_n^- Y_j^- + cY_n^- \Delta_{n-1}^-\right)$$

be a lift of $\mathcal{D}_3(\lambda)$, i.e., $\pi\bar{\mathcal{D}}_3(\lambda) = \mathcal{D}_3(\lambda)$. Moreover, assume that the map

$$\mathcal{A}(\mathfrak{g}_n) \otimes \mathbb{C}(\lambda-3) \ni T \otimes 1 \mapsto i(T)\bar{\mathcal{D}}_3(\lambda) \otimes 1 \in \mathcal{A}(\mathfrak{g}_{n+1}) \otimes \mathbb{C}(\lambda)$$

induces a homomorphism $\mathcal{N}_{\lambda-3}(\mathfrak{g}_n) \to \mathcal{N}_\lambda(\mathfrak{g}_{n+1})$ of semi-holonomic Verma modules. Then

$$a + b + c = 1,$$

$$(n-4+\lambda)b + (n-4+2\lambda)c = \frac{2}{3}(n-5+2\lambda).$$

Conversely, if (a, b, c) satisfies these conditions, then $\bar{\mathcal{D}}_3(\lambda)$ defines a rational family of homomorphisms of semi-holonomic Verma modules. The family $\mathcal{D}_3^T(\lambda)$ appears for $c = -\frac{1}{3}(n-5+2\lambda)$.

Proof. The condition $\pi\bar{\mathcal{D}}_3(\lambda) = \mathcal{D}_3(\lambda)$ is equivalent to $a + b + c = 1$ since $\mathfrak{n}^-$ is abelian. It is enough to analyze the conditions

$$[Y_i^+, \bar{\mathcal{D}}_3(\lambda)] \in J(\lambda) = \mathcal{A}(\mathfrak{g}_{n+1})(\mathfrak{m}_{n+1} \oplus \mathbb{C}(H_0-\lambda))$$

for $i = 1, \dots, n-1$. The arguments are similar to those in the proof of Theorem 5.1.1. However, it is now forbidden to use the relations $X \otimes Y - Y \otimes X = [X, Y]$ for $X, Y \in \mathfrak{n}^-$. It suffices to consider the case $i = 1$. The general case follows by a rotation as in the proof of Theorem 5.1.1. We determine the commutators

$$\left[Y_1^+, (Y_n^-)^3\right], \quad \left[Y_1^+, \Delta_{n-1}^- Y_n^-\right], \quad \left[Y_1^+, \sum_j Y_j^- Y_n^- Y_j^-\right], \quad \left[Y_1^+, Y_n^- \Delta_{n-1}^-\right].$$

First of all, we have in $\mathcal{A}(\mathfrak{g}_{n+1})$,

$$[Y_1^+, (Y_n^-)^3] = 2Y_1^- Y_n^- + 4Y_n^- Y_1^- + 6(Y_n^-)^2 M_{1n}. \tag{6.18.11}$$

Next, by Lemma 5.1.3,

$$[Y_1^+, \Delta_{n-1}^-] = -2Y_1^- + 4Y_1^- H_0 + \sum_{r=2}^{n-1} (2Y_1^- + 4Y_r^- M_{1r})$$

$$= 2(n-3)Y_1^- + 4Y_1^- H_0 + 4\sum_{r=2}^{n-1} Y_r^- M_{1r}. \tag{6.18.12}$$

Hence

$$[Y_1^+, \Delta_{n-1}^- Y_n^-] = [Y_1^+, \Delta_{n-1}^-]Y_n^- + \Delta_{n-1}^-[Y_1^+, Y_n^-]$$

$$= 2(n-3)Y_1^- Y_n^- + 4Y_1^- H_0 Y_n^- + 4\sum_{r=2}^{n-1} Y_r^- M_{1r} Y_n^- + 2\Delta_{n-1}^- M_{1n}$$

$$= 2(n-5)Y_1^- Y_n^- + 4Y_1^- Y_n^- H_0 + 4\sum_{r=2}^{n-1} Y_r^- Y_n^- M_{1r} + 2\Delta_{n-1}^- M_{1n}$$

$$\equiv 2(n-5+2\lambda)Y_1^- Y_n^- \tag{6.18.13}$$

mod $J(\lambda)$. Similarly, using (5.2.7) and

$$[M_{1n}, (Y_1^-)^2] = Y_1^-[M_{1n}, Y_1^-] + [M_{1n}, Y_1^-]Y_1^- = -(Y_n^- Y_1^- + Y_1^- Y_n^-),$$

we get

$$[Y_1^+, Y_n^- \Delta_{n-1}^-] = [Y_1^+, Y_n^-]\Delta_{n-1}^- + Y_n^-[Y_1^+, \Delta_{n-1}^-]$$

$$= 2M_{1n}\Delta_{n-1}^- + Y_n^- \left(2(n-3)Y_1^- + 4Y_1^- H_0 + 4\sum_{r=2}^{n-1} Y_r^- M_{1r}\right)$$

$$= 2M_{1n}(Y_1^-)^2 + 2\left((Y_2^-)^2 + \cdots + (Y_{n-1}^-)^2\right)M_{1n}$$

$$+ Y_n^- \left(2(n-3)Y_1^- + 4Y_1^- H_0 + 4\sum_{r=2}^{n-1} Y_r^- M_{1r}\right),$$

i.e.,

$$[Y_1^+, Y_n^- \Delta_{n-1}^-]$$

$$= 2(n-4)Y_n^- Y_1^- - 2Y_1^- Y_n^- + 4Y_n^- Y_1^- H_0 + 4Y_n^- \sum_{r=2}^{n-1} Y_r^- M_{1r} + 2\Delta_{n-1}^- M_{1n}$$

$$\equiv 2(n-4+2\lambda)Y_n^- Y_1^- - 2Y_1^- Y_n^- \tag{6.18.14}$$

mod $J(\lambda)$. Finally, we find

$$
\begin{aligned}
\left[Y_1^+, Y_j^- Y_n^- Y_j^-\right] &= [Y_1^+, Y_j^-]Y_n^- Y_j^- + Y_j^-[Y_1^+, Y_n^- Y_j^-] \\
&= 2M_{1j}Y_n^- Y_j^- + 2Y_j^-(M_{1n}Y_j^- + Y >_n^- M_{1j}) \\
&= 2Y_n^- Y_1^- + 2(Y_n^- Y_j^- + Y_j^- Y_n^-)M_{1j} + 2(Y_j^-)^2 M_{1n}
\end{aligned}
$$

for $j > 1$ and

$$
\begin{aligned}
\left[Y_1^+, Y_1^- Y_n^- Y_1^-\right][Y_1^+, Y_1^-]Y_n^- Y_1^- &+ Y_1^-[Y_1^+, Y_n^- Y_1^-] \\
&= 2H_0 Y_n^- Y_1^- + 2Y_1^-(M_{1n}Y_1^- + Y_n^- H_0) \\
&= -4Y_n^- Y_1^- - 2Y_1^- Y_n^- + 2(Y_n^- Y_1^- + Y_1^- Y_n^-)H_0.
\end{aligned}
$$

Hence

$$
\left[Y_1^+, \sum_1^{n-1} Y_j^- Y_n^- Y_j^-\right] \equiv 2(n-4+\lambda)Y_n^- Y_1^- + 2(\lambda-1)Y_1^- Y_n^-
$$

mod $J(\lambda)$. These results imply that $[Y_1^+, \bar{\mathcal{D}}_3(\lambda)]$ is of the form

$$
\begin{aligned}
-\frac{1}{3}(n-5+2\lambda)\left(2Y_1^- Y_n^- + 4Y_n^- Y_1^-\right) & \\
+ 2[a(n-5+2\lambda) + b(n-4+\lambda)Y_n^- Y_1^- &+ b(\lambda-1)Y_1^- Y_n^- \\
&+ c(n-4+2\lambda)Y_n^- Y_1^- - cY_1^- Y_n^-],
\end{aligned}
$$

i.e.,

$$
\begin{aligned}
2\left\{-\frac{1}{3}(n-5+2\lambda) + a(n-5+2\lambda) + b(\lambda-1) - c\right\} Y_1^- Y_n^- & \\
+ 2\left\{-\frac{2}{3}(n-5+2\lambda) + b(n-4+\lambda) + c(n-4+2\lambda)\right\} Y_n^- Y_1^-. &
\end{aligned}
$$

Now the condition $\frac{1}{3}(n-5+2\lambda) = a(n-5+2\lambda) + b(\lambda-1) - c$ is a consequence of

$$
\frac{2}{3}(n-5+2\lambda) = b(n-4+\lambda) + c(n-4+2\lambda)
$$

and $a + b + c = 1$. The proof is complete. $\square$

Next, we discuss the fourth-order family $\Theta_4(\lambda) : \mathcal{N}_{\lambda-4}(\mathfrak{g}_n) \to \mathcal{N}_\lambda(\mathfrak{g}_{n+1})$ constructed by translation of $\Theta_2(\lambda)$.

Theorem 6.18.5. *The homomorphism* $\Theta_4(\lambda) : \mathcal{N}_{\lambda-3}(\mathfrak{g}_n) \to \mathcal{N}_\lambda(\mathfrak{g}_{n+1})$ *is induced by right multiplication with*

$$
3\frac{1}{(n-9+2\lambda)(n-7+2\lambda)(n-4+2\lambda)(n-2+2\lambda)}(n-5+\lambda)(n-4+\lambda)\mathcal{D}_4^T(\lambda),
$$

where

$$
\mathcal{D}_4^T(\lambda) = \frac{1}{3}(n-5+2\lambda)(n-7+2\lambda)\left((Y_n^-)^4 + (Y_n^-)^2\Delta_{n-1}^-\right) - 2(n-5+2\lambda)\Delta_{n-1}^-(Y_n^-)^2
$$
$$
+ \frac{1}{3}\frac{n-7+2\lambda}{n-5+\lambda}\left(-(n-5+2\lambda)(\Delta_{n-1}^-)^2 + (n-2+2\lambda)L_{4,n-1}\right) - \frac{\lambda-2}{n-5+\lambda}(\Delta_{n-1}^-)^2
$$
$$
+ \frac{1}{3}\frac{(n-7+2\lambda)(n-5+2\lambda)}{n-5+\lambda}\left((\lambda+3)\Delta_{n-1}^-Y_n^{-2} - (n-2+2\lambda)\sum_1^{n-1}Y_j^-(Y_n^-)^2Y_j^-\right).
$$

Note that $\pi\mathcal{D}_4^T(\lambda) = \mathcal{D}_4(\lambda)$. In other words, $\mathcal{D}_4^T(\lambda)$ is a lift of $\mathcal{D}_4(\lambda)$ which defines a homomorphism of semi-holonomic Verma modules. Although $\mathcal{D}_4(\lambda)$ is polynomial in λ, its lift has a simple pole at $\lambda = 5 - n$ with residue

$$
\text{Res}_{5-n}(\mathcal{D}_4^T) = -\frac{(n-3)(n-8)}{3}
$$
$$
\left([(\Delta_{n-1}^-)^2 - L_{4,n-1}] + (n-5)\left[\Delta_{n-1}^-(Y_n^-)^2 - \sum_1^{n-1}Y_j^-(Y_n^-)^2Y_j^-\right]\right)
$$

and value

$$
\mathcal{D}_4^T\left(-\frac{n-5}{2}\right) = \frac{1}{n-5}\left((n-1)(\Delta_{n-1}^-)^2 - 4L_{4,n-1}\right) \tag{6.18.15}
$$

for $n \neq 5$. We recall that $\mathcal{D}_4(-\frac{n-5}{2}) = (\Delta_{n-1}^-)^2$. The proof is a straightforward calculation and we omit the details.

(6.18.15) shows why the lift $\mathcal{D}_4^T(-\frac{n-5}{2})$ of $(\Delta_{n-1}^-)^2$ does not exist for $n = 5$. This is the case which is not covered by the translation in the category of Verma module: $(\Delta_4^-)^2$ can not be constructed from Δ_4^- by translation. Instead, we have the rational family

$$
\frac{2}{\lambda}\left[(\Delta_4^-)^2 - L_4\right]
$$
$$
+ \left[\frac{1}{3}(\Delta_4^-)^2 + \frac{2}{3}L_4 - 4\left(\Delta_4^-(Y_5^-)^2 - \sum_i Y_i^-(Y_5^-)^2Y_i^-\right)\right] + \cdots. \tag{6.18.16}
$$

We can write $\mathcal{D}_4^T(\lambda)$ in the form

$$
\frac{1}{3}(n-5+2\lambda)(n-7+2\lambda)(Y_n^-)^4 - 2(n-5+2\lambda)
$$
$$
\times \left\{a(Y_n^-)^2\Delta_{n-1}^- + bY_n^-\Delta_{n-1}^-Y_n^- + c\Delta_{n-1}^-(Y_n^-)^2 + d\sum_1^{n-1}Y_j^-(Y_n^-)^2Y_j^-\right\}
$$
$$
+ (1-\alpha)(\Delta_{n-1}^-)^2 + \alpha L_{4,n-1}, \tag{6.18.17}
$$

where the coefficients are given by

$$a = -\frac{1}{6}(n-7+2\lambda),$$

$$b = 0,$$

$$c = 1 - \frac{1}{6}\frac{n-7+2\lambda}{n-5+\lambda}(\lambda+3),$$

$$d = \frac{1}{6}\frac{n-7+2\lambda}{n-5+\lambda}(n-2+2\lambda),$$

$$\alpha = \frac{1}{2}\frac{n-7+2\lambda}{n-5+\lambda}(n-2+2\lambda) \ (= \frac{1}{2}d).$$

In analogy to Theorem 6.18.4, $\mathcal{D}_4^T(\lambda)$ is an element in a one-parameter family of lifts of $\mathcal{D}_4(\lambda)$.

Theorem 6.18.6. *Let $\bar{\mathcal{D}}_4(\lambda)$ be of the form (6.18.17). Assume that $\bar{\mathcal{D}}_4(\lambda)$ is a lift of*

$$\mathcal{D}_4(\lambda) = \frac{1}{3}(2\lambda+n-5)(2\lambda+n-7)(Y_n^-)^4 - 2(2\lambda+n-5)(Y_n^-)^2\Delta_{n-1}^- + (\Delta_{n-1}^-)^2.$$

Moreover, assume that the map

$$\mathcal{A}(\mathfrak{g}_n) \otimes \mathbb{C}(\lambda-4) \ni T \otimes 1 \mapsto i(T)\tilde{\mathcal{D}}_4(\lambda) \otimes 1 \in \mathcal{A}(\mathfrak{g}_{n+1}) \otimes \mathbb{C}(\lambda)$$

induces a homomorphism $\mathcal{N}_{\lambda-4}(\mathfrak{g}_n) \to \mathcal{N}_\lambda(\mathfrak{g}_{n+1})$ of $\mathcal{A}(\mathfrak{g}_n)$-modules. Then

$$a + b + c + d = 1,$$

$$\alpha = 2d,$$

$$(n-5+\lambda)\alpha + 2(n-5+2\lambda)a = n-7+2\lambda,$$

$$(n-6+2\lambda)b - 2a = \frac{1}{3}(n-7+2\lambda).$$

Conversely, if (a, b, c, d) and α satisfy these conditions, then $\bar{\mathcal{D}}_4(\lambda)$ defines a rational family of homomorphisms of semi-holonomic Verma modules. The family $\mathcal{D}_4^T(\lambda)$ appears for $b = 0$.

Proof. The condition $\pi\bar{\mathcal{D}}_4(\lambda) = \mathcal{D}_4(\lambda)$ is equivalent to $a+b+c+d = 1$ since $\mathfrak{n}^-$ is abelian. As in the proof of Theorem 6.18.4, it is enough to analyze the condition

$$[Y_1^+, \bar{\mathcal{D}}_4(\lambda)] \in J(\lambda) = \mathcal{A}(\mathfrak{g}_{n+1})(\mathfrak{m}_{n+1} \oplus \mathbb{C}(H_0-\lambda)).$$

We determine the structure of the commutators $[Y_1^+, (Y_n^-)^4]$,

$$[Y_1^+, (Y_n^-)^2\Delta_{n-1}^-], \quad \left[Y_1^+, \sum_j Y_j^-(Y_n^-)^2Y_j^-\right],$$

$$[Y_1^+, Y_n^-\Delta_{n-1}^-Y_n^-], \ [Y_1^+, \Delta_{n-1}^-(Y_n^-)^2]$$

and $\left[Y_1^+, (\Delta_{n-1}^-)^2\right]$, $\left[Y_1^+, L_{4,n-1}\right]$. Since

$$\left[Y_1^+, (Y_n^-)^2\right] = \left[Y_1^+, Y_n^-\right] Y_n^- + Y_n^- \left[Y_1^+, Y_n^-\right]$$
$$= 2M_{1n}Y_n^- + 2Y_n^- M_{1n} = 2Y_1^- + 4Y_n^- M_{1n}$$

and

$$\left[M_{1n}, (Y_n^-)^2\right] = \left[M_{1n}, Y_n^-\right] Y_n^- + Y_n^- \left[M_{1n}, Y_n^-\right] = Y_1^- Y_n^- + Y_n^- Y_1^-$$

we get

$$\left[Y_1^+, (Y_n^-)^4\right] = \left[Y_1^+, (Y_n^-)^2\right] (Y_n^-)^2 + (Y_n^-)^2 \left[Y_1^+, (Y_n^-)^2\right]$$
$$= \left(2Y_1^- + 4Y_n^- M_{1n}\right) (Y_n^-)^2 + (Y_n^-)^2 \left(2Y_1^- + 4Y_n^- M_{1n}\right)$$
$$= 2Y_1^- (Y_n^-)^2 + 4Y_n^- \left[M_{1n}, (Y_n^-)^2\right] + 2(Y_n^-)^2 Y_1^- + 8(Y_n^-)^3 M_{1n}$$
$$\equiv 2Y_1^- (Y_n^-)^2 + 4Y_n^- Y_1^- Y_n^- + 6(Y_n^-)^2 Y_1^- \qquad (6.18.18)$$

mod $J(\lambda)$. In order to determine the commutator $\left[Y_1^+, (Y_n^-)^2 \Delta_{n-1}^-\right]$, we use the identity $\left[M_{1n}, \Delta_{n-1}^-\right] = -(Y_n^- Y_1^- + Y_1^- Y_n^-)$ which follows from

$$\left[M_{1n}, (Y_r^-)^2\right] = \left[M_{1n}, Y_r^-\right] Y_r^- + Y_r^- \left[M_{1n}, Y_r^-\right] = \begin{cases} 0 & r \neq 1 \\ -(Y_n^- Y_1^- + Y_1^- Y_n^-) & r = 1 \end{cases}.$$

Hence (6.18.12) yields

$$\left[Y_1^+, (Y_n^-)^2 \Delta_{n-1}^-\right] = \left[Y_1^+, (Y_n^-)^2\right] \Delta_{n-1}^- + (Y_n^-)^2 \left[Y_1^+, \Delta_{n-1}^-\right]$$
$$= \left(2Y_1^- + 4Y_n^- M_{1n}\right) \Delta_{n-1}^-$$
$$+ (Y_n^-)^2 \left(2(n-3)Y_1^- + 4Y_1^- H_0 + 4\sum_{r=2}^{n-1} Y_r^- M_{1r}\right)$$
$$\equiv 2Y_1^- \Delta_{n-1}^- - 4Y_n^- Y_1^- Y_n^- + 2(n-5)(Y_n^-)^2 Y_1^- + 4\lambda(Y_n^-)^2 Y_1^-$$
$$(6.18.19)$$

mod $J(\lambda)$. Again, using (6.18.12), we see that

$$\left[Y_1^+, \Delta_{n-1}^-(Y_n^-)^2\right] = \left[Y_1^+, \Delta_{n-1}^-\right] (Y_n^-)^2 + \Delta_{n-1}^- \left[Y_1^+, (Y_n^-)^2\right]$$
$$= \left(2(n-3)Y_1^- + 4Y_1^- H_0 + 4\sum_{r=2}^{n-1} Y_r^- M_{1r}\right) (Y_n^-)^2$$
$$+ \Delta_{n-1}^-(2Y_1^- + 4Y_n^- M_{1n})$$
$$\equiv 2(n-7)Y_1^- (Y_n^-)^2 + 4\lambda Y_1^- (Y_n^-)^2 + 2\Delta_{n-1}^- Y_1^- \qquad (6.18.20)$$

mod $J(\lambda)$. Next, (6.18.12) and $[H_0, \Delta_{n-1}^-] = -2\Delta_{n-1}^-$ yield

$$\left[Y_1^+, (\Delta_{n-1}^-)^2\right] = \left[Y_1^+, \Delta_{n-1}^-\right]\Delta_{n-1}^- + \Delta_{n-1}^-\left[Y_1^+, \Delta_{n-1}^-\right]$$

$$= 2(n-3)Y_1^-\Delta_{n-1}^- + 4Y_1^- H_0\Delta_{n-1}^- + 4\sum_{r=2}^{n-1} Y_r^- M_{1r}\Delta_{n-1}^-$$

$$+ 2(n-3)\Delta_{n-1}^- Y_1^- + 4\Delta_{n-1}^- Y_1^- H_0 + 4\Delta_{n-1}^-\sum_{r=2}^{n-1} Y_r^- M_{1r}$$

$$\equiv 2(n-7)Y_1^-\Delta_{n-1}^- + 2(n-3)\Delta_{n-1}^- Y_1^- + 4\lambda\left(Y_1^-\Delta_{n-1}^- + \Delta_{n-1}^- Y_1^-\right)$$

$$\tag{6.18.21}$$

mod $J(\lambda)$. (6.18.14) and $[M_{1n}, Y_n^-] = Y_1^-$ imply

$$\left[Y_1^+, Y_n^-\Delta_{n-1}^- Y_n^-\right] = \left[Y_1^+, Y_n^-\Delta_{n-1}^-\right]Y_n^- + Y_n^-\Delta_{n-1}^-\left[Y_1^+, Y_n^-\right]$$

$$= \Big(2(n-4)Y_n^- Y_1^- - 2Y_1^- Y_n^-$$

$$+ 4Y_n^- Y_1^- H_0 + 4Y_n^-\sum_{r=2}^{n-1} Y_r^- M_{1r} + 2\Delta_{n-1}^- M_{1n}\Big)Y_n^-$$

$$+ 2Y_n^-\Delta_{n-1}^- M_{1n}$$

$$\equiv 2(n-6)Y_n^- Y_1^- Y_n^- - 2Y_1^-(Y_n^-)^2 + 4\lambda Y_n^- Y_1^- Y_n^- + 2\Delta_{n-1}^- Y_1^-$$

$$\tag{6.18.22}$$

mod $J(\lambda)$. It remains to determine the commutators

$$\left[Y_1^+, \sum_j Y_j^-(Y_n^-)^2 Y_j^-\right], \quad \left[Y_1^+, \sum_j Y_j^-\Delta_{n-1}^- Y_j^-\right].$$

We find

$$[Y_1^+, L_{4,n-1}]$$

$$\equiv (2n-6+2\lambda)\Delta_{n-1}^- Y_1^- + (2\lambda-4)Y_1^-\Delta_{n-1}^- + 2(n-5+2\lambda)\sum_j Y_j^- Y_1^- Y_j^-$$

mod $J(\lambda)$ and

$$\left[Y_1^+, \sum_j Y_j^-(Y_n^-)^2 Y_j^-\right]$$

$$\equiv 2(n-5+\lambda)(Y_n^-)^2 Y_1^- + (2\lambda-4)Y_1^-(Y_n-)^2 + 2\sum_j Y_j^- Y_1^- Y_j^-$$

mod $J(\lambda)$. Now it follows that the commutator $[Y_1^+, \bar{\mathcal{D}}_4(\lambda)]$ is a linear combination of

$$Y_1^-(Y_n^-)^2, \quad Y_n^- Y_1^- Y_n^-, \quad (Y_n^-)^2 Y_1^-, \quad Y_1^-\Delta_{n-1}^-, \quad \Delta_{n-1}^- Y_1^-$$

and $\sum_j Y_j^- Y_1^- Y_j^-$ modulo $J(\lambda)$. The vanishing of the resulting six coefficients implies the six conditions listed in Table 6.5. Here we use the notation

$$a_0(\lambda) = \frac{1}{3}(n-5+2\lambda)(n-7+2\lambda), \quad a_1(\lambda) = -2(n-5+2\lambda).$$

	term	condition
1	$Y_1^-(Y_n^-)^2$	$2a_0 + a_1(-2b + 2c(n-7+2\lambda) + 2d(\lambda-2)) = 0$
2	$Y_n^- Y_1^- Y_n^-$	$4a_0 + a_1(-4a + 2b(n-6+2\lambda)) = 0$
3	$(Y_n^-)^2 Y_1^-$	$6a_0 + a_1(2a(n-5+2\lambda) + 2d(n-5+\lambda)) = 0$
4	$Y_1^- \Delta_{n-1}^-$	$2a_1 a + 2(1-\alpha)(n-7+2\lambda) + 2\alpha(\lambda-2) = 0$
5	$\Delta_{n-1}^- Y_1^-$	$2a_1(b+c) + 2(1-\alpha)(n-3+2\lambda) = 0$
6	$\sum_j Y_j^- Y_1^- Y_j^-$	$2a_1 d + 2\alpha(n-5+2\lambda) = 0.$

Table 6.5: The vanishing conditions

We analyze the conditions in Table 6.5 together with $a+b+c+d=1$. The condition 6 means $2d = \alpha$. The first three and the last three conditions are linearly dependent if $a+b+c+d=1$; that can be seen immediately by adding these sets of equations. Now condition 4 is equivalent to

$$2a(n-5+2\lambda) + \alpha(n-5+\lambda) = (n-7+2\lambda), \quad (6.18.23)$$

which, in turn, is equivalent to condition 3 using $2d = \alpha$. Thus all conditions are satisfied if we choose $a+b+c+d=1$, $\alpha = 2d$, (6.18.23) and condition 2. □

It is worth emphasizing a *special case*. The relations

$$[Y_j^+, L_{4,n}] \equiv (2n-4+2\lambda)\Delta_n^- Y_j^- + (2\lambda-4)Y_j^- \Delta_n^- + 2(n-4+2\lambda)\sum_j Y_j^- Y_1^- Y_j^-$$

and

$$[Y_j^+, (\Delta_n^-)^2] \equiv 2(n-6+2\lambda)Y_j^- \Delta_n^- + 2(n-2+2\lambda)\Delta_n^- Y_j^-$$

mod $J(\lambda)$ for $j = 1, \ldots, n$ show that for $\lambda = -\frac{n}{2}+2$ the linear combination

$$B_{4,n} = n(\Delta_n^-)^2 - 4L_{4,n}$$

satisfies $[Y_j^+, B_{4,n}] \equiv 0 \mod J(-\frac{n}{2}+2)$. But $\pi B_{4,n} = (n-4)(\Delta_n^-)^2$, i.e., $B_{4,n}$ defines a homomorphism $\mathcal{N}_{-\frac{n}{2}-2} \to \mathcal{N}_{-\frac{n}{2}+2}$ which lifts the intertwining operator $\mathcal{M}_{-\frac{n}{2}-2} \to \mathcal{M}_{-\frac{n}{2}+2}$ of Verma modules defined by right multiplication with $(\Delta_n^-)^2$. It is this operator which is the translation of Δ_n^- in the semi-holonomic category. If $n \neq 4$, then $B_{4,n}$ induces the Paneitz operator $(n-4)P_{4,n}$ on $C^\infty(M^n)$. For $n = 4$, the product $(n-4)P_{4,n}$ vanishes although $B_{4,n}$ is non-trivial.

Theorem 6.18.7. *The family* $\Theta_{2N}(\lambda) : \mathcal{N}_{\lambda-2N}(\mathfrak{g}_n) \to \mathcal{N}_\lambda(\mathfrak{g}_{n+1})$ ($N \geq 0$) *is induced by right multiplication with*

$$c_{2N} \left\{ \prod_{j=N+1}^{2N} \frac{1}{n-1-2j+2\lambda} \prod_{j=1}^{N} \frac{1}{n-2j+2\lambda} \right\}$$

$$\times \prod_{j=1}^{N} (n-1-(j+N)+\lambda)\, \mathcal{D}_{2N}^T(\lambda), \quad (6.18.24)$$

where $\mathcal{D}_{2N}^T(\lambda)$ *is a rational family with the property*

$$\pi \circ \mathcal{D}_{2N}^T(\lambda) = \mathcal{D}_{2N}(\lambda),$$

i.e., $\mathcal{D}_{2N}^T(\lambda)$ *lifts* $\mathcal{D}_{2N}(\lambda)$. *Here* $c_{2N} = (2N-1)(2N-3)\cdots 1$. *The poles of* $\mathcal{D}_{2N}^T(\lambda)$ *are at the zeros of the product*

$$\prod_{j=2}^{N}(n-1-(j+N)+\lambda),$$

i.e., at $\lambda \in \{-n+N+3,\ldots,-n+1+2N\}$. *These* $N-1$ *poles are (at most) simple. In particular, for odd* n, *the critical family* $\mathcal{D}_{n-1}^T(\lambda)$ *has (at most) a simple pole at* $\lambda = 0$.

Note that in (6.18.24) the zeros of the denominator are disjoint from the zeros of the nominator. There is no pole at the zero of $(n-2-N+\lambda)$ since $\mathcal{D}_2^T(\lambda)$ has *no* pole. Since we do not give formulas for the residues, it is not excluded that some of these vanish. For an explicit example we refer to Theorem 6.18.5 and the discussion following it.

Notice that for $\lambda = -\frac{n-1}{2} + N$, the identity (6.18.24) reads

$$\Theta_{2N}\left(-\frac{n-1}{2}+N\right) = c_{2N}\frac{(-1)^N}{(2N)!} \prod_{j=1}^{N}\left(\frac{n-1}{2}-j\right) \mathcal{D}_{2N}^T\left(-\frac{n-1}{2}+N\right).$$

Hence using $\pi\mathcal{D}_{2N}^T(-\frac{n-1}{2}+N) = \mathcal{D}_{2N}(-\frac{n-1}{2}+N) = (\Delta_{n-1}^-)^N$, we find

$$\pi\Theta_{2N}\left(-\frac{n-1}{2}+N\right) = (2N-1)\cdots 3\cdot 1\frac{(-1)^N}{(2N)!} \prod_{j=1}^{N}\left(\frac{n-1}{2}-j\right)(\Delta_{n-1}^-)^N$$

$$= \frac{(-1)^N}{(2N)!} \prod_{j=1}^{N}(j-\frac{1}{2})(n-1-2j)(\Delta_{n-1}^-)^N \qquad (6.18.25)$$

for the Nth translation of $i : \mathfrak{g}_n \hookrightarrow \mathfrak{g}_{n+1}$ in the holonomic category. In particular, for odd n $\pi\Theta_{n-1}(0) = 0$. (6.18.25) should be compared with the formula

$$\prod_{j=2}^{N}(j-1)(n-1-2j)\Delta_{n-1}^N$$

for an analogous composition of operators in [116]. Since the latter product vanishes for $2N = n - 1$, it does *not* yield the *critical* GJMS-operator in dimension $n - 1$ even in the flat case.

Proof. We use induction on N. The assertion is valid for $N = 0$. In the following, we identify operators with multiplicators. We assume that

$$\pi\Theta_{2N}(\lambda)(1 \otimes 1)$$

$$= c_{2N} \left\{ \prod_{j=N+1}^{2N} \frac{1}{n-1-2j+2\lambda} \prod_{j=1}^{N} \frac{1}{n-2j+2\lambda} \right\} \prod_{j=1}^{N}(n-1-(j+N)+\lambda)\mathcal{D}_{2N}(\lambda).$$

We claim that $\pi\Theta_{2N+2}(\lambda)(1 \otimes 1)$ is of the form

$$\alpha_{N+1}(\lambda)(Y_n^-)^{2N+2} + \alpha_N(\lambda)(Y_n^-)^{2N}\Delta_{n-1}^- + \cdots + \alpha_0(\lambda)(\Delta_{n-1}^-)^{N+1}.$$

Since $\pi\Theta_{2N+2}(\lambda)$ is an intertwining operator of Verma modules, we can apply Theorem 5.1.1. It follows that $\pi\Theta_{2N+2}(\lambda)(1 \otimes 1)$ is the product of $\mathcal{D}_{2N+2}(\lambda)$ with a rational function in λ. It only remains to identify the latter rational multiplier.

In order to prove the claim, we first notice that

$$\pi\Theta_{2N+2}(\lambda)(1 \otimes 1) = p(\lambda)\pi\Theta_{2N}(\lambda-1)i(\lambda-(2N+1))(1 \otimes 1) \qquad \text{(by definition)}$$

$$= c_{2N} \left\{ \prod_{j=N+2}^{2N+1} \frac{1}{n-1-2j+2\lambda} \prod_{j=2}^{N+1} \frac{1}{n-2j+2\lambda} \right\} \prod_{j=1}^{N}(n-1-(j+N+1)+\lambda)$$

$$\times p(\lambda)\left(\mathcal{D}_{2N}(\lambda-1) \otimes I_n\right)i(\lambda-(2N+1))(1 \otimes 1).$$

Hence it is enough to prove that

$$p(\lambda)\left(\mathcal{D}_{2N}(\lambda-1) \otimes I_n\right)i(\lambda-(2N+1))(1 \otimes 1)$$

is of the form

$$\frac{1}{n-1-2(2N+2)+2\lambda}\frac{1}{n-2+2\lambda}\left[\alpha_{N+1}(\lambda)(Y_n^-)^{2N+2} + \cdots + \alpha_0(\lambda)(\Delta_{n-1}^-)^{N+1}\right],$$

$$(6.18.26)$$

where $i(\lambda)$ and $p(\lambda)$ are defined by

$$i(\lambda)(1 \otimes 1) = -\lambda(1 \otimes 1 \otimes v_-) + \sum_{1}^{n-1} Y_i^- \otimes 1 \otimes v_i - \frac{1}{2(n-3+2\lambda)}(\Delta_{n-1}^- \otimes 1 \otimes v_+)$$

and

$$p(\lambda)(1 \otimes 1 \otimes v_-) = \frac{1}{n-2+2\lambda}\left(\Delta_n^- \otimes 1\right),$$

$$p(\lambda)(1 \otimes 1 \otimes v_i) = Y_i^- \otimes 1 \ (i = 1, \ldots, n), \quad p(\lambda)(1 \otimes 1 \otimes v_+) = 2\lambda \otimes 1$$

(notice the different dimensions). Now $(\mathcal{D}_{2N}(\lambda-1) \otimes I_n)\, i(\lambda-(2N+1))(1 \otimes 1)$ is a linear combination of elements of the form

$$(Y_n^-)^{2N-2a}(\Delta_{n-1}^-)^a \otimes 1 \otimes v_-,$$

$$\sum_1^{n-1} Y_i^- (Y_n^-)^{2N-2a}(\Delta_{n-1}^-)^a \otimes 1 \otimes v_i, \quad (Y_n^-)^{2N-2a}(\Delta_{n-1}^-)^{a+1} \otimes 1 \otimes v_+.$$

These elements can be written as linear combinations of elements of the form

$$(Y_n^-)^{2N-2a}(\Delta_{n-1}^-)^a(1 \otimes 1 \otimes v_-), \quad \sum_1^{n-1} Y_i^- (Y_n^-)^{2N-2a}(\Delta_{n-1}^-)^a(1 \otimes 1 \otimes v_i),$$

$$(Y_n^-)^{2N-2a}(\Delta_{n-1}^-)^{a+1}(1 \otimes 1 \otimes v_+). \quad (6.18.27)$$

Now applying $p(\lambda)$, yields an element of the form (6.18.26).

Next, we prove that the proportionality coefficient is as asserted. We use induction on N. Assuming the validity of $(6.18.24)_{2N}$ for Θ_{2N}, it suffices to prove that the coefficient of $(Y_n^-)^{2N+2}$ in Θ_{2N+2} coincides with the corresponding coefficient given by $(6.18.24)_{N+1}$.

Now contributions of the form $(Y_n^-)^{2N+2}$ can arise only from contributions of $(Y_n^-)^{2N}(1 \otimes v_-)$ in

$$-(\lambda-2N-1)\mathcal{D}_{2N}(\lambda-1) \otimes 1 \otimes v_-$$

$$+\sum_{i=1}^{n-1} Y_i^- \mathcal{D}_{2N}(\lambda-1) \otimes 1 \otimes v_i - \frac{1}{2(n-3-2(2N+1)+2\lambda)}\Delta_{n-1}^- \mathcal{D}_{2N}(\lambda-1) \otimes 1 \otimes v_+.$$

$$(6.18.28)$$

In order to determine these contributions, we write each term as a linear combination of terms as in (6.18.27). The first term equals $-(\lambda-2N-1)\mathcal{D}_{2N}(\lambda-1)(1 \otimes 1 \otimes v_-)$, i.e., it contributes

$$-(\lambda-2N-1)A_{2N}(\lambda-1)(Y_n^-)^{2N}(1 \otimes 1 \otimes v_-),$$

where $A_{2N}(\lambda)$ is the coefficient of $(Y_n^-)^{2N}$ in $\mathcal{D}_{2N}(\lambda)$. The sum in (6.18.28) contributes

$$-(n-1)A_{2N}(\lambda-1)(Y_n^-)^{2N}(1 \otimes 1 \otimes v_-).$$

Finally, the third term in (6.18.28) yields the contribution

$$-\frac{2(n-1)}{2(n-3-2(2N+1)+2\lambda)}A_{2N}(\lambda-1)(Y_n^-)^{2N}(1 \otimes 1 \otimes v_-).$$

Thus we find

$$-\left[(n-2-2N+\lambda)+\frac{n-1}{n-5-4N+2\lambda}\right]A_{2N}(\lambda-1)(Y_n^-)^{2N}(1 \otimes 1 \otimes v_-)$$

$$=-\frac{1}{n-5-4N+2\lambda}(n-3-2N+\lambda)(n-3+2\lambda-4N)A_{2N}(\lambda-1)(Y_n^-)^{2N}(1 \otimes 1 \otimes v_-).$$

An easy calculation shows that (5.1.3) implies the relation

$$-\frac{1}{2N+1}(n-3+2\lambda-4N)A_{2N}(\lambda-1) = A_{2N+2}(\lambda).$$

Hence the coefficient of $(Y_n^-)^{2N+2}(1 \otimes 1 \otimes v_-)$ in (6.18.28) is

$$\frac{1}{n-5-4N+2\lambda}(2N+1)(n-3-2N+\lambda)A_{2N+2}(\lambda).$$

Now we apply $p(\lambda)$ and find that the coefficient of $(Y_n^-)^{2N+2}$ in $\pi \circ \Theta_{2N+2}(\lambda)(1 \otimes 1)$ is given by

$$(2N+1)c_{2N}\left\{\prod_{j=N+2}^{2N+2}\frac{1}{n-1-2j+2\lambda}\prod_{j=1}^{N+1}\frac{1}{n-2j+2\lambda}\right\}$$

$$\times \prod_{j=1}^{N+1}(n-1-(j+N+1)+\lambda)A_{2N+2}(\lambda).$$

Since $c_{2N+2} = (2N+1)c_{2N}$, this product coincides with the coefficient of $(Y_n^-)^{2N+2}$ in (6.18.24)$_{N+1}$. The proof is complete. $\qquad\square$

There is an analogous result for odd order families.

Theorem 6.18.8. *The family* $\Theta_{2N+1}(\lambda) : \mathcal{N}_{\lambda-2N-1}(\mathfrak{g}_n) \to \mathcal{N}_\lambda(\mathfrak{g}_{n+1})$ $(N \geq 0)$ *is induced by right multiplication with*

$$c_{2N+1}\left\{\prod_{j=N+1}^{2N}\frac{1}{n-3-2j+2\lambda}\prod_{j=1}^{N}\frac{1}{n-2j+2\lambda}\right\}\prod_{j=1}^{N}(n-2-(j+N)+\lambda)\mathcal{D}_{2N+1}^T(\lambda),$$

$$(6.18.29)$$

where $\mathcal{D}_{2N+1}^T(\lambda)$ *is a rational family with the property*

$$\pi \circ \mathcal{D}_{2N+1}^T(\lambda) = \mathcal{D}_{2N+1}(\lambda),$$

i.e., $\mathcal{D}_{2N+1}^T(\lambda)$ *lifts* $\mathcal{D}_{2N+1}(\lambda)$. *The poles of* $\mathcal{D}_{2N+1}^T(\lambda)$ *are at the zeros of the product* $\prod_{j=1}^{N}(n-2-(j+N)+\lambda)$, *i.e., at* $\lambda \in \{-n+N+3, \ldots, -n+2+2N\}$. *These* N *poles are (at most) simple. In particular, for even* n, *the critical family* $\mathcal{D}_{n-1}^T(\lambda)$ *has (at most) a simple pole at* $\lambda = 0$.

Theorem 6.18.3 and the discussion following it illustrate the general result.

6.19   From Verma modules to tractors

The projection $p(\lambda) : \mathcal{M}_{\lambda-1} \otimes F \to \mathcal{M}_\lambda$, which is given by

$$1 \otimes 1 \otimes v_- \mapsto \frac{1}{n-2+2\lambda}(\Delta_n^- \otimes 1), \quad 1 \otimes 1 \otimes v_i \mapsto Y_i^- \otimes 1, \quad 1 \otimes 1 \otimes v_+ \mapsto 2\lambda \otimes 1,$$

induces a left G-equivariant differential operator

$$D(\lambda) : C^\infty(G, \mathbb{C}(\lambda))^P \to (C^\infty(G, \mathbb{C}(\lambda-1)) \otimes F^*)^P \qquad (6.19.1)$$

by sending u to

$$F \ni f \mapsto p(\lambda)(1 \otimes 1 \otimes f)(u) \in C^\infty(G),$$

where $\mathcal{U}(\mathfrak{g})$ acts on $C^\infty(G)$ by the right regular representation R. In explicit terms, we find

$$D(\lambda) : u \mapsto \frac{1}{n-2+2\lambda} R(\Delta_n^-)u \otimes v_-^* + \sum_i R(Y_i^-)u \otimes v_i^* + 2\lambda u \otimes v_+^* \qquad (6.19.2)$$

using a dual basis $\{v_\pm^*, v_j^*\}$ of F^*.

It is instructive to verify the mapping properties of $D(\lambda)$ *directly*. The relation $\pi^*(m)D(\lambda)u(gm) = u(g)$ for $m \in M$ follows from $\mathrm{Ad}(m)\Delta_n^- = \Delta_n^-$, $\pi^*(m)v_\pm^* = v_\pm^*$ and the fact that $Y_j^- \mapsto v_j$ is an M-equivariant isomorphism $\mathfrak{n}^- \to V_0$. It remains to prove that

$$R(H_0)D(\lambda)u + \pi^*(H_0)D(\lambda)u = (\lambda-1)D(\lambda)u, \quad R(Y_j^+)D(\lambda)u + \pi^*(Y_j^+)D(\lambda)u = 0$$

for $j = 1, \ldots, n$. In order to prove the second set of conditions, we calculate the sum

$$\frac{1}{n-2+2\lambda} R(Y_j^+ \Delta_n^-)u \otimes v_-^* + \sum_i R(Y_j^+ Y_i^-)u \otimes v_i^* + 2\lambda R(Y_j^+)u \otimes v_+^*$$

$$+ \frac{1}{n-2+2\lambda} R(\Delta_n^-)u \otimes Y_j^+(v_-^*) + \sum_i R(Y_i^-)u \otimes Y_j^+(v_i^*) + 2\lambda u \otimes Y_j^+(v_+^*)$$

for $u \in C^\infty(G, \mathbb{C}(\lambda))^P$, i.e., $R(Y_j^+)u = 0$, $R(H_0)u = \lambda u$, $R(m)u = u$. The identities

$$Y_j^+(v_-^*) = 0, \; Y_j^+(v_+^*) = -v_j^*, \; Y_j^+(v_i^*) = -2\delta_{ij}v_-^*$$

show that the sum equals

$$\frac{1}{n-2+2\lambda} R([Y_j^+, \Delta_n^-])u \otimes v_-^* + \sum_i R([Y_j^+, Y_i^-])u \otimes v_i^* - 2R(Y_j^-)u \otimes v_-^* - 2\lambda u \otimes v_j^*.$$

Now the commutator relations

$$[Y_j^+, \Delta_n^-] = 2(n-2)Y_j^- + 4Y_j^- H_0 + 4\sum_{r \neq j} Y_r^- M_{jr}$$

(see (6.18.3)) and $[Y_j^+, Y_i^-] = 2\delta_{ij}H_0 + 2M_{ji}$ simplify that sum to

$$\frac{1}{n-2+2\lambda} \left(2(n-2)R(Y_j^-)u + 4\lambda R(Y_j^-)u\right) \otimes v_-^*$$

$$+ 2\lambda u \otimes v_j^* - 2R(Y_j^-)u \otimes v_-^* - 2\lambda u \otimes v_j^* = 0.$$

In order to prove the first set of conditions, we use $H_0(v_\pm^*) = \mp v_\pm^*$, $H_0(v_i^*) = 0$ and $[H_0, \Delta_n^-] = -2\Delta_n^-$, $[H_0, Y_i^-] = -Y_i^-$. It follows that $R(H_0)D(\lambda)u + \pi^*(H_0)D(\lambda)u$ equals

$$\frac{1}{n-2+2\lambda}(R(H_0\Delta_n^-)u + R(\Delta_n^-)u)\otimes v_-^* + \sum_i R(H_0Y_i^-)u\otimes v_i^* + 2\lambda(R(H_0)u - u)\otimes v_+^*,$$

i.e.,

$$\frac{\lambda-1}{n-2+2\lambda}R(\Delta_n^-)u\otimes v_-^* + (\lambda-1)\sum_i R(Y_i^-)u\otimes v_i^* + 2\lambda(\lambda-1)u\otimes v_+^*$$

$$= (\lambda-1)D(\lambda)u.$$

Similarly, the embedding $i(\lambda) : \mathcal{M}_{\lambda-1} \to \mathcal{M}_\lambda \otimes F$ defined by

$$1\otimes 1 \mapsto -\lambda\otimes 1\otimes v_- + \sum_i Y_i^- \otimes 1\otimes v_i - \frac{1}{2(n-2+2\lambda)}(\Delta_n^- \otimes 1\otimes v_+)$$

induces a left G-equivariant differential operator

$$C(\lambda) : (C^\infty(G, \mathbb{C}(\lambda)) \otimes F^*)^P \to C^\infty(G, \mathbb{C}(\lambda-1))^P \tag{6.19.3}$$

by $u \mapsto i(\lambda)(1\otimes 1)(u)$. Here $\mathcal{U}(\mathfrak{g})\otimes F$ acts on the tensor product module $C^\infty(G)\otimes F^*$ by

$$T\otimes f : u\otimes f^* \mapsto \langle T(u\otimes f^*), f\rangle,$$

where $C^\infty(G)$ is viewed as a right G-module. In order to determine an *explicit* formula, we first calculate how $Y_j^- \otimes v_j$ and $\Delta_n^- \otimes v_+$ act on

$$u = u_- \otimes v_-^* + \sum_i u_i \otimes v_i^* + u_+ \otimes v_+^* \in C^\infty(G)\otimes F^*.$$

We find

$$(Y_j^- \otimes v_j)(u_- \otimes v_-^*) = \langle Y_j^-(u_- \otimes v_-^*), v_j\rangle = \langle Y_j^-(u_-)\otimes v_-^* - u_- \otimes v_j^*, v_j\rangle = -u_-,$$

$$(Y_j^- \otimes v_j)(u_i \otimes v_i^*)$$
$$= \langle Y_j^-(u_i \otimes v_i^*), v_j\rangle = \langle Y_j^-(u_i)\otimes v_i^* - 2u_i \otimes \delta_{ij}v_+^*, v_j\rangle = \delta_{ij}Y_j^-(u_i)$$

and

$$(Y_j^- \otimes v_j)(u_+ \otimes v_+^*) = \langle Y_j^-(u_+ \otimes v_+^*), v_j\rangle = \langle Y_j^-(u_+)\otimes v_+^*, v_j\rangle = 0.$$

Here we have used

$$Y_j^-(v_-^*) = -v_j^*, \quad Y_j^-(v_i^*) = -2\delta_{ij}v_+^*, \quad Y_j^-(v_+^*) = 0.$$

Hence
$$\sum_j (Y_j^- \otimes v_j)(u) = -nu_- + \sum_i Y_i^-(u_i).$$

By definition,
$$(\Delta_n^- \otimes v_+)(u) = \langle \Delta_n^-(u_- \otimes v_-^*), v_+ \rangle + \sum_i \langle \Delta_n^-(u_i \otimes v_i^*), v_+ \rangle + \langle \Delta_n^-(u_+ \otimes v_+^*), v_+ \rangle.$$

Thus using
$$(Y_j^-)^2(u_- \otimes v_-^*) = (Y_j^-)^2(u_i) \otimes v_-^* - 2Y_j^-(u_-) \otimes v_j^* + 2u_- \otimes V_+^*,$$
$$(Y_j^-)^2(u_i \otimes v_i^*) = (Y_j^-)^2(u_i) \otimes v_i^* - 4Y_j^-(u_i) \otimes \delta_{ij}v_+^*,$$
$$(Y_j^-)^2(u_i \otimes v_+^*) = (Y_j^-)^2(u_i) \otimes v_+^*,$$

we find
$$(\Delta_n^- \otimes v_+)(u) = 2nu_- - 4\sum_i Y_i^-(u_i) + \Delta_n^-(u_+).$$

Finally, we have $(\lambda \otimes v_-)(u) = \lambda u_-$. Therefore,

$$C(\lambda)u = -\lambda u_- - nu_-$$
$$+ \sum_i Y_i^-(u_i) - \frac{1}{2(n-2+2\lambda)}\left(2nu_- - 4\sum_i Y_i^-(u_i) + \Delta_n^-(u_+)\right)$$
$$= \frac{1}{2(n-2+2\lambda)}\left\{-2(n+2\lambda)(n+\lambda-1)u_- + 2(n+2\lambda)\sum_i Y_i^-(u_i) - \Delta_n^-(u_+)\right\}.$$
$$\tag{6.19.4}$$

It is instructive to verify the mapping properties of $C(\lambda)$ *directly*. The P-invariance of $u \in C^\infty(G, \mathbb{C}(\lambda)) \otimes F^*$ means that

$$Y_j^+(u_-) \otimes v_-^* + \sum_i Y_j^+(u_i) \otimes v_i^* + Y_j^+(u_+) \otimes v_+^* - 2u_j \otimes v_-^* - u_+ \otimes v_j^* = 0,$$

i.e.,
$$Y_j^+(u_+) = 0, \ Y_j^+(u_i) = \delta_{ij}u_+, \ Y_j^+(u_-) = 2u_j.$$

Moreover,

$$H_0(u_-) \otimes v_-^* + \sum_i H_0(u_i) \otimes v_i^* + H_0(u_+) \otimes v_+^* + u_- \otimes v_-^* - u_+ \otimes v_+^*$$
$$= \lambda\left(u_- \otimes v_-^* + \sum_i u_i \otimes v_i^* + u_+ \otimes v_+^*\right)$$

i.e.,
$$H_0(u_\pm) = (\lambda \pm 1)u_\pm, \ H_0(u_i) = \lambda u_i$$

and $m(u_\pm) = u_\pm$, $M_{ji}(u_i) = u_j$.

Now using these results, we find that $2(n-2+2\lambda)Y_j^+C(\lambda)u$ equals

$$- 2(n+2\lambda)(n+\lambda-1)Y_j^+(u_-) + 2(n+2\lambda)\sum_i Y_j^+Y_i^-(u_i) - Y_j^+\Delta_n^-(u_+)$$

$$= -4(n+2\lambda)(n+\lambda-1)u_j + 2(n+2\lambda)\left(\sum_i [Y_j^+,Y_i^-](u_i) + Y_i^-Y_j^+(u_i)\right)$$
$$- [Y_j^+,\Delta_n^-](u_+)$$

i.e.,

$$- 4(n+2\lambda)(n+\lambda-1)u_j + 2(n+2\lambda)\left((n-1)u_j + 2\lambda u_j + Y_j^-(u_+)\right)$$
$$- 2(n-2)Y_j^-(u_+) - 4(\lambda+1)Y_j^-(u_+) = 0$$

using (6.18.3). Similarly, it follows that $H_0C(\lambda)u = (\lambda-1)C(\lambda)u$ and $mC(\lambda)u = C(\lambda)u$. In other words, we have proved $C(\lambda)u \in C^\infty(G,\mathbb{C}(\lambda-1))^P$.

$p(\lambda)$ and $i(\lambda)$ also induce *twisted versions* of $D(\lambda)$ and $C(\lambda)$ which involve an additional G-module (W,τ). Let

$$\nabla_X^W : C^\infty(G) \otimes W \to C^\infty(G) \otimes W, \; u \mapsto (R \otimes \tau)(X)(u)$$

and

$$\Delta^W = \sum_j \nabla_{Y_j^-}^W \circ \nabla_{Y_j^-}^W,$$

i.e.,

$$\Delta^W(u \otimes w) = R(\Delta_n^-)u \otimes w + 2\sum_j R(Y_j^-)u \otimes \tau(Y_j^-)w + u \otimes \tau(\Delta_n^-)w.$$

Lemma 6.19.1. *$p(\lambda)$ induces a left G-equivariant operator*

$$D^W(\lambda) : (C^\infty(G,\mathbb{C}(\lambda)) \otimes W)^P \to (C^\infty(G,\mathbb{C}(\lambda-1)) \otimes W \otimes F^*)^P$$

by $u \mapsto (F \in f \mapsto p(\lambda)(1 \otimes 1 \otimes f)u)$. In explicit terms it is given by

$$u \mapsto \frac{1}{n-2+2\lambda}\Delta^W u \otimes v_-^* + \sum_i \nabla_{Y_i^-}^W u \otimes v_i^* + 2\lambda u \otimes v_+^*$$

In order to verify the mapping properties of $D^W(\lambda)$, the same arguments as in the untwisted case apply by replacing the G-module $C^\infty(G)$ by $C^\infty(G) \otimes W$. In fact, the arguments only rest on the commutator relations.

In order to work out the case $W = F^*$, we write an element of $C^\infty(G) \otimes F^*$ in the form

$$u = u_- \otimes v_-^* + \sum_i u_i \otimes v_i^* + u_+ \otimes v_+^*.$$

We find

$$\nabla_{Y_j^-}^{F^*}(u) = Y_j^-(u_-) \otimes v_-^* + \sum_i Y_j^-(u_i) \otimes v_i^* + Y_j^-(u_+) \otimes v_+^* - u_- \otimes v_j^* - 2u_j \otimes v_+^*$$

and

$$\Delta^{F^*}(u) = \Delta_n^-(u_-) \otimes v_-^* + \sum_i \Delta_n^-(u_i) \otimes v_i^* + \Delta_n^-(u_+) \otimes v_+^*$$

$$- 2\sum_j Y_j^-(u_-) \otimes v_j^* - 4\sum_{i,j} Y_j^-(u_i) \otimes \delta_{ij} v_+^* + 2nu_- \otimes v_+^*.$$

It is convenient to write the latter formulas in terms of matrices as

$$\nabla_{Y_j^-}^{F^*} : \begin{pmatrix} u_- \\ \omega \\ u_+ \end{pmatrix} \mapsto \begin{pmatrix} Y_j^- & 0 & 0 \\ -e_j & Y_j^- & 0 \\ 0 & -2e_j^t & Y_j^- \end{pmatrix} \begin{pmatrix} u_- \\ \omega \\ u_+ \end{pmatrix}, \tag{6.19.5}$$

and

$$\Delta^{F^*} : \begin{pmatrix} u_- \\ \omega \\ u_+ \end{pmatrix} \mapsto \begin{pmatrix} \Delta_n^- & 0 & 0 \\ -2Y^- & \Delta_n^- & 0 \\ 2n & -4(Y^-)^t & \Delta_n^- \end{pmatrix} \begin{pmatrix} u_- \\ \omega \\ u_+ \end{pmatrix}, \tag{6.19.6}$$

where $\omega = \begin{pmatrix} u_1 \\ \vdots \\ u_n \end{pmatrix}$ and $Y^- = \begin{pmatrix} Y_1^- \\ \vdots \\ Y_n^- \end{pmatrix}$. In the following sections, these constructions will be recognized as the flat case of the *tractor connection* and its associated *tractor Laplacian* (see Definition 6.20.1 and formula (6.22.2)).

Corollary 6.18.1 has the following counterpart in terms of $C(\lambda)$ and $D(\lambda)$.

Lemma 6.19.2. $C(\lambda) = \mathrm{tr}_{F^*} \circ D^{F^*}(\lambda)$, where $\mathrm{tr}_{F^*} : F^* \otimes F^* \to \mathbb{C}$ denotes the contraction defined by the scalar product

$$\left(\begin{pmatrix} \tau_- \\ \tau \\ \tau_+ \end{pmatrix}, \begin{pmatrix} \mu_- \\ \mu \\ \mu_+ \end{pmatrix} \right) = -\frac{1}{2}(\tau_-\mu_+ + \tau_+\mu_-) + (\tau, \mu)$$

in terms of the basis $\{v_-^*, v_i^*, v_+^*\}$.

Proof. A routine calculation shows that the scalar product is induced by the scalar product of $F \simeq \mathbb{R}^{1,n+1}$. Now for $u \in (C^\infty(G, \mathbb{C}(\lambda)) \otimes F^*)^P$, we have

$$D^{F^*}(\lambda)u = \frac{1}{n-2+2\lambda}\Delta^{F^*}u \otimes v_-^* + \sum_i \nabla_{Y_i}^{F^*}u \otimes v_i^* + 2\lambda u \otimes v_+^*$$

by Lemma 6.19.1. Now (6.19.5) and (6.19.6) imply

$$\operatorname{tr}(D^{F^*}(\lambda)u) = -\frac{1}{2}\frac{1}{n-2+2\lambda}\left(2nu_- - 4\sum_j Y_j^-(\omega_j) + \Delta_n^- u_+\right)$$
$$+ \sum_j(-u_- + Y_j^-(\omega_j)) - \lambda u_- = C(\lambda)u$$

using (6.19.4). The proof is complete. □

The relation

$$p\left(-\frac{n}{2}-1\right) \circ (\Delta_n^- \otimes I) \circ i\left(-\frac{n}{2}+2\right) = -\frac{n-4}{8}(\Delta_n^-)^2, \qquad (6.19.7)$$

i.e., the fact that $(\Delta_n^-)^2$ is the Zuckerman translate of Δ_n^- (see Theorem 6.18.1), induces analogous relations for differential operators. In the flat case $(\mathbb{R}^n, g_c)$, let

$$D\left(-\frac{n}{2}+2\right)u = \begin{pmatrix} \frac{1}{2}\Delta_n u \\ du \\ (4-n)u \end{pmatrix}$$

and

$$C\left(-\frac{n}{2}-1\right)\begin{pmatrix} u_- \\ \omega \\ u_+ \end{pmatrix} = -\frac{1}{2}\left\{\left(\frac{n}{2}-2\right)u_- + 2\delta\omega - \frac{1}{4}\Delta_n u_+\right\}.$$

In order to write the analog of (6.19.7), we use the operator

$$\Delta_n^{\mathcal{T}} = \begin{pmatrix} \Delta_n & 0 & 0 \\ -2d & \Delta_n & 0 \\ 2n & -4\delta & \Delta_n \end{pmatrix} \qquad (6.19.8)$$

on the space $C^\infty(\mathbb{R}^n) \oplus \Omega^1(\mathbb{R}^n) \oplus C^\infty(\mathbb{R}^n)$ (this is the flat version of (6.22.2)).

Lemma 6.19.3.

$$C\left(-\frac{n}{2}-1\right) \circ \Delta_n^{\mathcal{T}} \circ D\left(-\frac{n}{2}+2\right) = -\frac{n-4}{8}\Delta_n^2.$$

Proof. We notice that

$$\Delta_n^{\mathcal{T}} \circ D\left(-\frac{n}{2}+2\right) = \begin{pmatrix} \frac{1}{2}\Delta_n^2 \\ 0 \\ 0 \end{pmatrix}.$$

Now the result follows by a direct calculation. □

Lemma 6.19.3 is the flat special case of a formula which yields the Paneitz operator P_4 as a composition of operators on tractors (Theorem 6.20.7, Corollary 6.20.3).

6.20   Some elements of tractor calculus

In the present section, we discuss some basic ingredients of tractor calculus. The results will be used in Section 6.21 for the construction of the *tractor families*, which are conformally covariant curved versions of the constructions in Section 6.19. The following presentation is self-contained. For more information on tractor calculus we refer to [17], [88], [109] and [54].

The main objects will be the conformally invariant tractor connection ∇^T and the conformally invariant tractor D-operator $D(\lambda)$. The latter is conformally invariant as an operator on sections of certain vector bundles. Both the operator and the bundles depend on the parameter λ. By choosing appropriate trivializations of the bundles, the operator corresponds to a conformally covariant family of operators which acts on functional spaces which are independent of λ.

We identify the pair consisting of

$$\begin{pmatrix} u_- \\ \omega \\ u_+ \end{pmatrix} \in C^\infty(M) \oplus \Omega^1(M) \oplus C^\infty(M)$$

and the metric g with the pair consisting of

$$T(g,\varphi) \begin{pmatrix} u_- \\ \omega \\ u_+ \end{pmatrix} \in C^\infty(M) \oplus \Omega^1(M) \oplus C^\infty(M)$$

and the metric $e^{2\varphi}g$. Here

$$T(g,\varphi) \stackrel{\text{def}}{=} \begin{pmatrix} e^{-\varphi} & & \\ & e^\varphi & \\ & & e^\varphi \end{pmatrix} \begin{pmatrix} 1 & (d\varphi,\cdot) & \frac{1}{4}|d\varphi|^2 \\ 0 & 1 & \frac{1}{2}d\varphi \\ 0 & 0 & 1 \end{pmatrix}, \tag{6.20.1}$$

where $(\cdot,\cdot)$ and $|\cdot|$ refer to g, of course. In view of

$$T(e^{2\psi}g,\varphi) \circ T(g,\psi) = T(g,\varphi+\psi),$$

this defines an equivalence relation. The equivalence classes are called *dual tractors*.

Alternatively, dual tractors can be defined as sections of the *dual standard tractor bundle* $\mathcal{T}^*M$ on M. $\mathcal{T}^*M$ is an invariant of the conformal class c containing g. It can be regarded as an associated vector bundle for the P-principal fibre bundle $\mathcal{G} \to M$ (associated to c). The defining representation is the restriction to P of the standard representation of G on F^*.

The choice of a metric $g \in c$ defines a trivialization of $\mathcal{T}^*M$ and an isomorphism

$$\Gamma(\mathcal{T}^*M) \simeq C^\infty(M) \oplus \Omega^1(M) \oplus C^\infty(M).$$

Such identifications will be used throughout without further notice and, abusing notation, $\Gamma(\mathcal{T}^*M)$ will also denote the direct sum.

The following result is a consequence of the transformation rule (2.5.7) for scalar curvature.

Theorem 6.20.1 (Tractor D-operator). *The rational family*

$$D_M(g;\lambda) : C^\infty(M) \ni u \mapsto \begin{pmatrix} \frac{1}{n-2+2\lambda}(\Delta_g + \lambda\mathsf{J}(g))u \\ du \\ 2\lambda u \end{pmatrix} \in \Gamma(\mathcal{T}^*M) \qquad (6.20.2)$$

is conformally covariant in the sense that

$$D_M(\hat{g};\lambda) \circ e^{\lambda\varphi} = e^{(\lambda-1)\varphi} \circ T_M(g,\varphi) \circ D_M(g;\lambda), \quad \hat{g} = e^{2\varphi}g \qquad (6.20.3)$$

for all $\varphi \in C^\infty(M)$.

Proof. We start by proving the important identity

$$e^{-(\lambda-2)\varphi}(\hat{\Delta} + \lambda\hat{\mathsf{J}})(e^{\lambda\varphi}u)$$

$$= (\Delta + \lambda\mathsf{J})u + (n-2+2\lambda)(du, d\varphi) + \lambda\frac{n-2+2\lambda}{2}|d\varphi|^2 u. \quad (6.20.4)$$

In fact, by Lemma 4.2.1 and (2.5.7), the left-hand side equals

$$\Delta u + \lambda\delta(ud\varphi) + (n-2+\lambda)(d\varphi, du) + \lambda(n-2+\lambda)|d\varphi|^2 u + \lambda(\mathsf{J} - \Delta\varphi - \frac{n-2}{2}|d\varphi|^2)u$$

$$= (\Delta + \lambda\mathsf{J})u + \lambda(\delta(ud\varphi) - u\Delta\varphi) + \lambda\frac{n-2+2\lambda}{2}|d\varphi|^2 u + (n-2+\lambda)(d\varphi, du)$$

$$= (\Delta + \lambda\mathsf{J})u + (n-2+2\lambda)(d\varphi, du) + \lambda\frac{n-2+2\lambda}{2}|d\varphi|^2 u.$$

Notice that, in contrast to Chapter 4, we use here the convention that $-\Delta$ is the non-negative Laplacian and $-\delta$ is adjoint to d. It follows that

$$e^{-(\lambda-2)\varphi}\frac{1}{n-2+2\lambda}(\hat{\Delta} + \lambda\hat{\mathsf{J}})(ue^{\lambda\varphi}) = \frac{1}{n-2+2\lambda}(\Delta + \lambda\mathsf{J})u + (d\varphi, du) + \frac{1}{2}\lambda|d\varphi|^2 u.$$

In other words, the first components of both sides of (6.20.3) coincide. For the second components, (6.20.3) asserts the obvious identity $d(e^{\lambda\varphi}u) = e^{\lambda\varphi}(du + \lambda u d\varphi)$. The proof is complete. $\qquad \square$

It is natural to interpret (6.20.3) as stating that $D_M(\lambda)$ defines a conformally invariant operator from densities to dual tractors coupled with densities. A disadvantage of such a formulation, however, is that the functional spaces depend on the parameter λ. The above choice of trivializations of all bundles avoids such complications.

(6.20.1) gives rise to a *representation* of the conformal group of (M, g). Let κ be a conformal diffeomorphism of g, i.e., $\kappa_*(g) = e^{2\Phi}g$ for some $\Phi \in C^\infty(M)$. We define the operator

$$\eta_\lambda(\kappa) = T^{-1}(g, \Phi) \circ e^{-\lambda\Phi} \circ \kappa_*, \quad \lambda \in \mathbb{C} \qquad (6.20.5)$$

on $C^\infty(M) \oplus \Omega^1(M) \oplus C^\infty(M)$. Then

$$\eta_\lambda(\kappa_1)\eta_\lambda(\kappa_2) = \eta_\lambda(\kappa_1\kappa_2).$$

In fact, let Φ_1 and Φ_2 correspond to κ_1 and κ_2. Then

$$\eta_\lambda(\kappa_1)\eta_\lambda(\kappa_2) = T^{-1}(g,\Phi_1)e^{-\lambda\Phi_1}(\kappa_1)_*T^{-1}(g,\Phi_2)e^{-\lambda\Phi_2}(\kappa_2)_*$$
$$= e^{-\lambda(\Phi_1+(\kappa_1)_*\Phi_2)}T^{-1}(g,\Phi_1)(\kappa_1)_*T^{-1}(g,\Phi_2)(\kappa_2)_*.$$

But using

$$T(\kappa_*(g),\kappa_*(\varphi)) = \kappa_* \circ T(g,\varphi) \circ \kappa^*,$$

we simplify the right-hand side to

$$e^{-\lambda(\Phi_1+(\kappa_1)_*\Phi_2)}T^{-1}(g,\Phi_1)T^{-1}((\kappa_1)_*(g),(\kappa_1)_*(\Phi_2))(\kappa_1\kappa_2)_*$$
$$= e^{-\lambda(\Phi_1+(\kappa_1)_*\Phi_2)}T^{-1}(g,\Phi_1+(\kappa_1)_*\Phi_2)(\kappa_1\kappa_2)_*$$
$$= \eta_\lambda(\kappa_1\kappa_2).$$

In particular, for $(M,g) = (S^n, g_c)$, (6.20.5) defines a representation η_λ^c of G^{n+1} on $C^\infty(S^n) \oplus \Omega^1(S^n) \oplus C^\infty(S^n)$. It corresponds to the left regular representation on the space $(C^\infty(G,\mathbb{C}(\lambda)) \otimes F^*)^P$. Therefore, Theorem 6.20.1 shows that on the sphere S^n the tractor D-operator defines an equivariant operator:

$$D_{S^n}(g_c;\lambda) \circ \pi_\lambda^c(g) = \eta_{\lambda-1}^c(g) \circ D_{S^n}(g_c;\lambda), \quad g \in G^{n+1}.$$

It corresponds to the D-operator (6.19.1). Similar comments apply to the following construction.

Theorem 6.20.2 (Tractor C-operator). *The rational family*

$$C_M(g;\lambda) : \Gamma(\mathcal{T}^*M) \ni \begin{pmatrix} u_- \\ \omega \\ u_+ \end{pmatrix} \mapsto \frac{1}{2(n-2+2\lambda)}$$
$$\times \{-2(n+2\lambda)(n+\lambda-1)u_- + 2(n+2\lambda)\delta_g(\omega) - \Delta_g u_+ + (n+\lambda-1)J(g)u_+\}$$
(6.20.6)

is conformally covariant in the sense that

$$C_M(\hat{g};\lambda) \circ e^{\lambda\varphi} \circ T_M(g,\varphi) = e^{(\lambda-1)\varphi} \circ C_M(g;\lambda), \quad \hat{g} = e^{2\varphi}g \qquad (6.20.7)$$

for all $\varphi \in C^\infty(M)$.

Proof. Note that

$$C_M(\lambda)\begin{pmatrix} u_- \\ \omega \\ u_+ \end{pmatrix}$$
$$= -\frac{1}{2(n-2+2\lambda)}(2nu_- - 4\delta\omega + \Delta u_+ - (n+\lambda-1)Ju_+) - (\lambda+n)u_- + \delta\omega$$

is equivalent to (6.20.6). In these terms, (6.20.7) asserts that

$$
-\frac{1}{2(n-2+2\lambda)}\Big[2n\big(u_-+(d\varphi,\omega)+\tfrac{1}{4}|d\varphi|^2u_+\big)e^{(\lambda-1)\varphi}
$$
$$
-4\hat{\delta}\big((\omega+\tfrac{1}{2}u_+d\varphi)e^{(\lambda+1)\varphi}\big)+\hat{\Delta}\big(e^{(\lambda+1)\varphi}u_+\big)\Big]
$$
$$
-(\lambda+n)\Big[\big(u_-+(d\varphi,\omega)+\tfrac{1}{4}|d\varphi|^2u_+\big)e^{(\lambda-1)\varphi}\Big]
$$
$$
+\hat{\delta}\big((\omega+\tfrac{1}{2}u_+d\varphi)e^{(\lambda+1)\varphi}\big)+\eta\hat{J}u_+e^{(\lambda+1)\varphi}
$$
$$
=-\frac{1}{2(n-2+2\lambda)}\left[2nu_--4\delta\omega+\Delta u_+\right]e^{(\lambda-1)\varphi}
$$
$$
-(\lambda+n)u_-e^{(\lambda-1)\varphi}+\delta\omega e^{(\lambda-1)\varphi}+\eta Ju_+e^{(\lambda-1)\varphi}
$$

for
$$
\eta\overset{\text{def}}{=}\frac{n+\lambda-1}{2(n-2+2\lambda)}.
$$

Lemma 4.2.1 shows that this identity is equivalent to

$$
\eta\frac{n-2}{2}|d\varphi|^2+\eta\Delta\varphi+\eta\hat{J}e^{2\varphi}=\eta J.
$$

But the latter identity is a consequence of (2.5.7). □

The following Lemma extends the identity $p(\lambda)\circ i(\lambda-1)=0$ to the curved case.

Lemma 6.20.1. *The composition $C_M(g;\lambda-1)\circ D_M(g;\lambda)$ on $C^\infty(M)$ vanishes identically.*

Proof. Direct calculation. □

Next, we define a connection on dual standard tractors.

Definition 6.20.1 (Tractor connection). *Let*

$$
\nabla_X^T=\begin{pmatrix}\nabla_X & (\mathsf{P}(X),\cdot) & 0\\ -X^\flat & \nabla_X & \tfrac{1}{2}\mathsf{P}(X)\\ 0 & -2\langle X,\cdot\rangle & \nabla_X\end{pmatrix},\quad X\in\mathcal{X}(M).\tag{6.20.8}
$$

Here the 1-form $\mathsf{P}(X)\in\Omega^1(M)$ is defined by P via the relation $\mathsf{P}(X,Y)=\langle\mathsf{P}(X),Y\rangle$.

In Definition 6.20.1, all constructions (∇_X, P etc.) are to be understood with respect to a chosen metric g. The basic property of ∇^T is its conformal invariance. For a detailed discussion of the relation to the normal conformal Cartan connection we refer to [22].

Theorem 6.20.3. *The connection ∇^T is conformally invariant, i.e.,*

$$T(g,\varphi) \circ \nabla_X^T = \hat{\nabla}_X^T \circ T(g,\varphi) \tag{6.20.9}$$

for all vector fields $X \in \mathcal{X}(M)$. Here ∇^T and $\hat{\nabla}^T$ denote the respective tractor connections for g and $\hat{g} = e^{2\varphi}g$.

Proof. The assertion follows from the conformal transformation rules for ∇_X and $\mathsf{P}(X)$. Since $X\langle \omega, Y\rangle = \langle \nabla_X(\omega), Y\rangle + \langle \omega, \nabla_X(Y)\rangle$, the rule (2.5.1) implies

$$\hat{\nabla}_X(\omega) = \nabla_X(\omega) - \omega\langle d\varphi, X\rangle - d\varphi\langle \omega, X\rangle + X^\flat(d\varphi, \omega). \tag{6.20.10}$$

Moreover, the identity $\hat{\mathsf{P}} = \mathsf{P} - \Xi$ (see (2.5.9)) yields

$$\hat{\mathsf{P}}(X) = \mathsf{P}(X) - \left[\nabla_X(d\varphi) - \langle d\varphi, X\rangle d\varphi + \frac{1}{2}|d\varphi|^2 X^\flat \right]. \tag{6.20.11}$$

Now we prove that the matrix entries $\cdot_{23}$ of both sides of (6.20.9) coincide. For the other entries, the arguments are analogous and we omit the details. The assertion is that

$$\frac{1}{2}e^\varphi \mathsf{P}(X)\cdot + \frac{1}{2}e^\varphi d\varphi \nabla_X(\cdot) = -\frac{1}{4}e^{-\varphi}\hat{X}^\flat|d\varphi|^2\cdot + \frac{1}{2}\hat{\nabla}_X(e^\varphi d\varphi\cdot) + \frac{1}{2}e^\varphi \hat{\mathsf{P}}(X)$$

as operators $C^\infty(M) \to \Omega^1(M)$. (6.20.10) and (6.20.11) show that the right-hand side equals

$$-\frac{1}{4}e^\varphi X^\flat|d\varphi|^2 u + \frac{1}{2}\left(\nabla_X \cdot - \langle d\varphi, X\rangle\cdot - d\varphi\langle X, \cdot\rangle + X^\flat(d\varphi, \cdot)\right)(e^\varphi u d\varphi)$$

$$+ \frac{1}{2}e^\varphi\left(\mathsf{P}(X) - \nabla_X(d\varphi) + \langle d\varphi, X\rangle d\varphi - \frac{1}{2}|d\varphi|^2 X^\flat\right)u,$$

i.e.,

$$-\frac{1}{4}e^\varphi X^\flat|d\varphi|^2 u + \frac{1}{2}e^\varphi\nabla_X(d\varphi)u + \frac{1}{2}e^\varphi d\varphi\langle d\varphi, X\rangle u + \frac{1}{2}e^\varphi d\varphi\nabla_X u$$

$$-\frac{1}{2}e^\varphi\langle d\varphi, X\rangle u d\varphi - \frac{1}{2}e^\varphi d\varphi\langle X, d\varphi\rangle u + \frac{1}{2}e^\varphi|d\varphi|^2 X^\flat u$$

$$+\frac{1}{2}e^\varphi \mathsf{P}(X)u - \frac{1}{2}e^\varphi\nabla_X(d\varphi)u + \frac{1}{2}e^\varphi\langle d\varphi, X\rangle u d\varphi - \frac{1}{4}e^\varphi|d\varphi|^2 X^\flat u.$$

But the latter sum simplifies to

$$\frac{1}{2}e^\varphi \mathsf{P}(X)u + \frac{1}{2}e^\varphi d\varphi\nabla_X u.$$

The proof is complete. $\qquad\qquad\qquad\qquad\qquad\qquad\qquad\qquad\qquad\qquad\square$

Next, the connection $\nabla^{\mathcal{T}}_X$ gives rise to the *tractor curvature* endomorphisms

$$R^{\mathcal{T}}(X,Y) = \nabla^{\mathcal{T}}_X \nabla^{\mathcal{T}}_Y - \nabla^{\mathcal{T}}_Y \nabla^{\mathcal{T}}_X - \nabla^{\mathcal{T}}_{[X,Y]}. \qquad (6.20.12)$$

A calculation yields the explicit formula

$$R^{\mathcal{T}}(X,Y) = \begin{pmatrix} 0 & (\mathcal{C}(X,Y),\cdot) & 0 \\ 0 & \mathsf{C}(X,Y) & \frac{1}{2}\mathcal{C}(X,Y) \\ 0 & 0 & 0 \end{pmatrix}, \qquad (6.20.13)$$

where the endomorphism $\mathsf{C}(X,Y) : \Omega^1(M) \to \Omega^1(M)$ is defined by the Weyl tensor C via

$$\langle \mathsf{C}(X,Y)\omega, Z \rangle = \mathsf{C}(X,Y,\omega^\sharp, Z),$$

and

$$\mathcal{C}(X,Y) = \nabla_X(\mathsf{P})(Y,\cdot) - \nabla_Y(\mathsf{P})(X,\cdot) \in \Omega^1(M) \qquad (6.20.14)$$

is the Cotton tensor (see Lemma 4.2.7).

In fact, we verify the central term in formula (6.20.13). We find

$$\nabla_X \nabla_Y - \nabla_Y \nabla_X - \nabla_{[X,Y]} - \mathsf{P}(X)\langle Y,\cdot\rangle + \mathsf{P}(Y)\langle X,\cdot\rangle - X^\flat(\mathsf{P}(Y),\cdot) + Y^\flat(\mathsf{P}(X),\cdot)$$
$$= R(X,Y) + (\mathsf{P} \oslash g)(X,Y) = \mathsf{C}(X,Y).$$

We omit the details for the remaining entries.

The conformal invariance of $R^{\mathcal{T}}$ is equivalent to the conformal transformation laws

$$\hat{\mathsf{C}}(X,Y) = \mathsf{C}(X,Y) \quad \text{and} \quad \hat{\mathcal{C}}(X,Y) = \mathcal{C}(X,Y) - \mathsf{C}(X,Y)(d\varphi). \qquad (6.20.15)$$

(6.20.13) and the relation (4.2.14) between $\mathcal{C}$ and C imply that the curvature $\mathcal{R}^{\mathcal{T}}$ vanishes iff $\mathsf{C} = 0$.

Moreover, the tractor connection preserves a scalar product on dual tractors.

Definition 6.20.2 (Tractor scalar product). *The tractor scalar product is defined by*

$$\left(\begin{pmatrix} u_- \\ \omega \\ u_+ \end{pmatrix}, \begin{pmatrix} v_- \\ \eta \\ v_+ \end{pmatrix} \right)^{\mathcal{T}}_g = -\frac{1}{2}(u_- v_+ + u_+ v_-) + (\omega, \eta)_g. \qquad (6.20.16)$$

Lemma 6.20.2. *The tractor scalar product is conformally invariant and preserved by the connection $\nabla^{\mathcal{T}}_X$, i.e.,*

$$(T(g,\varphi)u, T(g,\varphi)v)^{\mathcal{T}}_{\hat{g}} = (u,v)^{\mathcal{T}}_g$$

and

$$\nabla_X(u,v)^{\mathcal{T}}_g = \left(\nabla^{\mathcal{T}}_X(u), v \right)^{\mathcal{T}}_g + \left(u, \nabla^{\mathcal{T}}_X(v) \right)^{\mathcal{T}}_g.$$

Proof. The assertions follow by direct calculations. We omit the details. □

Remark 6.20.1. *We use trivializations of tractor bundles and a definition of the tractor connection which slightly differ from those used in the literature. The choice here is dictated by the choice of conventions in connection with the discussion of homomorphisms of Verma modules in Section 6.18. For the convenience of the reader, we relate the conventions explicitly. We use the diagonal matrix* $\mathrm{diag}(1,1,2)$ *to conjugate*

$$
\begin{pmatrix} e^{-\varphi} & & \\ & e^{\varphi} & \\ & & e^{\varphi} \end{pmatrix}
\begin{pmatrix} 1 & (d\varphi,\cdot) & \frac{1}{4}|d\varphi|^2 \\ 0 & 1 & \frac{1}{2}d\varphi \\ 0 & 0 & 1 \end{pmatrix},
\quad
\begin{pmatrix} \nabla_X & (\mathsf{P}(X),\cdot) & 0 \\ -X^\flat & \nabla_X & \frac{1}{2}\mathsf{P}(X) \\ 0 & -2\langle X,\cdot\rangle & \nabla_X \end{pmatrix}
$$

into

$$
\begin{pmatrix} e^{-\varphi} & & \\ & e^{\varphi} & \\ & & e^{\varphi} \end{pmatrix}
\begin{pmatrix} 1 & (d\varphi,\cdot) & \frac{1}{2}|d\varphi|^2 \\ 0 & 1 & d\varphi \\ 0 & 0 & 1 \end{pmatrix},
\quad
\begin{pmatrix} \nabla_X & (\mathsf{P}(X),\cdot) & 0 \\ -X^\flat & \nabla_X & \mathsf{P}(X) \\ 0 & -\langle X,\cdot\rangle & \nabla_X \end{pmatrix}.
$$

For standard tractors, Bailey, Eastwood and Gover ([17]) use the convention

$$
\begin{pmatrix} e^{-\varphi} & & \\ & e^{-\varphi} & \\ & & e^{\varphi} \end{pmatrix}
\begin{pmatrix} 1 & -(d\varphi,\cdot) & -\frac{1}{2}|d\varphi|^2 \\ 0 & 1 & \mathrm{grad}(\varphi) \\ 0 & 0 & 1 \end{pmatrix},
\quad
\begin{pmatrix} \nabla_X & -\langle\mathsf{P}(X),\cdot\rangle & 0 \\ X & \nabla_X & \mathsf{P}(X)^\sharp \\ 0 & -(X,\cdot) & \nabla_X \end{pmatrix}
$$

for trivializations and tractor connection. The diagonal matrix $\mathrm{diag}(1,-1,-1)$ *conjugates these into*

$$
\begin{pmatrix} e^{-\varphi} & & \\ & e^{-\varphi} & \\ & & e^{\varphi} \end{pmatrix}
\begin{pmatrix} 1 & (d\varphi,\cdot) & \frac{1}{2}|d\varphi|^2 \\ 0 & 1 & \mathrm{grad}(\varphi) \\ 0 & 0 & 1 \end{pmatrix},
\quad
\begin{pmatrix} \nabla_X & \langle\mathsf{P}(X),\cdot\rangle & 0 \\ -X & \nabla_X & \mathsf{P}(X)^\sharp \\ 0 & -(X,\cdot) & \nabla_X \end{pmatrix}.
$$

The tractor connection $\nabla^{\mathcal{T}} : \Gamma(\mathcal{T}^*) \to \Gamma(\mathcal{T}^* \otimes T^*)$ gives rise to the *tractor Bochner-Laplacian*

$$\Delta^{\mathcal{T}} = \mathrm{tr}(\nabla^{\mathcal{T}} \circ \nabla^{\mathcal{T}}) : \Gamma(\mathcal{T}^*) \to \Gamma(\mathcal{T}^*). \tag{6.20.17}$$

Here

$$\nabla^{\mathcal{T}} : \Gamma(\mathcal{T}^* \otimes T^*) \to \Gamma(\mathcal{T}^* \otimes T^* \otimes T^*)$$

couples $\nabla^{\mathcal{T}}$ with the Levi-Civita connection ∇^{LC} on Ω^1, i.e.,

$$\nabla^{\mathcal{T}}(u \otimes \omega) = \nabla^{\mathcal{T}}(u) \otimes \omega + u \otimes \nabla^{LC}(\omega),$$

and tr denotes the contraction $T^* \otimes T^* \to \mathbb{C}$ defined by g. A standard calculation yields

Lemma 6.20.3.

$$\Delta^{\mathcal{T}} = \sum_i \nabla^{\mathcal{T}}_{e_i} \nabla^{\mathcal{T}}_{e_i} - \nabla^{\mathcal{T}}_{\nabla^{LC}_{e_i} e_i} = \delta^{\mathcal{T}} \nabla^{\mathcal{T}}$$

in terms of a local orthonormal frame e_i.

The following result extends the conformal covariance of the Yamabe operator $P_2 = \Delta - (\frac{n}{2}-1)\mathsf{J}$.

Theorem 6.20.4. *The operator*

$$\Box^{\mathcal{T}} \overset{\text{def}}{=} \Delta^{\mathcal{T}} - \left(\frac{n}{2}-1\right)\mathsf{J} : \Gamma(\mathcal{T}^*) \to \Gamma(\mathcal{T}^*)$$

is conformally covariant in the sense that

$$\Box^{\mathcal{T}}_{\hat{g}} \circ e^{(-\frac{n}{2}+1)\varphi} \circ T(g,\varphi) = e^{-(\frac{n}{2}+1)\varphi} \circ T(g,\varphi) \circ \Box^{\mathcal{T}}_g.$$

Proof. We use Lemma 6.20.3 and Theorem 6.20.3. Since $\hat{e}_i = e^{-\varphi}e_i$, we get

$$\nabla^{LC}_{\hat{e}_i}\hat{e}_i = e^{-\varphi}\nabla^{LC}_{e_i}(e^{-\varphi}e_i) = e^{-2\varphi}\left(\nabla^{LC}_{e_i}e_i - \langle d\varphi, e_i\rangle e_i\right)$$

and the transformation rule (2.5.1) yields

$$\hat{\nabla}^{LC}_{\hat{e}_i}\hat{e}_i = \nabla^{LC}_{\hat{e}_i}\hat{e}_i + 2e^{-2\varphi}\langle d\varphi, e_i\rangle e_i - e^{-2\varphi}\,\mathrm{grad}\,\varphi$$
$$= e^{-2\varphi}\left(\nabla^{LC}_{e_i}e_i + \langle d\varphi, e_i\rangle e_i - \mathrm{grad}\,\varphi\right).$$

Hence

$$\hat{\nabla}^{\mathcal{T}}_{\hat{\nabla}^{LC}_{\hat{e}_i}\hat{e}_i} = e^{-2\varphi}\left(\hat{\nabla}^{\mathcal{T}}_{\nabla^{LC}_{\hat{e}_i}\hat{e}_i} + \langle d\varphi, e_i\rangle\hat{\nabla}^{\mathcal{T}}_{e_i} - \hat{\nabla}^{\mathcal{T}}_{\mathrm{grad}\,\varphi}\right).$$

Moreover, we find

$$\hat{\nabla}^{\mathcal{T}}_{\hat{e}_i}\hat{\nabla}^{\mathcal{T}}_{\hat{e}_i} = e^{-2\varphi}\left(\hat{\nabla}^{\mathcal{T}}_{e_i}\hat{\nabla}^{\mathcal{T}}_{e_i} - \langle d\varphi, e_i\rangle\hat{\nabla}^{\mathcal{T}}_{e_i}\right)$$
$$= e^{-2\varphi}T\left(\nabla^{\mathcal{T}}_{e_i}\nabla^{\mathcal{T}}_{e_i} - \langle d\varphi, e_i\rangle\nabla^{\mathcal{T}}_{e_i}\right)T^{-1},$$

where $T = T(g,\varphi)$. It follows that

$$T^{-1}e^{2\varphi}\Delta^{\mathcal{T}}_{\hat{g}}T = \Delta^{\mathcal{T}}_g - 2\sum_i\langle d\varphi, e_i\rangle\nabla^{\mathcal{T}}_{e_i} + n\nabla^{\mathcal{T}}_{\mathrm{grad}\,\varphi}$$
$$= \Delta^{\mathcal{T}}_g + (n-2)\nabla^{\mathcal{T}}_{\mathrm{grad}\,\varphi}.$$

On the other hand, a calculation shows that

$$\Delta^{\mathcal{T}}(e^{\lambda\varphi}u) = e^{\lambda\varphi}\left(\Delta^{\mathcal{T}}u + 2\lambda\nabla^{\mathcal{T}}_{\mathrm{grad}\,\varphi}u + \lambda u\Delta\varphi + \lambda^2|d\varphi|^2u\right).$$

Therefore,

$$\Delta^{\mathcal{T}}_{\hat{g}}(e^{\lambda\varphi}u) = e^{-2\varphi}T\left(\Delta^{\mathcal{T}}_g + (n-2)\nabla^{\mathcal{T}}_{\mathrm{grad}\,\varphi}\right)(T^{-1}(e^{\lambda\varphi}u))$$
$$+ e^{(\lambda-2)\varphi}T\left((n-2+2\lambda)\nabla^{\mathcal{T}}_{\mathrm{grad}\,\varphi} + \lambda\Delta_g(\varphi) + \lambda(n-2+\lambda)|d\varphi|^2\right)(T^{-1}u),$$

i.e.,

$$\Delta^{\mathcal{T}}_{\hat{g}}(e^{\lambda\varphi}u) - e^{(\lambda-2)\varphi}T\Delta^{\mathcal{T}}_g(T^{-1}u)$$
$$= e^{(\lambda-2)\varphi}T\left\{(n-2+2\lambda)\nabla^{\mathcal{T}}_{\mathrm{grad}\,\varphi} + \lambda\Delta_g(\varphi) + \lambda(n-2+\lambda)|d\varphi|^2\right\}(T^{-1}u).$$

Together with

$$e^{\lambda\varphi}\hat{\mathsf{J}} - e^{(\lambda-2)\varphi}\mathsf{J} = e^{(\lambda-2)\varphi}\left(-\Delta\varphi - \frac{n-2}{2}|d\varphi|^2\right)$$

(see (2.5.7)) we obtain

$$(\Delta_{\hat{g}}^{\mathcal{T}} + \lambda\hat{\mathsf{J}})(e^{\lambda\varphi}u) - e^{(\lambda-2)\varphi}T(\Delta_g^{\mathcal{T}} + \lambda\mathsf{J})(T^{-1}u)$$
$$= e^{(\lambda-2)\varphi}T\left\{(n-2+2\lambda)\nabla_{\mathrm{grad}\,\varphi}^{\mathcal{T}} + (\lambda+\frac{n-2}{2})\lambda|d\varphi|^2\right\}(T^{-1}u) = 0$$

for $\lambda = -\frac{n}{2} + 1$. The proof is complete.   □

The latter proof also shows that the tractor D-operator can be iterated. More precisely, we have

Theorem 6.20.5. *The tractor D-operator $D_M(\lambda) : C^\infty(M) \to \Gamma(\mathcal{T}^*M)$ extends to a conformally covariant family*

$$D_M(\lambda) : \Gamma(\mathcal{T}^*M) \to \Gamma(\mathcal{T}^*M \otimes \mathcal{T}^*M)$$

by

$$D_M(g;\lambda) : u \mapsto \begin{pmatrix} \frac{1}{n-2+2\lambda}(\Delta^{\mathcal{T}}+\lambda\mathsf{J})u \\ \nabla^{\mathcal{T}}u \\ 2\lambda u \end{pmatrix}. \qquad (6.20.18)$$

Here all constructions are to be understood with respect to g and we use the identification

$$\Gamma(\mathcal{T}^*M) \oplus \Gamma(\mathcal{T}^*M \otimes \mathcal{T}^*M) \oplus \Gamma(\mathcal{T}^*M) \simeq \Gamma(\mathcal{T}^*M \otimes \mathcal{T}^*M).$$

The conformal covariance means

$$D_M(\hat{g};\lambda) \circ e^{\lambda\varphi} \circ T_M(g,\varphi) = e^{(\lambda-1)\varphi} \circ T_M(g,\varphi) \circ D_M(g;\lambda), \qquad (6.20.19)$$

where T_M denotes the respective trivializations of $\mathcal{T}^*M$ and $\mathcal{T}^*M \otimes \mathcal{T}^*M$.

Proof. The proof is analogous to the proof of Theorem 6.20.1. It rests on the identity

$$e^{-(\lambda-2)\varphi}(\hat{\Delta}^{\mathcal{T}}+\lambda\hat{\mathsf{J}})(e^{\lambda\varphi}u)$$
$$= T(\Delta^{\mathcal{T}}+\lambda\mathsf{J})(T^{-1}u) + (n-2+2\lambda)T\nabla_{\mathrm{grad}\,\varphi}^{\mathcal{T}}T^{-1}u + \lambda(\lambda+\frac{n-2}{2})|d\varphi|^2u$$

which was established at the end of the proof of Theorem 6.20.4; compare also with (6.20.4). We divide by $(n-2+2\lambda)$. It follows that

$$e^{-(\lambda-1)\varphi}\frac{1}{n-2+2\lambda}(\hat{\Delta}^{\mathcal{T}}+\lambda\hat{J})(e^{\lambda\varphi}u)$$

$$= e^{-\varphi}\left(1,(d\varphi,\cdot),\frac{1}{4}|d\varphi|^2\right)\begin{pmatrix}\frac{1}{n-2+2\lambda}T(\Delta^{\mathcal{T}}+\lambda J)T^{-1}u\\T\nabla^{\mathcal{T}}T^{-1}u\\2\lambda u\end{pmatrix}.$$

In other words, the first components of both sides of (6.20.19) coincide. For the second components, the assertion is the obvious identity

$$\hat{\nabla}^{\mathcal{T}}(e^{\lambda\varphi}u)=e^{\lambda\varphi}\left(\lambda u d\varphi+T\nabla^{\mathcal{T}}T^{-1}u\right).$$

The proof is complete.   □

The extension of $D_M(\lambda)$ from $C^\infty(M)$ to $\Gamma(\mathcal{T}^*M)$ described in Theorem 6.20.5 can be continued, and yields conformally covariant tractor D-operators

$$D_M(\lambda):\Gamma(\underbrace{\mathcal{T}^*M\otimes\cdots\otimes\mathcal{T}^*M}_{r})\to\Gamma(\underbrace{\mathcal{T}^*M\otimes\cdots\otimes\mathcal{T}^*M}_{r+1}).$$

These are defined by (6.20.18) using the connection $\nabla^{\mathcal{T}}$ acting on the space $\Gamma(\underbrace{\mathcal{T}^*M\otimes\cdots\otimes\mathcal{T}^*M}_{r})$ and the tractor Bochner-Laplacian

$$\Delta^{\mathcal{T}}:\Gamma(\underbrace{\mathcal{T}^*M\otimes\cdots\otimes\mathcal{T}^*M}_{r})\to\Gamma(\underbrace{\mathcal{T}^*M\otimes\cdots\otimes\mathcal{T}^*M}_{r}).$$

The corresponding operator $\Box^{\mathcal{T}}\stackrel{\text{def}}{=}\Delta^{\mathcal{T}}-\frac{n-2}{2}J$ is conformally covariant. We omit the details.

The following result is the curved analog of Lemma 6.19.2. It can be used to give an alternative proof of Theorem 6.20.2.

Lemma 6.20.4. $C_M(\lambda)=\mathrm{tr}\circ D_M(\lambda)$, where $D_M(\lambda):\Gamma(\mathcal{T}^*M)\to\Gamma(\mathcal{T}^*M\otimes\mathcal{T}^*M)$ and the trace denotes the contraction induced by the tractor scalar product.

Proof. By definition,

$$D(\lambda)u=\begin{pmatrix}\frac{1}{n-2+2\lambda}(\Delta^{\mathcal{T}}+\lambda J)u\\\nabla^{\mathcal{T}}u\\2\lambda u\end{pmatrix}$$

(see (6.20.18)). Let $u = \begin{pmatrix} u_- \\ \omega \\ u_+ \end{pmatrix}$. The explicit formulas (6.20.8) and (6.22.2) imply

$$\mathrm{tr}(D(\lambda)u) = -\lambda u_- + (-nu_- + \delta\omega + \frac{1}{2}Ju_+)$$
$$- \frac{1}{2}\frac{1}{n-2+2\lambda}\{2nu_- - 4\delta\omega + (\Delta - J)u_+\} - \frac{1}{2}\frac{1}{n-2+2\lambda}\lambda Ju_+.$$

A calculation shows that the latter sum equals $C(\lambda)u$. The proof is complete. □

Similarly, the family $C_M(\lambda) : \Gamma(T^*M) \to C^\infty(M)$ extends to conformally covariant families

$$\Gamma(\underbrace{T^*M \otimes \cdots \otimes T^*M}_{r+1}) \to \Gamma(\underbrace{T^*M \otimes \cdots \otimes T^*M}_{r})$$

by $C_M(\lambda) \overset{\mathrm{def}}{=} \mathrm{tr}\, D_M(\lambda)$ using a contraction. In explicit terms,

$$C_M(\lambda)u = \frac{1}{2(n-2-2\lambda)}$$
$$\times \{-2(n+2\lambda)(n+\lambda-1)u_- + 2(n+2\lambda)\delta^T\omega - \Delta^T u_+ + (n+\lambda-1)Ju_+\} \quad (6.20.20)$$

for $u = \begin{pmatrix} u_- \\ \omega \\ u_+ \end{pmatrix}$ with

$$u_\pm \in \Gamma(\underbrace{T^*M \otimes \cdots \otimes T^*M}_{r}), \quad \omega \in \Gamma(\underbrace{T^*M \otimes \cdots \otimes T^*M}_{r} \otimes T^*M).$$

Here δ^T denotes the divergence which is defined by the connections.

Now for an oriented codimension one submanifold (i.e., a hypersurface) Σ in M, the family $D_1(\lambda) = \nabla_N - \lambda H : C^\infty(M) \to C^\infty(\Sigma)$ is conformally covariant (Theorem 6.2.1). An analogous family $D_1^T(\lambda) : \Gamma(T^*M) \to \Gamma(T^*\Sigma)$ can be defined by the composition of two conformally covariant families. The following result defines one factor. $D_1^T(\lambda)$ will be defined in Lemma 6.20.8.

Lemma 6.20.5. *The family*

$$D^T(\lambda) \overset{\mathrm{def}}{=} \nabla_N^T - \lambda H$$

is conformally covariant in the sense that

$$e^{(\lambda-1)\varphi} \circ T(g, \varphi) \circ D^T(g; \lambda) = D^T(\hat{g}; \lambda) \circ e^{\lambda\varphi} \circ T(g, \varphi).$$

Proof. We observe that the composition

$$T(g,\varphi)^{-1} \circ e^{-(\lambda-1)\varphi} \circ D^{\mathcal{T}}(\hat{g};\lambda) \circ e^{\lambda\varphi} \circ T(g,\varphi)$$

is *linear* in λ. Using $\hat{N} = e^{-\varphi}N$, Theorem 6.20.3 implies that the absolute coefficient is

$$T(g,\varphi)^{-1} \circ \hat{\nabla}_{\hat{N}}^{\mathcal{T}} \circ T(g,\varphi) = \nabla_{N}^{\mathcal{T}}.$$

Moreover, using $\hat{H} = e^{-\varphi}(H + \nabla_N(\varphi))$, we find that the linear coefficient is

$$T(g,\varphi)^{-1} \circ \left[-\varphi \circ \hat{\nabla}_{\hat{N}}^{\mathcal{T}} - e^{\varphi}H(\hat{g}) + \hat{\nabla}_{\hat{N}}^{\mathcal{T}} \circ \varphi \right] \circ T(g,\varphi)$$

$$= T(g,\varphi)^{-1} \circ [\nabla_N(\varphi) - (H(g) + \nabla_N(\varphi))] \circ T(g,\varphi) = -H(g).$$

The proof is complete. □

Next, we define a tractor analog of the unit normal vector field of a hypersurface.

Definition 6.20.3 (Normal tractor). *For a hypersurface Σ of M and a metric g, let*

$$N^{\mathcal{T}} \stackrel{\text{def}}{=} \begin{pmatrix} H \\ N^{\flat} \\ 0 \end{pmatrix},$$

where all quantities are to be understood with respect to g. $N^{\mathcal{T}}$ is called the normal tractor of Σ for the metric g.

Note that $N^{\mathcal{T}}$ has length 1. The following result says that the normal tractor is a section of the dual tractor bundle on M, defined on the hypersurface Σ.

Lemma 6.20.6. $\hat{N}^{\mathcal{T}} = T_M(g,\varphi)N^{\mathcal{T}}$, *where $N^{\mathcal{T}}$ and $\hat{N}^{\mathcal{T}}$ are the respective normal tractors for the metrics g and $\hat{g} = e^{2\varphi}g$.*

Proof. The assertion is equivalent to $\hat{N}^{\flat} = e^{\varphi}N^{\flat}$ and $\hat{H} = e^{-\varphi}(H + \langle d\varphi, N \rangle)$. □

As a consequence, we find

Corollary 6.20.1. $D_1(M,\Sigma;g;\lambda) = (D_M(g;\lambda), N_g^{\mathcal{T}})_g^{\mathcal{T}}.$

Corollary 6.20.1 re-proves the conformal covariance of $D_1(M,\Sigma;g;\lambda)$ (Theorem 6.2.1). In fact, Lemma 6.20.2, Theorem 6.20.1 and Lemma 6.20.6 imply

$$e^{(\lambda-1)\varphi}D_1(g;\lambda)u = e^{(\lambda-1)\varphi}(D(g;\lambda)u, N_g^{\mathcal{T}})_g^{\mathcal{T}}$$

$$= e^{(\lambda-1)\varphi}\left(T(g,\varphi)D(g;\lambda)u, T(g,\varphi)N_g^{\mathcal{T}}\right)_{\hat{g}}^{\mathcal{T}}$$

$$= (D(\hat{g};\lambda)e^{\lambda\varphi}u, N_{\hat{g}}^{\mathcal{T}})_{\hat{g}}^{\mathcal{T}}$$

$$= D_1(\hat{g};\lambda)e^{\lambda\varphi}u.$$

Similar arguments can be used to re-prove the conformal covariance of $D^{\mathcal{T}}(\lambda)$ (Lemma 6.20.5).

It will be important to have an invariant linear map from tractors on M to tractors on a hypersurface Σ. The following result provides such a map.

Theorem 6.20.6 (Projection operator). *The linear operator*

$$\Pi_\Sigma \begin{pmatrix} u_- \\ \omega \\ u_+ \end{pmatrix} \stackrel{\text{def}}{=} \begin{pmatrix} i^*(u_-) - H\langle \omega, N \rangle + \frac{1}{4} H^2 i^*(u_+) \\ i^*(\omega) \\ i^*(u_+) \end{pmatrix}$$

is conformally covariant in the sense that

$$\Pi_\Sigma(\hat{g}) \circ e^{\lambda \varphi} \circ T_M(g, \varphi) = e^{\lambda i^*(\varphi)} \circ T_\Sigma(i^*(g), i^*(\varphi)) \circ \Pi_\Sigma(g) \qquad (6.20.21)$$

for all $\varphi \in C^\infty(M)$.

Proof. It is enough to prove the assertion for $\lambda = 0$. Then for the first component, (6.20.21) claims that

$$e^{-\varphi} \left(u_- + (d\varphi, \omega) + \frac{1}{4} |d\varphi|^2 u_+ \right) + \frac{1}{4} e^\varphi \hat{H}^2 u_+ - e^\varphi \hat{H} \left\langle \omega + \frac{1}{2} u_+ d\varphi, \hat{N} \right\rangle$$
$$= e^{-\varphi} \left(\left(u_- + \frac{1}{4} H^2 u_+ - H\langle \omega, N \rangle \right) + (i^*(d\varphi), i^*(\omega)) + \frac{1}{4} |i^*(d\varphi)|^2 u_+ \right)$$

on Σ. The assertions for the remaining two components are trivial. The above identity is equivalent to

$$e^{-\varphi}(d\varphi, \omega) - e^\varphi \hat{H} \langle \omega, \hat{N} \rangle = e^{-\varphi} (i^*(d\varphi), i^*(\omega)) - e^{-\varphi} H\langle \omega, N \rangle$$

and

$$e^{-\varphi} |d\varphi|^2 + e^\varphi \hat{H}^2 - 2e^\varphi \hat{H} \langle d\varphi, \hat{N} \rangle = e^{-\varphi} H^2 + e^{-\varphi} |i^*(d\varphi)|^2.$$

Both identities follow from $e^\varphi \hat{H} = H + \nabla_N(\varphi)$. The proof is complete. □

We extend the map Π_Σ to a conformally covariant operator

$$\Gamma(\underbrace{\mathcal{T}^* M \otimes \cdots \otimes \mathcal{T}^* M}_{r}) \to \Gamma(\underbrace{\mathcal{T}^* \Sigma \otimes \cdots \otimes \mathcal{T}^* \Sigma}_{r})$$

by

$$\Pi_\Sigma(u \otimes \cdots \otimes v) = \Pi_\Sigma(u) \otimes \cdots \otimes \Pi_\Sigma(v).$$

Remark 6.20.2. *The projection operator Π_Σ does not coincide with the orthogonal projection π_Σ along $N^{\mathcal{T}}$. In fact, for the tractor $u = \begin{pmatrix} u_- \\ \omega \\ u_+ \end{pmatrix}$, we find*

$$\pi_\Sigma(u) = u - (u, N^{\mathcal{T}})^{\mathcal{T}} N^{\mathcal{T}} = \begin{pmatrix} u_- - H\langle \omega, N \rangle + \frac{1}{2} H^2 u_+ \\ \omega - \langle \omega, N \rangle N^\flat + \frac{1}{2} u_+ H N^\flat \\ u_+ \end{pmatrix}.$$

π_Σ and Π_Σ actually satisfy different transformation laws. In fact, it follows from the definitions and Lemmas 6.20.2 and 6.20.6 that

$$\pi_\Sigma(\hat{g}) \circ T_M(g, \varphi) = T_M(g, \varphi) \circ \pi_\Sigma(g).$$

In order to relate π_Σ and Π_Σ, we introduce the linear operator

$$I(g) : \begin{pmatrix} v_- \\ \eta \\ v_+ \end{pmatrix} \mapsto \begin{pmatrix} v_- + \frac{1}{4}H^2 v_+ \\ \eta + \frac{1}{2}HN^\flat v_+ \\ v_+ \end{pmatrix} \in N^T(g)^\perp.$$

It defines an isomorphism between dual standard tractors on Σ and dual standard tractors on M (supported on Σ) which are orthogonal to N^T. It is easy to verify that

$$I(\hat{g}) \circ T_\Sigma(i^*(g), i^*(\varphi)) = T_M(g, \varphi) \circ I(g),$$

i.e., I is conformally covariant. Moreover, we find

$$I(g) \circ \Pi_\Sigma(g) = \pi_\Sigma(g).$$

Finally, note that $I(g)$ reduces to the identity if $H = 0$.

The following simple result illustrates these constructions by identifying the trace-free second fundamental form L_0 as a composition of operators on tractors.

Lemma 6.20.7. *In the situation of Definition 6.20.3, let*

$$\sigma : \begin{pmatrix} u_- \\ \omega \\ u_+ \end{pmatrix} \mapsto \omega - \frac{1}{2}du_+.$$

Then

$$\sigma \circ T(g, \varphi) = e^\varphi \circ \sigma \qquad (6.20.22)$$

and the bilinear form

$$(X, Y) \mapsto \left\langle \sigma \Pi_\Sigma \nabla_X^{M,T}(N^T), Y \right\rangle \in C^\infty(\Sigma), \; X, Y \in \mathcal{X}(\Sigma)$$

coincides with L_0.

Proof. (6.20.22) is equivalent to the obvious identity

$$e^\varphi \left(\omega + \frac{1}{2}d\varphi u_+ \right) - \frac{1}{2}d(e^\varphi u_+) = e^\varphi \left(\omega - \frac{1}{2}du_+ \right).$$

Now since

$$\nabla_X^{M,T}(N^T) = \begin{pmatrix} \langle dH + \mathsf{P}(N), X \rangle \\ -HX^\flat + \nabla_X^M(N^\flat) \\ 0 \end{pmatrix},$$

we find

$$\Pi_\Sigma \nabla_X^{M,T}(N^T) = \begin{pmatrix} -HX^\flat + i^* \overset{*}{\nabla_X^M}(N^\flat) \\ 0 \end{pmatrix}.$$

Hence the composition is given by $-HX^\flat + L(X) = L_0(X)$. □

Lemma 6.20.7 implies that $L_0 = 0$ if N^T is parallel, i.e., $\nabla_X^T N^T = 0$ for all $X \in \mathcal{X}(\Sigma)$. Conversely, if $L_0 = 0$, then the proof of Lemma 6.20.7 shows that

$$\nabla_X^T(N^T) = \begin{pmatrix} \langle dH + P(N), X \rangle \\ 0 \\ 0 \end{pmatrix}$$

for all $X \in \mathcal{X}(\Sigma)$. But $L_0 = 0$ and Lemma 6.25.2 imply that $P(N) + dH = 0$, i.e., $\nabla_X^T(N^T) = 0$. Hence we have proved

Corollary 6.20.2. Σ *is totally umbilic iff* N^T *is parallel.*

This is Proposition 2.9 in [17]. These considerations also show that

$$\Pi_\Sigma \nabla_X^T(N^T) = \begin{pmatrix} \frac{1}{n-2}\delta(L_0)(X) \\ L_0(X) \\ 0 \end{pmatrix} \in \Gamma(T^*\Sigma).$$

Now Lemma 6.20.5 and Theorem 6.20.6 yield

Lemma 6.20.8. *The family*

$$D_1^T(M,\Sigma;g;\lambda) \overset{def}{=} \Pi_\Sigma D^T(M,\Sigma;g;\lambda) : \Gamma(T^*M) \to \Gamma(T^*\Sigma)$$

is conformally covariant, i.e.,

$$e^{(\lambda-1)\varphi} \circ T_\Sigma(g,\varphi) \circ D_1^T(M,\Sigma;g;\lambda) = D_1^T(M,\Sigma;\hat{g};\lambda) \circ e^{\lambda\varphi} \circ T_M(g,\varphi).$$

The latter extension of $D_1(\lambda)$ from $C^\infty(M)$ to $\Gamma(T^*M)$ can be continued, and yields conformally covariant operators

$$D_1^T(\lambda) : \Gamma(\underbrace{T^*M \otimes \cdots \otimes T^*M}_{r}) \to \Gamma(\underbrace{T^*\Sigma \otimes \cdots \otimes T^*\Sigma}_{r}).$$

Finally, we briefly consider the Paneitz operator from the point of view of tractor calculus. The following result is due to Eastwood and Gover ([109]).

Theorem 6.20.7. *On* M^n,

$$\left(\Delta^T - \left(\frac{n}{2}-1\right)J\right) \circ D\left(-\frac{n}{2}+2\right) = \frac{1}{2}\begin{pmatrix} P_{4,n} \\ 0 \\ 0 \end{pmatrix}.$$

Proof. The explicit formula (6.22.2) implies that the left-hand side is given by

$$\begin{pmatrix} \Delta - \frac{n}{2}\mathsf{J} & -(d\mathsf{J}, \cdot) + 2\delta(\mathsf{P}\#\cdot) & \frac{1}{2}|\mathsf{P}|^2 \\ -2d & \Delta + \frac{n-4}{2}(2\mathsf{P}\# - \mathsf{J}) & \frac{1}{2}d\mathsf{J} + \mathsf{P}\#d \\ 2n & -4\delta & \Delta - \frac{n}{2}\mathsf{J} \end{pmatrix} \begin{pmatrix} \frac{1}{2}(\Delta - (\frac{n}{2}-2)\mathsf{J})u \\ du \\ -(n-4)u \end{pmatrix}.$$

Here we made use of

$$\mathrm{Ric} - 2\mathsf{P} - \left(\frac{n}{2}-1\right)\mathsf{J}g = \frac{n-4}{2}(2\mathsf{P} - \mathsf{J}g).$$

Hence the upper entry of the resulting tractor is given by

$$\frac{1}{2}\left(\Delta - \frac{n}{2}\mathsf{J}\right)\left(\Delta - (\frac{n}{2}-2)\mathsf{J}\right)u + 2\delta(\mathsf{P}\#du) - (d\mathsf{J}, du) - \frac{1}{2}(n-4)|\mathsf{P}|^2 u.$$

The latter sum equals

$$\frac{1}{2}\Big\{\Delta^2 u - (n-2)\mathsf{J}\Delta u - (n-4)(d\mathsf{J}, du)$$
$$+ 4\delta(\mathsf{P}\#du) - 2(d\mathsf{J}, du) + \frac{n-4}{2}\left(\frac{n}{2}\mathsf{J}^2 - \Delta\mathsf{J} - 2|\mathsf{P}|^2\right)u\Big\} = \frac{1}{2}P_{4,n}u$$

(see (4.1.7) and (4.2.11)); note that the Laplacian here has the opposite sign. The vanishing of the remaining two components is easy to check. We omit the details. $\square$

The following consequence is the curved version of Theorem 6.18.1.

Corollary 6.20.3. *On* M^n,

$$C\left(-\frac{n}{2}-1\right) \circ \left(\Delta^T - \left(\frac{n}{2}-1\right)\mathsf{J}\right) \circ D\left(-\frac{n}{2}+2\right) = -\frac{n-4}{8}P_{4,n}.$$

Corollary 6.20.3 implies that on M^4 the composition

$$C(-3) \circ (\Delta^T - \mathsf{J}) \circ D(0)$$

vanishes, i.e., does not yields a construction of P_4. In this respect, Theorem 6.20.7 is more useful, and motivated analogous constructions of the higher order critical GJMS-operators ([116]).

6.21 The tractor families $D_N^T(M, \Sigma; g; \lambda)$

In the present section, we use tractor D-operators for the definition of a series of conformally covariant families of differential operators

$$D_N^T(M^n, \Sigma^{n-1}; g; \lambda) : C^\infty(M^n) \to C^\infty(\Sigma^{n-1})$$

of *any* order N (even and odd) for a hypersurface Σ of M and an arbitrary background metric g. The superscript T stands for (curved) translation. These families, in particular, generalize the families $D_N^c(\lambda) = D_N(S^n, S^{n-1}; g_c; \lambda)$. The idea is as follows. We know that $D_N^c(\lambda) : C^\infty(S^n) \to C^\infty(S^{n-1})$ is induced by the families $\mathcal{D}_N(\lambda)$ of homomorphisms of Verma modules. By Theorem 6.18.7, $\mathcal{D}_N(\lambda)$ is an iterated Zuckerman translate of either $\mathcal{D}_0(\lambda) = i$ or $\mathcal{D}_1(\lambda) = Y_n^-$ (depending on the parity of N); here we simplify notation by writing $\mathcal{D}_N(\lambda)$ for the normalized version $\mathcal{D}_N^0(\lambda)$. The building blocks of the curved version of Zuckerman translation are the families $D(\lambda)$ and $C(\lambda)$. Theorem 6.18.7 and Theorem 6.18.8 immediately suggest the definitions.

Now two basic questions concerning these families are:

1. Which parts of the construction only depend on the metric on the submanifold Σ?

2. What can be said about the relation to GJMS-operators and Q-curvature on the submanifold?

A closely related question concerns the relation between the tractor families and the residue families. In order to address these questions, we formulate a system of conjectures and prove the *holographic duality* for a conformally flat metric h.

The tractor families may also shed some light on the *non-existence* of certain conformally covariant powers of the Laplacian. For a manifold of even dimension $n \geq 4$, there exists no conformally covariant power of the Laplacian of order $\geq n+2$ ([118], [112]). That non-existence is reflected in the theory of the families $D_{2N}^T(\cdot, \Sigma^n; \cdot; \lambda)$ as follows. Although the tractor construction is holomorphic in the spectral parameter (apart from trivial rational coefficients), the operator families of interest, i.e., those which generalize the families $D_{2N}(S^n, S^{n-1}; \lambda)$, in general, will have simple poles. These poles arise precisely under those conditions on the parameters (dimension, order and conformal weight) for which the existence of a conformally covariant power of the Laplacian (for a general metric) is forbidden. Therefore, we regard the corresponding residues as *obstructions* to the existence of such operators. For certain metrics, the obstructions may vanish, however. In such cases, the value of the family at the corresponding parameter is well defined, and is a conformally covariant power of the Laplacian on the target space. In particular, on the conformally flat round sphere S^{n-1} there are conformally covariant powers of the Laplacian for all orders. That corresponds to the vanishing of all obstructions in that case, and is reflected by the existence of the *holomorphic* families $D_{2N}(S^{n+1}, S^n; \lambda)$ (for all N) with the property

$$D_{2N}\left(S^{n+1}, S^n; g_c; -\frac{n}{2}+N\right) = P_{2N}(S^n, g_c)i^*.$$

We continue with the details. First, observe that for even order $2N$, the definition of

$$\Theta_{2N}(\lambda) : \mathcal{M}_{\lambda-2N}(\mathfrak{g}_n) \to \mathcal{M}_\lambda(\mathfrak{g}_{n+1})$$

(see Definition 6.18.1) is equivalent to

$$\Theta_{2N}(\lambda)$$
$$= p(\lambda) \circ p(\lambda-1) \circ \cdots \circ p(\lambda-N+1) \circ (i \otimes I_n) \circ i(\lambda-N) \circ \cdots \circ i(\lambda-2N+1),$$

which realizes the composition

$$\mathcal{M}_{\lambda-2N}(\mathfrak{g}_n) \to \mathcal{M}_{\lambda-2N+1}(\mathfrak{g}_n) \otimes F_n \to \cdots \to \mathcal{M}_{\lambda-N}(\mathfrak{g}_n) \otimes \underbrace{F_n \otimes \cdots \otimes F_n}_{N \text{ factors}}$$

$$\xrightarrow{i \otimes I_n} \mathcal{M}_{\lambda-N}(\mathfrak{g}_{n+1}) \otimes \underbrace{F_{n+1} \otimes \cdots \otimes F_{n+1}}_{N \text{ factors}} \to \cdots \to \mathcal{M}_\lambda(\mathfrak{g}_{n+1}).$$

Here $i(\cdot)$ and $p(\cdot)$ denote the usual embedding and projection as well as all twisted versions. Note also that both maps are used for the Lie algebras $\mathfrak{g}_n$ and $\mathfrak{g}_{n+1}$. $i : \mathfrak{g}_n \hookrightarrow \mathfrak{g}_{n+1}$ embeds the Lie algebras, and $I_n : F_n \hookrightarrow F_{n+1}$ denotes the G^n-equivariant embedding of standard representations. For the induced embeddings of tensor products we use the same symbol.

Definition 6.21.1. *Let $\Sigma \subset M$ be a hypersurface and g a metric on M. For $N \geq 0$, set*

$$D_{2N}^T(M, \Sigma; g; \lambda) = C_\Sigma(g; \lambda-2N+1) \circ \cdots \circ C_\Sigma(g; \lambda-N)$$
$$\circ \Pi_\Sigma(g) \circ D_M(g; \lambda-N+1) \circ \cdots \circ D_M(g; \lambda). \quad (6.21.1)$$

The composition (6.21.1) is conformally covariant since all factors are. More precisely, we have

$$e^{-(\lambda-2N)\varphi} \circ D_{2N}^T(M, \Sigma; e^{2\varphi}g; \lambda) \circ e^{\lambda\varphi} = D_{2N}^T(M, \Sigma, g; \lambda)$$

for all $\lambda \in \mathbb{C}$ and all $\varphi \in C^\infty(M)$.

Similarly, the odd order versions are defined in

Definition 6.21.2. *Let $\Sigma \subset M$ be a hypersurface and g a metric on M. We fix a unit normal field $N = N(g)$. For $N \geq 0$, set*

$$D_{2N+1}^T(M, \Sigma; g; \lambda) = C_\Sigma(g; \lambda-2N) \circ \cdots \circ C_\Sigma(g; \lambda-N+1)$$
$$\circ D_1^T(M, \Sigma; g; \lambda-N) \circ D_M(g; \lambda-N+1) \circ \cdots \circ D_M(g; \lambda).$$

We recall that $D_1^T(\lambda) = \Pi_\Sigma \circ D^T(\lambda)$ (Lemma 6.20.8). Again, the composition is conformally covariant since all factors are. More precisely, we have

$$e^{-(\lambda-2N-1)\varphi} \circ D_{2N+1}^T(M, \Sigma; e^{2\varphi}g; \lambda) \circ e^{\lambda\varphi} = D_{2N+1}^T(M, \Sigma, g; \lambda)$$

for all $\lambda \in \mathbb{C}$ and all $\varphi \in C^\infty(M)$.

We rephrase these definitions by saying that $D_{2N}^{\mathcal{T}}(\lambda)$ and $D_{2N+1}^{\mathcal{T}}(\lambda)$ are the Nth curved translations of i^* and ∇_N, respectively.

In [40], Branson and Gover used the tractor families $D_N^{\mathcal{T}}(\lambda)$ (for specific values of the parameter λ), for the construction of conformally covariant elliptic self-adjoint boundary value problems on manifolds M with boundary Σ. Such boundary value problems can be used to construct conformally covariant pseudo-differential operators on Σ (see also [111]). The convention in [40] differ from those used here in two respects: trivializations of tractor bundles (see Remark 6.20.1) and normalizations of tractor D-operators.

In the composition which defines $D_{2N}^{\mathcal{T}}(\lambda)$, the projection Π_Σ operates on sections in

$$\Gamma(\underbrace{\mathcal{T}^*M \otimes \cdots \otimes \mathcal{T}^*M}_{N \text{ factors}}).$$

It follows that the mean curvature H appears in the family $D_{2N}^{\mathcal{T}}(\lambda)$ with powers up to $2N$. Similarly, in the composition defining $D_{2N+1}^{\mathcal{T}}(\lambda)$ the operator Π_Σ operates on

$$\Gamma(\underbrace{\mathcal{T}^*M \otimes \cdots \otimes \mathcal{T}^*M}_{N \text{ factors}}).$$

This yields powers of H up to $2N$. The additional factor $D^{\mathcal{T}}$ contributes one more power of H so that in the family $D_{2N+1}^{\mathcal{T}}(\lambda)$ the mean curvature H appears with powers up to $2N + 1$. In particular, the top power of the mean curvature in the family $D_N^{\mathcal{T}}(\lambda)$ always coincides with the order of the family.

Theorem 6.18.7 and Theorem 6.18.8 say which coefficients have to be removed in order to get the families of interest. There are actually *two* types of coefficients. The first two products trivially arise by the definition of the *rational* C and D families. However, the product

$$\prod_{j=1}^{N}(n-1-(j+N)+\lambda)$$

is of a different nature. In the context of (holonomic) Verma modules, it appears as an overall coefficient in the formula for the Nth Zuckerman translation. We have seen that in the semi-holonomic setting that coefficient has the effect that the families of interest are no longer polynomial but rational with possible poles at the zeros of the above product. In the curved case, the situation is analogous with the only difference that some residues possibly vanish.

Now in the following definition, we remove the irrelevant coefficients from $D_N^{\mathcal{T}}(\lambda)$.

Definition 6.21.3. *For the pair (M^n, Σ^{n-1}), the factorizations*

$$D_{2N}^T(M, \Sigma; g; \lambda) = c_{2N} \left\{ \prod_{j=N+1}^{2N} \frac{1}{n-1-2j+2\lambda} \prod_{j=1}^{N} \frac{1}{n-2j+2\lambda} \right\}$$

$$\times \prod_{j=1}^{N} (n-1-(j+N)+\lambda) D_{2N}^T(M, \Sigma; g; \lambda)$$

and

$$D_{2N+1}^T(M, \Sigma; g; \lambda) = c_{2N+1} \left\{ \prod_{j=N+1}^{2N} \frac{1}{n-3-2j+2\lambda} \prod_{j=1}^{N} \frac{1}{n-2j+2\lambda} \right\}$$

$$\times \prod_{j=1}^{N} (n-2-(j+N)+\lambda) D_{2N+1}^T(M, \Sigma; g; \lambda)$$

define rational families $D_N^T(M, \Sigma; g; \lambda)$ of conformally covariant differential operators. The poles of the even order family $D_{2N}^T(\lambda)$ are at the zeros of the product $\prod_{j=1}^{N}(n-1-(j+N)+\lambda)$, i.e., at

$$\lambda \in \{-n+N+2, \dots, -n+1+2N\}. \tag{6.21.2}$$

These N poles are (at most) simple. The poles of the odd order family $D_{2N+1}^T(\lambda)$ are at the zeros of the product $\prod_{j=1}^{N}(n-2-(j+N)+\lambda)$, i.e., at

$$\lambda \in \{-n+N+3, \dots, -n+2+2N\}. \tag{6.21.3}$$

These N poles are (at most) simple.

Theorem 6.18.7 says that, for even n, the *critical* family

$$\Theta_n(\lambda) : \mathcal{N}_{\lambda-n}(\mathfrak{g}_{n+1}) \to \mathcal{N}_\lambda(\mathfrak{g}_{n+2})$$

(notice the choice of Lie algebras) factorizes as

$$(\cdots) \prod_{j=1}^{\frac{n}{2}} \left(n-\left(j+\frac{n}{2}\right)+\lambda\right) \mathcal{D}_n^T(\lambda).$$

Similarly, we factorize $D_n^T(M, \Sigma; g; \lambda) : C^\infty(M^{n+1}) \to C^\infty(\Sigma^n)$ as

$$(\cdots) \prod_{j=1}^{\frac{n}{2}} \left(n-\left(j+\frac{n}{2}\right)+\lambda\right) D_n^T(M, \Sigma; g; \lambda).$$

Therefore, it is natural to expect poles of $D_n^T(\cdot, \Sigma^n; g; \lambda)$ (at most) at the points

$$\lambda \in \left\{ -\frac{n}{2}+1, \ldots, 0 \right\}. \tag{6.21.4}$$

$\mathcal{D}_n^T(\lambda)$ has *no* pole at $\lambda = -\frac{n}{2}+1$. We expect that the same is true for $D_n^T(\lambda)$. Much more important, however, is the behaviour at $\lambda = 0$. This is the critical case.

$\mathcal{D}_n^T(\lambda)$ $(n \geq 4)$ has a simple pole at $\lambda = 0$ with a *non-vanishing* residue. However, in the curved case we conjecture that the residue actually vanishes.

Conjecture 6.21.1. *For even n, the critical family $D_n^T(\cdot, \Sigma^n; g; \lambda)$ is regular at $\lambda = 0$, i.e.,*

$$D_n^T(\cdot, \Sigma^n; g; 0) = 0.$$

Note that the operator $B_n(\cdot, \Sigma; g) = D_n^T(\cdot, \Sigma^n; g; 0)$ is a conformally covariant operator

$$e^{n\varphi} B_n(\hat{g}) = B_n(g).$$

The regularity of $D_n^T(\cdot, \Sigma^n; g; \lambda)$ at $\lambda = 0$ has the consequence that the value

$$P_n^T(M, \Sigma; g) = D_n^T(M, \Sigma^n; g; 0) : C^\infty(M) \to C^\infty(\Sigma) \tag{6.21.5}$$

and the quantity

$$Q_n^T(M, \Sigma; g) = \dot{D}_n^T(M, \Sigma^n; g; 0)(1) \in C^\infty(\Sigma) \tag{6.21.6}$$

are *well defined*. Moreover,

Lemma 6.21.1. *If $D_n^T(\cdot, \Sigma^n; g; 0) = 0$, let*

$$\tilde{P}_n^T(g) \stackrel{\text{def}}{=} P_n^T(g) - P_n^T(g)(1).$$

Then the natural pair $(\tilde{P}_n^T(g), Q_n^T(g))$ satisfies the fundamental identity

$$e^{n\varphi} Q_n^T(\hat{g}) = Q_n^T(g) - \tilde{P}_n^T(g)(\varphi).$$

Proof. We differentiate the identity

$$e^{-(\lambda+n)\varphi} \circ D_N^T(\hat{g}; \lambda) \circ e^{\lambda\varphi} = D_n^T(g; \lambda)$$

at $\lambda = 0$ and apply the result to $u = 1$. Then

$$-\varphi e^{n\varphi} D_n^T(\hat{g}; 0)(1) + e^{n\varphi} \dot{D}_n^T(\hat{g}; 0)(1) + e^{n\varphi} D_n^T(\hat{g}; 0)(\varphi) = \dot{D}_n^T(g; 0)(1),$$

i.e.,

$$e^{n\varphi} \dot{D}_n^T(\hat{g}; 0)(1) + \left\{ D_n^T(\hat{g}; 0)(\varphi) - \varphi D_n^T(\hat{g}; 0)(1) \right\} = \dot{D}_n^T(g; 0)(1)$$

using $e^{n\varphi} D_n^T(\hat{g}; 0) = D_n^T(g; 0)$. In other words,

$$e^{n\varphi} Q_n^T(\hat{g}) + \tilde{P}_n^T(g)(\varphi) = Q_n^T(g).$$

The proof is complete. □

It follows that, if the operator $P_n^T(g)$ annihilates constants, then the pair

$$(P_n^T(g), Q_n^T(g))$$

satisfies the fundamental identity

$$e^{n\varphi} Q_n^T(\hat{g}) = Q_n^T(g) - P_n^T(g)(\varphi), \quad \hat{g} = e^{2\varphi} g.$$

For *even* n, the *subcritical* family $D_{2N}^T(\cdot, \Sigma^n; g; \lambda)$ $(2N < n)$ is regular at $\lambda = -\frac{n}{2} + N$. Hence the value

$$P_{2N}^T(\cdot, \Sigma; g) = D_{2N}^T\left(\cdot; \Sigma; g; -\frac{n}{2} + N\right) \tag{6.21.7}$$

is well defined. Similarly, for *odd* n and all N, the family $D_{2N}^T(\cdot, \Sigma; g; \lambda)$ is regular at $\lambda = -\frac{n}{2} + N$. Hence the value

$$P_{2N}^T(\cdot; \Sigma; g) = D_{2N}^T\left(\cdot; \Sigma; g; -\frac{n}{2} + N\right) \tag{6.21.8}$$

is well defined. Now the relation $\mathcal{D}_{2N}(-\frac{n}{2} + N) = (\Delta_n^-)^N$ suggests

Conjecture 6.21.2. *For $n \geq 3$, let $2N < n$ if n is even. Then*

$$D_{2N}^T\left(\cdot, \Sigma^n; g; -\frac{n}{2} + N\right) = \mathbf{P}_{2N}(\Sigma^n, i^*(g))i^*$$

for a conformally covariant differential operator $\mathbf{P}_{2N}$ on Σ with leading part Δ_g^N. In particular, the operator on the left-hand side does not depend on the embedding.

The critical case for *even* n behaves differently.

Conjecture 6.21.3 (Decomposition). *Assuming Conjecture 6.21.1, the operator*

$$P_n^T(\cdot, \Sigma^n; g)$$

(see (6.21.5)) admits a decomposition

$$P_n^T(\cdot, \Sigma; g) = \mathbf{P}_n(\Sigma, i^*(g))i^* + P_n^e(\cdot, \Sigma; g)$$

into conformally covariant differential operators

$$\mathbf{P}_n : C^\infty(\Sigma) \to C^\infty(\Sigma), \quad P_n^e : C^\infty(M) \to C^\infty(\Sigma),$$

where $\mathbf{P}_n$ has leading term $\Delta^{\frac{n}{2}}$. The operator $\mathbf{P}_n$ only depends on the metric $i^*(g)$. The operator P_n^e depends on the metric in a neighborhood of Σ but lives on Σ. Similarly, $Q_n^T(\cdot, \Sigma; g) \in C^\infty(\Sigma)$ (see (6.21.6)) admits a decomposition

$$-(-1)^{\frac{n}{2}} Q_n^T(\cdot, \Sigma; g) = \mathbf{Q}_n(\Sigma, i^*(g)) + Q_n^e(\cdot, \Sigma; g).$$

Moreover, the pairs $(\mathbf{P}_n, \mathbf{Q}_n)$ and (P_n^e, Q_n^e) satisfy the respective fundamental identities

$$e^{n\varphi} \mathbf{Q}_n(\hat{g}) = \mathbf{Q}_n(g) - \mathbf{P}_n(g)(\varphi)$$

on Σ and

$$e^{n\varphi} Q_n^e(\hat{g}) = Q_n^e(g) - P_n^e(g)(\varphi).$$

The point of these decompositions is that the *intrinsic* pair $(\mathbf{P}_n, \mathbf{Q}_n)$ only depends on the metric $i^*(g)$ on Σ.

For *odd* n, the situation for the critical families $D_n^T(g; 0)$ is the following. In that case, the value $D_n^T(\cdot, \Sigma^n; g; 0)$ is not well defined, in general, since the family $D_n^T(\cdot, \Sigma^n; g; \lambda)$ has a simple pole at $\lambda = 0$. In fact, Theorem 6.18.8 says that, for odd n, the critical family

$$\Theta_n(\lambda) : \mathcal{N}_{\lambda-n}(\mathfrak{g}_{n+1}) \to \mathcal{N}_\lambda(\mathfrak{g}_{n+2})$$

factorizes as

$$(\cdots) \prod_{j=1}^{\frac{n-1}{2}} \left(n - 1 - \left(j + \frac{n-1}{2} \right) + \lambda \right) \mathcal{D}_n^T(\lambda).$$

Similarly, we factorize $D_n^T(M, \Sigma; g; \lambda) : C^\infty(M) \to C^\infty(\Sigma)$ as

$$(\cdots) \prod_{j=1}^{\frac{n-1}{2}} \left(n - 1 - \left(j + \frac{n-1}{2} \right) + \lambda \right) D_n^T(M, \Sigma; g; \lambda).$$

Therefore, it is natural to expect poles of $D_n^T(\cdot, \Sigma; g; \lambda)$ (at most) at the points

$$\lambda \in \left\{ -\frac{n-1}{2} + 1, \dots, 0 \right\}.$$

In particular, there is a possible pole at $\lambda = 0$. Let

$$B_n(M, \Sigma; g) = \mathrm{Res}_0(D_n^T(M, \Sigma; g; \lambda)) : C^\infty(M) \to C^\infty(\Sigma)$$

be the residue. B_n is a conformally covariant differential operator, i.e.,

$$e^{n\varphi} B_n(\hat{g}) = B_n(g).$$

Problem 6.21.1. *Let n be odd. Determine the residue B_n. Characterize in geometric terms the metrics with $B_n = 0$. Does B_n vanish if $L_0 = 0$? Is B_n induced by the residue $\mathrm{Res}_0(\Theta_n^T(\lambda))$ of the semi-holonomic family?*

Note that it is *not* true that the semi-holonomic $\mathrm{Res}_0(\Theta_n^T(\lambda)) \in T(\mathfrak{n}_{n+2}^-)$ is tangential, i.e., does *not* contain components Y_{n+1}^-. As an example, compare for $n = 3$ the algebraic structure given in Theorem 6.18.3 with the geometric structure given in Theorem 6.25.1.

Definition 6.21.4. *Let n be odd and assume that $B_n(M, \Sigma; g) = 0$. Then the critical family $D_n^T(M, \Sigma^n; g; \lambda)$ is regular at $\lambda = 0$. We set*

$$P_n^T(M, \Sigma; g) = D_n^T(M, \Sigma; g; 0), \quad \tilde{P}_n^T(g) = P_n^T(g) - P_n^T(g)(1),$$
$$Q_n^T(M, \Sigma; g) = \dot{D}_n^T(M, \Sigma; g; 0)(1).$$

Lemma 6.21.2. *For odd* n, *the natural pair* $(\tilde{P}_n^T(g), Q_n^T(g))$ *satisfies the fundamental identity*

$$e^{n\varphi} Q_n^T(\hat{g}) = Q_n^T(g) - \tilde{P}_n^T(g)(\varphi).$$

The proof is the same as for Lemma 6.21.1. In particular, the behaviour of odd order Q-curvature $Q_n^T(M, \Sigma; g)$ under $g \to e^{2\varphi}g$ is described by a linear *differential* operator.

If for odd n the obstruction operator

$$\mathrm{Res}_0(D_n^T(\cdot, \Sigma^n; g; \lambda)) : C^\infty(M) \to C^\infty(\Sigma)$$

is non-trivial, then it comes together with a notion of Q-curvature. In fact, the usual family proof shows that the pair

$$\left(D_n^T(\cdot, \Sigma^n; g; 0), \dot{D}_n^T(\cdot, \Sigma^n; g; 0)(1) \right)$$

satisfies a fundamental identity. Note that $D_n^T(\cdot, \Sigma^n; g; 0)(1) = 0$ since $D_M(0)(1) = 0$.

Another notion of an *odd* order Q-curvature was introduced in [97]. We briefly recall the definition. Assume that Σ^n is the boundary of the manifold M, and that g is a Poincaré-Einstein metric with conformal infinity $[h]$. Let $S(h; \lambda) : C^\infty(\Sigma) \to C^\infty(\Sigma)$ be the scattering operator (as in Chapter 3). Then

$$S(h; n)(1) = (-1)^{\frac{n}{2}} c_{\frac{n}{2}} Q_n(h)$$

for *even* n and $S(h; n)(1) = 0$ for *odd* n. Set

$$Q_n^{FG}(g; h) = \dot{S}(h; n)(1).$$

Here the notation indicates that Q_n^{FG} does not depend only on h. Q_n^{FG} satisfies the fundamental identity

$$e^{n\varphi} Q_n^{FG}(g; e^{2\varphi}h) = Q_n^{FG}(g; h) - S(h; n)(\varphi).$$

The operator $S(h; n)$ is a *non-local* pseudo-differential operator of order n on Σ. In particular, it follows that both notions of odd order Q-curvature differ.

Note that in [97], Q_n^{FG} is defined by a slightly different normalization (similar to that for the even order Q_n).

The integral of $Q_n^{FG}(g; h)$ only depends on $[h]$ and we have

Theorem 6.21.1 ([97]). *Let* n *be odd. Then*

$$- \int_{\Sigma^n} Q_n^{FG}(g; h) \, \mathrm{vol}(h) = V,$$

where V *is the renormalized volume in*

$$\int_{r \geq \varepsilon} \mathrm{vol}(g) = c_0 \varepsilon^{-n} + c_2 \varepsilon^{-n+2} + \cdots + c_{n-1} \varepsilon^{-1} + V + o(1), \quad \varepsilon \to 0$$

(*compare with* (6.6.25)).

An interesting *problem* is to uncover the relation between Q_n^{FG} (as defined in [97]) and Q_n^T (as defined here). In the case $n = 3$, we have

Lemma 6.21.3. $Q_3^{FG}(M^4, \Sigma^3; g; h)$ *coincides with* $Q_3^T(M^4, \Sigma^3; e^{2v}g)$, *up to a constant multiple. Here* $v \in C^\infty(M)$ *is a certain solution of the equation* $\Delta_g v = -3$ ([97]) *which induces a conformal compactification* $e^{2v}g$ *of the Poincaré-Einstein metric* g.

Proof. In the following formulas, we suppress constant multiples. By [70] (Lemma 2.2),

$$Q_3^{FG}(M^4, \Sigma^3; g; h) = T(e^{2v}g),$$

where $T = \nabla_N \tau$ is the order 3 Chang-Qing Q-curvature for the metric $e^{2v}g$ (see Theorem 6.26.1). Here we use that $L = 0$. On the other hand, by Theorem 6.26.2,

$$Q_3^T(M^4, \Sigma^3; e^{2v}g) = T(e^{2v}g).$$

This proves the assertion.   □

Does this relation generalize?

Note also that the conformal compactification $e^{2v}g$ has the property ([70])

$$Q_n(M^n, e^{2v}g) = 0 \tag{6.21.9}$$

for even n. In order to prove the vanishing result (6.21.9), we apply Theorem 3.2.3. Since $\tau_g = -n(n-1)$, we find

$$P_n(g)(v) = \prod_{j=\frac{n}{2}}^{n-2} \left(\Delta_g - \frac{j(n-1-j)}{n(n-1)}\tau_g \right) \Delta_g(v) = \prod_{j=\frac{n}{2}}^{n-2} j(n-1-j)\Delta_g(v) = -(n-1)!$$

and $Q_n(g) = (-1)^{\frac{n}{2}}(n-1)!$. Hence

$$e^{nv}Q_n(M^n, e^{2v}g) = Q_n(M^n, g) + (-1)^{\frac{n}{2}}P_n(g)(v)$$
$$= (-1)^{\frac{n}{2}}(n-1)! - (-1)^{\frac{n}{2}}(n-1)!$$
$$= 0.$$

Problem 6.21.2. *Determine the relation between* (P_n, Q_n) *and* $(\mathbf{P}_n, \mathbf{Q}_n)$.

There is one important special class of background metrics for which we can prove a result in the direction of Problem 6.21.2. We recall that for conformally flat h, the Fefferman-Graham expansion terminates at the third term (see Lemma 6.14.1 and its proof).

Theorem 6.21.2 (Holographic duality). *Let* $n \geq 3$. *On* $M^{n+1} = (0, \varepsilon) \times \Sigma^n$ *we consider a Poincaré-Einstein metric* $r^{-2}(dr^2 + h_r)$ *with conformal infinity* h. *Let* h *be conformally flat. Then*

$$D_{2N}^T(dr^2 + h_r; \lambda) = D_{2N}^{res}(h; \lambda).$$

Proof. We work in local coordinates so that $h = e^{2\varphi}h_c$ with $h_c = \sum_i dx_i^2$. The proof rests on the identity

$$D_{2N}^{\text{res}}(h_c; \lambda) = D_{2N}^T(dr^2 + h_c; \lambda).$$

In fact, Theorem 6.18.7 implies that $D_{2N}^T(dr^2 + h_c; \lambda)$ is induced by $\mathcal{D}_{2N}(\lambda)$, i.e.,

$$D_{2N}^T(dr^2 + h_c; \lambda) = D_{2N}^{nc}(\lambda).$$

On the other hand, Theorem 5.2.5 and Definition 6.6.2 yield

$$D_{2N}^{\text{res}}(h_c; \lambda) = D_{2N}^{nc}(\lambda).$$

As in Section 6.11, let κ be the local diffeomorphism with the properties

$$\kappa^* \left(r^{-2}(dr^2 + \hat{h}_r) \right) = r^{-2}(dr^2 + h_r)$$

and

$$i^* \left(\frac{\kappa^*(r)}{r} \right) = e^\varphi, \quad \hat{h} = e^{2\varphi}h.$$

Then for $h = h_c$ we get

$$D_{2N}^{\text{res}}(h_c; \lambda) \circ \left(\frac{\kappa^*(r)}{r} \right)^{-\lambda} \circ \kappa^* = D_{2N}^T(dr^2 + h_c; \lambda) \circ \left(\frac{\kappa^*(r)}{r} \right)^{-\lambda} \circ \kappa^*.$$

Now we apply (6.11.4), the conformal covariance of $D_{2N}^T(g; \lambda)$ and the relation $i^*(\kappa^*(r)/r) = e^\varphi$. We find

$$e^{-(\lambda - 2N)\varphi} \circ D_{2N}^{\text{res}}(h; \lambda) = e^{-(\lambda - 2N)\varphi} \circ D_{2N}^T \left((\kappa^*(r)/r)^2 (dr^2 + h_c); \lambda \right) \circ \kappa^*$$
$$= e^{-(\lambda - 2N)\varphi} \circ D_{2N}^T(dr^2 + h_r; \lambda)$$

using $i^*\kappa^* = \text{id}$. $\qquad\square$

There is an analogous result for odd order families. We omit the details.

As consequences, we obtain the following tractor formulas for GJMS-operators and critical Q-curvature.

Theorem 6.21.3. *Let $n \geq 3$ be even and assume that h is conformally flat. Then*

$$P_{2N}^T(dr^2 + h_r) = P_{2N}(h)i^* \quad \text{and} \quad \dot{D}_n^T(dr^2 + h_r; 0)(1) = -(-1)^{\frac{n}{2}}Q_n(h).$$

Proof. We find

$$P_{2N}^T(dr^2 + h_r) = D_{2N}^T \left(dr^2 + h_r; -\frac{n}{2} + N \right) \qquad \text{(by (6.21.7))}$$
$$= D_{2N}^{\text{res}} \left(h; -\frac{n}{2} + N \right) \qquad \text{(by Theorem 6.21.2)}$$
$$= P_{2N}(h)i^* \qquad \text{(by (6.6.15))}$$

and

$$\dot{D}_n^T(dr^2+h_r;0)(1) = \dot{D}_n^{\text{res}}(0)(1) \qquad \text{(by Theorem 6.21.2)}$$
$$= -(-1)^{\frac{n}{2}}Q_n(h) \qquad \text{(by (6.6.2))}.$$

The proof is complete. $\square$

Theorem 6.21.3 shows that the conformal flatness of h implies that $D_n^T(dr^2+h_r;\lambda) = P_n(h)i^*$, i.e., the extrinsic part $P_n^e(dr^2+h_r)$ vanishes. Similarly, we have $Q_n^e(dr^2+h_r) = 0$. We expect that the same is true for any h. More precisely,

Conjecture 6.21.4. $P_n^e(dr^2+h_r) = 0$ and $Q_n^e(dr^2+h_r) = 0$.

The holographic duality very likely holds true for more general metrics h. In Section 6.22 we prove its validity for any h if the order of the family is ≤ 4 (Theorem 6.22.6). The proof rests on factorization identities. For a conformally flat h, factorization identities for residue families (Theorem 6.11.1) immediately imply factorization identities for tractor families using Theorem 6.21.2. More precisely,

Theorem 6.21.4. *For a conformally flat metric h, the tractor family*

$$D_{2N}^T(dr^2+h_r;\lambda) : C^\infty([0,\varepsilon) \times \Sigma) \to C^\infty(\Sigma)$$

factorizes for

$$\lambda \in \left\{ -\frac{n+1}{2}, \ldots, -\frac{n+1}{2}+N \right\} \cup \left\{ -\frac{n}{2}+N, \ldots, -\frac{n}{2}+2N \right\}$$

into products of lower order tractor families and GJMS-operators. More precisely, for $j = 1, \ldots, N$, the identities

$$D_{2N}^T\left(dr^2+h_r; -\frac{n}{2}+2N-j\right) = P_{2j}(h) \circ D_{2N-2j}^T\left(dr^2+h_r; -\frac{n}{2}+2N-j\right)$$

and

$$D_{2N}^T\left(dr^2+h_r; -\frac{n+1}{2}+j\right) = D_{2N-2j}^T\left(dr^2+h_r; -\frac{n+1}{2}-j\right) \circ P_{2j}(dr^2+h_r)$$

hold true.

Above we used an indirect argument to prove Theorem 6.21.4. It rests on Theorem 6.21.2, i.e., on the relation to residue families and their factorizations. For the convenience of the reader, we next present the argument in a way which is independent of residue families. It only rests on the conformal covariance of tractor families and their factorizations for the flat metric $h_c = \sum_i dx_i^2$. For that metric, the factorization identities follow from

$$D_{2N}^T(dr^2+h_c;\lambda) = D_{2N}^{nc}(\lambda)$$

and the factorization identities for $\mathcal{D}_{2N}(\lambda)$ (see the arguments in the proof of Theorem 6.21.2). In fact, we compose the identity

$$D_{2N}^T \left(dr^2 + h_c; -\frac{n}{2} + 2N - j \right) = P_{2j}(h_c) \circ D_{2N-2j}^T \left(dr^2 + h_r; -\frac{n}{2} + 2N - j \right)$$

with

$$\left(\frac{\kappa^*(r)}{r} \right)^{\frac{n}{2} - 2N + j} \circ \kappa^*,$$

where κ is the local diffeomorphism in the proof of Theorem 6.21.2, i.e.,

$$\kappa_* \left(\kappa^*(r)/r \right)^2 (dr^2 + h_c) = dr^2 + h_r.$$

Using conformal covariance, we find

$$e^{(\frac{n}{2} + j)\varphi} \circ D_{2N}^T \left((\kappa^*(r)/r)^2 (dr^2 + h_c); -\frac{n}{2} + 2N - j \right) \circ \kappa^*$$
$$= P_{2j}(h_c) \circ e^{(\frac{n}{2} - j)\varphi} \circ D_{2N-2j}^T \left((\kappa^*(r)/r)^2 (dr^2 + h_c); -\frac{n}{2} + 2N - j \right) \circ \kappa^*,$$

i.e.,

$$D_{2N}^T \left(dr^2 + h_r; -\frac{n}{2} + 2N - j \right)$$
$$= \left[e^{(\frac{n}{2} + j)\varphi} \circ P_{2j}(h_c) \circ e^{(\frac{n}{2} - j)\varphi} \right] \circ D_{2N-2j}^T \left(dr^2 + h_r; -\frac{n}{2} + 2N - j \right)$$

by naturality and $i^* \kappa^* = \text{id}$. Now the first system of factorization identities follows from

$$P_{2j}(h) = e^{-(\frac{n}{2} + j)\varphi} \circ P_{2j}(h_c) \circ e^{(\frac{n}{2} - j)\varphi}, \quad h = e^{2\varphi} h_c.$$

Similarly, in order to prove the second system we compose

$$D_{2N}^T \left(dr^2 + h_c; -\frac{n+1}{2} + j \right) = D_{2N-2j}^T \left(dr^2 + h_c; -\frac{n+1}{2} - j \right) \circ P_{2j}(dr^2 + h_c)$$

with

$$\left(\frac{\kappa^*(r)}{r} \right)^{\frac{n+1}{2} - j}$$

and find

$$D_{2N}^T \left((\kappa^*(r)/r)^2 (dr^2 + h_c); -\frac{n+1}{2} + j \right)$$
$$= D_{2N-2j}^T \left((\kappa^*(r)/r)^2 (dr^2 + h_c); -\frac{n+1}{2} - j \right) \circ P_{2j} \left((\kappa^*(r)/r)^2 (dr^2 + h_c) \right)$$

using conformal covariance. Finally, we compose with κ^*. Naturality and $i^*\kappa^* = \mathrm{id}$ imply

$$D_{2N}^T\left(dr^2 + h_r; -\frac{n+1}{2} + j\right) = D_{2N-2j}^T\left(dr^2 + h_r; -\frac{n+1}{2} - j\right) \circ P_{2j}\left(dr^2 + h_r\right).$$

The proof is complete. $\square$

Of course, it would be interesting to extend Theorem 6.21.4 beyond the conformally flat category.

In the situation of Theorem 6.21.2, the trace-free part L_0 of L vanishes. In fact, it is a well-known observation of LeBrun ([164], Proposition 2.8) that the Einstein condition $\mathrm{Ric}(g) + ng = 0$ implies $L_0 = 0$, i.e., Σ is totally umbilic; see also (6.16.10). In the case $n = 3$, we will see later (Theorem 6.25.1) that L_0 actually determines the obstruction B_3, i.e., the vanishing of L_0 implies the vanishing of the obstruction B_3.

For *even* n, the poles of the families

$$\mathcal{D}_{2N}^T(\lambda) : \mathcal{N}_{\lambda-2N}(\mathfrak{g}_{n+1}) \to \mathcal{N}_\lambda(\mathfrak{g}_{n+2})$$

and

$$D_{2N}^T(M, \Sigma^n; g; \lambda) : C^\infty(M^{n+1}) \to C^\infty(\Sigma^n)$$

at the zeros of the product

$$\prod_{j=2}^N (n - (j+N) + \lambda)$$

seem to be related to the *non-existence* of certain conformally covariant powers of the Laplacian. More precisely, we recall (Theorem 5.2.2) that for all n,

$$\mathcal{D}_{2N}\left(-\frac{n}{2} + N\right) = (\Delta_n^-)^N, \ N \geq 1. \tag{6.21.10}$$

Thus in order to induce a curved version of $(\Delta_n^-)^N$, we would use the semi-holonomic lift $\mathcal{D}_{2N}^T(\lambda)$ or the tractor construction $D_{2N}^T(M, \Sigma^n; g; \lambda)$ at $\lambda = -\frac{n}{2} + N$. But the family $D_{2N}^T(\cdot; \Sigma^n; g; \lambda)$ has a pole at $\lambda = -\frac{n}{2} + N$ if

$$-\frac{n}{2} + N \in \{-n+N+2, \ldots, -n+2N\},$$

i.e., for

$$n = 4, 6, \ldots, 2N.$$

For even n, *obstructions* arise as follows. In view of the above mentioned poles, the value $D_{2N}^T(\cdot, \Sigma^n; g; -\frac{n}{2} + N)$ is possibly not well defined if the order $2N$ exceeds the dimension n of the target manifold Σ, i.e., $2N \geq n$. In the extreme case $2N = n$, however, Conjecture 6.21.1 states that the value $D_n^T(\cdot; \Sigma^n; g; 0)$ *is* still well defined. Thus we are left with the cases $2N > n$ (n even). These are precisely the cases for which the GJMS-construction does not work for general metrics (because of the obstructed Fefferman-Graham expansion). Moreover, Graham [118] proved that on a four-manifold Σ there exists no natural conformally covariant operator of the form $\Delta_\Sigma^3 + LOT$. For more details we refer to the discussion in Section 6.12. More generally, Gover and Hirachi ([112]) proved that the condition $2N \le n$ in the construction of [124], i.e., in Theorem 3.1.1, is sharp, i.e., Graham's non-existence extends to all cases $2N > n$ (n even). From the above perspective, an *obstruction* to the construction of a curved analog of $\mathcal{D}_{2N}(-\frac{n}{2} + N) = (\Delta_n^-)^N$ is the residue

$$B_{2N}(g) \stackrel{\text{def}}{=} \text{Res}_{-\frac{n}{2}+N}(D_{2N}^T(M, \Sigma^n; g; \lambda)) : C^\infty(M) \to C^\infty(\Sigma).$$

The conformal covariance of the tractor family implies that the operator B_{2N} is conformally covariant, i.e.,

$$e^{(\frac{n}{2}+N)\varphi} \circ B_{2N}(\hat{g}) \circ e^{(-\frac{n}{2}+N)\varphi} = B_{2N}(g).$$

In Lemma 6.12.1, we have seen that on manifolds of dimension $n = 4$, the second-order operator

$$\mathcal{R} = -\delta(\mathcal{B}\#d) + (\mathcal{B}, \mathsf{P})$$

is an obstruction to the existence of a conformally covariant cube of the Laplacian. It would be interesting to interpret this result from the tractor point of view and to relate the obstructions.

Problem 6.21.3. *Characterize in geometric terms the conformally invariant condition $B_{2N}(g) = 0$, $2N > n$ on g. Does the vanishing only depend on the metric $i^*(g)$ on Σ? Is the operator $B_{2N}(g)$ induced (in a sense) by the residue*

$$\text{Res}_{-\frac{n}{2}+N}(\Theta_{2N}^T(\lambda))$$

of the semi-holonomic family?

If $B_{2N}(g) = 0$, then the family $D_{2N}^T(M, \Sigma; g; \lambda)$ is regular at the argument $\lambda = -\frac{n}{2} + N$ and the conformally covariant operator

$$P_{2N}^T(g) = D_{2N}^T\left(M, \Sigma; g; -\frac{n}{2} + N\right) : C^\infty(M) \to C^\infty(\Sigma)$$

is well defined. Examples of such situations are provided by the group-equivariant families considered in Chapter 5.

Example 6.21.1. *The simplest case concerns the order 6 family $D_6^T(\lambda)$. For $\mathcal{D}_6(\lambda)$, we have the relation $\mathcal{D}_6(-\frac{n}{2} + 3) = (\Delta_n^-)^3$. The semi-holonomic lift $D_6^T(\lambda)$ has poles at $\lambda = -n + 6$ and $\lambda = -n + 5$. Thus, using $D_6^T(\cdot, \Sigma^n; g; \lambda)$, a construction of a curved analog P_6 of Δ^3 is problematic if $-\frac{n}{2} + 3 = -n + 6$, i.e., $n = 6$ and if $-\frac{n}{2} + 3 = -n + 5$, i.e., $n = 4$. According to Conjecture 6.21.1, the case $n = 6$ is not problematic. Therefore, it remains to analyze the vanishing of the obstruction operator $B_6(g) : C^\infty(M^5) \to C^\infty(\Sigma^4)$. What is the relation between B_6 and $\mathcal{R}$?*

Finally, we introduce the tractor Q-polynomial. The following definition generalizes the Q-polynomial in Definition 6.6.3. We restrict here to the critical case.

Definition 6.21.5 (Tractor Q-polynomial). *For (M^{n+1}, Σ^n), we define the associated critical tractor Q-polynomial by*

$$Q_n^T(M, \Sigma; g; \lambda) = D_n^T(M, \Sigma; g; \lambda)(1).$$

We do not discuss here the quality of the function $Q_n^T(g; \lambda)$, i.e., whether it actually *is* a polynomial. It remains an open problem to embed the full theory of the Q-polynomials which are derived from the residue families into an analogous theory for tractor Q-polynomials.

We have used homomorphisms of semi-holonomic Verma modules to suggest the existence of curved versions in terms of tractor D-operators. However, we expect that there is also a *direct* induction machine so that the rational family

$$\Theta_N(\lambda) : \mathcal{N}_{\lambda-N}(\mathfrak{g}_n) \to \mathcal{N}_\lambda(\mathfrak{g}_{n+1})$$

induces

$$D_N^T(M^n, \Sigma^{n-1}; g; \lambda) : C^\infty(M) \to C^\infty(\Sigma).$$

Such an induction machine should rest on conformal Cartan connections together with reductions of structures from M to Σ. Note that the Willmore curvature $H^2 - K$ of a surface $\Sigma \subset \mathbb{R}^3$ appears in the curvature of the reduction of the Cartan connection of $(\mathbb{R}^3, g_c)$ to Σ ([216], p.314) as well as in the Q-polynomial $Q_2(\mathbb{R}^3, \Sigma^2; g_c; \lambda)$ (see 6.3.1). In order to define an induction machine, it would be crucial to describe the operator Π_Σ in terms of Cartan's principal bundles.

6.22   Some results on tractor families

In the present section, we illustrate the perspective outlined in Section 6.21 by a series of results.

First of all, we confirm Conjecture 6.21.1 in the cases $n = 2$ and $n = 4$. For $n = 2$, the regularity of $D_2^T(\cdot, \Sigma^2; g; \lambda)$ at $\lambda = 0$ follows from

$$D_2^T(\cdot, \Sigma^2; g; \lambda) = \frac{1}{(-2 + 2\lambda)(1 + 2\lambda)} \lambda D_2(\cdot, \Sigma^2; g; \lambda) \qquad (6.22.1)$$

(Theorem 6.24.1) with the *holomorphic* family $D_2(\lambda)$. For $n = 4$, the proof of the analogous result requires some more work.

Theorem 6.22.1. $D_4^T(M^5, \Sigma^4; g; 0) = 0.$

Proof. We have to prove that $C_\Sigma(-3) \circ \{C_\Sigma(-2)\Pi_\Sigma D_M(-1)\} \circ D_M(0) = 0$. First, we observe that, up to irrelevant non-vanishing coefficients,

$$C_\Sigma(-3) \begin{pmatrix} u_- \\ \omega \\ u_+ \end{pmatrix} = -4\delta_\Sigma \omega - \Delta_\Sigma u_+, \quad D_M(0)u = \begin{pmatrix} \frac{1}{3}\Delta_M u \\ du \\ 0 \end{pmatrix}$$

(see (6.20.6) and (6.20.2)). Hence

$$\Pi_\Sigma D_M(0)u = \begin{pmatrix} \frac{1}{3}\Delta_M u - HNu \\ d_\Sigma u \\ 0 \end{pmatrix}.$$

Now we use the fact that

$$C_\Sigma(-2) \circ \Pi_\Sigma \circ D_M(-1) = (\Delta_\Sigma^\mathcal{T} - J_\Sigma) \circ \Pi_\Sigma = \square_\Sigma^\mathcal{T} \circ \Pi_\Sigma,$$

where $\Delta_\Sigma^\mathcal{T}$ is the tractor Laplacian on Σ. This is a tractor generalization (Theorem 6.24.2) of Theorem 6.24.1. Next, we need an explicit formula for $\Delta_\Sigma^\mathcal{T}$. The definition of $\nabla^\mathcal{T}$ implies the following formula for $\nabla_X^\mathcal{T} \nabla_Y^\mathcal{T}$:

$$\begin{pmatrix} \nabla_X\nabla_Y - P(X,Y) & \nabla_X(P(Y),\cdot) + (P(X),\nabla_Y\cdot) & \frac{1}{2}(P(X),P(Y)) \\ -X^\flat\nabla_Y - \nabla_X(Y^\flat\cdot) & \nabla_X\nabla_Y - P(X)\langle Y,\cdot\rangle - X^\flat(P(Y),\cdot) & \frac{1}{2}(\nabla_X(P(Y)\cdot) + P(X)\nabla_Y\cdot) \\ 2(X,Y) & -2\langle X,\nabla_Y\cdot\rangle - 2\nabla_X\langle Y,\cdot\rangle & \nabla_X\nabla_Y - P(X,Y) \end{pmatrix}.$$

Now applying Lemma 6.20.3 and the Weitzenböck formula

$$\mathrm{tr}(\nabla^{LC}\nabla^{LC}) = \Delta + \mathrm{Ric}\,\#$$

for the Bochner-Laplacian on 1-forms, we find

$$\Delta^\mathcal{T} = \begin{pmatrix} \Delta - J & -(dJ,\cdot) + 2\delta(P\#\cdot) & \frac{1}{2}|P|^2 \\ -2d & \Delta + \mathrm{Ric}\,\# - 2P\# & \frac{1}{2}dJ + P\#d \\ 2n & -4\delta & \Delta - J \end{pmatrix}. \tag{6.22.2}$$

Since for $n = 4$, $\mathrm{Ric} - 2P = Jg$, we obtain

$$\Delta_\Sigma^\mathcal{T} = \begin{pmatrix} \Delta_\Sigma - J_\Sigma & * & \frac{1}{2}|P_\Sigma|^2 \\ -2d_\Sigma & \Delta_\Sigma + J_\Sigma & * \\ 8 & -4\delta_\Sigma & \Delta_\Sigma - J_\Sigma \end{pmatrix}.$$

Hence the desired composition equals

$$C_\Sigma(-3) \begin{pmatrix} * \\ -\frac{2}{3}d_\Sigma\Delta_M u + 2d_\Sigma(HNu) + \Delta_\Sigma d_\Sigma u \\ \frac{8}{3}\Delta_M u - 8HNu - 4\Delta_\Sigma u \end{pmatrix}$$

$$= \frac{8}{3}\Delta_\Sigma\Delta_M u - 8\Delta_\Sigma(HNu) - 4\delta_\Sigma\Delta_\Sigma d_\Sigma u$$

$$- \frac{8}{3}\Delta_\Sigma\Delta_M u + 8\Delta_\Sigma(HNu) + 4\Delta_\Sigma^2 u = 0.$$

The proof is complete. $\qquad\qquad\qquad\qquad\qquad\qquad\qquad\qquad\qquad\qquad\square$

Remark 6.22.1. *Corollary 6.20.3 says that*

$$C_\Sigma(-3)(\Delta_\Sigma^T - \mathsf{J}_\Sigma)D_\Sigma(0) = 0 \qquad\qquad (6.22.3)$$

*on a 4-manifold Σ. Although that identity resembles Theorem 6.22.1, it is impor-
tant to emphasize the different nature of both assertions. In fact, while Theorem
6.22.1 says that*

$$\Pi_\Sigma D_M(0)u = \begin{pmatrix} \frac{1}{3}\Delta_M u - HNu \\ d_\Sigma u \\ 0 \end{pmatrix} \in \ker C_\Sigma(-3)\square_\Sigma^T,$$

(6.22.3) states that

$$D_\Sigma(0)u = \begin{pmatrix} \frac{1}{2}\Delta_\Sigma u \\ d_\Sigma u \\ 0 \end{pmatrix} \in \ker C_\Sigma(-3)\square_\Sigma^T$$

which follows from

$$(\Delta_\Sigma^T - \mathsf{J}_\Sigma)\begin{pmatrix} \frac{1}{2}\Delta_\Sigma u \\ d_\Sigma u \\ 0 \end{pmatrix} = \begin{pmatrix} * & * & * \\ -2d_\Sigma & \Delta_\Sigma & * \\ 8 & -4\delta_\Sigma & * \end{pmatrix}\begin{pmatrix} \frac{1}{2}\Delta_\Sigma u \\ d_\Sigma u \\ 0 \end{pmatrix} = \begin{pmatrix} * \\ 0 \\ 0 \end{pmatrix}.$$

Remark 6.22.2. *The curved translation*

$$D_4^T(M,\Sigma;g;0) = C_\Sigma(-3)C_\Sigma(-2)\Pi_\Sigma D_M(-1)D_M(0)$$

*of $i^* : C^\infty(M) \to C^\infty(\Sigma)$ is the curved analog of $\Theta_4(0)$. Therefore, we consider the
vanishing $D_4^T(0) = 0$ (Theorem 6.22.1) as the statement that the non-vanishing
element*

$$\Theta_4(0) \simeq (\Delta_4^-)^2 - L_4 \in \ker \pi$$

*(see (6.18.16)) induces (via Cartan geometry) a vanishing operator $C^\infty(M^5) \to
C^\infty(\Sigma^4)$. On the other hand, $(\Delta_4^-)^2 - L_4$ also appears in the curved translation of
the Yamabe operator. In fact, according to Theorem 6.18.1, it is a multiple of the
Zuckerman translate of Δ_4^-. The curved version of that construction is*

$$C_\Sigma(-3)\square_\Sigma^T D_\Sigma(0).$$

*Its vanishing (Corollary 6.20.3) is to be considered as the statement that the non-
vanishing element $(\Delta_4^-)^2 - L_4$ induces (via Cartan geometry) a vanishing operator
$C^\infty(\Sigma^4) \to C^\infty(\Sigma^4)$. We emphasize that the two respective inductions are of different nature. While the latter one uses intrinsic geometry of Σ the former one
involves reductions from M to Σ. The observations in Remark 6.22.1 actually indicate the different nature of both vanishing results.*

Next, we confirm Conjecture 6.21.2 in the special case $N = 2$. In particular, assuming $n \neq 4$, we prove that $D_4^T(\cdot, \Sigma^n; g; -(n-4)/2)$ only depends on the pull-back of the metric to Σ. The critical case $n = 4$ is more complicated, and will be discussed below.

Theorem 6.22.2. $D_4^T(M^{n+1}, \Sigma^n; g; -\frac{n-4}{2}) = P_{4,n}(\Sigma^n; g)i^*$ if $3 \leq n \neq 4$.

Proof. We present the details for $n = 3$. The general case is analogous. We first prove that the left-hand side does not depend on the embedding. For that purpose, we have to analyze the composition

$$(-3+2\lambda)(-5+2\lambda)(2+2\lambda)2\lambda D_4^T(\lambda)$$
$$= (-5+2\lambda)(2+2\lambda)C_\Sigma(\lambda-3)\{(-3+2\lambda)2\lambda\, C_\Sigma(\lambda-2)\Pi_\Sigma D_M(\lambda-1)\}D_M(\lambda)$$
$$= 3\lambda(\lambda-1)D_4^T(\lambda)$$

at $\lambda = \frac{1}{2}$, i.e.,

$$D_4^T\left(\frac{1}{2}\right) = 8C_\Sigma\left(-\frac{5}{2}\right)D_2^T\left(M, \Sigma; g; -\frac{1}{2}\right)D_M\left(\frac{1}{2}\right)$$
$$= 8C_\Sigma\left(-\frac{5}{2}\right)\left(\Delta_\Sigma^T - \frac{1}{2}J_\Sigma\right)\Pi_\Sigma D_M\left(\frac{1}{2}\right).$$

Here we have used

$$D_2^T(\lambda-1) = (2\lambda-3)(2\lambda)\frac{1}{\lambda}D_2^T(\lambda-1) = (2\lambda-3)2\lambda\frac{1}{\lambda}C_\Sigma(\lambda-2)\Pi_\Sigma D_M(\lambda-1)$$

and Theorem 6.24.2. Now we apply the explicit formulas

$$D_M\left(\frac{1}{2}\right)u = \begin{pmatrix} \frac{1}{3}\left(\Delta_M + \frac{1}{2}J_M\right)u \\ d_M u \\ u \end{pmatrix}, \; u \in C^\infty(M),$$

$$(-8)C_\Sigma\left(-\frac{5}{2}\right)u = -2u_- - 4\delta_\Sigma\omega - \Delta_\Sigma u_+ - \frac{1}{2}J_\Sigma u_+, \; u \in \Gamma(T^*\Sigma)$$

and (using $\text{Ric} - 2\mathsf{P} = -\mathsf{P} + J$)

$$\Delta_\Sigma^T - \frac{1}{2}J_\Sigma = \begin{pmatrix} \Delta_\Sigma - \frac{3}{2}J_\Sigma & -(d_\Sigma J_\Sigma, \cdot) + 2\delta_\Sigma(\mathsf{P}_\Sigma\#\cdot) & \frac{1}{2}|\mathsf{P}_\Sigma|^2 \\ -2d_\Sigma & \Delta_\Sigma - \mathsf{P}_\Sigma + \frac{1}{2}J_\Sigma & \frac{1}{2}d_\Sigma J_\Sigma + \mathsf{P}_\Sigma\#d_\Sigma \\ 6 & -4\delta_\Sigma & \Delta_\Sigma - \frac{3}{2}J_\Sigma \end{pmatrix}.$$

Hence

$$\Pi_\Sigma D_M\left(\frac{1}{2}\right)u = i^* \begin{pmatrix} \frac{1}{3}\left(\Delta_M + \frac{1}{2}J_M\right)u - HNu + \frac{1}{4}H^2u \\ d_\Sigma u \\ u \end{pmatrix}.$$

Now observe that

$$\begin{pmatrix} * \\ 0 \\ 0 \end{pmatrix} \in \ker\left(C_\Sigma\left(-\frac{5}{2}\right)\left(\Delta_\Sigma^{\mathcal{T}} - \frac{1}{2}J_\Sigma\right)\right).$$ (6.22.4)

In fact, the operator in (6.22.4) is of the form

$$\left(-2\left(\Delta_\Sigma - \frac{3}{2}J_\Sigma\right) + 8\delta_\Sigma d_\Sigma - 6\left(\Delta_\Sigma + \frac{1}{2}J_\Sigma\right), *, *\right) = (0, *, *).$$

It follows that $D_4^T(\frac{1}{2})$ lives on Σ. It remains to calculate the composition of

$$-\begin{pmatrix} * & -(dJ, \cdot) + 2\delta(\mathsf{P}\#\cdot) & \frac{1}{2}|\mathsf{P}|^2 \\ * & \Delta - \mathsf{P} + \frac{1}{2}J & \frac{1}{2}dJ + \mathsf{P}\#d \\ * & -4\delta & \Delta - \frac{3}{2}J \end{pmatrix}\begin{pmatrix} 0 \\ du \\ u \end{pmatrix}, \; u \in C^\infty(\Sigma)$$

with

$$u \mapsto -2u_- - 4\delta\omega - \left(\Delta + \frac{1}{2}J\right)u_+, \; u \in \Gamma(\mathcal{T}^*\Sigma);$$

here all constructions are with respect to Σ. The above matrix product yields

$$\begin{pmatrix} -(dJ, du) + 2\delta(\mathsf{P}\#du) + \frac{1}{2}|\mathsf{P}|^2 u \\ \Delta du + \frac{1}{2}Jdu + \frac{1}{2}u(dJ) \\ -3\Delta u - \frac{3}{2}Ju \end{pmatrix}.$$

Thus we find

$$2(dJ, du) - 4\delta(\mathsf{P}\#du) - |\mathsf{P}|^2 u$$

$$- 4\delta(\Delta du) - 2\delta(udJ) - 2\delta(Jdu) + \left(\Delta + \frac{1}{2}J\right)\left(3\Delta + \frac{3}{2}J\right)u$$

$$= -\Delta^2 u - 4\delta(\mathsf{P}\#du) + (du, dJ) + J\Delta u - |\mathsf{P}|^2 u + \frac{3}{4}J^2 u - \frac{1}{2}u\Delta J$$

$$= -\Delta^2 u + \delta(Jg - 4\mathsf{P})\#du + \frac{1}{2}\left(\frac{3}{2}J^2 - 2|\mathsf{P}|^2 - \Delta J\right)u$$

$$= -P_4 u$$

by (4.1.7). The proof is complete.   □

Remark 6.22.3. *The extrinsic construction of the Paneitz operator $P_4(\Sigma^n; g)$ in Theorem 6.22.2 is an analog of the intrinsic construction*

$$C_\Sigma\left(-\frac{n}{2}-1\right) \circ \left(\Delta_\Sigma^{\mathcal{T}} - \left(\frac{n}{2}-1\right)J_\Sigma\right) \circ D_\Sigma\left(-\frac{n}{2}+2\right) : C^\infty(\Sigma) \to C^\infty(\Sigma)$$

by curved translation (Corollary 6.20.3). The individual factors in the latter composition are given by the formulas

$$D_\Sigma\left(-\frac{n}{2}+2\right) = \begin{pmatrix} \frac{1}{2}\left(\Delta_\Sigma - \left(\frac{n}{2}-2\right)\mathsf{J}_\Sigma\right) \\ d_\Sigma u \\ -(n-4)u \end{pmatrix},$$

$$C_\Sigma\left(-\left(\frac{n}{2}+1\right)\right) \sim 2(n-4)u_- - 4\delta\omega - \left(\Delta_\Sigma - \left(\frac{n}{2}-2\right)\mathsf{J}_\Sigma\right)u_+$$

and

$$\Delta_\Sigma^T - \left(\frac{n}{2}-1\right)\mathsf{J}_\Sigma = \begin{pmatrix} \Delta_\Sigma - \frac{n}{2}\mathsf{J}_\Sigma & -(d_\Sigma\mathsf{J}_\Sigma, \cdot) + 2\delta_\Sigma(\mathsf{P}_\Sigma\#\cdot) & \frac{1}{2}|\mathsf{P}_\Sigma|^2 \\ -2d_\Sigma & \Delta_\Sigma + \left(\frac{n}{2}-2\right)(2\mathsf{P}_\Sigma - \mathsf{J}_\Sigma) & \frac{1}{2}d_\Sigma\mathsf{J}_\Sigma + \mathsf{P}_\Sigma\#d_\Sigma \\ 2n & -4\delta_\Sigma & \Delta_\Sigma - \frac{n}{2}\mathsf{J}_\Sigma. \end{pmatrix}$$

Now a calculation (see the proof of Theorem 6.20.7) shows that the composition is a multiple of $(n-4)P_4$ on Σ^n. In particular, it vanishes on Σ^4. In contrast, the operator $D_4^T(\cdot; \Sigma^4; 0)$ does not vanish and reproduces $P_4 i^*$ (together with a Paneitz-type operator of the embedding). Finally, note that the composition

$$\left(\Delta_\Sigma - \left(\frac{n}{2}-1\right)\mathsf{J}_\Sigma\right)\Pi_\Sigma D_M\left(-\left(\frac{n}{2}-2\right)\right)$$

depends on the embedding.

Next, we consider the Q-curvatures $Q_2^T(\cdot, \Sigma^2; g)$ and $Q_4^T(\cdot, \Sigma^4; g)$ more closely. Theorem 6.24.1 implies that for $n \geq 3$,

$$D_2^T(\cdot, \Sigma^n; dr^2 + h_r; \lambda) = D_2(\cdot, \Sigma^n; dr^2 + h_r; \lambda)$$

for the family $D_2(\lambda)$ constructed in Section 6.4. On the other hand, by Theorem 6.7.1, the residue of

$$\int_0^\varepsilon \int_\Sigma r^\lambda u\varphi \operatorname{vol}(dr^2 + h_r), \quad -\Delta_{r^{-2}(dr^2 + h_r)}u = \mu(n-\mu)u$$

at $\lambda = -\mu - 3$ yields the family

$$\delta_2(h; \lambda) = -\frac{1}{2(n-2+2\lambda)}D_2(dr^2 + h_r; \lambda).$$

Renormalization as in (6.6.7) gives the residue family

$$D_2^{\mathrm{res}}(h; \lambda) = -2(n-2+2\lambda)\delta_2(h; \lambda) = D_2(dr^2 + h_r; \lambda),$$

i.e.,

$$D_2^{\mathrm{res}}(\Sigma; h; \lambda) = D_2^T(M, \Sigma; dr^2 + h_r; \lambda). \tag{6.22.5}$$

Note that

$$D_2\left(dr^2+h_r; -\frac{n}{2}+1\right) = \left(\Delta - \left(\frac{n}{2}-1\right)\mathsf{J}\right) i^* = P_2(h)i^*.$$

This is the non-critical case of the holographic duality for order 2 families with Σ of dimension $n \geq 3$. For $n = 2$, the tractor family $D_2^T(\lambda)$ is not defined but the above arguments show that in this case the family $D_2(\lambda)$ from Section 6.3 should be used as a substitute.

Example 6.22.1. *For $n = 2$, we have $D_2^T(\cdot, \Sigma^2; g; \lambda) = D_2(\cdot, \Sigma^2; g; \lambda)$ by (6.22.1). Hence*

$$\dot{D}_2^T(\cdot, \Sigma^2; g; 0)(1) = \dot{D}_2(\cdot, \Sigma^2; g; 0)(1) = Q_2(\Sigma^2; g)$$

by Corollary 6.3.1. In particular, $Q_2^e(\cdot, \Sigma^2; g) = 0$ for all metrics g.

Example 6.22.2. *Let $n = 4$. We determine the tractor Q-polynomial*

$$Q_4^T(M, \Sigma^4; g; \lambda)$$

(see Definition 6.21.5) if $M = (0, \varepsilon) \times \Sigma^4$ and $g = dr^2 + (h + r^2 h_{(2)} + r^4 h_{(4)})$ are the first terms in the Taylor series of the Poincaré-Einstein metric $r^{-2}(dr^2 + h_r)$ associated to h. The fact that $h_{(4)}$ is not fully determined by $h = h_{(0)}$ will have no influence. In fact, $\operatorname{tr}(h_{(4)})$ is determined, and this is the only information on $h_{(4)}$ which enters into $Q_4^T(g; \lambda)$. By Definition 6.21.1, we have

$$D_4^T(M^5, \Sigma^4; g; \lambda) = C_\Sigma(\lambda-3) \circ \{C_\Sigma(\lambda-2) \circ \Pi_\Sigma \circ D_M(\lambda-1)\} \circ D_M(\lambda).$$

Now $D_4^T(\cdot, \Sigma^4; g; \lambda)$ is determined by the factorization

$$3\frac{1}{(2\lambda-4)}\frac{1}{(2\lambda-2)}\frac{1}{(2\lambda+1)}\frac{1}{(2\lambda+3)}\lambda(\lambda+1)D_4^T(\cdot, \Sigma^4; g; \lambda)$$

(see Definition 6.21.3) of this family. We use the identity

$$C_\Sigma(\lambda-1)\Pi_\Sigma D_M(\lambda) = \frac{\lambda+2}{2\lambda(2\lambda+3)} D_2^T(M, \Sigma; g; \lambda)$$

with

$$D_2^T(M, \Sigma; g; \lambda) = -(2\lambda+2)i^*\left(\Delta_M^T - \frac{3}{2}\mathsf{J}_M\right) + (2\lambda+3)\left(\Delta_\Sigma^T - \mathsf{J}_\Sigma\right)i^*$$

(see (6.24.1)). Hence $6\lambda D_4^T(g; \lambda)$ is given by the composition of

$$\begin{pmatrix} u_- \\ \omega \\ u_+ \end{pmatrix} \mapsto -4\lambda(\lambda-1)u_- + 4(\lambda-1)\delta_\Sigma\omega - (\Delta_\Sigma-\lambda\mathsf{J}_\Sigma)u_+$$

and

$$-2\lambda i^* \left(\Delta_M^T - \frac{3}{2} J_M \right) + (2\lambda+1)\left(\Delta_\Sigma^T - J_\Sigma \right) i^*$$

acting on

$$\begin{pmatrix} (\Delta_M + \lambda J_M)u \\ (2\lambda+3)d_M u \\ 2\lambda(2\lambda+3)u \end{pmatrix}.$$

Thus $6\lambda D_4^T(g;\lambda)(1) = 6\lambda Q_4^T(g;\lambda)$ *is given by the application of*

$$-4\lambda(\lambda-1)u_- + 4(\lambda-1)\delta_\Sigma\omega - (\Delta_\Sigma - \lambda J_\Sigma)u_+$$

to

$$-2\lambda i^* \begin{pmatrix} \Delta_M - \frac{5}{2} J_M & * & \frac{1}{2}|P_M|^2 \\ -2d_M & * & \frac{1}{2}d_M J_M \\ 10 & * & \Delta_M - \frac{5}{2} J_M \end{pmatrix} \begin{pmatrix} \lambda J_M \\ 0 \\ 2\lambda(2\lambda+3) \end{pmatrix}$$
$$+ (2\lambda+1) \begin{pmatrix} \Delta_\Sigma - 2 J_\Sigma & * & \frac{1}{2}|P_\Sigma|^2 \\ -2d_\Sigma & * & \frac{1}{2}d_\Sigma J_\Sigma \\ 8 & * & \Delta_\Sigma - 2 J_\Sigma \end{pmatrix} \begin{pmatrix} \lambda i^* J_M \\ 0 \\ 2\lambda(2\lambda+3) \end{pmatrix}.$$

Here we have used formula (6.22.2) for the tractor Laplacian. Thus we find that $6Q_4^T(g;\lambda)$ *is given by*

$$-4\lambda(\lambda-1)\left\{ (-2\lambda)i^* \left[(\Delta_M - \frac{5}{2} J_M)J_M + (2\lambda+3)|P_M|^2 \right] \right.$$
$$\left. + (2\lambda+1)\left[(\Delta_\Sigma - 2J_\Sigma)i^* J_M + (2\lambda+3)|P_\Sigma|^2 \right] \right\}$$
$$+ 4(\lambda-1)\delta_\Sigma \left\{ -2\lambda(2\lambda+1)i^* d_M J_M + (2\lambda+1)(-2d_\Sigma(i^* J_M) + (2\lambda+3)d_\Sigma J_\Sigma) \right\}$$
$$- (\Delta_\Sigma - \lambda J_\Sigma)\left\{ 10\lambda(2\lambda+1)i^* J_M + (2\lambda+1)(8i^* J_M - 4(2\lambda+3)J_\Sigma) \right\}.$$

Now using $i^* J_M = J_\Sigma$ *(see Lemma 6.11.1), the latter sum simplifies to*

$$-4\lambda(\lambda-1)\left\{ (-2\lambda)i^* \left[(\Delta_M - \frac{5}{2} J_M)J_M + (2\lambda+3)|P_M|^2 \right] \right.$$
$$\left. + (2\lambda+1)\left[(\Delta_\Sigma - 2J_\Sigma)J_\Sigma + (2\lambda+3)|P_\Sigma|^2 \right] \right\}$$
$$+ 4(\lambda-1)(2\lambda+1)\Delta_\Sigma J_\Sigma - 2(2\lambda+1)(\lambda-2)(\Delta_\Sigma - \lambda J_\Sigma)J_\Sigma.$$

Hence the coefficients of λ^0 *and* λ^1 *are*

$$-4\Delta_\Sigma J_\Sigma + 4\Delta_\Sigma J_\Sigma = 0$$

and

$$4\left\{(\Delta_\Sigma-2J_\Sigma)J_\Sigma+3|P_\Sigma|^2\right\}-4\Delta_\Sigma J_\Sigma+6\Delta_\Sigma J_\Sigma-4J_\Sigma^2$$
$$=12|P_\Sigma|^2-12J_\Sigma^2+6\Delta_\Sigma J_\Sigma=-6Q_4(\Sigma;h),$$

respectively. The latter results mean that

$$Q_4^T(g;0)=0\quad and\quad -\dot{Q}_4^T(g;0)=Q_4(\Sigma;h).$$

We continue with the calculation of the remaining coefficients of $6Q_4^T(g;\lambda)$. First of all, the coefficient of λ^4 is

$$16|P_M|^2-16|P_\Sigma|^2=0$$

by Lemma 6.11.2. For the coefficients of λ^2 and λ^3, we find (using Lemma 6.11.2) the respective formulas

$$-8\ddot{J}_M+6J_\Sigma^2-4|P_\Sigma|^2$$

and

$$8(\ddot{J}_M-|P_\Sigma|^2),$$

where

$$\ddot{J}_M\overset{\text{def}}{=}(\partial/\partial r)^2\big|_{r=0}(J_M).$$

We omit the details. Hence we obtain the formula

$$-6Q_4^T(g;\lambda)=6\lambda Q_4(h)+\lambda^2\left(8\ddot{J}_M-6J_\Sigma^2+4|P_\Sigma|^2\right)+\lambda^3\left(-8\ddot{J}_M+8|P_\Sigma|^2\right)$$
$$=6Q_4^{\text{res}}(h;\lambda)+8\lambda^2(\lambda-1)\left(|P_\Sigma|^2-\ddot{J}_M\right),\tag{6.22.6}$$

where $Q_4^{\text{res}}(h;\lambda)$ is the Q-polynomial given by (6.6.44).

In order to complete the evaluation of (6.22.6), we apply Lemma 6.11.1. Thus we have completed the proof of

Theorem 6.22.3. *In terms of the notation in Example 6.22.2,*

$$-Q_4^T(g;\lambda)=Q_4^{\text{res}}(h;\lambda).$$

In other words, for the background metric $g=dr^2+h_r$ which corresponds to a Poincaré-Einstein metric $r^{-2}(dr^2+h_r)$, the tractor Q-polynomial $Q_4^T(g;\lambda)$ is given by the Q-polynomial $Q_4^{\text{res}}(h;\lambda)$ (see (6.6.44)).

The following result shows that for the theory of Q_4 the trivial embedding is less natural than the one given by the Poincaré-Einstein extension.

Lemma 6.22.1. *In the situation of Example 6.22.2, i.e., $n = 4$ and the background metric $g = dr^2 + h_r$ so that $r^{-2}(dr^2 + h_r)$ is Einstein, we have $Q_4^e = 0$, i.e., for such a background metric the Q-curvature $Q_4^T(g)$ only depends on the Q-curvature of h. On the other hand, for the trivial embedding, i.e., for $g = dr^2 + h$ on $\mathbb{R} \times \Sigma$, we have*

$$Q_4^e = -\Delta \mathsf{J}.$$

Proof. The first assertion is contained in Theorem 6.22.3. For the convenience of the reader, we repeat the direct argument. In fact, since $i^*(\mathsf{P}_M) = \mathsf{P}_\Sigma$ and $H = 0$ (by Lemma 6.11.2) we find $\mathsf{J} = 0$, i.e., $Q_4^e = 0$. For the trivial embedding, we find $L = 0$, $\mathrm{Ric}_g = \begin{pmatrix} 0 & 0 \\ 0 & \mathrm{Ric}_h \end{pmatrix}$ and $\tau_g = \tau_h$. Therefore,

$$i^*(\mathsf{P}_M) = \frac{1}{3}\left(\mathrm{Ric}_h - \frac{\tau_h}{8}h\right), \quad \mathsf{P}_\Sigma = \frac{1}{2}\left(\mathrm{Ric}_h - \frac{\tau_h}{6}h\right),$$

i.e.,

$$\mathcal{J} = i^*(\mathsf{P}_M) - \mathsf{P}_\Sigma = -\frac{1}{6}\,\mathrm{Ric}_h + \frac{1}{24}\tau_h h = -\frac{1}{3}\mathsf{P}_\Sigma + \frac{1}{12}\mathsf{J}_h h.$$

Lemma 4.2.7 implies

$$\delta\mathcal{J} = -\frac{1}{3}d\mathsf{J}_h + \frac{1}{12}d\mathsf{J}_h = -\frac{1}{4}d\mathsf{J}_h.$$

But since $\mathrm{tr}(\mathcal{J}) = 0$, by Lemma 6.22.4, we find

$$Q_4^e = 4\delta(\delta\mathcal{J}) - 2\Delta(\mathrm{tr}\,\mathcal{J}) = -\Delta\mathsf{J}.$$

The proof is complete. $\qquad\square$

The following result establishes factorization identities for $D_4^T(dr^2 + h_r; \lambda)$ for general h.

Theorem 6.22.4. *In the situation of Example 6.22.2,*

$$D_4^T(dr^2 + h_r; 0) = P_4(h)i^*,$$

$$D_4^T\left(dr^2 + h_r; -\frac{3}{2}\right) = D_2\left(dr^2 + h_r; -\frac{7}{2}\right) \circ P_2(dr^2 + h_r),$$

$$D_4^T(dr^2 + h_r; 1) = P_2(h) \circ D_2(dr^2 + h_r; 1).$$

Proof. We begin with the proof of the first identity. We recall from the discussion in Example 6.22.2 that the projection Π_Σ is given by pull-back i^* and the family $6\lambda D_4^T(g; \lambda)$ is the composition of

$$u \mapsto -4\lambda(\lambda - 1)u_- + 4(\lambda - 1)\delta_\Sigma \omega - (\Delta_\Sigma - \lambda\mathsf{J}_\Sigma)u_+$$

and

$$-2\lambda i^*\left(\Delta_M^T - \frac{3}{2}\mathsf{J}_M\right) + (2\lambda + 1)\left(\Delta_\Sigma^T - \mathsf{J}_\Sigma\right)i^* \tag{6.22.7}$$

acting on

$$\begin{pmatrix} (\Delta_M+\lambda J_M)u \\ (2\lambda+3)d_M u \\ 2\lambda(2\lambda+3)u \end{pmatrix}.$$

We have to prove that the coefficient of λ in this decomposition coincides with $6P_4(h)$ (recall that, by Theorem 6.22.1, the coefficient of λ^0 vanishes). For that purpose, we write (6.22.7) in the explicit form

$$- 2\lambda i^* \begin{pmatrix} \Delta_M-\frac{5}{2}J_M & * & \frac{1}{2}|P_M|^2 \\ -2d_M & \Delta_M+P_M\#-\frac{1}{2}J_M & * \\ 10 & -4\delta_M & \Delta_M-\frac{5}{2}J_M \end{pmatrix} \begin{pmatrix} (\Delta_M+\lambda J_M)u \\ (2\lambda+3)d_M u \\ 2\lambda(2\lambda+3)u \end{pmatrix}$$

$$+ (2\lambda+1) \begin{pmatrix} \Delta_\Sigma-2J_\Sigma & -(d_\Sigma J_\Sigma,\cdot)+2\delta_\Sigma(P_\Sigma\#\cdot) & \frac{1}{2}|P_\Sigma|^2 \\ -2d_\Sigma & \Delta_\Sigma & \frac{1}{2}d_\Sigma J_\Sigma+P_\Sigma\#d_\Sigma \\ 8 & -4\delta_\Sigma & \Delta_\Sigma-J_\Sigma \end{pmatrix}$$

$$i^* \begin{pmatrix} (\Delta_M+\lambda J_M)u \\ (2\lambda+3)d_M u \\ 2\lambda(2\lambda+3)u \end{pmatrix}.$$

Here we have used the relations

$$(\mathrm{Ric}_M-2P_M)\#-\frac{3}{2}J_M = P_M\#-\frac{1}{2}J_M \quad \text{and} \quad (\mathrm{Ric}_\Sigma-2P_\Sigma)-J_\Sigma = P_\Sigma$$

and formula (6.22.2) for the tractor Laplacian on page 419. For the coefficient of λ, we obtain the sum of

$$4\left(\Delta_\Sigma-2J_\Sigma, -(d_\Sigma J_\Sigma,\cdot)+2\delta_\Sigma(P_\Sigma\#\cdot), *\right) i^* \begin{pmatrix} \Delta_M u \\ 3d_M u \\ 0 \end{pmatrix}$$

$$+ 8\left(-2\Delta_\Sigma, \delta_\Sigma i^*(\Delta_M+P_M\#-\frac{1}{2}J_M), *\right) \begin{pmatrix} \Delta_M u \\ 3d_M u \\ 0 \end{pmatrix} \quad (6.22.8)$$

and the coefficient of λ in

$$4(\lambda-1)(2\lambda+1)\left(-2\Delta_\Sigma(\Delta_M+\lambda J_M)u\right.$$

$$+ (2\lambda+3)\delta_\Sigma\Delta_\Sigma d_\Sigma u + 2\lambda(2\lambda+3)\delta_\Sigma\left(\frac{1}{2}(d_\Sigma J_\Sigma)u + P_\Sigma\#d_\Sigma u\right)\Big)$$

$$- (\Delta_\Sigma-\lambda J_\Sigma)2\lambda\left[-10(\Delta_M+\lambda J_M)u + 4(2\lambda+3)\Delta_M u - 2\lambda(2\lambda+3)\left(\Delta_M-\frac{5}{2}J_M\right)u\right]$$

$$- (\Delta_\Sigma-\lambda J_\Sigma)(2\lambda+1)\left[8(\Delta_M+\lambda J_M)u - 4(2\lambda+3)\Delta_\Sigma u + 2\lambda(2\lambda+3)\left(\Delta_\Sigma-2J_\Sigma\right)u\right].$$

The latter sum yields the contributions

$$8\Delta_\Sigma\Delta_M u + 8\Delta_\Sigma(J_M u) - 20\Delta_\Sigma^2 u - 24\left(\frac{1}{2}\delta_\Sigma((d_\Sigma J_\Sigma)u) + \delta_\Sigma(P_\Sigma\#d_\Sigma u)\right)$$

$$+ 20\Delta_\Sigma\Delta_M u - 24\Delta_\Sigma\Delta_M u$$

$$- \Delta_\Sigma\left[16\Delta_M u + 8J_M u - 32\Delta_\Sigma u + 6(\Delta_\Sigma - \lambda J_\Sigma)u\right] + J_\Sigma(8\Delta_M u - 12\Delta_\Sigma u)$$

$$= -12\Delta_\Sigma\Delta_M u + 6\Delta_\Sigma^2 u - 24\delta_\Sigma(P_\Sigma\#d_\Sigma u)$$

$$- 12(\Delta_\Sigma J_\Sigma)u - 12(d_\Sigma J_\Sigma, du) + 12\Delta_\Sigma(J_\Sigma u) - 12J_\Sigma\Delta_\Sigma u + 8J_\Sigma\Delta_M u.$$

The contributions in (6.22.8) are

$$4\Delta_\Sigma\Delta_M u - 8J_\Sigma\Delta_M u + 12(-(d_\Sigma J_\Sigma, d_\Sigma u) + \delta_\Sigma(P_\Sigma\#d_\Sigma u))$$

$$- 16\Delta_\Sigma\Delta_M u + 24\Delta_\Sigma\Delta_M u + 24\delta_\Sigma\left(P_M - \frac{1}{2}J_M\right)d_M u.$$

Hence we find

$$6\Delta_\Sigma^2 u + 24\delta_\Sigma(P_\Sigma\#d_\Sigma u) - 12\delta_\Sigma(J_\Sigma d_\Sigma u)$$

$$= 6\left(\Delta_\Sigma^2 u - \delta_\Sigma(2J_\Sigma - 4P_\Sigma\#)d_\Sigma u\right) = 6P_4 u.$$

This proves the first relation. For the proof of the second factorization, we observe that the formulas in Example 6.22.2 show that $(-9)D_4^T(-\frac{3}{2})u$ is given by the composition of

$$\begin{pmatrix} u_- \\ \omega \\ u_+ \end{pmatrix} \mapsto -15u_- - 10\delta_\Sigma\omega - \left(\Delta_\Sigma + \frac{3}{2}J_\Sigma\right)u_+$$

and

$$3i^* \begin{pmatrix} \Delta_M - \frac{5}{2}J_M & * & * \\ -2d_M & * & * \\ 10 & * & * \end{pmatrix} - 2 \begin{pmatrix} \Delta_\Sigma - 2J_\Sigma & * & * \\ -2d_\Sigma & * & * \\ 8 & * & * \end{pmatrix} i^*$$

acting on

$$\begin{pmatrix} (\Delta_M - \frac{3}{2}J_M)u \\ 0 \\ 0 \end{pmatrix} = \begin{pmatrix} P_2(dr^2 + h_r)u \\ 0 \\ 0 \end{pmatrix}.$$

In order to verify the explicit formula, we observe that

$$-15\left(3i^*\left(\Delta_M - \frac{5}{2}J_M\right) - 2(\Delta_\Sigma - 2J_\Sigma)i^*\right) - 10\delta_\Sigma(-6i^*d_M + 4d_\Sigma i^*)$$

$$- 14\left(\Delta_\Sigma + \frac{3}{2}J_\Sigma\right) = (-9)\left(5i^*\frac{\partial^2}{\partial r^2} + \left(\Delta_\Sigma - \frac{7}{2}J_\Sigma\right)i^*\right).$$

Similarly, it follows that $6D_4^T(1)u$ is given by $-(\Delta_\Sigma - J_\Sigma) = -P_2(h)$ applied to the last component of

$$-2i^* \left(\Delta_M^T - \frac{3}{2} J_M \right) + 3 \left(\Delta_\Sigma^T - J_\Sigma \right) i^*$$

acting on

$$\begin{pmatrix} (\Delta_M + J_M)u \\ 5d_M u \\ 10u \end{pmatrix}.$$

A calculation shows that this component is given by

$$-6 \left(-4 \frac{\partial^2}{\partial r^2} + (\Delta_\Sigma + J_\Sigma) \right) = -6D_2(1).$$

The proof is complete.   □

        Moreover, we have

Theorem 6.22.5. *In the situation of Example 6.22.2, the family* $D_4^T(dr^2 + h_r; \lambda)$ *is polynomial of degree 2.*

Proof. $6\lambda D_4^T(dr^2 + h_r; \lambda)$ is given by the composition of

$$\begin{pmatrix} u_- \\ \omega \\ u_+ \end{pmatrix} \mapsto -4\lambda(\lambda-1)u_- + 4(\lambda-1)\delta_\Sigma\omega - (\Delta_\Sigma - \lambda J_\Sigma)u_+$$

and

$$-2\lambda i^* \begin{pmatrix} \Delta_M - \frac{5}{2}J_M & * & \frac{1}{2}|P_M|^2 \\ -2d_M & * & \frac{1}{2}d_M J_M \\ 10 & * & \Delta_M - \frac{5}{2}J_M \end{pmatrix} + (2\lambda+1) \begin{pmatrix} \Delta_\Sigma - 2J_\Sigma & * & \frac{1}{2}|P_\Sigma|^2 \\ -2d_\Sigma & * & \frac{1}{2}d_\Sigma J_\Sigma \\ 8 & * & \Delta_\Sigma - 2J_\Sigma \end{pmatrix} i^*$$

acting on

$$\begin{pmatrix} (\Delta_M + \lambda J_M)u \\ (2\lambda+3)d_M u \\ 2\lambda(2\lambda+3)u \end{pmatrix}.$$

Thus $D_4^T(\lambda)$ is polynomial of degree ≤ 4. For the coefficients of λ^5 and λ^4 in the product $6\lambda D_4^T(\lambda)u$, we find the respective formulas

$$4 \left(i^*(|P_M|^2 u) - |P_\Sigma|^2 i^*(u) \right) 4 = 0$$

and

$$- 2|\mathsf{P}_\Sigma|^2 4 i^*(u)$$

$$+ \mathsf{J}_\Sigma \left(-2i^* \left(\Delta_M - \frac{5}{2}\mathsf{J}_M \right) u + 2(\Delta_\Sigma - 2\mathsf{J}_\Sigma) i^*(u) \right) 4$$

$$- 4 \left(-2i^* \left(\Delta_M - \frac{5}{2}\mathsf{J}_M \right)(\mathsf{J}_M u) + 2(\Delta_\Sigma - 2\mathsf{J}_\Sigma) i^*(\mathsf{J}_M u) \right)$$

$$= -8|\mathsf{P}_\Sigma|^2 i^*(u) - 8 \left(\mathsf{J}_\Sigma i^* \frac{\partial^2}{\partial r^2} - \frac{1}{2}\mathsf{J}_\Sigma^2 i^* \right) u$$

$$+ 8 \left(i^* \Delta_M (\mathsf{J}_M) u - \Delta_\Sigma (\mathsf{J}_\Sigma) i^*(u) + \mathsf{J}_\Sigma i^*(\Delta_M u) - \mathsf{J}_\Sigma (\Delta_\Sigma i^*(u)) - \frac{1}{2}\mathsf{J}_\Sigma^2 i^*(u) \right)$$

$$= -8|\mathsf{P}_\Sigma|^2 i^*(u) - 8 \left(\mathsf{J}_\Sigma i^* \frac{\partial^2}{\partial r^2} - \frac{1}{2}\mathsf{J}_\Sigma^2 i^* \right) u + 8 \left(|\mathsf{P}_\Sigma|^2 i^* + \mathsf{J}_\Sigma i^* \frac{\partial^2}{\partial r^2} - \frac{1}{2}\mathsf{J}_\Sigma^2 i^* \right) u$$

$$= 0$$

using Lemma 6.11.2 and Lemma 6.11.1. $\qquad\square$

As a corollary of Theorem 6.22.5, we find

Theorem 6.22.6. *Let* $n = 4$, h *any metric and* $r^{-2}(dr^2 + h_r)$ *the (approximate) Poincaré-Einstein metric with conformal infinity* h:

$$h_r = h + r^2 h_{(2)} + r^4 h_{(4)}.$$

Then

$$D_4^T(dr^2 + h_r; \lambda) = D_4^{\mathrm{res}}(h; \lambda).$$

Proof. The tractor family $D_4^T(dr^2 + h_r; \lambda)$ is polynomial of degree 2 (Theorem 6.22.5). Hence it is completely determined by the three factorization identities in Theorem 6.22.4. On the other hand, $D_4^{\mathrm{res}}(h; \lambda)$ satisfies an analogous system of three factorization identities. The structure of these identities and the relation

$$D_2^{\mathrm{res}}(h; \lambda) = D_2^T(dr^2 + h_r; \lambda)$$

(see (6.22.5)) imply the assertion. $\qquad\square$

Theorem 6.22.4 implies that $D_4^T(dr^2 + h_r; 0)(1) = 0$. Although a formula for $D_4^T(g; 0)$ for general g is rather complicated, it is easy to prove the analogous vanishing result.

Lemma 6.22.2. $D_4^T(\cdot, \Sigma^4; g; 0)(1) = 0$.

Proof. We note that the quantity $D_4^T(\cdot; \Sigma^4; g; 0)(1)$ is given by

$$C_\Sigma(-3)(\Delta_\Sigma^T - \mathsf{J}_\Sigma)\Pi_\Sigma \begin{pmatrix} \mathsf{J}_M \\ 0 \\ 6 \end{pmatrix}.$$

But for $u = \begin{pmatrix} u_- \\ \omega \\ u_+ \end{pmatrix} \in \Gamma(\mathcal{T}^*(\Sigma))$ $C_\Sigma(-3)u$ is a multiple of $-4\delta_\Sigma\omega - \Delta_\Sigma u_+$. Moreover,

$$(\Delta_\Sigma^T - J_\Sigma) \begin{pmatrix} i^*J_M + \frac{3}{2}H^2 \\ 0 \\ 6 \end{pmatrix} = \begin{pmatrix} * \\ -2d_\Sigma(i^*J_M + \frac{3}{2}H^2) + 3d_\Sigma J_\Sigma \\ 8i^*J_M + 12H^2 - 12J_\Sigma \end{pmatrix}$$

using (6.22.2). Hence the desired composition is

$$8\Delta_\Sigma(i^*J_M + \frac{3}{2}H^2) - 12\Delta_\Sigma J_\Sigma - 8\Delta_\Sigma(i^*J_M) - 12\Delta_\Sigma H^2 + 12\Delta_\Sigma J_\Sigma = 0.$$

The proof is complete.   □

For a general background metric g, the extrinsic contribution $P_4^e(\cdot, \Sigma^4; g)$
does not vanish. However, it is not so easy to see that it is a differential operator
which is tangential to Σ, i.e., lives on Σ. In fact, we recall that translation yields
the rational family

$$\frac{2}{\lambda}\left[(\Delta_4^-)^2 - L_4\right] + \left[\frac{1}{3}(\Delta_4^-)^2 + \frac{2}{3}L_4 - 4\left(\Delta_4^-(Y_5^-)^2 - \sum_i Y_i^-(Y_5^-)^2 Y_i^-\right)\right] + \cdots$$

(see (6.18.16)). Assuming that the residue $(\Delta_4^-)^2 - L_4$ induces $Res_0(D_4^T) = 0$
(Theorem 6.22.1) (see also Remark 6.22.2), it is natural to expect that

$$(\Delta_4^-)^2 - 4\left(\Delta_4^-(Y_5^-)^2 - \sum_i Y_i^-(Y_5^-)^2 Y_i^-\right) \qquad (6.22.9)$$

induces $D_4^T(0)$. Since the second sum contains two normal derivatives $(Y_5^-)^2$, it
seems to be natural to expect that the extrinsic part of $D_4^T(0)$ contains (at most)
two normal derivatives. But note that

$$\Delta_4^-(Y_5^-)^2 - \sum_i Y_i^-(Y_5^-)^2 Y_i^- = \sum_i Y_i^-\left([Y_i^-, Y_5^-]Y_5^- + Y_5^-[Y_i^-, Y_5^-]\right).$$

This formula suggests that we should expect (at most) one normal derivative
(combined with tangential derivatives and curvature). We will see later that the
correct operator actually is tangential, i.e., of normal order 0.

Next, we discuss the decomposition

$$-Q_4^T = Q_4 + Q_4^e$$

under the additional assumption $H = 0$. In particular, the following result gives an
explicit formula for the extrinsic Q_4-curvature of (X^5, M^4, g) if $H = 0$. It implies
that $D_4^T(g; 0)$, acting on $\ker i^*\nabla_N$, is $tangential$ to Σ if $H = 0$ (Theorem 6.22.8),
i.e., $D_4^T(g; 0)$ has normal order ≤ 1.

Theorem 6.22.7. *We consider $\Sigma^4 \hookrightarrow M^5$ with the background metric g. Assume that $H = 0$. Then $Q_4^T(g) = \dot{D}_4^T(g;0)1$ is given by*

$$-Q_4^T(g) = Q_4(i^*(g)) + Q_4^e(g),$$

where

$$Q_4^e = 4\delta(\delta\theta) - 2\Delta_\Sigma(\operatorname{tr}\theta) = 4\delta(\delta\theta) - \frac{2}{3}\Delta_\Sigma(\operatorname{tr}\wedge^2 L_0)$$

and

$$\theta \stackrel{\text{def}}{=} i^*(\mathsf{P}_M) - \mathsf{P}_\Sigma. \tag{6.22.10}$$

In particular, $Q_4^e = 0$ if $\theta = 0$.

Theorem 6.22.7 re-proves that $Q_4^e = 0$ in Example 6.22.2 since in that case $L = 0$ and $\theta = 0$. Theorem 6.22.7 confirms Conjecture 6.21.3 in the case $n = 4$ and $H = 0$.

In the case (S^5, S^4) with the round metric, we have $L = 0$, $\mathsf{P}_{S^5} = \frac{1}{2}g_{S^5}$ and $\mathsf{P}_{S^4} = \frac{1}{2}g_{S^4}$. Hence $\theta = 0$. Thus $Q_4^e = 0$ by Theorem 6.22.7. This corresponds to the observation that the family $D_4^c(\lambda) : C^\infty(S^5) \to C^\infty(S^4)$ calculates $Q_4(S^4; g_c)$ (see Lemma 6.1.2).

For the proof of Theorem 6.22.7, we write the Gauß equation in the form

Lemma 6.22.3. *For (M^{n+1}, Σ^n, g),*

$$i^* \mathsf{J}_M - \mathsf{J}_\Sigma = i^* \mathsf{P}_M(N, N) + \frac{1}{n-1}\operatorname{tr}(\wedge^2 L).$$

Proof. The Gauß equation (see (6.4.13))

$$\tau_M = \tau_\Sigma + 2\operatorname{Ric}_{NN} + \operatorname{tr}(L^2) - \operatorname{tr}(L)^2$$

is equivalent to

$$2n\mathsf{J}_M = 2(n-1)\mathsf{J}_\Sigma + 2\left((n-1)\mathsf{P}_M(N, N) + \mathsf{J}_M\right) + \operatorname{tr}(L^2) - \operatorname{tr}(L)^2,$$

i.e.,

$$2(n-1)\mathsf{J}_M = 2(n-1)\mathsf{J}_\Sigma + 2(n-1)\mathsf{P}_M(N, N) + 2\operatorname{tr}(\wedge^2 L).$$

This yields the assertion. □

We continue with the *proof of Theorem 6.22.7.*

Proof. The discussion in Example 6.22.2 shows that $6Q_4^T(g; \lambda)$ is given by the application of

$$\begin{pmatrix} u_- \\ \omega \\ u_+ \end{pmatrix} \mapsto -4\lambda(\lambda-1)u_- + 4(\lambda-1)\delta_\Sigma\omega - (\Delta_\Sigma - \lambda\mathsf{J}_\Sigma)u_+ \tag{6.22.11}$$

to

$$D_2^T(\cdot, \Sigma^4; g; \lambda-1)\begin{pmatrix} \mathsf{J}_M \\ 0 \\ 2\lambda(2\lambda+3) \end{pmatrix},$$

where $D_2^T(g;\lambda)$ is given by

$$2\lambda(2\lambda+3)C_\Sigma(\lambda-1)\Pi_\Sigma D_M(\lambda) = (\lambda+2)D_2^T(g;\lambda).$$

Now the explicit form of $D_2^T(g;\lambda)$ is more complicated than in Example 6.22.2. In fact, we find

$$2(\lambda+2)D_2^T(g;\lambda)u = 4\lambda(2\lambda+3)C_\Sigma(\lambda-1)\Pi_\Sigma D_M(\lambda)$$
$$= -4(\lambda+1)(\lambda+2)i^*(\Delta_M^T + \lambda J_M)u + 4(\lambda+1)(2\lambda+3)\delta_\Sigma^T i^* \nabla^{M,T} u$$
$$- 2\lambda(2\lambda+3)(\Delta_\Sigma^T - (\lambda+2)J_\Sigma)i^* u.$$

Notice that

$$D_2^T(g;-1) = \left(\Delta_\Sigma^T - J_\Sigma\right)i^* \quad \text{and} \quad D_2^T\left(g;-\tfrac{3}{2}\right) = i^*\left(\Delta_M^T - \tfrac{3}{2}J_M\right).$$

These results are special cases of Theorem 6.24.2 for $n = 5$, $H = 0$. Now we rewrite $D_2^T(g;\lambda-1)$ in the form

$$2(\lambda+1)D_2^T(g;\lambda-1) = -4\lambda(\lambda+1)i^*(\Delta_M^T + (\lambda-1)J_M)u$$
$$+ 4\lambda(2\lambda+1)\mathcal{R}u + 4\lambda(2\lambda+1)\Delta_\Sigma^T i^* u$$
$$- 2(\lambda-1)(2\lambda+1)(\Delta_\Sigma^T - (\lambda+1)J_\Sigma)i^* u$$

with

$$\mathcal{R} \stackrel{\text{def}}{=} \delta_\Sigma^T\left(i^* \nabla^{M,T} - \nabla^{\Sigma,T} i^*\right) : \Gamma(T^*M) \to \Gamma(T^*\Sigma).$$

We obtain the formula

$$D_2^T(g;\lambda-1)$$
$$= \left[-2\lambda\Delta_M^T u + (2\lambda+1)\Delta_\Sigma^T u + (\lambda-1)(-2\lambda J_M + (2\lambda+1)J_\Sigma)u\right] + \frac{2\lambda(2\lambda+1)}{\lambda+1}\mathcal{R}u$$
$$= \left[-2\lambda\left(\Delta_M^T + (\lambda-1)J_M\right) + (2\lambda+1)\left(\Delta_\Sigma^T + (\lambda-1)J_\Sigma\right)\right]u + \frac{2\lambda(2\lambda+1)}{\lambda+1}\mathcal{R}u.$$

In particular, the tractor family $D_2^T(g;\lambda)$ is only rational in λ if $\mathcal{R} \neq 0$. Now applying the explicit formulas for the tractor Laplacians, we find

$$D_2^T(g;\lambda-1)\begin{pmatrix} J_M \\ 0 \\ 2(2\lambda+3) \end{pmatrix} = \left[-2\lambda\begin{pmatrix} \Delta_M - J_M & * & \frac{1}{2}|P_M|^2 \\ -2d_M & * & \frac{1}{2}d_M J_M \\ 10 & * & \Delta_M - J_M \end{pmatrix}\right.$$
$$+ (2\lambda+1)\begin{pmatrix} \Delta_\Sigma - J_\Sigma & * & \frac{1}{2}|P_\Sigma|^2 \\ -2d_\Sigma & * & \frac{1}{2}d_\Sigma J_\Sigma \\ 8 & * & \Delta_\Sigma - J_\Sigma \end{pmatrix}\right]\begin{pmatrix} J_M \\ 0 \\ 2(2\lambda+3) \end{pmatrix}$$
$$+ (\lambda-1)(-2\lambda J_M + (2\lambda+1)J_\Sigma)\begin{pmatrix} J_M \\ 0 \\ 2(2\lambda+3) \end{pmatrix} + \frac{2\lambda(2\lambda+1)}{\lambda+1}\mathcal{R}\begin{pmatrix} J_M \\ 0 \\ 2(2\lambda+3) \end{pmatrix}.$$

Composing with (6.22.11) yields the sum of

$$-4\lambda(\lambda-1)\Big\{(-2\lambda)\left[(\Delta_M-\mathsf{J}_M)\mathsf{J}_M + (2\lambda+3)|\mathsf{P}_M|^2\right]$$
$$+ (2\lambda+1)\left[(\Delta_\Sigma-\mathsf{J}_\Sigma)\mathsf{J}_M + (2\lambda+3)|\mathsf{P}_\Sigma|^2\right] + (\lambda-1)(-2\lambda\mathsf{J}_M + (2\lambda+1)\mathsf{J}_\Sigma)\mathsf{J}_M\Big\}$$
$$+ 4(\lambda-1)\delta_\Sigma\big\{-2(2\lambda+1)(\lambda+1)d_\Sigma\mathsf{J}_M + (2\lambda+1)(2\lambda+3)d_\Sigma\mathsf{J}_\Sigma\big\}$$
$$- (\Delta_\Sigma-\lambda\mathsf{J}_\Sigma)\big\{-4(\lambda+1)(\lambda-2)(2\lambda+1)\mathsf{J}_M + 2(2\lambda+1)(\lambda-2)(2\lambda+3)\mathsf{J}_\Sigma\big\}$$

and the composition of (6.22.11) with

$$\frac{2\lambda(2\lambda+1)}{\lambda+1}\mathcal{R}\begin{pmatrix}\mathsf{J}_M\\0\\2(2\lambda+3)\end{pmatrix}.$$

We determine the coefficients of λ^0 and λ^1 in the above sum. For λ^0, we find immediately

$$8\Delta_\Sigma\mathsf{J}_M - 12\Delta_\Sigma\mathsf{J}_\Sigma - \Delta_\Sigma(8\mathsf{J}_M - 12\mathsf{J}_\Sigma) = 0.$$

This re-proves the special case $H = 0$ of Lemma 6.22.2. The coefficient of λ in the above sum is given by

$$4\left[(\Delta_\Sigma - \mathsf{J}_\Sigma)\mathsf{J}_M + 3|\mathsf{P}_\Sigma|^2 - \mathsf{J}_\Sigma\mathsf{J}_M\right]$$
$$+ 8\Delta_\Sigma\mathsf{J}_M + 8\Delta_\Sigma\mathsf{J}_M - 20\Delta_\Sigma\mathsf{J}_\Sigma - \Delta_\Sigma(20\mathsf{J}_M - 26\mathsf{J}_\Sigma) + \mathsf{J}_\Sigma(8\mathsf{J}_M - 12\mathsf{J}_\Sigma)$$
$$= 6\Delta_\Sigma\mathsf{J}_\Sigma + 12(|\mathsf{P}_\Sigma|^2 - \mathsf{J}_\Sigma^2)$$
$$= -6Q_4(\Sigma^4; g).$$

This yields the intrinsic Q_4-curvature of (Σ^4, g). Finally, we evaluate the extrinsic contribution of $\mathcal{R}$. It does not contribute to the coefficient of λ^0 and its contribution to the coefficient of λ^1 is

$$2\mathcal{R}\begin{pmatrix}\mathsf{J}_M\\0\\6\end{pmatrix} = 2\delta_\Sigma^T\left(i^*\nabla^{M,T} - \nabla^{\Sigma,T}i^*\right)\begin{pmatrix}\mathsf{J}_M\\0\\6\end{pmatrix}.$$

In order to make it explicit, we first note that for $X \in \mathcal{X}(\Sigma)$ and $u \in \Gamma(\mathcal{T}^*M)$,

$$i^*(\nabla_X^{M,T}u) - \nabla_X^{\Sigma,T}(i^*u)$$
$$= i^*\begin{pmatrix}\nabla_X^M & (\mathsf{P}_M(X),\cdot) & 0\\-X^\flat & \nabla_X^M & \frac{1}{2}\mathsf{P}_M(X)\\0 & -2\langle X,\cdot\rangle & \nabla_X^M\end{pmatrix}\begin{pmatrix}u_-\\\omega\\u_+\end{pmatrix}$$
$$- \begin{pmatrix}\nabla_X^\Sigma & (\mathsf{P}_\Sigma(X),\cdot) & 0\\-X^\flat & \nabla_X^\Sigma & \frac{1}{2}\mathsf{P}_\Sigma(X)\\0 & -2\langle X,\cdot\rangle & \nabla_X^\Sigma\end{pmatrix}i^*\begin{pmatrix}u_-\\\omega\\u_+\end{pmatrix}$$
$$= \begin{pmatrix}(\mathsf{P}_M(X),\omega) - (\mathsf{P}_\Sigma(X), i^*(\omega))\\i^*(\nabla_X^M\omega) - \nabla_X^\Sigma i^*(\omega) + \frac{1}{2}(i^*\mathsf{P}_M(X) - \mathsf{P}_\Sigma(X))u_+\\0\end{pmatrix}.$$

Hence if $\omega = 0$, then

$$i^*(\nabla_X^{M,\mathcal{T}} u) - \nabla_X^{\Sigma,\mathcal{T}}(i^* u) = \frac{1}{2} \begin{pmatrix} 0 \\ (i^*(\mathsf{P}_M(X)) - \mathsf{P}_\Sigma(X))u_+ \\ 0 \end{pmatrix}. \tag{6.22.12}$$

Now let

$$\theta = i^*\mathsf{P}_M - \mathsf{P}_\Sigma, \quad \theta(X) = i^*(\mathsf{P}_M(X)) - \mathsf{P}_\Sigma(X) \in \Omega^1(\Sigma), \; X \in \mathcal{X}(\Sigma).$$

For a geodesic frame $\{X_i\}$ of Σ, we find

$$2\mathcal{R} \begin{pmatrix} \mathsf{J}_M \\ 0 \\ 6 \end{pmatrix} = 6 \sum_i \nabla_{X_i}^{\mathcal{T}} \begin{pmatrix} 0 \\ \theta(X_i) \\ 0 \end{pmatrix}$$

$$= 6 \sum_i \begin{pmatrix} \nabla_{X_i}^{\Sigma} & (\mathsf{P}_\Sigma(X_i), \cdot) & 0 \\ -X_i^\flat & \nabla_{X_i}^{\Sigma} & \frac{1}{2}\mathsf{P}_\Sigma(X_i) \\ 0 & -2\langle X_i, \cdot\rangle & \nabla_{X_i}^{\Sigma} \end{pmatrix} \begin{pmatrix} 0 \\ \theta(X_i) \\ 0 \end{pmatrix}$$

$$= 6 \sum_i \begin{pmatrix} * \\ \nabla_{X_i}^{\Sigma} \theta(X_i) \\ -2\theta(X_i, X_i) \end{pmatrix}.$$

Thus composing with (6.22.11) (for $\lambda = 0$) we find the contribution

$$-24\delta_\Sigma \left(\sum_i \nabla_{X_i}^{\Sigma} \theta(X_i) \right) + 12\Delta_\Sigma(\operatorname{tr}\theta).$$

Hence

$$6Q_4^T(g;\lambda) = -6Q_4(\Sigma^4; g) - 24\delta_\Sigma \left(\sum_i \nabla_{X_i}^{\Sigma} \theta(X_i) \right) + 12\Delta_\Sigma(\operatorname{tr}\theta),$$

i.e.,

$$-Q_4^T(g;\lambda) = Q_4(\Sigma^4; g) + 4\delta_\Sigma(\delta\theta) - 2\Delta_\Sigma(\operatorname{tr}\theta).$$

Here we have used the fact that for the geodesic frame $\{X_i\}$,

$$\delta\theta(Y) = \sum_i \nabla_{X_i}(\theta)(X_i, Y) = \sum_i \langle \nabla_{X_i}(\theta(X_i)), Y\rangle.$$

Finally, we note that, by Lemma 6.22.3,

$$\operatorname{tr}\theta = \sum_{i=1}^{4} \mathsf{P}_M(X_i, X_i) - \mathsf{P}_\Sigma(X_i, X_i) = \mathsf{J}_M - \mathsf{J}_\Sigma - \mathsf{P}_M(N, N) = \frac{1}{3}\operatorname{tr}(\wedge^2 L_0).$$

The proof is complete. $\square$

Remark 6.22.4. *The proof of Theorem 6.22.7 also shows that the family $D_2^T(g; \lambda)$: $\Gamma(T^*M) \to \Gamma(T^*\Sigma)$ on tractors has a a simple pole at $\lambda = -2$ with residue $2\mathcal{R}$. In other words, the operator*

$$\mathcal{R} = \delta_\Sigma^T (i^* \nabla^{M,T} - \nabla^{\Sigma,T} i^*)$$

defines a conformally invariant operator on weighted tractors ($\lambda = -2$).

Note that, in the situation of Theorem 6.22.7, for closed Σ^4,

$$\int_\Sigma Q_4^e \, \mathrm{vol} = 0.$$

Next, we prove the fundamental identity for the extrinsic Q-curvature Q_4^e.

Theorem 6.22.8. *In the situation of Theorem 6.22.7, assume that $i^*\nabla_N\varphi = 0$. Then $\hat{H} = 0$ and*

$$e^{4\varphi}\hat{Q}_4^e = Q_4^e + P_4^e(\varphi),$$

where

$$P_4^e = -4\delta(\theta \# d), \quad \theta = i^*(\mathsf{P}_M) - \mathsf{P}_\Sigma.$$

The operator P_4^e satisfies

$$e^{4\varphi}P_4^e(\hat{g}) = P_4^e(g)$$

for all φ as above.

Note that the structure $P_4^e = -4\delta(\cdot)$ is suggested by (6.22.9).

Proof. Since $Q_4^e = 4\delta(\delta\theta) - 2\Delta(\mathrm{tr}\,\theta)$, we have to determine

$$e^{4\varphi}\hat{\delta}(\hat{\delta}\hat{\theta}) = \left(e^{4\varphi}\hat{\delta}e^{-2\varphi}\right)\left(e^{2\varphi}\hat{\delta}\hat{\theta}\right) \qquad (6.22.13)$$

and

$$e^{4\varphi}\hat{\Delta}(\hat{\mathrm{tr}}\hat{\theta}) = \left(e^{4\varphi}\hat{\Delta}e^{-2\varphi}\right)\left(e^{2\varphi}\hat{\mathrm{tr}}\hat{\theta}\right). \qquad (6.22.14)$$

Now (2.5.9) implies

$$(\hat{\theta} - \theta)(X, Y)$$
$$= -g(\nabla_X^M(\mathrm{grad}_M(\varphi)), Y) + g(\nabla_X^\Sigma(\mathrm{grad}_\Sigma(\varphi)), Y) - \frac{1}{2}(\nabla_N(\varphi))^2 g(X, Y)$$

for $X, Y \in \mathcal{X}(\Sigma)$. But

$$g(\nabla_X^M(\mathrm{grad}_M(\varphi)), Y) = g(\nabla_X^M(\mathrm{grad}_M^T(\varphi) + \mathrm{grad}_M^\perp(\varphi)), Y)$$
$$= g(\nabla_X^M(\mathrm{grad}_\Sigma(\varphi)), Y) + g(\nabla_X^M(\nabla_N(\varphi)N), Y)$$
$$= g(\nabla_X^\Sigma(\mathrm{grad}_\Sigma(\varphi)), Y) + \nabla_N(\varphi)L(X, Y).$$

Thus

$$\hat{\theta} - \theta = -\nabla_N(\varphi)L - \frac{1}{2}(\nabla_N(\varphi))^2 g.$$

In particular,

$$\hat{\theta} - \theta = 0$$

if $i^*\nabla_N\varphi = 0$. Moreover, we claim that for any symmetric bilinear form b,

$$e^{2\varphi}\hat{\delta}(b) = \delta(b) - b\#d\varphi - \text{tr}(b)d\varphi. \qquad (6.22.15)$$

Now (6.22.15) implies

$$e^{2\varphi}\hat{\delta}\hat{\theta} = \delta\theta - \theta\#d\varphi - \text{tr}(\theta)d\varphi.$$

Therefore, by Lemma 4.2.1, the right-hand side of (6.22.13) equals

$$\delta\left(\delta\theta - \theta\#d\varphi - \text{tr}(\theta)d\varphi\right).$$

Moreover, by Lemma 4.2.1, the right-hand side of (6.22.14) simplifies to

$$\Delta\,\text{tr}(\theta) - 2\delta(\text{tr}(\theta)d\varphi).$$

Hence

$$e^{4\varphi}\hat{Q}_4^e = Q_4^e - 4\delta(\theta\#d\varphi) - 4\delta(\text{tr}(\theta)d\varphi) + 4\delta(\text{tr}(\theta)d\varphi).$$

It remains to verify (6.22.15). But the transformation formula

$$\hat{\nabla}_X\omega = \nabla_X\omega - X(\varphi)\omega - \langle\omega, X\rangle d\varphi + (d\varphi, \omega)X^\flat, \quad \omega \in \Omega^1$$

implies

$$\begin{aligned}
\hat{\delta}(b) &= \sum_i \hat{\nabla}_{\hat{X}_i}(b(\hat{X}_i)) \\
&= e^{-2\varphi}\sum_i \left(\nabla_{X_i}(b(X_i)) - 2X_i(\varphi)b(X_i) - b(X_i, X_i)d\varphi + (b(X_i), d\varphi)X_i^\flat\right) \\
&= e^{-2\varphi}\left(\delta(b) - b\#d\varphi - \text{tr}(b)d\varphi\right).
\end{aligned}$$

Finally, using $\hat{\theta} = \theta$ and $e^{4\varphi}\hat{\delta}e^{-2\varphi} = \delta$ (Lemma 4.2.1), we find

$$e^{4\varphi}\hat{\delta}(\hat{\theta}\#du) = \delta e^{2\varphi}(\theta\#e^{-2\varphi}du) = \delta(\theta\#du).$$

This proves the conformal covariance of P_4^e. The proof is complete. $\qquad \square$

Corollary 6.22.1. *Let $H = 0$. Then*

$$D_4^T(g;0) = P_4(i^*(g))i^* + P_4^e(g)$$

on $\ker i^*\nabla_N$.

Proof. By the covariance of $D_4^T(g; \lambda)$ and Lemma 6.22.2 we have

$$e^{4\varphi} \dot{D}_4^T(\hat{g}; 0)(1) = \dot{D}_4^T(g; 0)(1) - D_4^T(g; 0)(\varphi),$$

i.e.,

$$e^{4\varphi} Q_4^T(\hat{g}) = Q_4^T(g) - D_4^T(g; 0)(\varphi).$$

Theorem 6.22.7 yields

$$\left[e^{4\varphi} Q_4(\hat{g}) - Q_4(g)\right] + \left[e^{4\varphi} Q_4^e(\hat{g}) - Q_4^e(g)\right] = D_4^T(g; 0)(\varphi).$$

Hence, by Theorem 6.22.8 (and the fundamental identity for Q_4),

$$P_4(g)(\varphi) + P_4^e(g)(\varphi) = D_4^T(g; 0)(\varphi)$$

for $\varphi \in \ker i^* \nabla_N$. The proof is complete. □

Theorem 6.22.7 and Theorem 6.22.8 have analogs without the assumption $H = 0$. In the following, we only prove an extension of Theorem 6.22.8 in the sense that we find a fundamental pair (P_4^e, Q_4^e). The detailed proof that these data actually are derived from $D_4^T(\lambda)$ will be given elsewhere.

Theorem 6.22.9. *The operator*

$$P_4^e(g) : C^\infty(\Sigma^4) \ni u \mapsto -4\delta\left(\mathcal{J} \# du\right) \in C^\infty(\Sigma^4)$$

with

$$\mathcal{J} = \theta + HL - \frac{1}{2}H^2 g = \theta + HL_0 + \frac{1}{2}H^2 g, \quad \theta = i^*(\mathsf{P}_M) - \mathsf{P}_\Sigma \qquad (6.22.16)$$

is conformally covariant

$$e^{4\varphi} P_4^e(\hat{g}) = P_4^e(g).$$

Examples 6.22.1.
(1) $H = 0$. Then $P_4^e(g)$ reduces to $\delta(-4\theta \# d)$. It vanishes iff $\theta = 0$.
(2) $L_0 = 0$. Under this conformally invariant condition $P_4^e(g)$ reduces to $\delta((-4\theta - 2H^2 g)\# d)$. It vanishes iff $\theta + \frac{1}{2}H^2 g = 0$.

Moreover, we have

Theorem 6.22.10. *Let*

$$Q_4^e = 4\delta(\delta\mathcal{J}) - 2\Delta(\mathrm{tr}(\mathcal{J})) \in C^\infty(\Sigma^4),$$

where $\mathcal{J}$ is defined by (6.22.16). Then

$$e^{4\varphi} Q_4^e(\hat{g}) = Q_4^e(g) + P_4^e(g)(\varphi).$$

We continue with the *proof of Theorem 6.22.9*.

Proof. The identity

$$\hat{\theta} = \theta - \nabla_N(\varphi)L - \frac{1}{2}(\nabla_N(\varphi))^2 g \qquad (6.22.17)$$

(see the proof of Theorem 6.22.8) and the transformation formulas

$$e^{\varphi}\hat{H} = H + \nabla_N(\varphi)$$

and $e^{-\varphi}\hat{L}_0 = L_0$ imply (using $\hat{\#} = e^{-2\varphi}\#$)

$$e^{4\varphi}\hat{P}_4^e(\hat{g}) = e^{4\varphi}\hat{\delta}\left(-4\hat{\mathcal{J}}\hat{\#}d\right) = \delta\Big(-4\big(\theta - \nabla_N(\varphi)L - \frac{1}{2}(\nabla_N(\varphi))^2 g\big)$$

$$- 4(H + \nabla_N(\varphi))L_0 - 2(H^2 + 2H\nabla_N(\varphi) + (\nabla_N(\varphi))^2)g\Big)\#d.$$

Now $L = L_0 + Hg$ implies that the latter term simplifies to

$$\delta\left(-4\theta - 4HL_0 - 2H^2 g\right)\#d = -4\delta(\mathcal{J}\#du).$$

The proof is complete. □

Note that the above proof also shows the conformal invariance of $\mathcal{J}$ in all dimensions. We emphasize that observation in the following result.

Theorem 6.22.11. *For any isometric embedding $i : (\Sigma^n, g) \hookrightarrow (X^{n+1}, g)$ $(n \geq 3)$, the symmetric bilinear form*

$$\mathcal{J} \stackrel{\text{def}}{=} \theta + HL - \frac{1}{2}H^2 g \in \Gamma(S^2 T^*\Sigma), \quad \theta = i^*(\mathsf{P}_M) - \mathsf{P}_\Sigma \qquad (6.22.18)$$

is conformally invariant: $\hat{\mathcal{J}} = \mathcal{J}$.

For the sake of convenience, we rewrite the argument.

Proof. We use the formula $\mathcal{J} = \theta + HL_0 + \frac{1}{2}H^2 g$. Then

$$\hat{\mathcal{J}} = \hat{\theta} + \hat{H}\hat{L}_0 + \frac{1}{2}\hat{H}^2\hat{g}$$

$$= \left[\theta - \nabla_N(\varphi)L - \frac{1}{2}(\nabla_N(\varphi))^2 g\right] + (H + \nabla_N(\varphi))L_0 + \frac{1}{2}(H + \nabla_N(\varphi))^2 g$$

$$= \left(\theta + HL_0 + \frac{1}{2}H^2 g\right) - \nabla_N(\varphi)L + \nabla_N(\varphi)L_0 + H\nabla_N(\varphi)g$$

$$= \mathcal{J}.$$

The proof is complete. □

Example 6.22.3. *We calculate the invariant $\mathcal{J}$ for $(X, \Sigma) = (\mathbb{B}^{n+1}, S^n)$ with the Euclidean metric which restricts to the round metric g_c. In view of $\mathsf{P}_X = 0$ and $\mathsf{P}_\Sigma = \frac{1}{2}g_c$, we find $\theta = -\frac{1}{2}g_c$. Now $L_0 = 0$ and $H = 1$ yields $\mathcal{J} = 0$. The situation is conformally equivalent to (S^{n+1}, S^n) with the round metric. In that case $\theta = 0$ and $L = 0$, i.e., $\mathcal{J} = 0$.*

Some comments are in order. Lemma 6.23.2 implies that for a totally umbilic hypersurface in a conformally flat background metric the $\mathcal{J}$-invariant vanishes. Example 6.22.3 is a very special case. The natural second-order differential operator $P_4^e(g)$ is *tangential* to Σ. But it is *not* a natural operator when regarded as a differential operator on Σ. In fact, its definition uses the pull-back $i^*(\mathsf{P}_M)$ of the Schouten tensor of the metric g in a neighborhood of Σ (containing the curvature R^M_{iNNj}), i.e., $P_4^e(g)$ is not determined by $i^*(g)$. Q_4^e is an exact divergence, i.e., its integral over a closed Σ vanishes. Although $\mathcal{J}$ is an invariant in all dimensions $n \geq 3$, the operator $P_4^e(g)$ is conformally covariant only for $n = 4$. $P_4^e(g)$ and $Q_4^e(g)$ vanish if $\mathcal{J}(g) \equiv 0$. This leads to the problem of characterizing the conformal submanifolds $\Sigma \hookrightarrow X$ with vanishing $\mathcal{J}$. More generally, it is interesting to characterize in geometric terms the vanishing of the pair $(P_4^e(g), Q_4^e(g))$.

Now we prove Theorem 6.22.10.

Proof. The proof is identical to that of Theorem 6.22.8 using the conformal invariance of $\mathcal{J}$ instead of θ. $\qquad\square$

The following lemma shows that it is of interest to consider the trace of the linear map $\mathcal{J}^\sharp$ on $T(\Sigma)$ corresponding to the $\mathcal{J}$-tensor. It implies that $\mathcal{J}$ is non-trivial.

Lemma 6.22.4. $\mathrm{tr}(\mathcal{J}^\sharp) = \frac{n}{2}(H^2 - \tau_e) = \frac{1}{2(n-1)}|L_0|^2 \geq 0$. *Equality holds true at umbilic points.*

Proof. As usual we shall use the same symbol for $\mathcal{J}$ and $\mathcal{J}^\sharp$. By definition of $\mathcal{J}$,

$$\mathrm{tr}(\mathcal{J}) = \mathrm{tr}(\theta) + \frac{n}{2}H^2.$$

But

$$\mathrm{tr}(\theta) = i^*(\mathsf{J}_M) - \mathsf{J}_\Sigma - i^*\mathsf{P}_M(N, N) = \frac{1}{2(n-1)}\left(\mathrm{tr}(L^2) - (\mathrm{tr}(L))^2\right)$$

by Lemma 6.22.3. Hence

$$\mathrm{tr}(\mathcal{J}) = \frac{1}{2(n-1)}\left(\mathrm{tr}(L^2) - nH^2\right)$$

$$= \frac{1}{2(n-1)}\left(\sum_{i=1}^{n}\lambda_i^2 - \frac{1}{n}\left(\sum_{i=1}^{n}\lambda_i\right)^2\right)$$

$$= \frac{1}{2n}\sum_{i=1}^{n}\lambda_i^2 - \frac{1}{n(n-1)}\sum_{i<j}\lambda_i\lambda_j.$$

Here λ_i are the eigenvalues of $L^\sharp$, i.e., the principal curvatures. On the other hand

$$H^2 - \tau_e = \frac{1}{n^2} \left(\sum_{i=1}^n \lambda_i \right)^2 - \frac{2}{n(n-1)} \sum_{i<j} \lambda_i \lambda_j$$

$$= \frac{1}{n^2} \sum_{i=1}^n \lambda_i^2 - \frac{2}{n} \frac{1}{n(n-1)} \sum_{i<j} \lambda_i \lambda_j.$$

This proves the first assertion. Now

$$\mathrm{tr}(\mathcal{J}) = \frac{1}{2(n-1)} \left(\mathrm{tr}(L^2) - nH^2 \right) = \frac{1}{2(n-1)} \mathrm{tr}(L_0^2) = \frac{1}{2(n-1)} |L_0|^2 \geq 0$$

completes the proof.   □

The conformal invariance of $\mathcal{J}$ and Lemma 6.22.4 imply the conformal invariance of the symmetric bilinear form $(H^2 - \tau_e)g$ ([76]) (see also (6.4.12)). For $n \geq 3$, the conformally invariant *Chen-Willmore functional*

$$\mathcal{CW} = \int_\Sigma (H^2 - \tau_e)^{\frac{n}{2}} \mathrm{vol}$$

is just a multiple of

$$\int_\Sigma \mathrm{tr}(\mathcal{J}^\sharp)^{\frac{n}{2}} \mathrm{vol}, \qquad\qquad\qquad (6.22.19)$$

i.e., a multiple of the functional

$$\int_\Sigma |L_0|^n \mathrm{vol}$$

(see [240], p. 256 and the references there describing the history of that functional). The critical points or the minima of $\mathcal{CW}$ are called Chen-Willmore submanifolds. Recall that for $n = 2$, $|L_0|^2 = H^2 - \tau_e = H^2 - K + \overline{K}$ (see page 197).

$\mathcal{J}$ is not well defined for $n = 2$ since P_Σ is not. But $(H^2 - \tau_e)\mathrm{vol}$ is well defined and reduces to the Willmore integrand

$$(H^2 - \lambda_1\lambda_2)\mathrm{vol} = (H^2 - K + \overline{K})\mathrm{vol} = \left(\frac{\lambda_1 - \lambda_2}{2} \right)^2 \mathrm{vol}$$

on the surface Σ (see (6.3.6)).

The conformal invariant $\mathcal{J}$ can be used to introduce an analog of the Chen-Willmore functional.

Theorem 6.22.12. *Let* $i : (\Sigma^n, g) \hookrightarrow (X^{n+1}, g)$ *be an isometric embedding* $(n \geq 3)$. *Then the integral*

$$\mathcal{W}(g) = \int_\Sigma |\mathcal{J}(g)|^{\frac{n}{2}} \mathrm{vol}(g) \geq 0$$

is conformally invariant. $\mathcal{W}(g)$ *vanishes iff* $\mathcal{J}(g) \equiv 0$.

In fact, $e^{-\varphi}\hat{L}_0 = L_0$ implies $e^{n\varphi}|\hat{L}_0|^n = |L_0|^n$ and $\hat{\mathcal{J}} = \mathcal{J}$ implies $e^{n\varphi}|\hat{\mathcal{J}}|^{\frac{n}{2}} = |\mathcal{J}|^{\frac{n}{2}}$. We recall that the vanishing of $\mathcal{J}(g)$ for $n = 4$ is equivalent to the vanishing of $P_4^e(g)$.

Of course, $\mathcal{W}$ gives rise to an $SO(1, n+2)$-invariant (i.e., Möbius-invariant) functional for embeddings $i : \Sigma^n \hookrightarrow S^{n+1}$ of Σ ($n \geq 3$) into the Möbius space (S^{n+1}, g_c).

Similarly, we can define other conformally invariant functionals in terms of the characteristic polynomials of the conformally invariant bilinear forms L_0 and $\mathcal{J}$. For instance,

$$\int_{\Sigma^n} \mathrm{tr}(\wedge^{\frac{n}{2}}\mathcal{J}^\sharp)\,\mathrm{vol} \tag{6.22.20}$$

is a conformal invariant.

Remark 6.22.5. *For an isometric embedding $i : \Sigma^m \hookrightarrow M^n$ of higher codimension, there is an analogous conformally invariant symmetric bilinear form $\mathcal{J}$. Let*

$$\mathcal{L}(X, Y) = -\nabla_X^M(Y)^\perp$$

be the $T(\Sigma)^\perp$-valued second fundamental form. In terms of an orthonormal basis $\{v_i\}$ of $T(\Sigma)^\perp$ we decompose $\mathcal{L} = \sum_i L_i v_i$ and define the mean curvature vector

$$\mathcal{H} = \frac{1}{m}\sum_i \mathrm{tr}(L_i)v_i.$$

Then

$$\mathcal{J} = i^*(\mathsf{P}_M) - \mathsf{P}_\Sigma + (\mathcal{L}, \mathcal{H}) - \frac{1}{2}|\mathcal{H}|^2 g$$

is conformally invariant.

The above results suggest a series of questions. Is P_n^e tangential for $n \geq 6$ and self-adjoint? Is Q_n^e an exact divergence for $n \geq 6$? Can one characterize the vanishing of P_n^e in terms of tensors? For $n = 4$, what is the geometric meaning of $\mathcal{J} \equiv 0$ and $Q_4^e(g) \equiv 0$? Can one prove the conformal invariance of $\mathcal{J}$ more directly by using conformally invariant calculus? Which symmetric bilinear forms arise as $\mathcal{J}$? Which functions arise as $Q_4^e(g)$? Analyze the critical points of the functional $\mathcal{W}$!

For the convenience of the reader, we *summarize* the results for $(M^5, \Sigma^4; g)$.

- Let $H = 0$. Then the pairs

$$(P_4, Q_4) = \left(\Delta^2 + \delta(2\mathsf{J}g - 4\mathsf{P})\#d, 2(\mathsf{J}^2 - |\mathsf{P}|^2) - \Delta\mathsf{J}\right)$$

and

$$(P_4^e, Q_4^e) = (\delta(-4\theta\#d), 4\delta(\delta\theta) - 2\Delta(\mathrm{tr}\,\theta))$$

satisfy the fundamental identities

$$e^{4\varphi}\hat{Q}_4 = Q_4 + P_4(\varphi) \quad \text{and} \quad e^{4\varphi}\hat{Q}_4^e = Q_4^e + P_4^e(\varphi)$$

if $\varphi \in \ker i^*\nabla_N$ (Theorem 6.22.8).

- For general g, the fundamental identity holds true for the pair

$$(P_4^e, Q_4^e) = (\delta(-4\mathcal{J}\#d), 4\delta(\delta\mathcal{J}) - 2\Delta(\operatorname{tr}\mathcal{J}))$$

(Theorem 6.22.10), where

$$\mathcal{J} = i^*(\mathsf{P}_M) - \mathsf{P}_\Sigma + HL - \frac{1}{2}H^2 g = i^*(\mathsf{P}_M) - \mathsf{P}_\Sigma + HL_0 + \frac{1}{2}H^2 g.$$

- The family $D_4^T(g; \lambda)$ is regular at $\lambda = 0$ (Theorem 6.22.1) and

$$\begin{aligned}
D_4^T(g; 0) &= P_4(i^*(g)) + P_4^e(g) &&\text{(on } \ker \nabla_N \text{ by Corollary 6.22.1)}\\
-Q_4^T(g) &= Q_4(i^*(g)) + Q_4^e(g) &&\text{(Theorem 6.22.7).}
\end{aligned}$$

- Moreover, for the metric $g = dr^2 + h_r$ (as in Example 6.22.2)

$$\begin{aligned}
D_4^T(dr^2 + h_r; 0) &= P_4(h)i^* &&\text{(Theorem 6.22.4)}\\
-Q_4^T(dr^2 + h_r) &= Q_4(h; \lambda) &&\text{(Theorem 6.22.3).}
\end{aligned}$$

In this case, the extrinsic pair (P_4^e, Q_4^e) vanishes ($\mathcal{J} = 0$).

In particular, the pair (P_4, Q_4) can be extracted naturally

- from the family $D_4^T(dr^2 + h_r; \lambda)$ and
- from the eigenfunctions of the Laplacian for the Poincaré-Einstein metric $r^{-2}(dr^2 + h_r)$.

For a general background metric g, the pair $(D_4^T(g; 0), \dot{D}_4^T(g; 0)(1))$ naturally decomposes as the sum of the *intrinsic* pair (P_4, Q_4) and the *extrinsic* pair (P_4^e, Q_4^e) of the embedding. For the latter construction of (P_4, Q_4), it is not necessary to solve the Einstein equation. In Section 6.23, we shall prove a formula for $\mathcal{J}$ in terms of L_0 and the Weyl-tensor of the background.

If Σ is the boundary of a compact manifold M, the families $D_N^T(M, \Sigma; g; \lambda)$ can be used to *define* conformally covariant boundary value problems for GJMS-operators on M. More precisely, for such purposes, it is convenient to consider *boundary value problems* for the closely related operators

$$\Box_{2N}^T \overset{\text{def}}{=} \underbrace{C_M\left(-\frac{n}{2} - N + 1\right) \circ \cdots \circ C_M\left(-\frac{n}{2} - 1\right)}_{N-1}$$

$$\circ \Box^T \circ \underbrace{D_M\left(-\frac{n}{2} + 2\right) \circ \cdots \circ D_M\left(-\frac{n}{2} + N\right)}_{N-1} : C^\infty(M) \to C^\infty(M),$$

where $\Box^T$ is a tractor version of the Yamabe operator (see Section 6.20). $\Box_{2N}^T$ is conformally covariant, i.e.,

$$e^{(\frac{n}{2} + N)\varphi} \circ \hat{\Box}_{2N}^T \circ e^{(-\frac{n}{2} + N)\varphi} = \Box_{2N}^T,$$

and is of the form $c\Delta_M^N + LOT$ for a non-vanishing constant c except for $2N \geq n$ (n even).

The operators $\square_{2N}^T$ were studied systematically in [116]. See also [109]. However, for even $n \geq 4$, the operator $\square_n^T$ does *not* yield a *non-trivial* operator of that form. In the simplest case $N = 2$ (Corollary 6.20.3),

$$C_M\left(-\frac{n}{2}-1\right) \circ \square_M^T \circ D_M\left(-\frac{n}{2}+2\right) \simeq (n-4)P_4(M).$$

Hence the composition degenerates to 0 for $n = 4$. For conformally flat metrics, the operator $\square_{2N}^T$ yields the GJMS-operator P_{2N} ([116]). It suffices to verify that on $\mathbb{R}^n$ with the Euclidean metric the operator $\square_{2N}^T$ coincides with a multiple of Δ_n^N. The latter fact is equivalent to the result that the Nth translate of Δ_n^- is a multiple of $(\Delta_n^-)^N$. This is a standard multiplicity one theorem. More precisely, we have the formula ([116])

$$c\prod_{j=2}^{N}(n-2j)(\Delta_n^-)^N, \ N \geq 2, \ c \neq 0.$$

The latter formula should be compared with

$$\pi\Theta_{2N}\left(-\frac{n}{2}+N\right) = c\prod_{j=1}^{N}(n-2j)(\Delta_n^-)^N, \ N \geq 1, \ c \neq 0$$

(see (6.18.25)).

In [40], the boundary operators $D_L^T(M, \Sigma; g; \lambda): C^\infty(M^n) \to C^\infty(\Sigma^{n-1})$ for $\lambda = -\frac{n-1}{2}+k$, $k \geq 1$ were used to construct conformally covariant self-adjoint *elliptic* boundary value problems for the operator $\square_{2N}^T$, $2N < n$ on M^n. In particular, such boundary value problems give rise to conformally covariant operators on the boundary Σ.

The observation that for even n, the tractor construction $\square_n^T$ on M^n *fails* to produce the critical GJMS-operator $P_n(M^n)$ can be seen as a reflection of the Eastwood-Slovak theorem that $(\Delta_n^-)^{\frac{n}{2}}$ does *not* lift to a homomorphism of semi-holonomic Verma modules ([92]). In the case $n = 4$, we have seen in Theorem 6.18.1 that semi-holonomic Zuckerman translation of Δ_n^- yields a multiple of $4L_{4,n} - n(\Delta_n^-)^2$. This is a lift of $(\Delta_n^-)^2$ only for $n \neq 4$ since it obviously projects to the holonomic $(n-4)(\Delta_4^-)^2$. Hence translation can't be used to lift $(\Delta_4^-)^2$, and actually do not exist by [92]. On the other hand, Theorem 6.18.5 yields a semi-holonomic family $\mathcal{D}_4^T(\lambda)$ which lifts the family $\mathcal{D}_4(\lambda)$. The fact that $\mathcal{D}_4(0) = (\Delta_4^-)^2$ and $\mathcal{D}_4^T(\lambda)$ has a pole at $\lambda = 0$ reflects the Eastwood-Slovak non-existence.

6.23 $\mathcal{J}$ and Fialkow's fundamental forms

In Section 6.21, we introduced the conformally invariant bilinear form

$$\mathcal{J} \in \Gamma(S^2(T^*\Sigma))$$

which is associated to any hypersurface $i : \Sigma \hookrightarrow X$ of dimension ≥ 3. In the present section, we discuss its relation to Fialkow's ([100]) fundamental forms and find an explicit formula for $\mathcal{J}$ in terms of L_0 and the Weyl tensor of the background.

In the fundamental work [100], Fialkow characterizes conformally equivalent non-umbilic submanifolds of conformally flat space in terms of *conformal fundamental forms* (see also [101] for the case of surfaces). For a modern presentation of the geometrical content of [101] we refer to [157]. See also [216] (Chapter 7) for the relation between Fialkow's invariants and Cartan geometry.

We describe the definition of the fundamental forms in the case of a hypersurface $\Sigma^n \hookrightarrow X^{n+1}$. Fialkow starts with the observation that

$$\mathcal{F}_I = |L_0|^2 g$$

is a conformally invariant bilinear form on Σ. In [100], $\mathcal{F}_I$ is called the *conformal measure tensor*. The *first* fundamental form $\mathcal{F}_I$ is non-degenerate except at the umbilic points of Σ. In particular, it induces a conformally invariant connection which, in turn, gives rise to conformally invariant curvature etc. The *second* fundamental form is the conformally invariant bilinear form $\mathcal{F}_{II} = |L_0|L_0$. Moreover, Fialkow introduces the *deviation tensor*

$$\mathcal{E} \stackrel{\text{def}}{=} i^*(\mathsf{P}_X) + HL_0 + \frac{1}{2}H^2 g - \left(\mathrm{Hess}(v) - dv \otimes dv + \frac{1}{2}|dv|^2 g \right), \qquad (6.23.1)$$

where

$$v \stackrel{\text{def}}{=} \log |L_0| \qquad\qquad\qquad (6.23.2)$$

(see formulas (6.12), (7.1) and (13.9) in [100]). Note that in [100] v has a different normalization and the curvature tensor has opposite sign.

$\mathcal{E}$ is conformally invariant, i.e., $\hat{\mathcal{E}} = \mathcal{E}$. Since

$$\mathcal{E} = \mathcal{J} + \mathsf{P}_\Sigma - \left(\mathrm{Hess}(v) - dv \otimes dv + \frac{1}{2}|dv|^2 g \right),$$

the conformal invariance of $\mathcal{E}$ is a consequence of the conformal invariance of $\mathcal{J}$ and the following observation.

Lemma 6.23.1. *For v as in* (6.23.2),

$$\left(\widehat{\mathrm{Hess}}(\hat{v}) - d\hat{v} \otimes d\hat{v} + \frac{1}{2}|d\hat{v}|^2 \hat{g} \right) - \left(\mathrm{Hess}(v) - dv \otimes dv + \frac{1}{2}|dv|^2 g \right)$$

$$= -\mathrm{Hess}(\varphi) + d\varphi \otimes d\varphi - \frac{1}{2}|d\varphi|^2 g = \hat{\mathsf{P}}_\Sigma - \mathsf{P}_\Sigma.$$

Proof. The transformation formulas

$$\widehat{\mathrm{Hess}}(u) = \mathrm{Hess}(u) - du \otimes d\varphi - d\varphi \otimes du + (du, d\varphi)g$$

for $u \in C^\infty(\Sigma)$ and $\hat{v} = v - \varphi$, imply the first identity. The second identity is (2.5.9). $\qquad\square$

An alternative proof of the conformal invariance of $\mathcal{E}$ using invariant calculus was given in [228].

The conformal invariance of $\mathcal{J}$ implies that the linear map $\mathcal{J}^\sharp$ transforms as $e^{2\varphi}\hat{\mathcal{J}}^\sharp = \mathcal{J}^\sharp$. Hence the traces $\sigma_p(\mathcal{J}^\sharp)$ of the induced maps

$$\wedge^p(\mathcal{J}^\sharp) : \wedge^p T(\Sigma) \to \wedge^p T(\Sigma)$$

transform according to

$$e^{2p\varphi}\sigma_p(\hat{\mathcal{J}}^\sharp) = \sigma_p(\mathcal{J}^\sharp).$$

In particular, the volume form

$$\mathrm{tr}(\wedge^{\frac{n}{2}}\mathcal{J}^\sharp)\,\mathrm{vol}$$

is conformally invariant and gives rise to the conformally invariant functional (6.22.20).

For a submanifold $\Sigma^n \hookrightarrow S^{n+1}$ of the Möbius space (conformal sphere), we express $\mathcal{J}$ in terms of L_0.

Lemma 6.23.2. *For a hypersurface $i : \Sigma^n \hookrightarrow S^{n+1}$ $(n \geq 3)$ of the sphere (S^{n+1}, g_c), the invariant form $\mathcal{J}$ is given by*

$$\mathcal{J} = \frac{1}{n-2}\left(L_0^2 - \frac{1}{2(n-1)}|L_0|^2 i^*(g_c)\right).$$

The form in brackets is conformally invariant for all $n \geq 2$.

Here L_0^2, by abuse of notation, denotes the bilinear form which corresponds to the operator $(L_0^\sharp)^2$. The form $|L_0|^2 g_c$ on Σ is Fialkow's first fundamental form. In the framework of Möbius geometry, it is sometimes called the Möbius-form ([142]).

Proof. We use the formulas $\mathrm{Ric}_{S^{n+1}} = n g_c$, $\mathsf{P}_{S^{n+1}} = \frac{1}{2}g_c$, $G = i^*(g_c) = g_\Sigma$ in the Gauß equations

$$i^*(\mathrm{Ric}_M) = \mathrm{Ric}_\Sigma + G + L^2 - nHL,$$

$$i^*(J_M) = J_\Sigma + \mathsf{P}_M(N, N) + \frac{1}{2(n-1)}\left(\mathrm{tr}(L^2) - (\mathrm{tr}(L))^2\right)$$

(see Lemma 6.22.3) for a hypersurface $\Sigma \hookrightarrow X$, and find

$$\mathrm{Ric}_\Sigma = (n-1)g_\Sigma - L^2 + nHL,$$

$$J_\Sigma = \frac{n}{2} - \frac{1}{2(n-1)}\left(\mathrm{tr}(L^2) - \mathrm{tr}(L)^2\right).$$

Hence

$$\mathcal{J} = i^*(\mathsf{P}_M) - \mathsf{P}_\Sigma + HL - \frac{1}{2}H^2 g_\Sigma$$

$$= \frac{1}{2}g_\Sigma - \frac{1}{n-2}(\mathrm{Ric}_\Sigma - \mathsf{J}_\Sigma g_\Sigma) + HL - \frac{1}{2}H^2 g_\Sigma$$

$$= \frac{1}{2}g_\Sigma - \frac{1}{n-2}\left[(n-1)g_\Sigma - L^2 + nHL - \left(\frac{n}{2} - \frac{1}{2(n-1)}\left(\mathrm{tr}(L^2) - \mathrm{tr}(L)^2\right)\right)g_\Sigma\right]$$

$$+ HL - \frac{1}{2}H^2 g_\Sigma$$

$$= \frac{1}{n-2}\left[L^2 - nHL - \frac{1}{2(n-1)}\left(\mathrm{tr}(L^2) - \mathrm{tr}(L)^2\right)g_\Sigma\right] + HL - \frac{1}{2}H^2 g_\Sigma.$$

Now we decompose $L = L_0 + Hg_\Sigma$ and find

$$\mathcal{J} = \frac{1}{n-2}\left(L_0^2 - \frac{1}{2(n-1)}|L_0|^2 g_\Sigma\right).$$

The proof is complete.   □

Corollary 6.23.1. *For a hypersurface $\Sigma^n \hookrightarrow S^{n+1}$, the invariant $\mathcal{J}$ vanishes precisely at umbilic points. Thus $\mathcal{J} \equiv 0$ iff Σ is a totally umbilic hypersurface.*

Proof. By Lemma 6.23.2, $\mathcal{J} = 0$ is equivalent to the relations

$$\mu_i^2 = \frac{1}{2(n-1)}\sum_i^n \mu_i^2$$

for the eigenvalues μ_i of $L_0^\sharp$. Summation implies $\sum_i \mu_i^2 = 0$, i.e., $\mu_i = 0$, i.e., $L_0 = 0$, i.e., $L = Hg$.   □

The following result extends Lemma 6.23.2 to the general case.

Lemma 6.23.3. *For a hypersurface $i : \Sigma^n \hookrightarrow X^{n+1}$ $(n \geq 3)$, the invariant form $\mathcal{J}$ is given by*

$$\mathcal{J}_{ij} = \frac{1}{n-2}\left((L_0^2)_{ij} - \frac{1}{2(n-1)}|L_0|^2 g_{ij} + \mathsf{C}_{iNNj}\right).$$

The form in brackets is conformally invariant for all $n \geq 2$.

Proof. Similarly as in the proof of Lemma 6.23.2, we have

$$\mathrm{Ric}_\Sigma = i^*(\mathrm{Ric}_M) - c_{23}(R_M) - L^2 + nHL,$$

$$\mathsf{J}_\Sigma = i^*(\mathsf{J}_M) - \mathsf{P}_M(N, N) - \frac{1}{2(n-1)}\left(\mathrm{tr}(L^2) - (\mathrm{tr}(L))^2\right)$$

by the Gauß equation, where $c_{23}(R)_{ij} = R_{iNNj}$. Hence

$$
\begin{aligned}
\mathcal{J} &= i^*(\mathsf{P}_M) - \mathsf{P}_\Sigma + HL - \frac{1}{2}H^2 g_\Sigma \\
&= i^*(\mathsf{P}_M) - \frac{1}{n-2}(\mathrm{Ric}_\Sigma - \mathsf{J}_\Sigma g_\Sigma) + HL - \frac{1}{2}H^2 g_\Sigma \\
&= i^*(\mathsf{P}_M) - \frac{1}{n-2}[i^*(\mathrm{Ric}_M) - i^*(\mathsf{J}_M)g_\Sigma] + \frac{1}{n-2}[c_{23}(R) - \mathsf{P}_M(N,N)g_\Sigma] \\
&\quad + \frac{1}{n-2}\left[\mathrm{tr}(L_0^2) - \frac{1}{2(n-1)}|L_0|^2 g_\Sigma\right] \\
&= i^*(\mathsf{P}_M) - \frac{n-1}{n-2}i^*(\mathsf{P}_M) + \frac{1}{n-2}[c_{23}(R) - \mathsf{P}_M(N,N)g_\Sigma] \\
&\quad + \frac{1}{n-2}\left[\mathrm{tr}(L_0^2) - \frac{1}{2(n-1)}|L_0|^2 g_\Sigma\right].
\end{aligned}
$$

But

$$
c_{23}(\mathsf{C})_{ij} = \mathsf{C}_{iNNj} = R_{iNNj} + (\mathsf{P} \oslash g)_{iNNj} = R_{iNNj} - \mathsf{P}_M(N,N)g_{ij} - \mathsf{P}_{ij}
$$

implies that $\mathcal{J}$ is given by

$$
\begin{aligned}
\frac{1}{n-2}[-i^*(\mathsf{P}_M) + c_{23}(R) - \mathsf{P}_M(N,N)g_\Sigma] &+ \frac{1}{n-2}\left(\mathrm{tr}(L_0^2) - \frac{1}{2(n-1)}|L_0|^2 g\right) \\
&= \frac{1}{n-2}\left(\mathrm{tr}(L_0^2) - \frac{1}{2(n-1)}|L_0|^2 g + c_{23}(\mathsf{C})\right).
\end{aligned}
$$

The proof is complete. □

Note that Lemma 6.23.3 re-proves the relation (Lemma 6.22.4)

$$
\mathrm{tr}(\mathcal{J}) = \frac{1}{2(n-1)}|L_0|^2. \tag{6.23.3}
$$

In view of the analogy between the symmetric tensor

$$
\frac{1}{n-2}\left(L_0^2 - \frac{1}{2(n-1)}|L_0|^2 g\right)
$$

and the Schouten tensor

$$
\frac{1}{n-2}\left(\mathrm{Ric} - \frac{1}{2(n-1)}\tau g\right),
$$

(6.23.3) should be viewed as an analog of $\mathrm{tr}(\mathsf{P}) = \frac{1}{2(n-1)}\tau$.

6.24 $D_2(g; \lambda)$ as a tractor family

The following result relates the tractor construction $D_2^T(M, \Sigma; g; \lambda)$ to the family $D_2(M, \Sigma; g; \lambda)$ discussed in Section 6.4. In particular, we obtain an alternative proof of the conformal covariance of that family.

Theorem 6.24.1. *For $n \geq 4$, let $\Sigma^{n-1} \subset M^n$ be an (oriented) codimension 1 submanifold. Then the family*

$$D_2^T(M, \Sigma; g; \lambda) = C_\Sigma(g; \lambda - 1) \circ \Pi_\Sigma \circ D_M(g; \lambda) : C^\infty(M) \to C^\infty(\Sigma)$$

has the form

$$\frac{1}{(n-5+2\lambda)(n-2+2\lambda)}(n-3+\lambda)D_2(M, \Sigma; g; \lambda).$$

We regard the composition in Theorem 6.24.1 as being induced by the family $\Theta_2(\lambda)$ (Theorem 6.18.2).

Proof. It follows immediately from the definitions that Π_Σ sends

$$(n-2+2\lambda)D_M(\lambda)u = \begin{pmatrix} (\Delta_M + \lambda \mathsf{J}_M)u \\ (n-2+2\lambda)du \\ 2\lambda(n-2+2\lambda)u \end{pmatrix}$$

to

$$\begin{pmatrix} (\Delta_M + \lambda \mathsf{J}_M)u + \frac{1}{2}(n-2+2\lambda)\lambda H^2 u - (n-2+2\lambda)HNu \\ (n-2+2\lambda)i^*du \\ 2\lambda(n-2+2\lambda)u \end{pmatrix}.$$

Hence the composition $2(n-5+2\lambda)(n-2+2\lambda)D_2^T(\lambda)$ is given by

$$u \mapsto -2\lambda(n-2+2\lambda)\left(\Delta_\Sigma - (n-3+\lambda)\mathsf{J}_\Sigma\right)u + 2(n-2+2\lambda)(n-3+2\lambda)\Delta_\Sigma u$$

$$- 2(n-3+2\lambda)(n-3+\lambda)\left[(\Delta_M + \lambda \mathsf{J}_M) + \frac{1}{2}\lambda(n-2+2\lambda)H^2 - (n-2+2\lambda)HN\right]u$$

$$= 2(n-2+2\lambda)(n-3+\lambda)\left(\Delta_\Sigma + \lambda \mathsf{J}_\Sigma\right)u - 2(n-3+2\lambda)(n-3+\lambda)\left(\Delta_M + \lambda \mathsf{J}_M\right)u$$

$$+ 2(n-3+2\lambda)(n-3+\lambda)\left(-\frac{1}{2}(n-2+2\lambda)\lambda H^2 + (n-2+2\lambda)HN\right)u.$$

It suffices to verify that the latter sum coincides with the product of $2(n-3+\lambda)$ with $D_2(M, \Sigma; g; \lambda)$. Now using $\mathsf{J}_M = Q_2(M)$ and $P_2(M) = \Delta_M - \frac{n-2}{2}Q_2(M)$, we find

$$- (n-3+2\lambda)(\Delta_M+\lambda J_M) + (n-2+2\lambda)(\Delta_\Sigma+\lambda J_\Sigma)$$

$$+ (n-3+2\lambda)(n-2+2\lambda)\left(HN - \lambda\frac{1}{2}H^2\right)$$

$$= -(n-3+2\lambda)P_2(M) + (n-2+2\lambda)P_2(\Sigma) + (n-3+2\lambda)(n-2+2\lambda)HN$$

$$+ 2\left(\lambda+\frac{n-2}{2}\right)\left(\lambda+\frac{n-3}{2}\right)(Q_2(\Sigma) - Q_2(M) - \lambda H^2).$$

The proof is complete. $\square$

In Section 6.21, we have used the following tractor generalization of Theorem 6.24.1. It can be regarded as the appropriate system of factorization identities.

Theorem 6.24.2. *Let M and Σ be as in Theorem 6.24.1. The conformally covariant family $D_2^T(M,\Sigma;g;\lambda)$ defined by*

$$(n-3+\lambda)D_2^T(M,\Sigma;g;\lambda)$$
$$= (n-5+2\lambda)(n-2+2\lambda)\{C_\Sigma(\lambda-1)\Pi_\Sigma D_M(\lambda)\} : \Gamma(\mathcal{T}^*M) \to \Gamma(\mathcal{T}^*\Sigma)$$

has the values

$$\Pi_\Sigma\left(\Delta_M^T - \frac{n-2}{2}J_M\right), \quad \left(\Delta_\Sigma^T - \frac{n-3}{2}J_\Sigma\right)\Pi_\Sigma$$

at $\lambda = -\frac{n-2}{2}$ and $\lambda = -\frac{n-3}{2}$, respectively.

Proof. For $\lambda = -\frac{n}{2}+1$ and $u \in \Gamma(\mathcal{T}^*M)$, the product $(n-2+2\lambda)D_M(\lambda)u$ equals

$$\begin{pmatrix}\left(\Delta_M^T - (\frac{n}{2}-1)J_M\right)u \\ 0 \\ 0\end{pmatrix}$$

(see (6.20.2)). Π_Σ maps this tractor into

$$\begin{pmatrix}\Pi_\Sigma\left(\Delta_M^T - (\frac{n}{2}-1)J_M\right)u \\ 0 \\ 0\end{pmatrix} \in \Gamma(\mathcal{T}^*\Sigma \otimes \mathcal{T}^*\Sigma)$$

(Theorem 6.20.6). Now, by (6.20.20), the product $(n-5+2\lambda)C_\Sigma(\lambda-1)$ sends the latter term into

$$- (n-3+2\lambda)(n-3+\lambda)\Pi_\Sigma\left(\Delta_M^T - \left(\frac{n}{2}-1\right)J_M\right)u$$

$$= (n-3+\lambda)\Pi_\Sigma\left(\Delta_M^T - \left(\frac{n}{2}-1\right)J_M\right)u.$$

This proves the first assertion. For the proof of the second one we notice that, by (6.20.20),

$$C_\Sigma\left(-\frac{n-1}{2}\right)u = \frac{1}{4}\left(\Delta_\Sigma^{\mathcal{T}} - \frac{n-3}{2}\mathsf{J}_\Sigma\right)u_+,$$

where $u_+ \in \Gamma(\mathcal{T}^*\Sigma)$ is the last component of $u \in \Gamma(\mathcal{T}^*\Sigma \otimes \mathcal{T}^*\Sigma)$. Now the last component of the composition $\Pi_\Sigma D_M(-\frac{n-3}{2})u$ is $-(n-3)\Pi_\Sigma u \in \Gamma(\mathcal{T}^*\Sigma)$. Hence we find the expression

$$-\frac{n-3}{4}\left(\Delta_\Sigma^{\mathcal{T}} - \frac{n-3}{2}\right)\Pi_\Sigma u$$

for the composition $C_\Sigma(-\frac{n-1}{2})\Pi_\Sigma D_M(-\frac{n-3}{2})u$, i.e.,

$$\frac{n-3}{2}D_2^{\mathcal{T}}\left(M, \Sigma; -\frac{n-3}{2}\right)$$
$$= (-2)C_\Sigma\left(-\frac{n-1}{2}\right)\Pi_\Sigma D_M\left(-\frac{n-3}{2}\right) = \frac{n-3}{2}\left(\Delta_\Sigma^{\mathcal{T}} - \frac{n-3}{2}\right)\Pi_\Sigma.$$

The proof is complete.   □

Although we do not discuss here an explicit formula for $D_2^{\mathcal{T}}(M, \Sigma; g; \lambda)$ for general g, one special case is not hard to get. We recall that for the Poincaré-Einstein metric $r^{-2}(dr^2 + h_r)$ on $M^n = (0, \varepsilon) \times \Sigma^{n-1}$, the family $D_2(dr^2 + h_r; \lambda)$ takes the form

$$-(n-3+2\lambda)i^*\left(\Delta_M - \frac{n-2}{2}\mathsf{J}_M\right) + (n-2+2\lambda)\left(\Delta_\Sigma - \frac{n-3}{2}\mathsf{J}_\Sigma\right)i^*$$

with $i^*\mathsf{J}_M = \mathsf{J}_\Sigma$. In particular, it is linear in λ. This result extends to the tractor generalization

$$D_2^{\mathcal{T}}(M, \Sigma; g; \lambda) : \Gamma(\mathcal{T}^*M) \to \Gamma(\mathcal{T}^*\Sigma)$$

defined in Theorem 6.24.2. In fact, by Lemma 6.11.2, the Schouten tensor of $dr^2 + h_r$ pulls back to the Schouten tensor of h. Hence the tractor connections are compatible under pull-back. These observation imply that

$$D_2^{\mathcal{T}}(M, \Sigma; g; \lambda)$$
$$= -(n-3+2\lambda)\Pi_\Sigma\left(\Delta_M^{\mathcal{T}} - \frac{n-2}{2}\mathsf{J}_M\right) + (n-2+2\lambda)\left(\Delta_\Sigma^{\mathcal{T}} - \frac{n-3}{2}\mathsf{J}_\Sigma\right)\Pi_\Sigma, \quad (6.24.1)$$

where Π_Σ is given by pull-back. In particular, the family is *linear* in λ.

The following result provides an alternative tractor construction of the second-order family $D_2(M, \Sigma; g; \lambda)$.

Theorem 6.24.3.

$$(n-2+2\lambda)D_2(M,\Sigma; g; \lambda)u$$
$$= -(D^{\mathcal{T}}(g; \lambda-1)D_M(g; \lambda)u, N^{\mathcal{T}}) - \lambda(n-1)\left(\lambda+\frac{n-2}{2}\right)\left(H^2 - \tau_e\right)u.$$

Note that $H^2 - \tau_e$ is a local conformal invariant, i.e., $e^{2\varphi}(\hat{H}^2 - \hat{\tau}_e) = H^2 - \tau_e$ (see Section 6.4).

Proof. We use the formula

$$D^{\mathcal{T}}(\lambda-1)C(\lambda)u =$$
$$\begin{pmatrix} (\nabla_N - (\lambda-1)H)(\Delta+\lambda\mathsf{J})u + (n-2+2\lambda)(\mathsf{P}(N), du) \\ -N^\flat(\Delta_M + \lambda\mathsf{J}_M)u + (n-2+2\lambda)(\nabla_N - (\lambda-1)H)\,du + \lambda(n-2+2\lambda)\mathsf{P}(N)u \\ -2(n-2+2\lambda)\nabla_N u + 2\lambda(n-2+2\lambda)(\nabla_N - (\lambda-1)H)u \end{pmatrix}.$$

Hence for the scalar product we obtain

$$(D^{\mathcal{T}}(\lambda-1)D_M(\lambda)u, N^{\mathcal{T}})$$
$$= -(\Delta_M + \lambda\mathsf{J}_M)u + (n-2+2\lambda)\left(\langle N, \nabla_N du\rangle - (\lambda-1)HNu\right) + \lambda(n-2+2\lambda)\mathsf{P}(N,N)u$$
$$+ H\left[(n-2+2\lambda)\nabla_N u - \lambda(n-2+2\lambda)(\nabla_N - (\lambda-1)H)u\right].$$

Now let N be the geodesic normal field. Then we find for the scalar product

$$-\Delta_M u + (n-2+2\lambda)N^2 u - 2(\lambda-1)(n-2+2\lambda)HNu$$
$$+ \left\{-\lambda\mathsf{J}_M + \lambda(n-2+2\lambda)\mathsf{P}(N,N) + \lambda(\lambda-1)(n-2+2\lambda)H^2\right\}u$$
$$= (n-3+2\lambda)N^2 u - \Delta_\Sigma u - [2(\lambda-1)(n-2+2\lambda)+n-1]HNu + \{\cdots\}$$
$$= (n-3+2\lambda)N^2 u - \Delta_\Sigma u - (n-3+2\lambda)(2\lambda-1)HNu + \{\cdots\}.$$

The latter sum can be reformulated as

$$(n-3+2\lambda)\Delta_M u - (n-2+2\lambda)\Delta_\Sigma u - (n-3+2\lambda)(n-2+2\lambda)HNu + \{\cdots\}$$
$$= (n-3+2\lambda)P_2(M)u - (n-2+2\lambda)P_2(\Sigma)u - (n-3+2\lambda)(n-2+2\lambda)HNu$$
$$+ (n-3+2\lambda)\frac{n-2}{2}\mathsf{J}_M u - (n-2+2\lambda)\frac{n-3}{2}\mathsf{J}_\Sigma u + \{\cdots\}.$$

That result differs from $-D_2(M,\Sigma; g; \lambda)$ by the zeroth-order operator

$$-\lambda\mathsf{J}_M + \lambda(n-2+2\lambda)\mathsf{P}(N,N) + \lambda(\lambda-1)(n-2+2\lambda)H^2$$
$$+\frac{n-2}{2}(n-3+2\lambda)\mathsf{J}_M u - \frac{n-3}{2}(n-2+2\lambda)\mathsf{J}_\Sigma u$$
$$+ 2\left(\lambda+\frac{n-2}{2}\right)\left(\lambda+\frac{n-3}{2}\right)\left(Q_2(M) - Q_2(\Sigma) - \lambda H^2\right).$$

The latter sum is a polynomial of order 3 in λ with absolute coefficient

$$\frac{n-2}{2}(n-3)\mathsf{J}_M - \frac{n-3}{2}(n-2)\mathsf{J}_\Sigma + \frac{(n-2)(n-3)}{2}\left(Q_2(M) - Q_2(\Sigma)\right) = 0$$

and leading coefficient 0. It remains to determine the respective coefficients α and β of λ^2 and λ. α and β satisfy the relations

$$
\begin{aligned}
- \mathsf{J}_M + (n-2)\mathsf{P}(N,N) &- (n-2)H^2 + (n-2)\mathsf{J}_M - (n-3)\mathsf{J}_\Sigma \\
&+ (2n-5)\left(Q_2(\Sigma) - Q_2(M)\right) + \frac{(n-2)(n-3)}{2}H^2 = \beta \quad (6.24.2)
\end{aligned}
$$

and

$$2\mathsf{P}(N,N) + (n-4)H^2 + 2\left(Q_2(\Sigma) - Q_2(M)\right) - (2n-5)H^2 = \alpha. \qquad (6.24.3)$$

But (6.24.2) is equivalent to

$$-(n-2)\mathsf{J}_M + (n-2)\mathsf{J}_\Sigma - \frac{(n-1)(n-2)}{2}H^2 + \mathrm{Ric}_N - \mathsf{J}_M = \beta,$$

i.e.,

$$-\frac{1}{2}\tau_M + \frac{1}{2}\tau_\Sigma + \mathrm{Ric}_N - \frac{(n-1)(n-2)}{2}H^2 = \beta,$$

i.e.,

$$\beta = -\frac{(n-1)(n-2)}{2}\left(H^2 - \tau_e\right)$$

using $\tau_M = \tau_\Sigma + 2\,\mathrm{Ric}_N - (n-1)(n-2)\tau_e$ (see Section 6.4). Similarly, we find that (6.24.3) is equivalent to

$$2\,\mathrm{Ric}_N - 2\mathsf{J}_M + \tau_\Sigma - 2(n-2)\mathsf{J}_M - (n-1)(n-2)H^2 = (n-2)\alpha,$$

i.e.,

$$2\,\mathrm{Ric}_N + \tau_\Sigma - \tau_M - (n-1)(n-2)H^2 = (n-2)\alpha,$$

i.e.,

$$(n-1)\left(\tau_e - H^2\right) = \alpha.$$

Hence the zeroth operator is given by

$$-(n-1)\left(\lambda^2 + \lambda\frac{n-2}{2}\right)\left(H^2 - \tau_e\right) = -\lambda(n-1)\left(\lambda + \frac{n-2}{2}\right)\left(H^2 - \tau_e\right).$$

The proof is complete $\square$

6.25 The family $D_3^T(M, \Sigma; g; \lambda)$

In the present section, we analyze the tractor family $D_3^T(\lambda)$ of order 3. The results will be used in Section 6.26 to relate this family to results of Chang and Qing [68].

The first result concerns the obstruction.

Theorem 6.25.1. *The family*

$$D_3^T(M, \Sigma; g; \lambda) \stackrel{\text{def}}{=} C_\Sigma(g; \lambda-2) \circ \Pi_\Sigma \circ D^T(g; \lambda-1) \circ D_M(g; \lambda) : C^\infty(M^4) \to C^\infty(\Sigma^3)$$

has the form

$$\frac{3}{(2\lambda-3)(2\lambda+2)} \lambda D_3^T(M, \Sigma; g; \lambda)$$

for a rational family $D_3^T(\cdot; \lambda)$ with a simple pole at $\lambda = 0$. Its residue is the operator

$$R \stackrel{\text{def}}{=} \frac{2}{3} \delta_\Sigma \left(L_0 \# d_\Sigma \right) i^* : C^\infty(M^4) \to C^\infty(\Sigma^3).$$

R is conformally covariant:

$$e^{3\varphi} R(e^{2\varphi} g) = R(g).$$

Since $D_3^T(M, \Sigma; g; \lambda)$ corresponds to $\mathcal{D}_3^T(\lambda)$ (Theorem 6.18.3), it is natural to regard the residue R as being induced by the residue (see (6.18.9))

$$\text{Res}_0(\mathcal{D}_3^T(\lambda)) = -\frac{2}{3} \sum_1^3 Y_i^- \left(Y_4^- Y_i^- - Y_i^- Y_4^- \right).$$

Proof. We first determine the composition $D^T(g; \lambda-1) \circ D_M(g; \lambda)$ for general n. The definitions give

$$(n-2+2\lambda) D^T(\lambda-1) D_M(\lambda) u = \left(\nabla_N^T - (\lambda-1)H \right) \begin{pmatrix} (\Delta_M + \lambda \mathsf{J}_M) u \\ (n-2+2\lambda) du \\ 2\lambda(n-2+2\lambda) u \end{pmatrix}$$

$$= \begin{pmatrix} \nabla_N - (\lambda-1)H & (\mathsf{P}(N), \cdot) & 0 \\ -N^\flat & \nabla_N - (\lambda-1)H & \frac{1}{2}\mathsf{P}(N) \\ 0 & -2\langle N, \cdot \rangle & \nabla_N - (\lambda-1)H \end{pmatrix} \begin{pmatrix} (\Delta_M + \lambda \mathsf{J}_M) u \\ (n-2+2\lambda) du \\ 2\lambda(n-2+2\lambda) u \end{pmatrix}$$

$$= \begin{pmatrix} (\nabla_N - (\lambda-1)H)(\Delta_M + \lambda \mathsf{J}_M) u + (n-2+2\lambda)(\mathsf{P}(N), du) \\ -N^\flat(\Delta_M + \lambda \mathsf{J}_M) u + (n-2+2\lambda)(\nabla_N - (\lambda-1)H) du + \lambda(n-2+2\lambda)\mathsf{P}(N) u \\ -2(n-2+2\lambda)\nabla_N u + 2\lambda(n-2+2\lambda)(\nabla_N - (\lambda-1)H) u \end{pmatrix}.$$

The second ingredient is a formula for the composition $C_\Sigma(g; \lambda) \circ \Pi_\Sigma$ for general n.

The definitions yield

$$
2(n-3+2\lambda)C_\Sigma(\lambda)\Pi_\Sigma
\begin{pmatrix} u_- \\ \omega \\ u_+ \end{pmatrix}
= 2(n-3+2\lambda)C_\Sigma(\lambda)
\begin{pmatrix} u_- - H\langle \omega, N\rangle + \frac{1}{4}H^2 u_+ \\ i^*(\omega) \\ u_+ \end{pmatrix}
$$

$$
= -2(n-1+2\lambda)(n-2+\lambda)\left\{ u_- - H\langle \omega, N\rangle + \frac{1}{4}H^2 u_+ \right\}
$$

$$
+ 2(n-1+2\lambda)\delta_\Sigma i^*(\omega) - \Delta_\Sigma u_+ + (n-2+\lambda)J_\Sigma u_+,
$$

i.e.,

$$
2(n-7+2\lambda)C_\Sigma(\lambda-2)\Pi_\Sigma
\begin{pmatrix} u_- \\ \omega \\ u_+ \end{pmatrix}
$$

$$
= -2(n-5+2\lambda)(n-4+\lambda)\left\{ u_- - H\langle \omega, N\rangle + \frac{1}{4}H^2 u_+ \right\}
$$

$$
+ 2(n-5+2\lambda)\delta_\Sigma i^*(\omega) - \Delta_\Sigma u_+ + (n-4+\lambda)J_\Sigma u_+.
$$

It follows that the composition $D_3^{\mathcal{T}}(\lambda)$ has the structure

$$
2(n-2+2\lambda)(n-7+2\lambda)D_3^{\mathcal{T}}(\lambda) = (n-4+\lambda)\{\cdots\}
$$
$$
+ 2(n-5+2\lambda)(n-2+2\lambda)\delta_\Sigma \left(i^* \left(\nabla_N - (\lambda-1)H\right) du + \lambda i^* \mathsf{P}(N)u \right)
$$
$$
- 2(\lambda-1)(n-2+2\lambda)\Delta_\Sigma i^*(\nabla_N - \lambda H)u.
$$

For $n = 4$, we find

$$
D_3^{\mathcal{T}}(\lambda) = \frac{\lambda}{4(2\lambda-3)(\lambda+1)}\{\cdots\}
$$
$$
+ \frac{2\lambda-1}{2\lambda-3}\delta_\Sigma \left(i^* \left(\nabla_N - (\lambda-1)H\right) du + \lambda i^* \mathsf{P}(N)u \right) - \frac{\lambda-1}{2\lambda-3}\Delta_\Sigma i^*(\nabla_N - \lambda H)u.
$$

Now write $D_3^{\mathcal{T}}(\lambda)$ in the form

$$
D_3^{\mathcal{T}}(\lambda) = \frac{3}{(2\lambda-3)(2\lambda+2)}\lambda D_3^{T}(\lambda)
$$

with a rational $D_3^{T}(\lambda)$. Then $D_3^{T}(\lambda)$ has a simple pole at $\lambda = 0$ with residue

$$
\frac{2}{3}\delta_\Sigma \left\{ d_\Sigma i^* \nabla_N - i^* \nabla_N d - H d_\Sigma i^* \right\}.
$$

In view of Lemma 6.25.1, the residue equals

$$
\frac{2}{3}\left[\delta_\Sigma(L\#d_\Sigma) - \delta_\Sigma(Hg\#d_\Sigma)\right]i^* = \frac{2}{3}\delta_\Sigma(L_0\#d_\Sigma)i^*.
$$

The proof is complete. $\square$

Lemma 6.25.1. $d_\Sigma i^* \nabla_N u - i^* \nabla_N du = L \# d_\Sigma i^* u.$

Proof. Let N be the unit geodesic normal and $X \in \mathcal{X}(\Sigma)$. X will also denote an arbitrary extension of X to a neighborhood of Σ. Then

$$\langle \nabla_N du, X \rangle = -\langle du, \nabla_N X \rangle + N \langle du, X \rangle$$
$$= -\langle du, \nabla_X N \rangle - \langle du, [N, X] \rangle + N \langle du, X \rangle$$
$$= -\langle du, \nabla_X N \rangle + \langle d(\nabla_N u), X \rangle.$$

Hence on Σ we have

$$\langle \nabla_N du, X \rangle - \langle d(\nabla_N u), X \rangle = -\langle du, \nabla_X N \rangle = -\langle d_\Sigma u, \nabla_X N \rangle$$
$$= -L(X, (d_\Sigma u)^\sharp) = -\langle L \# d_\Sigma u, X \rangle.$$

The proof is complete. $\qquad\square$

If the residue of $D_3^T(\lambda)$ at $\lambda = 0$ vanishes, then its value $P_3^T = D_3^T(0)$ and the curvature quantity $Q_3^T = \dot{D}_3^T(0)(1)$ are well defined. Next, we determine these data. We begin with Q_3^T.

Theorem 6.25.2. *Assume that* $n = 4$ *and* $L_0 = 0$. *Then for the critical Q-polynomial*

$$Q_{3,T}(g; \lambda) \overset{\text{def}}{=} \frac{3}{2} D_3^T(g; \lambda)(1)$$

we find $Q_{3,T}(g; 0) = 0$ *and*

$$Q_{3,T} \overset{\text{def}}{=} \dot{Q}_{3,T}(0) = \frac{1}{2} N J_M + J_M H + J_\Sigma H - HP(N, N) - \Delta_\Sigma H + \frac{1}{2} H^3.$$

$Q_{3,T}$ is a convenient renormalization of Q_3^T (Definition 6.21.4).

Proof. The formula in the proof of Theorem 6.25.1 implies

$$(2\lambda + 2) D^T(\lambda - 1) D_M(\lambda)(1) = \lambda \begin{pmatrix} \nabla_N - (\lambda - 1) H) J_M \\ -N^\flat J_M + 2(\lambda + 1) P(N) \\ -4(\lambda - 1)(\lambda + 1) H \end{pmatrix}$$

Now $2(2\lambda - 3) C_\Sigma(\lambda - 2) \Pi_\Sigma$ maps this tractor into the scalar

$$\lambda \Big\{ -2(2\lambda - 1)\lambda \left[(\nabla_N - (\lambda - 1)H) J_M - (\lambda^2 - 1)H^3 + J_M H - 2(\lambda + 1)HP(N, N) \right]$$
$$+ 4(2\lambda - 1)(\lambda + 1)\delta_\Sigma i^* P(N) + (\lambda^2 - 1)\Delta_\Sigma H - 4\lambda(\lambda^2 - 1) J_\Sigma H \Big\}.$$

It follows that $Q_{3,T}(0)$ is proportional to

$$-4\delta_\Sigma i^* P(N) - 4\Delta_\Sigma H.$$

By Lemma 6.25.2, that sum vanishes. This proves $Q_{3,T}(0) = 0$. Next, using the relation

$$2(2\lambda+2)(2\lambda-3)D_3(\lambda)(1) = 6\lambda D_3^T(\lambda)(1),$$

we read off $\frac{3}{2}\dot{D}_3^T(0)(1)$ from the above formula. We obtain

$$\frac{3}{2}\dot{D}_3^T(0)(1) = \frac{1}{4}\left\{2\left[(\nabla_N+H)\mathsf{J}_M+H^3+\mathsf{J}_MH-2H\mathsf{P}(N,N)\right]+4\delta_\Sigma i^*\mathsf{P}(N)+4\mathsf{J}_\Sigma H\right\}$$

$$= \frac{1}{2}N\mathsf{J}_M+\mathsf{J}_MH+\mathsf{J}_\Sigma H - H\mathsf{P}(N,N)+\delta_\Sigma i^*\mathsf{P}(N)+\frac{1}{2}H^3.$$

In order to complete the proof, we apply Lemma 6.25.2.   □

Lemma 6.25.2. $i^*\mathsf{P}(N) = \frac{1}{n-2}\delta(L_0) - d_\Sigma H$.

Proof. An application of the Codazzi-Mainardi equation

$$(R(X,Y)Z, N) = \nabla_Y(L)(X, Z) - \nabla_X(L)(Y, Z)$$

yields

$$(n-2)\langle i^*\mathsf{P}(N), Y\rangle = \mathrm{Ric}(N, Y)$$
$$= \sum_i (R(Y, e_i)e_i, N)$$
$$= \sum_i \nabla_{e_i}(L)(Y, e_i) - \nabla_Y(L)(e_i, e_i)$$
$$= \sum_i \nabla_{e_i}(L_0)(Y, e_i) + \sum_i \nabla_{e_i}(gH)(Y, e_i) - \nabla_Y(gH)(e_i, e_i)$$
$$= \delta(L_0)(Y) + \sum_i \nabla_{e_i}(H)g(Y, e_i) - \nabla_Y(H)\sum_i g(e_i, e_i)$$
$$= \delta(L_0)(Y) + \nabla_Y(H) - (n-1)\nabla_Y(H)$$
$$= \delta(L_0)(Y) - (n-2)\langle d_\Sigma H, Y\rangle.$$

The proof is complete.   □

Theorem 6.25.3. *Let $n = 4$. Assume that $L_0 = 0$. Then*

$$\frac{3}{2}D_3^T(0)$$

$$= \left\{\frac{1}{2}N\Delta_M+\Delta_\Sigma N\right\} + H\Delta_M - H\langle\nabla_N d\cdot, N\rangle + (\mathsf{P}(N), d\cdot) - \mathsf{J}_\Sigma N - \frac{3}{2}H^2Nu.$$

Proof. Similarly as in the proof of Theorem 6.25.1, we first find an explicit formula for the composition $2(n-2+2\lambda)(n-7+2\lambda)D_3^T(\lambda)$. It is the sum of

$$-2(n-5+2\lambda)(n-4+\lambda)\Big\{(\nabla_N-(\lambda-1)H)(\Delta_M+\lambda\mathsf{J}_M)u + (n-2+2\lambda)(\mathsf{P}(N), du)$$

$$+ \frac{n-2+2\lambda}{2}(\lambda-1)H^2(\nabla_N-\lambda H)u$$

$$+ H\big((\Delta_M+\lambda\mathsf{J}_M)u - (n-2+2\lambda)\langle\nabla_N-(\lambda-1)H)du, N\rangle$$

$$- \lambda(n-2+2\lambda)\mathsf{P}(N,N)u\big)\Big\} \tag{6.25.1}$$

and

$$2(n-5+2\lambda)(n-2+2\lambda)\delta_\Sigma\left\{i^*(\nabla_N-(\lambda-1)H)du + \lambda i^*\mathsf{P}(N)u\right\}$$

$$- 2(n-2+2\lambda)(\lambda-1)\Delta_\Sigma(\nabla_N-\lambda H)u \tag{6.25.2}$$

$$+ 2(n-4+\lambda)(n-2+2\lambda)(\lambda-1)\mathsf{J}_\Sigma(\nabla_N-\lambda H)u.$$

For $n = 4$, Theorem 6.25.1 shows that this is a polynomial in λ with vanishing absolute coefficient. We determine the coefficient of λ. The above formula implies that this coefficient is the sum of

$$2\left(\nabla_N\Delta_M u + 2(\mathsf{P}(N), du) - H^2\nabla_N u + 2H\Delta_M u - 2H((\nabla_N + H)du, N)\right)$$

$$= 2\nabla_N\Delta_M u + 4H\Delta_M u - 4H(((\nabla_N + H)du, N) + 4(\mathsf{P}(N), du) - 6H^2 Nu,$$

$$-4\left(-\delta_\Sigma(Hd_\Sigma u) + \delta_\Sigma(i^*\mathsf{P}(N)u)\right) + 4\delta_\Sigma(\nabla_N du + Hdu), \tag{6.25.3}$$

$-4\Delta_\Sigma(Hu)$ and $-4\mathsf{J}_\Sigma Nu$. Now by Lemma 6.25.2, the first two terms in (6.25.3) add up to $4\Delta_\Sigma(Hu)$. This sum cancels with $-4\Delta_\Sigma(Hu)$. But by Lemma 6.25.1, the third term in (6.25.3) simplifies to $4\Delta_\Sigma Nu$ using $L_0 = 0$. Thus we obtain

$$4\left(\frac{1}{2}N\Delta_M u + \Delta_\Sigma Nu + H\delta_M u - H(\nabla_N du, N) + (\mathsf{P}(N), du) - \mathsf{J}_\Sigma Nu - \frac{3}{2}H^2 Nu\right).$$

In view of the relation $2(2\lambda + 2)(2\lambda - 3)D_3^T(\lambda) = 6\lambda D_3^T(\lambda)$, the asserted formula for $\frac{3}{2}D_3^T(0)$ follows. $\qquad\square$

For general dimension, i.e., in the non-critical case $n \neq 4$, the situation is similar. Then we can write

$$D_3^T(M^n, \Sigma^{n-1}; g; \lambda) = (\cdots)(n-4+\lambda)D_3^T(M^n, \Sigma^{n-1}; g; \lambda)$$

for a rational family $D_3^T(M^n, \Sigma^{n-1}; g; \lambda)$ with a simple pole at $\lambda = -n + 4$ and a generally non-vanishing residue

$$\mathrm{Res}_{-n+4}(D_3^T(M, \Sigma; g; \lambda)) : C^\infty(M^n) \to C^\infty(\Sigma^{n-1}).$$

However, for $n \neq 4$, the residue is more complicated than for $n = 4$. In particular, we have (up to a constant multiple)

$$D_3^T \left(\cdot; -\frac{n}{2}+2 \right) = \mathrm{Res}_{-n+4}(D_3^T(\cdot; \lambda)) + \left(\frac{n}{2}-2 \right)(\cdots).$$

Grant ([129]) found a conformally covariant operator $\mathcal{L}_n$ of the form

$$\mathcal{L}_n = \mathrm{Res}_{-n+4}(D_3^T) + \left(\frac{n}{2}-2 \right)(\cdots).$$

$\mathcal{L}_n$ vanishes if $L_0 = 0$. It follows that

$$D_3^T \left(-\frac{n}{2}+2 \right) - \mathcal{L}_n = (n-4)P_3$$

for a conformally covariant third-order operator P_3 of the form

$$\left\{ \frac{1}{2}N\Delta_M + \Delta_\Sigma N \right\} + LOT.$$

P_3 acts on the same space as the Paneitz operator P_4 and defines a conformally covariant boundary condition for P_4. For $n = 4$ and $L_0 \neq 0$ the relation between Grant's P_3 and the Chang-Qing operator P_3 is not clear. However, for $n = 4$ and $L_0 = 0$, we will clarify the relation in Section 6.26.

An interesting observation is that both *factorization identities* for $\mathcal{D}_3(\lambda)$ and $\mathcal{D}_3^T(\lambda)$ continue to hold true for $D_3^T(\lambda)$. More precisely,

Lemma 6.25.3.

$$D_3^T(M, \Sigma; g; \lambda) = D_1(M, \Sigma; g; \lambda-2) \circ P_2(M, g) \quad for \ \lambda = -\frac{n-2}{2}$$

and

$$D_3^T(M, \Sigma; g; \lambda) = P_2(\Sigma, g) \circ D_1(M, \Sigma; g; \lambda) \quad for \ \lambda = -\frac{n-5}{2}.$$

Proof. For $\lambda = -\frac{n-2}{2}$, the formula for $D_3^T(\lambda)$ which was derived during the proof of Theorem 6.25.3 implies

$$2(n-2+2\lambda)(n-7+2\lambda)D_3^T(\lambda) = 6(n-4+\lambda)D_3^T(\lambda)$$
$$= -2(n-5+2\lambda)(n-4+\lambda)\left\{ (\nabla_N - (\lambda-1)H)(\Delta_M + \lambda J_M) + H(\Delta_M + \lambda J_M) \right\}$$
$$= 6(n-4+\lambda)D_1(\lambda-2)P_2(M)$$

using $D_1(\lambda) = \nabla_N - \lambda H$. Similarly, for $\lambda = -\frac{n-5}{2}$, the same formula yields

$$2(n-2+2\lambda)(n-7+2\lambda)D_3^T(\lambda) = 6(n-4+\lambda)D_3^T(\lambda)$$
$$= 2(n-2+2\lambda)(\lambda-1)\left\{ -\Delta_\Sigma(\nabla_N - \lambda H) + \frac{n-3}{2}J_\Sigma(\nabla_N - \lambda H) \right\}$$
$$= 6(n-4+\lambda)P_2(\Sigma)D_1(\lambda).$$

This completes the proof.   □

Remark 6.8.2 states the analogous factorization identities for order 3 residue families.

Theorem 6.25.4. *Let $n = 4$. Then*

$$D_3^T(dr^2 + h_r; \lambda) = D_3^{\mathrm{res}}(h; \lambda).$$

Proof. The formulas (6.25.1) and (6.25.2) in the proof of Theorem 6.25.3 show that for the metric $dr^2 + h_r$,

$$6\lambda D_3^T(\lambda) = 2(2\lambda+2)(2\lambda-3)D_3^T(\lambda)$$
$$= -2(2\lambda-1)\lambda \nabla_N(\Delta_M + \lambda J_M)u + 4\lambda(\lambda+1)\Delta_\Sigma(\nabla_N u) + 4\lambda(\lambda+1)(\lambda-1)J_\Sigma \nabla_N u.$$

Here we have used that $H = 0$. Hence

$$6D_3^T(\lambda) = \lambda^2 \{-4\nabla_N(J_M u) + 4J_\Sigma \nabla_N u\}$$
$$+ \lambda\{-4\nabla_N(\Delta_M u) + 2\nabla_N(J_M u) + 4\Delta_\Sigma(\nabla_N u)\}$$
$$+ \{2\nabla_N(\Delta_M u) + 4\Delta_\Sigma(\nabla_N u) - 4J_\Sigma(\nabla_N u)\}.$$

Now $i^*(J_M) = J_\Sigma$ (Lemma 6.11.1) and $i^*\nabla_N(J_M) = 0$ imply that the coefficient of λ^2 vanishes. For the coefficient of λ, we find

$$-4\left(\frac{\partial^3}{\partial r^3} - J_\Sigma \frac{\partial^2}{\partial r^2} + \Delta_\Sigma \frac{\partial}{\partial r}\right) + 2J_\Sigma \frac{\partial}{\partial r} + 4\Delta_\Sigma \frac{\partial}{\partial r} = -4\frac{\partial^3}{\partial r^3} + 6J_\Sigma \frac{\partial}{\partial r}$$

using arguments as in the proof of Lemma 6.8.1. Similar arguments yield

$$2\nabla_N \Delta_M + 4\Delta_\Sigma \nabla_N - 4J_\Sigma \nabla_N$$
$$= \left(2\frac{\partial^3}{\partial r^3} - 2J_\Sigma \frac{\partial}{\partial r} + 2\Delta_\Sigma \frac{\partial}{\partial r}\right) + 4\Delta_\Sigma \frac{\partial}{\partial r} - 4J_\Sigma \frac{\partial}{\partial r}$$
$$= 6\left(\frac{1}{3}\frac{\partial^3}{\partial r^3} + (\Delta_\Sigma - J_\Sigma)\frac{\partial}{\partial r}\right)$$

for the coefficient of λ^0. Hence

$$D_3^T(\lambda) = \lambda\left(-\frac{2}{3}\frac{\partial^3}{\partial r^3} + J_\Sigma \frac{\partial}{\partial r}\right) + \left(\frac{1}{3}\frac{\partial^3}{\partial r^3} + (\Delta_\Sigma - J_\Sigma)\frac{\partial}{\partial r}\right).$$

Since this formula coincides with (6.8.4), the proof is complete. $\square$

Remark 6.25.1. *Theorem 6.25.4 continues to hold true for $n \geq 4$. In fact, on $(0, \varepsilon) \times \Sigma^{n-1}$ we have*

$$D_3^{\mathrm{res}}(\lambda) = -\frac{1}{3}(n-5+2\lambda)\frac{\partial^3}{\partial r^3} + (\Delta_\Sigma + (\lambda-1)J_\Sigma)\frac{\partial}{\partial r}$$
$$= \lambda\left(-\frac{2}{3}\frac{\partial^3}{\partial r^3} + J_\Sigma \frac{\partial}{\partial r}\right) + \left(-\frac{n-5}{3}\frac{\partial^3}{\partial r^3} + (\Delta_\Sigma - J_\Sigma)\frac{\partial}{\partial r}\right)$$

by (6.8.4). On the other hand, (6.25.1) and (6.25.2) yield

$$6D_3^T(\lambda) = -2(n-5+2\lambda)\nabla_N(\Delta_M + \lambda \mathsf{J}_M)$$
$$+ 2(n-2+2\lambda)\Delta_\Sigma \nabla_N + 2(n-2+2\lambda)(\lambda-1)\mathsf{J}_\Sigma \nabla_N.$$

Similar arguments as in the proof of Theorem 6.25.4 yield the assertion. Note that the crucial point in the proof are the identities

$$i^*(\mathsf{J}_M) = \mathsf{J}_\Sigma \quad and \quad i^*\nabla_N(\mathsf{J}_M) = 0.$$

Now by construction the pair

$$(P_{3,T}(g), Q_{3,T}(g)) \stackrel{\text{def}}{=} \frac{3}{2}\left(D_3^T(g;0), \dot{D}_3^T(g;0)(1)\right) \tag{6.25.4}$$

satisfies the fundamental identity

$$e^{3\varphi}Q_{3,T}(e^{2\varphi}g) = Q_{3,T}(g) - P_{3,T}(g)(\varphi).$$

Thus we have proved

Theorem 6.25.5. *The natural pair*

$$P_{3,T} = \left\{\frac{1}{2}N\Delta_M + \Delta_\Sigma N\right\} + H\Delta_M - H(\nabla_N d\cdot, N) + (\mathsf{P}(N), d\cdot) - \mathsf{J}_\Sigma N - \frac{3}{2}H^2 N$$

and

$$Q_{3,T} = \frac{1}{2}N\mathsf{J}_M + H\mathsf{J}_M + \mathsf{J}_\Sigma H - H\mathsf{P}(N,N) - \Delta_\Sigma H + \frac{1}{2}H^3$$

satisfies the fundamental identity

$$e^{3\varphi}\hat{Q}_{3,T} = Q_{3,T} - P_{3,T}.$$

Example 6.25.1. *Let S^3 be the equator of S^4 with the round metric g_c. Then $\tau_{S^4} = 12$, $\mathsf{J}_{S^4} = 2$ and $\tau_{S^3} = 6$, $\mathsf{J}_{S^3} = \frac{3}{2}$. Hence, by Theorem 6.25.3,*

$$P_{3,T} = \frac{3}{2}D_3^T(0) = \left\{\frac{1}{2}N\Delta_{S^4} + \Delta_{S^3}N\right\} + (\mathsf{P}(N), d\cdot) - \frac{3}{2}N.$$

Now $\mathsf{P} = \frac{1}{2}(\mathrm{Ric} - 2g) = \frac{1}{2}g$, i.e., $\mathsf{P}(N) = \frac{1}{2}N^\flat$. Therefore,

$$D_3^T(0) = \frac{1}{3}N\Delta_{S^4} + \frac{2}{3}\Delta_{S^3}N - \frac{2}{3}N.$$

This formula coincides with $D_3^c(0)$ for $n = 4$.

Example 6.25.2. *Let S^3 be the unit sphere in $\mathbb{R}^4$ with the flat metric. For the inner normal $N = -\partial/\partial r$, we have $H = -1$. Thus*

$$P_{3,T} = \frac{3}{2}D_3^T(0) = -\frac{1}{2}\frac{\partial}{\partial r}\Delta_{\mathbb{R}^4} - \Delta_{S^3}\frac{\partial}{\partial r} - \Delta_{S^3}.$$

Therefore, $D_3^T(0) = D_3^\flat(0)$ (see (5.4.14)).

In Section 6.26, we will relate the latter result to results of Branson-Gilkey ([38]) and Chang-Qing ([68]). See also Corollary 5.4.1.

6.26 The pair (P_3, Q_3)

Let (M^4, g) be a smooth compact four-manifold with boundary $\Sigma^3 = \partial M^4$. In [68], Chang and Qing introduced a natural pair (P_3, T)

$$P_3(M, \Sigma; g) : C^\infty(M) \to C^\infty(\Sigma), \quad T(M, \Sigma; g) \in C^\infty(\Sigma)$$

which satisfies the fundamental identity

$$e^{3\varphi} T(M, \Sigma; e^{2\varphi} g) = T(M, \Sigma; g) + P_3(M, \Sigma; g)(\varphi). \tag{6.26.1}$$

In the present section, we discuss that pair under the conformally invariant assumption that the trace-free part L_0 of the second fundamental form vanishes, i.e., $L = Hg$, i.e., the boundary is totally umbilic. In particular, we shall relate the pair (P_3, T) to the tractor constructions in Section 6.25.

Theorem 6.26.1 ([68]). *Let N be the inner unit normal of ∂M. Assume that $L_0(g) = 0$. Then*

$$P_3(M, \Sigma; g) \overset{\text{def}}{=} \left\{ \frac{1}{2} N \Delta_M + \Delta_\Sigma N \right\} + H \Delta_\Sigma + (F - 2\mathsf{J}_M) N - (dH, d\cdot) \tag{6.26.2}$$

and the scalar curvature quantity

$$T(M, \Sigma; g) \overset{\text{def}}{=} -\frac{1}{2} N \mathsf{J}_M - 3\mathsf{J}_M H + \mathrm{Ric}_N H - 3H^3 + \frac{1}{3} \mathrm{tr}(L^3) + \Delta_\Sigma H \tag{6.26.3}$$

satisfy the fundamental identity (6.26.1). *Notice that P_3 annihilates constants. All quantities are to be understood with respect to g, of course; we have suppressed the obvious restrictions i^* and, as before, N denotes also the normal derivative ∇_N. Finally, F is given by* (2.5.19).

In Theorem 6.26.1, all terms are written so that a direct comparison with [68] is easy. Notice, however, that our conventions differ from those in [68]. In particular, in [68] the signs of Laplacians are opposite and their pair (N, L) satisfies $e^\varphi \hat{L} = L - N(\varphi) g$.

In [68], the proof of Theorem 6.26.1 uses the explicit formulas for the conformal variation of local invariants of order 3 given in [38].

Remark 6.26.1. *In [68], there is also an analogous pair (P_3, T) if no condition on L is posed. It is given by*

$$P_3 = \left\{ \frac{1}{2} N \Delta_M + \Delta_\Sigma N \right\} + \left\{ 2H \Delta_\Sigma - (L, \mathrm{Hess}_\Sigma) \right\} + (F - 2\mathsf{J}_M) N - (dH, d\cdot)$$

and

$$T = -\frac{1}{2} N \mathsf{J}_M - 3\mathsf{J}_M H + (G, L) - 3H^3 + \frac{1}{3} \mathrm{tr}(L^3) + \Delta_\Sigma H.$$

The following result relates Theorem 6.26.1 to the tractor family of order 3.

Theorem 6.26.2. *Assume that $L_0 = 0$. Then*

$$P_3 = P_{3,T}$$

and

$$T = -Q_{3,T} = -\frac{1}{2}N\mathsf{J}_M - 2H\mathsf{J}_\Sigma + \Delta_\Sigma H + H^3.$$

Proof. Lemma 6.25.2 implies that

$$(\mathsf{P}(N), du) = \mathsf{P}(N, N)Nu + (i^*\mathsf{P}(N), i^*du) = \mathsf{P}(N, N)Nu - (dH, du).$$

Thus we have to prove that (6.26.2) coincides with

$$P_{3,T}u = \left\{ \frac{1}{2}N\Delta_M u + \Delta_\Sigma Nu \right\} + H\Delta_M u$$

$$- H\langle \nabla_N du, N\rangle + \mathsf{P}(N, N)Nu - (dH, du) - \mathsf{J}_\Sigma Nu - \frac{3}{2}H^2 Nu.$$

We assume that N is the geodesic unit normal field. Then $\Delta_M u = \langle \nabla_N du, N\rangle + 3HNu + \Delta_\Sigma u$. Hence

$$P_{3,T}u = \left\{ \frac{1}{2}N\Delta_M u + \Delta_\Sigma Nu \right\} + H\Delta_M u$$

$$+ \frac{3}{2}H^2 Nu + \frac{1}{2}(\mathrm{Ric}_N - \mathsf{J}_M)Nu - (dH, du) - \mathsf{J}_\Sigma Nu.$$

Therefore, it suffices to verify the relation

$$\frac{1}{2}\left(\mathrm{Ric}_N - \frac{1}{6}\tau_M \right) - \frac{1}{4}\tau_\Sigma + \frac{3}{2}H^2 = \mathrm{Ric}_N - \frac{1}{3}\tau_M. \qquad (6.26.4)$$

But the Gauß equation and $L_0 = 0$ imply

$$\tau_M = \tau_\Sigma + 2\,\mathrm{Ric}_N - \mathrm{tr}(L)^2 + |L|^2 = \tau_\Sigma + 2\,\mathrm{Ric}_N - 6H^2. \qquad (6.26.5)$$

Hence the left-hand side of (6.26.4) equals

$$\frac{1}{2}\,\mathrm{Ric}_N - \frac{1}{12}\tau_M - \frac{1}{4}\tau_M + \frac{1}{2}\,\mathrm{Ric}_N = \mathrm{Ric}_N - \frac{1}{3}\tau_M.$$

This proves (6.26.4). The relation $T = -Q_{3,T}$ follows from the validity of the corresponding fundamental identities. An alternative direct argument is as follows. We recall (Theorem 6.25.2) that

$$-Q_{3,T} = -\frac{1}{2}N\mathsf{J}_M - \mathsf{J}_M H - \mathsf{J}_\Sigma H + \Delta_\Sigma H + \mathsf{P}(N, N)H - \frac{1}{2}H^3.$$

Hence, for the proof of the relation $T = -Q_{3,T}$, it is enough to prove that

$$H(P(N,N) - J_\Sigma) - \frac{1}{2}H^3 = -2J_M H + \mathrm{Ric}_N H - 3H^3 + \frac{1}{3}\mathrm{tr}(L^3).$$

In view of $\mathrm{tr}(L^3) = 3H^3$, that identity is equivalent to

$$H\left(\frac{1}{2}\mathrm{Ric}_N - \frac{1}{12}\tau_M - \frac{1}{4}\tau_\Sigma\right) - \frac{1}{2}H^3 = -2J_M H + H\,\mathrm{Ric}_N - 2H^3. \qquad (6.26.6)$$

Now (6.26.5) implies that the left-hand side of (6.26.6) equals

$$H\left(\mathrm{Ric}_N - \frac{1}{3}\tau_M\right) - 2H^3.$$

This proves (6.26.6) and hence $T = -Q_{3,T}$. Finally, we find

$$\begin{aligned}
T &= -\frac{1}{2}NJ_M + H(\mathrm{Ric}_N - 3J_M - 2H^2) + \Delta_\Sigma H \\
&= -\frac{1}{2}NJ_M + H\left(-\frac{1}{2}\tau_\Sigma + H^2\right) + \Delta_\Sigma H
\end{aligned}$$

using (6.26.5). The proof is complete. $\qquad\qquad\qquad\qquad\qquad\qquad\square$

If L vanishes, the formulas for P_3 and T further simplify to

$$P_3(M, \Sigma; g) = \left\{\frac{1}{2}N\Delta_M + \Delta_\Sigma N\right\} + \left(\mathrm{Ric}_N - \frac{1}{3}\tau_M\right)N \qquad (6.26.7)$$

and

$$T(M, \Sigma; g) = -\frac{1}{12}N\tau_M. \qquad (6.26.8)$$

The renormalization

$$\frac{2}{3}P_3(M, \Sigma; g) = \left\{\frac{1}{3}N\Delta_M + \frac{2}{3}\Delta_\Sigma N\right\} + \left(\frac{2}{3}\mathrm{Ric}_N - \frac{2}{9}\tau_M\right)N$$

fits better with the normalizations used in the present work. In the light of Theorem 6.4.1, an obvious question is whether it is more natural to write this operator in the alternative form

$$\frac{1}{3}NP_2(M, g) + \frac{2}{3}P_2(\Sigma, g)N + LOT.$$

The following calculation shows that this is actually the case. We have to determine the lower order terms. We first recall that $P_2(M) = \Delta_M - \frac{1}{6}\tau_M$ and $P_2(\Sigma) = \Delta_\Sigma - \frac{1}{8}\tau_\Sigma$. Hence

$$LOT = \left[\frac{1}{18}\tau_M + \frac{1}{12}\tau_\Sigma - \frac{2}{9}\tau_M + \frac{2}{3}\mathrm{Ric}_N\right]N + \frac{1}{18}N\tau_M.$$

The assumption $L = 0$ implies $\tau_M = \tau_\Sigma + 2\operatorname{Ric}_N$ on Σ. Therefore, the above formula for LOT simplifies to

$$\left[-\frac{1}{12}\tau_\Sigma + \frac{1}{3}\operatorname{Ric}_N\right] N + \frac{1}{18}N\tau_M.$$

Now we observe that

$$Q_2(M) - Q_2(\Sigma) = \frac{1}{6}\tau_M - \frac{1}{4}\tau_\Sigma = -\frac{1}{12}\tau_\Sigma + \frac{1}{3}\operatorname{Ric}_N.$$

In other words, we have proved

Lemma 6.26.1. *If $L = 0$, then the operator $\frac{2}{3}P_3(M,\Sigma;g)$ can be written in the form*

$$\left[\frac{1}{3}NP_2(M;g) + \frac{2}{3}P_2(\Sigma;g)N\right] + (Q_2(M;g) - Q_2(\Sigma;g))N + Q_3(M,\Sigma;g),$$

where $Q_3(M,\Sigma;g) = -\frac{2}{3}Q_{3,T} = \frac{1}{3}N\mathsf{J}_M$.

Lemma 6.26.1 is similar in spirit to the formula for the family $D_2(g;\lambda)$ in Theorem 6.4.1. In both cases, the algebraic theory (Verma modules) determines the top degree part by replacing the Lie-algebraic Laplacians Δ^- by Yamabe operators. Then the lower orders terms are naturally formulated in terms of Q-curvatures and mean curvature.

The condition $L = 0$ is rather restrictive and, in particular, not conformally invariant. Therefore, it is of interest to have the following generalization of Lemma 6.26.1.

Theorem 6.26.3. *Assume that $L_0 = 0$. Then*

$$\frac{2}{3}P_3(M,\Sigma;g) = \left[\frac{1}{3}NP_2(M;g) + \frac{2}{3}P_2(\Sigma;g)N\right] + \frac{2}{3}\left(HP_2(\Sigma;g) - (dH,d\cdot)\right)$$
$$+ \left(Q_2(M;g) - Q_2(\Sigma;g) + 2H^2\right)N$$
$$+ Q_3(M,\Sigma;g) + \frac{2}{3}\left((P_2(\Sigma;g) - Q_2(\Sigma;g))H + H^3\right),$$

where $Q_3(M,\Sigma;g) = -\frac{2}{3}Q_{3,T}$.

Proof. We first rewrite (6.26.2) in terms of $P_2(M)$ and $P_2(\Sigma)$. We get

$$P_3(M,\Sigma;g) = \left\{\frac{1}{2}NP_2(M,g) + P_2(\Sigma,g)N\right\} + HP_2(\Sigma,g) - (dH,d\cdot)$$
$$+ \left(\operatorname{Ric}_N - \frac{1}{3}\tau_M + \frac{1}{12}\tau_M + \frac{1}{8}\tau_\Sigma\right)N + \frac{1}{12}N\tau_M + \frac{1}{8}H\tau_\Sigma.$$

Now (6.26.5) implies

$$\text{Ric}_N - \frac{1}{4}\tau_M + \frac{1}{8}\tau_\Sigma = \frac{1}{2}\text{Ric}_N - \frac{1}{8}\tau_\Sigma + \frac{3}{2}H^2$$

and

$$Q_2(M) - Q_2(\Sigma) = \frac{1}{6}\tau_M - \frac{1}{4}\tau_\Sigma = \frac{1}{6}\left(\tau_\Sigma + 2\,\text{Ric}_N - 6H^2\right) - \frac{1}{4}\tau_\Sigma$$
$$= \frac{2}{3}\left(\frac{1}{2}\text{Ric}_N - \frac{1}{8}\tau_\Sigma\right) - H^2.$$

Hence

$$\text{Ric}_N - \frac{1}{3}\tau_M + \frac{1}{8}\tau_\Sigma = \frac{3}{2}\left(Q_2(M) - Q_2(\Sigma)\right) + 3H^2.$$

Therefore,

$$P_3(M, \Sigma; g) = \left\{\frac{1}{2}NP_2(M, g) + P_2(\Sigma, g)N\right\} + HP_2(\Sigma, g) - (dH, d\cdot)$$
$$+ \frac{3}{2}\left(Q_2(M) - Q_2(\Sigma)\right) + 2H^2)N + \frac{1}{12}N\tau_M + \frac{1}{8}H\tau_\Sigma.$$

Thus it only remains to prove that

$$\frac{1}{12}N\tau_M + \frac{1}{8}H\tau_\Sigma = -Q_{3,T} + (P_2(\Sigma) - Q_2(\Sigma))H + H^3.$$

But this is a consequence of the formula for $Q_{3,T}$ in Theorem 6.26.2. The proof is complete. □

It it not known whether the curvature quantity T also has an interpretation in the spirit of a holographic formula for even order Q-curvature. Moreover, it would be interesting to remove the assumption $L_0 = 0$ in Theorem 6.26.2.

We finish with the description of the relation between Q_4, Q_3 and the Euler characteristic of a four-manifold with a totally umbilic boundary.

Theorem 6.26.4. *Assume that the boundary Σ^3 of M^4 is totally umbilic for g, i.e., $L_0(g) = 0$. Then*

$$\chi(M^4) = \frac{1}{32\pi^2}\int_M \left(|C|^2 + 4Q_4\right)\text{vol}(g) - \frac{1}{4\pi^2}\frac{3}{2}\int_\Sigma Q_3\,\text{vol}_\Sigma(g).$$

Corollary 6.26.1. *If $L_0(g) = 0$, then the sum*

$$\int_M Q_4\,\text{vol}(g) - 3\int_\Sigma Q_3\,\text{vol}_\Sigma(g)$$

is conformally invariant.

Theorem 6.26.4 shows that for $L_0 = 0$, the boundary integral in terms of Q_3 is *directly* related to the boundary integral in the Gauß-Bonnet formula. If L_0 does *not* vanish, then the boundary integral in Gauß-Bonnet differs from the integral of Q_3 by the integral of a pointwise conformally invariant quantity.

Proof. We use the formula ([38], formula (4.3))

$$\chi(M) = \frac{1}{32\pi^2} \int_M \left(\tau_M^2 - 4|\operatorname{Ric}|^2 + |R|^2 \right) \operatorname{vol}$$
$$- \frac{1}{24\pi^2} \int_\Sigma \left(9\tau_M H - 18FH - 6(T,L) + 54H^3 - 18H|L|^2 + 4\operatorname{tr}(L^3) \right) \operatorname{vol}_\Sigma .$$

Note that our conventions differ from those in [38]: their Laplacian has opposite sign and L is the negative of ours. We recall from Section 6.14 that the first integrand equals $8\operatorname{Pf}_4$, i.e., the first integral is just

$$\frac{1}{(2\pi)^2} \int_M \operatorname{Pf}_4 .$$

We have also seen in Section 6.14 that

$$Q_4 = 2\operatorname{Pf}_4 - \frac{1}{4}|C|^2 - \Delta J_M$$

(here $-\Delta$ denotes the non-negative Laplacian). Thus the first integral can be written in the form

$$\frac{1}{32\pi^2} \int_M \left(|C|^2 + 4Q_4 + 4\Delta J_M \right) \operatorname{vol} .$$

Next, we consider the boundary term. It simplifies under the assumption $L_0 = 0$. For the second integrand, we obtain

$$9\tau_M H - 18H\operatorname{Ric}_N - 6H(\tau_M - 2\operatorname{Ric}_N) + 12H^3 = 3H\tau_M - 6H\operatorname{Ric}_N + 12H^3.$$

Using the relation $\tau_M = \tau_\Sigma + 2\operatorname{Ric}_N - 6H^2$, that term simplifies to $3H\tau_\Sigma - 6H^3$. Thus

$$\chi(M) = \frac{1}{32\pi^2} \int_M \left(|C|^2 + 4Q_4 + 4\Delta J_M \right) \operatorname{vol} - \frac{1}{4\pi^2} \int_\Sigma \left(\frac{1}{2}H\tau_\Sigma - H^3 \right) \operatorname{vol}_\Sigma .$$

Now Green's formula and Theorem 6.26.2 yield

$$\chi(M) = \frac{1}{32\pi^2} \int_M \left(|C|^2 + 4Q_4 \right) \operatorname{vol} - \frac{1}{4\pi^2} \int_\Sigma \left(\frac{1}{2}N J_M + \frac{1}{2}H\tau_\Sigma - H^3 \right) \operatorname{vol}_\Sigma$$
$$= \frac{1}{32\pi^2} \int_M \left(|C|^2 + 4Q_4 \right) \operatorname{vol} + \frac{1}{4\pi^2} \int_\Sigma Q_{3,T} \operatorname{vol}_\Sigma .$$

The relation $Q_{3,T} = -\frac{3}{2}Q_3$ completes the proof.   □

Bibliography

[1] O. Aharony, S. Gubser, J. Maldacena, H. Ooguri, and Y. Oz. Large N field theories, string theory and gravity. *Phys. Rep.*, 323(3-4):183–386, 2000. `arXiv:hep-th/9905111v3`.

[2] P. Albin. Renormalizing curvature integrals on Poincaré-Einstein manifolds. `arXiv:math/0504161v1`.

[3] S. Alexakis. The decomposition of global conformal invariants: On a conjecture of Deser and Schwimmer. `arXiv:0711.1685v1`.

[4] S. Alexakis. On conformally invariant differential operators. `arXiv:math/0608771v3`.

[5] S. Alexakis. On the decomposition of global conformal invariants I. *Annals of Math.*, 2006. `arXiv:math/0509571v3`.

[6] S. Alexakis. On the decomposition of global conformal invariants. II. *Adv. Math.*, 206(2):466–502, 2006. `arXiv:math/0509572v3`.

[7] S. Alexakis and R. Mazzeo. Renormalized area and properly embedded minimal surfaces in hyperbolic 3-manifolds. `arXiv:math/0802.2250v2`.

[8] E. Álvarez, J. Conde, and L. Hernández. The Dirichlet obstruction in AdS/CFT. *Internat. J. Modern Phys. D*, 12(8):1415–1429, 2003. `arXiv:hep-th/0303164v2`.

[9] E. Álvarez, J. Conde, and L. Hernández. Rudiments of holography. *Internat. J. Modern Phys. D*, 12(4):543–582, 2003. `arXiv:hep-th/0205075v5`.

[10] M. Anderson. L^2 curvature and volume renormalization of AHE metrics on 4-manifolds. *Math. Res. Lett.*, 8(1-2):171–188, 2001. `arXiv:math/0011051v2`.

[11] M. Anderson. Geometric aspects of the AdS/CFT correspondence. In *AdS/CFT correspondence: Einstein metrics and their conformal boundaries*, volume 8 of *IRMA Lect. Math. Theor. Phys.*, pages 1–31. Eur. Math. Soc., Zürich, 2005. `arXiv:hep-th/0403087v2`.

[12] D. Anselmi. Quantum irreversibility in arbitrary dimension. *Nuclear Phys. B*, 567(1-2):331–359, 2000. `arXiv:hep-th/9905005v3`.

[13] M.F. Atiyah and R. Bott. A Lefschetz fixed point formula for elliptic complexes. I. *Ann. of Math.* (2), 86:374–407, 1967.

[14] M.F. Atiyah and R. Bott. A Lefschetz fixed point formula for elliptic complexes. II. Applications. *Ann. of Math.* (2), 88:451–491, 1968.

[15] I.G. Avramidi. *Covariant methods for the calculation of the effective action in quantum field theory and investigation of higher-derivative quantum gravity*. PhD thesis, Moscow State Lomonosov University, 1986. `arXiv: hep-th/9510140v3`.

[16] P. Bäcklund. Families of equivariant differential operators and Anti-de Sitter spaces. 2007. `arXiv:0806.4508v1`.

[17] T.N. Bailey, M. Eastwood, and A.R. Gover. Thomas's structure bundle for conformal, projective and related structures. *Rocky Mountain J. Math.*, 24(4):1191–1217, 1994.

[18] F. Bastianelli, S. Frolov, and A. Tseytlin. Conformal anomaly of $(2,0)$ tensor multiplet in six dimensions and AdS/CFT correspondence. *J. High Energy Phys.*, (2):Paper 13, 21, 2000. `arXiv:hep-th/0001041v2`.

[19] R. Baston. Verma modules and differential conformal invariants. *J. Differential Geom.*, 32(3):851–898, 1990.

[20] R. Baston and M. Eastwood. Invariant operators. In *Twistors in mathematics and physics*, volume 156 of *London Math. Soc. Lecture Note Ser.*, pages 129–163. Cambridge Univ. Press, 1990.

[21] H. Bateman and A. Erdelyi. *Higher transcendental functions*, volume I. 1953.

[22] H. Baum. The conformal analog of Calabi-Yau manifolds. In *Handbook of Pseudo-Riemannian Geometry and Supersymmetry*, IRMA-Series. Publishing House of EMS, 2007. (to appear).

[23] W. Beckner. Sharp Sobolev inequalities on the sphere and the Moser-Trudinger inequality. *Ann. of Math.* (2), 138(1):213–242, 1993.

[24] A. Besse. *Einstein manifolds*, volume 10 of *Ergebnisse der Mathematik und ihrer Grenzgebiete*. Springer-Verlag, 1987.

[25] O. Biquard and R. Mazzeo. Parabolic geometries as conformal infinities of Einstein metrics. *Arch. Math.* (*Brno*), 42(suppl.):85–104, 2006.

[26] C. Bishop and P. Jones. Hausdorff dimension and Kleinian groups. *Acta Math.*, 179(1):1–39, 1997.

[27] M. Bonk, J. Heinonen, and E. Saksman. Logarithmic potentials, quasiconformal flows, and Q-curvature. *Duke Math. J.*, 142(2):197–239, 2008.

[28] D. Borthwick. *Spectral theory of infinite-area hyperbolic surfaces*, volume 256 of *Progress in Mathematics*. Birkhäuser Boston Inc., 2007.

[29] R. Bousso. The holographic principle. *Rev. Modern Phys.*, 74(3):825–874, 2002.

[30] T.P. Branson. Differential operators canonically associated to a conformal structure. *Math. Scand.*, 57(2):293–345, 1985.

[31] T.P. Branson. *The functional determinant*, volume 4 of *Lecture Notes Series*. Seoul National University Research Institute of Mathematics Global Analysis Research Center, 1993.

[32] T.P. Branson. Sharp inequalities, the functional determinant, and the complementary series. *Trans. Amer. Math. Soc.*, 347(10):3671–3742, 1995.

[33] T.P. Branson. An anomaly associated with 4-dimensional quantum gravity. *Comm. Math. Phys.*, 178(2):301–309, 1996.

[34] T.P. Branson. Spectral theory of invariant operators, sharp inequalities, and representation theory. In *The Proceedings of the 16th Winter School "Geometry and Physics" (Srní,* 1996), number 46, pages 29–54, 1997.

[35] T.P. Branson. *Q*-curvature and spectral invariants. *Rend. Circ. Mat. Palermo (2) Suppl.*, 75:11–55, 2005.

[36] T.P. Branson. *Q*-curvature, spectral invariants, and representation theory. *SIGMA Symmetry Integrability Geom. Methods Appl.*, 3:Paper 090, 31, 2007. arXiv:0709.2471v1.

[37] T.P. Branson, Sun-Yung A. Chang, and P. Yang. Estimates and extremals for zeta function determinants on four-manifolds. *Comm. Math. Phys.*, 149(2):241–262, 1992.

[38] T.P. Branson and P. Gilkey. The functional determinant of a four-dimensional boundary value problem. *Trans. Amer. Math. Soc.*, 344(2):479–531, 1994.

[39] T.P. Branson, P. Gilkey, and J. Pohjanpelto. Invariants of locally conformally flat manifolds. *Trans. Amer. Math. Soc.*, 347(3):939–953, 1995.

[40] T.P. Branson and A.R. Gover. Conformally invariant non-local operators. *Pacific J. Math.*, 201(1):19–60, 2001.

[41] T.P. Branson and A.R. Gover. Origins, applications and generalizations of the *Q*-curvature. The American Institute of Mathematics, 2003. http://www.aimath.org/WWN/confstruct/confstruct.pdf.

[42] T.P. Branson and A.R. Gover. Conformally invariant operators, differential forms, cohomology and a generalisation of *Q*-curvature. *Comm. Partial Differential Equations*, 30(10-12):1611–1669, 2005. arXiv:math/0309085v2.

[43] T.P. Branson and A.R. Gover. Origins, applications and generalisations of the *Q*-curvature. *Acta Appl. Math.*, 102(2-3):131–146, 2008.

[44] T.P. Branson and A.R. Gover. Variational status of a class of fully nonlinear curvature prescription problems. *Calc. Var. Partial Differential Equations*, 32(2):253–262, 2008. arXiv:math/0610773v1.

[45] T.P. Branson and B. Ørsted. Conformal indices of Riemannian manifolds. *Compositio Math.*, 60(3):261–293, 1986.

[46] T.P. Branson and B. Ørsted. Conformal deformation and the heat operator. *Indiana Univ. Math. J.*, 37(1):83–110, 1988.

[47] T.P. Branson and B. Ørsted. Conformal geometry and global invariants. *Diff. Geom. Appl.*, 1:279–308, 1991.

[48] T.P. Branson and B. Ørsted. Explicit functional determinants in four dimensions. *Proc. Amer. Math. Soc.*, 113(3):669–682, 1991.

[49] S. Brendle. Global existence and convergence for a higher order flow in conformal geometry. *Ann. of Math.* (2), 158(1):323–343, 2003.

[50] U. Bunke and M. Olbrich. Group cohomology and the singularities of the Selberg zeta function associated to a Kleinian group. *Ann. of Math.* (2), 149(2):627–689, 1999.

[51] U. Bunke and M. Olbrich. The spectrum of Kleinian manifolds. *J. Funct. Anal.*, 172(1):76–164, 2000.

[52] F. Burstall and D. Calderbank. Submanifold geometry in generalized flag manifolds. *Rend. Circ. Mat. Palermo* (2) *Suppl.*, 72:13–41, 2004.

[53] P. Buser. *Geometry and spectra of compact Riemann surfaces*, volume 106 of *Progress in Mathematics*. Birkhäuser Boston Inc., 1992.

[54] A. Čap and A.R. Gover. Tractor calculi for parabolic geometries. *Trans. Amer. Math. Soc.*, 354(4):1511–1548 (electronic), 2002.

[55] A. Čap and A.R. Gover. Standard tractors and the conformal ambient metric construction. *Ann. Global Anal. Geom.*, 24(3):231–259, 2003. arXiv: math/0207016v1.

[56] A. Čap, A.R. Gover, and V. Souček. Conformally invariant operators via curved Casimirs: examples. arXiv:0808.1978v1.

[57] A. Čap, J. Slovák, and V. Souček. Invariant operators on manifolds with almost Hermitian symmetric structures. I. Invariant differentiation. *Acta Math. Univ. Comenian.* (*N.S.*), 66(1):33–69, 1997.

[58] A. Čap, J. Slovák, and V. Souček. Invariant operators on manifolds with almost Hermitian symmetric structures. II. Normal Cartan connections. *Acta Math. Univ. Comenian.* (*N.S.*), 66(2):203–220, 1997.

[59] A. Čap, J. Slovák, and V. Souček. Invariant operators on manifolds with almost Hermitian symmetric structures. III. Standard operators. *Differential Geom. Appl.*, 12(1):51–84, 2000.

[60] A. Čap, J. Slovák, and V. Souček. Bernstein-Gelfand-Gelfand sequences. *Ann. of Math.* (2), 154(1):97–113, 2001.

[61] Cheng-Hung Chang and D. Mayer. The transfer operator approach to Selberg's zeta function and modular and Maass wave forms for PSL(2, **Z**). In *Emerging applications of number theory* (*Minneapolis, MN, 1996*), volume 109 of *IMA Vol. Math. Appl.*, pages 73–141. Springer, 1999.

[62] Sun-Yung A. Chang. On zeta functional determinant. In *Partial differential equations and their applications* (*Toronto, ON, 1995*), volume 12 of *CRM Proc. Lecture Notes*, pages 25–50. Amer. Math. Soc., 1997. With notes taken by Jie Qing.

[63] Sun-Yung A. Chang. *Non-linear elliptic equations in conformal geometry.* Zürich Lectures in Advanced Mathematics. European Mathematical Society (EMS), Zürich, 2004. (reviewed by R. Mazzeo in Bull. AMS 44 (2) (2007), 323–330).

[64] Sun-Yung A. Chang. Conformal invariants and partial differential equations. *Bull. Amer. Math. Soc. (N.S.)*, 42(3):365–393 (electronic), 2005.

[65] Sun-Yung A. Chang and Hao Fang. A class of variational functionals in conformal geometry. *Intern. Math. Research Notices*, 008, 2008. `arXiv: 0803.0333v1`.

[66] Sun-Yung A. Chang, M. Gursky, and P. Yang. An equation of Monge-Ampère type in conformal geometry, and four-manifolds of positive Ricci curvature. *Ann. of Math.* (2), 155(3):709–787, 2002.

[67] Sun-Yung A. Chang, M. Gursky, and P. Yang. A conformally invariant sphere theorem in four dimensions. *Publ. Math. Inst. Hautes Études Sci.*, pages 105–143, 2003.

[68] Sun-Yung A. Chang and J. Qing. The zeta functional determinants on manifolds with boundary. I. The formula. *J. Funct. Anal.*, 147(2):327–362, 1997.

[69] Sun-Yung A. Chang and J. Qing. The zeta functional determinants on manifolds with boundary. II. Extremal metrics and compactness of isospectral set. *J. Funct. Anal.*, 147(2):363–399, 1997.

[70] Sun-Yung A. Chang, J. Qing, and P. Yang. On the renormalized volumes for conformally compact Einstein manifolds. *ESI*, 2004. (preprint 1513).

[71] Sun-Yung A. Chang, J. Qing, and P. Yang. On the topology of conformally compact Einstein 4-manifolds. In *Noncompact problems at the intersection of geometry, analysis, and topology*, volume 350 of *Contemp. Math.*, pages 49–61. Amer. Math. Soc., 2004.

[72] Sun-Yung A. Chang and P. Yang. Extremal metrics of zeta function determinants on 4-manifolds. *Ann. of Math.* (2), 142(1):171–212, 1995.

[73] Sun-Yung A. Chang and P. Yang. On a fourth order curvature invariant. In *Spectral problems in geometry and arithmetic (Iowa City, IA, 1997)*, volume 237 of *Contemp. Math.*, pages 9–28. Amer. Math. Soc., 1999.

[74] Sun-Yung A. Chang and P. Yang. Non-linear partial differential equations in conformal geometry. In *Proceedings of the International Congress of Mathematicians, Vol. I (Beijing, 2002)*, pages 189–207, 2002.

[75] I. Chavel. *Eigenvalues in Riemannian geometry*, volume 115 of *Pure and Applied Mathematics*. Academic Press Inc., 1984.

[76] B. Chen. Some conformal invariants of submanifolds and their applications. *Boll. Un. Mat. Ital.* (4), 10:380–385, 1974.

[77] B. Chow and D. Knopf. *The Ricci flow: an introduction*, volume 110 of *Mathematical Surveys and Monographs*. American Mathematical Society, 2004.

[78] A. Connes. Trace formula in noncommutative geometry and the zeros of the Riemann zeta function. *Selecta Math. (N.S.)*, 5(1):29–106, 1999.

[79] G. D'Ambra and M. Gromov. Lectures on transformation groups: geometry and dynamics. In *Surveys in differential geometry (Cambridge, MA, 1990)*, pages 19–111. Lehigh Univ., 1991.

[80] S. de Haro, K. Skenderis, and S. Solodukhin. Holographic reconstruction of spacetime and renormalization in the AdS/CFT correspondence. *Comm. Math. Phys.*, 217(3):595–622, 2001. `arXiv:hep-th/0002230v3`.

[81] S. Deser and A. Schwimmer. Geometric classification of conformal anomalies in arbitrary dimensions. *Phys. Lett. B*, 309(3-4):279–284, 1993. `arXiv:hep-th/9302047v1`.

[82] E. D'Hoker and D. Freedman. Supersymmetric gauge theories and the ADS/CFT correspondence. In *Strings, branes and extra dimensions. TASI 2001*, pages 3–158. World Sci. Publ., River Edge, NJ, 2004. `arXiv:hep-th/0201253v2`.

[83] Z. Djadli. Opérateurs géométriques et géométrie conforme. In *Actes de Séminaire de Théorie Spectrale et Géométrie. Vol. 23. Année 2004–2005*, volume 23 of *Sémin. Théor. Spectr. Géom.*, pages 49–103. Univ. Grenoble I, 2005.

[84] Z. Djadli and A. Malchiodi. Existence of conformal metrics with constant Q-curvature. *Ann. of Math. (2)*, 168(3):813–858, 2008. `arXiv:math/0410141v3`.

[85] M. do Carmo. *Riemannian geometry*. Birkhäuser Boston Inc., 1992.

[86] H. Donnelly. Heat equation and the volume of tubes. *Invent. Math.*, 29(29):239–243, 1975.

[87] M.J. Duff. Twenty years of the Weyl anomaly. *Classical Quantum Gravity*, 11(6):1387–1403, 1994.

[88] M. Eastwood. Notes on conformal differential geometry. In *The Proceedings of the 15th Winter School "Geometry and Physics" (Srní, 1995)*, volume 43, pages 57–76, 1996.

[89] M. Eastwood. Variations on the de Rham complex. *Notices Amer. Math. Soc.*, 46(11):1368–1376, 1999.

[90] M. Eastwood and J. Rice. Conformally invariant differential operators on Minkowski space and their curved analogues. *Comm. Math. Phys.*, 109(2):207–228, 1987.

[91] M. Eastwood and M. Singer. A conformally invariant Maxwell gauge. *Physics Letters*, 107A(2):73–74, 1985.

[92] M. Eastwood and J. Slovák. Semiholonomic Verma modules. *J. Algebra*, 197(2):424–448, 1997.

[93] Khalil El Mehdi. Prescribing Q-curvature on higher dimensional spheres. *Ann. Math. Blaise Pascal*, 12:259–295, 2005.

[94] C. Falk. Konforme Berandung vollständiger Anti-deSitter- Mannigfaltigkeiten. Master's thesis, Humboldt-Universität Berlin, 2007.

[95] C. Falk and A. Juhl. Universal recursive formulas for Q-curvature. `arXiv: math/0804.2745v2`.

[96] C. Fefferman and C.R. Graham. The ambient metric. `arXiv:0710.0919v1`.

[97] C. Fefferman and C.R. Graham. Q-curvature and Poincaré metrics. *Math. Res. Lett.*, 9(2-3):139–151, 2002. `arXiv:math/0110271v1`.

[98] C. Fefferman and K. Hirachi. Ambient metric construction of Q-curvature in conformal and CR geometries. *Math. Res. Lett.*, 10(5-6):819–831, 2003. `arXiv:math/0303184v2`.

[99] C. Feffermann and C.R. Graham. Conformal invariants. *Astérisque*, Numero Hors Serie:95–116, 1985. The mathematical heritage of Élie Cartan (Lyon, 1984).

[100] A. Fialkow. Conformal differential geometry of a subspace. *Trans. Amer. Math. Soc.*, 56:309–433, 1944.

[101] A. Fialkow. Conformal classes of surfaces. *Amer. J. Math.*, 67:583–616, 1945.

[102] C. Frances. *Géométrie et dynamique Lorentziennes conformes*. PhD thesis, ENS Lyon, 2002. `http://www.umpa.ens-lyon.fr/~cfrances/`.

[103] C. Frances. The conformal boundary of anti-de Sitter space-times. In *AdS/CFT correspondence: Einstein metrics and their conformal boundaries*, volume 8 of *IRMA Lect. Math. Theor. Phys.*, pages 205–216. Eur. Math. Soc., Zürich, 2005.

[104] D. Fried. The zeta functions of Ruelle and Selberg. I. *Ann. Sci. École Norm. Sup.* (4), 19(4):491–517, 1986.

[105] I.M. Gelfand and G.E. Shilov. *Generalized functions. Vol. 1.* Academic Press, 1964 [1977]. Translated from the Russian by Eugene Saletan.

[106] P. Gilkey. The spectral geometry of a Riemannian manifold. *J. Differential Geometry*, 10(4):601–618, 1975.

[107] P. Gilkey. *Invariance theory, the heat equation, and the Atiyah-Singer index theorem*. Studies in Advanced Mathematics. CRC Press, Boca Raton, FL, second edition, 1995.

[108] A.R. Gover. q curvature prescription; forbidden functions and the GJMS null space. `arXiv:0810.5604v1`.

[109] A.R. Gover. Aspects of parabolic invariant theory. *Rend. Circ. Mat. Palermo* (2) *Suppl.*, 59:25–47, 1999. The 18th Winter School "Geometry and Physics" (Srní, 1998).

[110] A.R. Gover. Laplacian operators and Q-curvature on conformally Einstein manifolds. *Math. Ann.*, 336(2):311–334, 2006. arXiv:math/0506037v3.

[111] A.R. Gover. Conformal Dirichlet-Neumann maps and Poincaré-Einstein manifolds. *SIGMA Symmetry Integrability Geom. Methods Appl.*, 3:Paper 100, 21, 2007. arXiv:0710.2585v2.

[112] A.R. Gover and K. Hirachi. Conformally invariant powers of the Laplacian— a complete nonexistence theorem. *J. Amer. Math. Soc.*, 17(2):389–405, 2004. arXiv:math/0304082v2.

[113] A.R. Gover and P. Leitner. A class of compact Poincaré-Einstein manifolds: properties and construction. arXiv:0808.2097v1.

[114] A.R. Gover and P. Leitner. A sub-product construction of Poincaré-Einstein metrics. *Intern. J. Math.*, 2009. arXiv:math/0608044v2.

[115] A.R. Gover and P. Nurowski. Obstructions to conformally Einstein metrics in n dimensions. *J. Geom. Phys.*, 56(3):450–484, 2006. arXiv:math/0405304v3.

[116] A.R. Gover and L. Peterson. Conformally invariant powers of the Laplacian, Q-curvature, and tractor calculus. *Comm. Math. Phys.*, 235(2):339–378, 2003. arXiv:math-ph/0201030v3.

[117] C.R. Graham. Extended obstruction tensors and renormalized volume coefficients. arXiv:0810.4203v1.

[118] C.R. Graham. Conformally invariant powers of the Laplacian. II. Nonexistence. *J. London Math. Soc.* (2), 46(3):566–576, 1992.

[119] C.R. Graham. Volume and area renormalizations for conformally compact Einstein metrics. In *The Proceedings of the 19th Winter School "Geometry and Physics" (Srní, 1999)*, volume 63, pages 31–42, 2000. arXiv:math/9909042v1.

[120] C.R. Graham. Conformal powers of the Laplacian via stereographic projection. *SIGMA Symmetry Integrability Geom. Methods Appl.*, 3:Paper 121, 4, 2007. arXiv:0711.4798v2.

[121] C.R. Graham. Jet isomorphism for conformal geometry. *Arch. Math. (Brno)*, 43(5):389–415, 2007. arXiv:0710.1671v2.

[122] C.R. Graham and K. Hirachi. The ambient obstruction tensor and Q-curvature. In *AdS/CFT correspondence: Einstein metrics and their conformal boundaries*, volume 8 of *IRMA Lect. Math. Theor. Phys.*, pages 59–71. Eur. Math. Soc., Zürich, 2005. arXiv:math/0405068v1.

[123] C.R. Graham and K. Hirachi. Inhomogeneous ambient metrics. In *Symmetries and overdetermined systems of partial differential equations*, volume 144 of *IMA Vol. Math. Appl.*, pages 403–420. Springer, New York, 2008.

[124] C.R. Graham, R. Jenne, L.J. Mason, and G.A. J. Sparling. Conformally invariant powers of the Laplacian. I. Existence. *J. London Math. Soc.* (2), 46(3):557–565, 1992.

[125] C.R. Graham and A. Juhl. Holographic formula for Q-curvature. *Adv. Math.*, 216(2):841–853, 2007.

[126] C.R. Graham and J.M. Lee. Einstein metrics with prescribed conformal infinity on the ball. *Adv. Math.*, 87(2):186–225, 1991.

[127] C.R. Graham and E. Witten. Conformal anomaly of submanifold observables in AdS/CFT correspondence. *Nuclear Phys. B*, 546(1-2):52–64, 1999. `arXiv:hep-th/9901021v1`.

[128] C.R. Graham and M. Zworski. Scattering matrix in conformal geometry. *Invent. Math.*, 152(1):89–118, 2003. `arXiv:math/0109089v1`.

[129] D. Grant. A conformally invariant third order Neumann-type operator for hypersurfaces. Master's thesis, University of Auckland, 2003.

[130] A. Gray. *Tubes*, volume 221 of *Progress in Mathematics*. Birkhäuser Verlag, second edition, 2004.

[131] C. Guillarmou. Generalized Krein formula, determinants and Selberg zeta function in even dimension. `arXiv:math/0512173v3`.

[132] C. Guillarmou. Resonances and scattering poles on asymptotically hyperbolic manifolds. *Math. Res. Lett.*, 12(1):103–119, 2005. `arXiv:math/0403545v1`.

[133] C. Guillarmou and F. Naud. Wave 0-trace and length spectrum on convex co-compact hyperbolic manifolds. *Commun. Anal. Geom.*, 14(5):945–967, 2006. `arXiv:math.DG/0606223v1`.

[134] M. Gursky and J. Viaclovsky. Fully nonlinear equations on Riemannian manifolds with negative curvature. *Indiana Univ. Math. J.*, 52(2):399–419, 2003. `arXiv:math/0210303v2`.

[135] M. Gursky and J. Viaclovsky. Volume comparison and the σ_k-Yamabe problem. *Adv. Math.*, 187(2):447–487, 2004. `arXiv:math/0210302v2`.

[136] S. Hawking. Zeta function regularization of path integrals in curved space-time. *Comm. Math. Phys.*, 55:133–148, 1977.

[137] S. Hawking. The path-integral approach to quantum gravity. In S. Hawking and W. Israel, editors, *General relativity. An Einstein centenary survey*. Cambridge University Press, 1979.

[138] H. Hecht and W. Schmid. Characters, asymptotics and n-homology of Harish-Chandra modules. *Acta Math.*, 151(1-2):49–151, 1983.

[139] D.A. Hejhal. *The Selberg trace formula for* PSL$(2, R)$. *Vol. I.* Springer-Verlag, Berlin, 1976. Lecture Notes in Mathematics, Vol. 548.

[140] S. Helgason. *Groups and geometric analysis*, volume 113 of *Pure and Applied Mathematics*. Academic Press Inc., 1984.

[141] M. Henningson and K. Skenderis. The holographic Weyl anomaly. *JHEP*, 7:Paper 23, 12 pp. (electronic), 1998. `arXiv:hep-th/9806087v2`.

[142] U. Hertrich-Jeromin. *Introduction to Möbius differential geometry*, volume 300 of *London Mathematical Society Lecture Note Series*. Cambridge University Press, 2003.

[143] M. Herzlich. A remark on renormalized volume and Euler characteristic for ACHE 4-manifolds. *Differential Geom. Appl.*, 25(1):78–91, 2007. `arXiv:math/0305134v2`.

[144] P. Hislop, P. Perry, and S.-H. Tang. CR-invariants and the scattering operator for complex manifolds with CR-boundary. *C. R. Math. Acad. Sci. Paris*, 342(9):651–654, 2006. `arXiv:0709.1103v1`.

[145] J. Holland and G. Sparling. Conformally invariant powers of the ambient Dirac operator. `arXiv:math/0112033v2`.

[146] G.T. Horowitz and J. Polchinski. Gauge/gravity duality. `arXiv:gr-qc/0602037v3`.

[147] C. Imbimbo, A. Schwimmer, S. Theisen, and S. Yankielowicz. Diffeomorphisms and holographic anomalies. *Classical Quantum Gravity*, 17(5):1129–1138, 2000. Strings '99 (Potsdam).

[148] H. Iwaniec. *Spectral methods of automorphic forms*, volume 53 of *Graduate Studies in Mathematics*. American Mathematical Society, Providence, RI, second edition, 2002.

[149] R.W. Jenne. *A construction of conformally invariant differential operators*. PhD thesis, University of Washington, 1988.

[150] A. Juhl. Invariant extension of automorphic functions, differential intertwining operators, and Jacobi polynomials. (in preparation).

[151] A. Juhl. *Cohomological theory of dynamical zeta functions*, volume 194 of *Progress in Mathematics*. Birkhäuser Verlag, 2001.

[152] M. Kashiwara, A. Kowata, K. Minemura, K. Okamoto, T. Ōshima, and M. Tanaka. Eigenfunctions of invariant differential operators on a symmetric space. *Ann. of Math. (2)*, 107(1):1–39, 1978.

[153] A. Katok and B. Hasselblatt. *Introduction to the modern theory of dynamical systems*, volume 54 of *Encyclopedia of Mathematics and its Applications*. Cambridge University Press, 1995.

[154] A. Kholodenko. Boundary conformal field theories, limit sets of Kleinian groups and holography. *J. Geom. Phys.*, 35(2-3):193–238, 2000.

[155] S. Kichenassamy. On a conjecture of Fefferman and Graham. *Adv. Math.*, 184(2):268–288, 2004.

[156] A. Knapp. *Representation theory of semisimple groups*, volume 36 of *Princeton Mathematical Series*. Princeton University Press, 1986. An overview based on examples.

[157] O. Kowalski. Partial curvature structures and conformal geometry of submanifolds. *J. Differential Geometry*, 8:53–70, 1973.

[158] C. Kozameh, E. Newman, and P. Nurowski. Conformal Einstein equations and Cartan conformal connection. *Classical Quantum Gravity*, 20(14):3029–3035, 2003.

[159] C. Kozameh, E. Newman, and K. Tod. Conformal Einstein spaces. *Gen. Relativity Gravitation*, 17(4):343–352, 1985.

[160] K. Krasnov. Holography and Riemann surfaces. *Adv. Theor. Math. Phys.*, 4:929–979, 2000. `arXiv:hep-th/0005106v2`.

[161] K. Krasnov and J.-M. Schlenker. On the renormalized volume of hyperbolic 3-manifolds. *Comm. Math. Phys.*, 279(3):637–668, 2008. `arXiv:math/0607081v3`.

[162] R. Kulkarni. Conformal structures and Möbius structures. In *Conformal geometry (Bonn, 1985/1986)*, Aspects Math., E12, pages 1–39. Vieweg, 1988.

[163] P.D. Lax and R.S. Phillips. The asymptotic distribution of lattice points in Euclidean and non-Euclidean spaces. *J. Funct. Anal.*, 46(3):280–350, 1982.

[164] C. LeBrun. $\mathcal{H}$-space with a cosmological constant. *Proc. Roy. Soc. London Ser. A*, 380:171–185, 1982.

[165] J.M. Lee and T.H. Parker. The Yamabe problem. *Bull. Amer. Math. Soc. (N.S.)*, 17(1):37–91, 1987.

[166] T. Leistner and P. Nurowski. Ambient metrics for n-dimensional pp-waves. `arXiv:0810.2903v2`.

[167] J. Lewis and D. Zagier. Period functions for Maass wave forms. I. *Ann. of Math. (2)*, 153(1):191–258, 2001.

[168] M. Listing. Conformally invariant Cotton and Bach tensors in n dimensions. `arXiv:math/0408224v1`.

[169] M. Lyubich and Y. Minsky. Laminations in holomorphic dynamics. *J. Differential Geom.*, 47(1):17–94, 1997.

[170] A. Malchiodi. Conformal metrics with constant Q-curvature. *SIGMA Symmetry Integrability Geom. Methods Appl.*, 3:Paper 120, 11, 2007. `arXiv:0712.2123v1`.

[171] J. Maldacena. The large N limit of superconformal field theories and supergravity. *Adv. Theor. Math. Phys.*, 2(2):231–252, 1998. `arXiv:hep-th/9711200v3`.

[172] J. Maldacena. Tasi 2003 lectures on AdS/CFT. 2004. `arXiv:hep-th/0309246v5`.

[173] N. Mandouvalos. Scattering operator, Eisenstein series, inner product formula and "Maass-Selberg" relations for Kleinian groups. *Mem. Amer. Math. Soc.*, 78:iv+87, 1989.

[174] Y. Manin and M. Marcolli. Holography principle and arithmetic of algebraic curves. *Adv. Theor. Math. Phys.*, 5(3):617–650, 2001.

[175] R. Manvelyan, K. Mkrtchyan, and R. Mkrtchyan. Conformal invariant powers of the Laplacian, Fefferman-Graham ambient metric and Ricci gauging. *Phys. Lett. B*, 657(1-3):112–119, 2007. `arXiv:0707.1737v3`.

[176] R. Manvelyan and D.H. Tchrakian. Conformal coupling of the scalar field with gravity in higher dimensions and invariant powers of the Laplacian. *Phys. Lett. B*, 644(5-6):370–374, 2007. `arXiv:hep-th/0611077v2`.

[177] C. Margerin. Géométrie conforme en dimension 4: ce que l'analyse nous apprend. *Astérisque*, 307:Exp. No. 950, ix–x, 415–468, 2006. Séminaire Bourbaki. Vol. 2004/2005.

[178] K. Matsuzaki and M. Taniguchi. *Hyperbolic manifolds and Kleinian groups*. Oxford Mathematical Monographs. The Clarendon Press Oxford University Press, 1998.

[179] R. Mazzeo. The Hodge cohomology of a conformally compact metric. *J. Differential Geom.*, 28(2):309–339, 1988.

[180] R. Mazzeo and R.B. Melrose. Meromorphic extension of the resolvent on complete spaces with asymptotically constant negative curvature. *J. Funct. Anal.*, 75(2):260–310, 1987.

[181] R. Melrose. *Geometric scattering theory*. Stanford Lectures. Cambridge University Press, 1995.

[182] S. Miller and W. Schmid. Automorphic distributions, *L*-functions, and Voronoi summation for GL(3). *Ann. of Math. (2)*, 164(2):423–488, 2006. `arXiv:math/0408100v3`.

[183] S. Morosawa, Y. Nishimura, M. Taniguchi, and T. Ueda. *Holomorphic dynamics*, volume 66 of *Cambridge Studies in Advanced Mathematics*. Cambridge University Press, Cambridge, 2000.

[184] D. Mumford, C. Series, and D. Wright. *Indra's pearls*. Cambridge University Press, 2002. The vision of Felix Klein.

[185] S. Nayatani. Patterson-Sullivan measure and conformally flat metrics. *Math. Z.*, 225(1):115–131, 1997.

[186] C. Ndiaye. Curvature flows on four manifolds with boundary. `arXiv:0708.2029v1`.

[187] C.B. Ndiaye. Constant Q-curvature metrics in arbitrary dimension. *J. Funct. Anal.*, 251(1):1–58, 2007.

[188] P. Nicholls. *The ergodic theory of discrete groups*, volume 143 of *London Mathematical Society Lecture Note Series*. Cambridge University Press, 1989.

[189] K. Okikiolu. Critical metrics for the determinant of the Laplacian in odd dimensions. *Ann. of Math. (2)*, 153(2):471–531, 2001.

[190] M. Olbrich. Cohomology of convex cocompact groups and invariant distributions on limit sets. Habilitation Göttingen, `arXiv:math/0207301`, 2001.

[191] B. Osgood, R. Phillips, and P. Sarnak. Extremals of determinants of Laplacians. *J. Funct. Anal.*, 80(1):148–211, 1988.

[192] S. Paneitz. A quartic conformally covariant differential operator for arbitrary pseudo-Riemannian manifolds (summary). *SIGMA Symmetry Integrability Geom. Methods Appl.*, 4:Paper 036, 3, 2008.

[193] I. Papadimitriou and K. Skenderis. AdS/CFT correspondence and geometry. In *AdS/CFT correspondence: Einstein metrics and their conformal boundaries*, volume 8 of *IRMA Lect. Math. Theor. Phys.*, pages 73–101. Eur. Math. Soc., Zürich, 2005. `arXiv:hep-th/0404176v2`.

[194] T. Parker and S. Rosenberg. Invariants of conformal Laplacians. *J. Differential Geom.*, 25(2):199–222, 1987.

[195] S. Patterson and P. Perry. The divisor of Selberg's zeta function for Kleinian groups. *Duke Math. J.*, 106(2):321–390, 2001. Appendix A by Charles Epstein.

[196] S.J. Patterson. Lectures on measures on limit sets of Kleinian groups. In *Analytical and geometric aspects of hyperbolic space (Coventry/Durham, 1984)*, volume 111 of *London Math. Soc. Lecture Note Ser.*, pages 281–323. Cambridge Univ. Press, 1987.

[197] S.J. Patterson. *An introduction to the theory of the Riemann zeta-function*, volume 14 of *Cambridge Studies in Advanced Mathematics*. Cambridge University Press, 1988.

[198] S. Paycha and S. Rosenberg. Conformal anomalies via canonical traces. In *Analysis, geometry and topology of elliptic operators*, pages 263–294. World Sci. Publ., Hackensack, NJ, 2006.

[199] R. Penrose. Conformal treatment of infinity. In *Relativité, Groupes et Topologie (Lectures, Les Houches, 1963 Summer School of Theoret. Phys., Univ. Grenoble)*, pages 565–584. Gordon and Breach, 1964.

[200] P. Perry. The Laplace operator on a hyperbolic manifold. II. Eisenstein series and the scattering matrix. *J. Reine Angew. Math.*, 398:67–91, 1989.

[201] P. Perry. The spectral geometry of geometrically finite hyperbolic manifolds. In *Spectral theory and mathematical physics: a Festschrift in honor of Barry Simon's 60th birthday*, volume 76 of *Proc. Sympos. Pure Math.*, pages 289–327. Amer. Math. Soc., 2007.

[202] L. Peterson. Future directions of research in geometry: a summary of the panel discussion at the 2007 Midwest Geometry Conference. *SIGMA Symmetry Integrability Geom. Methods Appl.*, 3:Paper 081, 7, 2007. `http://www.emis.de/journals/SIGMA/MGC2007.html`.

[203] A.M. Polyakov. Quantum geometry of bosonic strings. *Phys. Lett. B*, 103(3):207–210, 1981.

[204] R. Ponge. Logarithmic singularities of Schwartz kernels and local invariants of conformal and CR structures. `arXiv:0710.5783v1`.

[205] J. Qing and D. Raske. Compactness for conformal metrics with constant Q-curvature on locally conformally flat manifolds. *Calc. Var. Partial Differential Equations*, 26(3):343–356, 2006.

[206] J. Ratcliffe. *Foundations of hyperbolic manifolds*, volume 149 of *Graduate Texts in Mathematics*. Springer, second edition, 2006.

[207] D.B. Ray and I.M. Singer. R-torsion and the Laplacian on Riemannian manifolds. *Advances in Math.*, 7:145–210, 1971.

[208] R. Riegert. A nonlocal action for the trace anomaly. *Phys. Lett. B*, 134(1-2):56–60, 1984.

[209] D. Ruelle. Zeta-functions for expanding maps and Anosov flows. *Invent. Math.*, 34(3):231–242, 1976.

[210] W. Schmid. Automorphic distributions for $SL(2, \mathbb{R})$. In *Conférence Moshé Flato 1999, Vol. I (Dijon)*, volume 21 of *Math. Phys. Stud.*, pages 345–387. Kluwer Acad. Publ., 2000.

[211] A. Schwimmer and S. Theisen. Diffeomorphisms, anomalies and the Fefferman-Graham ambiguity. *JHEP*, 8:Paper 32, 16, 2000. `arXiv:hep-th/0008082v1`.

[212] A. Schwimmer and S. Theisen. Universal features of holographic anomalies. *JHEP*, 10:001, 15 pp. (electronic), 2003. `arXiv:hep-th/0309064v1`.

[213] A. Selberg. Harmonic analysis and discontinuous groups in weakly symmetric Riemannian spaces with applications to Dirichlet series. *J. Indian Math. Soc. (N.S.)*, 20:47–87, 1956.

[214] N. Seshadri. Approximate Einstein ACH metrics, volume renormalization, and an invariant for contact manifolds. `arXiv:0707.0597v2`.

[215] N. Seshadri. Volume renormalization for complete Einstein-Kähler metrics. *Differential Geom. Appl.*, 25(4):356–379, 2007. `arXiv:math/0404455v2`.

[216] R.W. Sharpe. *Differential geometry*, volume 166 of *Graduate Texts in Mathematics*. Springer-Verlag, 1997. Cartan's generalization of Klein's Erlangen program, with a foreword by S.S. Chern.

[217] Wei-Min Sheng, N. Trudinger, and Xu-Jia Wang. The Yamabe problem for higher order curvatures. *J. Differential Geom.*, 77(3):515–553, 2007. `arXiv:math/0505463v1`.

[218] K. Skenderis. Asymptotically anti-de Sitter spacetimes and their stress energy tensor. In *Strings 2000. Proceedings of the International Superstrings Conference (Ann Arbor, MI)*, volume 16, pages 740–749, 2001. `arXiv:hep-th/0010138v1`.

[219] K. Skenderis. Lecture notes on holographic renormalization. *Classical Quantum Gravity*, 19(22):5849–5876, 2002. European Winter School on the Quantum Structure of Spacetime (Utrecht, 2002).

[220] K. Skenderis and S. Solodukhin. Quantum effective action from the AdS/CFT correspondence. *Phys. Lett. B*, 472(3-4):316–322, 2000. `arXiv: hep-th/9910023v1`.

[221] J. Slovák. *Natural operators on conformal manifolds*. PhD thesis, Brno, 1993.

[222] J. Slovák. Parabolic geometries, 1997. Habilitation, Masaryk University.

[223] H.M. Smith. *Principles of Holography*. Wiley-Interscience, 1969.

[224] D. Sullivan. The density at infinity of a discrete group of hyperbolic motions. *Inst. Hautes Études Sci. Publ. Math.*, 50:171–202, 1979.

[225] D. Sullivan. Discrete conformal groups and measurable dynamics. *Bull. Amer. Math. Soc. (N.S.)*, 6(1):57–73, 1982.

[226] Dennis Sullivan. Conformal dynamical systems. In *Geometric dynamics (Rio de Janeiro, 1981)*, volume 1007 of *Lecture Notes in Math.*, pages 725–752. Springer, Berlin, 1983.

[227] L. Susskind and J. Lindesay. *An Introduction to Black Holes, Information and the String Theory Revolution: The Holographic Universe*. World Scientific Publishing, 2004.

[228] K. Takano and T. Imai. Note on the conformal theory of subspaces. *Tensor (N.S.)*, 3:108–118, 1954.

[229] L. Takhtajan and Lee-Peng Teo. Liouville action and Weil-Petersson metric on deformation spaces, global Kleinian reciprocity and holography. *Comm. Math. Phys.*, 239(1-2):183–240, 2003. `arXiv:math/0204318v2`.

[230] E.C. Titchmarsh. *The theory of the Riemann zeta-function*. The Clarendon Press Oxford University Press, New York, second edition, 1986. Edited and with a preface by D.R. Heath-Brown.

[231] W. Ugalde. A construction of critical GJMS operators using Wodzicki's residue. *Comm. Math. Phys.*, 261(3):771–788, 2006.

[232] M. Urbański. Measures and dimensions in conformal dynamics. *Bull. Amer. Math. Soc. (N.S.)*, 40(3):281–321 (electronic), 2003.

[233] E. van den Ban and H. Schlichtkrull. Asymptotic expansions and boundary values of eigenfunctions on Riemannian symmetric spaces. *J. Reine Angew. Math.*, 380:108–165, 1987.

[234] D.V. Vassilevich. Heat kernel expansion: user's manual. *Phys. Rep.*, 388(5-6):279–360, 2003.

[235] J. Viaclovsky. Conformal geometry, contact geometry, and the calculus of variations. *Duke Math. J.*, 101(2):283–316, 2000.

[236] D. Vogan. *Representations of real reductive Lie groups*, volume 15 of *Progress in Mathematics*. Birkhäuser Boston, 1981.

[237] X. Wang. Holography and the geometry of certain convex cocompact hyperbolic 3-manifolds. `arXiv:math/0210164v2`.

[238] X. Wang. On conformally compact Einstein manifolds. *Math. Res. Lett.*, 8(5-6):671–688, 2001.

[239] T.J. Willmore. *Riemannian geometry*. Oxford Science Publications. The Clarendon Press Oxford University Press, New York, 1993.

[240] T.J. Willmore. Surfaces in conformal geometry. *Ann. Global Anal. Geom.*, 18(3-4):255–264, 2000. Special issue in memory of Alfred Gray (1939–1998).

[241] E. Witten. Anti de Sitter space and holography. *Adv. Theor. Math. Phys.*, 2(2):253–291, 1998. `arXiv:hep-th/9802150v2`.

[242] V. Wünsch. On conformally invariant differential operators. *Math. Nachr.*, 129:269–281, 1986.

[243] G. Zuckerman. Tensor products of finite and infinite dimensional representations of semisimple Lie groups. *Ann. Math. (2)*, 106(2):295–308, 1977.

Index

AdS_n anti de Sitter space, 6
$B_{2N}(g)$ obstruction operator, 417
$C^\infty(S^m)_\lambda$, 72
$C_M(\lambda)$ tractor C-operator, 390
D^W, 385
$D^T(\lambda)$, 398
$D_N^b(\lambda)$, 172, 173
$D_2(g;\lambda)$, 196, 201
$D_M(\lambda)$ tractor D-operator, 389
$D_N^T(g;\lambda)$ tractor family, 407
$D_N^T(g;\lambda)$, 405
$D_N^c(\lambda)$ intertwining family, 133
$D_N^{\mathrm{res}}(h;\lambda)$ residue family, 217
$D_N^{nc}(\lambda)$ intertwining family, 133
$D_n^{\mathrm{res}}(h;\lambda)$ critical residue family, 215
E_{2n} Euler form, 333
F, 86
F_n standard representation, 48
G G-tensor, 86
G^n, 63
H mean curvature, 46
H^+ upper hemisphere, 68
H_0, 64
I_λ intertwining operator, 3
$I_\lambda(\mathfrak{g}_n)$ ideal defined by ξ_λ, 130
K^n, 64
L second fundamental form, 46
L total holographic anomaly, 225
L_0 trace-free second fundamental form, 53
M^n, 64
M_{ij}, 64
N^T normal tractor, 399
$N^\pm$, 65

P parabolic subgroup, 65
P_3 Chang-Qing operator, 463
P_3^T, 457
P_4 Paneitz operator, 11, 107
P_4^e, 53
P_c^- contracting part of Poincaré map, 4
P_n critical GJMS-operator, 88
P_n^e, 409
$P_{2N} = P_{2N,n}$ GJMS-operator, 88
$P_{4,n}$ Paneitz operator, 107
Q-curvature
    critical, 90
    extrinsic, 409
    for Einstein metrics, 343
    fundamental identity, 90
    in dimension 4, 106
    odd order, 410
    subcritical, 90
Q-polynomial, 218
Q_3^T, 457
Q_4, 20, 106, 230
Q_4^e, 53
Q_6, 28, 36, 255, 257
Q_n critical Q-curvature, 90
$Q_n^T(g;\lambda)$ tractor Q-polynomial, 418
$Q_n^T(g)$ tractor Q-curvature, 408
Q_n^e extrinsic Q-curvature, 409
Q_n^{FG}, 411
$Q_{2N} = Q_{2N,n}$ Q-curvature, 90
$Q_{2N}^{\mathrm{res}}(\lambda)$ Q-polynomial, 218
$Q_{3,T}$, 457
$Q_{4,n}$, 107
$Q_{6,n}$, 261

R Riemannian curvature, 83
R^T tractor curvature, 393
S Einstein tensor, 115
T-curvature, 463
$T(g,\varphi)$ trivialization, 388
$V(g_E, h)$ renormalized volume, 309
W Weyl group, 64
$X^\flat$, 106
Y Yamabe operator, 11
$Y_j^\pm$, 64
Z_Γ Selberg zeta function, 4
Z_Γ^R Ruelle zeta function, 4
$\#$, 106
$\mathcal{A}(\mathfrak{g})$ semi-holonomic universal enveloping algebra, 357
$\mathcal{B}$ Bach tensor, 240
C Weyl tensor, 84
Δ Laplacian, 87
Δ^W, 385
Δ^T, 394
Δ^T tractor Laplacian, 394
Δ_n^T, 387
Δ_n^- Laplacian, 117
Ein_n, 6
$\mathcal{G}_M$ ray bundle, 88
$\mathcal{H}$ mean curvature vector, 443
$\mathcal{H}_0$ height, 67
Hess   covariant Hessian form, 83
J, 84
$\Lambda(\Gamma)$ limit set, 3
$\mathcal{M}_\lambda$ Verma module, 15
$\mathcal{M}(V_\sigma)$ Verma module, 357
$\mathcal{N}(V_\sigma)$ semi-holonomic Verma module, 357
$\mathcal{N}_\lambda$ semi-holonomic Verma module, 360
$\Omega(\Gamma)$ proper set, 3
$\mathcal{P}_\lambda$ Poisson transformation, 72, 76
Π_Σ projection, 400
P Schouten tensor, 84
$\mathsf{P}(X)$, 391
Ric   Ricci tensor, 83
Ric_N, 86

$\mathcal{S}(\lambda)$ scattering operator, 94
$\mathcal{T}^*M$ dual tractor bundle, 388
$\mathcal{T}_j(h; \lambda)$, 215
$\Theta_N(\lambda)$, 367
$\mathcal{U}(\mathfrak{g})$ universal enveloping algebra, 356
$\mathcal{X}(M)$ vector fields on M, 83
$\mathcal{C}$ Cotton tensor, 393
$\mathfrak{a}$, 64
α restricted root, 129
α_0, 64
$\mathbb{B}^n$ ball model, 2, 66
$\mathbb{C}(\lambda)$, 15, 129
$\delta_N(h; \lambda)$, 218
$\delta(\Gamma)$ critical exponent, 4
ext_λ extension operator, 179
$\mathfrak{g}_n$, 63
$\mathbf{g}$, 88
$\mathbb{H}^n$ hyperbolic space, 68
$\mathcal{J}$, 439
κ, 220
κ stereographic projection, 67
$\mathfrak{m}_n$, 64
μ_{PS} Patterson-Sullivan measure, 5
$\mathfrak{n}_n^\pm$, 64
∇ Levi-Civita connection, 82
∇^W, 385
∇^T tractor connection, 391
ω Cartan connection, 358
$\omega^\sharp$, 106
$\mathfrak{p}$ parabolic subalgebra, 15, 129
π, 357
π_Σ, 400
π_λ^c compact model of principal series, 72
π_λ^{nc} non-compact model of principal series, 75
$\mathrm{res}_{\Omega(\Gamma)}$ restriction to $\Omega(\Gamma)$, 180
$\sharp$, 106
$\square^T$, 395
$\mathcal{D}_N^0(\lambda)$, 118, 128
$\mathcal{D}_N(\lambda)$, 118, 127
$\mathcal{D}_{2N}^T(\lambda)$, 378

τ scalar curvature, 84

τ_e extrinsic scalar curvature, 207

θ, 433

$\tilde{\Theta}_N(\lambda)$, 367

$\tilde{g}$ ambient metric, 89

$\mathcal{CW}$ Chen-Willmore functional, 442

ξ_λ character, 15, 129

ζ Riemann zeta function, 4

$c(\lambda)$ c-function, 96

g-trace, 85

g_N Nayatani metric, 59

$i(\lambda)$, 364

$p(\lambda)$, 364

v_n holographic anomaly, 24

v_{2j} holographic coefficients, 24

w, 64

AdS/CFT duality, 28, 336, 343

ambient metric, 89

asymptotic volume, 309

Atiyah-Bott-Lefschetz formula, 8

automorphic distributions, 179

Bach tensor, 240

Bianchi identities, 112

Bochner-Weitzenböck formula, 112

Bruhat decomposition, 69

Cartan connection, 358

Chang-Qing operator, 463

Chen-Willmore functional, 442

Chen-Willmore quantity, 190

Chern-Gauß-Bonnet theorem, 333

Codazzi equation, 458

conformal

   compactification, 92

   index density, 30

   infinity, 92

Cotton tensor, 112, 393

critical exponent, 4

critical GJMS-operator, 88

critical residue family, 215

curvature tensor, 83

curved translation principle, 47

Deser-Schwimmer classification, 334

Einstein's static universe, 6

Einstein-Hilbert action, 29

Eisenstein series, 103

Euler characteristic, 27

Euler form, 333

extension operator, 179

extrinsic scalar curvature, 207

factorization identities

   for intertwining families, 139, 148

   for residue families, 266

Fefferman-Graham expansion, 22

Fefferman-Graham obstruction, 240, 249

Fialkow's fundamental forms, 446

fundamental identity, 90

fundamental pair, 191

G-tensor, 86

Gauß equation, 207

Gegenbauer polynomial, 118

geodesic flow, 7

GJMS-operators, 88

Gover's formula, 343

Harish-Chandra's c-function, 96

Henningson-Skenderis test, 249, 342

Hessian, 83

holographic

   anomaly, 24

   coefficients, 24

   duality, 412

   renormalization, 29

holographic formula

   for Q_4, 230

   for Q_6, 254

   for critical Q-curvature, 26, 225

   for subcritical Q-curvature, 300

hyperbolic space

   as symmetric space, 65

   ball model, 66

   upper half space model, 68

induced representation, 131
Iwasawa decomposition, 65

Koszul formula, 82
Kulkarni-Nomizu product, 84

Langlands decomposition, 15
Laplacian, 87
Lefschetz fixed point formula, 8
Lichnerowicz Laplacian, 242
light cone, 65
limit set, 81

mean curvature, 46
mean curvature vector, 443

Nayatani metric, 59, 81
Newton's relations, 246
normal tractor, 399

Paneitz
    curvature, 20, 106
    operator, 11, 107
    quantity, 106
Patterson-Sullivan measure, 5
PBH transformations, 25
Pfaffian, 333
Poincaré-Einstein metric, 93
Poisson transformation, 72, 76
Polyakov formula, 30, 32, 340
principal series representation, 72,
    75
proper set, 3

renormalized volume, 309
residue families, 217
Ricci tensor, 83

scalar curvature, 84
scattering operator, 94
Schouten tensor, 84
second fundamental form, 46
semi-holonomic Verma module, 357
spherical metric, 68

standard tractor bundle, 388
stereographic projection, 67
subcritical GJMS-operator, 88

T-tensor, 86
tractor
    C-operator, 390
    D-operator, 389
    Q-polynomial, 418
    Bochner-Laplacian, 394
    connection, 391
    curvature endomorphism, 393
    scalar product, 393
tractor families, 407
triangle decomposition, 15

universal enveloping algebra, 356

Verma module, 356

Weyl tensor, 84
Weyl's tube formula, 336
Willmore functional, 197

Yamabe
    equation, 91
    operator, 11

zeta function
    Riemann, 4
    Ruelle, 4
    Selberg, 4
Zuckerman translation, 360

Progress in Mathematics (PM)

Edited by
Hyman Bass, University of Michigan, USA
Joseph Oesterlé, Institut Henri Poincaré, Université Paris VI, France
Alan Weinstein, University of California, Berkeley, USA

Progress in Mathematics is a series of books intended for professional mathematicians and scientists, encompassing all areas of pure mathematics. This distinguished series, which began in 1979, includes research level monographs, polished notes arising from seminars or lecture series, graduate level textbooks, and proceedings of focused and refereed conferences. It is designed as a vehicle for reporting ongoing research as well as expositions of particular subject areas.

Progress in Mathematics (PM)

Edited by
Hyman Bass, University of Michigan, USA
Joseph Oesterlé, Institut Henri Poincaré, Université Paris VI, France
Alan Weinstein, University of California, Berkeley, USA

Progress in Mathematics is a series of books intended for professional mathematicians and scientists, encompassing all areas of pure mathematics. This distinguished series, which began in 1979, includes research level monographs, polished notes arising from seminars or lecture series, graduate level textbooks, and proceedings of focused and refereed conferences. It is designed as a vehicle for reporting ongoing research as well as expositions of particular subject areas.

PM 257: Lakshmibai, V. / Littelmann, P. / Seshadri, C.S. (Eds.)
Schubert Varieties (due 2009)
ISBN 978-0-8176-4153-5

PM 256: Borthwick, D.
Spectral Theory of Infinite-Volume Hyperbolic Surfaces (2007). ISBN 978-0-8176-4524-3

PM 255: Kobayashi, T. / Schmid, W. / Yang, J.-H. (Eds.)
Representation Theory and Automorphic Forms (2007). ISBN 978-0-8176-4505-2

PM 254: Ma, X. / Marinescu, G.
Holomorphic Morse Inequalities and Bergman Kernels (2007)
Winner of the Ferran Sunyer i Balaguer Prize 2006.
ISBN 978-3-7643-8096-0

PM 253: Ginzburg, V. (Ed.)
Algebraic Geometry and Number Theory (2006).
ISBN 978-0-8176-4471-7

PM 252: Maeda, Y. / Michor, P. / Ochiai, T. / Yoshioka, A. (Eds.)
From Geometry to Quantum Mechanics (2007).
ISBN 978-0-8176-4512-8

PM 251: Boutet de Monvel, A. / Buchholz, D. / Iagolnitzer, D. / Moschella, U. (Eds.)
Rigorous Quantum Field Theory (2006). ISBN 978-3-7643-7433-4

PM 250: Unterberger, A.
The Fourfold Way in Real Analysis (2006). ISBN 978-3-7643-7544-7

PM 249: Kock, J. / Vainsencher, I.
An Invitation to Quantum Cohomology. Kontsevich's Formula for Rational Plane Curves (2006).
ISBN 978-0-8176-4456-7

PM 248: Bartholdi, L. / Ceccherini-Silberstein, T. / Smirnova-Nagnibeda, T. / Zuk, A. (Eds.) Infinite Groups: Geometric, Combinatorial and Dynamical Aspects (2006)
ISBN 978-3-7643-7446-4

PM 247: Baues, H.-J.
The Algebra of Secondary Cohomology Operations (2006)
ISBN 978-3-7643-7448-8

PM 246: Dragomir, S. / Tomassini, G., Differential Geometry and Analysis on CR Manifolds (2006).
ISBN 978-0-8176-4388-1

PM 245: Fels, G. / Huckleberry, A.T. / Wolf, J.A.
Cycle Spaces of Flag Domains. A Complex Geometric Viewpoint (2006). ISBN 978-0-8176-4391-1

PM 244: Etingof, P. / Retakh, V. / Singer, I.M. (Eds.)
The Unity of Mathematics (2006)
ISBN 978-0-8176-4076-7

PM 243: Bernstein, J. / Hinich, V. / Melnikov, A. (Eds.)
Studies in Lie Theory (2006)
ISBN 0-8176-4342-3

PM 242: Dufour, J.-P. / Zung, N.T.
Poisson Structures and their Normal Forms (2005)
ISBN 978-3-7643-7334-4

PM 241: Seade, J.
On the Topology of Isolated Singularities in Analytic Spaces

Winner of the Ferran Sunyer i Balaguer Prize 2005
ISBN 978-3-7643-7322-1

PM 240: Ambrosetti, A. / Malchiodi, A. Perturbation Methods and Semilinear Elliptic Problems in R^n (2005). *Winner of the Ferran Sunyer i Balaguer Prize 2005.*
ISBN 978-3-7643-7321-4

PM 239: van der Geer, G. / Moonen, B.J.J. / Schoof, R. (Eds.)
Number Fields and Function Fields – Two Parallel Worlds (2005)
ISBN 978-0-8176-4397-3

PM 238: Sabadini, I. / Struppa, D.C. / Walnut, D.F. (Eds.)
Harmonic Analysis, Signal Processing, and Complexity (2005).
ISBN 978-0-8176-4358-4

PM 237: Kulish, P.P. / Manojlovic, N. / Samtleben, H. (Eds.)
Infinite Dimensional Algebras and Quantum Integrable Systems (2005). ISBN 978-3-7643-7215-6

PM 236: Hotta, R. / Takeuchi, K. / Tanisaki, T.
D-Modules, Perverse Sheaves, and Representation Theory (due 2007)
ISBN 978-0-8176-4363-8

PM 235: Bogomolov, F. / Tschinkel, Y. (Eds.)
Geometric Methods in Algebra and Number Theory (2005).
ISBN 978-0-8176-4349-2

PM 234: Kowalski, O. / Musso, E. / Perrone, D. (Eds.)
Complex, Contact and Symmetric Manifolds. (2005)
ISBN 978-0-8176-3850-4

BIRKHÄUSER